# Fish Physiology

Conservation Physiology for the Anthropocene – Issues and Applications

Volume 39B

This is Volume 39B in the

FISH PHYSIOLOGY series
Edited by Anthony P. Farrell, Colin J. Brauner and Erika J. Eliason
Honorary Editors: William S. Hoar and David J. Randall

*A complete list of books in this series appears at the end of the volume*

# Fish Physiology

## Conservation Physiology for the Anthropocene – Issues and Applications

Volume 39B

Edited by


**Nann A. Fangue**
Department of Wildlife, Fish and Conservation Biology,
University of California - Davis, Davis, CA, United States

**Steven J. Cooke**
Fish Ecology and Conservation Physiology Laboratory,
Department of Biology, Carleton University, Ottawa, ON, Canada

**Anthony P. Farrell**
Department of Zoology, and Faculty of Land and Food Systems,
The University of British Columbia, Vancouver, British Columbia, Canada

**Colin J. Brauner**
Department of Zoology, The University of British Columbia,
Vancouver, British Columbia, Canada

**Erika J. Eliason**
Department of Ecology, Evolution, and Marine Biology,
University of California - Santa Barbara, Santa Barbara, CA, United States

Academic Press is an imprint of Elsevier
50 Hampshire Street, 5th Floor, Cambridge, MA 02139, United States
525 B Street, Suite 1650, San Diego, CA 92101, United States
The Boulevard, Langford Lane, Kidlington, Oxford OX5 1GB, United Kingdom
125 London Wall, London, EC2Y 5AS, United Kingdom

First edition 2022



ISBN: 978-0-12-824268-1
ISSN: 1546-5098

For information on all Academic Press publications
visit our website at https://www.elsevier.com/books-and-journals

*Publisher:* Zoe Kruze
*Acquisitions Editor:* Sam Mahfoudh
*Editorial Project Manager:* Jhon Michael Peñano
*Production Project Manager:* Sudharshini Renganathan
*Cover Designer:* Christian Bilbow

Typeset by STRAIVE, India



**Cover images**: The Nassau grouper (Epinephelus striatus) is considered Endangered by the International Union for the Conservation of Nature. They face diverse threats that necessitate adopting a systems approach to understanding threats and identifying effective conservation strategies. From endocrinology studies to understand the basis for sex change and to enable captive breeding to studies of sensory physiology to understand how they find spawning aggregations and studies on the physiological consequences of fisheries interactions, Nassau grouper exemplify the need for and promise of Conservation Physiology - Dr. Sean J. Landsman.

# Contents

# Contributors

**Vera Maria Fonseca de Almeida-Val** (209), Laboratory of Ecophysiology and Molecular Evolution, Brazilian National Institute for Research of the Amazon, Manaus, Brazil

**W. Gary Anderson** (39), Department of Biological Sciences, University of Manitoba, Winnipeg, MB, Canada

**Melinda R. Baerwald** (1), Division of Integrated Science & Engineering, California Department of Water Resources, West Sacramento, CA, United States

**Nolan N. Bett** (429), Department of Forest and Conservation Sciences, University of British Columbia, Vancouver, BC, Canada

**Kim Birnie-Gauvin** (581), National Institute of Aquatic Resources, Technical University of Denmark, Silkeborg, Denmark

**Derek Campos** (209), Laboratory of Ecophysiology and Molecular Evolution, Brazilian National Institute for Research of the Amazon, Manaus, Brazil

**Lauren J. Chapman** (273), Department of Biology, McGill University, Montréal, QC, Canada

**Steven J. Cooke** (175), Fish Ecology and Conservation Physiology Laboratory, Department of Biology and Institute of Environmental and Interdisciplinary Science, Carleton University, Ottawa, ON, Canada

**James A. Crossman** (39), Fish and Aquatics, BC Hydro, Castlegar, BC, Canada

**Brittany E. Davis** (1), Division of Integrated Science & Engineering, California Department of Water Resources, West Sacramento, CA, United States

**Gudrun De Boeck** (111), ECOSPHERE, Department of Biology, University of Antwerp, Antwerp, Belgium

**Margaret F. Docker** (489), Department of Biological Sciences, University of Manitoba, Winnipeg, MB, Canada

**Rafael Mendonça Duarte** (209), Biosciences Institute, São Paulo State University (UNESP), Coastal Campus, São Paulo, Brazil

**Nann A. Fangue** (1), Department of Wildlife, Fish, and Conservation Biology, University of California, Davis, CA, United States

**Anthony P. Farrell** (429), Department of Zoology, University of British Columbia, Vancouver, BC, Canada

**Vincent Fugère** (273), Département des Sciences de l'Environnement, Université du Québec à Trois-Rivières, Trois-Rivières, QC, Canada

**Scott G. Hinch** (429), Department of Forest and Conservation Sciences, University of British Columbia, Vancouver, BC, Canada

**Jack Hollins** (175), University of Windsor, Department of Integrative Biology, Windsor, ON, Canada

**Björn Illing** (325), ARC Centre of Excellence for Coral Reef Studies, James Cook University, Townsville, QLD, Australia; Thünen Institute of Fisheries Ecology, Bremerhaven, Germany

**Nicholas S. Johnson** (489), U.S. Geological Survey, Great Lakes Science Center, Hammond Bay Biological Station, Millersburg, MI, United States

**Shaun S. Killen** (175), Institute of Biodiversity, Animal Health and Comparative Medicine, College of Medical, Veterinary and Life Sciences, Graham Kerr Building, University of Glasgow, Glasgow, United Kingdom

**Barbara Koeck** (175), Institute of Biodiversity, Animal Health and Comparative Medicine, College of Medical, Veterinary and Life Sciences, Graham Kerr Building, University of Glasgow, Glasgow, United Kingdom

**Lisa M. Komoroske** (581), Department of Environmental Conservation, University of Massachusetts Amherst, Amherst, MA, United States

**Robert J. Lennox** (175), Laboratory for Freshwater Ecology and Inland Fisheries (LFI) at NORCE, Norwegian Research Centre, Nygårdsporten 112, Bergen, Norway

**Elizabeth A. Nyboer** (273), Department of Biology, Carleton University, Carleton Technology and Training Centre, Ottawa, ON, Canada

**Essie Rodgers** (111), School of Biological Sciences, University of Canterbury, Christchurch, New Zealand

**Jodie L. Rummer** (325), ARC Centre of Excellence for Coral Reef Studies; College of Science and Engineering, James Cook University, Townsville, QLD, Australia

**Andrea Schreier** (39), Department of Animal Science, College of Agricultural and Environmental Sciences, University of California, Davis, CA, United States

**Ted R. Sommer** (1), Division of Integrated Science & Engineering, California Department of Water Resources, West Sacramento, CA, United States

**Katherine K. Strailey** (393), Program in Ecology, Evolution, and Conservation Biology, University of Illinois at Urbana-Champaign, Champaign, IL, United States

**Cory D. Suski** (393), Department of Natural Resources and Environmental Sciences, University of Illinois at Urbana-Champaign, Champaign, IL, United States

**Raewyn M. Town** (111), ECOSPHERE, Department of Biology, University of Antwerp, Antwerp, Belgium

**Adalberto Luis Val** (209), Laboratory of Ecophysiology and Molecular Evolution, Brazilian National Institute for Research of the Amazon, Manaus, Brazil

**Michael P. Wilkie** (489), Department of Biology and Laurier Institute for Water Science, Wilfrid Laurier University, Waterloo, ON, Canada

**Yuzo R. Yanagitsuru** (1), Department of Wildlife, Fish, and Conservation Biology, University of California, Davis, CA, United States

# Preface

The world is changing.

Indeed, scholars have now declared that we have entered the Anthropocene epoch—a period distinct from the Holocene. The Anthropocene is defined by the manifold effects of human activities on planet Earth. By any and all measures, humans are changing environmental conditions and contributing to the loss of biodiversity.

Yet, all is not lost. There remains optimism that it is possible to address or mitigate some environmental threats and change our relationship with nature. Doing so could lead to opportunities for restoring biodiversity. In fact, the United Nations just launched (in 2021) the "Decade of Ecosystem Restoration," emphasizing that it is time to act.

Whether it be environmental change or direct interactions between humans (or human infrastructure) and wildlife, organisms can be influenced at various levels of biological organization. Although resource managers and conservation practitioners tend to focus on populations, communities, and ecosystems, it is often processes that play out at the level of the individual and are driven by physiological systems that influence these higher levels of biological organization. For that reason, the discipline of "conservation physiology" has emerged in recognition that physiological tools, knowledge, and concepts can be used to understand and solve conservation problems. Although conservation physiology is relevant to all taxa (including plants and microbes), there has been a particular focus on vertebrates, and more specifically, on fish.

Fish stocks are generally in poor condition around the globe—from the high seas to inland waters. The threats facing fish are many—from overfishing and bycatch, to fragmentation of migration corridors, to existing and emerging environmental pollutants, habitat loss, invasive species, climate change, and so on. In an attempt to stem this trend, a conservation physiology "toolbox" is increasingly being used to help understand the potential mechanisms by which these threats and stressors are influencing fish populations. A firm understanding of the mechanisms at play informs the potential solutions that may mitigate these threats. Thus, to directly address these conservation problems, conservation physiology is being used to develop solutions—from bycatch reduction strategies to restoration plans for fish passage. In other words, conservation physiology extends from problem identification to real-life solutions. To date, however, a synthesis related to conservation physiology specific to wild fish is lacking.

Consequently, the aim of Volume 39 of the *Fish Physiology* series is to generate a synthesis related to the physiology of fish in the Anthropocene. Specifically, in Volume 39A, we consider the ways in which different physiological systems (e.g., sensory physiology, cardiorespiratory) are relevant to conservation physiology. In Volume 39B, we present case studies to explore the ways in which physiology has been or can be used to understand or solve conservation problems (e.g., bycatch, habitat alteration, noise pollution). The first chapter in Volume 39A is written by an established group of scientists and sets the stage for what is meant by a systems approach to conservation. The last chapter of Volume 39B is written by early career researchers and is intended to be a forward-looking perspective on the state and future of conservation physiology of fishes.

Collectively, this volume provides an integrated synthesis that celebrates the successes achieved so far while identifying opportunities to further benefit the management and conservation of wild fish populations. It is our hope that this volume will serve as a resource for learners, fish physiologists, conservation practitioners, and fisheries managers.

We are grateful for the generosity of the many referees who provided peer reviews and thoughtful comments on the various chapters prior to their publication. We are especially appreciative of the authors who generated high-quality chapters during a period of uncertainty and challenges related to the global COVID-19 pandemic. The ideas shared in this book are truly at the frontier of applied fish physiology and conservation. We are fortunate to be part of a caring and supportive community and hope that this volume will serve to further build connections among those working on the conservation physiology of fishes. The team at Elsevier assisted with advancing this project. This book is dedicated to all of the scientists, practitioners, and community members who devote their professional and personal lives to generating and applying evidence to protect and restore fish populations and aquatic ecosystems around the globe.

**Nann A. Fangue**
**Steven J. Cooke**
**Anthony P. Farrell**
**Colin J. Brauner**
**Erika J. Eliason**

# Abbreviations

| | |
|---|---|
| **[Hb]** | hemoglobin concentration |
| **μatm** | microatmospheres |
| **$\mu mol\,g^{-1}\,h^{-1}$** | μmoles per gram per hour |
| **$^{14}$C-TFM** | carbon-14 labeled TFM |
| **2,3DPG** | 2,3 diphosphoglycerate |
| **AAS** | absolute aerobic scope, the difference between the MMR and SMR |
| **AChE** | acetylcholinesterase |
| **ACTH** | adrenocorticotropic hormone |
| **ADCP** | acoustic Doppler current profiler |
| **ADME** | absorption, distribution, metabolism, excretion |
| **AEP** | auditory evoked potential |
| **AHA** | Aldrich humic acid |
| **AHH** | aryl hydrocarbon hydroxylase |
| **AHR** | Aryl Hydrocarbon Receptors |
| **Al** | aluminum |
| **ALP** | alkaline phosphatase |
| **ALT** | alanine aminotransferase |
| **AO** | adverse outcome, often an impact on survival, growth, reproduction or population size |
| **AOP Network** | sets of AOPs sharing at least one common MIE, KE or AO |
| **AOP** | adverse outcome pathway, a pathway of events resulting in an Adverse Outcome |
| **ARE** | antioxidant response elements |
| **ARR** | acclimation response ratio |
| **AS** | aerobic scope |
| **ASA** | aerobic scope for activity |
| **ASR** | aquatic surface respiration |
| **AST** | aspartate aminotransferase |
| **ATP** | adenosine triphosphate |
| **BAM** | intracellular bioactive metal pool |
| **BC** | British Columbia |
| **BChE** | butyrylcholinesterase |
| **BIM** | intracellular bioinactive metal pool |
| **BIPOC** | Black, Indigenous, and people of color |

**BLM** biotic ligand model
**BPA** bisphenol A
**C:N ratio** carbon:nitrogen ratio
**$Ca^{2+}$** calcium ion
**$CaNO_3$** calcium nitrate
**CAT** catalase
**CCO** cytochrome c oxidase
**Cd** cadmium
**CDFW** California Department of Fish and Wildlife
**CESA** California Endangered Species Act
**CFD** computational fluid dynamics
**CFTR** cystic transmembrane conductance regulator
**$CH_4$** methane
**CK** creatinine kinase
**$Cl^-$** chloride ion
**Co** cobalt
**$CO_2$** carbon dioxide
**COSEWIC** Committee on the Status of Endangered Wildlife in Canada
**COSSARO** Committee on the Status of Species at Risk in Ontario
**Cr** chromium
**CRF** corticotropin releasing factor
**CRISPR/Cas9** Clustered Regularly Interspaced Short Palindromic Repeats/Cas9
**CS** citrate synthase
**$CT_{max}$** critical thermal maxima
**$CT_{min}$** critical thermal minima
**CTX** ciguatoxin
**Cu** copper
**CVP** Central Valley Project
**CYP1A** enzyme of the cytochrome P450 family
***D*** diffusion coefficients of the various species
***d*** diffusion layer thickness
**DBP** deltamethrin
**DDT** dichlorodiphenyl trichloroethane
**DEB** dynamic energy budget
**DEIJ** diversity, equity, inclusion and justice
**DNA** deoxyribonucleic acid
**D-nets** drift nets
**DNP** 2,4-dinitrophenylhydrazine
**DO** dissolved oxygen
**DOC** dissolved organic carbon
**dph** days post-hatch

**dsRNA** double-stranded RNA
**E2** 17β-estradiol
**$EC_{50}$** the exposure concentration at which 50% of the test population showed an effect
**ECD** endocrine disruptor
**$ED_{50}$** the internalized dose at which 50% of the test population showed an effect
**eDNA** environmental DNA
**EDP** environmental, physiological and demography
**EEA** European Environment Agency
**ELISA** enzyme-linked immunosorbent assay
**EMG** electromyogram
**ENM** ecological niche model
**EPA** Environmental Protection Agency
**EPOC** excess post-exercise oxygen consumption
**EQS** environmental quality standards
**EROD** ethoxyresorufin-*O*-deethylase
**ETS** electron transport system
**F1** first filial, the first filial generation of animal offspring
**FA** fulvic acids
**FAO** Food and Agricultural Organization of the United Nations
**FAS** factorial aerobic scope
**FCCL** Fish Conservation and Culture Laboratory
**Fe** iron
**FELS** fish early life stage
**FEOW** Freshwater Ecoregions of the World
**FESA** Federal Endangered Species Act
**FET** fish embryo tests
**FI** fluorescence index
**FIE** fisheries-induced evolution
**FRAP Assay** ferric reducing antioxidant power assay
**GA** glucuronic acid
**GABA** γ-aminobutyric acid
**GBR** Great Barrier Reef
**GBT** gas bubble trauma
**GLANSIS** Great Lakes Aquatic Nonindigenous Species Information System
**GLFC** Great Lakes Fishery Commission
**GnRH** gonadotropin-releasing hormone
**GOLT** Gill-oxygen limitation theory
**GPx** glutathione peroxidase
**GR** glutathione reductase
**GSH** oxidized glutathione
**GSSG** reduced glutathione

| | |
|---|---|
| **GST** | glutathione-*S*-transferase |
| **GTP** | guanosine triphosphate |
| **GWAS** | genome-wide association study |
| **$H^+$** | hydrogen |
| **$H^+$-ATPase** | proton ATPase pump |
| **$H_2O_2$** | hydrogen peroxide |
| **$H_2S$** | hydrogen sulfide |
| **HA** | humic acids |
| **Hb** | hemoglobin |
| **$HC_5$** | concentration at which 5% of the biological species are affected |
| **$HCO_3^-$** | bicarbonate |
| **Hg** | mercury |
| **HIF-1$\alpha$** | hypoxia inducing factor |
| **$HO^{\bullet}$** | hydroxyl radical |
| **HPA** | hypothalamic–pituitary–adrenal |
| **HPI axis** | hypothalamic–pituitary–interrenal axis |
| **HPLC-MS-MS** | high pressure liquid chromatography with tandem mass spectrometry |
| **HS** | humic substances |
| **HSI** | hepatosomatic index |
| **HSP** | heat shock protein |
| **Ht** | hematocrit |
| **HVA** | homeoviscous adaptation |
| **Hz** | hertz |
| **IBI** | index of biological integrity |
| **IBM** | individual based model |
| **IgM** | immunoglobulins |
| **ILOS** | incipient lethal oxygen saturation where LOE occurs |
| **INPA** | National Institute for Research of the Amazon |
| **IPCC** | Intergovernmental Panel on Climate Change |
| **IPLC** | Indigenous peoples and local communities |
| **IPP** | inositol pentaphosphate |
| **IUCN** | International Union for the Conservation of Nature |
| **$J_{Amm}$** | ammonia excretion rate |
| **$J_{diff}$** | diffusive flux of the bioreactive free metal ion from the bulk medium toward the biointerface |
| **$J_{in}$** | uptake rate |
| **$J_{max}$** | maximum uptake rate |
| **$J_{net}$** | net flux rate |
| **$J_{out}$** | efflux rate |
| **$J_u$** | biouptake flux |
| **K** | Fulton's condition factor |
| **$K^+$** | potassium ion |

| | |
|---|---|
| **$K_a$** | affinity constant |
| **$k_a$** | association rate constant |
| **$k_{ads}$** | adsorption rate constants |
| **$K_{app}$** | apparent stability constant |
| **KCl** | potassium chloride |
| **$k_d$** | dissociation rate constant |
| **$k_{des}$** | desorption rate constants |
| **KE** | key event |
| **KER** | key event relationships |
| **$K_m$** | Michaelis-Menten constant |
| **$K_{ow}$** | octanol-water partition coefficient |
| **KPa** | kilopascal |
| **$l$** | reaction layer thickness |
| **LC/MS** | liquid chromatography with mass spectrometry |
| **$LC_{50}$** | the exposure concentration lethal to 50% of the test population |
| **LC99.9** | concentration of a toxicant that causes 99.9 % mortality to a population of test organisms |
| **$LD_{50}$** | the internalized concentration or dose lethal to 50% of the test population |
| **LDH** | lactate dehydrogenase |
| **LGBTQIA+** | Lesbian, gay, bisexual, transgender, queer, intersex, asexual and other gender identities |
| **LOE** | loss of equilibrium |
| **LOEC** | lowest-observed-effect-concentration |
| **LPO** | Lipid Peroxidation |
| **LPS** | lipopolysaccharides |
| **LVB Ecoregion** | Lake Victoria Basin Ecoregion |
| **MCH** | mean corpuscular hemoglobin |
| **MCHC** | mean corpuscular hemoglobin concentration |
| **MDA** | malonic dialdehyde |
| **MFO** | mixed function oxygenase |
| **$mg\,L^{-1}$** | milligrams per liter |
| **$Mg^{2+}$** | magnesium ion |
| **MHW** | marine heatwave |
| **MIE** | molecular initiating event, the direct interaction of the chemical with a molecular target |
| **MLC** | minimum lethal concentration |
| **MMR** | maximal aerobic metabolic rate, metabolic rate during maximal sustainable aerobic exercise, maximum metabolic rate |
| **Mn** | manganese |
| **Mo** | molybdenum |

| | |
|---|---|
| **$\dot{M}_{O2}$** | oxygen consumption rate |
| **$\dot{M}O_{2max}$** | maximum oxygen uptake rate |
| **$\dot{M}O_{2rest}$** | resting oxygen uptake rate |
| **MOA** | mode-of-action |
| **MRC** | mitochondria rich cells |
| **mRNA** | messenger ribonucleic acid |
| **MT** | metallothionein |
| **MYA** | million years ago |
| **$Na^+$** | sodium ion |
| **$Na^+$,$K^+$, ATPase** | sodium potassium pump |
| **NaCl** | sodium chloride |
| **NBT** | nitro blue tetrazolium |
| **NCC** | $Na^+$:$Cl^-$ cotransporter |
| **NCE** | nonconsumptive effect |
| **Ne** | effective population size |
| **NGO** | nongovernmental organization |
| **$NH_4^+$** | ammonium |
| **Ni** | nickel |
| **NKA** | $Na^+$/$K^+$-ATPase |
| **NKCC1** | $Na^+$:$K^+$:2$Cl^-$ cotransporter |
| **$NO_2^-$** | nitrite |
| **NOAA** | National Oceanic and Atmospheric Administration |
| **NOEC** | no-observed-effect-concentration |
| **NOM** | natural organic matter |
| **NTU** | nephelometric turbidity units |
| **$O_2$** | oxygen |
| **$O_2^{\cdot-}$** | superoxide anion radical |
| **OCLTT** | oxygen- and capacity-limited thermal tolerance |
| **OECD** | Organisation for Economic Co-operation and Development |
| **OMZ** | oxygen minimum zone |
| **ORAC assay** | oxygen radical absorbance capacity assay |
| **PAHs** | polycyclic aromatic hydrocarbons |
| **Pb** | lead |
| **PBB** | polybrominated biphenyl |
| **PBDE** | polybrominated diphenyl ether |
| **PBI** | proton binding index |
| **PBT** | parentage based tagging |
| **PBTK** | physiologically based toxicokinetic model |
| **PCB** | polychlorinated biphenyl |
| **$pCO_2$** | partial pressure of carbon dioxide |
| **PCr** | phosphocreatine |
| **$P_{crit}$** | critical oxygen tension |

| | |
|---|---|
| **PDA** | public data archiving |
| **PFK** | phosphofructokinase |
| **pH** | negative log concentration of $H^+$ in an aqueous solution |
| **PIT** | passive integrated transponder |
| **PNEC** | predicted no effect concentrations |
| **$pO_2$** | partial pressure of oxygen |
| **$Po_{2crit}$, $P_{crit}$ or $O_{2crit}$** | critical minimum oxygen pressure where regulation of oxygen uptake is compromised |
| **POD** | pelagic organism decline |
| **PPM** | parts per million |
| **ppt** | parts per thousand |
| **PSU** | practical saline units |
| **$Q_{10}$** | temperature quotient |
| **QSAR** | quantitative structure–activity relationship |
| **QTL** | quantitative trait locus |
| **RBC** | red blood cell |
| **RCPs** | representative concentration pathways |
| **REACH** | Registration, Evaluation, Authorisation, and Restriction of Chemicals |
| **RIA** | radioimmunoassay |
| **RMR** | routine or resting metabolic rate |
| **RNA** | ribonucleic acid |
| **RNAi** | RNA interference |
| **RNA-Seq** | ribonucleic acid sequencing |
| **RNS** | reactive nitrogen species |
| **ROMS** | ROS-modified substances |
| **ROS** | reactive oxygen species |
| **RT-qPCR** | real-time quantitative polymerase chain reaction |
| **SAC** | Sportfishing Association of California |
| **$SAC_{340}$** | specific absorption coefficient at 340 nm |
| **SARA** | Species at Risk Act (Canada) |
| **SDA** | specific dynamic action |
| **SDM** | species distribution model |
| **SEP** | Salmon Enhancement Program |
| **SFE** | San Francisco Estuary |
| **SGC** | São Gabriel da Cachoeira |
| **SIMONI** | smart integrated monitoring strategy |
| **siRNAs** | short interfering, double-stranded RNA |
| **SMR** | standard metabolic rate |
| **Sn** | tin |
| **SNP** | single nucleotide polymorphism |
| **SOD** | superoxide dismutase |
| **SSD** | species sensitivity distribution |
| **s-SDH** | Serum-Sorbitol Dehydrogenase |

| | |
|---|---|
| **STEM** | Science, technology, engineering and mathematics |
| **SW** | seawater |
| **SWP** | State Water Project |
| **SWRCB** | State Water Resource Control Board |
| **T** | temperature |
| **TBA** | thiobarbituric acid |
| **TBARS** | TBA-reactive substances |
| $T_{crit}$ | critical temperature |
| $T_{crit,min}$ | lower critical thermal limit |
| $T_{critmax}$ | upper critical thermal limit |
| **TD** | toxicodynamics |
| **TDGS** | Total dissolved gas supersaturation |
| **TEK** | Traditional ecological knowledge |
| **TEP** | transepithelial potential |
| **TEPP** | tetraethyl pyrophosphate |
| **TFM** | 3-trifluoromethyl-4-nitrophenol |
| **TJ** | tight junctions |
| **TK** | toxicokinetics |
| **TKTD** | toxicokinetic-toxicodynamic model |
| $T_{opt}$ | thermal optima |
| $T_{optA}$ | optimal temperature for maximum aerobic scope |
| $T_{optAS}$ | optimal aerobic scope temperature window |
| **TSM** | thermal safety margin |
| **TSR** | temperature size rule |
| **TSS** | total suspended solids |
| **U/S ratio** | ratio of unsaturated to saturated fatty acids |
| $U_{crit}$ | critical swimming speed |
| **UDP-GT** | uridine diphosphate (UDP)-glucuronyl transferase |
| **USEPA** | United States Environmental Protection Agency |
| **USFWS** | United States Fish and Wildlife Service |
| **UV** | ultraviolet |
| **V-ATPase** | vacuolar type $H^+$-ATPase |
| **VEGFC** | vascular endothelial growth factor |
| **VTG** | vitellogenin |
| **WHAM** | Windermere humic aqueous model |
| **WWF** | World Wildlife Fund |
| **Zn** | zinc |

Chapter 1

# Using physiology to recover imperiled smelt species

**Yuzo R. Yanagitsuru[a], Brittany E. Davis[b], Melinda R. Baerwald[b], Ted R. Sommer[b], and Nann A. Fangue[a,*]**

[a]*Department of Wildlife, Fish, and Conservation Biology, University of California, Davis, CA, United States*

[b]*Division of Integrated Science & Engineering, California Department of Water Resources, West Sacramento, CA, United States*

[*]*Corresponding author: e-mail: nafangue@ucdavis.edu*

## Chapter Outline



Delta smelt is an endemic and critically endangered fish species inhabiting the San Francisco Estuary that has become entangled in the conflict between environmental and human needs in California, United States. Their inordinately sensitive nature has reputed them as an indicator species for the health of this estuary ecosystem, and many natural resource management actions center around delta smelt conservation. Decades of physiological studies defining the environmental tolerances of the species have

**Fish Physiology, Vol. 39B. https://doi.org/10.1016/bs.fp.2022.04.012**

informed many of these conservation actions and have also culminated in a genetically managed conservation hatchery to be used for supplementing wild populations. Despite substantial advancements in our knowledge of delta smelt, the wild population continues to decline, and the species is now threatened with extinction. It is thus timely to synthesize our current state of knowledge of delta smelt conservation physiology. In this chapter, we will review physiological studies on delta smelt, how they have informed conservation efforts, and highlight how physiological studies will continue to inform conservation and management actions.

## 1 Introduction

Delta smelt (*Hypomesus transpacificus*, family: Osmeridae) (Fig. 1) is a once abundant, but now critically endangered, forage fish species endemic to the San Francisco Estuary (SFE). Delta smelt have become central to ongoing natural resource conflicts between environmental and human needs in California, United States (Moyle et al., 2018). Their imperiled status has garnered substantial academic attention (Moyle et al., 2016) and physiological data on their environmental tolerances has played a significant role in shaping conservation actions in the SFE. However, despite decades of major conservation efforts to protect delta smelt, the population has continued to decline, and is now threatened with extinction (Fig. 2; Tempel et al., 2021; Hobbs et al., 2017; Moyle et al., 2018).

Delta smelt conservation is uniquely difficult due to their seemingly unavoidable direct competition with humans for water resources. Unlike other imperiled California species such as the salmonids that only utilize the SFE seasonally, delta smelt are year-round residents and their protection methods can limit the diversion of water resources from the SFE for human needs. As a result, delta smelt have gained a poor reputation with the urban and agricultural sectors and this negative public perception continues to detract from conservation efforts (Moyle et al., 2018). As the frequency and severity of droughts in California increase due to climate change, and water resources become ever scarcer, these conflicts will only heighten and make delta smelt conservation even more challenging ( Jeffries et al., 2016). As wild populations continue to decline even with management efforts, a glimmer of hope for the species lies in the conservation aquaculture of delta smelt at facilities like the UC Davis Fish Conservation and Culture Laboratory (FCCL). In this section, we provide a brief overview of the issues facing the SFE and the delta smelt. In Section 2, we highlight how physiological approaches have been used to understand the factors affecting the decline of delta smelt, and in Section 3, we discuss how conservation and management actions have been influenced by physiological studies. The delta smelt provides a particularly compelling case study for the importance of conservation physiology in the Anthropocene.

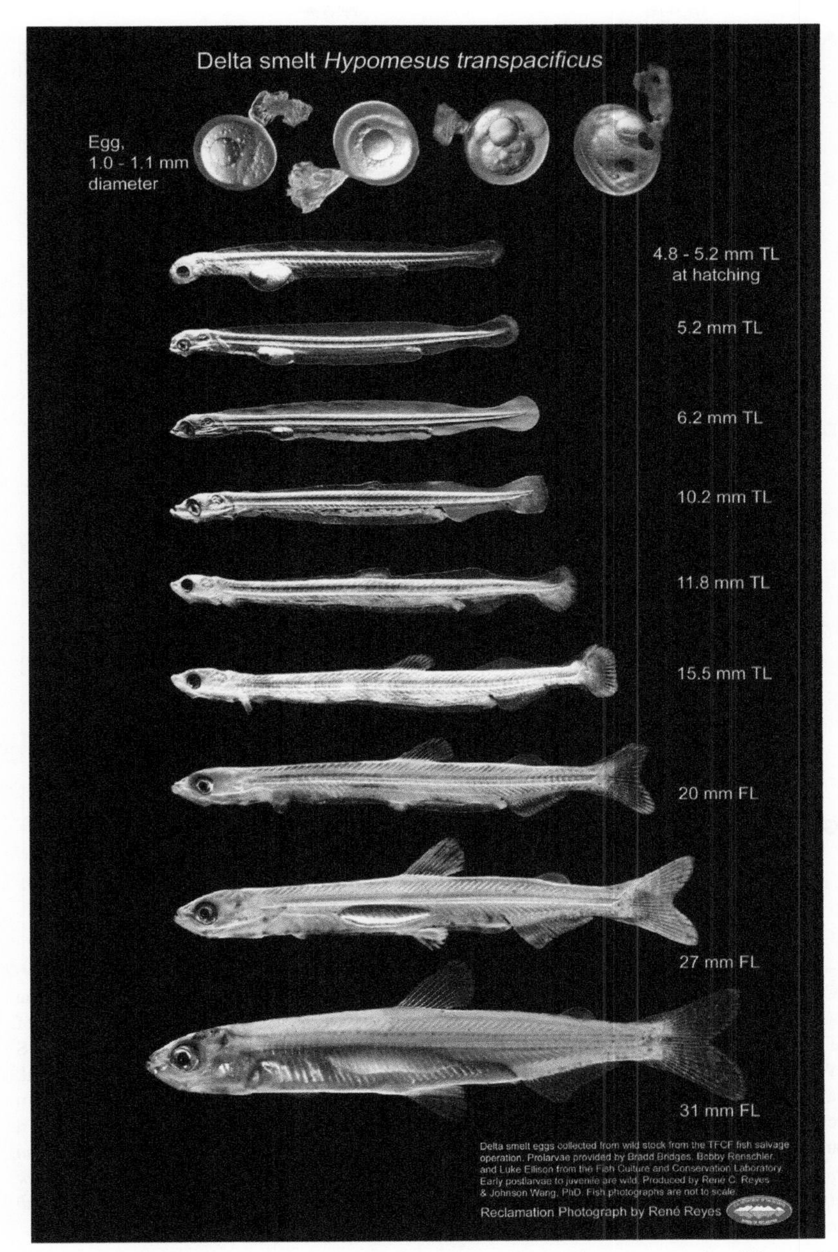


**FIG. 1** Delta smelt throughout development from egg to adult. *Photo courtesy: René Reyes, U.S. Bureau of Reclamation.*

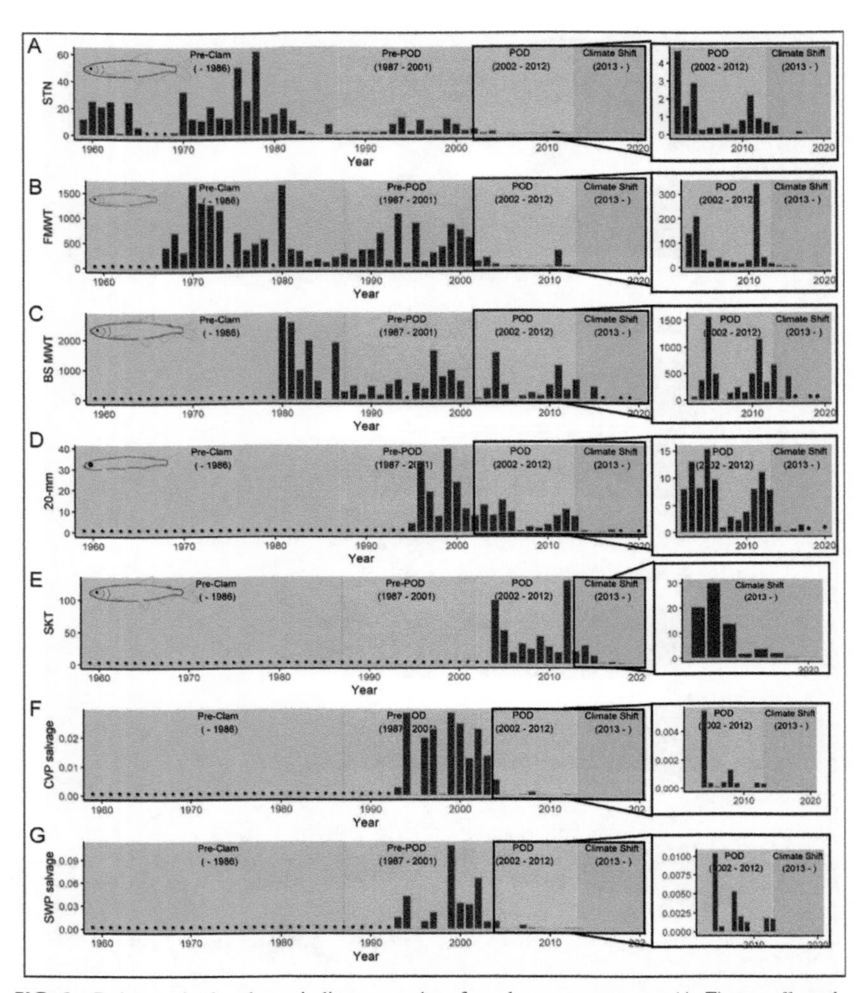


**FIG. 2** Delta smelt abundance indices over time from long term surveys (A–E) as well as the CVP and SWP pump salvages (F and G; calculated in catch per acre foot) show the decline of the delta smelt population. STN, summer townet survey; FMWT, fall midwater trawl; BS MWT, San Francisco Bay Study Midwater Trawl; 20-mm, 20-mm Survey; SKT, Spring Kodiak Trawl; Pre-Clam, before introduction of overbite clam; Pre-POD, before pelagic organism decline; POD, pelagic organism decline; climate shift, regime change marking a shift toward warmer temperatures after the historic California drought between 2012 and 2016 began. *From Tempel, T.L., Malinich, T.D., Burns, J., Barros, A., Burdi, C.E., Hobbs, J.A., 2021. The value of long-term monitoring of the San Francisco Estuary for Delta Smelt and Longfin Smelt. Calif. Fish Game 107, 148–171.*

## 1.1 San Francisco Estuary: History of human development and restructuring of delta smelt habitat

The SFE is the third largest estuary in the United States and the largest on the Pacific Coast of the Americas. Broadly, the SFE is divided into four segments: San Francisco Bay, San Pablo Bay, Suisun Bay, and the Sacramento-San Joaquin Delta (Delta from hereon) (Fig. 3). With its diverse suite of habitats encompassing marine bays, brackish marshes, and tidal freshwater wetlands, the SFE supports over 750 species of plants and animals including many endemic and endangered species (Healey et al., 2016). Of particular biological and political importance is the Delta, formed by the confluence of the Sacramento and San Joaquin Rivers, which enter Suisun Bay and ultimately drains outward to the San Pablo and San Francisco Bays. In addition to providing

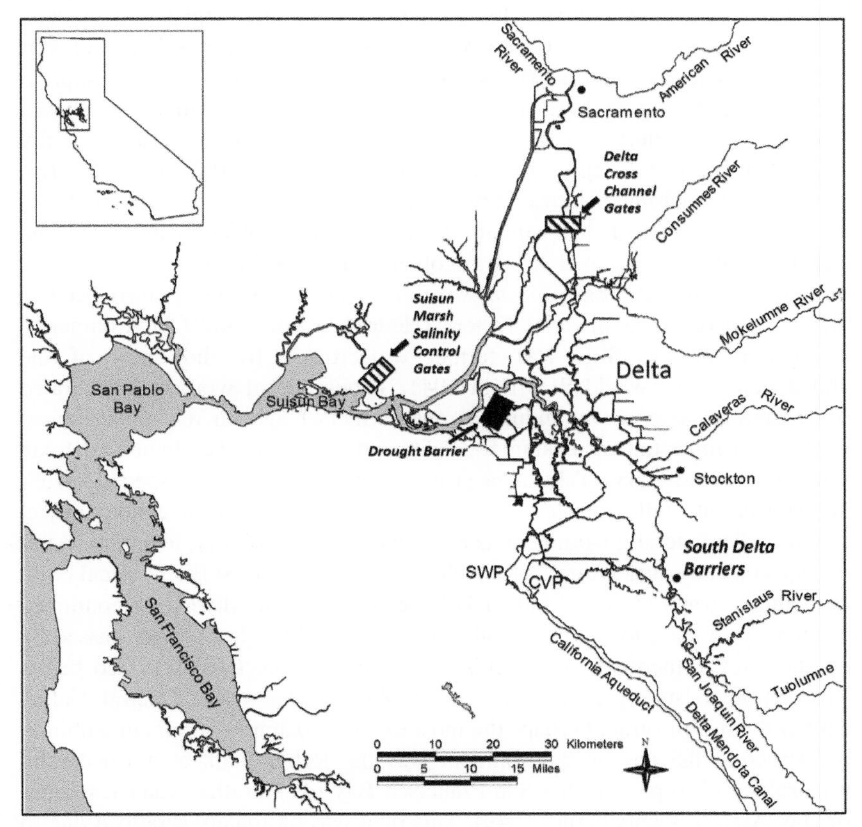


**FIG. 3** Map of the San Francisco Estuary including some of the major water management features. *From Sommer, T., 2020. How to respond? An introduction to current bay-delta natural resources management options. San Franc. Estuary Watershed Sci. 18, 1, reprinted with permission.*

year-round nursery and feeding habitats for fish and wildlife, the Delta supplies water to over 25 million California residents and a multi-billion dollar agriculture industry (Brown et al., 2013). As a result, the distribution of freshwater from the Delta to meet environmental and human demands is an ongoing natural resource conflict in California.

Many of the ecological issues surrounding the SFE today stem from the history of human development, which is marked by more than a century of diking and draining tributaries by local landowners and ambitious civil works projects that exploited and restructured the natural environment for human use (Nichols et al., 1986; Takekawa et al., 2006). Extensive human development around the SFE began in 1848 when hundreds of thousands of people, attracted by the idea of striking it rich, immigrated to the area during the California gold rush. It was during this time that prospectors discovered the agricultural value of the Central Valley region with its fertile soils and plentiful water (Dilsaver, 1985). This discovery stimulated legislation (federal Arkansas Act of 1850, state Green Act of 1850) to convert grassland and wetland habitats to agricultural fields and to develop infrastructure to support agriculture in the region. Large networks of levees, large rim dams (e.g., Shasta, Oroville, Trinity, etc.), and water diversions were constructed across virtually every tributary in the region during this time, blocking 90% of historic fish spawning and rearing habitats (Yoshiyama et al., 2001) and reducing flows and sediment transport downstream. As the human population increased and agricultural lands developed, urban runoff increased, polluting the local waters. Furthermore, striped bass (*Morone saxatilis*) were intentionally introduced in 1879 to support a commercial fishery (Hedrick et al., 1987). Environmental degradation of the SFE was further exacerbated by the entry of the United States into World War II in 1941. Thousands of workers immigrated to the area to support war industries and farmers began to rely on chemical practices to meet the increasing food demand, which further increased toxic runoff into the estuary (Takekawa et al., 2006). The war also sped up work on the Central Valley Project (CVP), a large-scale water diversion project passed by the federal government as a national defense priority, to supply water to Central Valley agriculture, but which greatly altered the SFE's physical environment and water regime (Madera Tribune, 1941). Water diversions continued to intensify in the 1960s when another large water diversion project passed by the state government, the California State Water Project (SWP), also began construction to supply agriculture and drinking water to the Central Valley and Southern California. Perhaps the most extreme example of human ambition to restructure the natural environment was the Reber Plan, an unsuccessful proposal to fill in parts of the San Francisco Bay to centralize water resources for Bay Area residents (Cole, 1949). This plan was ultimately scrapped due to technological infeasibilities in the early 1960s, marking the end of an era of grandiose civil works projects in the region (Jackson and Paterson, 1977).

The inordinate scale by which the natural environment was exploited has left the SFE with only 3% of its historical (pre-1850) tidal wetland habitats (Whipple et al., 2012). It has also developed the region into the nexus of California's water distribution network, which is relied upon by over 70% of California residents and the agriculture industry responsible for producing over half of the fruits, nuts, and vegetables grown in the United States (California Department of Food and Agriculture, 2021). Even with environmental regulations in place, this enormous human demand for water consumes between 35% and 65% of the annual freshwater outflow from the Delta, depending on time of year (Sommer et al., 2011). In addition to reduced freshwater input, invasive species such as the overbite clam (*Potamocorbula amurensis*), changes to the food web, and contaminants have also upended the biological system of the SFE (Feyrer et al., 2003; Fong et al., 2016; MacNally et al., 2010; Fig. 2). The culmination of this cascade of changes has been the decline of an entire assemblage of fish species in the SFE beginning in the 1980s in a phenomenon known as the Pelagic Organism Decline (POD) (Fig. 2; Sommer et al., 2007). Of the several SFE fish species that are currently in jeopardy (i.e., longfin smelt (*Spirinchus thaleichthys*), winter and spring-run Chinook salmon (*Oncorhynchus tshawytscha*), and green sturgeon (*Acipenser medirostris*), among others), the delta smelt has experienced a particularly catastrophic decline and has thus gained special attention.

## 1.2 Delta smelt

The delta smelt is a small, pelagic, semi-anadromous, forage fish species endemic to the upper SFE that were historically abundant throughout the entire Delta and to the Western extent of San Pablo Bay (Erkkila et al., 1950; Ganssle, 1966; Moyle et al., 1992; Radtke, 1966; Sweetnam et al., 1993; Wang, 1991). While some two-year old fish have been observed in the wild (Moyle, 2002), most delta smelt have an annual life cycle (Lewis et al., 2021). This, paired with their relatively low fecundity (1200–2600 embryos/female), makes them particularly susceptible to rapid population decline as a single poor year could devastate recruitment for the following year (Moyle, 2002). Historically, these declines were temporary, and the population rebounded, but since the POD in the 1980s, their abundance has continually decreased (Sommer et al., 2007). As a result, they were listed as threatened under both the California Endangered Species Act (CESA) and Federal Endangered Species Act (FESA) in 1993 (U.S. Office of the Federal Register, 1993). As their population continued to plummet, they were up-listed as endangered under the CESA in 2009 (California Fish and Game Commission, 2009). However, for the FESA, they have remained listed only as threatened; since 2010, they have failed to be up-listed and instead have been considered warranted but precluded for endangered status (U.S. Office of the Federal Register, 2010, 2020). Finally, in 2014, they were classified as critically endangered by the International Union for Conservation in Nature Red List of Threatened Species (International Union for Conservation of Nature, 2014). In 2015, the California Department of Fish and Wildlife's (CDFW) annual Summer Townet Survey recorded their first zero delta smelt abundance index (a comparative measure of relative abundance) since monitoring began in 1959. In response, the U.S. Fish and Wildlife Service's (USFWS) Enhanced Delta Smelt Monitoring Program (EDSM) was established in 2016 to monitor all life stages of delta smelt with higher spatiotemporal resolution than CDFW's historical surveys. The delta smelt's population collapse has continued to be documented in CDFW's historical surveys as well as with the newer EDSM surveys. Zero indices were recorded between 2018 and 2021 for CDFW's Summer Townet Survey, Fall Midwater Trawl, and 20mm survey (California Department of Fish and Wildlife, 2021). Similarly, EDSM surveys showed a steep decline in delta smelt abundance from 365 individuals caught in 2017 to 22 individuals caught in 2020 (U.S. Fish and Wildlife Service et al., 2022). There is no question that delta smelt are now ecologically extinct and if current trends continue, they are projected to become extinct in the wild within the next decade (Hobbs et al., 2017; Moyle et al., 2018).

Delta smelt are highly sensitive and serve as an indicator of the health of the SFE ecosystem. As such, delta smelt have become a politically important species with studies on their physiology and ecology influencing environmental water management policies (California Natural Resources Agency, 2005, 2016; U.S. Bureau of Reclamation, 2019). Over the last three decades, these studies have implicated a multitude of factors stemming from human disruption in the decline of delta smelt such as: habitat loss, water diversions/exports, pollution, invasive species, reduction of turbidity, limited food sources, climate change, and modern droughts (Fig. 2; Hobbs et al., 2017; Moyle et al., 2016, 2018; Sommer et al., 2007). Furthermore, these studies have resulted in the establishment of a genetically managed captive refuge population (Baskerville-Bridges et al., 2005b; Lindberg et al., 2013), informed tidal habitat restoration efforts (Sherman et al., 2017), and modified water management (e.g., salinity-field management; Sommer, 2020). While there are still knowledge gaps for the best practices for the conservation of delta smelt, the work conducted thus far has equipped resource managers with a suite of tools that can be applied in the recovery of delta smelt.

## 2 Using physiology to understand the factors affecting the decline of delta smelt

Most early studies on delta smelt were field studies that correlated environmental conditions to their abundance (Moyle et al., 1992; Nobriga et al., 2000; Stevens and Miller, 1983). While these studies were critical for generating hypotheses for which factors may be affecting delta smelt in the wild, they were less useful for disentangling the extent to which each factor affected delta smelt abundance. Laboratory based studies filled, and continue to fill, this knowledge gap and have been crucial for revealing the mechanisms that are contributing to delta smelt decline. Here, we focus on mechanistic studies of key abiotic environmental factors crucial for informing conservation efforts for delta smelt: temperature, salinity, turbidity, and anthropogenic contaminants.

## 2.1 Temperature

Increasing temperatures due to climate change, water diversions, droughts, or anomalous warming events pose a significant threat to delta smelt. Initial studies focused on identifying the maximum temperature that wild adult delta smelt can tolerate using critical thermal methodology, the results of which are used for informing water management in the Delta today. Adult delta smelt have mean upper critical thermal tolerance values ($CT_{Max}$) between 24 °C and 28 °C depending on acclimation temperature and life stage (Davis et al., 2019; Jeffries et al., 2016; Komoroske et al., 2014; Swanson et al., 2000), corroborating long term field data that showed decreased delta smelt abundance at temperatures above 24 °C (Nobriga et al., 2008). These results revealed that delta smelt had lower upper thermal tolerances compared to other fish species in the Delta (i.e., Sacramento splittail (*Pogonichthys macrolepidotus*), green sturgeon, and some populations of Chinook salmon), and that they are living close to their thermal maximum (Mayfield and Cech, 2004; Sardella et al., 2008; Young and Cech, 1996; Zillig et al., 2020). In fact, temperatures in the Delta today can exceed delta smelt $CT_{Max}$ (up to 29 °C) in some locations during the summer (Jeffries et al., 2016). Historically, the number of days within a year when maximum surface water temperatures exceeded delta smelt $CT_{Max}$ in the Delta has been relatively low (between 2 and 25 days annually between 1969 and 2008) (Brown et al., 2013). However, with climate change, the number of thermally unsuitable days for delta smelt is projected to approximately double by 2050, and the range of delta smelt habitat is thus expected to decrease (Brown et al., 2013).

While extreme high temperature events can be acutely lethal for delta smelt, the effects of sustained or acute exposures to sublethal high temperatures may also contribute to their population decline. Transcriptomic responses of delta smelt acutely exposed to temperatures 4–6 °C below their $CT_{Max}$ (between 22 °C and 24 °C) suggested that delta smelt likely experience physiological stress at temperatures well below their $CT_{Max}$ at all life stages (Komoroske et al., 2015). Their reduced capacity to modify gene expression for molecular chaperones when exposed to these sublethal high temperatures even after acclimation for 3 weeks to different temperatures suggest that delta smelt do not have the molecular machinery to cope with periods of thermal stress. Additionally, sublethal high temperature results in few changes in gene expression for metabolic processes, indicating that they cannot modify their metabolic physiology to match the energetic demand necessary for tolerating stressful high temperature environments (Jeffries et al., 2016; Komoroske et al., 2015). Furthermore, while adult delta smelt (200–250 days post-hatch [dph]) can tolerate 19 °C, individuals acclimated to this temperature for three weeks had increased expression of genes associated with macromolecular damage but did not have heightened thermal tolerance compared to those acclimated to 16 °C, suggesting that delta smelt lack the ability to truly acclimate to temperatures as low as 19 °C and that this temperature can impart negative effects on organismal performance (Komoroske et al., 2014, 2015).

The inability of delta smelt to fully acclimate to higher temperatures is further supported by their swimming performance and behavior. Critical swimming speed of wild adult delta smelt is not influenced by acclimation temperatures as commonly observed in other fish species (Swanson et al., 1998). Despite this, sub-adult (160–200 dph) delta smelt chronically exposed to 21 °C have routine swimming velocities 50% higher than fish at 17 °C, indicating that routine activity at higher temperatures account for a higher percentage of their maximal swimming capacity and thus delta smelt at higher temperatures incur a higher locomotor energetic cost. This is further pronounced when fish at 21 °C are exposed to conspecific alarm or predator cues, which induce even higher swimming velocities averaging 85% of their critical swimming speeds that do not return to baseline levels within 20 min as observed in lower temperatures (Davis et al., 2019). Considering that Delta water temperatures can remain above 20 °C for extended durations during the summer, temperature may be contributing to delta smelt habitat compression and the species may already be living under chronic thermal stress.

In addition to the direct consequences that high temperatures have on delta smelt, warming waters can also generate favorable conditions for invasive species that compete with or predate upon delta smelt, or engineer ecosystems into habitats unsuitable for the species. Wakasagi (*Hypomesus nipponensis*) and Mississippi silverside (*Menidia audens*, synonym of *Menidia beryllina*) are small, non-native forage fish species that occupy a similar ecological niche to delta smelt. Like delta smelt, all three species feed primarily on zooplankton and thus directly compete for food resources. Unlike delta smelt, they have a comparatively high thermal tolerance and flexible transcriptional responses that allow them to acclimate to high temperatures, which may confer a competitive advantage over delta smelt in high temperature conditions (Davis et al., 2019; Komoroske et al., 2020; Swanson et al., 2000). In addition to competition for food resources, gut content analyses indicate that Mississippi silversides also consume larval delta smelt in offshore habitats. Although a relatively small proportion of Mississippi silversides were detected with delta smelt DNA within their guts, the highly abundant and pervasive nature of the species potentially adds a considerable predation pressure on delta smelt (Baerwald et al., 2012; Schreier et al., 2016).

Higher temperatures can also increase the growth rate and spread of the invasive plant, Brazilian waterweed (*Egeria densa*). This spread simultaneously reduces the pelagic habitats preferred by delta smelt and increases predation risk by introduced predatory species like largemouth bass (*Micropterus salmoides*) that use vegetated areas as cover to ambush pelagic open-water prey (Conrad et al., 2016; Ferrari et al., 2014). While only largemouth bass were experimentally found to increase predation on delta smelt in vegetated areas, it is conceivable that predation risk by other species documented with delta smelt in their guts (Schreier et al., 2016) that prefer vegetated areas, such as bluegill (*Lepomis macrochirus*), could also increase (Conrad et al., 2016; Ferrari et al., 2014). Coupled with the increased risk of predation due to reduced group cohesiveness of delta smelt in warmer waters, higher temperatures could significantly increase predation pressure on the species (Davis et al., 2019).

## 2.2 Salinity

The risk of increased salinity within the Delta posed by climate change, drought, anomalous warming events, and water diversions is a major concern for conservation managers. Controlled freshwater releases by conservation managers have maintained relatively stable salinity conditions in the SFE but as sea level rises and more saltwater pushes inland, the demand for limited freshwater resources to control salinity may exceed what is available. Long-term field studies have demonstrated the importance of controlling salinity for delta smelt as delta smelt abundance drops when salinities increase beyond 3 ppt. However, laboratory studies found that delta smelt appear to be fully capable of acclimating to a wide range of salinities by restructuring their gills (Kammerer et al., 2016) and employing a suite of molecular mechanisms that are commonly used by other euryhaline fish (Komoroske et al., 2016). Delta smelt euryhaline physiology is reflected in their ability to quickly regulate their internal osmolality to baseline levels within 6 h during an acute exposure to 18.5 ppt water. While internal osmolality did not return to baseline levels within the 14 day experimental period after acute seawater (34 ppt) exposure, internal osmolality had a continual decrease toward baseline levels during this experimental period, indicating that fish were headed toward recovery (Komoroske et al., 2016). This translates to a high tolerance to a wide range of salinities. For example, even the embryonic life stage, which is the most sensitive to salinity, experiences the negative effects of reduced hatching success and yolk fraction (i.e. yolk volume normalized to egg volume) only when exposed to salinities higher than 10 ppt (Romney et al., 2019). Mid-larval delta smelt (30–32 dph) experience no mortality when exposed to 18.5 ppt water, the highest recorded salinity in the Delta, for at least 14 days and nearly all (99% survival) juveniles (140–164 dph) and adults (200–250 dph) can survive seawater exposure for at least 21 days when salinity is raised gradually (Komoroske et al., 2014, 2016). One potential consequence of inhabiting high salinity environments is the energetically expensive mechanisms (e.g., energetic cost of gill $Na^{+}/K^{+}$ ATPase activity and protein synthesis; Milligan and McBride, 1985; Pannevis and Houlihan, 1992) involved with acclimating to high salinity environments. However, there is limited evidence to suggest that delta smelt experience an energy deficit in higher salinities; oxygen consumption rates do not differ for adult delta smelt acclimated to salinities between freshwater and 12ppt (Hammock et al., 2017), which suggests that energetic costs are not prohibitive for delta smelt inhabiting salinities typical of the Delta. Overall, it is unlikely that salinity directly impacts the distribution of delta smelt.

One potential mechanism for the discrepancy between field and laboratory salinity tolerance is variation in food availability. The low salinity regions of the Delta typically have higher zooplankton abundance than higher salinity areas, and delta smelt captured in these regions have high stomach fullness and body conditions (Hammock et al., 2017). For example, *Eurytemora affinis*, one of the calanoid copepods that constitutes most of the delta smelt diet (now replaced by the invasive *Pseudodiaptomus forbesi*) (Moyle et al., 2016) has reduced survival, and feeding performance in higher salinities when its food is limited (Hammock et al., 2016). Additionally, the non-native and filter feeding overbite clam increases feeding rates at higher salinities likely to compensate for the higher metabolic costs of inhabiting these areas, which may exacerbate the decrease in zooplankton abundance at higher salinities (Paganini et al., 2010). While physiological studies of salinity showed limited evidence of a direct negative impact, delta smelt abundance may be affected by indirect mechanisms such as alterations in food availability.

## 2.3 Turbidity

Delta smelt abundance in the wild has a curvilinear relationship with water clarity, a multifactorial measure affected by turbidity (water cloudiness) and light intensity; delta smelt abundance increases as water clarity lowers up to a point (i.e., optimal turbidity), thereafter abundance decreases as water clarity continues to lower (Hamilton and Murphy, 2020). Of the two aspects of water clarity, there has been a larger emphasis on understanding the effects of turbidity on delta smelt because areas of higher turbidity are associated with higher food availability, which provides a potential mechanism behind the correlation between delta smelt abundance and water clarity (Jassby and Cloern, 2000; Kimmerer, 2002; Sobczak et al., 2002), and because turbidity has decreased in the SFE following a 36% decrease in suspended solid concentration between 1998 and 1999 (Schoellhamer, 2011). Furthermore, infrastructure exists that allows managers to alter turbidity within the Delta by increasing mixing via water releases, making turbidity a feasible environmental condition to control.

Turbidity has been demonstrated to play an important role in improving feeding performance for delta smelt larvae. Laboratory studies have shown that elevated turbidities are critical for the feeding success and thus survival of first-feeding larvae and improve feeding performance for larvae up to 60 dph (Baskerville-Bridges and Lindberg, 2004; Hasenbein et al., 2013; Tigan et al., 2020). This is consistent with larvae of several other fish species, which also require turbid waters to feed (Faulk and Holt, 2005; Naas and Harboe, 1992; Stuart and Drawbridge, 2011). Although this has not been empirically demonstrated for delta smelt, the mechanism by which turbidity improves larval feeding is largely accepted to be due to the improved visual contrast of prey against the water column (Baskerville-Bridges and Lindberg, 2004; Baskerville-Bridges et al., 2005b). Contrast is particularly important for feeding upon largely translucent prey such as rotifers, which are provided as feed for larval delta smelt at the FCCL, or natural prey like the calanoid copepod: *Eurytemora affinis*. An alternative hypothesis is that higher turbidities scatter light more evenly across the water column, promoting a more uniform dispersion of phototactic smelt larvae, which then reduces competition for food between individuals (Rieger and Summerfelt, 1997). Though this hypothesis has not been directly tested, the finding that light intensity only affects feeding performance for delta smelt when turbidity was $\geq$9 NTU (nephelometric turbidity units) (Baskerville-Bridges and Lindberg, 2004) provides some support for it. The positive relationship between higher turbidities and feeding performance appears to weaken and potentially reverse as delta smelt age; juvenile delta smelt (120 dph) have a weak but negative correlation with feeding performance and turbidities between 0 and 250 NTU (Hasenbein et al., 2013). This is consistent with successful culture practices at the FCCL where juveniles are reared at lower turbidities than larvae. Finally, while the effects of elevated turbidity on adult delta smelt have not been investigated mechanistically, the success of the FCCL in rearing adults in clear water demonstrates that adult feeding success is not contingent upon turbid waters as it is for larvae (Baskerville-Bridges et al., 2005b).

In addition to improvements to feeding performance, turbidity has been hypothesized to improve physiological performance for delta smelt by reducing stress. However, there is currently limited evidence to support this. Larvae (60 dph) showed a trend whereby there was decreased cortisol gene expression when held in 25 and 80 NTU water for 24 h but these results were not statistically significant (Hasenbein et al., 2016). Similarly, juvenile (120 dph) delta smelt showed no differences in whole-body cortisol levels between 0 and 250 NTU turbidities when exposed for 2 h (Hasenbein et al., 2013). Though these studies showed no significant differences, the relatively short exposure periods and the highly sensitive nature of delta smelt may have blurred any turbidity effect on the stress response.

There is more support for turbidity benefiting delta smelt by reducing predation risk. As an open-water species with relatively weak swimming performance, delta smelt rely on turbid waters for cover from predators. Laboratory studies showed that delta smelt are less likely to be predated upon by largemouth bass in turbid waters (Ferrari et al., 2014). It is conceivable that the increased cover from higher turbidity water may provide protection from other visual predators in the SFE such as striped bass as well. Similarly, field studies indicate that delta smelt experience lower predation pressure from Mississippi silverside in turbid conditions despite there being a likely high overlap between the species in waters of higher turbidity (Mahardja et al., 2016; Schreier et al., 2016). Finally, higher turbidities may also indirectly decrease predation pressure from largemouth bass and other predators that prefer vegetated areas by reducing the spread of Brazilian waterweed (Ferrari et al., 2014).

## 2.4 Anthropogenic contaminants

A suite of industrial, agricultural, and urban pollutants regularly drains into the Delta with particularly pronounced pulses of contaminants added to the ecosystem during heavy rains and storms (Weston et al., 2014, 2015a, 2019). As a result, the SFE has been identified as impaired for aquatic life by several specific contaminants on the United States Environmental Protection Agency (EPA) 2010 List of Impaired Water Bodies (U.S. Environmental Protection Agency, 2011). Histopathological studies of field-caught juvenile delta smelt show that these contaminants are already affecting wild delta smelt and are considered an important stressor in the Delta (Hammock et al., 2015). Studies on contaminant toxicity have used model organisms such as zebrafish (*Danio rerio*), fathead minnows (*Pimephales promelas*), and Mississippi silversides to assess both acute and chronic toxicity of these contaminants to determine the potential effects that organisms in the Delta may be experiencing (Beggel et al., 2011; Cole et al., 2016; Frank et al., 2018, 2019). While useful for understanding the mechanisms by which these contaminants act on organisms, they are less useful for determining the lethal and sublethal concentrations for delta smelt; delta smelt generally experience sublethal and lethal effects at contaminant concentrations lower than those experienced by these model species (Jeffries et al., 2015).

Pyrethroids and organophosphates are some of the most common pesticides used in agricultural fields surrounding the Delta. Both are neurotoxic pesticides, but their modes of action differ. While the primary mode of action of pyrethroids is to block the closure of voltage gated ion channels and prevent the repolarization of neurons (Ray and Fry, 2006), organophosphates inhibit acetylcholinesterase, ultimately causing neurons to continually stimulate (Hsieh et al., 2001). Additionally, exposure to pyrethroid pesticides can result in endocrine disruption (Brander et al., 2013, 2016a,b) with long-term population consequences (White et al., 2017). Laboratory studies assessing the toxicity of these pesticides on delta smelt have focused on larvae, which are considered the most sensitive life stage of fishes due to it being a critical stage of development (Hjort, 1914; Sifa and Mathias, 1987). Additionally, the lipophilic nature of these pesticides results in higher uptake in larvae via their yolk-sac, further increasing their vulnerability to exposure (Mundy et al., 2020, 2021). The commonly used pyrethroid pesticides: permethrin, bifenthrin, and esfenvalerate all affect the expression of genes related to neurological function and development in delta smelt at concentrations as low as 1.4, 0.002, and 0.06 $\mu g L^{-1}$, respectively (Connon et al., 2009; Jeffries et al., 2015; Mundy et al., 2020). Correlating with the neurological effects, permethrin, bifenthrin, esfenvalerate, and the organophosphate, chlorpyrifos, cause behavioral changes that may make larvae more susceptible to predation such as hyperactivity and freezing behaviors at concentrations as low as 0.05, 0.002, 0.06, and 0.05 $\mu g L^{-1}$, respectively (Connon et al., 2009; Mundy et al., 2020, 2021; Segarra et al., 2021). Considering that all four of these chemicals have been measured in the Delta at concentrations equal to or exceeding these sublethal concentrations, and that hundreds of additional compounds have been detected in individual Delta water samples (Moschet et al., 2017), it is likely that contaminant mixture exposure on the early life stages of delta smelt are directly affecting their survival and recruitment into the population (Bacey et al., 2005; Connon et al., 2019; Fong et al., 2016; Weston et al., 2014, 2015b, 2019).

Herbicides are some of the most frequently detected contaminants in aquatic ecosystems, but they have historically been considered of low concern for animals because they target biochemical pathways absent in animals. This belief is reflected in the widespread use of herbicides in the Delta such as: penoxsulam, imazamox, fluridone, and glyphosate to control invasive weeds such as Brazilian waterweed to aid delta smelt habitat recovery. However, penoxsulam, imazamox, and fluridone act similarly to chlorpyrifos by inhibiting acetylcholinerase. In adult delta smelt exposed to these herbicides for 6h, there were no alterations in swimming behavior though the potential for altered behavior with longer exposures still exists and the effects of these chemicals on larvae are unknown (Jin et al., 2018). Fluridone and glyphosate are estrogen mimics, a class of contaminants that are known to adversely affect reproductive performance. Endocrine disruption by estrogen mimics is measured with changes in concentration of the biomarker: 17β-estradiol (E2). Fluridone raises the concentration of the female hormone, E2, in both male and female delta smelt at concentrations as low as 0.21 μM and glyphosate raises the concentration of E2 in males at concentrations as low as 0.46 μM (Jin et al., 2018). Impairments to reproductive performance by these herbicides have not been assessed in delta smelt. However, the altered E2 concentrations are alarming because endocrine disruption has reduced sperm concentrations in other fish species (Aravindakshan et al., 2004; Wang et al., 2019), which could affect fertilization rates and thus recruitment of delta smelt.

## 2.5 Synthesis

The diversity of stressors and their direct and indirect effects on delta smelt make protecting the species a gargantuan task. We reviewed the current state of knowledge on just four abiotic factors that have been extensively studied for delta smelt. However, there are certainly other environmental disturbances that are affecting delta smelt but that we have limited information for. For example, dissolved oxygen is a major environmental factor that affects the distribution of many aquatic organisms that has not been studied for delta smelt and is assumed to be a non-issue as most regions of the SFE do not experience low dissolved oxygen levels (Feyrer et al., 2007). Furthermore, there are likely interactive effects of consequence that only add to the complexities for conservation. Relatively few studies have addressed how multiple stressors can affect delta smelt thus far (Davis et al., 2019; Hasenbein et al., 2013, 2016; Segarra et al., 2021), which limits our ability to predict how delta smelt will respond to management strategies that often focus on single stressors. Conducting the studies necessary to fully describe the interactive effects of every stressor delta smelt face is impractical due to the sheer diversity of stressors present in the system. However, this should not discourage further studies evaluating the impacts of the interactions of select stressors. Particularly in the case of contaminants, some of which affect organisms differently under different temperatures and salinities, unveiling the negative impacts that a contaminant has on delta smelt can influence policies surrounding their use.

## 3 Conservation efforts and management actions influenced by physiological studies

Section 2 illustrated how physiological studies have improved our understanding of the fundamental biology and life cycle of delta smelt and some of the key factors influencing their decline. Below, we describe some of the major management applications of this information. We focus on the development of culture methods, but also include examples of how the information has been used for other management applications of natural resources such as regulations (e.g., water rights, species-listings and take permits), water infrastructure (e.g., reservoir releases, gates, barriers), habitat conservation (e.g., tidal wetland restoration, sediment supplementation, aquatic weed removal), and biological management (e.g., supplementation; Sommer, 2020).

### 3.1 Development and optimization of a captive culture for delta smelt

Research began in 1992 to develop methods for the captive culture of delta smelt, and the UC Davis affiliated FCCL was formally established in 1996 to maintain a captive delta smelt population (Baskerville-Bridges et al., 2005b). The ever-decreasing abundance of source fish from the wild, near-annual life cycle, low fecundity, high sensitivity to handling stress, and extended larval period made these fish particularly difficult to culture. As a result, it took many years to close the life cycle of delta smelt and develop a culture protocol robust enough to produce adequate numbers of study specimens for laboratory studies (Baskerville-Bridges et al., 2005a,b). In 2008, as the wild population continued to decline, a genetic management plan for the captive delta smelt population was implemented to maintain a refuge population to be used for supplementation efforts (Fisch et al., 2013; Lindberg et al., 2013). A small backup population of the spawning cohort was also established in 2008 at USFWS Livingston Stone National Fish Hatchery should an unpredicted catastrophe happen to the FCCL population. FCCL is currently the only source of delta smelt for researchers and conservation managers. However, since 2020, managers have been working together to expand the Livingston Stone population to become a second genetic refuge population.

Extreme sensitivity by wild broodfish to handling was the first major bottleneck for the development of the delta smelt captive culture. Early attempts in 1993 to transport wild broodstock in rectangular coolers resulted in less than 10% survival within 72h post-collection. While transportation methods quickly improved to yield approximately 75% survival with the use of cylindrical polyethylene bags, supplementation with pure oxygen, salt, and the water conditioning polymer, NovAqua (Kordon Co. Hayward, California) (Swanson et al., 1996), it took another decade for collection methods to be optimized. Today, broodstock are transported with 120-L black carboys filled to the brim to avoid air pockets, pads underneath carboys to dampen vibrations, and stocking densities of no more than 4 fish $L^{-1}$. Broodstock collections now incorporate water temperature (<12°C), depth (7.5–9m), and time of day (morning) for collections as well to capture the healthiest fish to ensure higher survival and quality of wild broodstock (Baskerville-Bridges et al., 2005b). Although collection and transportation protocols have improved for delta smelt, the inherent sensitivity of the species to handling continues to be a challenge as exemplified by the decreased survival of fish when handled by inexperienced staff (Baskerville-Bridges et al., 2005b). This is an important factor to consider for delta smelt hatchery operations as the seasonal hiring that is often practiced at hatcheries for other fish species such as salmon is impractical for delta smelt.

By optimizing maintenance of broodstock, the FCCL was able to improve the quality and quantity of eggs and maximize production from the limited broodstock (see Chapter 6, Volume 39A: Bernier and Alderman, 2022). At FCCL, delta smelt were found to have improved health in dark tanks and to have faster gonadal development and improved egg quality when housed outdoors under natural lighting compared to indoors under fluorescent lighting. Delta smelt broodstock are now housed in black tanks outdoors and only brought indoors to manipulate spawning timing via alterations in photoperiod. Wild broodstock transitioned well to dry feed (95–99% transition) and an optimal feeding protocol dependent on season and size of fish was developed. Finally, while broodstock spawned volitionally within tanks, egg production was infrequent, fertilization and hatching rates were low, and the adhesive eggs were often covered in food and feces upon collection. Considerable effort was thus placed on developing strip-spawning methods for the tiny and sensitive delta smelt, which confers the advantage of controlled spawn timing, higher fertilization and hatching success, and cleaner embryos. FCCL has since transitioned to spawning broodstock exclusively via stripping and is capable of spawning almost year-round, albeit with lower spawning efficiency during the summer (Baskerville-Bridges et al., 2005b).

Similar to many other cultured fish species, optimizing larviculture proved to be the most challenging and demanding aspect of developing the delta smelt captive culture. Larvae are highly sensitive to abiotic conditions and indoor recirculating systems with precisely controlled environmental conditions are used instead of the flow-through systems used for juveniles and adults (Lindberg et al., 2000). Larvae also had improved survival in larger tank sizes (120 L) and higher stocking densities (40 fish $L^{-1}$). This necessitated multiple spawns from a limited supply of broodfish to fill a single larval tank. Furthermore, larvae resorb yolk within 5 days and feed exclusively on live zooplankton until they reach the juvenile stage between 60 and 80 dph. This required an intensive food culture program to produce enough nutritious zooplankton to feed the early life stages (Lindberg et al., 1999). Furthermore, the use of green water by suspending algae paste in water to increase turbidity is necessary to stimulate larval feeding, which results in higher biofouling in larval tanks and in turn reduced water quality. Removing detritus by siphon proved too disruptive for the sensitive larvae and two-ridge ramshorn snails (*Helisoma anceps*) are now employed to feed on the accumulating biomass. These snails have proven to be highly effective at reducing biofouling within larval tanks and are easily produced with minimal maintenance (Hung et al., 2018). Overcoming larviculture bottlenecks such as developing appropriate housing and feeding requirements was a critical success for closing the delta smelt's life cycle as establishing a conservation hatchery for delta smelt is required to even conceive of future efforts to supplement wild populations.

Cultured juveniles and adults are comparatively easier to rear, and the few challenges for culturing later life stages had simple solutions. Early juveniles (~20 mm) are especially sensitive to handling, and water-to-water transport is crucial when moving fish to new systems. However, this period of particularly high sensitivity is transient and by the time juveniles reach 30 mm in size, they can tolerate brief air exposures and handling with a soft brine shrimp net (Baskerville-Bridges et al., 2005b). Maintenance of water quality becomes easier as delta smelt age as juveniles and adults can tolerate the fluctuating Delta water conditions from filtered and UV-sterilized flow-through systems, disturbances by siphons for removing detritus, and transition easily to artificial feeds that do not foul the water as quickly relative to live feed. Perhaps the most challenging aspect of rearing juvenile and adult delta smelt is preventing the outbreak of disease. Rearing the later life stages at high stocking densities proved infeasible due to the high likelihood of a disease outbreak and are thus reared at lower densities (1 fish $L^{-1}$). Due to this, maintaining large populations of juveniles and adults requires a large number of tanks, which increases the costs of operations and the land required at the hatchery (Baskerville-Bridges et al., 2005b).

The FCCL faced the major challenge of optimizing a culture protocol for a very sensitive species, with very few fish to experiment with, in a short time span. As a result, experimental studies conducted to inform the development of delta smelt culture methods had low replication. Some of these challenges are expected and common for developing culture methods (e.g., larviculture, disease outbreaks), while others presented unexpected problems requiring novel solutions often taken for granted when working with more robust species (e.g., require experienced full time staff, no corners in containers, cushions during transportation). Now that the delta smelt culture has been established, many of the methods are being revisited to further optimize delta smelt culture methods. For example, studies that have direct relevance for the continued improvement of the delta smelt culture such as: examining feeding responses to different turbidity and lighting conditions, immune responses to common diseases, stress responses to different salinities, and testing novel filtration systems, have all been recently conducted (Castillo et al., 2014; Frank et al., 2017; Hasenbein et al., 2016; Hung and Piedrahita, 2011; Tigan et al., 2020).

## 3.2 Genetic management

Captive and wild populations can rapidly genetically diverge from each other without gene flow, either from genetic drift or domestication (see Chapter 8, Volume 39A: Jeffries et al., 2022; Waters et al., 2015). The primary goal of genetic management at conservation hatcheries is to minimize this divergence by maintaining genetic variation of the refuge population as representative of the wild population through the generations, in captivity (Lindberg et al., 2013). Although delta smelt have three distinct life history strategies: semi-anadromous, freshwater resident, and brackish resident (Hobbs et al., 2019), population genetics studies did not find any distinct population segments. Thus, delta smelt are managed as a single panmictic population (Fisch et al., 2011; Trenham et al., 1998).

Delta smelt in captivity are particularly susceptible to rapid genetic divergence from wild populations due to difficulty in maintaining the species in captivity, their annual life cycle, and the shrinking wild population. However, by working closely with the UC Davis Genomic Variation Laboratory to genotype all broodfish, using informed single pair crosses to minimize kinship, family equalizations to remove reproductive variation among families, rearing in multi-family groups to minimize tank effects, and annual genetic supplementation with wild broodstock (Lindberg et al., 2013), the FCCL has been largely successful for maintaining genetic variation and minimizing differentiation between captive and wild populations but there is still evidence for increasing domestication within the refuge population (Finger et al., 2018).

Some families inherently survive in a hatchery setting better than others, resulting in an inevitable hatchery effect (Allendorf, 1993; Frankham, 2008). Family size is equalized at the embryo and larval stage to minimize reproductive variance but the intermixing of families within larval tanks may lead to competition between families that result in unequal representation of certain families. Similar to other hatcheries, these effects are observed quickly. For example, survival and growth of the third filial generation ($F_3$) delta smelt were higher compared to previous generations and $F_1$ (wild × wild) fish had reduced stress response markers compared to $F_0$ (wild fish) fish after transportation (Afentoulis et al., 2013; Lindberg et al., 2013). This has resulted in higher recovery of offspring from pair crosses with higher hatchery ancestry over time. Given these circumstances, it is perhaps unsurprising that the refuge population is showing genetic evidence of adaptation to hatchery conditions (Finger et al., 2018).

Because of their annual life cycle and the difficulty of maintaining this species in captivity, which can impose strong selection, delta smelt are particularly susceptible to rapid domestication. The FCCL is currently at the $F_{13}$ and with the increasing difficulty of capturing wild delta smelt to genetically supplement the refuge population, it is becoming exceedingly difficult to maintain genetic similarity between the captive and wild fish. This is exemplified in recent years where the FCCL has struggled to collect wild adult delta smelt to genetically supplement the refuge population. Currently, the FCCL is permitted to capture 100 wild adult delta smelt annually to supplement their broodstock. Historically, the FCCL staff have easily collected this take limit but in the 2018–19 season, were only able to collect 28 wild broodfish. In the 2019–20 season, FCCL had to double their effort to collect 93 wild broodfish. In the 2020–21 season, FCCL was only able to capture two wild broodfish despite extensive collection efforts by FCCL and agency partners. Collection of adult delta smelt across all monitoring programs occurring in the Delta is now extremely uncommon and future broodstock collection efforts are unlikely to produce many, if any, broodfish. Moreover, delta smelt are known to hybridize with other smelt species such as longfin smelt (*Spirinchus thaleichthys*) and the invasive wakasagi (*Hypomesus nipponensis*). Though hybridization occurs at low frequencies, this still contributes to the decrease in availability of the already limited wild broodfish (Fisch et al., 2014).

The observed domestication within the refuge population is indicative of genetic adaptation to captive conditions (Finger et al., 2018), which can lower fitness upon reintroduction to the wild (Christie et al., 2012). This knowledge brings a sense of urgency to the implementations of tools to supplement the wild population (i.e., releasing captive raised animals into the wild). A key tenant of supplementation is to avoid introducing captive individuals that are increasingly maladaptive to wild conditions, and therefore less likely to survive and reproduce. It is imperative that supplementation commence as soon as possible now that the refuge population is at a high risk of becoming closed.

### 3.3 Future directions: Supplementation of wild delta smelt populations

Planning for the supplementation of the wild delta smelt populations is underway and reliant on the wealth of knowledge and experience gained from past physiological studies defining the delta smelt's environmental tolerances, swimming ability, and ability to evade predators (Fig. 4). Supplementation is identified as a strategy in the latest Delta Smelt Biological Opinion issued to U.S. Bureau of Reclamation and Department of Water Resources (U.S. Bureau of Reclamation, 2019) and plans are underway to experimentally release fish into the SFE in late 2021 to early 2022. However, developing methods for supplementation still presents several major questions and challenges: How can we increase production of expensive delta smelt to meet the demands of supplementation? Will released delta smelt be resilient to current and future conditions? What location and life stage provides the best chances of a successful release? How can we transport and release highly sensitive delta smelt in mass quantities safely? How can we minimize predation upon released fish? Moreover, successful delta smelt supplementation will be complicated by legal constraints and permitting requirements that must be considered (Lessard et al., 2018). Several planning teams consisting of fisheries agency scientists and academics have formed to design and coordinate each step of release logistics. The rapid development of these teams and their monumental task to meet expedited goals of release has relied heavily on physiological research to inform transport and release strategies to minimize stress and maximize survival (Swanson et al., 1996). For example, research has informed life-stage sensitivities and site locations for release, transport methodologies to reduce stress and maximize survival, acclimation requirements and understanding post-release responses in physiology and behavior. Predator control remains at pilot stage for fish management but evaluation of the potential benefits of this topic are informed by bioenergetic modeling of specific predators (Loboschefsky et al., 2012), as well as focused laboratory research on delta smelt predation (Davis et al., 2019).

Releasing captive fish into the wild is controversial and raises many concerns about the effectiveness of this strategy. Supplementation may divert attention and funding from other restoration efforts that some argue have more promise. There is also a question over the efficacy of releasing delta smelt into the highly altered and potentially unsuitable SFE habitat. Furthermore, there is concern over the rising domestication index of the delta smelt refuge population (Finger et al., 2018) and the potential of supplementation to "genetically pollute" the wild population with hatchery adapted genes (Glover et al., 2012; Karlsson et al., 2016). But the intense genetic management of the refuge population has, to date, minimized genetic divergence between the refuge and wild populations, and the risk of swamping the wild population with hatchery adapted genes is considered low (Evans et al., 2019; U.S. Fish and Wildlife Service et al., 2022). Doing nothing and

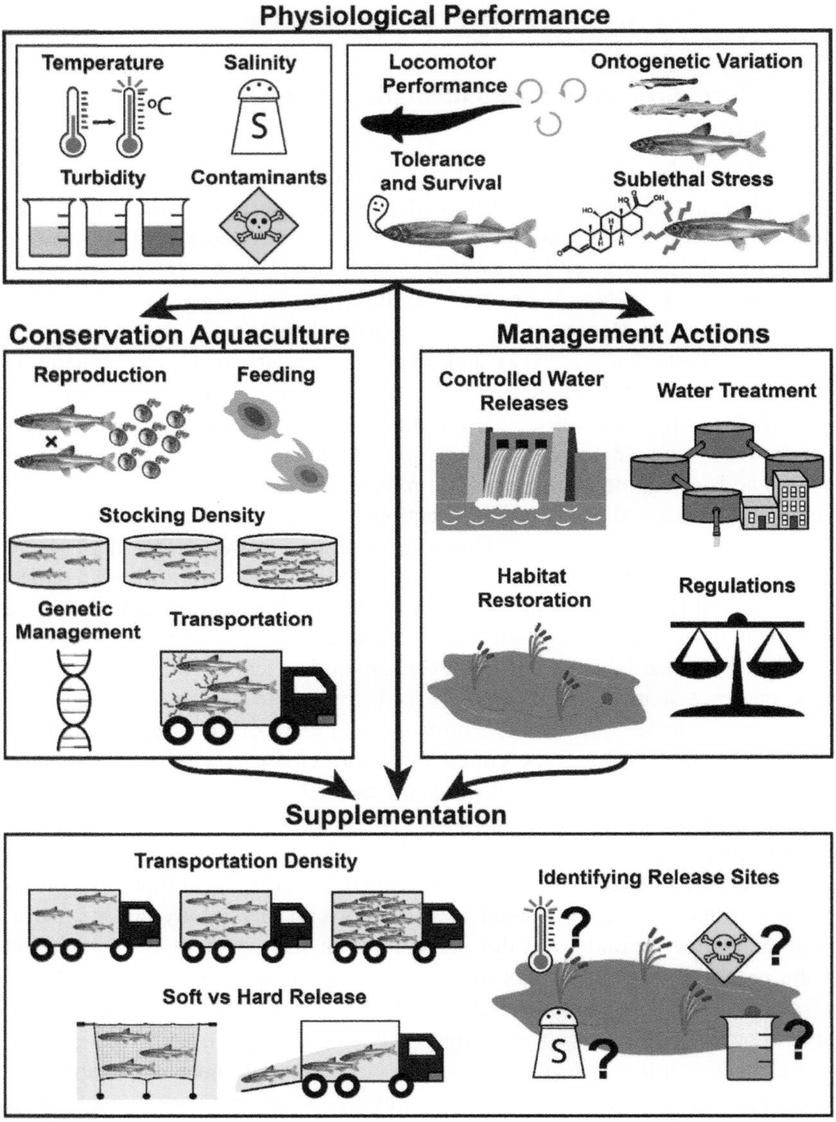

**FIG. 4** Physiology underpins effective conservation strategies for delta smelt. Understanding the physiological responses of delta smelt to environmental conditions has informed the development of a conservation aquaculture program and regulatory applications (see Section 2). Refining delta smelt conservation aquaculture required additional physiology research to develop methods for reproductive control, rearing (e.g., feeding and stocking density), genetic management, and transportation (see Sections 3.1 and 3.2). Delta smelt environmental tolerances defined the conditions necessary to maintain in the San Francisco Estuary for the effective use of water infrastructure (i.e., water releases from dams and reservoirs, and water treatment plants) and for informing habitat restoration plans (see Section 3.4). This accumulated knowledge is informing the implementation of the next major conservation action: supplementation of wild populations with fish from conservation hatcheries (see Section 3.3). Physiology research will continue to aid this conservation strategy by informing methods for the mass-transportation of delta smelt, identifying release sites, and developing effective release methods (i.e., soft vs hard release). *Photos of delta smelt embryo, larvae, and adults courtesy of René Reyes.*

allowing the wild population to go extinct is now considered the greater risk (Hobbs et al., 2017). Nevertheless, the effects of rising domestication indices on the physiology and behavior of delta smelt remain sources of concern that are being investigated. Additionally, even though captive and wild populations have remained genetically similar, it is unknown whether the refuge population can tolerate future warming conditions. It is possible that the genetic management plan could shift its focus from neutral variation to allelic variants of genes that may confer a selective advantage in a changing climate. However, there is evidence that suggests delta smelt are already living at the upper end of their thermal capacity (Davis et al., 2019; Komoroske et al., 2014; Swanson et al., 2000) so there may not be much more genetic variation upon which selection can act.

The resource intensive nature of delta smelt rearing and the resultant low production of refuge fish is a potential bottleneck for successful supplementation. A Biological Opinion issued an estimate that 125,000 adult delta smelt (>200 dph) must be released annually for supplementation to have a meaningful impact (U.S. Bureau of Reclamation, 2019). In 2020, FCCL produced 32,000, 200 dph adults of which 6600 are necessary to regenerate the refuge population and the backup population at Livingston Stone National Fish Hatchery. This leaves only 18,800 fish or 15% of the recommended number of fish for supplementation purposes. Reaching the Biological Opinion supplementation goals, with current resources, is infeasible. Though FCCL has begun to increase production of delta smelt for supplementation efforts with the goal of increasing their production of 120 dph fish to 50,000 individuals over the next 3 years, meeting the requirements written in the Biological Opinion will require a substantial expansion of delta smelt facilities. To address this, efforts are underway to construct a Fish Technology Center that would serve as another delta smelt conservation hatchery to greatly expand the refuge population and support delta smelt research and supplementation efforts (Lessard et al., 2018).

Even with these facility expansions, the exorbitant costs of raising delta smelt may hinder efforts to adequately increase production to meet the demands for supplementation. For example, FCCL produced 32,000 adult delta smelt in 2020 with a $2.7 million operating budget, amounting to approximately $84 per fish in 2020 (U.S. Department of the Interior, 2020). In contrast, salmonids of conservation concern produced at California hatcheries cost between $0.23 and $0.96 per smolt, or between 0.4% and 1.1% of the cost of an adult delta smelt (Cavallo et al., 2009; U.S. Department of the Interior, 2020). Similarly, the endangered Rio Grande silvery minnow which has been supplemented by captive fish since 2002 cost only $3.48 per fish produced or 4.1% of the cost of a delta smelt in 2015 (Middle Rio Grande Endangered Species Collaborative Program, 2015). It is likely that alternative, cost-efficient methods for rearing delta smelt will be necessary to meet the supplementation requirements in the future. A particularly enticing direction is rearing delta smelt in low maintenance naturalized mesocosms that have natural populations of zooplankton as it may reduce domestication by rearing them in an environment more similar to the Delta (Hung et al., 2019).

## 3.4 The contribution of physiological data to additional management actions

Beyond biological interventions such as large-scale supplementation efforts built on decades of experimental and field research, additional aquatic resource management tools have been developed and brought to bear on these complex problems. In general, there are three categories that we summarize below with examples: Regulatory applications, water infrastructure, and habitat restoration.

Regulatory Applications: Regulatory approaches are often an important first step in implementing major changes in water or ecosystem management. For delta smelt, the three most relevant categories of regulations are: species listings, water quality, and water rights. As a result of numerous declines in species such as delta smelt, there is a major focus on providing protections for sensitive species through listings under the FESA and CESA. The process for species listings is difficult and substantial, requiring basic information about the species status and its stressors. For delta smelt, physiological results described above represented some of the core information used to identify basic environmental requirements, and how habitat modifications create stressors (U.S. Bureau of Reclamation, 2019; U.S. Fish and Wildlife Service, 2008).

The primary regulatory organizations that deal with water-quality constituents in the SFE are the United States EPA and several entities in the California EPA. Among the various physiological metrics, the most relevant for water quality management in the SFE is delta smelt salinity tolerance and field occurrence thresholds relative to salinity (Brown et al., 2013; Davis et al., 2019; Hasenbein et al., 2013; Kimmerer et al., 2009; Komoroske et al., 2014), which influences water diversion. The State Water Resources Control Board (SWRCB) has the primary responsibility to manage water rights for much of the year through a unique metric: X2, which is the distance from the Golden Gate Bridge at which daily average salinity is 2 practical saline units (PSU; roughly equivalent to ppt). This metric is controlled by outflow, tides, and meteorological inputs, the former of which is a major focus of Delta management (see below). Though the efficacy of this practice is controversial (Feyrer et al., 2011; Kimmerer, 2002; Murphy and Weiland, 2019; Nobriga et al., 2008), the low salinity habitats created by X2 are used as a surrogate indicator for delta smelt habitat and management of the monthly average location of X2 is intended to maintain suitable habitat for the species (U.S. Fish and Wildlife Service, 2008).

Water Infrastructure: The SFE likely has the most complicated and frequently used water management infrastructure on the planet (Sommer, 2020). The infrastructure tools include five major groups: (1) reservoirs; (2) water diversion; (3) gates; (4) barriers; and (5) water treatment. These tools are used to meet the environmental water quality criteria and endangered species requirements set by regulatory agencies. As noted above, much of the focus is on salinity management, which is guided by delta smelt salinity tolerances and field occurrence thresholds (Brown et al., 2013; Davis et al., 2019; Hasenbein et al., 2013; Kimmerer et al., 2009; Komoroske et al., 2014). However, water treatment plants manage for a broader suite of constituents that are also informed by physiological data such as heavy metals (mercury, copper, selenium) pathogens, sediments, polychlorinated biphenyls, nutrients (nitrogen), and numerous pesticides (e.g., organophosphates [orthophosphates, diazonin, and chlorpyrifos]). These contaminants are a major issue to consider as they have been demonstrated to impact neurodevelopment and behavior (Mundy et al., 2020, 2021; Segarra et al., 2021), bioaccumulate through the food web (Derby et al., 2021; Fuller et al., 2021), and can have transgenerational effects (DeCourten et al., 2020; Major et al., 2020).

Physiology also informs one of the most prominent components of water management: diversions from the large SWP and CVP facilities. For example, water exports are managed to maintain daily average turbidity at a level <12NTU for key locations, with greater than 12NTU considered a "turbidity bridge" correlated with the behavioral migration of delta smelt to spawn (Grimaldo et al., 2009). If 12NTU is exceeded at key locations during the spawning season, water exports are reduced in an effort to minimize entrainment of delta smelt at SWP and CVP pumping facilities. Additionally, the operations management season for delta smelt is ended when temperatures in the South Delta exceed 25°C, the delta smelt $CT_{Max}$ (Komoroske et al., 2014; Swanson et al., 2000), indicating delta smelt are no longer in the entrainment zone and pumping operations can proceed without delta smelt regulations.

Habitat Restoration: Given the extremely high degree of habitat loss and alteration, another major focus of resource management in the SFE is habitat management. This includes several major areas that are strongly guided by basic physiological criteria such as salinity, temperature, turbidity, and oxygen levels. The primary tools include: (1) floodplain and riparian restoration; (2) tidal wetland restoration; (3) aquatic weed removal; (4) sediment supplementation; and (5) temperature management. In addition, management of salinity (via X2; see above) is also considered an approach to habitat management (U.S. Fish and Wildlife Service, 2008). However, the large scale required for meaningful impacts makes habitat management difficult and, in some cases, infeasible. For example, many areas of the SFE are unsuitable for habitat restoration as the physicochemical parameters of the water are outside of the delta smelt tolerance range. Temperature in particular is a parameter that poses issues for habitat restoration as the range of areas within the Delta where temperatures exceed the thermal tolerance of delta smelt have been increasing (Halverson et al., 2021). The extensive spread of aquatic weeds has limited habitat restoration success to localized regions and only for floating varieties. Furthermore, the effects of large-scale sediment supplementation on water quality are currently unknown. Finally, temperature is largely dictated by air temperature and the cold freshwater releases from reservoirs can only transiently cool the Delta water due to limitations in available freshwater resources (Sommer, 2020). These complications have thus far impeded the success of conservation actions in the Delta.

Together, water quality standards, resiliency strategies, and understanding of physiology and behavior of delta smelt have shaped a more recent regulatory framework for adaptive management of water and species. The X2 standard was carried forward in the most recent USFWS Biological Opinion (U.S. Bureau of Reclamation, 2019) and CDFW Incidental Take Permit (California Department of Fish and Wildlife, 2020) with additional management tools including flexibility in spring outflow, additional summer flows, and holding of water for subsequent years, all of which aim to improve habitat conditions and water quality for species such as delta smelt. Many of the resiliency strategies including reoperation of the Suisun Marsh Control Gates and the North Delta Food Subsidies project (flow intervention to transport food) have transitioned to mitigation and conservation actions to offset effects of long-term water operations of the CVP and SWP in the SFE (California Department of Fish and Wildlife, 2020; U.S. Bureau of Reclamation, 2019) aiming to increase habitat suitability, connectivity, and food availability for delta smelt. Additionally, management of CVP and SWP exports during winter and spring are regulated by a number of delta smelt thresholds, informed by physiology and behavior studies, and a life cycle model of delta smelt (Smith et al., 2021).

## 4 Concluding remarks

Researchers have collected a tremendous amount of knowledge about the delta smelt's physiological requirements for survival and this research has translated to many of the management practices that are used to protect the Delta ecosystem and its inhabitants such as delta smelt today (Fig. 4). Conservation of delta smelt is an ongoing effort and management practices are imperfect with many questions left to answer and challenges to face. However, physiology will continue to play a critical role in addressing these questions and be used to solve many of the new issues that arise. Annotated transcriptomes and genomes have been assembled, investigations into producing genetically resilient delta smelt have begun (see Section 3.3), methods for transporting small numbers (up to 12,000 individuals at a time) of delta smelt are being developed, and preliminary trials for holding subadult delta smelt in cages within the Delta have been successful. However, many knowledge gaps remain before successful supplementation can be implemented and physiological studies will be crucial for addressing them. For example: how does stocking density affect stress in delta smelt in transportation containers and in cages deployed within the Delta? Do delta smelt need an acclimation period in cages in the Delta prior to release, or can they be released directly? How can early life stages of delta smelt be reared in cages within the Delta? What are the best environmental conditions for releasing delta smelt? How will contaminants impact the released populations? These are just a few of the questions that physiological studies will help to answer toward this effort.

While we have focused on delta smelt in this chapter, it is important both to remember that a whole assemblage of fish species is declining in the SFE and to recognize the interconnectedness of the SFE to the rest of the world. As a sentinel species, the decline of delta smelt acts as a harbinger for the conservation crises that may come. Other SFE species such as Chinook salmon, green sturgeon, longfin smelt, and even the introduced striped bass are also in severe decline. Longfin smelt have garnered special attention recently as they have experienced a particularly precipitous decline with their abundances dropping to approximately 1% of their historic abundances in the SFE, threatening them with extirpation, and have thus far eluded development of a captive culture (Hobbs et al., 2017; Nobriga and Rosenfield, 2016; Yanagitsuru et al., 2021). With each species having a unique life history and physiological requirements for survival that may sometimes conflict with one another, conservation of fish species in the SFE is a monumental task. But this is a task worthy of our efforts. The urgency to protect fish species in the SFE extends beyond just the region. The decline of anadromous fish species in the SFE could directly impact marine ecosystems even as far North as the Aleutian Islands, Pacific Northwest (Saglam et al., 2021). Though we are experiencing an extinction crisis across the globe (Ceballos et al., 2017; see Chapter 1, Volume 39A: Cooke et al., 2022), there is reason for hope. Research has already laid the groundwork for implementing effective conservation plans and there is infrastructure in place or being built for the conservation of fish species. Furthermore, public support for environmental protection in the United States has trended upwards (Kim and Urpelainen, 2018). While there are major environmental challenges in the SFE, progress in fish physiology and other scientific areas will continue to fuel innovation in species management. Hence, we are optimistic that management actions such as supplementation and habitat restoration will lead to substantial progress in the status of native fishes. At the very least, these actions will help support resilience, a critical need to help buffer fish populations from future issues.

## References

Afentoulis, V., DuBois, J., Fujimura, R., Region, B.D., 2013. Stress Response of Delta Smelt, Hypomesus transpacificus, in the Collection, Handling, Transport, and Release Phase of Fish Salvage at the John E. Skinner Delta Fish Protective Facility. Calif. Fish Game, Bay Delta Region.

Allendorf, F.W., 1993. Delay of adaptation to captive breeding by equalizing family size. Conserv. Biol. 7, 416–419.

Aravindakshan, J., Paquet, V., Gregory, M., Dufresne, J., Fournier, M., Marcogliese, D.J., Cyr, D.G., 2004. Consequences of xenoestrogen exposure on male reproductive function in spottail shiners (Notropis hudsonius). Toxicol. Sci. 78, 156–165.

Bacey, J., Spurlock, F., Branch, E.M., 2005. Biological Assessment of Urban and Agricultural Streams in the California Central Valley (Fall 2002 Through Spring 2004). CalEPA, CalDPR, Sacramento, CA, USA.

Baerwald, M.R., Schreier, B.M., Schumer, G., May, B., 2012. Detection of threatened Delta Smelt in the gut contents of the invasive Mississippi Silverside in the San Francisco Estuary using TaqMan assays. Trans. Am. Fish. Soc. 141, 1600–1607.

Baskerville-Bridges, B., Lindberg, C., 2004. The effect of light intensity, alga concentration, and prey density on the feeding behavior of delta smelt larvae. Am. Fish. Soc. Symp. 39, 219–227.

Baskerville-Bridges, B., Lindberg, J.C., Van Eenennaam, J., Doroshov, S.I., 2005a. Delta Smelt Culture and Research Program Final Report: 2003–2005. University of California Davis. Report to CALFED Bay-Delta Program. ERP-02-P31.

Baskerville-Bridges, B., Lindberg, J.C., Doroshov, S.I., 2005b. Manual for the Intensive Culture of Delta Smelt (*Hypomesus transpacificus*). University of California-Davis, Sacramento (CA). Report to CALFED Bay-Delta Program. ERP-02-P31.

Beggel, S., Connon, R., Werner, I., Geist, J., 2011. Changes in gene transcription and whole organism responses in larval fathead minnow (*Pimephales promelas*) following short-term exposure to the synthetic pyrethroid bifenthrin. Aquat. Toxicol. 105, 180–188.

Bernier, N.J., Alderman, S.L., 2022. Applied aspects of fish endocrinology. Fish Physiol. 39A, 253–320.

Brander, S.M., Connon, R.E., He, G., Hobbs, J.A., Smalling, K.L., Teh, S.J., White, J.W., Werner, I., Denison, M.S., Cherr, G.N., 2013. From 'omics to otoliths: responses of an estuarine fish to endocrine disrupting compounds across biological scales. PLoS One 8, e74251.

Brander, S.M., Gabler, M.K., Fowler, N.L., Connon, R.E., Schlenk, D., 2016a. Pyrethroid pesticides as endocrine disruptors: molecular mechanisms in vertebrates with a focus on fishes. Environ. Sci. Technol. 50, 8977–8992.

Brander, S.M., Jeffries, K.M., Cole, B.J., DeCourten, B.M., White, J.W., Hasenbein, S., Fangue, N.A., Connon, R.E., 2016b. Transcriptomic changes underlie altered egg protein production and reduced fecundity in an estuarine model fish exposed to bifenthrin. Aquat. Toxicol. 174, 247–260.

Brown, L.R., Bennett, W.A., Wagner, R.W., Morgan-King, T., Knowles, N., Feyrer, F., Schoellhamer, D.H., Stacey, M.T., Dettinger, M., 2013. Implications for future survival of delta smelt from four climate change scenarios for the Sacramento–San Joaquin Delta, California. Estuar. Coasts 36, 754–774.

California Department of Fish and Wildlife, 2020. Long-term Operation of the State Water Project in the Sacramento San Joaquin Delta. California Department of Fish and Wildlife.

California Department of Fish and Wildlife, 2021. Publically Available Survey Data. https://www.dfg.ca.gov/delta/data/.

California Department of Food and Agriculture, 2021. County Agricultural Commissioner's Reports Crop Year 2018–2019. California Department of Food and Agriculture.

California Fish and Game Commission, 2009. Office of Administrative Law's Notice ID #Z2009-0106-06. Office of Administrative Law's File ID#2009-1119-01S. Section 670.5, Title 14, CCR, Uplist Delta Smelt from Threatened to Endangered Species Status. California Fish and Game Commission.

California Natural Resources Agency, 2005. Delta Smelt Action Plan. California Department of Water Resources and California Department of Fish and Game, Sacramento, USA.

California Natural Resources Agency, 2016. Delta Smelt Resiliency Strategy. California Natural Resources Agency.

Castillo, G., Morinaka, J., Fujimura, R., DuBois, J., Baskerville-Bridges, B., Lindberg, J., Tigan, G., Ellison, L., Hobbs, J., 2014. Evaluation of calcein and photonic marking for cultured delta smelt. N. Am. J. Fish Manag. 34, 30–38.

Cavallo, B., Brown, R., Lee, D., 2009. Hatchery and Genetic Management Plan for Feather River Hatchery Spring-Run Chinook Salmon Program. Feather River Hatchery and Genetic Management Plan.

Ceballos, G., Ehrlich, P.R., Dirzo, R., 2017. Biological annihilation via the ongoing sixth mass extinction signaled by vertebrate population losses and declines. Proc. Natl. Acad. Sci. 114, 6089–6096.

Christie, M.R., Marine, M.L., French, R.A., Waples, R.S., Blouin, M.S., 2012. Effective size of a wild salmonid population is greatly reduced by hatchery supplementation. Heredity 109, 254–260.

Cole, C.F., 1949. A solution to some San Francisco bay area problems-the reber plan. J. Geogr. 48, 112–120.

Cole, B.J., Brander, S.M., Jeffries, K.M., Hasenbein, S., He, G., Denison, M.S., Fangue, N.A., Connon, R.E., 2016. Changes in *Menidia beryllina* gene expression and in vitro hormone-receptor activation after exposure to estuarine waters near treated wastewater outfalls. Arch. Environ. Contam. Toxicol. 71, 210–223.

Cooke, S.J., Fangue, N.A., Bergman, J.N., Madliger, C.L., Cech Jr. J.J., Eliason, E.J., Brauner, C. J., Farrell, A.P., 2022. Conservation physiology and the management of wild fish populations in the Anthropocene. Fish Physiol. 39A, 1–31.

Connon, R.E., Geist, J., Pfeiff, J., Loguinov, A.V., D'Abronzo, L.S., Wintz, H., Vulpe, C.D., Werner, I., 2009. Linking mechanistic and behavioral responses to sublethal esfenvalerate exposure in the endangered delta smelt; Hypomesus transpacificus (Fam. Osmeridae). BMC Genomics 10, 1–18.

Connon, R.E., Hasenbein, S., Brander, S.M., Poynton, H.C., Holland, E.B., Schlenk, D., Orlando, J.L., Hladik, M.L., Collier, T.K., Scholz, N.L., Incardona, J.P., 2019. Review of and recommendations for monitoring contaminants and their effects in the San Francisco Bay – Delta. San Franc. Estuary Watershed Sci. 17, 2.

Conrad, J.L., Bibian, A.J., Weinersmith, K.L., De Carion, D., Young, M.J., Crain, P., Hestir, E.L., Santos, M.J., Sih, A., 2016. Novel species interactions in a highly modified estuary: association of Largemouth Bass with Brazilian waterweed Egeria densa. Trans. Am. Fish. Soc. 145, 249–263.

Davis, B.E., Hansen, M.J., Cocherell, D.E., Nguyen, T.X., Sommer, T., Baxter, R.D., Fangue, N.A., Todgham, A.E., 2019. Consequences of temperature and temperature variability on swimming activity, group structure, and predation of endangered delta smelt. Freshw. Biol. 64, 2156–2175.

DeCourten, B.M., Forbes, J.P., Roark, H.K., Burns, N.P., Major, K.M., White, J.W., Li, J., Mehinto, A.C., Connon, R.E., Brander, S.M., 2020. Multigenerational and transgenerational effects of environmentally relevant concentrations of endocrine disruptors in an estuarine fish model. Environ. Sci. Technol. 54, 13849–13860.

Derby, A.P., Fuller, N.W., Hartz, K.E.H., Segarra, A., Connon, R.E., Brander, S.M., Lydy, M.J., 2021. Trophic transfer, bioaccumulation and transcriptomic effects of permethrin in inland silversides, *Menidia beryllina*, under future climate scenarios. Environ. Pollut. 275, 116545.

Dilsaver, L.M., 1985. After the gold rush. Geogr. Rev. 75, 1–18.

Erkkila, L.F., Moffett, J.W., Cope, O.B., Smith, B.R., Nelson, R.S., 1950. Sacramento–San Joaquin Delta Fishery Resources: Effects of Tracy Pumping Plant and Delta Cross Channel. U.S. Fish and Wildlife Service Special Scientific Report Fisheries, p. 5.

Evans, M.L., Hard, J.J., Black, A.N., Sard, N.M., O'Malley, K.G., 2019. A quantitative genetic analysis of life-history traits and lifetime reproductive success in reintroduced Chinook salmon. Conserv. Genet. 20, 781–799.

Faulk, C.K., Holt, G.J., 2005. Advances in rearing cobia Rachycentron canadum larvae in recirculating aquaculture systems: live prey enrichment and greenwater culture. Aquaculture 249, 231–243.

Ferrari, M.C., Ranåker, L., Weinersmith, K.L., Young, M.J., Sih, A., Conrad, J.L., 2014. Effects of turbidity and an invasive waterweed on predation by introduced largemouth bass. Environ. Biol. Fish 97, 79–90.

Feyrer, F., Herbold, B., Matern, S.A., Moyle, P.B., 2003. Dietary shifts in a stressed fish assemblage: consequences of a bivalve invasion in the San Francisco Estuary. Environ. Biol. Fish 67, 277–288.

Feyrer, F., Nobriga, M.L., Sommer, T.R., 2007. Multidecadal trends for three declining fish species: habitat patterns and mechanisms in the San Francisco Estuary, California, USA. Can. J. Fish. Aquat. Sci. 64, 723–734.

Feyrer, F., Newman, K., Nobriga, M., Sommer, T., 2011. Modeling the effects of future outflow on the abiotic habitat of an imperiled estuarine fish. Estuar. Coasts 34, 120–128.

Finger, A.J., Mahardja, B., Fisch, K.M., Benjamin, A., Lindberg, J., Ellison, L., Ghebremariam, T., Hung, T.C., May, B., 2018. A conservation hatchery population of delta smelt shows evidence of genetic adaptation to captivity after 9 generations. J. Hered. 109, 689–699.

Fisch, K.M., Henderson, J.M., Burton, R.S., May, B., 2011. Population genetics and conservation implications for the endangered delta smelt in the San Francisco Bay-Delta. Conserv. Genet. 12, 1421–1434.

Fisch, K.M., Ivy, J.A., Burton, R.S., May, B., 2013. Evaluating the performance of captive breeding techniques for conservation hatcheries: a case study of the delta smelt captive breeding program. J. Hered. 104, 92–104.

Fisch, K.M., Mahardja, B., Burton, R.S., May, B., 2014. Hybridization between delta smelt and two other species within the family Osmeridae in the San Francisco Bay-Delta. Conserv. Genet. 15, 489–494.

Fong, S., Louie, S., Werner, I., Davis, J., Connon, R.E., 2016. Contaminant effects on California Bay–Delta species and human health. San Franc. Estuary Watershed Sci. 14, 5.

Frank, D.F., Hasenbein, M., Eder, K., Jeffries, K.M., Geist, J., Fangue, N.A., Connon, R.E., 2017. Transcriptomic screening of the innate immune response in delta smelt during an *Ichthyophthirius multifiliis* infection. Aquaculture 473, 80–88.

Frank, D.F., Miller, G.W., Harvey, D.J., Brander, S.M., Geist, J., Connon, R.E., Lein, P.J., 2018. Bifenthrin causes transcriptomic alterations in mTOR and ryanodine receptor-dependent signaling and delayed hyperactivity in developing zebrafish (*Danio rerio*). Aquat. Toxicol. 200, 50–61.

Frank, D.F., Brander, S.M., Hasenbein, S., Harvey, D.J., Lein, P.J., Geist, J., Connon, R.E., 2019. Developmental exposure to environmentally relevant concentrations of bifenthrin alters transcription of mTOR and ryanodine receptor-dependent signaling molecules and impairs predator avoidance behavior across early life stages in inland silversides (*Menidia beryllina*). Aquat. Toxicol. 206, 1–13.

Frankham, R., 2008. Genetic adaptation to captivity in species conservation programs. Mol. Ecol. 17, 325–333.

Fuller, N., Hartz, K.E.H., Johanif, N., Magnuson, J.T., Robinson, E.K., Fulton, C.A., Poynton, H.C., Connon, R.E., Lydy, M.J., 2021. Enhanced trophic transfer of chlorpyrifos from resistant *Hyalella azteca* to inland silversides (*Menidia beryllina*) and effects on acetylcholinesterase activity and swimming performance at varying temperatures. Environ. Pollut. 291, 118217.

Ganssle, D., 1966. Fishes and decapods of San Pablo and Suisun bays. Fish. Bull. 133, 64–94.

Glover, K.A., Quintela, M., Wennevik, V., Besnier, F., Sørvik, A.G., Skaala, Ø., 2012. Three decades of farmed escapees in the wild: a spatio-temporal analysis of Atlantic salmon population genetic structure throughout Norway. PLoS One 7, e43129.

Grimaldo, L.F., Sommer, T., Van Ark, N., Jones, G., Holland, E., Moyle, P.B., Herbold, B., Smith, P., 2009. Factors affecting fish entrainment into massive water diversions in a tidal freshwater estuary: can fish losses be managed? N. Am. J. Fish Manag. 29, 1253–1270.

Halverson, G.H., Lee, C.M., Hestir, E.L., Hulley, G.C., Cawse-Nicholson, K., Hook, S.J., Bergamaschi, B.A., Acuña, S., Tufillaro, N.B., Radocinski, R.G., Rivera, G., 2021. Decline in thermal habitat conditions for the endangered delta smelt as seen from Landsat satellites (1985–2019). Environ. Sci. Technol. 56, 185–193.

Hamilton, S.A., Murphy, D.D., 2020. Use of affinity analysis to guide habitat restoration and enhancement for the imperiled delta smelt. Endanger. Species Res. 43, 103–120.

Hammock, B.G., Hobbs, J.A., Slater, S.B., Acuña, S., Teh, S.J., 2015. Contaminant and food limitation stress in an endangered estuarine fish. Sci. Total Environ. 532, 316–326.

Hammock, B.G., Lesmeister, S., Flores, I., Bradburd, G.S., Hammock, F.H., Teh, S.J., 2016. Low food availability narrows the tolerance of the copepod Eurytemora affinis to salinity, but not to temperature. Estuar. Coasts 39, 189–200.

Hammock, B.G., Slater, S.B., Baxter, R.D., Fangue, N.A., Cocherell, D., Hennessy, A., Kurobe, T., Tai, C.Y., Teh, S.J., 2017. Foraging and metabolic consequences of semi-anadromy for an endangered estuarine fish. PLoS One 12, e0173497.

Hasenbein, M., Komoroske, L.M., Connon, R.E., Geist, J., Fangue, N.A., 2013. Turbidity and salinity affect feeding performance and physiological stress in the endangered delta smelt. Integr. Comp. Biol. 53, 620–634.

Hasenbein, M., Fangue, N.A., Geist, J.P., Komoroske, L.M., Connon, R.E., 2016. Physiological stress biomarkers reveal stocking density effects in late larval Delta Smelt (*Hypomesus transpacificus*). Aquaculture 450, 108–115.

Healey, M., Goodwin, P., Dettinger, M., Norgaard, R., 2016. The state of Bay–Delta science 2016: an introduction. San Franc. Estuary Watershed Sci. 14, 5.

Hedrick, R.P., McDowell, T., Groff, J., 1987. Mycobacteriosis in cultured striped bass from California. J. Wildl. Dis. 23, 391–395.

Hjort, J., 1914. Fluctuations in the Great Fisheries of Northern Europe Viewed in the Light of Biological Research. ICES.

Hobbs, J., Moyle, P.B., Fangue, N., Connon, R.E., 2017. Is extinction inevitable for Delta Smelt and Longfin Smelt? An opinion and recommendations for recovery. San Franc. Estuary Watershed Sci. 15, 2.

Hobbs, J.A., Lewis, L.S., Willmes, M., Denney, C., Bush, E., 2019. Complex life histories discovered in a critically endangered fish. Sci. Rep. 9, 1–12.

Hsieh, B.H., Deng, J.F., Ger, J., Tsai, W.J., 2001. Acetylcholinesterase inhibition and the extrapyramidal syndrome: a review of the neurotoxicity of organophosphate. Neurotoxicology 22, 423–427.

Hung, T.C., Piedrahita, R.H., 2011. The performance and impact of a bubble-wash bead filter in a recirculating green water larval culture system for delta smelt (*Hypomesus transpacificus*). Aquac. Eng. 45, 60–65.

Hung, T.C., Stevenson, T., Sandford, M., Ghebremariam, T., 2018. Temperature, density and ammonia effects on growth and fecundity of the ramshorn snail (*Helisoma anceps*). Aquac. Res. 49, 1072–1079.

Hung, T.C., Rosales, M., Kurobe, T., Stevenson, T., Ellison, L., Tigan, G., Sandford, M., Lam, C., Schultz, A., Teh, S., 2019. A pilot study of the performance of captive-reared delta smelt *Hypomesus transpacificus* in a semi-natural environment. J. Fish Biol. 95, 1517–1522.

International Union for Conservation of Nature, 2014. Red List of Threatened Species. Version 2014.3. www.iucnredlist.org. Downloaded on 27 May 2015.

Jackson, W.T., Paterson, A.M., 1977. The Sacramento-San Joaquin Delta: The Evolution and Implementation of Water Policy: An Historical Perspective. University of California Water Resources Center, UC Berkeley.

Jassby, A.D., Cloern, J.E., 2000. Organic matter sources and rehabilitation of the Sacramento–San Joaquin Delta (California, USA). Aquat. Conserv. 10, 323–352.

Jeffries, K.M., Komoroske, L.M., Truong, J., Werner, I., Hasenbein, M., Hasenbein, S., Fangue, N.A., Connon, R.E., 2015. The transcriptome-wide effects of exposure to a pyrethroid pesticide on the Critically Endangered delta smelt *Hypomesus transpacificus*. Endanger. Species Res. 28, 43–60.

Jeffries, K.M., Connon, R.E., Davis, B.E., Komoroske, L.M., Britton, M.T., Sommer, T., Todgham, A.E., Fangue, N.A., 2016. Effects of high temperatures on threatened estuarine fishes during periods of extreme drought. J. Exp. Biol. 219, 1705–1716.

Jeffries, K.M., Jeffrey, J.D., Holland, E.B., 2022. Applied aspects of gene function for the conservation of fishes. Fish Physiol. 39A, 389–433.

Jin, J., Kurobe, T., Ramírez-Duarte, W.F., Bolotaolo, M.B., Lam, C.H., Pandey, P.K., Hung, T.C., Stillway, M.E., Zweig, L., Caudill, J., Lin, L., 2018. Sub-lethal effects of herbicides penoxsulam, imazamox, fluridone and glyphosate on Delta Smelt (*Hypomesus transpacificus*). Aquat. Toxicol. 197, 79–88.

Kammerer, B.D., Hung, T.C., Baxter, R.D., Teh, S.J., 2016. Physiological effects of salinity on Delta Smelt, *Hypomesus transpacificus*. Fish Physiol. Biochem. 42, 219–232.

Karlsson, S., Diserud, O.H., Fiske, P., Hindar, K., Grant, S.W., 2016. Widespread genetic introgression of escaped farmed Atlantic salmon in wild salmon populations. ICES J. Mar. Sci. 73, 2488–2498.

Kim, S.E., Urpelainen, J., 2018. Environmental public opinion in US states, 1973–2012. Environ. Pollut. 27, 89–114.

Kimmerer, W.J., 2002. Physical, biological, and management responses to variable freshwater flow into the San Francisco Estuary. Estuaries 25, 1275–1290.

Kimmerer, W.J., Gross, E.S., MacWilliams, M.L., 2009. Is the response of estuarine nekton to freshwater flow in the San Francisco Estuary explained by variation in habitat volume? Estuar. Coasts 32, 375–389.

Komoroske, L.M., Connon, R.E., Lindberg, J., Cheng, B.S., Castillo, G., Hasenbein, M., Fangue, N.A., 2014. Ontogeny influences sensitivity to climate change stressors in an endangered fish. Conserv. Physiol. 2, cou008.

Komoroske, L.M., Connon, R.E., Jeffries, K.M., Fangue, N.A., 2015. Linking transcriptional responses to organismal tolerance reveals mechanisms of thermal sensitivity in a mesothermal endangered fish. Mol. Ecol. 24, 4960–4981.

Komoroske, L.M., Jeffries, K.M., Connon, R.E., Dexter, J., Hasenbein, M., Verhille, C., Fangue, N.A., 2016. Sublethal salinity stress contributes to habitat limitation in an endangered estuarine fish. Evol. Appl. 9, 963–981.

Komoroske, L.M., Jeffries, K.M., Whitehead, A., Roach, J.L., Britton, M., Connon, R.E., Verhille, C., Brander, S.M., Fangue, N.A., 2020. Transcriptional flexibility during thermal challenge corresponds with expanded thermal tolerance in an invasive compared to native fish. Evol. Appl. 14, 931–949.

Lessard, J., Cavallo, B., Anders, P., Sommer, T., Schreier, B., Gille, D., Schreier, A., Finger, A., Hung, T.C., Hobbs, J., May, B., 2018. Considerations for the use of captive-reared delta smelt for species recovery and research. San Franc. Estuary Watershed Sci. 16, 3.

Lewis, L.S., Denney, C., Willmes, M., Xieu, W., Fichman, R.A., Zhao, F., Hammock, B.G., Schultz, A., Fangue, N., Hobbs, J.A., 2021. Otolith-based approaches indicate strong effects of environmental variation on growth of a Critically Endangered estuarine fish. Mar. Ecol. Prog. Ser. 676, 37–56.

Lindberg, J.C., Baskerville-Bridges, B., Van Eenennaam, J.P., Doroshov, S.I., 1999. Development of Delta Smelt Culture Techniques: Year-End Report 1999. California Department of Water Resources report (DWR B-81581), Sacramento, California.

Lindberg, J., Baskerville-Bridges, B., Doroshov, S., 2000. Update on delta smelt culture with an emphasis on larval feeding behavior. IEP Newsl. 13, 45–49.

Lindberg, J.C., Tigan, G., Ellison, L., Rettinghouse, T., Nagel, M.M., Fisch, K.M., 2013. Aquaculture methods for a genetically managed population of endangered Delta Smelt. N. Am. J. Aquac. 75, 186–196.

Loboschefsky, E., Benigno, G., Sommer, T., Rose, K., Ginn, T., Massoudieh, A., Loge, F., 2012. Individual-level and population-level historical prey demand of San Francisco Estuary Striped Bass using a bioenergetics model. San Franc. Estuary Watershed Sci. 10, 1.

MacNally, R., Thomson, J.R., Kimmerer, W.J., Feyrer, F., Newman, K.B., Sih, A., Bennett, W.A., Brown, L., Fleishman, E., Culberson, S.D., Castillo, G., 2010. Analysis of pelagic species in decline in the upper San Francisco Estuary using multivariate autoregressive modeling (MAR). Ecol. Appl. 20, 1417–1430.

Madera Tribune, 1941. Valley Project to be Speeded Will Become Part of National Defense Steps, Declares Director. Madera Tribune. 28 July 1941. LXXVIII (47). Accessed 25 April 2021.

Mahardja, B., Conrad, J.L., Lusher, L., Schreier, B., 2016. Abundance trends, distribution, and habitat associations of the invasive Mississippi Silverside (*Menidia audens*) in the Sacramento–San Joaquin Delta, California, USA. San Franc. Estuary Watershed Sci. 14, 2.

Major, K.M., DeCourten, B.M., Li, J., Britton, M., Settles, M.L., Mehinto, A.C., Connon, R.E., Brander, S.M., 2020. Early life exposure to environmentally relevant levels of endocrine disruptors drive multigenerational and transgenerational epigenetic changes in a fish model. Front. Mar. Sci. 7, 471.

Mayfield, R.B., Cech Jr., J.J., 2004. Temperature effects on green sturgeon bioenergetics. Trans. Am. Fish. Soc. 133, 961–970.

Middle Rio Grande Endangered Species Collaborative Program, 2015. Fiscal Year 2015 Annual Report. Bureau of Reclamation Contract #R16PX00944. U.S. Geological Survey.

Milligan, L.P., McBride, B.W., 1985. Energy costs of ion pumping by animal tissues. J. Nutr. 115, 1374–1382.

Moschet, C., Lew, B.M., Hasenbein, S., Anumol, T., Young, T.M., 2017. LC-and GC-QTOF-MS as complementary tools for a comprehensive micropollutant analysis in aquatic systems. Environ. Sci. Technol. 51, 1553–1561.

Moyle, P.B., 2002. Inland Fishes of California: Revised and Expanded. University of California Press.

Moyle, P.B., Herbold, B., Stevens, D.E., Miller, L.W., 1992. Life history and status of delta smelt in the Sacramento-San Joaquin Estuary, California. Trans. Am. Fish. Soc. 121, 67–77.

Moyle, P.B., Brown, L.R., Durand, J.R., Hobbs, J.A., 2016. Delta smelt: life history and decline of a once-abundant species in the San Francisco estuary. San Franc. Estuary Watershed Sci. 14, 6.

Moyle, P.B., Hobbs, J.A., Durand, J.R., 2018. Delta Smelt and water politics in California. Fisheries 43, 42–50.

Mundy, P.C., Carte, M.F., Brander, S.M., Hung, T.C., Fangue, N., Connon, R.E., 2020. Bifenthrin exposure causes hyperactivity in early larval stages of an endangered fish species at concentrations that occur during their hatching season. Aquat. Toxicol. 228, 105611.

Mundy, P.C., Hartz, K.E.H., Fulton, C.A., Lydy, M.J., Brander, S.M., Hung, T.C., Fangue, N., Connon, R.E., 2021. Exposure to permethrin or chlorpyrifos causes differential dose-and time-dependent behavioral effects at early larval stages of an endangered teleost species. Endanger. Species Res. 44, 89–103.

Murphy, D.D., Weiland, P.S., 2019. The low-salinity zone in the San Francisco Estuary as a proxy for delta smelt habitat: a case study in the misuse of surrogates in conservation planning. Ecol. Indic. 105, 29–35.

Naas, K.E., Harboe, T., 1992. Enhanced first feeding of halibut larvae (*Hippoglossus hippoglossus* L.) in green water. Aquaculture 105, 143–156.

Nichols, F.H., Cloern, J.E., Luoma, S.N., Peterson, D.H., 1986. The modification of an estuary. Science 231, 567–573.

Nobriga, M.L., Rosenfield, J.A., 2016. Population dynamics of an estuarine forage fish: disaggregating forces driving long-term decline of Longfin Smelt in California's San Francisco Estuary. Trans. Am. Fish. Soc. 145, 44–58.

Nobriga, M., Hymanson, Z., Oltmann, R., 2000. Environmental factors influencing the distribution and salvage of young delta smelt: a comparison of factors occurring in 1996 and 1999. IEP Newsl. 13, 55–65.

Nobriga, M.L., Sommer, T.R., Feyrer, F., Fleming, K., 2008. Longterm trends in summertime habitat suitability for delta smelt (*Hyposmesus transpacificus*). San Franc. Estuary Watershed Sci. 6, 1.

Paganini, A., Kimmerer, W.J., Stillman, J.H., 2010. Metabolic responses to environmental salinity in the invasive clam *Corbula amurensis*. Aquat. Biol. 11, 139–147.

Pannevis, M.C., Houlihan, D.F., 1992. The energetic cost of protein synthesis in isolated hepatocytes of rainbow trout (Oncorhynchus mykiss). J. Comp. Physiol. B. 162, 393–400.

Radtke, L.D., 1966. Distribution of Smelt, Juvenile Sturgeon, and Starry Flounder in the Sacramento-San Joaquin Delta. California Department of Fish and Game. Fish Bull., 136.

Ray, D.E., Fry, J.R., 2006. A reassessment of the neurotoxicity of pyrethroid insecticides. Pharmacol. Ther. 111, 174–193.

Rieger, P.W., Summerfelt, R.C., 1997. The influence of turbidity on larval walleye, Stizostedion vitreum, behavior and development in tank culture. Aquaculture 159, 19–32.

Romney, A.L., Yanagitsuru, Y.R., Mundy, P.C., Fangue, N.A., Hung, T.C., Brander, S.M., Connon, R.E., 2019. Developmental staging and salinity tolerance in embryos of the delta smelt, *Hypomesus transpacificus*. Aquaculture 511, 634191.

Saglam, I.K., Hobbs, J.A., Baxter, R., Lewis, L.S., Benjamin, A., Finger, A.J., 2021. Genome-wide analysis reveals regional patterns of drift, structure, and gene flow in longfin smelt (*Spirinchus thaleichthys*) in the northeastern Pacific. Can. J. Fish. Aquat. Sci. 78, 1793–1804.

Sardella, B.A., Sanmarti, E., Kültz, D., 2008. The acute temperature tolerance of green sturgeon (Acipenser medirostris) and the effect of environmental salinity. J. Exp. Zool. A Ecol. Genet. Physiol. 309, 477–483.

Schoellhamer, D.H., 2011. Sudden clearing of estuarine waters upon crossing the threshold from transport to supply regulation of sediment transport as an erodible sediment pool is depleted: San Francisco Bay, 1999. Estuar. Coasts 34, 885–899.

Schreier, B.M., Baerwald, M.R., Conrad, J.L., Schumer, G., May, B., 2016. Examination of predation on early life stage Delta Smelt in the San Francisco estuary using DNA diet analysis. Trans. Am. Fish. Soc. 145, 723–733.

Segarra, A., Mauduit, F., Amer, N.R., Biefel, F., Hladik, M.L., Connon, R.E., Brander, S.M., 2021. Salinity changes the dynamics of pyrethroid toxicity in terms of behavioral effects on newly hatched delta smelt larvae. Toxics 9, 40.

Sherman, S., Hartman, R., Contreras, D., 2017. Effects of Tidal Wetland Restoration on Fish: A Suite of Conceptual Models. Interagency Ecological Program Technical Reports. pp. 1–358.

Sifa, L., Mathias, J.A., 1987. The critical period of high mortality of larvae fish—a discussion based on current research. Chin. J. Oceanol. Limnol. 5, 80–96.

Smith, W.E., Polansky, L., Nobriga, M.L., 2021. Disentangling risks to an endangered fish: using a state-space life cycle model to separate natural mortality from anthropogenic losses. Can. J. Fish. Aquat. Sci. 99, 1–22.

Sobczak, W.V., Cloern, J.E., Jassby, A.D., Müller-Solger, A.B., 2002. Bioavailability of organic matter in a highly disturbed estuary: the role of detrital and algal resources. Proc. Natl. Acad. Sci. 99, 8101–8105.

Sommer, T., 2020. How to respond? An introduction to current bay-delta natural resources management options. San Franc. Estuary Watershed Sci. 18, 1.

Sommer, T., Armor, C., Baxter, R., Breuer, R., Brown, L., Chotkowski, M., Culberson, S., Feyrer, F., Gingras, M., Herbold, B., Kimmerer, W., 2007. The collapse of pelagic fishes in the upper San Francisco Estuary: El colapso de los peces pelagicos en la cabecera del Estuario San Francisco. Fisheries 32, 270–277.

Sommer, T., Mejia, F.H., Nobriga, M.L., Feyrer, F., Grimaldo, L., 2011. The spawning migration of delta smelt in the upper San Francisco estuary journal issue. San Franc. Estuary Watershed Sci. 9, 2.

Stevens, D.E., Miller, L.W., 1983. Effects of river flow on abundance of young Chinook salmon, American shad, longfin smelt, and delta smelt in the Sacramento-San Joaquin River system. N. Am. J. Fish Manag. 3, 425–437.

Stuart, K.R., Drawbridge, M., 2011. The effect of light intensity and green water on survival and growth of cultured larval California yellowtail (*Seriola lalandi*). Aquaculture 321, 152–156.

Swanson, C., Mager, R.C., Doroshov, S.I., Cech, J.J., Jr., 1996. Use of salts, anesthetics, and polymers to minimize handling and transport mortality in delta smelt. Trans. Am. Fish. Soc. 125, 326–329.

Swanson, C., Young, P.S., Cech, J.J., 1998. Swimming performance of delta smelt: maximum performance, and behavioral and kinematic limitations on swimming at submaximal velocities. J. Exp. Biol. 201, 333–345.

Swanson, C., Reid, T., Young, P.S., Cech, J.J., Jr., 2000. Comparative environmental tolerances of threatened delta smelt (*Hypomesus transpacificus*) and introduced wakasagi (*H. nipponensis*) in an altered California estuary. Oecologia 123, 384–390.

Sweetnam, D.A., Stevens, D.E., Chadwick, H.K., Petrovich, A., 1993. Report to the Fish and Game Commission: Status Review of the Delta Smelt (*Hypomesus Transpacificus*) in California. Department of Fish and Game, State of California, The Resources Agency.

Takekawa, J.Y., Woo, I., Spautz, H., Nur, N., Grenier, J.L., Malamud-Roam, K., Nordby, J.C., Cohen, A.N., Malamud-Roam, F., La Cruz, S.E.W., 2006. Environmental threats to tidal-marsh vertebrates of the San Francisco Bay estuary. Stud. Avian Biol. 32, 176.

Tempel, T.L., Malinich, T.D., Burns, J., Barros, A., Burdi, C.E., Hobbs, J.A., 2021. The value of long-term monitoring of the San Francisco Estuary for Delta Smelt and Longfin Smelt. Calif. Fish Game 107, 148–171.

Tigan, G., Mulvaney, W., Ellison, L., Schultz, A., Hung, T.C., 2020. Effects of light and turbidity on feeding, growth, and survival of larval Delta Smelt (*Hypomesus transpacificus*, Actinopterygii, Osmeridae). Hydrobiologia 847, 2883–2894.

Trenham, P.C., Shaffer, H.B., Moyle, P.B., 1998. Biochemical identification and assessment of population subdivision in morphologically similar native and invading smelt species (Hypomesus) in the Sacramento–San Joaquin estuary, California. Trans. Am. Fish. Soc. 127, 417–424.

U.S. Bureau of Reclamation, 2019. Biological Opinion for the Reinitiation of Consultation on the Coordinated Operations of the Central Valley Project and State Water Project. National Marine Fisheries Service, National Oceanic and Atmospheric Administration, U.S. Department of Commerce.

U.S. Bureau of Reclamation, 2019. Reclamation Managing Water in the West. Environmental Assessment. Delta Smelt Fall Habitat Action in 2019. U.S. Department of the Interior.

U.S. Department of the Interior, 2020. Fiscal Year 2020 Obligation Plan for CVPIA Authorities. Central Valley Project, California. Interior Region 10—California-Great Basin. U.S. Bureau of Reclamation.

U.S. Environmental Protection Agency, 2011. EPA Decision Concerning California's 2008–2010 Clean Water Act Section 303(d) List. U.S. Environmental Protection Agency.

U.S. Fish and Wildlife Service, 2008. Formal Endangered Species Act Consultation on the Proposed Coordinated Operations of the Central Valley Project (CVP) and State Water Project (SWP). United States Department of the Interior.

U.S. Fish and Wildlife Service, Senegal, T., Mckenzie, R., Speegle, J., Perales, B., Bridgman, D., Erly, K., Staiger, S., Arrambide, A., Gilbert, M., 2022. Delta smelt supplementation strategy. In: Interagency Ecological Program and US Fish and Wildlife Service: San Francisco Estuary Enhanced Delta Smelt Monitoring Program Data, 2016-2021 ver 8. Environmental Data Initiative.

U.S. Office of the Federal Register, 1993. Endangered and threatened wildlife and plants: determination of threatened status for the Delta Smelt. Fed. Regist. 58 42 (5 March 1993), 12854–12864.

U.S. Office of the Federal Register, 2010. Endangered and threatened wildlife and plants; 12-month finding on a petition to reclassify the delta smelt from threatened to endangered throughout its range. Fed. Regist. 75 66 (April 7,2010), 17667–17680.

U.S. Office of the Federal Register, 2020. Endangered and threatened wildlife and plants; review of domestic species that are candidates for listing as endangered or threatened; annual notification of findings on resubmitted petitions; annual descriptions of progress on listing actions. Fed. Regist. 85 221 (16 November 2020), 73164–73179.

Wang, J.C., 1991. Early Life Stages and Early Life History of the Delta Smelt, Hypomesus transpacificus, in the Sacramento-San Joaquin Estuary, With Comparison of Early Life Stages of the Longfin Smelt, Spirinchus thaleichthys. Interagency Ecological Study Program for the Sacramento-San Joaquin Estuary.

Wang, Y.Q., Li, Y.W., Chen, Q.L., Liu, Z.H., 2019. Long-term exposure of xenoestrogens with environmental relevant concentrations disrupted spermatogenesis of zebrafish through altering sex hormone balance, stimulating germ cell proliferation, meiosis and enhancing apoptosis. Environ. Pollut. 244, 486–494.

Waters, C.D., Hard, J.J., Brieuc, M.S., Fast, D.E., Warheit, K.I., Waples, R.S., Knudsen, C.M., Bosch, W.J., Naish, K.A., 2015. Effectiveness of managed gene flow in reducing genetic divergence associated with captive breeding. Evol. Appl. 8, 956–971.

Weston, D.P., Asbell, A.M., Lesmeister, S.A., Teh, S.J., Lydy, M.J., 2014. Urban and agricultural pesticide inputs to a critical habitat for the threatened delta smelt (*Hypomesus transpacificus*). Environ. Toxicol. Chem. 33, 920–929.

Weston, D.P., Chen, D., Lydy, M.J., 2015a. Stormwater-related transport of the insecticides bifenthrin, fipronil, imidacloprid, and chlorpyrifos into a tidal wetland, San Francisco Bay, California. Sci. Total. Environ. 527, 18–25.

Weston, D.P., Schlenk, D., Riar, N., Lydy, M.J., Brooks, M.L., 2015b. Effects of pyrethroid insecticides in urban runoff on Chinook salmon, steelhead trout, and their invertebrate prey. Environ. Toxicol. Chem. 34, 649–657.

Weston, D., Moschet, C., Young, T., Johanif, N., Poynton, H., Major, K., Connon, R., Hasenbein, S., 2019. Chemical and toxicological impacts to Cache Slough following storm-driven contaminant inputs. San Franc. Estuary Watershed Sci. 17, 3.

Whipple, A.A., Grossinger, R.M., Rankin, D., Stanford, B., Askevold, R.A., 2012. Sacramento-San Joaquin Delta Historical Ecology Investigation: Exploring Pattern and Process. San Francisco Estuary Institute-Aquatic Science Center, Richmond, CA.

White, J.W., Cole, B.J., Cherr, G.N., Connon, R.E., Brander, S.M., 2017. Scaling up endocrine disruption effects from individuals to populations: outcomes depend on how many males a population needs. Environ. Sci. Technol. 51, 1802–1810.

Yanagitsuru, Y.R., Main, M.A., Lewis, L.S., Hobbs, J.A., Hung, T, Connon, R.E., Fangue, N.A., 2021. Effects of temperature on hatching and growth performance of embryos and yolk-sac larvae of a threatened estuarine fish: longfin smelt (Spirinchus thaleichthys). Aquaculture 537, 736502.

Yoshiyama, R.M., Gerstung, E.R., Fisher, F.W., Moyle, P.B., 2001. Historical and present distribution of Chinook salmon in the Central Valley drainage of California. Fish. Bull. 179, 71–176.

Young, P.S., Cech J.J., Jr., 1996. Environmental tolerances and requirements of splittail. Trans. Am. Fish. Soc. 125, 664–678.

Zillig, K.W., Cocherell, D.E., Fangue, N.A., 2020. Interpopulation Variation Among Juvenile Chinook Salmon From California and Oregon. The United States Environmental Protection Agency Region 9—Pacific Southwest Region, San Francisco, CA. https://www.epa.gov/sfbay-delta/2020-chinook-salmon-interpopulation-thermal-tolerance-investigation.

Chapter 2

# Conservation aquaculture—A sturgeon story

W. Gary Anderson[a,*], Andrea Schreier[b], and James A. Crossman[c]
[a]Department of Biological Sciences, University of Manitoba, Winnipeg, MB, Canada
[b]Department of Animal Science, College of Agricultural and Environmental Sciences, University of California, Davis, CA, United States
[c]Fish and Aquatics, BC Hydro, Castlegar, BC, Canada
*Corresponding author: e-mail: Gary.Anderson@umanitoba.ca

## Chapter Outline



Sturgeon are found circumglobally throughout the northern hemisphere. Longevity, late age to sexual maturation and infrequent spawning are typical life history strategies that have supported their evolution over millennia but at the same time have made them vulnerable to extirpation or extinction when population sizes have been reduced as a result of overharvest, habitat degradation and/or pollution. As a consequence sturgeons are among the most at risk species as listed by the International Union for the Conservation of Nature. Aquaculture has been used as a conservation strategy to arrest declines or

Fish Physiology, Vol. 39B. https://doi.org/10.1016/bs.fp.2022.04.013

maintain current population levels for several sturgeon species. Here we describe the development of this conservation strategy in sturgeons, examining the importance of understanding genotype and environment in development of appropriate rearing strategies toward improved conservation practice for this imperiled group of fishes. Recent research indicates a significant effect of environment on phenotype, with early life history being particularly relevant; evidence suggests that repatriation should be the preferred approach where possible. We suggest that while aquaculture can be a valuable conservation and recovery tool, it is particularly effective when the impacts on phenotypic development during early life history are considered and combined with effective post-release monitoring and implementation of additional restoration efforts.

## 1 Introduction

Since the late 20th century the aquaculture industry has frequently been viewed with skepticism by conservation practitioners (Brannon, 1993; Goodman, 1990; Hillborn, 1992). There is substance to this view as industrial aquaculture operations may play a role in the extirpation of wild populations through, among other things, disruption of native habitat (Islam, 2005); introduction of non-native species (Fisher et al., 2014); dilution of wild genetic integrity (Allendorf and Phelps, 1980; Naylor et al., 2005); introduction of parasitic corridors (Larsen and Vormedal, 2021); and removal of broodstock (Williot, 2011). In recent times, practices have somewhat altered the perception of aquaculture's environmental impact. Continual advances in technology alongside an ever-increasing understanding and implementation of best practices to protect and preserve the genetic integrity of wild stocks means we are now in a position where focus on effective mitigation of environmental impacts created by the global aquaculture industry is critically important. As a benefit, improvements in the aquaculture industry have, by extension, led to significant improvements in the use of hatcheries toward conservation of wild populations (Edwards, 2015) as well as a sustainable source of fish for human consumption that alleviates pressure on declining wild populations.

Conservation aquaculture can be defined broadly as the intent to rear aquatic organisms for the enhancement, maintenance or repatriation of wild stocks as a natural resource (Froehlich et al., 2017) and is by no means fish specific. However, for the purposes of this chapter we will apply this definition to fish and in particular sturgeons. At face value conservation aquaculture uses aquaculture techniques to conserve or protect endangered fish populations (Anders, 1998; Flagg and Nash, 1999), but important distinctions should be made regarding the ultimate intent of conservation aquaculture: to preserve the fish or fishery? While the two are often not mutually exclusive the ultimate goals of either are quite different and have been defined in the Ramsar declaration for global sturgeon conservation (2006), where the expressed intent to preserve the fish must be accompanied by a variety of additional factors such as habitat restoration and maintenance of genetic integrity. That said, a successful conservation aquaculture program that conserves or restores a species or population could ultimately be a victim of its own success, possibly supporting a fishery in perpetuity. Thus, the initial objectives of a conservation aquaculture program become integral to its long-term success and implementation. However, much of our discussion in this chapter will focus on the role and evolution of conservation aquaculture toward conserving the species and not necessarily supporting a fishery, but it is important to recognize the importance of aquaculture in restoring or maintaining a fishery; a practice that is conducted globally to great effect. To that end, the history of conservation aquaculture for salmonids in the Pacific Northwest serves as an excellent case study toward understanding and development of best practices in using aquaculture to conserve multiple species and fisheries that might be considered as threatened or endangered (Hillborn, 1992; MacKinlay et al., 2004).

Hatchery programs began in British Columbia (BC), Canada in the late 19th/early 20th century primarily to mitigate the adverse impact mining operations had on habitat, but these programs had little impact on existing populations (Foerster, 1968). However, in response to rapidly declining salmon stocks in the 1970's, the Salmon Enhancement Program (SEP) was established in BC, Canada to improve freshwater survival (Larkin, 1974). This heralded a dramatic shift in the production and release of hatchery-reared fish as one tool used by fisheries managers to arrest these declines (MacKinlay et al., 2004).

In simplistic terms conservation aquaculture relies on the collection of gametes from wild caught broodstock native to the river system of concern. Although historically some programs have used gametes from non-native broodstock, this practice is now much less common and only considered if the outlook for a native population is particularly bleak. These gametes are then fertilized, embryos are incubated and at hatch fish are reared within a controlled environment until release. For salmon this would normally be the alevin or fry stage, but timing of release will vary depending on management strategy and/or known population bottlenecks (MacKinlay et al., 2004). Importantly, conservation aquaculture programs will control some or all of these steps, typically to optimize certain aspects of the species biology (e.g., survival, growth). Thus, at some point during early rearing fish will spend a reasonable amount of time in an environment that is different from the natural systems in which the species has evolved. Further, hatchery rearing can lead to highly selective processes through the breeding, rearing, and stocking stages—all of which may be contrary to natural processes (e.g., mate selection, exploratory behavior, predator avoidance) that otherwise might occur uninhibited or unmanipulated in the wild environment (Osborne et al., 2020).

Upon inception of the approaches used in conservation aquaculture it was often erroneously considered that the more fish that were released the greater the chances of success (Hillborn, 1992), in particular since life stage specific survival in the wild was not fully understood. Since the establishment of the SEP, billions of salmon fry have been released up and down the west coast of North America with the intent of increasing productivity and maximizing long-term stock viability of Pacific salmon populations (MacKinlay et al., 2004). The approach has had some degree of success, but it has also highlighted some significant failures in conservation aquaculture programs (Anders, 1998; Brannon, 1993; Hillborn, 1992), one of which being the negative implications of overstocking. Nonetheless, significant improvements have been made in the adaptive management of programs (Walters, 2007) to both respond to new information and develop alternate approaches in the rearing process. Regardless of the species, results consistently demonstrate the physiological and behavioral traits of hatchery-reared salmon are different from their wild conspecifics (Chittenden et al., 2008; Greene, 1952; Shrimpton et al., 1994). Furthermore, despite actions to maximize or maintain wild genotypes (inherited genetic makeup) of fish reared in conservation aquaculture (MacKinlay et al., 2004), it has been shown the hatchery rearing environment has a significant impact on phenotype (expression of physiological and/or behavioral traits) of reared fish when compared to wild conspecifics (Chittenden et al., 2010), underscoring the critical nature of understanding longer-term impacts of early rearing environments in conservation aquaculture on population persistence.

## 1.1 The sturgeon story

Sturgeons are one of the most imperiled groups of animals currently on the Red List of Threatened Species with the International Union of Conservation of Nature (IUCN), with up to 85% of all sturgeon at risk of extinction (Table 1). The recovery of sturgeon species has been hindered by their unique life history characteristics evolved over millions of years that include longevity in excess of 100 years, sexually dimorphic age to maturation, and intermittent spawning, with males reproducing more frequently than females (Billard and Lecointre, 2001; Haxton et al., 2016). While these characteristics allow for adaptations to change over a long time period, sturgeon have faced significant anthropogenic impacts occurring within a few generations for most species. Long-term adaptation to historic conditions such as migration distance, habitat requirements, water quality and the close link between flow

**TABLE 1** Conservation aquaculture programs for sturgeon, with status, location of stocking (population), date started, date ended (if known), and source of progeny.

| | IUCN Status | Population | Date started | Date ended | Progeny source | References |
|---|---|---|---|---|---|---|
| Adriatic sturgeon (*Acipenser naccarii*) | CE | Lombardy Basin | 1991 | | CB | Arlati and Poliakova (2009), Boscari and Congiu (2014), Paschos et al. (2008) |
| | | Vemeto Basin | 1999 | | CB | |
| | | Kalamas River | 2000 | | CB | |
| | | Po River | ? | | CB | |
| Amur sturgeon (*Acipenser schrencki*) | CE | Amur River Basin | 1988 | | WB | Chebanov and Billard (2001), Wei et al. (2004) |
| Atlantic sturgeon (*Acipenser oxyrinchus oxyrinchus*) | NT, EXW (Baltic) | Hudson River | 1994 | | WB | Butkauskas et al. (2019), J. Gessner (IGB Berlin, pers. comm.) Henderson et al. (2004), Kolman et al. (2011), Mohler et al. (2012), Purvina and Medne (2018), St. Pierre (1999), Secor et al. (2000a) |
| | | Nanticoke River | 1996 | | WB? | |
| | | Vistula drainage | 2006 | | WB, CB[a] | |
| | | Oder drainage | 2006 | 2009 | WB, CB[a] | |
| | | Neris River | 2011 | 2020 | WB, CB | |
| | | Šventoji River | 2011 | 2020 | WB, CB | |
| | | Daugava River | 2013 | | WB, CB | |
| | | Narva River | 2015 | | WB, CB | |
| | | Parnü River | 2021 | | WB, CB | |
| | | Volkhov River | 2022 | | WB, CB | |

Continued

**TABLE 1** Conservation aquaculture programs for sturgeon, with status, location of stocking (population), date started, date ended (if known), and source of progeny.—Cont'd

| | IUCN Status | Population | Date started | Date ended | Progeny source | References |
|---|---|---|---|---|---|---|
| Beluga | CE | Caspian Sea basin | 1955 | | WB | Abdelhay and Tahori (2006), Antognazza et al. (2021), Chebanov et al. (2002), M. Chebanov (State Center for Sturgeon Conservation; pers. comm.), Holostenco et al. (2019), Secor et al. (2000b), Smederevac-Lalić et al. (2011), Williot et al. (2002a) |
| (*Huso huso*) | | Don River | 1950s | 1999 | WB | |
| | | Kuban River | 1976 | 1999[b] | WB | |
| | | Danube River | 1990s | | WB, CB | |
| | | Po River | 2019 | 2020 | CB | |
| Chinese sturgeon (*Acipenser sinensis*) | CE | Yangtze River | 1983 | | WB, CB | Wei et al. (1997), Wei et al. (2004) |
| European sturgeon (*Acipenser sturio*) | CE | Gironde-Garonne-Dordogne basin | 1995[c] | | WB, CB | Brevé et al. (2014), Gessner et al. (2011a), Gessner et al. (2011b), J. Gessner (IGB Berlin, pers. comm.), Kirschbaum et al. (2011),Williot, 2011 |
| | | Elbe River | 1996[d] | | CB | |
| | | Oste River | 1996[d] | | CB | |
| | | Stör River | 1996[d] | | CB | |
| | | Rhine River | 2012 | | CB | |
| | | Havel River | | | | |
| | | Mulde River | | | | |

| | | | | | | |
|---|---|---|---|---|---|---|
| Gulf sturgeon | NT | Suwannee River | 1992 | | WB | Neidig et al. (2002), Sulak |
| (*Acipenser oxyrinchus desotoi*) | | Hillsborough River | 2000 | 2001 | WB | et al. (2014) |
| Kaluga (*Huso dauricus*) | CE | Amur River Basin | 1992 | | WB, CB | Chebanov and Billard (2001), Koshelev et al. (2014), Li et al. (2013), Wei et al. (2004) |
| Lake sturgeon (*Acipenser fulvescens*) | LC | Red Cedar River | 1979 | 1982 | WB | Baker and Scribner (2017), Berkman et al. (2020), Bezold and Peterson (2008), Bruch et al. (2021), Chalupnicki et al. (2011), Dieterman et al. (2010), Dittman et al. (2015), Drauch and Rhodes (2007), Holst and Zollweg-Horan (2018), Ganus et al. (2018), Kampa et al. (2014), Lyons et al. (2000), McDougall et al. (2014), Runstrum et al. (2002), Smith (2009), Walker and Alford (2016), Welsh and Jackson (2014), Welsh et al. (2019), Welsh et al. (2020) |
| | | Menominee River | 1982 | 2006 | WB | |
| | | St. Louis River | 1983 | 2000[e] | WB | |
| | | Mississippi River | 1984 | 2015 | WB | |
| | | Lake Superior | 1989 | 2002 | WB | |
| | | Wisconsin River | 1990s | | WB | |
| | | Cheatham Reservoir | 1992 | | WB | |
| | | Clinch River | 1992 | | WB | |
| | | Missouri River | 1992 | 2015 | WB | |
| | | Grasse River | 1993 | | WB | |
| | | Black Lake (NY) | 1993 | | WB | |
| | | Manitowish River | 1994 | | WB | |
| | | Oswegatchie River | 1994 | | WB | |
| | | Wolf River system | 1994 | | WB | |
| | | Manitowoc River | 1994 | 2007 | WB | |
| | | Oneida Lake | 1995 | | WB | |
| | | Cayuga Lake | 1995 | | WB | |
| | | Yellow River | 1995 | | WB | |
| | | Clam River | 1995 | 2003 | WB | |
| | | Nelson River | 1990s | | WB | |
| | | St. Lawrence River | 1996 | | WB | |

*Continued*

**TABLE 1** Conservation aquaculture programs for sturgeon, with status, location of stocking (population), date started, date ended (if known), and source of progeny.—Cont'd

| | IUCN Status | Population | Date started | Date ended | Progeny source | References |
|---|---|---|---|---|---|---|
| | | Red Lake Basin | 1997 | 2007 | WB | |
| | | Detroit River (MN) | 1998 | 2007 | WB | |
| | | Otter Tail River | 1998 | 2007 | WB | |
| | | Biron Flowage | 1998 | 2003 | WB | |
| | | St. Regis River | 1998 | | WB | |
| | | Lake Wisconsin | 2000 | 2002 | WB | |
| | | Castle Rock Lake | 2000 | 2004 | WB | |
| | | French Broad River | 2000 | | WB | |
| | | Upper Tennessee River | 2000 | | WB | |
| | | Holston River | 2001 | | WB | |
| | | Black Lake (MI) | 2001 | | WB | |
| | | Eau Claire River | 2002 | 2003 | WB | |
| | | Buffalo River | 2002 | 2007 | WB | |
| | | Fort Louden Reservoir | 2002 | | WB | |
| | | Coosa River | 2002 | | | |
| | | Genesee River | 2003 | | WB | |
| | | Milwaukee River | 2003 | | WB | |
| | | Roseau River | 2004 | 2007 | WB | |
| | | Round Lake | 2004 | 2007 | WB | |
| | | Eastman River | 2004 | 2008 | WB | |

| | | | | | | |
|---|---|---|---|---|---|---|
| | | Raquette River | 2004 | | WB | |
| | | Manistee River | 2004 | | WB | |
| | | Flambeau River | 2005 | | WB | |
| | | Branch River | 2005 | 2007 | WB | |
| | | Saskatchewan River | 2005 | 2008 | WB | |
| | | Cumberland River | 2006 | | WB | |
| | | Salmon River | 2012 | | WB | |
| | | Lake Ontario | 2014 | 2015 | WB | |
| | | Minnesota River | ? | | WB | |
| | | Red River | ? | | WB | |
| | | Namekagon River | ? | | WB | |
| | | Winnipeg River | | | WB | |
| | | | | | WB | |
| Pallid sturgeon (*Scaphirhynchus albus*) | E | Atchafalaya River | 2004 | | WB | Saltzgiver et al. (2012), Steffensen et al. (2019), R. Wilson (USFWS, pers. comm.) |
| | | Mississippi River | 1992[f] | | WB | |
| | | Missouri River | 1992[f] | | WB, CB | |
| | | Platte River | 1997 | | WB | |
| | | Yellowstone River | 1998 | | WB, CB | |
| Persian sturgeon (*Acipenser persicus*) | CE | Caspian Sea basin | 1970s | | WB, CB | Abdelhay and Tahori (2006) |
| Russian sturgeon (*Acipenser gueldenstaedtii*) | CE | Caspian Sea basin | 1955 | | WB | Abdelhay and Tahori (2006), Chebanov and Billard (2001), Chebanov et al. (2002), Holostenco et al. (2019), Secor et al. (2000b), Smederevac-Lalić et al., 2011), Williot et al. (2002a) |
| | | Kuban River and reservoirs | 1950s | | WB, CB | |
| | | Don River | 1956 | | WB, CB | |
| | | Danube River | 1990s | | WB, CB | |
| | | Ural Sea basin | ? | | ? | |

*Continued*

**TABLE 1** Conservation aquaculture programs for sturgeon, with status, location of stocking (population), date started, date ended (if known), and source of progeny.—Cont'd

| | IUCN Status | Population | Date started | Date ended | Progeny source | References |
|---|---|---|---|---|---|---|
| Sakhalin sturgeon (*Acipenser mikadoi*) | CE | Tumnin River | 2005 | | WB, CB | J. Gessner (IGB Berlin, pers. comm.), Koshelev et al. (2012) |
| | | Lake Tunaicha | 2005 | 2009[b] | | |
| Ship sturgeon (*Acipenser nudiventris*) | CE | Caspian Sea basin | 1970s | | WB | Abdelhay and Tahori (2006), M. Chebanov (State Center for Sturgeon Conservation, pers. comm.), Lagutov and Lagutov (2008), Vasilyeva et al. (2019) |
| | | Ural River | ? | | WB | |
| | | Lake Baikal | 1970s | 1980s | CB | |
| | | Kuban River | 2017 | | CB | |
| | | Don River | 2017 | | CB | |
| Shortnose sturgeon (*Acipenser brevirostrum*) | V | Savannah River | 1984 | 1992 | WB | Quattro et al. (2002), Smith and Dingley (1984), Smith et al. (2002) |
| | | Cooper River | 1983 | | WB | |
| Shovelnose sturgeon (*Scaphirhynchus platyorhynchus*) | LC | Bighorn River | 1996 | 2020 | WB | Hogberg et al. (2021), Koch and Quist (2010) |
| | | Scioto River | ? | | ? | |
| | | Ohio River | ? | | ? | |
| Siberian sturgeon (*Acipenser baerii*) | E | Siberian Rivers | 1999 | | ? | Chebanov and Billard (2001), Korentovich and Litvenenko (2018), Q. Wei (Chinese Academy of Fishery Sciences, pers. comm.) |
| | | Lake Baikal | 1999 | | ? | |
| | | Selenga River | 2000 | 2020 | CB | |
| | | Ob River | 2007 | | ? | |

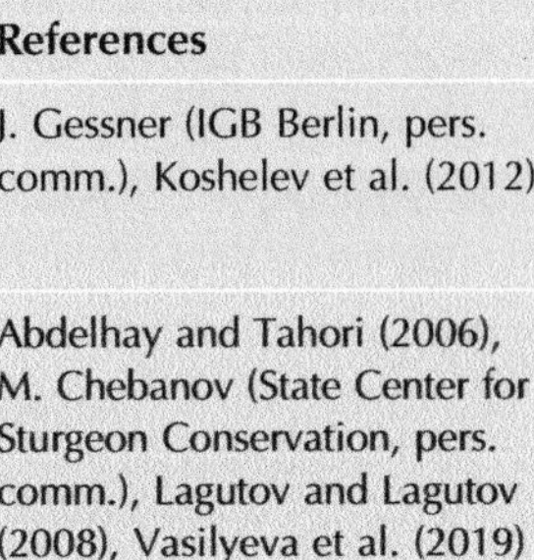

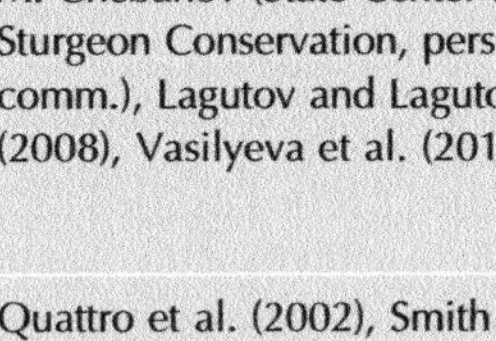

| | | | | | | |
|---|---|---|---|---|---|---|
| Sterlet (*Acipenser ruthenus*) | V | Dniester River | 1970s | | WB, CB | Friedrich et al. (2019), J. Gessner (IGB Berlin, pers. comm.), Holostenco et al. (2019), Kolman et al. (2016), Pegasov (2009) |
| | | Dnieper River | 1970s | | WB, CB | |
| | | Don River | 1970s | | WB, CB | |
| | | Kuban River | 1970s | | WB, CB | |
| | | Danube River basin | 1989[g] | | WB, CB | |
| | | Moscow River basin | ? | | ? | |
| Stellate sturgeon (*Acipenser stellatus*) | CE | Caspian Sea basin | 1955 | | WB | Abdelhay and Tahori (2006), Chebanov and Billard (2001), Chebanov et al. (2002), J. Gessner (IGB Berlin, pers. comm.) Holčik et al. (2006), Holostenco et al. (2019), Secor et al. (2000b), Smederevac-Lalić et al., 2011), Williot et al. (2002a) |
| | | Kuban River and reservoirs | 1950s | 2000[b] | WB | |
| | | Don River | 1956 | 1999[b] | WB | |
| | | Aral Sea basin | 1960s | | WB | |
| | | Danube River | 1990s | 2008[b] | WB | |
| | | Ural Sea basin | ? | | ? | |
| White sturgeon (*Acipenser transmontanus*) | LC | Columbia River | 2001 | | WB, R | Hildebrand et al. (2016), Ward et al. (2002) |
| | | Kootenai River | 1990 | | WB | |
| | | Snake River | 1989 | | WB, R | |
| | | Nechako River | 2015 | | WB, R | |
| Yangtze sturgeon (*Acipenser dabryanus*) | CE | Yangtze River | 1983 | | CB | Q. Wei (Chinese Academy of Fishery Sciences, pers. comm.), Wu et al. (2014), Zhang et al. (2011) |

CE=critically endangered, E=endangered LC=least concern, NT=near threatened, V=vulnerable. WB=wild broodstock, CB=captive broodstock, R=repatriation.
[a] *Captive broodstock in development.*
[b] *Stocking did not occur every year in this range.*
[c] *Recovery attempts for* Acipenser sturio *in France began in the mid-1980s, broodstock stated to be established in 1992 but releases did not occur until 1995.*
[d] *Culture of* Acipenser sturio *for release in Germany began in 1996 but releases did not occur until 2008.*
[e] *Progeny result of fertilized egg collection but not from St. Louis River adults so can't be classified as repatriation, per se.*
[f] *Culture of pallid sturgeon began in 1992 but releases did not occur until 1994.*
[g] *Culture of sterlet for conservation began as early as 1989 but releases did not occur until the early 1990s.*

and temperature as the key triggers for spawning have rendered the species vulnerable to recent anthropogenic changes.

A number of factors have been identified as causal in the ongoing decline in sturgeons with construction of dams blocking migratory corridors and disrupting habitat, overfishing, and pollution being the top threats regardless of region or species (Peterson et al., 2007; Williot et al., 2002a). However, realization of the decline of these fishes is not recent. Research into rearing and growing sturgeon in captivity first began in Russia in 1869 with the sterlet sturgeon, *Acipenser ruthenus* (Milshtein, 1969) and was quickly followed in 1876 by live culturing of lake sturgeon, *Acipenser fulvescens* in North America (Post, 1890). Conservation hatcheries with the intent to stock sturgeon have been running for quite some time but the most intense research focus on conservation aquaculture of sturgeon began in Russia and included components still relevant today such as the addition of silt to de-adhese eggs (Borodin, 1898); use of live food culturing techniques (reviewed in Vedrasco et al., 2002) and promotion of egg maturation and ovulation with hypophysial hormones (reviewed in: Barannikova, 1987). By the mid-late 20th century hundreds of millions of juveniles were being released to enhance existing stocks (Chebanov and Billard, 2001). However, as techniques were developed and our understanding of sturgeon biology improved, it became apparent that success of sturgeon conservation hatcheries was not simply a numbers game. This consideration is still relevant today where many thousands of fall-released young of the year lake sturgeon have a very poor survival rate relative to hundreds of spring released yearlings (McDougall et al., 2014). Therefore, as with Pacific salmon, it is evident a clearer understanding of rearing environment and the resulting effect on phenotype is required to maximize success, that is, to produce a fish with a wild genetic pedigree that when stocked into the environment is likely to survive and contribute to the next generation.

In this chapter we detail the developments in aquaculture of sturgeon with the intent of describing species and life history-specific knowledge to inform conservation and recovery actions. We begin with a description of the different methods and reasoning behind collection of progeny and/or broodstock for conservation purposes (Fig. 1). We provide a detailed examination of the impact of abiotic and biotic factors in early rearing on phenotypic development, understanding species differences in life history and the association they may have in the development of rearing environments most appropriate for a species-specific wild phenotype. We then discuss stocking approaches that can inform best practices and risks that must be considered. Many of the challenges in developing novel procedures in conservation aquaculture depend on measuring the success or failure of these approaches in the wild and we provide some examples of methods used to assess the effectiveness

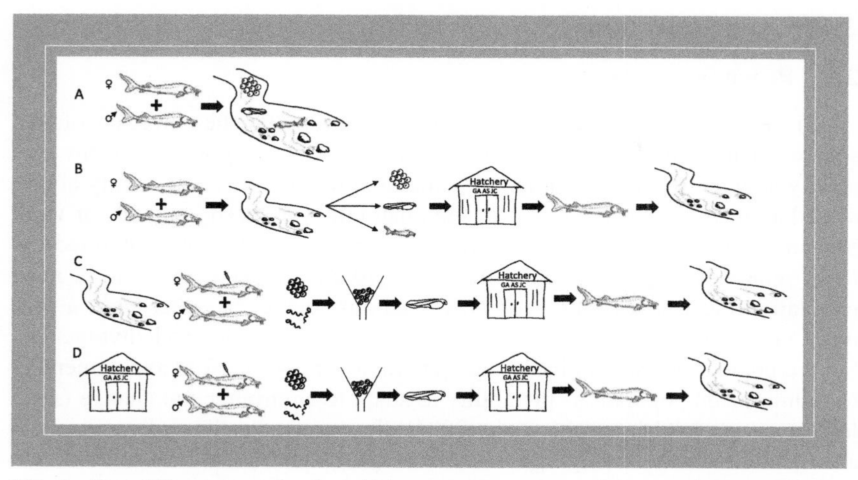


**FIG. 1** The different scenarios by which progeny may repopulate an existing population. (A) Represents a wild-type scenario where males and females will mate, embryo's hatch and larvae and juveniles grow-out. (B) Illustrates repatriation methods, where males and females mate in the wild and progeny are removed from the source location at one of three life history stages, embryogenesis, yolk-sac larvae or free-swimming and feeding larvae; grown-out in the hatchery before stocking at the source location as 0+ fingerlings in the fall or yearlings the following spring/early summer. (C) Illustrates capture of wild broodstock adults. Gametes can be collected from fish in the act of spawning or adults are transported and held in captivity where promotion of gamete maturation and ovulation occurs through hormone injection. In this example we assume that fish are captured and held in captivity for no longer then days to weeks. It is important to note that administration of GnRH is not always conducted as wild caught spawning males will frequently release milt for collection if captured at the appropriate time. What is less likely is the coordinated release of eggs from females so there is often a requirement to hold females for a period of time following hormone administration prior to collection of eggs. Eggs are then fertilized and embryo's incubated in a MacDonald hatching jar after de-adhesion. Fish are grown-out in the hatchery prior to restocking as fall fingerlings or spring yearlings. Wild adults are released near their capture location following spawning. (D) Illustrates captive broodstock where the process is similar to that described in C with the exception that adults are held in captivity as broodstock for months to years. Hormone administration is nearly always required for both males and females as broodstock adults will normally have been maintained within the hatchery an extended period of time, eggs are fertilized in the hatchery and fish are grown-out in the hatchery until release. Adults raised exclusively in a hatchery environment and used for broodstock would fall under category D. Use of such individuals for conservation purposes is not recommended unless it is a last resort due to the potential deleterious effects of domestication (see text for details). Finally, sperm can be sourced from cryopreserved samples to use in particularly dire situations, however, the resultant impact on phenotypic development in sturgeons using cryopreserved sperm are unknown.

of sturgeon conservation aquaculture programs. We close the chapter with a best-practices approach recognizing that there is no "one size fits all" solution to the conservation of these enigmatic fishes found in aquatic habitats across the northern hemisphere.

# 2 Progeny selection

## 2.1 Progeny source

When developing a conservation aquaculture program, the selection of the source population(s) is critical. For supplementation programs, progeny are ideally sourced from the recipient population, provided it is sufficiently abundant that adequate numbers of breeders, naturally produced gametes, or wild progeny can be collected and natural reproduction would not be harmed by these removals (Fig. 1). A reintroduction program seeking to create a new population must necessarily seek an exogenous source. In either case, a primary consideration in source selection is genetic diversity and divergence. Source populations should have high genetic diversity to afford supplemented or reintroduced populations the ability to adapt to environmental change (e.g., lake sturgeon; Welsh and McClain, 2004). Several studies have demonstrated an association between high genetic diversity and individual post-release survival (Scott et al., 2020) and population establishment (Zeisset and Beebee, 2012). Genetic differentiation between potential sources and recipient populations should also be considered for supplementation programs. Outbreeding depression resulting from crosses between genetically divergent populations has been reported for aquatic species (Byrne and Silla, 2020; Huff et al., 2011) though the risk of outbreeding may be minimized by following recommendations proposed by Frankham et al. (2011).

Once a source population(s) has been selected, hatchery practices can maximize genetic diversity in released cohorts. First, for programs using broodstock, numerous individuals can be crossed to produce progeny for release and full and partial factorial mating schemes can increase a cohort's effective population size (Ne) (Busack and Knudsen, 2007; Fiumera et al., 2004). Cohorts with high Ne will experience less genetic drift, retaining more genetic diversity than similar cohorts with low Ne. Minimizing the variance in family size within and among years so no one family is numerically overrepresented also increases Ne in a released cohort (Fisch et al., 2015). Another practice that increases genetic diversity preservation is using unique wild adults or progeny to establish cohorts each year rather than relying on a captive broodstock (Fig. 1). Captive broodstock can be defined as either:

**(1)** Wild caught fish held in a hatchery either temporarily or until the subsequent spawning season with the express intent of using these fish for gamete production and eventual release of the adults back to the wild, or;

**(2)** As progeny reared within the hatchery environment until adulthood, again with the express intent of using these individuals for gamete production. In this case adults may remain captive for multiple spawning cycles and never be released into the wild.

Importantly, broodstock raised from progeny born in captivity and broodstock maintained long-term (years) in the hatchery may well experience domestication selection, or unnatural selection pressures in a hatchery that may reduce the fitness of their progeny once released in the wild. Indeed, reduced fitness resulting from domestication selection has been observed in salmonids (Araki et al., 2007; Evans et al., 2014) and other fishes (Finger et al., 2018) and thus maintaining captive broodstock is not recommended for sturgeon conservation programs.

Generally, sturgeon aquaculture programs for conservation in North America have relied on wild adult sturgeon as a progeny source rather than establishing captive broodstocks (Table 1; Fig. 1). The Upper Missouri River pallid sturgeon (*Scaphirhynchus albus*) captive breeding program, which temporarily brings wild adults into the Gavins Point National Fish Hatchery, South Dakota, USA for spawning, also maintains a captive broodstock but their progeny are only used to supply juvenile pallid sturgeon for experimental stocking (Saltzgiver et al., 2012, E. Heist, Southern Illinois University, pers. comm.). Sturgeon supplementation programs typically source progeny from the recipient population (e.g., Ireland et al., 2002; St. Pierre, 1999). However, reintroduction programs have tended to source progeny from populations with abundant adults (e.g., Lake Winnebago lake sturgeon), nearby populations likely to be genetically similar, or both (e.g., Berkman et al., 2020; Welsh et al., 2019, 2020).

## 2.2 Progeny collection

There are a number of ways that wild adult sturgeon can be used to source progeny (Fig. 1). First, sexually mature adults can be captured on the spawning grounds and transported into a hatchery for spawning when ripe (e.g., St. Pierre, 1999), where hormonal induction of final maturation and ovulation of eggs in females and sperm in males can be conducted (Bayunova et al., 2006; Doroshov et al., 1997; Genz et al., 2014; Williot et al., 2002b) (Fig. 1). In extreme cases where the Ne is dangerously low or there is a critical need to increase the number of contributing males, cryopreserved sperm can be used to fertilize the eggs. However, this approach is not common in sturgeon conservation programs and to the authors' knowledge there are no long-term studies examining phenotypic development in sturgeons using cryopreserved sperm. Alternatively, wild adults in the act of spawning can be captured on the spawning grounds for gamete collection and egg fertilization can be conducted in the field (Forsythe et al., 2013) prior to transport to the hatchery or gametes can be fertilized at the hatchery (e.g., Crossman et al., 2011a). Lastly, following wild spawning events, fertilized eggs and drifting larvae may be collected with egg mats or drift nets (D-nets) at or downstream of spawning sites and transported to a hatchery for rearing (Auer and Baker, 2002; Crossman et al., 2011a; Jay et al., 2014; Thorstensen et al., 2019). The latter approach, initially applied in lake sturgeon, is referred to as repatriation (Dowling et al., 2004) or head-starting (Heppell et al., 1996) (Fig. 1).

An important and vital distinction between broodstock-based and repatriation approaches is the length of time individuals experience the wild environment during early life history. Because wild spawning allows for natural mate selection and spawning location, eggs and larvae collected for repatriation experience natural selective pressures at early developmental stages. Further, in some populations, multiple wild progeny sources may be available. The Upper Columbia River white sturgeon population transitioned from a wild broodstock approach to a repatriation program to increase genetic diversity (Jay et al., 2014). In addition to improvements in genetic diversity, collection of wild-origin progeny from natural spawning events represented significantly more spawning events across the full spawning distribution compared to the broodstock approach (Fig. 2). Reflecting natural spawning time in conservation aquaculture programs is important as repeatability for spawn time has been shown in some species (e.g., female lake sturgeon; Forsythe et al., 2012). While methods for broodstock spawning have improved over time, transport of wild adults to a hatchery for artificially induced spawning resulted in later spawning in early years compared to natural spawn timing in the wild (Fig. 2). The environment x genotype interaction for individuals fertilized in the hatchery is entirely different from those captured in the wild and subsequently reared in a hatchery prior to repatriation. Indeed, the initial rearing approach in broodstock-based aquaculture may involve incubation in McDonald jars which has been shown to induce stress and affect gene expression (Earhart et al., 2020a) and/or can significantly affect movement and survival in lake sturgeon (Crossman et al., 2011a). In addition, mechanical shock to eggs during the artificial fertilization process in a hatchery setting can cause spontaneous autopolyploidy, or triploidization due to retention of the egg's second polar body (Gille et al., 2015). Repatriation can also represent more wild adults (and therefore more genetic diversity) than broodstock-based approaches (Crossman et al., 2011b; Jay et al., 2014; Thorstensen et al., 2019). Thus far, repatriation has been used successfully for white sturgeon (*Acipenser transmontanus*) in the Upper Columbia (Canadian and American programs; Crossman and Korman, 2021) and Middle Snake Rivers. Repatriation is not possible for sturgeon in all systems, however. To employ this technique successfully, spawning sites must be known and occur in areas where gear can be deployed safely to capture eggs/larvae efficiently. Deep, turbid or poorly characterized systems do not lend themselves to successful repatriation, and poor representation of a wild cohort due to inefficient sampling would not realize the genetic diversity benefits of repatriation.

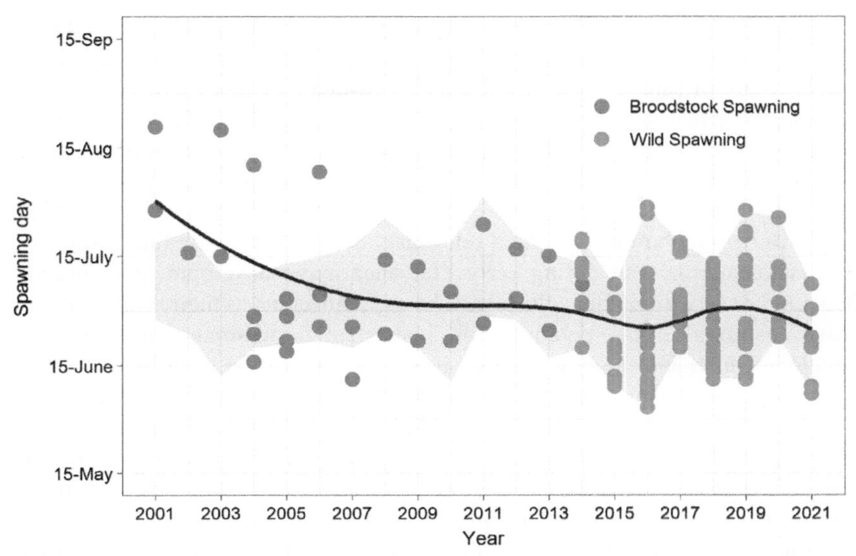


**FIG. 2** Lessons learned from 20 years of conservation aquaculture for white sturgeon in the Upper Columbia River population, Canada, 2001 to 2021. Comparison of in-hatchery spawn timing of broodstock to spawning events documented in the wild demonstrating improvements to spawn timing and representation of a more natural distribution of spawning events following a transition to incorporating wild-origin progeny. Broodstock spawning points represent individual spawning events for adults captured in the river and transported to the hatchery. Wild spawning points represent progeny collected as embryos or larvae from natural spawning events in the river. The solid line is a Loess smoothing function to visualize the trend of spawning events over time. The distribution of spawning events in the wild is shown in the gray ribbon and represents the earliest and latest document spawning events within each year.

When sturgeon conservation aquaculture programs are limited to a broodstock-based approach, releases can be performed over multiple years to increase genetic diversity and the number of parents represented (Drauch and Rhodes, 2007; Welsh and Jackson, 2014).

# 3 Influence of rearing environment on phenotypic development

## 3.1 Environment/phenotype interactions

The notion of the environment shaping phenotype is not new, with initial observations that different rearing environments result in development of different forms in animals first documented by Aristotle (350 BC; see Burggren, 2020). In comparison to natural environments, conservation hatcheries by default create an artificial environment with reduced predation pressure, pathogen exposure and environmental fluctuations alongside abundant resources to maximize growth and survival of fish during the first weeks to months of life when they are most vulnerable to mortality in the wild. In essence Darwin's notion of "survival of the fittest" is substantially more relaxed in comparison to wild fish where there is significantly greater selection pressure during early life history (Osborne et al., 2020). While improved survival and growth in conservation aquaculture is of concern, the fact that fish grown in captivity have had little to no experience of the environment into which they will be transferred is of equal relevance. Specifically, with little exposure to a wild environment during early life, phenotypic development of individuals in conservation aquaculture will be very different to their wild counterparts, resulting in the potential for a substantial mis-match between the developing fish and the environment into which it will be stocked.

Life history theory speculates that if early rearing environment matches the adult environment, the individual may show maximum fitness (Monaghan, 2008); therefore, one might imagine with the environmental mis-match between hatcheries and the wild that fish raised under artificial means of any sort would be maladapted and the process would be doomed to failure. However, it is also generally accepted that animals reared in a favorable environment during early life will do better than conspecifics raised in a resource poor environment (Monaghan, 2008; Reid et al., 2006) and will demonstrate improved fitness as adults regardless of whether the adult environment is resource rich or poor (Grafen, 1988; Lindstrom, 1999). This premise suggests that fish raised in a conservation aquaculture program might have better survival and improved fitness when stocked in the wild, although timing of the stocking event is critical (Crossman et al., 2011a; McDougall et al., 2014).

A major determinant in the fate and fitness of a hatchery reared individual is found in intraspecific variation and, in particular, the plasticity of the phenotype being examined (McKenzie et al., 2020). There are a number of definitions related to the general notion of phenotypic plasticity (Wright and Turko, 2016), all related to how the timing in the animal's life cycle and duration of environmental change may influence development trajectory on physiological and behavioral responses of an organism to a change in the environment. Transgenerational and developmental plasticity are perhaps the most relevant concepts when considering the role of conservation aquaculture programs that are often reliant on artificial selection of a mate, which in itself is a highly adaptive process (reviewed in Neff et al., 2011), and at some point an artificial early life history rearing environment. Transgenerational plasticity is where progeny phenotype is influenced by the parent's environmental experiences and is facilitated by epigenetic processes that can be heritable over a number of generations and/or result in immediate plasticity in the offspring (Donelson et al., 2018; Moran et al., 2016). Sturgeons do not lend themselves to transgenerational studies given their longevity and late age to sexual maturity. However, one might reasonably assume that use of captive broodstock and domestication would strongly influence the phenotype of individuals used for stock enhancement. Indeed, a growing body of literature and the focus of this section has led to a greater understanding within sturgeons of the impact of the environment on developmental plasticity, defined in this context as an irreversible change in adult phenotype as a result of an environmental perturbation during early development (McKenzie et al., 2020; Wright and Turko, 2016).

It is generally accepted that development plasticity occurs in response to changes in the environment during discrete developmental windows, a concept first identified in the early 20th century (Stockard, 1921). In essence, if an individual is subjected to a change in the environment during a specific period, the developmental trajectory is permanently modified or "switched" (i.e., fixed) for that individual (Beaman et al., 2016; Burggren and Mueller, 2015; Noh et al., 2017; Wilson and Franklin, 2002). However, the fixed nature of the phenotype is also related to the measured trait and while there are numerous examples of irreversible phenotype switching during development, there are also a number of examples where the organism resumes a normal adult phenotype following environmental perturbation during development (Burggren, 2020; Utz et al., 2014). As with transgenerational plasticity, the generation time for sturgeons make them less than ideal models to examine the ultimate consequences of developmental phenotypic plasticity. However, proximate consequences have been frequently examined and are very relevant to the management strategies employed in conservation aquaculture and indeed conservation strategies in general for these imperiled fishes. Normally, a single stressor—single time approach is taken to examine the effect of a given environmental parameter on developmental plasticity (Melendez and Mueller, 2021); however, the natural environment is multidimensional and so inclusion of dose dependency of the stressor alongside timing allows for examination of the more subtle impacts on phenotypic development (Dammerman et al., 2015; Mueller et al., 2016).

In the following sections we examine how changes in early-rearing--environment have influenced phenotypic development in sturgeon. We consider timing and duration of environmental impacts throughout development and proximate consequences of these changes across a broad range of sturgeons. We use this information to develop an understanding of best practices in sturgeon conservation aquaculture to build a framework for future approaches in this field.

## 3.2 Typical life-history characteristics of sturgeons

To develop an appropriate framework for hatchery management in a conservation setting, knowledge of the natural rearing environment in which the species of interest has evolved is critical. For sturgeons, fast flowing water and a rocky/cobble substrate is a universal requirement for spawning (Osborne et al., 2020; Peterson et al., 2007; Williot et al., 2002a) with water depth and temperature during the time of spawning being system and population specific. Sturgeon eggs are demersal and once fertilized become sticky and adhere to substrate surfaces within or immediately downstream from spawning sites (Finley et al., 2018; Siddique et al., 2016). The stickiness is caused by transformation of a jelly coat of mucopolysaccharides surrounding the chorion (Monaco and Doroshov, 1983) and secures the egg to substrate relatively free of sedimentation during incubation. Given this is an evolved mechanism it is reasonable to assume that egg stickiness may be associated with optimal embryogenesis and hatch in the wild as they adhere to substrates at locations with physical parameters (e.g., water velocity) proximal to and selected by spawning adults.

Time to hatch varies between species but unsurprisingly temperature is positively correlated to developmental rate in all species examined thus far (Dettlaff et al., 1993; Duong et al., 2011; Gisbert and Williot, 2002b; Kappenman et al., 2013; Wang et al., 1985). Sturgeons hatch as eleutheroembryos or yolk sac larvae, and are often called free swimming larvae at this stage. A significant amount of development remains to be completed as yolk is absorbed prior to the onset of exogenous feeding. The length of time to exogenous feeding is dependent on temperature but maternal investment in yolk reserves and/or genotype can be of critical importance (Dammerman et al., 2015). Upon emergence, sturgeon will drift and settle on favorable substrate for foraging and grow-out. The length of drift is largely dependent on topography and river characteristics (Auer and Baker, 2002) ranging from expansive distances (hundreds of km) (Braaten et al., 2008) to relatively short distances (tens of meters) (McDougall et al., 2017), but can also be impacted by family, prior thermal history (Hastings et al., 2013), attributes and location of the spawning site and salinity. Following the initial drift phase, embryos consume yolk sac resources and toward the end of the yolk sac phase start drifting again to reach productive river sections when they start exogenous feeding. Once settled, fish begin active foraging, initially feeding on small prey items before transitioning to larger invertebrates as they grow, with an increasing gape facilitating capture of larger prey. It is this particular life stage that is the least described in many species of sturgeon and yet arguably one of the most vulnerable, particularly for north-temperate species, as sufficient resources must be obtained in order to survive protracted periods of low resource availability in the upcoming winter. In subsequent years, sturgeon will continue to forage and grow and eventually reach sexual maturity.

Best practices in conservation aquaculture should be focussed primarily on early life history parameters as these are the factors that will ultimately shape the phenotype of the fish destined for stocking, which typically occurs within the first 6 months of life, prior to the first winter, or the following year as spring yearlings when fish are of a sufficient size to cope with the stressors of the natural environment.

### 3.3 Timing of intervention

Undoubtedly the most vulnerable time in any fish's life is as an embryo or free-swimming larva, thus successful conservation programs that employ hatchery rearing as a component of their management strategy will intervene at one or both of these life stages. We also know that both genetic and environmental factors influence developmental rate, time to hatch and subsequent dispersal in lake sturgeon (Duong et al., 2011) thus it is of no surprise that differences in phenotypic development may be observed between artificially spawned individuals and those collected during larval drift and subsequently returned or "repatriated" to the site of collection (Crossman et al., 2014). The cause of differences in phenotypic development in fish raised from egg in a hatchery environment versus those captured as free-swimming larvae then raised in a hatchery environment centres on the developmental windows concept (Burggren, 2020). In a hatchery environment with a purely conservation focus, the goal is to recreate a wild phenotype but this may not be entirely possible depending on how plastic a given trait might be. However, as many of the environmental effects on phenotypic development occur during embryogenesis and yolk-sac larvae, many traits may already have developed as wild-type in those fish captured as drifting or free-swimming larvae, grown in the hatchery and then repatriated to their natural environment. Growth and survival at key stages during early life history were assessed over a two-year period in two cohorts of lake sturgeon (Crossman et al., 2014). Three methods of progeny collection were tested and fish were either reared in a streamside rearing facility or traditional hatchery environment. In essence the study demonstrated rearing dependent variation in growth and survival and suggested that conservation aquaculture programs could and perhaps should incorporate alternate hatchery rearing environments with a variable environment, such as streamside rearing to promote phenotypic variation as typically observed in natural systems (Crossman et al., 2014). However, it is important to note that not all programs may lend themselves to stream-side rearing facilities as significant infrastructure may not exist. Thus it remains central to the success of sturgeon conservation aquaculture programs to understand the impact of specific and/or combinations of abiotic and biotic factors on phenotypic development in sturgeons.

# 4 Factors affecting phenotypic development in sturgeon

## 4.1 Temperature

### 4.1.1 Growth and mortality

Sturgeons are found circum-globally, and although confined to the northern hemisphere, a number of individual species demonstrate a substantial latitudinal range (Power and McKinley, 1997). While it is recognized that numerous biotic and abiotic factors influence population specific growth rates across a species' range (McDougall et al., 2018), it remains that temperature is the most pervasive environmental factor that directly or indirectly influences growth rate in fishes. As a consequence, it is of no surprise that a number of studies have examined the impact of temperature on a variety of traits in developing sturgeon, with examination of development and growth rate during early life history perhaps being the most prevalent.

Wang et al. (1985) compared the developmental rate during embryogenesis of lake and white sturgeon between 10 and 26 °C. They reported optimal survival temperatures of 14–17 °C and a significant increase in mortality and developmental abnormalities beyond 20 °C. In Siberian sturgeon (*Acipenser baerii*), a positive relationship between temperature (12–20 °C) and growth was reported, with a decrease in survival and growth between 20 and 24 °C (Aidos et al., 2017; Park et al., 2013). Furthermore, a linear relationship between temperature (8.3–12.0 °C) and cleavage divisions during early embryogenesis was reported in Siberian sturgeon. However, this relationship was increasingly curvilinear between incubation temperatures of 13–23 °C with temperatures exceeding 20 °C detrimental to development (Gisbert and Williot, 2002b). The positive relationship between development and growth continued to the yolk sac stage where a similar effect of temperature was seen in yolk absorption rates in Siberian sturgeon raised between 16 and 22 °C (Aidos et al., 2017). Similarly, increased temperature resulted in a decreased period of embryogenesis in Atlantic sturgeon (*Acipenser oxyrinchus*) but unlike Siberian sturgeon, the effect of temperature on yolk absorption was less pronounced (Kolman et al., 2018) at least within the range tested.

In green sturgeon (*Acipenser medirostris*), there was again a positive relationship between developmental rate and temperature reported during embryogenesis in temperatures ranging between 11 and ~17.5 °C, after which there was a decline in hatch success and an increase in developmental abnormalities to the upper temperature tested of 24 °C (Van Eenennaam et al., 2005). Similar results were reported for green sturgeon by Linares-Casenave et al. (2013), as skeletal deformities increased significantly beyond 20 °C and survival significantly decreased when temperatures exceeded 26 °C. A positive linear relationship between temperature and growth has also been reported for both pallid (*Scaphirhynchus albus*) and shovelnose (*S. platorhynchus*) sturgeon

(12–20 °C), however, the relationship once again became increasingly curvilinear as temperatures exceeded 20 °C (Kappenman et al., 2013). In lake sturgeon, temperature effects on development from fertilization to exogenous feeding were indexed using developmental milestones in a hatchery setting. Models predicted an approximate 2.3% daily change in development at 10 °C and 8.3% daily change in development at 19.9 °C (Eckes et al., 2015). Such data can be used to accurately control developmental rate in the hatchery or determine time of emergence in wild populations which can have significant implications on year class strength. For example, the reduced growth rate and increased developmental rate at lower temperatures in pallid sturgeon may lead to an increase in average drift time post-emergence, a situation that could lead to a mis-match between environment and resources required as the fish begin to feed exogenously (Mrnak et al., 2020).

Interestingly, temperature had no effect on body mass or mortality in Siberian sturgeon raised at three distinct temperatures (16, 19 and 22 °C) during the yolk sac stage, although yolk was absorbed the fastest in the 22 °C treatment group. However, growth was not assessed when fish began to feed exogenously (Aidos et al., 2020) which is a common occurrence in many hatchery-based studies examining the effects of temperature on growth in sturgeons. Research that is less frequently reported are longer-term studies that examine the impact of early rearing environment months, or even years, later. This is of significance as manipulation of environmental temperature during early life history in fish has been shown to have a fixed effect on a variety of measured traits ( Johnston, 2006; Maynard et al., 1996; Wiley et al., 1993). Unlike Siberian sturgeon, there was a positive relationship between growth and temperature (10–18 °C) in lake sturgeon at the yolk sac phase (Wassink et al., 2019), and this positive relationship seems to persist throughout the first year of life when fish are exposed to increased rearing temperatures (Yoon et al., 2019a,b). However, if increases in temperature were applied post-yolk sac absorption, the effect of temperature on the growth phenotype was transient (Yoon et al., 2022). Recently, developmental plasticity in growth was demonstrated in lake sturgeon where fish raised at higher temperatures during the first three weeks post-hatch showed an accelerated growth rate for the first year of life when placed in a common garden setup with fish raised at lower temperatures over the same time period (Brandt et al., 2022). Furthermore, this accelerated growth was shown to be fixed later in life as the same group of fishes displayed higher growth rate post-winter almost a year later (Brandt et al., 2021), a pattern in growth previously indicated from mark recapture data in the field (McDougall et al., 2020).

The impact of temperature on development and growth pervades a host of physiological factors. For example beluga (*Huso huso*) and Persian sturgeon (*Acipenser persicus*) showed a positive relationship between temperature and growth during the yolk sac larval stage and the improved growth observed at higher temperatures was paralleled with an increase in whole-body protease activity indicative of increased absorption and assimilation of the high protein concentration in yolk (Hashemi et al., 2018). Furthermore, temperature may differentially impact growth of specific organs during early life history. In juvenile sterlet sturgeon, temperature was positively correlated with growth at an optimum of 20 °C but declined at 24 °C and furthermore, rearing sterlet sturgeon at 24 °C resulted in a male biased population in the hatchery (Havasi et al., 2018), something that conservation practitioners would want to avoid. Finally, in white sturgeon, measured traits such as body length and mass were predictably larger in fish raised in warmer water (17 °C) compared with fish raised in cooler water (12.5 °C), however, total gill filament area was larger in fish raised in cooler waters, indicative of temperature dependent trade-offs in how resources are allocated ( Jay et al., 2020).

A common hatchery practice is to grow-out fish through the first winter of life at elevated temperatures to maximize growth and chance of survival when fish are stocked the following spring as yearlings. However, depending on the ecosystem, large daily fluctuations in temperature can be present - an environment that hatchery-raised fish would be naive to. Few studies have examined the effect of daily variations in temperature on sturgeon biology and those that have, have centered on their effect on growth. Interestingly, green sturgeon exposed to widely variable daily temperatures (+/− 10 °C daily variation) at 47 days post hatch resulted in accelerated growth compared to stable (15 °C) or narrowly variable temperatures (+/− 4 °C daily variation) (Rodgers et al., 2018). Temperature effects are also additive in many systems and while there is generally a positive relationship between temperature and growth, increasing temperatures can have negative consequences if there is an environmental mis-match. For example, juvenile green sturgeon fed optimal rations showed improved growth performance at 19 °C compared to cooler rearing temperatures but reduced growth performance when fed reduced rations at 19 °C compared to cooler temperatures (Poletto et al., 2018).

From a management perspective, understanding the physiological challenges for fish in relation to the timing of when they are released from conservation aquaculture has clear consequences on the success of the program and by default, hatchery operations. Temperature manipulation can be used as a tool to increase thermotolerance in fish (Kültz, 2005) and therefore prepare fish for stocking in the wild. In Atlantic sturgeon, early short term exposure to increased environmental temperature led to a 2–4 fold lower incidence of gene dysregulation in the liver transcriptome of fish compared with naive fish, indicative of a prior thermal history influencing the physiological preparedness of fish transferred to the wild (Yebra-Pimentel et al., 2020). In two latitudinally distinct populations of lake sturgeon, thermal history during early rearing and population source had a significant effect on growth and survival with the more southern population demonstrating a higher critical thermal maximum than the northern population at warmer (24 °C vs 16 °C) rearing temperatures. However, the northern population demonstrated an increased mass over time in comparison to the southern population at 16 and 20°C (Bugg et al., 2020) which may indicate latitudinal countergradient variation with northern fish growing faster at those temperatures to take advantage of a shorter growing season (Fangue et al., 2009). Population and rearing temperature also had a significant effect on cold adaptation in lake sturgeon with the southern population mounting a more vigorous cellular stress response compared to their northern counterparts when acutely exposed to a cold shock (Bugg et al., 2021b). Importantly the two source populations in these studies have evolved under distinctly different thermal regimes (Bugg et al., 2020, 2021b) and as a consequence, this has impacted the thermal biology and physiological response of these fish to changes in rearing and stocking temperatures. Essentially conservation aquaculture programs must be mindful of population specific environmental norms when establishing rearing environments in addition to the shifting baseline of environmental temperatures in more recent times as a result of global climate change.

#### *4.1.2 Whole-body and cellular stress response*

The endocrine stress response is critically important to the evolutionary success of vertebrates (Baker, 2019; Denver, 2009). Therefore, understanding the relationship between temperature and development of the cortisol response in sturgeons provides significant insight into how hatchery reared fishes may cope with exposure to a stressor when stocked into the inherently stressful wild environment. In fish there is a reliance on maternal investment of cortisol in the egg during early development which is succeeded by development of the necessary molecular machinery required for de novo synthesis (Nesan and Vijayan, 2013, 2016). De novo synthesis of cortisol is normally present post hatch although the exact timing is species dependent. This is frequently followed by a period of hyporesponsiveness where the fish does not, or is incapable of, mounting an appropriate stress response. Such a period has been demonstrated in white (Simontacchi et al., 2009); Persian (Falahatkar et al., 2014) and lake sturgeon (Earhart et al., 2020b; Zubair et al., 2012).

In Siberian sturgeon, there was a positive relationship between whole-body cortisol and temperature during the larval yolk sac stage which the authors attributed to increased metabolic demands at higher temperature, as yolk absorption rates were accelerated in warmer water (Aidos et al., 2020). Temperature was also positively correlated with whole-body cortisol in white and lake sturgeon (Bates et al., 2014; Zubair et al., 2012). However, lake sturgeon raised in warmer water (18°C) and then exposed to a stressor showed reduced increases in whole-body cortisol in response to that stressor compared to those raised at 10°C, but no difference in baseline levels were reported (Wassink et al., 2019). In addition, those fish raised at 18°C had significantly higher activity and survival rates during behavioral and predatory trials, suggesting that exposure to high stress environments (18 °C) in early life may be adaptive to high stress predatory environments later in life (Wassink et al., 2019).

Cellular markers of stress are also frequently used to assess the responsiveness of fish to a stressor (Akbarzadeh et al., 2018). Expression patterns of heat shock proteins (hsp's) are often used as the primary target to examine cellular stress (Fangue et al., 2006; Komoroske et al., 2015) particularly in light of global climate change. To assess recovery potential following acute exposure to increased temperatures, newly hatched green sturgeon larvae were exposed to 26 °C for 3 days from 16 °C and heat shock protein expression was assessed by western blot. There was a robust cellular response to this acute heat shock and expression levels of the measured hsp's remained elevated for up to 9 days post-acute-heat-shock (Werner et al., 2007). Increases in expression of hsp 90a and 90b were observed in northern and southern populations of lake sturgeon following an acute exposure to a heat stress with no difference between populations (Bugg et al., 2020). However, pre-acclimation of both populations to 24 °C resulted in a more rapid response by the southern population in hsp70 expression post-acute-heat-shock, which may be one reason why the southern population adapted to 24 °C had a higher critical thermal maximum (CTmax) in comparison to the northern population adapted to 24 °C (Bugg et al., 2020). Exposure of juvenile shortnose sturgeon (*Acipenser brevirostrum*) held at 15 °C to acute heat stress at different rates did not yield any difference in CTmax nor did it effect changes in expression of hsp70 or 90a (Zhang et al., 2017), suggesting that the single population of shortnose sturgeon used in the study may be capable of mounting a robust cellular response to acute thermal stress.

### 4.1.3 Swimming and metabolism

Sturgeon are not known for their swimming ability; indeed with a heterocercal tail and presence of scutes increasing drag they are considered relatively poor swimmers in relation to most teleosts (Webb, 1986). However, swimming remains critically important for ultimate fitness and survival of individuals. Thus, understanding the relationship between rearing environment and swimming performance would aid in development of hatchery programs for a given species. A standard swimming test used routinely as a physiological performance metric in fishes is the critical swimming test or *U*crit (Brett, 1964). In juvenile Chinese sturgeon (*Acipenser sinensis*), *U*crit was significantly greater in fish acclimated to 15 °C as opposed to 10 and 25 °C (Yuan et al., 2017) and a similar positive relationship between temperature and *U*crit was reported in both shovelnose and pallid sturgeon, though there was no species difference in swimming performance which likely reflected the very similar habitat the two species share within their natural range (Adams et al., 2003). Deslauriers and Kieffer (2012) demonstrated a steady and significant increase in *U*crit as temperature increased from 5 to 25 °C in shortnose sturgeon. This increase in swimming performance was mirrored with a significant effect of temperature on metabolic rate in shortnose sturgeon acclimated to 15, 20 and 25 °C with a maximum oxygen consumption obtained in fish acclimated to 25 °C and the highest metabolic scope at the intermediate temperature of 20 °C (Zhang and Kieffer, 2017). Neither burst swimming speed nor a volitional downward swimming behavior were affected by different rearing temperatures in the first three weeks of life in lake sturgeon demonstrating the plastic nature of the two measured swimming performance metrics and underscoring the importance of this trait in survival. However, there was an effect of age likely due to seasonal temperature variation (Brandt et al., 2022). A similar example of plasticity in measured traits, regardless of environment, is development of metabolic phenotypes in sturgeon. In a series of experiments manipulating temperature, substrate and dissolved oxygen, routine metabolic rate was found to be relatively consistent regardless of rearing environment and any differences observed as a result of the rearing environment were largely transient in nature, although some subtle variations existed (Yoon et al., 2019a,b, 2021).

### *4.1.4 Homeoviscous adaptation*

A critical aspect in the operation of conservation aquaculture programs is achieving the best time for release. Survival of the fish released must be balanced with operational costs and logistics of the hatchery. One key component related to survival in north temperate fish species is the ability to maintain normal cellular function in very low temperatures. Fatty acid metabolism is of critical importance in this process, not just as a source of energy during protracted periods of limited resources in the winter months, but in a process known as homeoviscous adaptation (HVA). HVA is undertaken by animals when adapting to a colder environment; essentially the cell membrane becomes more fluid as the ratio between unsaturated and saturated (U/S ratio) fatty acids increases. Rearing temperature has been shown to influence fatty acid metabolism in white (Buddington et al., 1993), Siberian (Vasconi et al., 2018) and lake sturgeon (Yoon et al., 2022). White sturgeon raised at cold temperatures during incubation showed no effect on U/S ratio when exposed to cold temperatures, however, post hatch larvae raised in cooler temperatures did show an increase in the U/S ratio when exposed to colder temperatures later in life, suggesting the ability for homeoviscous adaptation was acquired at hatch (Buddington et al., 1993). In lake sturgeon, a decrease in temperature at approximately 5 months post hatch led to a significant increase in both mono- and polyunsaturated fatty acids in phospholipids and triglycerides, indicative of a homeovsicous response (Yoon et al., 2022). Dietary sources alongside endogenous enzymatic conversion of fatty acids aid in preparation of the fish for cold environments (see below).

### 4.1.5 Additional traits

The immune response in sturgeons is very understudied, but what has been described suggests that the immune system is comparatively slow to develop and thus may mean sturgeons could be at a high risk of pathogenic infection as they rely on their innate immune system at a very young age (Gradil et al., 2014a,b). This may be particularly problematic in a hatchery environment where there is a greater risk of infection spreading. However, in lake sturgeon exposed to increasing concentrations of the bacterial endotoxins—lipopolysaccharides (LPS), molecular evidence suggested that the fish were capable of mounting an innate immune response at least at 30 days post fertilization (Bugg et al., 2021a). Further, a positive relationship between temperature and immune response was reported in 7 month old shortnose sturgeon held at 11 and 20 °C (Gradil et al., 2014a).

Microbial communities are recognized as increasingly important in the appropriate development of fish for conservation purposes. It is reasonable to expect a significant interaction between environment and phenotype when considering microbial colonization of the fish gut during early life history. Interestingly, however, only modest differences in community were observed in the gut of developing lake sturgeon raised on flow through untreated river water vs filtered groundwater leading the authors to suggest evidence of selectivity between the microbial community within the gastrointestinal tract and the host (Razak and Scribner, 2020).

The inner ear bones, or otoliths in fishes, acts as a key organ for orientation and balance in teleosts (Campana, 1999). In teleosts, early rearing in a hatchery environment results in a significantly different otolith phenotype in comparison to wild fishes (Chittenden et al., 2010; Reimer et al., 2017), which can have negative consequences on balance and orientation later in life (Oxman et al., 2007). In lake sturgeon, increased rearing temperature resulted in not only different growth rates but larger otoliths with a greater proportion of calcite deposits (Loeppky et al., 2021). In the same study fish were also raised in decreased pH reflecting predicted climate change scenarios to test the hypothesis that a more acidic environment would impede the deposition of calcium carbonate crystals in the otolith. Interestingly, the otoliths of fish raised in decreased environmental pH did not differ from control fish, nor did the interaction between temperature and pH have an effect, suggesting that the lake sturgeon, much like the white sturgeon, has the physiological capacity to maintain internal pH in the face of large fluctuations of pH in the external environment (Shartau et al., 2017).

## 4.2 Hypoxia

How fish respond to hypoxic conditions is relevant in both an aquaculture and wild setting. Intensive culture can often result in decreased dissolved oxygen (DO) in holding tanks and so called "dead zones" of low oxygen content in the natural environment can influence survival of natural stocks, particularly if the areas of low oxygen content overlap with spawning grounds. It has been shown in a number of species that sturgeon are relatively sensitive to reductions in oxygen. In white sturgeon, environmental hypoxia caused a decrease in growth in fish raised at 15, 20 and 25 °C and was particularly severe at 25 °C where a reduction in activity of the fish at this temperature suggested metabolic suppression under hypoxic and elevated temperature conditions (Cech et al., 1984). This notion was supported in the same species where measurement of metabolic rate and swimming activity under normoxic and hypoxic conditions at three life stages all showed a decrease in oxygen consumption and swimming under hypoxic conditions (Crocker and Cech, 1997). Presumably the hypometabolic response would aid in promoting survival under protracted periods of environmental hypoxia.

Similar findings were reported for European sturgeon (*Acipenser sturio*) exposed to varying hypoxic conditions at one of two temperatures (20 or 26 °C) throughout early life history (Delage et al., 2014) with critical oxygen tension estimated to be below 30% DO at 20 °C 2.7 $mg.L^{-1}$) and 40% DO at 26 °C (3.2 $mg.L^{-1}$). These estimated oxygen thresholds at relevant environmental temperatures are valuable tools in the development of appropriate management strategies for European sturgeon (Delage et al., 2020), Atlantic and shortnose sturgeon (Secor and Gunderson, 1998; Secor and Niklitschek, 2001) particularly in the light of global climate change. In shortnose sturgeon there was a positive relationship between DO and temperature with LC50 values of 26% (2.2 $mg.L^{-1}$) and 28% (2.3 $mg.L^{-1}$) oxygen saturation at 25 and 26 °C, respectively increasing to 42% (3.2 $mg.L^{-1}$) oxygen saturation at 30 °C (Campbell and Goodman, 2004). Growth rate is perhaps the most frequently examined trait in response to manipulation of oxygen content; however, metabolic rate is also of interest as this will have implications for energy balance as the fish develops, particularly in the first year of life. Lake sturgeon reared in mild hypoxia (80% DO; 7.8 $mg.L^{-1}$) at 16 °C during early life history were shown to compromise both forced maximum metabolic rate (a metabolic rate estimate determined after an individual is forced to exercise to presumptive exhaustion by gentle chasing within a confined space) and metabolic scope post-winter suggesting a longer term impact of hypoxia exposure during early rearing on development of metabolic phenotypes (Yoon et al., 2019b). Furthermore, a more proximate impact was demonstrated in that those fish raised in 80% DO had a higher mortality rate during acclimation to a colder environment (Yoon et al., 2019b).

Conversely, positive effects of increased oxygen availability have been reported in the sterlet sturgeon raised for commercial purposes, where fish raised in hyperoxic conditions (>100% DO) had a faster growth rate and reduced food conversion efficiency compared to fish raised in normoxic conditions (Ineno et al., 2018). However, it is less likely that fish in the wild would be exposed to hyperoxic conditions for prolonged periods and thus the relevance of this approach for conservation purposes may be questioned.

## 4.3 Salinity

In sturgeons, three types of life history have been suggested in regard to salinity, each with increasing levels of salinity tolerance (Nelson et al., 2013; Rochard et al., 1990). Stenohaline species such as lake sturgeon may have tolerance for a narrow range of salinities but typically inhabit a single salinity throughout their life. Euryhaline and anadromous species such as Adriatic sturgeon survive in a wide range of salinities, spawn in freshwater, migrate to brackish/estuarine water as juveniles, grow-out and then return to freshwater to spawn as adults. Amphihaline and diadromous species such as green sturgeon survive in freshwater and seawater, spawn in freshwater and migrate to full-strength seawater to grow-out prior to returning to freshwater to spawn as adults. Importantly, many of these species exhibit substantial plasticity in that at least two or indeed all three life history traits or ecomorphs can be found within a single species such as shortnose sturgeon (Rochard et al., 1990). Of the 24 species of extant sturgeon, only a few exhibit a stenohaline life history and so understanding timing of migration to the sea and extent of salinity tolerance is informative for conservation aquaculture in regard to growing and release strategies, as a euryhaline or amphihaline life history may require a period of pre-adaptation to a saline environment prior to release.

In the euryhaline Persian sturgeon, hatch success was high at 2 parts per thousand (ppt) but declined at salinities of 4 ppt and higher suggesting mildly brackish water rearing may be a viable solution for that species if access to freshwater for hatchery operations was limited (Khatooni et al., 2012). However, salinity acclimation studies at 20 days post hatch suggested limited tolerance to brackish water (~4 ppt) continued (Khatooni et al., 2011). In Adriatic sturgeon (*Acipenser naccarii*), complete development of the kidneys, gut and gills was required for salinity tolerance, at approximately 36 days of age. However, even at 150 days old the maximum salinity tolerated by these fish was approximately two thirds full strength seawater (Cataldi et al., 1999) in comparison to complete tolerance to full strength seawater at 1 to 1.5 years of age (Cataldi et al., 1995). In shortnose sturgeon, tolerance to two thirds strength seawater was observed at 1 year but the same age class was relatively intolerant to full strength seawater (Downie and Kieffer, 2016a) and in 3 month old ship sturgeon (*Acipenser nudiventris*), tolerance to salinity beyond brackish water was severely limited (Shahkar et al., 2015). In Chinese sturgeon, 8 month old fish were able to successfully adapt to approximately two thirds seawater suggesting moderate salinity tolerance at this life stage (Zhao et al., 2011) and in the amphihaline green sturgeon, tolerance to full strength seawater was achieved at 134 days post-hatch and matched a steady increase in the capacity of key osmoregulatory organs to maintain an internal sodium and water balance in the face of an increasingly saline environment (Allen et al., 2011).

In Atlantic sturgeon, growth rate significantly improved in higher saline waters, with age indicative of an age/size related tolerance to increasing salinity (Niklitschek and Secor, 2009). The effect of size and age on salinity tolerance was also examined in white sturgeon, where 1 year old fish at 10 or 30 g were acutely exposed to increased salinities and the authors concluded a moderate effect of size on salinity tolerance as the degree of internal perturbations was less in the larger fish (Amiri et al., 2009). Juvenile green sturgeon may be found in fresh or brackish water environments at any age or size but there was a distinct relationship between age/size and salinity tolerance of full strength seawater as green sturgeon typically migrate to full strength seawater at approximately 1.5 years of age and 75 cm total length (Allen and Cech, 2007). Culturing Siberian sturgeon in brackish water did not result in improved growth or survival despite the presence of wild juveniles in brackish water environments approaching one third strength seawater (Rodriguez et al., 2002). Finally, ion concentration will influence growth rate in presumptive stenohaline sturgeons such as lake sturgeon that have limited tolerance to increased salinity (LeBreton and Beamish, 1998). Interestingly, a single population raised in two different freshwater systems had distinct differences in growth rate that may have been related to the reduced availability of calcium ions in the system showing a lower growth rate (Genz and Hicks, 2021).

As with temperature, salinity will influence performance in several physiological processes. Indeed, in some fishes the pre-adaptation to increased salinity or smoltification can result in a transient effect on swimming performance which may be size/life stage dependent (Graham et al., 1996). However, in the juvenile Chinese sturgeon and shortnose sturgeon there was no effect of length at age on swimming performance at a time and size when both species are capable of tolerating increased salinity (He et al., 2013; Penny and Kieffer, 2019). Conversely, Allen et al. (2006) reported a transient decrease in *U*crit as green sturgeon acclimated to increased salinity, and a negative relationship between salinity acclimation and *U*crit was reported in Adriatic sturgeon (McKenzie et al., 2001). It is possible that greater preparatory steps have evolved for salinity tolerance in the green and Adriatic sturgeon analogous to smoltification in salmonids.

## 4.4 Substrate

As described, sturgeon naturally spawn over rocky substrate, something that is normally entirely absent from traditional hatcheries. However, substantial research has demonstrated significant benefits of substrate during early development in several sturgeon species. For example, in white sturgeon, yolk-sac larvae reared over artificial substrate until 16 days post hatch demonstrated a significant increase in growth compared to those raised in a bare tank, and the combined effect of temperature and substrate resulted in higher growth rate in white sturgeon raised over substrate at 17.5 °C at the yolk-sac stage compared to those raised over no substrate at 13.5 °C. Interestingly, temperature had no effect on yolk absorption rate whereas substrate did (Boucher et al., 2014; McAdam, 2011). Further, routine metabolic rates were lower and whole-body glycogen content was higher in gravel reared individuals suggesting those fish raised in a bare tank were diverting energy stores to something other than somatic growth (Boucher et al., 2018). Rearing with coarse hard substrate at the yolk-sac stage in white sturgeon tended to result in reduced whole-body cortisol at the larval life stage (Bates et al., 2014) and a positive effect on growth, whole-body lipid concentrations and gut development which persisted for at least 1 month of development when substrate was removed post-yolk-sac absorption (Baker et al., 2014). Similarly, the presence of substrate during embryogenesis and the yolk-sac life stage in lake sturgeon led to a more robust whole-body cortisol response compared to those fish raised over no substrate (Zubair et al., 2012). Conversely substrate (sand, gravel, no substrate) was not shown to have a significant effect on cortisol levels in free swimming yolk-sac larvae of Atlantic sturgeon, despite there being a pronounced positive effect on growth in those fish raised over substrate as yolk-sac larvae when the fish in each treatment began feeding (Gessner et al., 2009). The effect of substrate on swimming performance was also examined in juvenile shortnose sturgeon where there was no clear effect on *U*crit (Downie and Kieffer, 2016b) nor was there an influence of substrate and flow on growth in lake sturgeon (Merks and Anderson, unpublished). However, flow was shown to increase both *U*crit and endurance in Dabry's sturgeon (*Acipenser dabryanus*) when the fish were exposed to step controlled increases in water velocity from 15 days post hatch (dph) to 40 dph (Du et al., 2014).

In summary, those studies that report a positive effect of substrate presence during rearing on length and mass-at-age have typically recommended inclusion of either natural or artificial substrates in hatchery rearing programs provided it can be removed at the onset of exogenous feeding (Yoon et al., 2019a). Accordingly, both natural and artificial substrates (e.g., sinking Bio-Balls) that can be easily cleaned and maximize interstitial spaces during early rearing of yolk-sac larvae have been adopted in many hatcheries ( Jay et al., 2020). The exact mechanism behind this adaptive response to substrate is unknown but may be related to reduced energy expenditure during yolk-sac absorption with greater investment in development prior to drift and the onset of exogenous feeding. A similar effect has been observed during embryogenesis in lake sturgeon where individuals hatching from eggs allowed to adhere to egg mats had improved survival and growth for at least 70 dph compared to those eggs incubated in a MacDonald hatching jar (Earhart et al., 2020a). Lake sturgeon embryos have also been successfully reared on artificial surfaces in heath trays (Crossman et al., 2014) which also allows for improved control of microbial infections which can occur at higher rates in hatcheries (Fujimoto et al., 2013). Further, green sturgeon embryos incubated in a

MacDonald hatching jar illustrated 68–99% higher mortality at hatch compared to fish developing in upwelling incubators (Van Eenennaam et al., 2008). These experimental assessments at the embryogenesis stage align with observations in the wild on the importance of substrate during early life (Finley et al., 2018) and, by default, the selection of spawning habitats by wild spawning adults.

## 4.5 Maternal investment

Despite shortnose and Atlantic sturgeon occupying equivalent environmental niches, Atlantic sturgeon were reported to have accelerated embryogenesis compared to shortnose sturgeon at similar temperatures (11–15 °C) (Hardy and Litvak, 2004). The authors attributed these findings to an evolved earlier spawning time and larger egg size of shortnose sturgeon leading to a protracted period of embryogenesis in comparison to Atlantic sturgeon. These data also imply a greater maternal investment by shortnose sturgeon. Why this may be the case is unknown and research on maternal investment in sturgeon reproductive success and subsequent phenotypic development is relatively understudied. Measurement of maternal contribution through parentage analysis has been shown to explain a significant amount of the variation in the embryonic and larval development until dispersal in lake sturgeon (Duong et al., 2011). Furthermore, female spawning site selection was shown to influence offspring phenotype at hatch (length at age) (Dammerman et al., 2015). However, while in-stream microhabitat during egg incubation influenced some traits during early development, the environmental effect on phenotypic development was surpassed by the intrinsic maternal genetic input as larvae aged to the exogenous feeding stage and beyond (Dammerman et al., 2020).

Measurement of cortisol in non-fertilized and fertilized eggs has been conducted in sturgeons using both competitive binding assays (Earhart et al., 2020a; Simontacchi et al., 2009) and liquid chromatography tandem mass spectrometry (Bussy et al., 2017). Assessing values of cortisol in the egg facilitates understanding potential variation in female investment in the egg and/or the possible transgenerational impacts of maternal stress on resulting phenotype of the progeny. In lake sturgeon, egg cortisol concentration was shown to vary between females (Wassink et al., 2019) indicative of variation in maternal investment. To examine the possible effect on offspring phenotype, lake sturgeon eggs were exposed to high or low cortisol to mimic maternal stress. Subsequent exposure of the offspring to a stressor or not post hatch suggested a strong potential for maternal effects on physiological and behavioral phenotypes of the resulting offspring (Wassink et al., 2020). One such maternal effect may be the input of thiamine (vitamin B1) into the egg by the mother. Thiamine deficient eggs are known to result in suboptimal individuals at hatch in sterlet sturgeon (Ghiasi et al., 2017). Interestingly, thiamine concentration in the egg of lake sturgeon was shown to be significantly higher in early spawning females compared to eggs from late spawning females, suggesting that timing of spawn may negatively impact female investment into the egg, perhaps due to differential dietary intake by early and late spawning individuals (Larson et al., 2021).

Often closely associated with the influence of maternal investment and phenotypic development is the effect of epigenetics. This is where environmental variables may affect fish developmental plasticity. In essence one way in which fish respond to their environments is through epigenetic regulation. Epigenetic regulation refers to modulation of gene expression caused by interactions between the environment and the genome. Indeed, environmental conditions can turn genes "on" and "off" through DNA methylation, histone acetylation, and transcription of non-coding RNA that binds to and disables gene transcripts. Little work has been conducted on any sturgeon species to examine the effects of epigenetics, but it is reasonable to assume these factors will influence phenotypic development given the growing body of literature from epigenetics research conducted on teleosts (see Chapter 8, Volume 39A: Jeffries et al., 2022).

## 4.6 Diet

During the first few months of life there are well-documented bottlenecks in survival for all sturgeons that are related to the transition between diets. The cause of the increased mortality during dietary transitions is hypothesized to be a lack of ingestion and/or digestion (Sánchez-Hernández et al., 2019). In lake sturgeon a transient increase in whole-body cortisol has been observed during diet switching (Earhart et al., 2020b) indicative of a stressor, but may also be an important regulator of gut development in the fish as inhibition of cortisol synthesis at that time resulted in significant increases in mortality (Earhart et al., 2020b). Timing of the first transition from yolk to exogenous feeding is species and temperature dependent with *Artemia* spp. being a typical starter diet at this stage (Gisbert and Williot, 2002a; Vedrasco et al., 2002). Once sufficient growth has occurred there is a second transition from *Artemia* to an *Oligochaete* sp. and or a formulated diet for ongrowing. Timing of diet switching, frequency of feeding, enrichment and type of diet have all been shown to influence growth and survival of sturgeon in culture (Bauman et al., 2015, 2016; Falahatkar et al., 2013; Kelly and Arnold, 1999; Klassen and Peake, 2007, 2008; Kolman and Kapusta, 2018; Lindberg and Doroshov, 1986; Valentine et al., 2017; Vedrasco et al., 2002; Yoon et al., 2020). In essence, presentation of food to yolk-sac larvae a few days in advance of complete yolk-sac absorption promotes survival, enrichment with fatty acids tends to have a positive effect on growth, and gradual introduction of the diet used after *Artemia* will generally result in improved survival and growth.

Few studies have examined the effect of diet on traits other than growth and mortality, but it has been shown that enrichment of diet with unsaturated fatty acids resulted in improved hypoxia tolerance in Adriatic sturgeon (Randall et al., 1992) with the causes of improved tolerance likely from whole-body reductions in standard metabolic rate and increased cardiac performance in low oxygen levels (McKenzie et al., 1999). Furthermore, in lake sturgeon fed live *Artemia* there was a reduction in survival and whole-body energy density at 32 days post-fertilization compared to those fed dead *Artemia*. However, by the time the fish were 133 days post-fertilization, the fish fed live *Artemia* earlier in life demonstrated higher maximum metabolic rates suggesting a proximate benefit of a live diet earlier in life (Yoon et al., 2020). In addition to promoting hypoxia tolerance in sturgeons, enrichment of fatty acids in the diet may also promote homeoviscous adaptation in white sturgeon (Buddington et al., 1993) and survival in lake sturgeon (Yoon et al., unpublished).

## 4.7 Rearing density

In fish with a yolk-sac phase, reduced rearing density has been shown to increase total length at emergence and increase the likelihood of successful first feeding (Cushing, 1972; Yufera and Darias, 2007) and the same is evident in sturgeon. Beginning immediately post-hatch, rearing density has a significant effect on survival and growth. In Siberian sturgeon, free-swimming yolk-sac larvae reared at 30 individuals $L^{-1}$ were significantly heavier and longer at first feed than fish reared at 80 or 150 individuals $L^{-1}$ (Aidos et al., 2018). Likewise, in lake sturgeon, fish reared at 9688 individuals per $m^{-3}$ were significantly longer and had better survival than fish reared at 19,375 and 32,292 individuals.$m^3$ at time of emergence (Bauman et al., 2015). Further, those reared at higher density tended to emerge earlier than those reared at a lower density ( Jay et al., 2017). It is evident, therefore, that rearing density is an environmental stressor in sturgeons and this has been empirically demonstrated in shortnose sturgeon where fish stocked at moderate (20 kg.$m^3$) and high densities (40 kg.$m^3$) showed a significant increase in circulating levels of cortisol compared to those reared at low densities (10 kg.$m^3$) (Wuertz et al., 2006). However, social context plays a role in regulating both the whole-body cortisol response and cellular hsp response in sturgeon. One year old juvenile lake sturgeon held in isolation had a significantly prolonged cortisol stress response compared to those held in groups (Allen et al., 2009a) and 3 month old lake sturgeon demonstrated improved recovery in cellular and whole-body cortisol responses to an acute heat stress when exposed to the stressor in a group as opposed to in isolation (Yusishen et al., 2020). Thus, appropriate rearing densities may not only result in improved growth performance and success at first feed but may also influence other physiological traits such as the cortisol stress response as fish age.

## 5 Stocking techniques and prescriptions

Conservation aquaculture programs may undertake various rearing and stocking strategies to improve transition to the wild environment and ultimately increase post-release growth or survival. Methods to prepare fish for changes in temperature and salinity prior to release are described above. Furthermore, several studies in non-sturgeon fishes have shown that addition of enrichments such as substrate or structure to tanks, replication of natural conditions such as flow and temperature variation and feeding of live prey can promote development and behaviors likely to increase survival and minimize dispersal post-release (Braithewaite and Salvanes, 2005, Oldenberg et al., 2011; Roberts et al., 2011; Trippel et al., 2018; but see Fast et al., 2008). Predator conditioning, or controlled exposure to predators in the hatchery, and acclimation to simulated natural shelters in the hatchery have been shown to improve post-release survival in a number of fish species (D'Anna et al., 2012; Kawabata et al., 2011; Trippel et al., 2018). In lake sturgeon, naive age-0 hatchery-reared individuals demonstrated fairly rapid recognition of predators (Crossman et al., 2018) and can be trained to recognize predator odors when paired with whole-body homogenates of conspecifics (Sloychuk et al., 2016).

Several sturgeon conservation aquaculture programs provide environmental enrichments to increase post-release survival and promote imprinting on natal rivers. Streamside rearing facilities, where eggs and larvae are reared in water sourced from their natal rivers, are used for some Atlantic sturgeon (Jörn Gessner, personal communication), lake sturgeon (Crossman et al., 2014; Holtgren et al., 2007), white sturgeon (Hildebrand et al., 2016) and sterlet programs (Friedrich et al., 2019). Mann et al. (2011) found that age-0 lake sturgeon released from a streamside rearing facility in the Little Manistee River exhibited similar growth, movement patterns, and substrate preferences as wild age-0 lake sturgeon in the river. Advantages of streamside rearing in Upper Black River lake sturgeon included higher daily egg survival, greater diversity in incubation time and size at hatch, and higher recapture rates, the latter indicating higher post-release survival (Crossman et al., 2011a, 2014). However, fluctuating temperatures at the Upper Black River streamside rearing facility and exposure to native pathogens increased microbial infection rates relative to traditional hatchery facilities (Crossman et al., 2014). Use of streamside facilities has been tested experimentally in other sturgeon species. Streamside rearing of the European sturgeon was compared to traditional rearing and diel behaviors differed between the two groups for the first three days post release, although survival and growth rates were similar (Carrera-García et al., 2017).

Another strategy that may improve post-release survival of fishes is soft release, or initial stocking into an enclosure in the release area that allows fish to acclimate to the natural environment and recover from transport stress before being exposed to predators (Brown and Day, 2002). Soft release has been used with equivocal outcomes in birds, mammals and reptiles (de Miliano et al., 2016; DeGregorio et al., 2020; Mitchell et al., 2011; Richardson et al., 2015; Sasmal et al., 2015; Tennant and Germano, 2017). Although fewer studies have evaluated the effects of soft release on fishes, a more rapid adoption of normal behaviors and higher post-release survival have been reported for fish experiencing on-site acclimation relative to non-acclimated fish (Brennan et al., 2006; Hervas et al., 2010; Jonssonn et al., 1999; Walsh et al., 2013). Soft released brown trout (*Salmo trutta*) exhibited greater growth and higher condition factor 2 months post-release compared to individuals released immediately ( Jonssonn et al., 1999). However, managers must consider potential deleterious effects of soft release, such as the attraction of predators to acclimation cages (Fairchild et al., 2008). To date, there are no reports of soft release strategies used in sturgeon conservation aquaculture.

Biological characteristics of fish at the time of release can also affect survival and post-release behavior. For example, larger fish have been reported to experience higher post-release survival (Brown and Day, 2002; Leber, 1995; Sparrevohn and Strøttrup, 2007; Weber et al., 2020). Lake sturgeon that were larger at the time of stocking into the Black River had higher survival than individuals smaller at release (Baker and Scribner, 2017), but a similar size - survival relationship was not observed for white sturgeon in the Kootenai River (Ireland et al., 2002). Sturgeon that are older at the time of stocking tend to exhibit higher survival than younger individuals (Crossman et al., 2011a; McDougall et al., 2014; Steffensen et al., 2019). McDougall et al. (2014) reported that lake sturgeon stocked into Nelson River at age 1 year exhibited 17.7–130 fold increase post-release survival compared to individuals stocked at less than 1 year of age.

The timing and location of release can also influence post-release survival, behavior, or growth. If predation is a significant threat to individuals shortly after release, timing stocking events to avoid peak predator activity could improve survival (Crossman et al., 2011a; Olla et al., 1998). While releases of large numbers of fish may occur at locations with easy access to the water's edge, discussion on release location should include aligning important habitats with the life stage being released. This will likely minimize post-release dispersal behaviors that could increase encounters with predators. Seasonality of releases should also be considered and maximizing growth opportunities prior to the first winter is a common objective. Hervas et al. (2010) found that hatchery-reared white seabass (*Atractoscion nobilis*) survived best when released in spring and worst when released in winter. Kootenai River white sturgeon cohorts released in spring generally survived better than cohorts released in other seasons (Dinsmore et al., 2015, cited in Schreier et al., 2015).

Like all anthropogenic activities, conservation aquaculture is not without risk to extant wild populations. One concern is that hatchery-reared sturgeon not properly imprinted to their natal river may stray into other river systems.

Like salmonids, diadromous sturgeon exhibit natal philopatry, returning to their natal river to spawn, and natural straying rates are low (Grunwald et al., 2008; Moser et al., 2016). Unlike salmonids, diadromous sturgeon such as Atlantic, Gulf, and green sturgeon may enter non-natal rivers and estuaries during non-reproductive times to forage (Dugo et al., 2004; Israel et al., 2009; Waldman et al., 2013). This natural foraging migration is not problematic if individuals don't compete with the resident population for food or habitat. However, a problem can arise if individuals stray into a non-natal river during spawning, increasing gene flow to unnaturally high levels which reduces interpopulation genetic diversity. Straying of hatchery-reared individuals has been reported for several sturgeon species. Shortnose sturgeon stocked into the Savannah River were later captured in the Ogeechee and Edisto Rivers, making up 7.4% and 10.6% of adults in those rivers, respectively (Smith et al., 2002). Some individuals from the Savannah River releases were recaptured in the Edisto River in spawning condition, suggesting that movements into the Edisto River were for reproduction, not foraging. In lake sturgeon, Lake Winnebago origin individuals stocked into the St. Louis River were found within the spawning runs of other Lake Superior tributaries (Welsh et al., 2019). To minimize the risk of straying, managers can expose early life stage sturgeon to waters of their natal rivers, such as through a streamside rearing program. While specific timing of imprinting is uncertain, it is assumed to occur very early in life (Boiko, 1999; Boiko and Grigor'yan, 2002). Releasing sturgeon at early life stages may also reduce the risk of straying, though it may come at a tradeoff with higher post-release mortality.

In some regions, reducing straying risk also decreases the risk of interspecific hybridization among species inhabiting different rivers. Hybridization occurs readily among sturgeons, even across ploidy levels, and hybrids with even chromosome counts are often fertile (Havelka and Arai, 2019). Fertile interspecific hybrids can backcross with parental species, resulting in introgression of non-native genes into native gene pools, which can lead to outbreeding depression. Interspecific hybridization occurred in the Upper Danube River when Siberian sturgeon accidentally released from commercial aquaculture operations mated with a threatened sterlet population (Ludwig et al., 2009). Hybridization occurring independent of aquaculture operations can also impact conservation aquaculture programs. High levels of hybridization between pallid and shovelnose sturgeon in some shared portions of their range have led to the development of genetic methods to distinguish between species and broodstock collection for pallid conservation aquaculture is restricted to regions where hybridization rates are negligible ( Jordan et al., 2019).

Risks related to both sturgeon conservation aquaculture programs and recovery of wild populations are density-dependent impacts on growth and survival on wild individuals from increasing abundances of hatchery fish. These impacts also occur within and between hatchery year classes as hatchery-reared fish are often released at ages/sizes that are susceptible to density-dependent mortality (Justice et al., 2009; Lorenzen and Camp, 2019). Conservation aquaculture programs must balance uncertainties related to year class survival with the risk of exceeding the carrying capacity of the habitat (Hildebrand et al., 2016). Increased densities of sturgeon in the wild may not only impact survival of younger age classes but may result in decreased growth and condition of all ages, leading to reproductive delays in mature adults. In populations where natural recruitment may occur at low levels (e.g., Upper Columbia River, Fisheries and Oceans, 2014), density-dependent impacts may disproportionately affect wild recruits in their first year compared to hatchery-reared fish that are grown to larger sizes. Depending on the abundance of hatchery-origin fish in the system, these impacts on wild recruits may go undetected in post-release monitoring. Another related concern is unequal post-release survival among families, resulting in the overrepresentation of some families relative to others and a reduction in the genetic and phenotypic variability conserved by conservation aquaculture, stressing the importance of post-release genetic monitoring (Crossman and Korman, 2021; Schreier et al., 2015).

A final risk that conservation aquaculture programs for sturgeon must consider is spontaneous autopolyploidy. Although spontaneous autopolyploidy has been documented in the wild, it occurs at a higher rate in aquaculture (Schreier et al., 2021). Spontaneous autopolyploid sturgeon are viable and morphologically similar to normal ploidy sturgeon. Fertility has been observed in male and female spontaneous autopolyploid white and Siberian sturgeon and mating between normal and spontaneous autopolyploid individuals produces viable offspring of intermediate ploidy (Drauch Schreier et al., 2011; Havelka et al., 2014). Preliminary evidence suggests that intermediate ploidy females may experience delayed reproductive maturation and reduced fecundity, both of which hinder conservation goals (Schreier and Van Eenennaam, 2019, as cited in Van Eenennaam et al., 2020). While further data in the wild is needed to evaluate long-term risks, management of conservation aquaculture programs need to consider precautionary approaches that reduce the frequency that spontaneous autopolyploids occur. Very low-levels have been documented in repatriation programs that rely on wild-origin progeny (Schreier et al., 2021; James Crossman, unpublished data). Broodstock-based conservation aquaculture programs can reduce production and release of spontaneous autopolyploids by stirring eggs gently during de-adhesion (Van Eenennaam et al., 2020) and screening the ploidy of each maternal family prior to release (Schreier et al., 2021).

## 6 Measuring success

To measure and achieve success, recovery programs need to have clearly defined objectives related to the use of aquaculture for conservation. The overarching goal of many conservation aquaculture programs is largely attributed to replacing or restoring natural recruitment, therefore leading to an end point where aquaculture is not required. More specific objectives may include preventing extirpation, preserving genetic integrity, or restoring a resource that was once available. In the absence of clear objectives that support a long-term action plan for recovery, decision making around conservation aquaculture can become complex (Anders, 1998) and releases of hatchery-origin fish may continue in spite of signals in the wild population that may call for revaluation of the program. Therefore, robust post-release monitoring programs are required to provide information along a timeframe that can support decision making. Importantly, measuring success requires the ability to distinguish hatchery-origin fish in the wild to ensure post-release monitoring programs can be effectively implemented.

## 6.1 Marking techniques to assess success

In order to evaluate success, post-release monitoring is required at a level that can inform program objectives. This requires a tagging method to discriminate hatchery from wild fish with each method providing a different level of assessment of program effectiveness. Size at release can dictate which tagging method is used as older/larger fish can carry a larger electronic or external tag, whereas tagging of very young fish may be limited to chemical and/or genetic methods of identification. The former method allows for both short and long-term identification of individuals whereas the latter is limited to identification of released families or cohorts.

Typically to monitor conservation aquaculture program success, fish are marked just prior to release. Furthermore, monitoring success will be greatly improved with more fish marked. Given that most hatcheries will release cohorts of fish as age 0+ or age 1+ individuals and aim to mark as many fish as possible, attachment of larger telemetry tags externally or internally is usually either impractical or cost-prohibitive and may only be applied to a subset of fish for research purposes. Therefore, knowledge of fish movement, habitat use, spawning migrations and general activity is usually gained from separate telemetry studies tagging individuals that are large enough that the size of the tag is not expected to have negative impacts on the individual's condition. Passive Integrated Transponder (PIT) tags are a commonly used method once fish achieve a size where the tag can be implanted (e.g., fall age 0+ or spring age 1+ releases). PIT tags present a number of advantages including high retention rates (>95% in some cases), identification of individual fish (Hamel et al., 2012; Rien et al., 1994), and operation over a long time period which is critical for sturgeon. However, they are substantially greater in cost than conventional external "T bar" or anchor tags but this cost is offset by the reduced retention of conventional external tags (Rien et al., 1994) and the significant opportunity for collecting information post-release. A PIT tag can last for the life of the individual as it does not require internal power and remains dormant until activated by an electromagnetic field produced by a hand-held reader. Detection of PIT tags is usually achieved through direct handling of individuals or, if the system is sufficiently small, PIT tag receiver arrays. The latter has been shown to be extremely effective for assessing movement of lake sturgeon in the Black Lake System in Northern Michigan, USA. For fish that are too small to receive PIT tags at the time of release, scute removal may be performed. In both pallid (Steffensen et al., 2019) and white sturgeon (Fisheries and Oceans, 2014), removal of scutes is a common practice and depending on which scute is removed one can use this method to identify year class of stocked individuals or it can serve as a secondary mark in the event a PIT tag fails and it is critical that hatchery-origin fish are identifiable upon capture.

Chemical signatures are another way to "batch mark" large cohorts of fish and the utility of microchemical signatures deposited in the fish hard structures (e.g., fin ray or otolith) has received increasing attention over the last decade (Bakhshalizadeh et al., 2021; Loeppky et al., 2019; Loewen et al., 2016; Nelson et al., 2013; Smith and Whitledge, 2011). One important consideration for chemical signatures is the length of time a "mark" may persist within the hard structure of the fish. As otoliths are deemed largely inert, once deposited the elemental signature may be there for the fish's lifetime (Campana, 1999). Fin rays are a much more dynamic tissue with material being resorbed and deposited throughout the fish's lifetime and therefore, the signature may not persist for as long. This holds especially true for age-0 fish where signal persistence can be as low as 70% over a 3 month period (Lochet et al., 2011).

However, even with relatively short immersion times, chemical signatures have been shown to persist in fin rays for years, as measured using stable isotope analyses (Carriere et al., 2016; Smith and Whitledge, 2011; Veinott and Evans, 1999). The technique relies on elements from the environment being taken up by the fish and ultimately being deposited into the hard structure. Where a fish may experience large fluctuations in water chemistry throughout its life cycle, such as the diadromous green sturgeon, interrogation of the elemental signatures of the fin ray will determine when these migrations occurred (Allen et al., 2009b). More recently, improved measurement resolution and understanding of environmental factors that may influence deposition of elements into the fin ray have facilitated a finer scale understanding of habitat use of sturgeons within freshwater systems (Allen et al., 2018; Loeppky and Anderson, 2020). Importantly for conservation aquaculture, hatchery water is often taken from a different source to where fish will be released, as a consequence the natural element signature of the fin ray will be different between the hatchery and wild environment facilitating relatively straightforward identification of the source of an individual with no advanced marking required (Bakhshalizadeh et al., 2021; Loeppky et al., 2019).

One final marking method for programs relying on releases of early life stages is parentage based tagging (PBT) (Steele et al., 2019), where all hatchery parents and recaptured offspring are genotyped and parentage analysis determines whether individuals originate from hatchery families or natural production. PBT using single nucleotide polymorphism (SNP) markers has been shown to be a highly accurate means for identifying individual origin in salmonid fisheries (Beacham et al., 2019; Jensen et al., 2021; Steele et al., 2013). This approach lends itself well to sturgeon conservation aquaculture programs where adults are transported to a hatchery environment or all gametes are taken from adults in the field. PBT cannot be used for repatriation programs where adult genotypes are unknown and requires that adults contributing to a year class are not able to spawn together in the wild, which would confound hatchery and wild production.

## 6.2 Post release monitoring

For conservation aquaculture programs focused on recovery, understanding success in the wild is critical to ensuring ongoing refinements to the program that can be made to optimize hatchery techniques, stocking prescriptions, or the overall recovery process. This may require an adaptive monitoring program that is developed with specific decision points and timelines, either based on population targets or results from specific monitoring. There has been significant investment in post-release monitoring of hatchery-origin sturgeon for many species (Table 2), in particular for programs that initiated conservation aquaculture in the past few decades (Table 1). In some cases, large-scale releases of small sturgeon from hatcheries were both logistically challenged to apply marks to all individuals or stocking occurred prior to development of advanced marking techniques (e.g., Ponto-Caspian region; Chebanov and Billard, 2001). Therefore, while significant monitoring occurs in these regions, results from ongoing studies may not differentiate hatchery fish from wild fish.

Several post-release monitoring objectives are more commonly applied than others, with efforts to document growth, survival, and habitat use in highest frequency (Table 2). Post-release monitoring of sturgeon has largely followed the traditional framework of fisheries assessments, relying on direct capture and subsequent recapture of individuals. However, capture probabilities are variable depending on the species, life stage, and river system of interest and, in some cases, there is a significant lag between release and recapture. This can result in a long time-span prior to having sufficient information to inform management of aquaculture programs and other recovery actions. Accordingly, telemetry has been a leading method to study movements and habitat use, informing release locations that align with preferred habitats, or providing information on survival that can be difficult to obtain at younger ages through conventional sampling methods.

**TABLE 2 Common objectives and measures evaluated in post-release monitoring programs of hatchery-origin sturgeon to determine effectiveness of conservation aquaculture programs.**

| Objective | Measures | Species specific studies |
|---|---|---|
| Survival | Direct:<br>Mark-recapture studies to assess survival | Atlantic sturgeon (Mohler et al., 2012)<br>Gulf sturgeon (Sulak et al., 2014)<br>White sturgeon (Crossman and Korman, 2021; Ireland et al., 2002; Justice et al., 2009)<br>Pallid sturgeon (Eder et al., 2015; Steffensen et al., 2010, 2012, 2016)<br>Lake sturgeon (Crossman et al., 2011b; McDougall et al., 2014, 2020) |
| | Indirect:<br>Telemetry | Atlantic sturgeon (Secor et al., 2000)<br>European sturgeon (Carrera-García et al., 2017)<br>Lake sturgeon (Crossman et al., 2009) |
| Growth | Somatic growth | Dabry's sturgeon (Wu et al., 2014)<br>Lake sturgeon (Dittman et al., 2015; McDougall et al., 2020)<br>White sturgeon (Hildebrand et al., 2016; Hildebrand and Parsley, 2013; Ireland et al., 2002; Paragamian et al., 2005)<br>Pallid sturgeon (Gerrity, 2005) |
| | Relative condition factor[a] | Atlantic sturgeon (Mohler et al., 2012)<br>Pallid sturgeon (Steffensen and Mestl, 2016)<br>White sturgeon (Hildebrand and Parsley, 2013; Hildebrand et al., 2016; Maskill, 2020; Crossman and Korman, 2021) |
| | Relative weight | Shovelnose sturgeon (Quist et al., 1998)<br>White sturgeon (Beamesderfer, 1993; Hildebrand et al., 2016; Hildebrand and Parsley, 2013; Ireland et al., 2002; Paragamian et al., 2005) |
| Abundance | Abundance and density | Kaluga sturgeon (Koshelev et al., 2014)<br>Amur sturgeon (Koshelev et al., 2014)<br>White sturgeon (Crossman and Korman, 2021; Justice et al., 2009) |

*Continued*

**TABLE 2 Common objectives and measures evaluated in post-release monitoring programs of hatchery-origin sturgeon to determine effectiveness of conservation aquaculture programs.—Cont'd**

| Objective | Measures | Species specific studies |
|---|---|---|
| Sources of mortality | Predation | Lake sturgeon (Caroffino et al., 2010; Duong et al., 2011; Forsythe et al., 2013, 2018; Crossman et al., 2018; Waraniak et al., 2018a,b; |
| | Contaminants | White sturgeon (Babey et al., 2020) |
| | | European sturgeon (Acolas et al., 2020) |
| | | White sturgeon (Kruse and Scarnecchia, 2002) |
| Habitat use/ movements | Telemetry | Beluga sturgeon (Doukakis et al., 2008) |
| | | Chinese sturgeon (Wu et al., 2018) |
| | | Dabrys sturgeon (Li et al., 2021) |
| | | European sturgeon (Acolas et al., 2012, 2017; Carrera-García et al., 2017) |
| | | Lake sturgeon (Mann et al., 2011; Dittman et al., 2015; Lacho et al., 2020; Hegna et al., 2020) |
| | | Pallid sturgeon (Eder et al., 2015; Gerrity et al., 2008; Jordan et al., 2006) |
| | Microchemistry | Shortnose sturgeon (Smith et al., 2002; Trested et al., 2011) |
| | | White sturgeon (Neufeld and Rust, 2009; Young and Scarnecchia, 2005) |
| | | Russian sturgeon (Korneev and Luzhnyak, 2012) |
| | | Atlantic sturgeon (Balazik et al., 2012) |
| | | Lake sturgeon (Allen et al., 2018; Loeppky 2021) |
| | | Pallid sturgeon (Phelps et al., 2012) |
| | | Shovelnose sturgeon (Phelps et al., 2012) |
| | | Shortnose sturgeon (Altenritter et al., 2015) |
| | | White sturgeon (Sellheim et al., 2017) |
| | | Russian sturgeon (Arai and Miyazaki, 2001) |

**TABLE 2 Common objectives and measures evaluated in post-release monitoring programs of hatchery-origin sturgeon to determine effectiveness of conservation aquaculture programs.—Cont'd**

| Objective | Measures | Species specific studies |
|---|---|---|
| Genetic diversity | Direct:<br>Known family pedigrees | Lake sturgeon (Crossman et al., 2011b)<br>White sturgeon (Crossman and Korman, 2021; Schreier et al., 2015) |
| | Indirect:<br>Molecular analyses, parentage, relatedness, effective population size, | Review (Anders et al., 2012)<br>Pallid sturgeon (DeHaan et al., 2008)<br>White sturgeon (Schreier et al., 2015)<br>European sturgeon (Roques et al., 2018) |
| Diet | Direct:<br>Taxonomic identification of prey found in stomach contents | Atlantic sturgeon (Secor et al., 2000a); Bogacka-Kapusta et al., 2011)<br>Pallid sturgeon (Braaten et al., 2012; Gerrity, 2005; Gerrity et al., 2006)<br>White sturgeon (Crossman et al., 2016)<br>Russian sturgeon (Korneev and Luzhnyak, 2012) |
| | Indirect:<br>Stable isotopes<br>Genetic | Gulf sturgeon (Gu et al., 2001)<br>Lake sturgeon (Smith et al., 2015, 2016)<br>Pallid sturgeon (Andvik et al., 2010)<br>Shovelnose sturgeon (French et al., 2013) |
| Sex and stage of maturity | Direct:<br>Histological analysis of gonadal tissue<br>Endoscopy<br>Ultrasonography | Review by Webb et al. (2019)<br>Pallid sturgeon (Holmquist et al., 2019; Cox, 2020)<br>White sturgeon (Maskill et al., 2022)<br>Lake sturgeon (McGuire et al., 2019) |
| | Indirect:<br>Molecular analyses,<br>Plasma sex steroids, | Review by Webb et al. (2019)<br>Pallid sturgeon (Holmquist et al., 2019; Cox, 2020)<br>White sturgeon (Maskill et al., 2022)<br>Genetic sex ID for multiple species (Kuhl et al., 2021) |

Species specific examples are provided from the literature either from post-release monitoring studies or, where applicable, studies that highlight approaches that could be applied to address the objective. Measures are categorized as either direct or indirect where possible, with direct being measured from the individual at the time of capture and indirect being applied though a different method or process of a sample.

[a]*The mass of an individual relative to the average mass of others of the same length.*

More recently, programs have focused on advancing monitoring to assign sex and stage of maturity in hatchery-origin fish (Maskill et al., 2022) as well as describe post-release genetic diversity (Crossman and Korman, 2021; Schreier et al., 2015). As hatchery-origin fish approach an age where reproduction may begin, documenting reproductive indices (e.g., reproductive structure: the proportion of the population that is spawning in any given year) is essential to gauging the success of the program, through the successful replacement of breeding adults. While there are few published studies assigning sex and stage of maturity to post-release hatchery-origin sturgeon, many are underway and the methods and tools have been well developed on both cultured and wild sturgeons (Webb et al., 2019). While measuring and maximizing genetic diversity in the hatchery is a common goal, it is documented less frequently following release, in part due to challenges in capturing sufficient numbers of progeny in the wild. Importantly, sturgeon have a long-life span and while recaptures over a few years may not address questions related to diversity, low levels of monitoring applied consistently through time can result in significant learning (e.g., Crossman and Korman, 2021).

## 7 Conclusions—Uncertainties and areas of study critically required

Despite significant advancement in conservation aquaculture for sturgeon, domestication in hatchery reared fishes and its effect in the wild remain poorly understood. Fitness reduction in salmonids associated with hatchery rearing is well documented but is challenging to study in sturgeon given delayed maturation and intermittent spawning and requires more research. Several studies have looked at relative reproductive success of wild sturgeon and monitoring to document reproductive indices in hatchery-origin fish is ongoing. Epigenetic changes occur after hatchery rearing but studies just show associations and don't demonstrate causality. It is recommended that programs strive to follow best practices including:

- **(i)** development of clear objectives that support program decision making
- **(ii)** provision of environmental conditions during rearing, in particular early life stage rearing, that are reflective of the natural system
- **(iii)** selection of progeny that best represent genetic diversity and adaptations (e.g., spawning time) of the wild population
- **(iv)** reduce rearing time in the hatchery by optimizing release strategies
- **(v)** develop post-release monitoring that informs decision making and the effectiveness of the program in meeting stated objectives.

For recovery of many sturgeon populations, restoring natural recruitment to a level that supports population persistence is an overarching goal and the focus of restoration efforts beyond just conservation aquaculture. In many cases, releases from aquaculture have extended time to extirpation and allowed for concurrent investigations into population declines or recruitment failure to occur. While releases of hatchery-origin fish might help restore a natural age class structure and prevent extirpation, large abundances of hatchery-origin fish may complicate detecting improvements made to natural recruitment from other recovery actions and may increase interest in restoring fishery resources prior to recovery being fully addressed. Recent recovery actions are focusing on habitat restoration, either to improve spawning habitat (McAdam et al., 2018), ecosystem function (Kootenai Tribe of Idaho, 2009), or connectivity within rivers (Cooke et al., 2020). Determining the success of these actions in stimulating natural recruitment may be confounded by increased abundance of hatchery fish and complicate decisions around when to cease operating a hatchery supplementation program. The human dimensions component to these decisions can be challenging, with investment into infrastructure and employment having been significant for decades. In other cases, fishery resources may have been restored and or are being maintained through releases of hatchery fish. Accordingly, it's important that recovery programs critically evaluate aquaculture as an ongoing conservation action to ensure both short and long-term recovery objectives are being met and that aquaculture is implemented in parallel with other restoration activities.

## Acknowledgments

The authors thank Jörn Gessner (IGB Leibniz-Institute of Freshwater Ecology and Inland Fisheries), Molly Webb, (US Fish and Wildlife Service), Dawn Dittman (US Geological Survey) and Tim Haxton (Ontario Ministry of Natural Resources) for providing unpublished information pertinent to this review. Ed Heist, Jörn Gessner, Gwang-Seok Rex Yoon and several anonymous reviewers provided input that undoubtedly improved the scope and clarity of this chapter, the North American Sturgeon and Paddlefish Society members and World Sturgeon Conservation Society members for their persistent and unrelenting pressure on legislative bodies to enhance and preserve this enigmatic group of fishes. WGA is supported by a Natural Sciences and Engineering Research Council of Canada (NSERC) grant number RGPIN-2020-05328 and NSERC/Manitoba Hydro Industrial Research Chair in Conservation Aquaculture of Lake Sturgeon.

## References

Abdelhay, H.A., Tahori, H.B., 2006. Fingerling production and release for stock enhancement of sturgeon in the Southern Caspian Sea: an overview. J. Appl. Ichthyol. 22 (Suppl. 1), 125–131.

Acolas, M.L., Rochard, E., Le Pichon, C., Rouleau, E., 2012. Downstream migration patterns of one-year-old hatchery-reared European sturgeon (*Acipenser sturio*). J. Exp. Mar. Biol. Ecol. 430, 68–77.

Acolas, M.L., Le Pichon, C., Rochard, E., 2017. Spring habitat use by stocked one year old European sturgeon *Acipenser sturio* in the freshwater-oligohaline area of the Gironde estuary. Estuar. Coast. Shelf Sci. 196, 58–69.

Acolas, M.L., Davail, B., Gonzalez, P., Jean, S., Clérandeau, C., Morin, B., Gourves, P.Y., Daffe, G., Labadie, P., Perrault, A., Lauzent, M., 2020. Health indicators and contaminant levels of a critically endangered species in the Gironde estuary, the European sturgeon. Environ. Sci. Pollut. Res. 27 (4), 3726–3745.

Adams, S.R., Adams, G.L., Parson, G.R., 2003. Critical swimming speed and behaviour of juvenile shovelnose sturgeon and pallid sturgeon. Trans. Am. Fish. Soc. 132, 392–397.

Aidos, L., Pinheiro Valente, L.M., Sousa, V., Lanfranchi, M., Domeneghini, C., Giancamillo, A.D., 2017. Effects of different rearing temperatures on muscle development and stress response in the early larval stages of *Acipenser baerii*. Eur. J. Histochem. 61, 287–294.

Aidos, L., Vasconi, M., Abbate, F., Valente, L.M.P., Lanfranchi, M., Giancamillo, A.D., 2018. Effects of stocking density on reared Siberian sturgeon (*Acipenser baerii*) larval growth, muscle development and fatty acids composition in a recirculating aquaculture system. Aquacult. Res. 50, 588–598.

Aidos, L., Cafison, A., Berlotto, D., Bazzocchi, C., Radaelli, G., Giancamillo, A.D., 2020. How different rearing temperatures affect growth and stress status of Siberian sturgeon, *Acipenser baerii* larvae. J. Fish Biol. 96, 913–924.

Akbarzadeh, A., Gunther, O.P., Houde, A.L., Lee, S., Ming, T.J., Jeffries, K.M., Hinch, S.G., Miller, K.M., 2018. Developing specific molecular biomarkers for thermal stress in salmonids. BMC Genomics 19, 749.

Allen, P.J., Hodge, B., Werner, I., Cech, J.J., 2006. Effects of ontogeny, season, and temperature on the swimming performance of juvenile green sturgeon (*Acipenser medirostris*). Can. J. Fish. Aquat. Sci. 63, 1360–1369.

Allen, P.J., Cech, J.J., 2007. Age/size effects on juvenile green sturgeon, *Acipenser medirostris*, oxygen consumption, growth and osmoregulation in saline environments. Environ. Biol. Fishes 79, 211–219.

Allen, P.J., Barth, C.C., Peake, S.J., Abrahams, M.V., Anderson, W.G., 2009a. Cohesive social behaviour shortens the stress response: the effects of conspecifics on the stress response in lake sturgeon *Acipenser fulvescens*. J. Fish Biol. 74, 90–104.

Allen, P.J., Hobbs, J.A., Cech, J.J., Van Eenennaam, J.P., Doroshov, S.I., 2009b. Using trace elements in pectoral fin rays to assess life history movements in sturgeon: estimating age at initial seawater entry in Klamath River green sturgeon. Trans. Am. Fish. Soc. 138, 240–250.

Allen, P.J., McEnroe, M., Forostyan, T., Cole, S., Nicholl, M.M., Hodge, B., Cech, J.J., 2011. Ontogeny of salinity tolerance and evidence for seawater-entry preparation in juvenile green sturgeon, *Acipenser medirostris*. J. Comp. Physiol. 181B, 1045–1062.

Allen, P.J., DeVries, R.J., Fox, D.A., Gabitov, R.I., Anderson, W.G., 2018. Trace element and strontium isotopic analysis of Gulf Sturgeon fin rays to assess habitat use. Environ. Biol. Fishes 101, 469–488.

Allendorf, F.W., Phelps, S.R., 1980. Loss of genetic variation in a hatchery stock of cutthroat trout. Trans. Am. Fish. Soc. 109, 537–543.

Altenritter, M.E., Kinnison, M.T., Zydlewski, G.B., Secor, D.H., Zydlewski, J.D., 2015. Assessing dorsal scute microchemistry for reconstruction of shortnose sturgeon life histories. Environ. Biol. Fishes 98 (12), 2321–2335.

Amiri, B.M., Baker, D.W., Morgan, J.M., Brauner, C.J., 2009. Size dependent early salinity tolerance in two sizes of juvenile white sturgeon, *Acipenser transmontanus*. Aquaculture 286, 121–126.

Anders, P., 1998. Conservation aquaculture and endangered species: can objective science prevail over risk anxiety? Fisheries 23, 28–31.

Andvik, R.T., VanDeHey, J.A., Fincel, M.J., French, W.E., Bertrand, K.N., Chipps, S.R., Klumb, R.A., Graeb, B.D.S., 2010. Application of non-lethal stable isotope analysis to assess feeding patterns of juvenile pallid sturgeon *Scaphirhynchus albus*: a comparison of tissue types and sample preservation methods. J. Appl. Ichthyol. 26 (6), 831–835.

Antognazza, C.M., Vanetti, I., De Santis, V., Bellani, A., Di Francesco, M., Puzzi, C.M., Casoni, A.G., Zaccara, S., 2021. Genetic investigation of four beluga sturgeon (*Huso huso*, L.) broodstocks for its reintroduction in the Po River. Environments 8, 25.

Arai, T., Miyazaki, N., 2001. Use of otolith microchemistry to estimate the migratory history of the Russian sturgeon, *Acipenser gueldenstaedtii*. J. Mar. Biolog. Assoc. U.K. 81 (4), 709–710.

Araki, H., Cooper, B., Blouin, M., 2007. Genetic effects of captive breeding cause a rapid, cumulative fitness decline in the wild. Science 318, 100–103.

Arlati, G., Poliakova, L., 2009. Restoration of Adriatic sturgeon (*Acipenser naccarii*) in Italy: situation and perspectives. In: Carmona, R., Domezain, A., García-Gallego, M., Hernando, J.A., Rodríguez, F., Ruíz-Rejón, M. (Eds.), Biology, Conservation and Sustainable Development of Sturgeons. Springer Science + Business Media B. V, pp. 237–245.

Auer, N.A., Baker, E.A., 2002. Duration and drift of larval lake sturgeon in the Sturgeon River, Michigan. J. Appl. Ichthyol. 18, 557–564.

Babey, C.N., Gantner, N., Williamson, C.J., Spendlow, I.E., Shrimpton, J.M., 2020. Evidence of predation of juvenile white sturgeon (*Acipenser transmontanus*) by North American river otter (*Lontra canadensis*) in the Nechako River, British Columbia, Canada. J. Appl. Ichthyol. 36 (6), 780–784.

Baker, D.W., Mcadam, D.S.O., Boucher, M., Huynh, K.T., Brauner, C.J., 2014. Swimming performance and larval quality are altered by rearing substrate at early life phases in white sturgeon, *Acipenser transmontanus* (Richardson, 1836). J. Appl. Ichthyol. 30, 1461–1472.

Baker, E., Scribner, K.T., 2017. Cohort-specific estimates of first-year survival are positively associated with size at stocking for lake sturgeon *Acipenser fulvescens* (Rafinesque 1817) stocked in Black Lake, Michigan, USA. J. Appl. Ichthyol. 33 (5), 892–897.

Baker, M.E., 2019. Steroid receptors and vertebrate evolution. Mol. Cell. Endocrinol. 496, 110526.

Bakhshalizadeh, S., Tchaikovsky, A., Bani, A., Prohaska, T., 2021. Using fin ray microchemistry to discriminate hatchery reared juvenile age-0 Persian sturgeon by their origin in the Southern Caspian sea region using split stream ICP-MS/MC ICP-MS. Fish. Res. 243, 106093.

Balazik, M.T., McIninch, S.P., Garman, G.C., Fine, M.L., Smith, C.B., 2012. Using energy dispersive x-ray fluorescence microchemistry to infer migratory life history of Atlantic sturgeon. Environ. Biol. Fishes 95 (2), 191–194.

Barannikova, I.A., 1987. Review of sturgeon farming in the Soviet Union. J. Appl. Ichthyol. 27, 62–67.

Bates, L.C., Boucher, M.A., Shrimpton, J.M., 2014. Effect of temperature and substrate on whole body cortisol and size of larval white sturgeon (*Acipenser transmontanus* Richardson, 1836). J. Appl. Ichthyol. 30, 1259–1263.

Bauman, J.M., Baker, E.A., Marsh, T.L., Scribner, K.T., 2015. Effects of rearing density on total length and survival of lake sturgeon free embryos. N. Am. J. Aquac. 77, 444–448.

Bauman, J.M., Woodward, B.M., Baker, E.A., Marsh, T.L., Scribner, K.T., 2016. Effects of family, feeding frequency, and alternate food type on body size and survival of hatchery-produced and wild-caught lake sturgeon larvae. N. Am. J. Aquac. 78, 136–144.

Bayunova, L., Canario, A.V.M., Semenkova, T., Dyubin, V., Sverdlova, O., Trenkler, I., Barannikova, I., 2006. Sex steroids and cortisol levels in the blood of stellate sturgeon (*Acipenser stellatus*, Pallas) during final maturation induced by LH-RH-analogue. J. Appl. Ichthyol. 22, 334–339.

Beacham, T.D., Wallace, C., Jonsen, K., McIntosh, B., Candy, J.R., Willis, D., Lynch, C., Moore, J.-S., Bernatchez, L., Withler, R.E., 2019. Comparison of coded-wire tagging with parentage-based tagging and genetic stock identification in a large-scale coho salmon fisheries application in British Columbia, Canada. Evol. Appl. 12 (2), 230–254.

Beaman, J.E., White, C.R., Seebacher, F., 2016. Evolution of plasticity: mechanistic link between development and reversible acclimation. Trends Ecol. Evol. 31, 237–249.

Beamesderfer, R.C., 1993. A standard weight (Ws) equation for white sturgeon. Calif. Fish Game 79, 63–69.

Berkman, L.K., Anderson, M.R., Herzog, D.P., Moore, T.L., Eggert, L.S., 2020. A genetic assessment of Missouri's lake sturgeon after 30 years of restoration releases. N. Am. J. Fish. Manag. 40 (3), 700–712.

Bezold, J., Peterson, D.L., 2008. Assessment of lake sturgeon reintroduction in the Coosa River system, Georgia-Alabama. Am. Fish. Soc. Symp. 62, 571–586.

Billard, R., Lecointre, G., 2001. Biology and conservation of sturgeon and paddlefish. Rev. Fish Biol. Fish. 10, 355–392.

Bogacka-Kapusta, E., Wiszniewski, G., Duda, A., Kapusta, A., 2011. Feeding of hatchery-reared juvenile Atlantic sturgeon, *Acipenser oxyrinchus* Mitchill, released into the Drwęca River. Fish Aqua. Life. 19 (2), 113–117. https://doi.org/10.2478/v10086-011-0013-8.

Boiko, N.E., 1999. Formation of olfaotory imprinting and thyroid hormone metabolism in early ontogonesis of Russian sturgeon. In: Rosenthal, H., et al. (Eds.), 1999: Proceedings of the 3rd International Symposium on Sturgeon. Piacenza, Italy. 1997. J. Appl. lchthyol., 15 (4-5): 287.

Boiko, N.E., Grigor'yan, R.A., 2002. Effect of thyroid hormones on imprinting of chemical signals at early ontogenesis of the sturgeon *Acipenser gueldenstaedtii*. J. Evol. Biochem. Physiol. 38 (2), 218–222.

Borodin, N.A., 1898. Experimental of Artificial Insemination of Sturgeon Eggs and Other Biological Observations Constructed on Ural River in Spring 1897—Vastnik Ryboptomy-Schelennosti St Petersburg. Vol. 6–7.

Boscari, E., Congiu, L., 2014. The need for genetic support in restocking activities and ex situ conservation programmes: the case of the Adriatic sturgeon (*Acipenser naccarii* Bonaparte, 1836) in the Ticino River Park. J. Appl. Ichthyol. 30 (6), 1416–1422.

Boucher, M.A., McAdam, S.O., Shrimpton, J.M., 2014. The effect of temperature and substrate on the growth, development and survival of larval white sturgeon. Aquaculture 430, 139–148.

Boucher, M.A., Baker, D.W., Brauner, C.J., Shrimpton, J.M., 2018. The effect of substrate rearing on growth, aerobic scope and physiology of larval white sturgeon, *Acipenser transmontanus*. J. Fish Biol. 92, 1731–1746.

Braaten, P.J., Fuller, D.B., Holte, L.D., Lott, R.D., Viste, W., Brandt, T.F., Legare, R.G., 2008. Drift dynamics of larval pallid sturgeon and shovelnose sturgeon in a natural side channel of the upper Missouri River, Montana. N. Am. J. Fish. Manag. 28, 808–826.

Braaten, P.J., Fuller, D.B., Lott, R.D., Haddix, T.M., Holte, L.D., Wilson, R.H., Bartron, M.L., Kalie, J.A., DeHaan, P.W., Ardren, W.R., Holm, R.J., 2012. Natural growth and diet of known-age pallid sturgeon (*Scaphirhynchus albus*) early life stages in the upper Missouri River basin, Montana and North Dakota. J. Appl. Ichthyol. 28 (4), 496–504.

Braithewaite, V.A., Salvanes, A.G.V., 2005. Environmental variability in early rearing environment generates behaviourally flexible cod: implications for rehabilitating wild populations. Proc. R. Soc. B Biol. Sci. 272, 1107–1113.

Brandt, C., Groening, L., Klassen, C., Anderson, W.G., 2021. Effects of rearing temperature on volitional and escape response swimming performance in Lake Sturgeon, *Acipenser fulvescens*, from hatch to age1. Environ. Biol. Fishes 104, 737–750.

Brandt, C., Groening, L., Klassen, C., Anderson, W.G., 2022. Effects of rearing temperature on yolksac volume and growth rate in Lake Sturgeon, *Acipenser fulvescens*, from hatch to age-1. Aquaculture 546. https://doi.org/10.1016/j.aquaculture.2021.737352.

Brannon, E.L., 1993. The perpetual oversight of hatchery programs. Fish. Res. 18, 19–27.

Brennan, N.P., Darcy, M.C., Leber, K.M., 2006. Predator-free enclosures improve post-release survival of stocked common snook. J. Exp. Mar. Biol. Ecol. 335, 302–311.

Brett, J.R., 1964. The respiratory metabolism and swimming performance of young Sockeye Salmon. J. Fish. Res. Board Canada 21, 1183–1226.

Brevé, N.W.P., Vis, H., Houben, B., de Laak, G.A.J., Breukelaar, A.W., Acolas, M.L., de Bruijn, Q.A.A., Spierts, I., 2014. Exploring the possibilities of seaward migrating juvenile European sturgeon *Acipenser sturio* L., in the Dutch part of the Rhine River. J. Coast. Conserv. 18, 131–143.

Brown, C., Day, R.L., 2002. The future of stock enhancements: lessons for hatchery practice from conservation biology. Fish Fish. 3, 79–94.

Bruch, R.M., Eggold, B.T., Schiller, A., Wawrzyn, W., 2021. Projected lake sturgeon recovery in the Milwaukee River, Wisconsin, USA. J. Appl. Ichthyol. 37 (5), 643–654. https://doi.org/10.1111/jai.14238.

Buddington, R.K., Hazel, J.R., Doroshov, S.I., Van Eenennaam, J.P., 1993. Ontogeny of the capacity for homeoviscous adaptation in white sturgeon (*Acipenser transmontanus*). J. Exp. Zool. 265, 18–28.

Bugg, W.S., Yoon, G.R., Schoen, A.N., Laluk, A., Brandt, C., Anderson, W.G., Jeffries, K., 2020. Effect of acclimation temperature on the thermal physiology in two geographically distinct populations of age-0 Lake Sturgeon (*Acipenser fulvescens*). Conserv. Physiol. 08 (01), coaa087. https://doi.org/10.1093/conphys/coaa087.

Bugg, W.S., Jeffries, K.M., Anderson, W.G., 2021a. Survival and gene expression responses in immune challenged larval lake sturgeon. Fish Shellfish Immunol 112, 1–7.

Bugg, W.S., Yoon, G.R., Brandt, C., Earhart, M.L., Anderson, W.G., Jeffries, K.M., 2021b. The effects of population and thermal acclimation on the growth, condition, and cold responsive mRNA expression of age-0-lake sturgeon (*Acipenser fulvescens*). J. Fish Biol. 99 (6), 1912–1927. https://doi.org/10.1111/jfb.14897 (accepted).

Burggren, W.W., Mueller, C.A., 2015. Developmental critical windows and sensitive periods as three-dimensional constructs in time and space. Physiol. Biochem. Zool. 88, 91–102.

Burggren, W.W., 2020. Phenotypic switching resulting from developmental plasticity: fixed or reversible? Front. Physiol. 10, 1634.

Busack, C., Knudsen, C.M., 2007. Using factorial mating designs to increase the effective number of breeders in fish hatcheries. Aquaculture 273, 24–32.

Bussy, U., Wassink, L., Scribner, K.T., Li, W., 2017. Determination of cortisol in lake sturgeon (*Acipenser fulvescens*) eggs by liquid chromatography tandem mass spectrometry. J. Chromatogr. 1040B, 162–168.

Butkauskas, D., Pilinkovskij, A., Ragauskas, A., Kesminas, V., Fopp-Bayat, D., 2019. Genetic characterization of Atlantic sturgeon stocking material used in Lithuania to restore the Baltic Sea population. Acta Ichthyol. Piscat. 49, 251–256.

Byrne, P.G., Silla, A.J., 2020. An experimental test of the genetic consequences of population augmentation in an amphibian. Conserv. Sci. Pract. 2, e194.

Campana, S.E., 1999. Chemistry and composition of fish otoliths: pathways, mechanisms and applications. Mar. Ecol. Prog. Ser. 188, 263–297.

Campbell, J.G., Goodman, R.R., 2004. Acute sensitivity of juvenile shortnose sturgeon to low dissolved oxygen concentrations. Trans. Am. Fish. Soc. 133, 772–776.

Caroffino, D.C., Sutton, T.M., Elliott, R.F., Donofrio, M.C., 2010. Predation on early life stages of lake sturgeon in the Peshtigo River, Wisconsin. Trans. Am. Fish. Soc. 139 (6), 1846–1856.

Carrera-García, E., Rochard, E., Acolas, M., 2017. Effects of rearing practice on post-release young-of-the-year behavior: *Acipenser sturio* early life in freshwater. Endanger. Species Res. 34, 269–281.

Cataldi, E., Ciccotti, E., Di Marco, P., Di Santo, O., Bronzi, P., Cataudella, S., 1995. Acclimation trials of juvenile Italian sturgeon to different salinities: morpho-physiological descriptors. J. Fish Biol. 47, 609–618.

Carriere, B., Gillis, D., Halden, N., Anderson, W.G., 2016. Strontium metabolism in the juvenile Lake Sturgeon, *Acipenser fulvescens*, and further evaluation of the isotope as a marking tool for stock discrimination. J. Appl. Ichthyol. 32, 258–266.

Cataldi, E., Barzaghi, C., Di Marco, P., Boglionel, C., Dini, L., McKenzie, D.J., Bronzi, P., Cataudella, S., 1999. Some aspects of osmotic and ionic regulation in Adriatic sturgeon *Acipenser naccarii*. I: ontogenesis of salinity tolerance. J. Appl. Ichthyol. 15, 57–60.

Cech, J.J., Mitchell, S.J., Wragg, T.E., 1984. Comparative growth of juvenile white sturgeon and striped bass: effects of temperature and hypoxia. Estuaries 7, 12–18.

Chalupnicki, M.A., Dittman, D.E., Carlson, D.M., 2011. Distribution of lake sturgeon in New York: 11 years of restoration management. Am. Midl. Nat. 165, 364–371.

Chebanov, M., Billard, R., 2001. The culture of sturgeons in Russia: production of juveniles for stocking and meat for human consumption. Aquat. Living Resour. 14, 375–381.

Chebanov, M.S., Karnaukhov, G.I., Galich, E.V., Chmir, Y.N., 2002. Hatchery stock enhancement and conservation of sturgeon, with an emphasis on the Azov Sea populations. J. Appl. Ichthyol. 18, 463–469.

Chittenden, C.M., Sura, S., Butterworth, K.G., Cubitt, K.F., Plantalech-Menella, N., Balfry, S., Oakland, F., McKinely, R.S., 2008. Riverine, estuarine and marine migratory behaviour and physiology of wild and hatchery-reared coho salmon (*Oncorhynchus kisutch*) smolts descending the Campbell River, BC. J. Fish Biol. 72, 614–628.

Chittenden, C.M., Biagi, C.A., Davidsen, J.G., Davidsen, A.G., Kondo, H., McKnight, A., Pedersen, O.P., Raven, P.A., Rikardsen, A.H., Shrimpton, J.M., Zuehlke, B., 2010. Genetic versus rearing-environment effects on phenotype: hatchery and natural rearing effects on hatchery-and wild-born coho salmon. PLoS One 5 (8), e12261.

Cooke, S.J., Cech, J.J., Glassman, D.M., Simard, J., Louttit, S., Lennox, R.J., Cruz-Font, L., O'Connor, C.M., 2020. Water resource development and sturgeon (Acipenseridae): state of the science and research gaps related to fish passage, entrainment, impingement and behavioural guidance. Rev. Fish Biol. Fish. 30 (2), 219–244.

Cox, T.L., 2020. Reproductive Ecology of Hatchery-Origin Pallid Sturgeon Upstream of Fort Peck Reservoir, Montana. Master's Thesis, Montana State University. 88 pp.

Crocker, C.E., Cech, J.J., 1997. Effects of environmental hypoxia on oxygen consumption rate and swimming activity in juvenile white sturgeon, *Acipenser transmontanus*, in relation to temperature and life intervals. Environ. Biol. Fishes 50, 383–389.

Crossman, J.A., Forsythe, P.S., Baker, E.A., Scribner, K.T., 2009. Overwinter survival of stocked age-0 lake sturgeon. J. Appl. Ichthyol. 25 (5), 516–521.

Crossman, J.A., Forsythe, P.S., Scribner, K.T., Baker, E.A., 2011a. Hatchery rearing environment and age affect survival and movements of stocked juvenile lake sturgeon. Fish. Manag. Ecol. 18, 132–144.

Crossman, J.A., Scribner, K.T., Thuy Yen, D., Davis, C.A., Forsythe, P.S., Baker, E.A., 2011b. Gamete and larval collection methods and hatchery rearing environments affect levels of genetic diversity in early life stages of lake sturgeon (*Acipenser fulvescens*). Aquaculture 310, 312–324.

Crossman, J.A., Scribner, K.T., Davis, C.A., Forsythe, P.S., Baker, E.A., 2014. Survival and growth of lake sturgeon during early life stages as a function of rearing environment. Trans. Am. Fish. Soc. 143, 104–116.

Crossman, J.A., Scribner, K.T., Forsythe, P.S., Baker, E.A., 2018. Lethal and non-lethal effects of predation by native fish and an invasive crayfish on hatchery-reared age-0 lake sturgeon (*Acipenser fulvescens Rafinesque*, 1817). J. Appl. Ichthyol. 34 (2), 322–330.

Crossman, J.A., Jay, K.J., Hildebrand, L.R., 2016. Describing the diet of juvenile white sturgeon in the upper Columbia River Canada with lethal and nonlethal methods. N. Am. J. Fish. Manag. 36, 421–432. https://doi.org/10.1080/02755947.2015.1125976.

Crossman, J.A., Korman, J., 2021. Evaluating the Risk of Family Representation on the Long-Term Recovery of White Sturgeon in the Transboundary Reach of the Columbia River. Report Submitted to the Department of Fisheries and Oceans, Vancouver British Columbia, Canada. 90 pp.

Cushing, D.H., 1972. The production cycle and numbers of marine fish. Symp. Zool. Soc. Lond. 29, 213–232.

D'Anna, G., Giacalone, V.M., Fernández, T.V., Vaccaro, A.M., Pipitone, C., Mirto, S., Mazzola, S., Badalamenti, F., 2012. Effects of predator and shelter conditioning on hatchery-reared white seabream *Diplodus sargus* (L., 1758) released at sea. Aquaculture 356–357, 91–97.

Dammerman, K.J., Steibel, J.P., Scribner, K.T., 2015. Genetic and environmental components of phenotypic and behavioral trait variation during lake sturgeon (*Acipenser fulvescens*) early ontogeny. Environ. Biol. Fishes 98, 1659–1670.

Dammerman, K.J., Steibel, J.P., Scribner, K.T., 2020. Relative influences of microhabitat incubation conditions and genetic parentage effects on lake sturgeon (*Acipenser fulvescens*) offspring traits during early ontogeny. Environ. Biol. Fish. 103, 1565–1581.

de Miliano, J., Stefano, J.D., Courtney, P., Temple-Smith, P., Coulson, G., 2016. Soft-release versus hard-release for reintroduction of an endangered species: an experimental comparison using eastern barred bandicoots (*Perameles gunnii*). Wild. Res. 43, 1–12.

DeGregorio, B., Moody, R., Myers, H., 2020. Soft release translocation of Texas horned lizards (*Phrynosoma cornutum*) on an urban military installation in Oklahoma, United States. Animals 10, 1358.

DeHaan, P.W., Jordan, G.R., Ardren, W.R., 2008. Use of genetic tags to identify captive-bred pallid sturgeon (*Scaphirhynchus albus*) in the wild: improving abundance estimates for an endangered species. Conserv. Genet. 9 (3), 691.

Delage, N., Cachot, J., Rocahrd, E., Fraty, R., Jatteau, P., 2014. Hypoxia tolerance of European sturgeon (*Acipenser sturio* L., 1758) young stages at two temperatures. J. Appl. Ichthyol. 30, 1195–1202.

Delage, N., Couturier, B., Jatteau, P., Larcher, T., Ledevin, M., Goubin, H., Cachot, J., Rochard, E., 2020. Oxythermal window drastically constrains the survival and development of European sturgeon early life phases. Environ. Sci. Pollut. Res. 27, 3651–3660.

Denver, R.J., 2009. Structural and functional evolution of vertebrate neuroendocrine stress systems. Ann. N. Y. Acad. Sci. 1163, 1–16.

Deslauriers, D., Kieffer, J.D., 2012. The effects of temperature on swimming performance of juvenile shortnose sturgeon (*Acipenser brevirostrum*). J. Appl. Ichthyol. 28, 176–181.

Dettlaff, T.A., Ginsburg, A.S., Schmalhausen, O.I., 1993. Sturgeon Fishes Developmental Biology and Aquaculture. Springer-Verlag, Berlin, Heidelberg, New-York, pp. 1–300.

Dieterman, D.J., Frank, J., Painovich, N., Staples, D.F., 2010. Lake sturgeon population status and demography in the Kettle River, Minnesota, 1992–2007. N. Am. J. Fish. Manag. 30, 337–351.

Dinsmore, S.J., Rust, P., Hardy, R., Ross, T.J., Stephenson, S., Young, S., 2015. Kootenai River juvenile white sturgeon population analyses. In: Technical Report to the Kootenai Tribe of Idaho, Bonners Ferry, p. 37.

Dittman, D.E., Chalupnicki, M.A., Johnson, J.H., Snyder, J., 2015. Reintroduction of lake sturgeon (*Acipenser fulvescens*) into the St. Regis River, NY: post-release assessment of habitat use and growth. Northeast. Nat. 22, 704–716.

Donelson, J.M., Salinas, S., Munday, P.L., Shama, L.N.S., 2018. Transgenerational plasticity and climate change experiments: Where do we go from here? Glob. Chang. Biol. 24, 13–34.

Doroshov, S.I., Moberg, G.P., Van Eenennaam, J.P., 1997. Observations on the reproductive cycle of cultured white sturgeon, *Acipenser transmontanus*. Environ. Biol. Fishes 48, 265–278.

Doukakis, P., Erickson, D., Baimukhanov, M., Bokova, Y., Erbulekov, S., Nimatov, A., Pikitch, E.K., 2008. Field and genetic approaches to enhance knowledge of Ural River sturgeon biology. In: Lagutov, V. (Ed.), Rescue of Sturgeon Species in the Ural River basin. Springer, Amsterdam, pp. 277–292.

Dowling, T.E., Marsh, P.C., Kelsen, A.T., Tibbets, C.A., 2004. Genetic monitoring of wild and repatriated populations of endangered razorback sucker (*Xyrauchen texanus*, Catostomidae, Teleostei) in Lake Mohave, Arizona-Nevada. Mol. Ecol. 14, 123–135.

Downie, A.T., Kieffer, J.T., 2016a. THe physiology of juvenile shortnose sturgeon (*Acipenser brevirostrum*) during acute salinity challenge. Can. J. Zool. 94, 677–683.

Downie, A.T., Kieffer, J.T., 2016b. A split decision: the impact of substrate type on the swimming behaviour, substrate preference and UCrit of juvenile shortnose sturgeon (*Acipenser brevirostrum*). Environ. Biol. Fishes 100, 17–25.

Drauch, A.M., Rhodes, O.E., 2007. Genetic evaluation of the lake sturgeon reintroduction program in the Mississippi and Missouri Rivers. N. Am. J. Fish. Manag. 27, 434–442.

Drauch Schreier, A., Gille, D., Mahardja, B., May, B., 2011. Neutral microsatellite markers confirm the octoploid origin and reveal spontaneous autopolyploidy in white sturgeon, *Acipenser transmontanus*. J. Appl. Ichthyol. 27, 24–33.

Du, H., Wei, Q.W., Xie, X., Shi, L.L., Wu, J.M., Qiao, X.M., Liu, Z.G., 2014. Improving swimming capacity of juvenile Dabry's sturgeon, (*Acipenser dabryanus* Dumeril, 1869) in current-enriched culture tanks. J. Appl. Ichthyol. 30, 1445–1450.

Dugo, B.M.A., Kreiser, B.R., Ross, S.T., Slack, W.T., Heise, R.J., Bowen, B.R., 2004. Conservation and management implications of fine scale genetic structure of Gulf sturgeon in the Pascagoula River, Mississippi. J. Appl. Ichthyol. 20, 243–251.

Duong, T.Y., Scribner, K.T., Crossman, J.A., Forsythe, P.S., Baker, E.A., 2011. Environmental and maternal effects on embryonic and larval developmental time until dispersal of lake sturgeon (*Acipenser fulvescens*). Can. J. Fish. Aquat. Sci. 68, 643–654.

Earhart, M.L., Bugg, W.S., Wiwchar, C., Kroeker, J., Jeffries, K.M., Anderson, W.G., 2020a. Shaken, rattled and rolled: The effects of hatchery-rearing techniques on endogenous cortisol production, stress-related gene expression, growth and survival in larval Lake Sturgeon, *Acipenser fulvescens*. Aquaculture 522, 735116.

Earhart, M.L., Ali, J.L., Bugg, W.S., Jeffries, K.M., Anderson, W.G., 2020b. Endogenous cortisol production and its relationship with feeding transitions in larval Lake Sturgeon (*Acipenser fulvescens*) Comp. Biochem. Physiol. 249A, 110777.

Eckes, O.T., Aliosi, D.B., Sandheinrich, M.B., 2015. Egg and larval development index for Lake Sturgeon. N. Am. J. Aquac. 77, 211–216.

Eder, B.L., Steffensen, K.D., Haas, J.D., Adams, J.D., 2015. Short-term survival and dispersal of hatchery-reared juvenile pallid sturgeon stocked in the channelized Missouri River. J. Appl. Ichthyol. 31 (6), 991–996.

Edwards, P., 2015. Aquaculture environment interactions: past, present and likely future trends. Aquaculture 447, 2–14.

Evans, M.L., Wilke, N.F., O'Reilly, P.T., Fleming, I.A., 2014. Transgenerational effects of parental rearing environment influence the survivorship of captive-born offspring in the wild. Conserv. Lett. 7, 371–379.

Fairchild, E.A., Rennels, N., Howell, W.H., 2008. Predators are attracted to acclimation cages used for winter flounder stock enhancement. Rev. Fish. Sci. 16, 262–268.

Falahatkar, B., Akhaven, S.R., Efatpanah, I., Meknatkhah, B., 2013. Effect of winter feeding and starvation on the growth performance of young-of-year (YOY) great sturgeon *Huso huso*. J. Appl. Ichthyol. 29, 26–30.

Falahatkar, B., Akhavan, S.R., Ghaedi, G., 2014. Egg cortisol response to stress at early stages of development in Persian sturgeon *Acipenser persicus*. Aquac. Int. 22, 215–223.

Fangue, N.A., Hofmeister, M., Schulte, P.M., 2006. Intraspecific variation in thermal tolerance and heat shock protein gene expression in common Killifish, *Fundulus heteroclitus*. J. Exp. Biol. 209, 2859–2872.

Fangue, N.A., Podrabsky, J.E., Crawshaw, L.I., Schulte, P.M., 2009. Countergradient variation in temperature preference in populations of killifish *Fundulus heteroclitus*. Physiol. Biochem. Zool. 82, 776–786.

Fast, D.E., Neeley, D., Lind, D.T., Johnston, M.V., Strom, C.R., Bosch, W.J., Knudsen, C.M., Schroder, S.L., Watson, B.D., 2008. Survival comparison of spring Chinook salmon reared in a production hatchery under optimum conventional and seminatural conditions. Trans. Am. Fish. Soc. 137, 1507–1518.

Finger, A.J., Mahardja, B., Fisch, K.M., Benjamin, A., Lindberg, J., Ellison, L., Ghebremariam, T., Hung, T., May, B., 2018. A conservation hatchery population of delta smelt shows evidence of genetic adaptation to captivity after 9 generations. J. Hered. 109, 689–699.

Finley, A.O., Forsythe, P.S., Crossman, J.A., Baker, E.A., Scribner, K.T., 2018. Assessing impact of exogenous features on biotic phenomena in the presence of strong spatial dependence: a lake sturgeon case study in natural stream settings. PLoS One 13 (12), e0204150.

Fisch, K.M., Kozfkay, C.C., Ivy, J.A., Ryder, O.A., Waples, R.S., 2015. Fish hatchery genetic management techniques: integrating theory with implementation. N. Am. J. Aquac. 77, 343–357.

Fisher, A.C., Volpe, J.P., Fisher, J.T., 2014. Occupancy dynamics of escaped farmed Atlantic salmon in Canadian Pacific coastal salmon streams: implications for sustained invasions. Biol. Invasions 16, 2137–2146.

Fisheries and Oceans, 2014. Recovery strategy for white sturgeon (Acipenser transmontanus) in Canada [Final]. Species at Risk Act Recovery Strategy Series. Fisheries and Oceans Canada, Ottawa, pp. 1–252.

Fiumera, A.C., Porter, B.A., Looney, G., Asmussen, M.A., Avise, J.A., 2004. Maximizing offspring production while maintaining genetic diversity in supplemental breeding programs of highly fecund managed species. Conserv. Biol. 18, 94–101.

Flagg, T.A., Nash, C.E., 1999. A conceptual framework for conservation hatchery strategies for Pacific salmonids. In: NOAA Technical Memo NMFS-NWFCS-38. https://www.webapps.nwfsc.noaa.gov/assets/25/5377_06172004_101351_tm38.pdf.

Foerster, R.E., 1968. The sockeye salmon, *Oncorhynchus nerka*. Bull. Fish. Res. Board Can. 162, 422.

Forsythe, P.S., Crossman, J.A., Bello, N.M., Baker, E.A., Scribner, K.T., 2012. Individual-based analyses reveal high repeatability in timing and location of reproduction in lake sturgeon (*Acipenser fulvescens*). Can. J. Fish. Aquat. Sci. 69 (1), 60–72.

Forsythe, P.S., Scribner, K.T., Crossman, J.A., Ragavendran, A., Baker, E.A., 2013. Experimental assessment of the magnitude and sources of lake sturgeon egg mortality. Trans. Am. Fish. Soc. 142 (4), 1005–1011.

Forsythe, P.S., Crossman, J.A., Firkus, C.P., Scribner, K.T., Baker, E.A., 2018. Effects of crayfish density, body size and substrate on consumption of lake sturgeon (Acipenser fulvescens Rafinesque, 1817) eggs by invasive rusty crayfish [(Orconectes rusticus (Girard, 1852)]. J. Appl. Ichthyol., 314–321. https://doi.org/10.1111/jai.13562.

Frankham, R., Ballou, J.D., Eldridge, M.D.B., Lacy, R.C., Ralls, K., Dudash, M.R., Fenster, C.B., 2011. Predicting the probability of outbreeding depression. Conserv. Biol. 25, 465–475.

French, W.E., Graeb, B.D., Bertrand, K.N., Chipps, S.R., Klumb, R.A., 2013. Size-dependent trophic patterns of pallid sturgeon and shovelnose sturgeon in a large river system. J. Fish Wildl. Manag. 4 (1), 41–52.

Friedrich, T., Reinartz, R., Gessner, J., 2019. Sturgeon re-introduction in the Upper and Middle Danube River Basin. J. Appl. Ichthyol. 35, 1059–1068.

Froehlich, H.E., Gentry, R.R., Halper, B.S., 2017. Conservation aquaculture: shifting the narrative and paradigm of aquaculture's role in resource management. Biol. Conserv. 215, 162–168.

Fujimoto, M., Crossman, J.A., Scribner, K.T., Marsh, T.L., 2013. Microbial community assembly and succession on lake sturgeon egg surfaces as a function of simulated spawning stream flow rate. Microb. Ecol. 66 (3), 500–511.

Ganus, J.E., Mullen, D.M., Miller, B.T., Cobb, V.A., 2018. Quantification of emigration and habitat use inform stocking rates of lake sturgeon (*Acipenser fulvescens, Rafinesque*, 1817) in the Cumberland River, Tennessee, USA. J. Appl. Ichthyol. 34 (2), 331–340.

Genz, J., McDougall, C.A., Burnett, D., Arcinas, L., Kheeto, S., Anderson, W.G., 2014. Induced spawning of wild-caught Lake Sturgeon (*Acipenser fulvescens*): assessment of hormonal and stress responses, gamete quality and survival. J. Appl. Ichthyol. 30, 1565–1577.

Genz, J., Hicks, R.N., 2021. Response in growth, scute development and whole body ion content of *Acipenser fulvescens* reared in water of differing chemistries. Animals 11, 1419.

Gerrity, P.C., 2005. Habitat Use, Diet, and Growth of Hatchery-Reared Juvenile Pallid Sturgeon and Indigenous Shovelnose Sturgeon in the Missouri River above Fort Peck Reservoir. Master's Thesis, Department of Ecology, Montana State University.

Gerrity, P.C., Guy, C.S., Gardner, W.M., 2006. Juvenile pallid sturgeon are piscivorous: a call for conserving native cyprinids. Trans. Am. Fish. Soc. 135 (3), 604–609.

Gerrity, P.C., Guy, C.S., Gardner, W.M., 2008. Habitat use of juvenile pallid sturgeon and shovelnose sturgeon with implications for water-level management in a downstream reservoir. N. Am. J. Fish. Manag. 28 (3), 832–843.

Gessner, J., Kamerichs, C.M., Kloas, W., Wuertz, S., 2009. Behavioural and physiological responses in early life phases of Atlantic sturgeon (*Acipenser oxyrinchus* Mitchill 1815) towards different substrates. J. Appl. Ichthyol. 25, 83–90.

Gessner, J., Tautenhahn, M., Spratte, S., Arndt, G.M., von Nordheim, H., 2011a. Development of a German Action Plan for restoration of the European sturgeon *Acipenser sturio* L.—implementing international commitments on a national scale. J. Appl. Ichthyol. 27, 192–198.

Gessner, J., Spratte, S., Kirschbaum, F., 2011b. Historic overview on the status of the European sturgeon (*Acipenser studio*) and its fishery in the North Sea and its tributaries with a focus on German waters. In: Williot, P.W., Rochard, E., Desse-Berset, N., Kirschbaum, F., Gessner, J. (Eds.), Biology and Conservation of the European Sturgeon *Acipenser sturio* L. 1758: The Reunion of the European and Atlantic Sturgeons. Springer Verlag, Berlin, pp. 195–220.

Ghiasi, S., Falahatkar, B., Arslan, M., Dabrowski, K., 2017. Physiological changes and reproductive performance of Sterlet sturgeon, *Acipenser ruthenus* injected with thiamine. Anim. Reprod. Sci. 178, 23–30.

Gille, D.A., Famula, T.R., May, B.P., Schreier, A.D., 2015. Evidence for a maternal origin of spontaneous autopolyploidy in cultured white sturgeon (*Acipenser transmontanus*). Aquaculture 435, 467–474.

Gisbert, E., Williot, P., 2002a. Advances in the larval rearing of Siberian sturgeon. J. Fish Biol. 60, 1071–1092.

Gisbert, E., Williot, P., 2002b. Duration of synchronous egg cleavage cycles at different temperatures in Siberian sturgeon (*Acipenser baerii*). J. Appl. Ichthyol. 18, 271–274.

Goodman, M.L., 1990. Preserving the genetic diversity of salmonid stocks: a call for federal regulation of hatchery programs. Environ. Law 20, 111–116.

Gradil, A.M., Wright, G.M., Speare, D.J., Wadowska, D.W., Purcell, S., Fast, M.D., 2014a. The Effects of Temperature and Body Size on Immunological Development and Responsiveness in juvenile Shortnose Sturgeon (*Acipenser brevirostrum*). Fish Shellfish Immunol. 40, 545–555.

Gradil, A.M., Wright, G.M., Wadowska, D.W., Fast, M.D., 2014b. Ontogeny of the immune system in Acipenserid juveniles. Dev. Comp. Immunol. 44, 303–314.

Grafen, A., 1988. On the uses of data on lifetime reproductive success. In: Clutton-Brock, T. (Ed.), Reproductive Success. University of Chicago Press, Chicago, IL, pp. 454–471.

Graham, W.D., Thorpe, J.E., Metcalfe, N.B., 1996. Seasonal current holding performance of juvenile Atlantic salmon in relation to temperature and smolting. Can. J. Fish. Aquat. Sci. 53, 80–86.

Greene, C.W., 1952. Results from stocking brook trout of wild and hatchery strains at Stillwater Pond. Trans. Am. Fish. Soc. 81, 43–52.

Grunwald, C., Maceda, L., Waldman, J., Stabile, J., Wirgin, I., 2008. Conservation of Atlantic sturgeon *Acipenser oxyrinchus oxyrinchus*: delineation of stock structure and distinct population segments. Conserv. Genet. 9, 1111–1124.

Gu, B., Schell, D.M., Frazer, T., Hoyer, M., Chapman, F.A., 2001. Stable carbon isotope evidence for reduced feeding of Gulf of Mexico sturgeon during their prolonged river residence period. Estuar. Coast. Shelf Sci. 53 (3), 275–280.

Hamel, M.J., Hammen, J.J., Pegg, M.A., 2012. Tag retention of T-bar anchor tags and passive integrated transponder tags in Shovelnose sturgeon. N. Am. J. Fish. Manag. 32, 533–538.

Hardy, R.S., Litvak, M.K., 2004. Effects of temperature on the early development, growth, and survival of shortnose sturgeon, *Acipenser brevirostrum*, and Atlantic sturgeon, *Acipenser oxyrhynchus*, yolk-sac larvae. Environ. Biol. Fishes 70, 145–154.

Hashemi, S., Ghomi, M.R., Sohrabnezhad, N., 2018. Effect of water temperature on growth, survival and digestive enzyme activities of Beluga (*Huso huso*) and Persian sturgeon (*Acipenser persicus*) larvae from hatching to yolk sac absorption stage. J. Appl. Ichthyol. 34, 1324–1330.

Hastings, R.P., Bauman, J.M., Baker, E.A., Scribner, K.T., 2013. Post-hatch dispersal of lake sturgeon (*Acipenser fulvescens*, Rafinesque, 1817) yolk-sac larvae in relation to substrate in an artificial stream. J. Appl. Ichthyol. 29, 1208–1213.

Havasi, M., Lefler, K.K., Takacs, D., Ronyai, A., 2018. How do long-term effects of temperature influence sex ratio, somatic and gonadal development in juvenile sterlet (*Acipenser ruthenus* L.)—an additional climate change consequence? Aquacult. Res. 49, 3577–3585.

Havelka, M., Hulák, Ráb, P., Rábova, M., Lieckfeldt, D., Ludwing, A., Rodina, M., Gela, D., Pšenička, M., Bytyutskyy, D., Flajšhans, M., 2014. Fertility of a spontaneous hexaploid male Siberian sturgeon, *Acipenser baerii*. BMC Genet. 15, 5.

Havelka, M., Arai, K., 2019. Hybridization and polyploidization in sturgeon. In: Wang, H., Piferrer, F., Chen, S., Shen, Z. (Eds.), Sex Control in Aquaculture. vol. II. Wiley, Hoboken, pp. 669–687.

Haxton, T.J., Sulak, K., Hildebrand, L., 2016. Status of scientific knowledge of North American sturgeon. J. Appl. Ichthyol. 32, 5–10.

He, X., Lu, S., Liao, M., Zhu, X., Zhnag, M., Li, S., You, X., Chen, J., 2013. Effects of age and size on critical swimming speed of juvenile Chinese sturgeon, *Acipenser sinensis*, at seasonal temperatures. J. Fish Biol. 82, 1047–1105.

Hegna, J., Scribner, K., Baker, E., 2020. Movements, habitat use, and entrainment of stocked juvenile lake sturgeon in a hydroelectric reservoir system. Can. J. Fish. Aquat. Sci. 77 (3), 611–624.

Henderson, A.P., Spidle, A.P., King, T.L., 2004. Genetic diversity, kinship analysis, and broodstock management of captive Atlantic sturgeon for population restoration. Am. Fish. Soc. Symp. 44, 621–633.

Heppell, S.S., Crowder, L.B., Crouse, D.T., 1996. Models to evaluate headstarting as a management tool for long-lived turtles. Ecol. Appl. 6, 556–565.

Hervas, S., Lorenzen, K., Shane, M.A., Drawbridge, M.A., 2010. Quantitative assessment of a white seabass (*Atractoscion nobilis*) stock enhancement program in California: post-release dispersal, growth, and survival. Fish. Res. 105, 237–243.

Hildebrand, L.R., Parsley, M., 2013. Upper Columbia White Sturgeon Recovery Plan—2012 Revision. Unpubl. Report Prepared for the Upper Columbia White Sturgeon Recovery Initiative. pp. 129 + 1 app. Available at www.uppercolumbiasturgeon.org. (Accessed on 11 October 2021).

Hildebrand, L., Drauch Schreier, A., Lepla, K., McAdam, S.O., McLellan, J., Parsley, M.J., Paragamian, V.L., Young, S.P., 2016. Status of the white sturgeon (*Acipenser transmontanus* Richardson, 1863) throughout the species range, threats to survival, and prognosis for the future. J. Appl. Ichthyol. 32 (Suppl. 1), 261–312.

Hillborn, R., 1992. Hatcheries and the future of salmon in the Northwest. Fisheries 17, 5–8.

Hogberg, N.P., Skorupski, J.A., Hochhalter, S.J., 2021. Natural recruitment potential of a reintroduced shovelnose sturgeon population in the Bighorn River, Wyoming. N. Am. J. Fish. Manag. 41 (5), 1288–1298.

Holčik, J., Klindová, A., Masár, J., Mézáros, J., 2006. Sturgeons in the Slovakian rivers of the Danube River basin: an overview of their current status and proposal for their conservation and restoration. J. Appl. Ichthyol. 22 (Suppl. 1), 17–22.

Holmquist, L.M., Guy, C.S., Tews, A., Webb, M.A., 2019. First maturity and spawning periodicity of hatchery-origin pallid sturgeon in the upper Missouri River above Fort Peck Reservoir, Montana. J. Appl. Ichthyol. 35, 138–148. https://doi.org/10.1111/jai.13751.

Holostenco, D.N., Ciorpac, M., Paraschiv, M., Iani, M., Hont, Ş., Taflan, E., Suciu, R., Rişnoveanu, G., 2019. Overview of the Romanian sturgeon supportive stocking programme in the Lower Danube River system. Sci. Ann. Danube Delta Inst. 24, 21–30.

Holst, L., Zollweg-Horan, E., 2018. Lake sturgeon recovery plan 2018–2024. Technical Report Issued by the New York State Department of Environmental Conservation., p. 46.

Holtgren, J.M., Ogren, S.A., Paquet, A.J., Fajfer, S., 2007. Design of a portable streamside rearing facility for lake sturgeon. N. Am. J. Aquac. 69, 317–323.

Huff, D.D., Miller, L.M., Chizinski, C.J., Vondracek, B., 2011. Mixed-source reintroductions lead to outbreeding depression in second-generation descendants of a native North American fish. Mol. Ecol. 20, 4246–4258.

Ineno, T., Kodama, R., Taguchi, T., Yamada, K., 2018. Growth and maturation in sterlet *Acipenser ruthenus* under high concentrations of dissolved oxygen. Fish. Sci. 84, 605–612.

Ireland, S.C., Beamesderfer, R.C.P., Paragamian, V.L., Wakkinen, V.D., Siple, J.T., 2002. Success of hatchery-rearing juvenile white sturgeon (*Acipenser transmontanus*) following release in the Kootenai River, Idaho, USA. J. Appl. Ichthyol. 18, 642–650.

Islam, M.S., 2005. Nitrogen and phosphorus budget in coastal and marine cage aquaculture and impacts of effluent loading on ecosystem: review and analysis towards model development. Mar. Pollut. Bull. 50, 48–61.

Israel, J.A., Bando, K.J., Anderson, E.C., May, B., 2009. Polyploid microsatellite data reveal stock complexity among estuarine North American green sturgeon (*Acipenser medirostris*). Can. J. Fish. Aquat. Sci. 66, 1491–1504.

Jay, K.J., Crossman, J.A., Scribner, K.T., 2014. Estimates of effective number of breeding adults and reproductive success for white sturgeon. Trans. Am. Fish. Soc. 143 (5), 1204–1216.

Jay, K.J., McGuire, J.M., Scribner, K.T., 2017. Ecological conditions affect behavioral and morphological trait variability of lake sturgeon (*Acipenser fulvescens* Rafinesque, 1817) yolk-sac larvae. J. Appl. Ichthyol. 34, 341–347.

Jay, K.J., Crossman, J.A., Scribner, K.T., 2020. Temperature affects transition timing and phenotype between key developmental stages in white sturgeon *Acipenser transmontanus* yolk-sac larvae. Environ. Biol. Fishes 103, 1149–1162.

Jeffries, K.M., Jeffrey, J.D., Holland, E.B., 2022. Applied aspects of gene function for the conservation of fishes. Fish Physiol. 39A, 389–433.

Jensen, A.J., Schreck, C.B., Hess, J.E., Bohn, S., O'Malley, K.G., Peterson, J.T., 2021. Application of genetic stock identification and parentage-based tagging in a mixed stock recreational Chinook salmon fishery. N. Am. J. Fish. Manag. 41 (1), 130–141.

Johnston, I.A., 2006. Environment and plasticity of myogenesis in teleost fish. J. Exp. Biol. 209, 2249–2264.

Jonssonn, S., Brännäs, E., Lundqvist, H., 1999. Stocking of brown trout, *Salmo trutta* L.: effects of acclimatization. Fish. Manag. Ecol. 6, 459–473.

Jordan, G.R., Klumb, R.A., Wanner, G.A., Stancill, W.J., 2006. Poststocking movements and habitat use of hatchery-reared juvenile pallid sturgeon in the Missouri River below Fort Randall Dam, South Dakota and Nebraska. Trans. Am. Fish. Soc. 135 (6), 1499–1511.

Jordan, G.R., Heist, E.J., Kuhajda, B.R., Moyer, G.R., Hartfield, P., Piteo, M.S., 2019. Morphological identification overestimates the number of pallid sturgeon in the Lower Mississippi River due to extensive introgressive hybridization. Trans. Am. Fish. Soc. 148, 1004–1023.

Justice, C., Pyper, B.J., Beamesderfer, R.C.P., Paragamiaŋ, V.L., Rust, P.J., Neufeld, M.D., Ireland, S.-C., 2009. Evidence of density- and size-dependent mortality in hatchery-reared juvenile white sturgeon (*Acipenser transmontanus*) in the Kootenai River. Can. J. Fish. Aquat. Sci. 66, 802–815.

Kampa, J., Hatzenbeler, G., Jennings, M., 2014. Status and management of lake sturgeon (*Acipenser fulvescens* Rafinesque, 1817) in the upper St. Croix River and Namekagon River, Wisconsin, USA. J. Appl. Ichthyol. 30, 1387–1392.

Kappenman, K.M., Webb, M.A.H., Greenwood, M., 2013. The effect of temperature on embryo survival and development in pallid sturgeon *Scaphirhynchus albus* (Forbes & Richardson 1905) and shovelnose sturgeon *S. platorynchus* (Rafinesque, 1820). J. Appl. Ichthyol. 29, 1193–1203.

Kawabata, Y., Asami, K., Kobayashi, M., Sato, T., Okuzawa, K., Yamada, H., Yoseda, K., Arai, N., 2011. Effect of shelter acclimation on the post-release survival of hatchery-reared black-spot tuskfish *Choerodon schoenleinii*: laboratory experiments using the reef-resident predator white-streaked grouper *Epinephelus ongus*. Fish. Sci. 77, 79–85.

Kelly, J.L., Arnold, D.E., 1999. Effects of ration and temperature on growth of age-0 Atlantic sturgeon. N. Am. J. Aquac. 61, 51–57.

Khatooni, M.M., Amiri, B.M., Hoseinifar, S.H., Jafari, V., Makhdomi, N., 2011. Acclimation potential of *Acipenser persicus* post-larvae to abrupt or gradual increase in salinity. J. Appl. Ichthyol. 27, 528–532.

Khatooni, M.M., Amiri, B.M., Mirvaghefi, A., Jafari, V., Hoseinifar, S.H., 2012. The effects of salinity on the fertilization rate and rearing of the Persian sturgeon (*Acipenser persicus*) larvae. Aquacult. Int. 20, 1097–1105.

Kirschbaum, F., Williot, P., Fredrich, F., Tiedemann, R., Gessner, J., 2011. Restoration of the European sturgeon Acipenser sturio in Germany. In: Williot, P.W., Rochard, E., Desse-Berset, N., Kirschbaum, F., Gessner, J. (Eds.), Biology and Conservation of the European Sturgeon *Acipenser sturio* L. 1758: The Reunion of the European and Atlantic Sturgeons. Springer Verlag, Berlin, pp. 309–334.

Klassen, C.N., Peake, S.J., 2007. The use of black fly larvae as a food source for hatchery-reared lake sturgeon. N. Am. J. Aquac. 69, 223–228.

Klassen, C.N., Peake, S.J., 2008. Effect of diet switch timing and food source on survival and growth of lake sturgeon. J. Appl. Ichthyol. 24, 527–533.

Koch, J.D., Quist, M., 2010. Current status and trends in shovelnose sturgeon (*Scaphirhynchus platorynchus*) management and conservation. J. Appl. Ichthyol. 26, 491–498.

Kolman, R., Kapusta, A., 2018. Food characteristics and feeding management on sturgeon with a special focus on the Siberian sturgeon. In: Williot, P., Nonnotte, G., Chebanov, M. (Eds.), The Siberian Sturgeon (*Acipenser baerii*, Brandt, 1869) Volume 2—Farming. Springer, Cham.

Kolman, R., Kapusta, A., Duda, A., Wiszniewski, G., 2011. Review of the current status of the Atlantic sturgeon *Acipenser oxyrinchus oxyrinchus* Mitchill 1815, in Poland: principles, previous experience, and results. J. Appl. Ichthyol. 27, 186–191.

Kolman, R., Hudĩi, A., Zubcov, E., 2016. Endangered species of sturgeon require active protection—restitution sterlet population in the Dniestr. In: Proceedings of "Sustainable Use, Protection of Animal World and Forest Management in the Context of Climate Change.", pp. 209–210.

Kolman, R., Khudyi, C., Kushniryk, O., Khuda, L., Prusinska, M., Wiszniewski, G., 2018. Influence of temperature and Artemia enriched with x-3 PUFAs on the early ontogenesis of Atlantic sturgeon, *Acipenser oxyrinchus* Mitchill, 1815. Aquacult. Res. 49, 1740–1751.

Komoroske, L.M., Connon, R.E., Jeffries, K.M., Fangue, N.A., 2015. Linking transcriptional response to organismal tolerance reveals mechanisms of thermal sensitivity in a mesothermal endangered fish. Mol. Ecol. 24, 4960–4981.

Kootenai Tribe of Idaho, 2009. Kootenai River Habitat Restoration Project Master Plan: A Conceptual Feasibility Analysis and Design Framework. Kootenai Tribe of Idaho, Bonners Ferry, ID., pp. 1–386.

Korentovich, M., Litvenenko, A., 2018. Artificial production of Siberian sturgeon fingerlings for restocking of the Siberian Ob'-Irtysh basin: a synthesis. In: Williot, P., Nonnotte, G., Chebanov, M. (Eds.), The Siberian Sturgeon (*Acipenser baerii*, Brandt, 1869) Volume 2—Farming. Springer, Cham.

Korneev, A.A., Luzhnyak, V.A., 2012. Biological peculiarities of the industrially bred juvenile Russian sturgeon *Acipenser gueldenstaedtii* (*Acipenseridae, Acipenseriformes*) in the Don. J. Appl. Ichthyol. 52 (4), 268–276.

Koshelev, V.N., Mikodina, E.V., Mironova, T.N., Presnyakov, A.V., Novosadov, A.G., 2012. New data on biology and distribution of Sakhalin sturgeon *Acipenser mikadoi*. J. Ichthyol. 52, 619–627.

Koshelev, V., Shmigirilov, A., Ruban, G., 2014. Current status of feeding stocks of the kaluga sturgeon *Huso dauricus* Georgi, 1775, and Amur sturgeon *Acipenser schrenckii* Brandt, 1889, in Russian waters. J. Appl. Ichthyol. 30, 1310–1318.

Kruse, G.O., Scarnecchia, D.L., 2002. Assessment of bioaccumulated metal and organochlorine compounds in relation to physiological biomarkers in Kootenai River white sturgeon. J. Appl. Ichthyol. 18 (4–6), 430–438.

Kuhl, H., Guiguen, Y., Höhne, C., Kreuz, E., Du, K., Klopp, C., Lopez-Roques, C., Yebra-Pimentel, E.S., Ciorpac, M., Gessner, J., Holostenco, D., 2021. A 180 Myr-old female-specific genome region in sturgeon reveals the oldest known vertebrate sex determining system with undifferentiated sex chromosomes. Philos.Trans. R. Soc. B 376 (1832), 20200089.

Kültz, D., 2005. Molecular and evolutionary basis of the cellular stress response. Annu. Rev. Physiol. 67, 225–257.

Lacho, C.D., McDougall, C.A., Nelson, P.A., Legge, M.M., Gillespie, M.A., Michaluk, Y., Klassen, C.N., Macdonald, D., 2020. Evaluation of a deepwater release method for hatchery-reared lake sturgeon. N. Am. J. Fish. Manag. 40 (4), 828–839.

Lagutov, V., Lagutov, V., 2008. The Ural River sturgeons: population dynamics, catch, and reasons for decline. In: Lagutov, V. (Ed.), Rescue of Sturgeon Species in the Ural River. Springer Science + Business Media B. V, Dordrecht, pp. 193–276.

Larkin, P.A., 1974. Play it again Sam—an essay on salmon enhancement. J. Fish. Res. Board Can. 31 (8), 1433–1459.

Larsen, M.L., Vormedal, I., 2021. The environmental effectiveness of sea lice regulation: Compliance and consequences for farmed and wild salmon. Aquaculture 532, 736000.

Larson, D., Scribner, K.T., Drabowski, K., Lee, B.-J., Crossman, J., 2021. Egg lipid and thiamine vary between early and late spawning lake sturgeon. J. Appl. Ichthyol. 37, 655–663.

Leber, K.M., 1995. Significance of fish size-at-release on enhancement of striped mullet fisheries in Hawaii. J. World Aquacult. Soc. 26, 143–153.

LeBreton, G.T.O., Beamish, F.W.H., 1998. The influence of salinity on ionic concentrations and osmolarity of blood serum in lake sturgeon, *Acipenser fulvescens*. Environ. Biol. Fishes 52, 477–482.

Li, Y.H., Kynard, B., Wei, Q.W., Zhang, H., Du, H., Li, Q.K., 2013. Effects of substrate and water velocity on migration by early-life stages of kaluga, *Huso dauricus* (Georgi, 1775): an artificial stream study. J. Appl. Ichthyol. 29, 713–720.

Li, J., Wang, C., Pan, W., Du, H., Zhang, H., Wu, J., Wei, Q., 2021. Migration and distribution of adult hatchery reared Yangtze sturgeons (*Acipenser dabryanus*) after releasing in the upper Yangtze River and its implications for stock enhancement. J. Appl. Ichthyol. 37, 3–11.

Linares-Casenave, J., Werner, I., Van Eenennaam, J.P., Doroshov, S.I., 2013. Temperature stress induces notochord abnormalities and heat shock proteins expression in larval green sturgeon (*Acipenser medirostris* Ayres 1854). J. Appl. Ichthyol. 29, 958–967.

Lindberg, J.C., Doroshov, S.I., 1986. Effect of diet switch between natural and prepared foods on growth and survival of White Sturgeon juveniles. Trans. Am. Fish. Soc. 115, 166–171.

Lindstrom, J., 1999. Early development and fitness in birds and mammals. Trends Ecol. Evol. 14, 343–348.

Lochet, A., Jatteau, P., Gessner, J., 2011. Detection of chemical marks for stocking purposes in sturgeon species. J. Appl. Ichthyol. 27, 444–449.

Loeppky, A.R., McDougall, C.M., Anderson, W.G., 2019. Identification of hatchery-reared Lake Sturgeon, *Acipenser fulvescens*, using natural elemental signatures and stable isotope marking of fin rays. N. Am. J. Fish. Manag. 40, 61–74.

Loeppky, A.R., Anderson, W.G., 2020. Environmental influences on uptake kinetics and partitioning of strontium in age-0 Lake Sturgeon, *Acipenser fulvescens*: effects of temperature and ambient calcium concentrations. Can. J. Fish. Aquat. Sci. 78, 612–622.

Loeppky, A.R., Belding, L., Quijada-Rodriguez, A.R., Morgan, F., Pracheil, B.M., Chakoumakos, B.C., Anderson, W.G., 2021. Otolith polymorph composition in sturgeons: influence of ontogenetic development and environmental conditions. Sci. Rep. 11, 1–10.

Loewen, T.N., Carriere, B., Reist, J.D., Halden, N.M., Anderson, W.G., 2016. Review: Linking physiology and biomineralization processes to ecological inferences on the life history of fishes. Comp. Biochem. Physiol. 202A, 123–140.

Lorenzen, K., Camp, E.V., 2019. Density-dependence in the life history of fishes: when is a fish recruited? Fish. Res. 217, 5–10.

Ludwig, A., Lippold, S., Debus, L., Reinartz, R., 2009. First evidence of hybridization between endangered sterlets (*Acipenser ruthenus*) adn exotic Siberian sturgeons (*Acipenser baerii*) in the Danube River. Biol. Invasions 11, 753–760.

Lyons, J., Cochran, P.A., Fago, D., 2000. Wisconsin Fishes 2000: Status and Distribution. Publication WISCU-B-00-001. University of Wisconsin Sea Grant Institute, Madison.

MacKinlay, D.D., Lehmann, S., Bateman, J., Cook, R., 2004. Pacific Salmon Hatcheries in British Columbia. Am. Fish. Soc. Symp. 44, 54–75.

Mann, K.A., Holtgren, J.M., Auer, N.A., Ogren, S.A., 2011. Comparing size, movement, and habitat selection of wild and streamside reared lake sturgeon. N. Am. J. Fish. Manag. 31, 305–314.

Maskill, P.A.C., 2020. Description of the reproductive structure, size, growth, and condition of hatchery-origin white sturgeon in the lower Columbia River, British Columbia, Canada. Montana State University-Bozeman, College of Letters & Science, Bozeman, MT. Doctoral dissertation.

Maskill, P.A., Crossman, J.A., Webb, M.A., Marrello, M.M., Guy, C.S., 2022. Accuracy of histology, endoscopy, ultrasonography, and plasma sex steroids in describing the population reproductive structure of hatchery-origin and wild white sturgeon. J. Appl. Ichthyol. 38, 3–16. https://doi.org/10.1111/jai.14280.

Maynard, D.J., Flagg, T.A., Mahnken, C.V.W., Schroder, S.L., 1996. Natural rearing technologies for increasing post-release survival of hatchery-reared salmon. Bull. Natl. Res. Inst. Aquacult. Suppl. 2, 71–77.

McAdam, S.O., 2011. Effects of substrate condition on habitat use and survival by white sturgeon (*Acipenser transmontanus*) larvae and potential implications for recruitment. Can. J. Fish. Aquat. Sci. 68, 812–822.

McAdam, S.O., Crossman, J.A., Williamson, C., St-Onge, I., Dion, R., Manny, B.A., Gessner, J., 2018. If you build it, will they come? Spawning habitat remediation for sturgeon. J. Appl. Ichthyol. 34 (2), 258–278.

McDougall, C.A., Welsh, A.B., Gosselin, T., Anderson, W.G., Nelson, P.A., 2017. Rethinking the influence of hydroelectric development on gene flow in a long-lived fish, the Lake Sturgeon, *Acipenser fulvescens*. PLoS One 12 (3), e0174269.

McDougall, C.A., Nelson, P.A., Barth, C.C., 2018. Extrinsic factors influencing somatic growth of Lake Sturgeon. Trans. Am. Fish. Soc. 147, 459–479.

McDougall, C.A., Nelson, P.A., Aiken, J.K., Burnett, D.C., Barth, C.C., MacDonell, D.S., Michaluk, Y., Klassen, C.N., Macdonald, D., 2020. Hatchery rearing of lake sturgeon to age 1 prior to stocking: a path forward for species recovery in the Upper Nelson River, Manitoba, Canada. N. Am. J. Fish. Manag. 40 (4), 807–827.

McDougall, C.A., Pisiak, D.J., Barth, C.C., Blanchard, M.A., MacDonell, D.S., Macdonald, D., 2014. Relative recruitment success of stocked age-1 vs age-0 lake sturgeon (*Acipenser fulvescens* Rafinesque, 1817) in the Nelson River, northern Canada. J. Appl. Ichthyol. 30, 1451–1460.

McGuire, J.M., Bello-Deocampo, D., Bauman, J., Baker, E., Scribner, K.T., 2019. Histological characterization of gonadal development of juvenile Lake sturgeon (*Acipenser fulvescens*). Environ. Biol. Fishes 102 (7), 969–983.

McKenzie, D.J., Cataldi, E., Romano, P., Owen, S.F., Taylor, E.W., Bronzi, P., 2001. Effects of acclimation to brackish water on the growth, respiratory metabolism, and swimming performance of young-of-the-year Adriatic sturgeon (*Acipenser naccarii*). Can. J. Fish. Aquat. Sci. 58, 1104–1112.

McKenzie, D.J., Piraccini, G., Agnisola, C., Steffensen, J.F., Bronzi, P., Bolis, C.L., Tota, B., Taylor, E.W., 1999. The influence of dietary fatty acid composition on the respiratory and cardiovascular physiology of Adriatic sturgeon (*Acipenser naccarii*): a review. J. Appl. Ichthyol. 15, 265–269.

McKenzie, D.J., Zhang, Y., Eliason, E., Schulte, P., Claireaux, G., Blasco, F., Nati, J.J.H., Farrell, A.P., 2020. Intraspecific variation in tolerance of warming in fishes. J. Fish Biol. 98, 1536–15555.

Melendez, C.L., Mueller, C.A., 2021. Effect of increased embryonic temperature during developmental windows on survival, morphology and oxygen consumption of rainbow trout, *Oncorhynchus mykiss*. Comp. Biochem. Physiol. 252, 110834.

Milshtein, V.V., 1969. 100th anniversary of sturgeon farming. J. Appl. Ichthyol. 9, 271–273.

Mitchell, A.M., Wellicome, T.I., Brodie, D., Cheng, K.M., 2011. Captive-reared burrowing owls show higher site-affinity, survival, and reproductive performance when reintroduced using a soft-release. Biol. Conserv. 144, 1382–1391.

Mohler, J.W., Sweka, J.A., Kahnle, A., Hattala, K., Higgs, A., DuFour, M., 2012. Growth and survival of hatchery-produced Atlantic sturgeon released as young-of-year into the Hudson River, New York. J. Fish. Wild. Manag. 3, 23–32.

Monaco, G., Doroshov, S.I., 1983. Mechanical de-adhesion and incubation of white sturgeon eggs (*Acipenser transmontanous* Richardson) in jar incubators. Aquaculture 35, 117–123.

Monaghan, P., 2008. Early growth conditions, phenotypic development and environmental change. Philos. Soc. R. Soc. B 363, 1635–1645.

Moran, E.V., Hartig, F., Bell, D.M., 2016. Intraspecific trait variation across scales: implications for understanding global change responses. Glob. Chang. Biol. 22, 137–150.

Moser, M.L., Israel, J.A., Neuman, M., Lindley, S.T., Erickson, D.L., McCovey Jr., B.W., Klimley, A.P., 2016. Biology and life history of green sturgeon (*Acipenser medirostris* Ayres, 1854): state of the science. J. Appl. Ichthyol. 32, 67–86.

Mrnak, J.T., Heironimus, L.B., James, D.A., Chipps, S.R., 2020. Effect of water velocity and temperature on energy use, behaviour and mortality of pallid sturgeon *Scaphirhynchus albus*. J. Fish Biol. 97, 1690–1700.

Mueller, C.A., Willis, C., Burggren, W.W., 2016. Salt sensitivity of the morphometry of *Artemia franciscana* during development: a demonstration of 3-D critical windows. J. Exp. Biol. 219, 571–581.

Naylor, R., Hindar, K., Fleming, I.A., Goldburg, R., Williams, S., Volpe, J., Whoriskey, F., Eagle, J., Kelso, D., Mangel, M., 2005. Fugitive salmon: assessing the risks of escaped fish from net-pen aquaculture. BioScience 55, 427–437.

Neff, B.D., Garner, S.R., Pitcher, T.E., 2011. Conservation and enhancement of wild fish populations: preserving genetic quality versus genetic diversity. Can. J. Fish. Aquat. Sci. 68, 1139–1154.

Neidig, C.L., Leber, K.M., Varga, D., Graves, S., Roberts, D.E., 2002. Movements, habitat preferences, growth rates, and survival of hatchery-reared sub-adult Gulf of Mexico sturgeon in two reaches of the Hillsborough River, FL. Mote Marine Laboratory, St Petersburg, FL. USA, pp. 1–13. FMRI Report No. MML-840.

Nelson, T.C., Doukakis, P., Lindley, S.T., Schreier, A.D., Hightower, J.E., Hildebrand, L.R., Whitlock, R.E., Webb, M.A., 2013. Research tools to investigate movements, migrations, and life history of sturgeons (Acipenseridae), with an emphasis on marine-oriented populiations. PLoS One 8 (8), e71552.

Nesan, D., Vijayan, M.M., 2013. Role of glucocorticoid in developmental programming, evidence from zebrafish. Gen. Comp. Endocrinol. 181, 35–44.

Nesan, D., Vijayan, M., 2016. Maternal cortisol mediates hypothalamus-pituitary-interrernal axis development in zebrafish. Sci. Rep. 6, 22582.

Neufeld, M.D., Rust, P.J., 2009. Using passive sonic telemetry methods to evaluate dispersal and subsequent movements of hatchery-reared white sturgeon in the Kootenay River. J. Appl. Ichthyol. 25, 27–33.

Niklitschek, E.J., Secor, D.H., 2009. Dissolved oxygen, temperature and salinity effects on the ecophysiology and survival of juvenile Atlantic sturgeon in estuarine waters: I. Laboratory results. J. Exp. Mar. Biol. Ecol. 381, S150–S160.

Noh, S., Everman, E.R., Berger, C.M., Morgan, T.J., 2017. Seasonal variation in basal and plastic cold tolerance: adaptation is influenced by both long- and short-term phenotypic plasticity. Ecol. Evol. 7, 5248–5257.

Oldenberg, E.W., Guy, C.S., Cureton, E.S., Webb, M.A.H., Gardner, W.M., 2011. Effects of acclimation on poststocking dispersal and physiological condition of age-1 pallid sturgeon. J. Appl. Ichthyol. 27, 436–443.

Olla, B.L., Davis, M.W., Ryer, C.H., 1998. Understanding how the hatchery environment represses or promotes the development of behavioral survival skills. Bull. Mar. Sci. 62, 531–550.

Osborne, M.J., Dowling, T.E., Scribner, K.T., Turner, T.F., 2020. Wild at heart: programs to diminish negative ecological and evolutionary effects of conservation hatcheries. Biol. Conserv. 251, 108768.

Oxman, D.S., Barnett-Johnson, R., Smith, M.E., Coffin, A., Miller, D.L., Josephson, R., Popper, A.N., 2007. The effect of vaterite deposition on sound reception, otolith morphology, and inner ear sensory epithelia in hatchery-reared chinook salmon (*Oncorhynchus tshawytscha*). Can. J. Fish. Aquat. Sci. 64 (11), 1469–1478.

Paragamian, V.L., Beamesderfer, R.C., Ireland, S.C., 2005. Status, population dynamics, and future prospects of the endangered Kootenai River white sturgeon population with and without hatchery intervention. Trans. Am. Fish. Soc. 134 (2), 518–532.

Park, C., Lee, S.Y., Kim, D.S., Nam, Y.K., 2013. Effects of incubation temperature on egg development, hatching and pigment plug evacuation in farmed Siberian sturgeon *Acipenser baerii*. Fish. Aquat. Sci. 16, 25–31.

Paschos, I., Perdikaris, C., Guava, E., Nathanailides, C., 2008. Sturgeons in Greece: a review. J. Appl. Ichthyol. 24, 131–137.

Pegasov, V.A., 2009. The ecological problems of introductions and reintroductions of sturgeon. In: Carmona, R., Domezain, A., García-Gallego, M., Hernando, J.A., Rodríguez, F., Ruíz-Rejón, M. (Eds.), Biology, Conservation and Sustainable Development of Sturgeons. Springer Science + Business Media B. V, pp. 339–343.

Penny, F.M., Kieffer, J.D., 2019. Lack of change in swimming capacity (Ucrit) following acute salinity exposure in juvenile shortnose sturgeon (*Acipenser brevirostrum*). Fish Physiol. Biochem. 45, 1167–1175.

Peterson, D., Vecsei, P., Jennings, C., 2007. Ecology and biology of the lake sturgeon: A synthesis of current knowledge of a threatened North American Acipenseridae. Rev. Fish Biol. Fish. 17 (1), 59–76. https://doi.org/10.1007/s11160-006-9018-6.

Phelps, Q.E., Whitledge, G.W., Tripp, S.J., Smith, K.T., Garvey, J.E., Herzog, D.P., Ostendorf, D.E., Ridings, J.W., Crites, J.W., Hrabik, R.A., Doyle, W.J., 2012. Identifying river of origin for age-0 *Scaphirhynchus* sturgeons in the Missouri and Mississippi rivers using fin ray microchemistry. Can. J. Fish. Aquat. Sci. 69 (5), 930–941.

Poletto, J.B., Martin, B., Danner, E., Baird, S.E., Cocherell, D.E., Hamda, N., Cech, J.J., Fangue, N.A., 2018. Assessment of multiple stressors on the growth of larval green sturgeon *Acipenser medirostris*: implications for recruitment of early life-history stages. J. Fish Biol. 93, 952–960.

Post, H., 1890. The sturgeon; some experiments in hatching. Trans. Am. Fish. Soc. 19, 36–40.

Power, M., McKinley, R.S., 1997. Latitudinal variation in lake sturgeon size as related to the thermal opportunity for growth. Trans. Am. Fish. Soc. 126, 549–558.

Purvina, S., Medne, R., 2018. Reintroduction of sturgeon, *Acipenser oxyrinchus*, in the Gulf of Riga, East-Central Baltic Sea. Arch. Pol. Fish. 26, 39–46.

Quattro, J.M., Grieg, T.W., Coykendall, D.K., Bowen, B.W., Baldwin, J.D., 2002. Genetic issues in aquatic species management: the shortnose sturgeon (*Acipenser brevirostrum*) in the southeastern United States. Conserv. Genet. 3, 155–166.

Quist, M.C., Guy, C.S., Braaten, P.J., 1998. Standard weight (Ws) equation and length categories for shovelnose sturgeon. N. Am. J. Fish. Manag. 18 (4), 992–997.

Randall, D.J., McKenzie, D.J., Abrami, G., Bondiolotti, G.P., Natiello, F., Bronzi, P., Bolis, L., Agradi, E., 1992. Effects of diet on responses to hypoxia in sturgeon (*Acipenser naccarii*). J. Exp. Biol. 170, 113–125.

Razak, S.A., Scribner, K.T., 2020. Ecological and ontogenetic components of larval lake sturgeon gut microbiota assembly, successional dynamics, and ecological evaluation of neutral community processes. Appl. Environ. Microbiol. 86, e02662-19.

Reid, J.M., Bignal, E.M., Bignal, S., McCracken, D.I., Monaghan, P., 2006. Spatial variation in demography and population growth rate: the importance of natal location. J. Anim. Ecol. 75, 1201–1211.

Reimer, T., Dempster, T., Wargelius, A., Fjelldal, P.G., Hansen, T., Glover, K.A., Swearer, S.E., 2017. Rapid growth causes abnormal vaterite formation in farmed fish otoliths. J. Exp. Biol. 220, 2965–2969.

Richardson, K., Castro, I.C., Brunton, D.H., Armstrong, D.P., 2015. Not so soft? Delayed release reduces long-term survival in a passerine reintroduction. Oryx 49, 535–541.

Rien, T.A., Beamesderfer, R.C.P., Foster, C.A., 1994. Retention, recognition, and effects on survival of several tags and marks for white sturgeon. Calif. Fish Game 80, 161–170.

Roberts, L.J., Taylor, J., Garcia de Leaniz, C., 2011. Environmental enrichment reduces maladaptive risk-taking behavior in salmon reared for conservation. Biol. Conserv. 144, 1972–1979.

Rochard, E., Castelnaud, G., Lepage, M., 1990. Sturgeon (Pisces: Acipenseridae): threats and prospects. J. Fish Biol. 37 (Suppl. A), 123–132.

Rodgers, E.M., Cocherell, D.E., Nguyen, T.X., Todgham, A.E., Fangue, N.A., 2018. Plastic responses to diel thermal variation in juvenile green sturgeon, *Acipenser medirostris*. J. Therm. Biol. 76, 147–155.

Rodriguez, A., Gallardo, M.A., Gisbert, E., Santilariz, S., Ibarz, A., Sanchez, J., Catello-Orvay, F., 2002. Osmoregulation in juvenile Siberian sturgeon (*Acipenser baerii*). Fish Physiol. Biochem. 26, 345–354.

Roques, S., Berrebi, P., Rochard, E., Acolas, M.L., 2018. Genetic monitoring for the successful re-stocking of a critically endangered diadromous fish with low diversity. Biol. Conserv. 221, 91–102.

Runstrum, A., Bruch, R.M., Reiter, D., Cox, D., 2002. Lake sturgeon (*Acipenser fulvescens*) on the Menominee Indian Reservation: an effort toward co-management and population restoration. J. Appl. Ichthyol. 18, 481–485.

Saltzgiver, M.J., Heist, E.J., Hedrick, P.W., 2012. Genetic evaluation of the initiation of a captive population: the general approach and a case study in the endangered pallid sturgeon (*Scaphirhynchus albus*). Conserv. Genet. 13, 1381–1391.

Sánchez-Hernández, J., Nunn, A.D., Adams, C.E., Amundsen, P.A., 2019. Causes and consequences of ontogenetic dietary shifts: a global synthesis using fish models. Biol. Rev. 94, 539–554.

Sasmal, I., Honness, K., Bly, K., McCaffery, M., Kunkel, K., Jenks, J.A., Phillips, M., 2015. Release method evaluation for swift fox reintroduction at Bad River Ranches in South Dakota. Res. Ecol. 23, 491–498.

Schreier, A., Stephenson, S., Rust, P., Young, S., 2015. The case of the endangered Kootenai River white sturgeon (*Acipenser transmontanus*) highlights the importance of post-release genetic monitoring in captive and supportive breeding programs. Biol. Conserv. 192, 74–81.

Schreier, A., Van Eenennaam, J.P., 2019. 2018 Genetic Diversity Monitoring of White Sturgeon in the Kootenai Tribe of Idaho Native Fish Conservation Aquaculture Program and Study of 10N Reproductive Development. Technical Report to the Kootenai Tribe of Idaho, Bonners Ferry., p. 19.

Schreier, A.D., Van Eenennaam, J.P., Anders, P., Young, S., Crossman, J., 2021. Spontaneous autopolyploidy in the Acipenseriformes, with recommendations for management. Rev. Fish Biol. Fish. 31, 159–180.

Scott, P.A., Allison, L.J., Field, K.J., Averill-Murray, R.C., Shaffer, H.B., 2020. Individual heterozygosity predicts translocation success in threatened desert tortoises. Science 370, 1086–1089.

Secor, D.H., Gunderson, T.E., 1998. Effects of hypoxia and temperature on survival, growth, and respiration of juvenile Atlantic sturgeon, *Acipenser oxyrhynchus*. Fish. Bull. 96, 603–613.

Secor, D.H., Niklitschek, E.J., Stevenson, J.T., Gunderson, T.E., Minkkinen, S.P., Richardson, B., Florence, B., Mangold, M., Skjeveland, J., Henderson-Arzapalo, A., 2000a. Dispersal and growth of yearling Atlantic sturgeon, *Acipenser oxyrinchus*, released into Chesapeake Bay. Fish. Bull. 98, 800–810.

Secor, D.H., Arefjev, V., Nikolaev, A., Sharov, A., 2000b. Restoration of sturgeons: lessons from the Caspian Sea sturgeon ranching program. Fish Fish. 1, 215–230.

Secor, D.H., Niklitschek, E.J., 2001. Hypoxia and sturgeons. In: Report to the Chesapeake Bay Program Dissolved Oxygen Criteria Team. Chesapeake Biological Laboratory, Technical Report Series Number TS-314-01-CBL, Solomons, Maryland.

Sellheim, K., Willmes, M., Hobbs, J.A., Glessner, J.J.G., Jackson, Z.J., Merz, J.E., 2017. Validating fin ray microchemistry as a tool to reconstruct the migratory history of White Sturgeon. Trans. Am. Fish. Soc. 146 (5), 844–857.

Shahkar, E., Kim, D.-J., Mohseni, M., Yun, H., Bai, S.C., 2015. Effects of salinity changes on hematological responses in juvenile ship sturgeon, *Acipenser nudiventris*. Fish. Aquat. Sci. 18, 45–50.

Shartau, R.B., Baker, D.W., Brauner, C.J., 2017. White sturgeon (*Acipenser transmontanus*) acid–base regulation differs in response to different types of acidoses. J. Comp. Physiol. 187B, 985–994.

Shrimpton, J.M., Bernier, N.J., Randall, D.J., 1994. Changes in cortisol dynamics in wild and hatchery reared juvenile coho salmon (*Oncorhynchus kisutch*) during smoltification. Can. J. Fish. Aquat. Sci. 51, 2179–2187.

Siddique, M.A.M., Psenicka, M., Cosson, J., Dzyuba, B., Rodina, M., Golpour, A., Linhart, O., 2016. Egg stickiness in artificial reproduction of sturgeon: and overview. Rev. Aquacult. 8, 18–29.

Simontacchi, C., Negrato, E., Pazzaglia, M., Bertotto, D., Poltronieri, C., Radaelli, G., 2009. Whole-body concentrations of cortisol and sex steroids in white sturgeon (*Acipenser transmontanus*, Richardson 1836) during early development and stress response. Aquacult. Int. 17, 7–14.

Sloychuk, J.R., Chivers, D.P., Ferrari, M.C.O., 2016. Juvenile lake sturgeon go to school: life skills training for hatchery fish. Trans. Am. Fish. Soc. 145, 287–294.

Smederevac-Lalić, M., Jaríc, I., Višnjić, Ž., Skorić, S., Cvijaniović, G., Gačić, Z., Lenhardt, M., 2011. Management approaches and aquaculture of sturgeons in the Lower Danube region countries. J. Appl. Ichthyol. 27 (Suppl. 3), 94–100.

Smith, T.I.J., Dingley, E.K., 1984. Review of the biology and culture of Atlantic (*Acipenser oxyrinchus*) and shortnose (*A. brevirostrum*). J. World Maric. Soc. 15, 210–218.

Smith, T.I.J., McCord, J.W., Collines, M.R., Post, W.C., 2002. Occurrence of stocked shortnose sturgeon *Acipenser brevirostrum* in non-target rivers. J. Appl. Ichthyol. 18, 470–474.

Smith, A.L., 2009. Lake sturgeon (*Acipenser fulvescens*) stocking in North America. In: Technical Report, Ontario Ministry of Natural Resources, Peterborough. pp 17 p + app.

Smith, K.T., Whitledge, G.W., 2011. Evaluation of a stable-isotope labelling technique for mass marking fin rays of age-0 lake sturgeon. Fish. Manag. Ecol. 18, 168–175.

Smith, A., Marty, J., Power, M., 2015. Non-lethal sampling of lake sturgeon for stable isotope analysis: comparing pectoral fin-clip and dorsal muscle for use in trophic studies. J. Great Lakes Res. 41 (1), 292–297.

Smith, A., Smokorowski, K., Marty, J., Power, M., 2016. Stable isotope characterization of Rainy River, Ontario, lake sturgeon diet and trophic position. J. Great Lakes Res. 42 (2), 440–447.

Sparrevohn, C.R., Strøttrup, J.G., 2007. Post-release survival and feeding in reared turbot. J. Sea Res. 57 (2–3 Spec. Iss), 151–161.

St. Pierre, R.A., 1999. Restoration of Atlantic sturgeon in the northeastern USA with special emphasis on culture and restocking. J. Appl. Ichthyol. 15, 180–182.

Steele, C.A., Anderson, E.C., Ackerman, M.W., Hess, M.A., Campbell, N.R., Narum, S.R., Campbell, M.R., 2013. A validation of parentaqe-based tagging using hatchery steelhead in the Snake River basin. Can. J. Fish. Aquat. Sci. 70 (7), 1046–1054.

Steele, C.A., Hess, M., Narum, S., Campbell, M., 2019. Parentage-based tagging: reviewing the implementation of a new tool for an old problem. Fisheries 44, 412–422.

Steffensen, K.D., Powell, L.A., Koch, J.D., 2010. Assessment of hatchery-reared pallid sturgeon survival in the lower Missouri River. N. Am. J. Fish. Manag. 30 (3), 671–678.

Steffensen, K.D., Powell, L.A., Pegg, M.A., 2012. Population size of hatchery-reared and wild pallid sturgeon in the lower Missouri River. N. Am. J. Fish. Manag. 32 (1), 159–166.

Steffensen, K.D., Mestl, G.E., 2016. Assessment of pallid sturgeon relative condition in the upper channelized Missouri River. J. Freshwater Ecol. 31 (4), 583–595.

Steffensen, K.D., Powell, L.A., Stukel, S.M., Winders, K.R., Doyle, W.J., 2016. Updated assessment of hatchery-reared pallid sturgeon (Forbes & Richardson, 1905) survival in the lower Missouri River. J. Appl. Ichthyol. 32 (1), 3–10.

Steffensen, K.D., Hamel, M.J., Spurgeon, J.J., 2019. Post-stocking pallid sturgeon *Scaphirhychus albus* growth, dispersal, and survival in the lower Missouri River. J. Appl. Ichthyol. 35, 117–127.

Stockard, C.R., 1921. Developmental rate and structural expression: an experimental study of twins, 'double monsters' and single deformities, and the interaction among embryonic organs during their origin and development. Am. J. Anat. 28, 115–275.

Sulak, K.J., Randall, M.T., Clugston, J.P., 2014. Survival of hatchery Gulf sturgeon (*Acipenser oxyrinchus desotoi* Mitchill, 1815) in the Suwannee River, Florida: a 19-year evaluation. J. Appl. Ichthyol. 30, 1428–1440.

Tennant, E.N., Germano, D.J., 2017. Survival of translocated Heermann's kangaroo rats (*Dipodomys heermanni*) in the San Joaquin Desert of California using hard and soft release methods. West. Wildl. 4, 1–11.

Thorstensen, M., Bates, P., Lepla, K., Schreier, A., 2019. To breed or not to breed? Maintaining genetic diversity in white sturgeon supplementation programs. Conserv. Genet. 20, 997–1007.

Trested, D.G., Ware, K., Bakal, R., Isely, J.J., 2011. Microhabitat use and seasonal movements of hatchery-reared and wild shortnose sturgeon in the Savannah River, South Carolina–Georgia. J. Appl. Ichthyol. 27 (2), 454–461.

Trippel, N.A., Porak, W.F., Leone, E., 2018. Post-stocking survival of conditioned and pond-reared compared to indoor pellet-reared advanced fingerling Florida bass. N. Am. J. Fish. Manag. 38, 1039–1049.

Utz, M., Jeschke, J.M., Loeschcke, V., Gabriel, W., 2014. Phenotypic plasticity with instantaneous but delayed switches. J. Theor. Biol. 340, 60–72.

Valentine, S.A., Bauman, J.M., Scribner, K.T., 2017. Effects of alternative food types on body size and survival of hatchery-reared lake sturgeon larvae. N. Am. J. Aquac. 79, 275–282.

Van Eenennaam, J.P., Linares-Casenave, J., Deng, X., Doroshov, S.I., 2005. Effect of incubation temperature on green sturgeon embryos, *Acipenser medirostris*. Environ. Biol. Fishes 72, 145–154.

Van Eenennaam, J.P., Linares-Casenave, J., Muget, J.B., Doroshov, S.I., 2008. Induced spawning, artificial fertilization and egg incubation techniques for green sturgeon. N. Am. J. Aquac. 70, 434–445.

Van Eenennaam, J.P., Fiske, J.A., Leal, M.J., Cooley-Rieders, C., Todgham, A.E., Conte, F.S., Schreier, A.D., 2020. Mechanical shock during egg de-adhesion and post-ovulatory ageing contribute to spontaneous autopolyploidy in white sturgeon culture (*Acipenser transmontanus*). Aquaculture 515, 734530.

Vasconi, M., Aidos, L., Giancamillo, A.D., Bellagamba, F., Domeneghini, C., Moretti, V.M., 2018. Effect of temperature on fatty acid composition and development of unfed Siberian sturgeon (*A.baerii*) larvae. J. Appl. Ichthyol. 35, 296–302.

Vasilyeva, L., Elhetawy, A.I.G., Sudakova, N.V., Astafyeva, S.S., 2019. History, current status and prospects of sturgeon aquaculture in Russia. Aquacult. Res. 50, 979–993.

Vedrasco, A.V., Lobchenko, I.P., Billard, R., 2002. The culture of live food for sturgeon juveniles, a mini review of the Russian literature. Int. Rev. Hydrobiol. 87, 569–575.

Veinott, G.I., Evans, R.D., 1999. An examination of elemental stability in the fin ray of the white sturgeon with laser ablation-sampling-inductively coupled plasma mass spectrometry (LAS-ICP-MS). Trans. Am. Fish. Soc. 128, 352–361.

Waldman, J.R., King, T., Savoy, T., Maceda, L., Grunwald, C., Wirgin, I., 2013. Stock origins of subadult and adult Atlantic sturgeon, *Acipenser oxyrinchus*, in a non-natal estuary, Long Island Sound. Estuar. Coast. 36, 257–267.

Walker, D.J., Alford, J.B., 2016. Mapping lake sturgeon spawning habitat in the Upper Tennessee River using side-scan sonar. N. Am. J. Fish. Manag. 36, 1097–1105.

Walsh, M.L., Fujimoto, H., Yamamoto, T., Yamada, T., Takahashi, Y., Yamashita, Y., 2013. Post-release performance and assessment of cage-conditioned Japanese flounder, *Paralichthys olivaceus*, in Wakasa Bay, Japan. Rev. Fish. Sci. 21, 247–257.

Walters, C.J., 2007. Is adaptive management helping to solve fisheries problems? Ambio 36 (4), 304–307.

Wang, Y.L., Binkowski, F.P., Doroshov, S.I., 1985. Effect of temperature on early development of white and lake sturgeon, *Acipenser transmontanus* and *A. fulvescens*. Environ. Biol. Fishes 14, 43–50.

Waraniak, J.M., Baker, E.A., Scribner, K.T., 2018a. Molecular diet analysis reveals predator–prey community dynamics and environmental factors affecting predation of larval lake sturgeon *Acipenser fulvescens* in a natural system. J. Fish Biol. 93 (4), 616–629.

Waraniak, J.M., Blumstein, D.M., Scribner, K.T., 2018b. Barcoding PCR primers detect larval lake sturgeon (*Acipenser fulvescens*) in diets of piscine predators. Conserv. Genet. Res. 10 (2), 259–268.

Ward, D., Holmes, J., Kern, J., Hughes, M., Rien, T., Langness, O., Gilliland, D., Cady, B., James, B., DeVore, J., Gadomski, D., Parsley, M., Kofoot, P., Kappenman, K., Parker, B., Webb, M., Anthony, C., Schreck, C., Fitzpatrick, M., 2002. White sturgeon mitigation and restoration in the Columbia and Snake Rivers upstream from Bonneville Dam. In: Report submitted to the Bonneville Power Administration, Portland, Oregon, p. 167.

Wassink, L., Bussy, U., Li, W., Scribner, K.T., 2019. High-stress rearing temperature in *Acipenser fulvescens* affects physiology, behaviour and predation rates. Anim. Behav. 157, 153–165.

Wassink, L., Huerta, B., Li, W., Scribner, K.T., 2020. Interaction of egg cortisol and offspring experience influences stress-related behaviour and physiology in lake sturgeon. Anim. Behav. 161, 49–59.

Webb, P.W., 1986. Kinematics of lake sturgeon, *Acipenser fulvescens*, at cruising speeds. Can. J. Zool. 64, 2137–2141.

Webb, M.A.H., Van Eenennaam, J.P., Crossman, J.A., Chapman, F.A., 2019. A practical guide for assigning sex and stage of maturity in sturgeons and paddlefish. J. Appl. Ichthyol. 35 (1), 169–186.

Weber, M.J., Weber, R.E., Ball, E.E., Meerbeek, J.R., 2020. Using radiotelemetry to evaluate post-stocking survival and behavior of large fingerling walleye in three Iowa, USA, lakes. N. Am. J. Fish. Manag. 40, 48–60.

Wei, Q., He, J., Yang, D., Zheng, W., Li, L., 2004. Status of sturgeon aquaculture and sturgeon trade in China: a review based on two recent nationwide surveys. J. Appl. Ichthyol. 20, 321–332.

Wei, Q., Ke, F., Zhang, J., Zhuang, P., Luo, J., Zhou, R., Yang, W., 1997. Biology, fisheries, and conservation of sturgeons and paddlefish in China. Environ. Biol. Fishes 48, 241–255.

Welsh, A., McClain, J.R., 2004. Development of a management plan for lake sturgeon within the Great Lakes basin based on population genetics structure. In: Final Project Report. Great Lakes Fishery Trust Project, (2001.75).

Welsh, A.B., Carlson, D.M., Schlueter, S.L., Jackson, J.R., 2020. Tracking stocking success in a long-lived species through genetics and demographics: evidence of natural reproduction in lake sturgeon after twenty-two years. Trans. Am. Fish. Soc. 149, 121–130.

Welsh, A.B., Jackson, J.R., 2014. The effect of multi-year vs single-year stocking on lake sturgeon (*Acipenser fulvescens* Rafinesque, 1817) genetic diversity. J. Appl. Ichthyol. 30, 1524–1530.

Welsh, A.B., Schumacher, L., Quinlan, H.R., 2019. A reintroduced lake sturgeon population comes of age: a genetic evaluation of stocking success in the St. Louis River. J. Appl. Ichthyol. 35, 149–159.

Werner, I., Linares-Casenave, J., Van Eenennaam, J.P., Doroshov, S.I., 2007. The effect of temperature stress on development and heat-shock protein expression in larval green sturgeon (*Acipenser medirostris*). Environ. Biol. Fishes 79, 191–200.

Wiley, R.W., Whaley, R.A., Satake, J.B., Fowden, M., 1993. An evaluation of the potential for training trout to increase poststocking survival in streams. N. Am. J. Fish. Manag. 13, 171–177.

Williot, P., Arlati, G., Chebanov, M., Gulyas, T., Kasimov, R., Kirschbaum, F., Patriche, N., Pavlovskaya, L.P., Poliakova, L., Pourkazemi, M., Kim, Y., Zhuang, P., Zholdasova, I.M., 2002a. Status and management of Eurasian sturgeon: an overview. Int. Rev. Hydrobiol. 87, 483–506.

Williot, P., Gulyas, T., Ceapa, C., 2002b. An analogue of GnRH is effective for induction of ovulation and spermiation in farmed Siberian sturgeon, *Acipenser baerii* Brandt. Aquacult. Res. 33, 735–737.

Williot, P., 2011. Some Ex-Situ-Related Approaches for Assessing the Biological Variability of Acipenser sturio. Biology and Conservation of the European Sturgeon Acipenser sturio L. 1758. Springer, Berlin, Heidelberg, pp. 635–646.

Wilson, R.S., Franklin, C.E., 2002. Testing the beneficial acclimation hypothesis. Trends Ecol. Evol. 17, 66–70.

Wright, P.A., Turko, A.J., 2016. Amphibious fishes evolution and phenotypic plasticity. J. Exp. Biol. 219, 2245–2259.

Wu, J.M., Wei, Q.W., Du, H., Wang, C.Y., Zhang, H., 2014. Initial evaluation of the release programme for Dabry's sturgeon (*Acipenser dabryanus* Duméril, 1868) in the upper Yangtze River. J. Appl. Ichthyol. 30 (6), 1423–1427.

Wu, C., Chen, L., Gao, Y., Jiang, W., 2018. Seaward migration behavior of juvenile second filial generation Chinese sturgeon *Acipenser sinensis* in the Yangtze River, China. Fish. Sci. 84 (1), 71–78.

Wuertz, S., Lutz, I., Gessner, J., Loeschau, P., Hogans, B., Kirschbaum, F., Kloas, W., 2006. The influence of rearing density as environmental stressor on cortisol response of shortnose sturgeon (*Acipenser brevirostrum*). J. Appl. Ichthyol. 22, 269–273.

Yebra-Pimentel, E.S., Reis, B., Gessner, J., Wuertz, S., Dirks, R.P.H., 2020. Temperature training improves transcriptional homeostasis after heat shock in juvenile Atlantic sturgeon (*Acipenser oxyrinchus*). Fish Physiol. Biochem. 46, 1653–1664.

Yoon, G.R., Deslauriers, D., Enders, E.C., Treberg, J.R., Anderson, W.G., 2019a. Effects of temperature, dissolved oxygen and substrate on the development of metabolic phenotypes in age-0 lake sturgeon, *Acipenser fulvescens*: implications for overwintering survival. Can. J. Fish. Aquat. Sci. 76, 1596–1607.

Yoon, G.R., Deslauriers, D., Anderson, W.G., 2019b. Influence of a dynamic rearing environment on development of metabolic phenotypes in age-0 lake sturgeon, *Acipenser fulvescens*. Conserv. Physiol. 7, coz055.

Yoon, G.R., Deslauriers, D., Anderson, W.G., 2020. Influence of prey condition and incubation method on mortality, growth and metabolic rate during early life history in lake sturgeon, *Acipenser fulvescens*. J. Appl. Ichthyol. 36, 759–767.

Yoon, G.R., Earhart, M., Wang, Y., Suh, M., Anderson, W.G., 2021. Effects of temperature and food availability on liver fatty acid composition and plasma cortisol concentration in age-0 lake sturgeon: Support for homeoviscous adaptation. Comp. Biochem. Physiol. A 261, 111056.

Yoon, G.R., Groening, L., Klassen, C.N., Brandt, C., Anderson, W.G., 2022. Long-term effects of temperature on growth, energy density, whole-body composition and aerobic scope of age-0 Lake sturgeon (*A. fulvescens*). Aquaculture 547, 737505.

Young, W.T., Scarnecchia, D.L., 2005. Habitat use of juvenile white sturgeon in the Kootenai River, Idaho and British Columbia. Hydrobiologia 537 (1), 265–271.

Yuan, X., Zhou, Y.H., Huang, Y.P., Guo, W.T., Johnson, D., Jiang, Q., Jing, J.J., Tu, Z.Y., 2017. Effects of temperature and fatigue on the metabolism and swimming capacity of juvenile Chinese sturgeon (*Acipenser sinensis*). Fish Physiol. Biochem. 43 (5), 1279–1287.

Yufera, M., Darias, M.J., 2007. The onset of exogenous feeding in marine fish larvae. Aquaculture 268, 53–63.

Yusishen, M.E., Yoon, G.R., Bugg, W., Jeffries, K.M., Currie, S., Anderson, W.G., 2020. Love thy neighbour: Social buffering following exposure to an acute thermal stressor in a gregarious fish, the lake sturgeon (*Acipenser fulvescens*). Comp. Biochem. Physiol. 243A, 110686.

Zeisset, I., Beebee, T.J.C., 2012. Donor population size rather than local adaptation can be a key determinant of amphibian translocation success. Anim. Conserv. 16, 359–366.

Zhang, H., Wei, Q.W., Du, H., Li, L.X., 2011. Present status and risk for extinction of the Dabry's sturgeon (*Acipenser dabryanus*) in the Yangtze River watershed: a concern for intensified rehabilitation needs. J. Appl. Ichthyol. 27, 181–185.

Zhang, Y., Kieffer, J.D., 2017. The effect of temperature on the resting and post-exercise metabolic rates and aerobic metabolic scope in shortnose sturgeon *Acipenser brevirostrum*. Fish Physiol. Biochem. 43, 1245–1252.

Zhang, Y., Loughery, J.R., Martyniuk, C.J., Kieffer, J.D., 2017. Physiological and molecular responses of juvenile shortnose sturgeon (*Acipenser brevirostrum*) to thermal stress. Comp. Biochem. Physiol. 203A, 314–321.

Zhao, F., Qu, L., Zhuang, P., Zhang, L., Liu, J., Zhang, T., 2011. Salinity tolerance as well as osmotic and ionic regulation in juvenile Chinese sturgeon (*Acipenser sinensis* Gray, 1835) exposed to different salinities. J. Appl. Ichthyol. 27, 231–234.

Zubair, S., Peake, S.J., Hare, J.F., Anderson, W.G., 2012. The effect of temperature and substrate on the development of the cortisol stress response in the lake sturgeon, *Acipenser fulvescens*, Rafinesque (1817). Environ. Biol. Fishes 93, 577–587.

Chapter 3

# Using ecotoxicology for conservation: From biomarkers to modeling

Gudrun De Boeck[a,*], Essie Rodgers[b], and Raewyn M. Town[a]
[a]*ECOSPHERE, Department of Biology, University of Antwerp, Antwerp, Belgium*
[b]*School of Biological Sciences, University of Canterbury, Christchurch, New Zealand*
[*]*Corresponding author: e-mail: gudrun.deboeck@uantwerpen.be*

## Chapter Outline



Fish Physiology, Vol. 39B. https://doi.org/10.1016/bs.fp.2022.06.002

An endless list of new chemicals are entering nature, which makes it an impossible task to assess all possible mixture combinations at all possible concentrations and conditions that are leading to the ubiquitous anthropogenic impacts on the aquatic environment resulting from deteriorating water quality. Therefore, ecotoxicology is moving more toward a mechanistic understanding of toxicological processes, using trait-based approaches and sublethal molecular and physiological endpoints to understand the mode of action of pollutants and the adverse outcomes at the organismal and population level. These molecular and physiological endpoints can be used as biomarkers, applicable in the field. This brings ecotoxicological research much closer to conservation physiology. Understanding the relationships between chemical reactivity in the water and in organisms, and assessing the consequences at higher levels, allows conservation physiologists and managers to take the right restoration measures for an optimal improvement of the aquatic habitats of concern. In this chapter we discuss the role which the promising approach of mechanistic-based Adverse Outcome Pathways (AOPs) can play in ecotoxicological research. It studies a pathway of events, from the direct interaction of a chemical with a molecular target, through subsequent intermediate events at cellular, tissue, organ and individual organism levels which then result in an Adverse Outcome (AO) relevant to ecotoxicological risk assessment and regulatory decision-making. In this context, we also discuss the importance of modeling, including bioavailability based and effect based models. Finally, we reflect on the possibilities that meta-analysis has to offer to detect unifying physiological processes, as well as interesting outliers.

# 1 Introduction

## 1.1 Ecotoxicology: The need to combine ecology and basic toxicology

Environmental pollution can have both natural (forest fires, volcanic action, ...) and anthropogenic causes (urban waste, industry, agriculture, ...), and is described as the addition of any substance or any form of energy (e.g., thermal pollution) to the environment at a rate faster than it can be dispersed, diluted, decomposed, recycled, or stored in some harmless form. Therefore, anthropogenic pollution has been around since humankind started to live in larger groups. It was problematic even in ancient cities that struggled to handle their waste streams and generally provided a poor living environment. However, where early pollution was limited to such point sources, the situation was severely exacerbated by the industrial revolution when pollution became more widespread. Because aquatic ecosystems concentrate urban runoff from large catchment areas (Fig. 1), aquatic organisms are often confronted with a multitude of pollutants and stressors threatening their habitat (Schwarzenbach et al., 2006). Aquatic ecotoxicology is a multidisciplinary field which combines toxicology with ecology and aims to integrate the effects of pollutants across all levels of biological organization: from the molecular and subcellular level to whole aquatic communities and ecosystems.

Whereas the toxicological side of ecotoxicology was traditionally based on classic exposure tests of model organisms, the ecological side was largely built on traditional biomonitoring. Exposure tests consisted of constructing

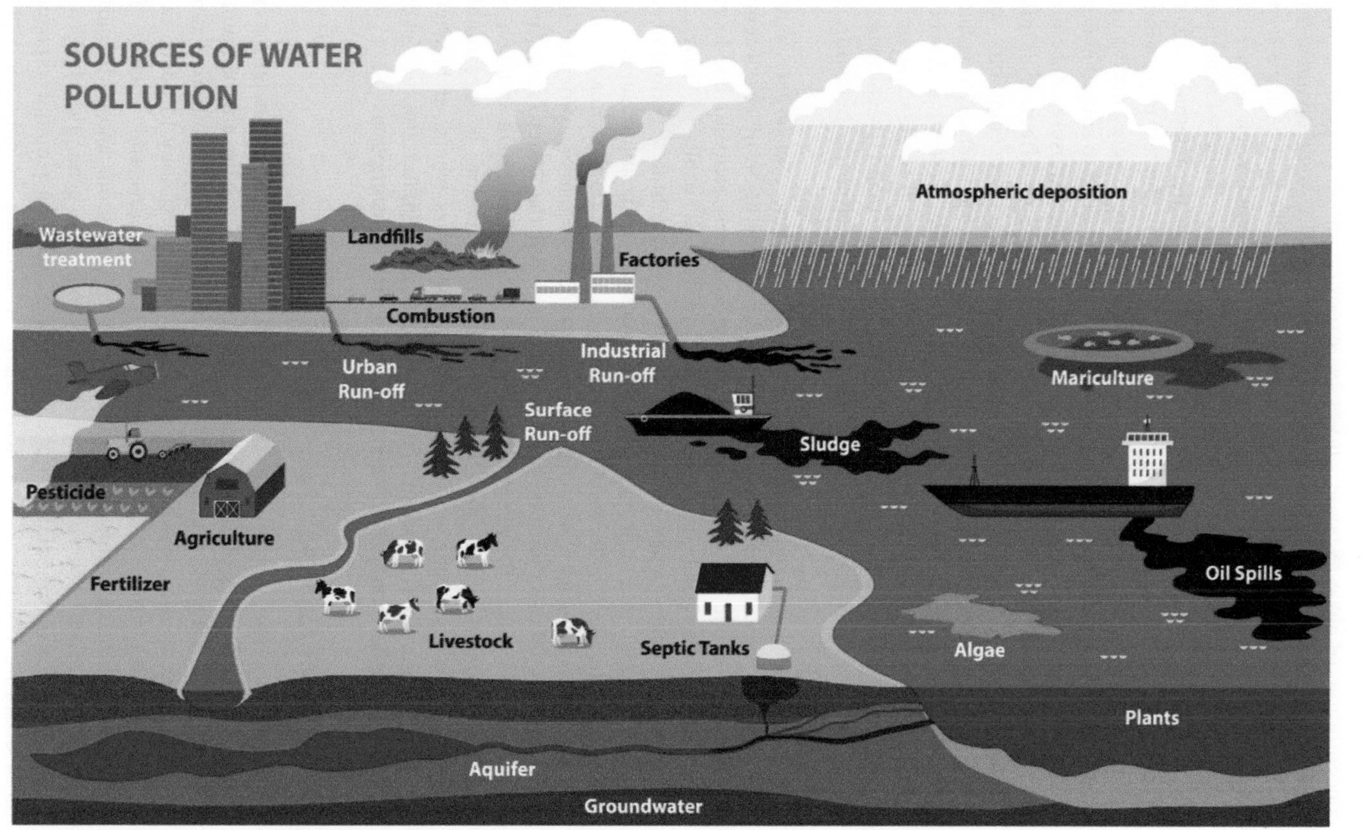


**FIG. 1** Different types of pollution from large catchment areas accumulate in the aquatic environment. *From VectorStock.*

concentration or dose-response curves by exposing model organisms (e.g., algae, invertebrates, fish) to a concentration gradient of a single compound in a well-defined water composition at a set temperature and could be acute (typically 24–96h) or chronic toxicity tests (>7 days). Endpoints such as survival, growth and reproduction were recorded. The most common output of acute tests was an $LC_{50}$ or $LD_{50}$, i.e., the exposure concentration or the internalized concentration or dose, respectively, that was lethal to 50% of the test population.

When looking at growth and reproduction, tests were necessarily long-term, ideally life-long, and the organisms were exposed to sublethal concentration gradients generating an $EC_{50}$ or $ED_{50}$, i.e., the exposure concentration or dose at which 50% of the population showed an effect. This also allowed calculation of the NOEC (no-observed-effect-concentration) and LOEC (lowest-observed-effect-concentration). Such tests are heavily regulated and protocols are described by the EPA (Environmental Protection Agency, USA: https://www.epa.gov/test-guidelines-pesticides-and-toxic-substances) and OECD (Organisation for Economic Co-operation and Development: https://www.oecd.org/chemicalsafety/testing/oecdguidelinesforthetestingofchemicals.htm).

On the other hand, biomonitoring described abundances of target species or community descriptors to assess the ecological condition of ecosystems, and can also include measurements of pollutant residues, elements and metabolites as pollution indicators (Derocles et al., 2018). It results in biological water quality indices (again based on, e.g., algae, invertebrates, fish numbers and health status), which provide direct measures of the health of an ecosystem's fauna and flora. Additionally, it gives information on the bioaccumulation and biomagnification of toxicants in the aquatic organisms provided the concentrations of the toxicants are measured in both the water (sometimes sediments) and biota. Bioaccumulation occurs when a pollutant is taken up at a faster rate than it is metabolized or excreted leading to measurable residues in the organism, while biomagnification occurs whenever bioaccumulation is amplified through the food chain, i.e., higher trophic levels accumulate higher pollutant concentrations via prey consumption.

## 1.2 Acclimatization vs adaptation

Acclimatization and adaptation to pollutants play pivotal roles in determining the long-term effects of pollutants on fish populations. Both are compensatory responses, conferring increased resistance to pollutants but at different time scales. Acclimatization is where fish remodel their physiology to counteract pollutant effects within a single generation, a rapid, reversible change. In contrast, genetic adaptation occurs over multiple generations and particularly when pollutants affect survival and/or reproduction. Transgenerational plasticity (TGP) is an intermediate process that occurs when a parental exposure alters offspring phenotypes via non-genetic inheritance or epigenetic processes (Ho and Burggren, 2010). Thus, acclimatization and TGP can be effective mechanisms to buffer fish populations from pollutants, "buying time" for genetic adaptation to take hold. Consequently, besides causing lethal or sublethal effects, pollutants need to be viewed as potential drivers of natural selection.

Increased pollutant resistance generally arises from enhanced detoxification or from metabolic biotransformation mechanisms, and lowered sensitivity of toxicokinetic and toxicodynamic pathways (Spurgeon et al., 2020). Most ecotoxicological studies have focused on short-term, within-generation responses to pollutants that lack an evolutionary perspective. However, evolutionary toxicology is a rapidly expanding field and the importance of examining multi-generational responses to persistent pollutants is becoming increasingly recognized (Brady et al., 2017; Klerks et al., 2011; Rodríguez-Romero et al., 2021). Therefore, the risks of pollution in a rapidly changing world need to be accurately predicted from multi-generation studies (rather than just single-generation studies) which can examine all types of compensation, including within-generation acclimatization, parental and carry-over effects, and selection effects, along with the cumulative outcome of all these processes (Decourten et al., 2020). Indeed, multi-generation studies have shown pollutant resistance can increase when parents beneficially program offspring phenotypes to enhance fitness, e.g., parental zebrafish (*Danio rerio*) exposed to dietary crude oil increased survival of F1-offspring raised in oiled environments and these changes were linked to epigenetic inheritance (i.e., histone modification and global DNA methylation; Bautista et al., 2020). Conversely, the negative effects of pollutants can worsen over successive generations, but fish are poorly represented in multi-generation studies (Loria et al., 2019) because of the need for rapid breeding and easy culture, as with invertebrates, particularly Daphnia spp. Well suited for future multi-generation studies are short-lived fishes such as zebrafish, western mosquitofish (*Gambusia affinis*), fathead minnow (*Pimephales promela*) and three-spined stickleback (*Gasterosteus aculeatus*) (Thoré et al., 2021a). Turquoise killifish (*Nothobranchius furzeri*), for example, are among the shortest-lived fishes (Thoré et al., 2021a,b) with a life-span of just 5–6 months, a generation time of 3 months in captivity and females producing approximately 20–50 eggs per day. Furthermore, their eggs can be stored in a dormant state, eliminating the need to maintain labor-intensive, continuous cultures (Philippe et al., 2018). Moreover, the physiology of turquoise killifish is well-characterized and species-specific genome editing tools are available (Thoré et al., 2021a).

For genetic adaptation to occur, genetically based variation in pollutant-resistance must already exist in a fish population. Selection experiments are essential for evaluating the heritability of pollution resistance. For example, cadmium (Cd) resistance increased in least killifish (*Heterandria formosa*) (Xie and Klerks, 2003) after six generations of selection, with survival time to a Cd exposure being three times longer compared to control lines; realized heritability for Cd-resistance was estimated as 0.50. Another common approach is to compare pollutant-resistance of populations either in exposed and unexposed (control) areas (e.g., Nacci et al., 2010), or along a pollution gradient. For example, resistance to lead (Pb) of western mosquitofish populations living in a highly contaminated habitat (Bayou Trepagnier, USA) was higher than that of a nearby control stream (Klerks and Lentz, 1998). However, the discovery that this Pb resistance disappeared after rearing in clean water for 34 days suggests that population differences were due to acclimatization rather than genetic adaptation. Common-garden experiments, where fish from different populations are reared under identical conditions, are therefore essential in disentangling the contributions of phenotypic plasticity from true genetic adaptation acting on variation already present in the population.

Population comparisons increasingly incorporate molecular and "omics-based" tools (Whitehead et al., 2017) to shed mechanistic insight into the underlying population divergence. For example, genome-wide microarrays have uncovered differences between pollutant-tolerant and pollutant-sensitive Atlantic killifish (*Fundulus heteroclitus*) populations. Moreover, these differences were consistent among tolerant populations (Bozinovic and Oleksiak, 2010). Also, a genome-wide association study (GWAS) with three-spined stickleback living across 21 locations of a historical mercury (Hg) pollution gradient revealed 28,450 single nucleotide polymorphisms (SNPs) associated with accumulated levels of Hg; identified ten SNPs associated with Hg in muscle, and a significant association with at least one locus on Chromosome 4 (Calboli et al., 2021). Candidate gene, SNP and QTL (quantitative trait locus) mapping approaches can then offer insight into which molecular markers could be targets of selection in pollutant-tolerant populations. At a coarser scale, phylogenetic analyses can be useful in exploring genes and detoxification pathways associated with pollution resistance. For example, phylogenetic approaches have revealed that nitrite ($NO_2^-$) toxicity in fish has a strong phylogenetic signature, associated with taxon-specific patterns in chloride channels (Brady et al., 2017).

Whether adaptive evolution can keep pace with emerging pollutants and climate change is an unanswered question under intense investigation. Obviously, the extent of pre-existing genetically based variation for pollutant-resistance will greatly influence the efficacy of evolutionary rescue, as will the specific nature of the mechanism. Because genetic variation tends to be inversely correlated with population size, evolutionary rescue may be less likely in small populations (Bell and Gonzalez, 2009). The pace and severity of pollutant exposure or environmental change will also limit the types of compensatory responses that can contribute to survival (Whitehead et al., 2017). Likewise, a rapid and a severe pollution may limit acclimatization responses and adaptational responses in species with long generation times or late-in-life maturation (a longer exposure before reproduction). Severe levels of pollution that cause population declines with insufficient time for adaptive genetic variants to increase in frequency are also problematic.

Pollution or environmental change can interact through diverse physiological challenges that require evolution of diverse pathways to maintain fitness (Whitehead et al., 2017). Adaptation to one type of stressor or pollutant may increase a species' susceptibility to other stressors due to fitness trade-offs (Whitehead et al., 2017). Indeed, Atlantic killifish from a highly polluted river with a heightened pollutant-tolerance came at the cost of reduced hypoxia tolerance and poorer overall condition (Meyer and Di Giulio, 2003). The potential trade-offs associated with adaptive responses therefore need to be explored to provide a holistic view of how species will fare moving into the future. Running multi-generation studies with multiple pollutants/stressors can provide enhanced ecological realisms, but these studies are rare. Integrating evolutionary perspectives into ecotoxicology is a promising way forward to gain much needed insight into how the physiology and fitness of fishes will be shaped by pollutants and global change.

## 1.3 Adverse outcome pathways

An endless list of new chemicals occurring in nature, plus a range of other stressors, present an impossible task to assess all possible mixture combinations at all possible concentrations and conditions (Altenburger et al., 2004; Bozinovic and Pörtner, 2015; Nikinmaa, 2013). Therefore, ecotoxicology is moving more toward a mechanistic understanding of toxicological processes, using trait-based approaches and sublethal molecular and physiological endpoints to understand the mode of action of pollutants (Ankley et al., 2010). Functional traits are then defined as "morphological, biochemical, physiological, structural, phenological, or behavioral characteristics that are expressed in phenotypes of individual organisms and are considered relevant to the response of such organisms to the environment and/or their effects on ecosystem properties" (Violle et al., 2007). They affect the fitness of an organism by altering survival, growth and/or reproduction, and therefore indirectly have an impact on populations and ecosystems. This brings ecotoxicological research much closer to ecophysiology and conservation physiology, and inherently involves the need to understand the relationships between chemical reactivity in the exposure media and within organisms, especially when dealing with polluted areas.

A promising new approach in this regard is based on lab-driven (Groh et al., 2015a,b) mechanistic-based Adverse Outcome Pathways (AOPs) (Ankley et al., 2010; Garcia-Reyero and Perkins, 2011) or AOP Networks (Knapen et al., 2015). An AOP provides a pathway of events (Fig. 2) starting from a molecular initiating event (MIE; the direct interaction of the chemical with a molecular target) through subsequent intermediate events at cellular, tissue, organ and individual organism levels which then result in an Adverse Outcome (AO), one that is relevant to ecotoxicological risk assessment and regulatory decision-making (Ankley et al., 2010; Groh et al., 2015a). An AO is often an impact on survival, growth and reproduction when used in an ecotoxicological context, but can also describe a population decline. Intermediate events are assigned to be Key Events (KEs) described by qualitative and quantitative Key Event Relationships (KERs) when they result in a biological change that is necessary for the AO to occur and is measurable.

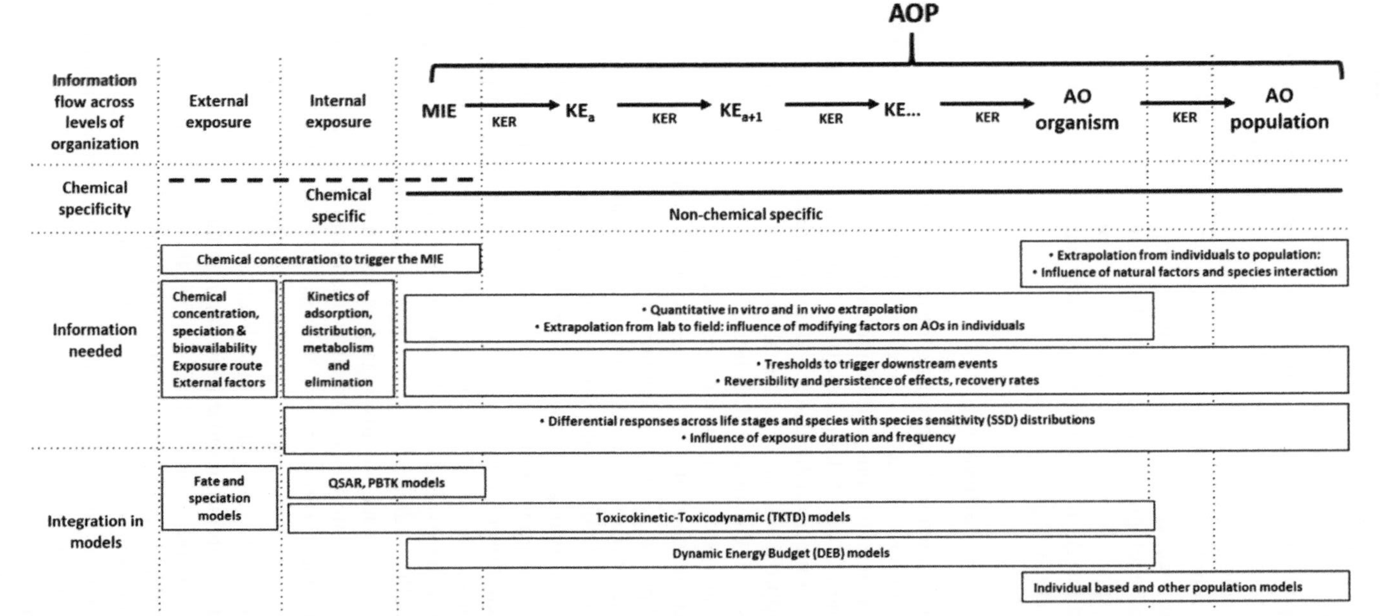


**FIG. 2** Information flow in ecotoxicological risk assessment: Top panel depicts the information flow during risk assessment, indicating the AOP position in the process, followed by chemical specificity, information that needs to be obtained and existing uncertainties, and how this can be integrated in models (discussed below). Abbreviations: AOP, adverse outcome pathway; MIE, molecular initiating event; KER, key event relationship; KE, key event; AO, adverse outcome; SSD, species sensitivity distributions; QSAR, quantitative structure-activity relationship; PBTK, physiologically based toxicokinetic models; TKTD, toxicokinetic-toxicodynamic models; DEB, dynamic energy budget models. *Adapted from Groh, K.J., Carvalho, R.N., Chipman, J.K., Denslow, N.D., Halder, M., Murphy C.A., Roelofs, D., Rolaki, A., Schirmer, K., Watanabe, K.H., 2015a. Development and application of the adverse outcome pathway framework for understanding and predicting chronic toxicity: I. Challenges and research needs in ecotoxicology. Chemosphere 120, 764–777.*

However, a single AOP may not capture all interactions of complex webs of molecular and biochemical perturbations. Therefore, sets of AOPs sharing at least one common MIE, KE or AO, an AOP Network, can reinforce the capacity for a realistic risk assessment. A few such studies have successfully adopted an AOP Network for field studies of long-term on endocrine disruption in a lake (Kidd et al., 2007), and near a papermill effluent in Jackfish Bay, Canada (Miller et al., 2015). In Jackfish Bay, individual responses of fish were used to predict the ongoing impacts as well as recovery of fish populations after stressor mitigation. The European City Fish project, aimed at developing a generic methodology for ecological risk assessment of urban rivers (Van der Oost et al., 2020), likewise identified indirect links between pollutant exposure and adverse effects on individuals (e.g., energy reallocation showed to be a significant "generic" KE) although some relevant variables were still lacking to analyze all interactions among pollutant exposure, MIEs, KEs and AOs. As such, fish biomarkers and a bioanalytical monitoring strategy for SIMONI Risk Indication (Smart Integrated Monitoring strategy, Van der Oost et al., 2017a,b) seemed to be valuable indicators of risks to macroinvertebrate and fish populations due to micropollutants (Van der Oost et al., 2020). AOP use is further facilitated in online communities by recording the AOP's in the Collaborative Adverse Outcome Pathway Wiki or AOP-Wiki (aopwiki.org) and visualize and integrating them in AOP Xplorer.

Predicting potential hazards with AOPs and AOP Networks could reduce the need for whole-organism toxicity tests and field studies. Furthermore, by starting with molecular events, AOP development and validation offers potential for highly desirable *in silico* (e.g., quantitative structure–activity relationship or QSAR models, structural alerts), *in chemico* (e.g., receptor binding assays), *in vitro* (e.g., stem cell assays) and *in vivo* tests using non-protected taxa or life stages (e.g., fish embryo tests), for 21st century toxicity assessment (Krewski et al., 2010).

## 2 Molecular initiating events, key events and their use as biomarkers

Organisms strive toward internal homeostasis, or a stable internal environment in a steady state, to survive. Organisms, however, are not completely shut off from the outside world despite the protection of cell membranes and external epithelia. The obligatory epithelial exchange surfaces for respiratory gasses, water, food and waste products, i.e., gill, gut and skin are in the case of fish, are seeded with transport channels and proteins. Pollutants can either interfere with these epithelia and their processes, inducing an MIE within these cells, or can use epithelia as sites of entry and disturb homeostasis elsewhere. Such disturbances can be followed by additional KEs and MIEs, which can be measured for use as a biomarker: "a naturally occurring molecule, gene, or characteristic by which a particular pathological or physiological process, disease, etc., can be identified" (Oxford Languages).

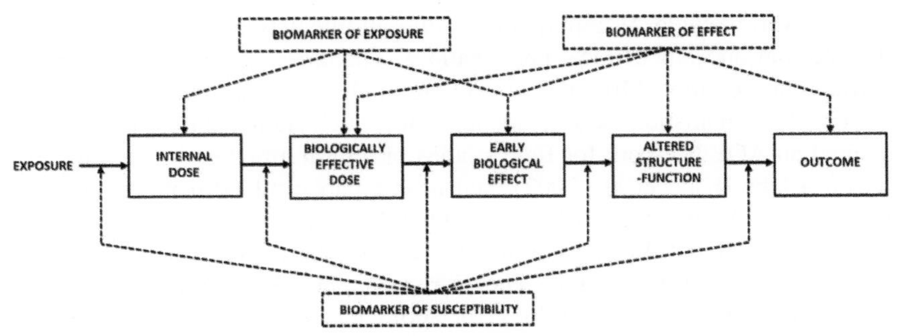


**FIG. 3** Simplified flow chart of classes of biological markers (indicated by boxes) which represent a continuum of changes and might not always be distinct. Solid arrows indicate progression, if it occurs, to the next class of marker. Dashed arrows indicate that the individual biomarker influences and/or indicates the rate of progression. *Adapted from Committee on Biological Markers of the National Research Council, 1987. Biological markers in environmental health research. Environ. Health Perspect. 74, 3–9.*

Life history traits such as survival, growth, reproduction and ultimately changes in population abundance are the AO and are the most ecologically relevant biomarkers but take a long time to evaluate. Physiological and molecular MIEs and KEs, in contrast, are often fast responding biomarkers that act as early warning signals. Early warning signals are valuable if susceptible to the effects of an exposure vary with life stage, species or population (Fig. 3).

Biomarkers can be classified in other ways, e.g., for exposure (e.g., internal dose), for effect (e.g., altered function) or for susceptibility (e.g., presence of uptake or defence mechanisms). An extensive literature review for Gladstone Harbor (Queensland, Australia), for example, revealed the most suitable biomarkers for exposure (e.g., CYP1A, EROD, SOD, HSP, MT, DNA strand breaks, lipid peroxidation, micronuclei, apoptosis) and for effect (e.g., histopathology, lipid storage index) (Kroon et al., 2017). That said, an AOP framework would classify most of these biomarkers as KEs leading to adverse outcomes.

Importantly, not all measured biomarkers indicate harm. Internal changes are continuously sensed by cells, organs and/or organisms and negative feedback loops are used to restore homeostasis, an active process called allostasis (McEwen and Wingfield, 2010; Romero et al., 2009). Stress responses resulting in allostasis are called eustress. A negative impact or an allostatic overload called distress, however, occurs when either the capacity to restore homeostasis is exceeded, or it is energetically too costly, or it is irreversible. Setting homeostasis to another sustainable and reversible level could be acclimation. Regardless of being a bioindicator of eustress or distress, the biomarker response is an early warning signal, perhaps creating opportunities to react before irreversible damage occurs (Brosset et al., 2021; Schreck and Tort, 2016).

The next section considers some commonly used MIE and KE biomarkers. Their adoption usually reflects an ease of measurement, either in field or in lab settings, and their reliability and ecological relevance (Adams and Ham, 2011; Brosset et al., 2021; Jeffrey et al., 2015; Madliger et al., 2016; McKenzie et al., 2016). Brosset et al. (2021) developed a flow chart to decide on the use of physiological biomarkers in fisheries management, which could easily be extended to other field studies as well.

## 2.1 Stress hormones

Stress hormones such as adrenaline and cortisol (glucocorticosteroid hormone) communicate homeostatic disturbances. Adrenaline is released rapidly (s to min) and will rapidly mobilize a fight-or-flight (-or-freeze) response and sugars from glycogen stores. A slower cortisol release (min to h) aims to restore pre-stress conditions by adjusting energy allocation to new demands. These two stress hormones are under control of the corticotropin releasing factor (CRF), the principal hypothalamic factor for controlling the stress axis. CRF in turn stimulates release of the adrenocorticotropic hormone (ACTH) from the pituitary gland *pars distalis*, which then triggers adrenaline release from the chromaffin tissue and cortisol release from the cortisol-producing interrenal cells in the head kidney, i.e., the hypothalamic-pituitary-interrenal (HPI) axis (Flik et al., 2006; Gorissen and Flik, 2016).

Traditionally, the release of these hormones is measured in plasma/blood via an ELISA or RIA assay, which can give a snap-shot picture of the level of stress exposure. The blood collection procedure is almost always a confounding factor for the quickly released adrenaline, which is consequently less suitable than cortisol as a biomarker since a fish must be netted to collect blood. If blood collection is rapid (5–15 min) the initial rise in cortisol can be detected. Cortisol is a glucocorticoid and a mineralocorticoid hormone and can signal general or osmoregulatory stresses. Acute elevations of cortisol may last hours and are used as an indication of the duration before reaching homeostasis. However, cortisol can be chronically elevated, and this would be worrisome (Schreck and Tort, 2016; Yada and Tort, 2016).

Several methods are used to measure cortisol (reviewed by Sadoul and Geffroy, 2019) using water or mucus samples (Carbajal et al., 2019; Ellis et al., 2004; Guardiola et al., 2016; Oliveira et al., 2018), or biopsies of gill tissue (Gesto et al., 2015), gut contents (Bertotto et al., 2010), faeces (Cao et al., 2017; Lupica and Turner, 2009), scales (Aerts et al., 2015) and fins (Bertotto et al., 2010). Some are less invasive (e.g., water, mucus, faeces) and some give a picture of cumulative cortisol release over time (e.g., faeces, scales), providing a reliable indicator of chronic stress but perhaps not the intensity of the stress, which is possible using blood samples. Extraction techniques for cortisol are complex and time-consuming (Sadoul and Geffroy, 2019). HPLC-MS–MS offers very low detection limits and the needed sensitivity and selectivity for small and/or low concentration samples. HPLC-MS–MS profiles glucocorticoid: (i) the dominant hormone (cortisol); (ii) direct precursors

(17α-hydroxyprogesterone and 11-deoxycortisol) that lead to cortisol production); (iii) phase I endogenous metabolites (cortisone and 20β-dihydrocortisone); and (iv) the most abundant, more polar phase I metabolites for excretion (tetrahydrocortisol and tetrahydrocortisone) to establish if exogenous glucocorticoids present in the environment (e.g., from water) or anthropogenic derived glucocorticoids (e.g., from hands) may have influenced the results (Aerts, 2018).

Beyond traditional stress biomarkers, genomic tools are now identifying candidate genes and unique signatures of them that are associated with stress responses which could become reliable early biomarkers of stressors (reviewed by Eissa and Wang, 2014). Even metabolomics has expanded beyond routine identification of simple biomarkers toward the exploration of molecular mechanisms of pollutants on specific metabolic pathways ( Johnson et al., 2016), and it can highlight involvement of metabolic pathways that were not considered previously (Lardon et al., 2013). When available, this should be the primary choice when using invasive techniques, both for studies in the laboratory and in the field (Morrison et al., 2007). Metabolomics using Nuclear Magnetic Resonance and Mass Spectrometry is an expensive promising, elaborate, sensitive and detailed screening of small metabolites (for a recent review see Zhang et al., 2021 or book see Álvarez-Muñoz and Farré, 2020). By analyzing the complete metabolomic profile, contaminant-induced metabolism alterations and up/down-regulation of metabolic pathways are revealed rapidly and in detail at a cellular level.

## 2.2 Blood and tissue metabolites

Other blood and tissue measurements are used as biomarkers. Reference values and the effects of pollutants are given for different fish species for blood electrolytes (sodium, potassium, chloride, magnesium, calcium), for enzymes including aspartate aminotransferase (AST), alanine aminotransferase (ALT), alkaline phosphatase (ALP), creatinine kinase (CK), and lactate dehydrogenase (LDH), and for metabolites such as glucose, albumin, total cholesterol, fatty acids, triglycerides, total protein, creatinine (based on a review of >350 publications in Folmar, 1993). Also hematological parameters such as hematocrit, hemoglobin and methemoglobin, hormones such as insulin, growth hormone, prolactin, cortisol, thyroid hormones, and sex steroid hormones such as androgens, estrogens, and progestogens as well as vitellogenin are given in this review. Such data bases, which consider age, sex and season of the year, are invaluable when comparing fish collected from reference and contaminated sites, or when sampling the same location over time before and after remediation (Ahmed et al., 2020; Bojarski and Witeska, 2020). Some parameters (e.g., glucose, lactate, protein, hematocrit) are now much easier to measure in the field because handheld point-of-care technologies, if rigorous calibration and validation of these units for use with fish is performed (Andrewartha et al., 2016; Clark et al., 2008; Stoot et al., 2014).

In the lab, new automated techniques such as flow cytometry and blood cell counting are being used, as shown in a study on micro- and nano-plastics where a complete blood count was used for the evaluation of hematopoiesis (Guerrera et al., 2021). Even lab analytical tools need to be calibrated for fish use (Harter et al., 2014).

Understanding changes in energy stores and metabolism is crucial, as they impact the functioning of organisms, populations and ecosystems (Cooke et al., 2013; Kearney et al., 2021). Indeed, increased energy metabolism and decreased whole-body energy stores are expected from the glycolytic and gluconeogenic actions of cortisol. Not only carbohydrate metabolism, but also protein and lipid metabolism are affected as energy fluxes between these metabolic pathways play an important role (De Boeck et al., 2001; Liew et al., 2013). Measuring whole body or liver and muscle energy stores is a sensitive indictor of the metabolic cost of a chronic exposure, provided total body mass and hepatosomatic index are included to differentiate between energy stores on a per weight basis and on a total availability basis.

Guerrera et al. (2021) for example, reported decreased liver glycogen stores and effects on lipid tissues. Simple condition indices such as Fulton's condition factor ($K=100*(Weight/Length^3)$) and the hepatosomatic index ($HSI=100*(Weight_{liver}/Weight)$), as well as the measurement of energy reserves, are used as to assess population health (Claireaux et al., 2004; Jeffrey et al., 2015). In three-spined stickleback, K and HSI were significant predictors of energy reserves (lipid, protein, glycogen and total energy), but the proportion of variance accounted for was small and the relationship highly variable with the season (Chellapa et al., 1995). Whereas for some studies K but not HSI correlated with metal accumulation in a pollution gradient (De Jonge et al., 2015), in others it did not (Bervoets and Blust, 2003) even though a threshold for metal accumulation above which K was always low could be determined. Therefore, K and HSI seem useful, although insensitive field biomarkers because they are highly dependent on the season. Handheld fat meters have been routinely used with migratory salmon in the field (Cooke et al., 2005, 2006; Crossin and Hinch, 2005).

## 2.3 Energy metabolism and challenge tests

Non-invasive techniques can measure whole organism metabolic rates and capacities as biomarkers for environmental stress. Direct calorimetry is rarely attempted to directly measure metabolic rates in poikilothermic aquatic organisms. Instead respirometry to measure the rate of oxygen removal from water (oxygen uptake rate) can indirectly measure aerobic energy metabolism. If $CO_2$ and nitrogen releases to the water are concurrently measured, instantaneous fuel use can be stoichiometrically calculated as carbohydrate, protein and lipid fractions (Alsop et al., 1999; De Boeck et al., 2001; Kieffer et al., 1998; Lauff and Wood, 1996). Stress increases oxygen uptake and switches fuel usage, and pollutants can limit the capacity for oxygen uptake.

Aerobic scope represents the capacity for oxygen uptake, a concept developed by Fry about 75 years ago (Fry, 1947; Fry and Hart, 1948). This capacity can be expressed either as an absolute (AAS) difference between the standard metabolic rate (SMR) at rest (for review Chabot et al., 2016) and the maximal aerobic metabolic rate (MMR) of that organism (for review Little et al., 2020) at the same temperature, or as a factorial aerobic scope (FAS) of MMR and SMR. Both AAS and FAS are used in different ways to characterize the capacity of a fish to increase its aerobic performance to forage, escape predators and to grow and reproduce. AAS was used to develop the oxygen and capacity-limited thermal tolerance (OCLTT) theory, which hypothesizes that organisms or populations of organisms cannot persist in habitats where their aerobic scope deviates too far from an optimum when the capacity to supply oxygen to tissues becomes limited at temperature extremes, such as those expected with climate change (Pörtner, 2001, 2010; Pörtner and Farrell, 2008). The healthy discussion of the universality of the OCLTT across species indirectly led to improved guidelines on how to measure oxygen uptake in fishes (Chabot et al., 2016, 2021; Clark et al., 2013; Farrell, 2016; Halsey et al., 2018; Pörtner et al., 2017). A reduction in AAS or FAS is a useful indicator of stress in ecotoxicology (Callaghan et al., 2016; Mager and Grosell, 2011; McKenzie et al., 2007). Yet care is needed with interpretations because cross-tolerance with increased temperature can elevate AAS (Gomez Isaza et al., 2020a; Opinion et al., 2020, 2021).

More general response to stress is increased SMR, or an increase in the related routine metabolic rate (RMR, metabolic rate during normal activity) (Farrell et al., 1998; McGeer et al., 2000; McKenzie et al., 2007; Metcalfe et al., 2016; Rodgers and Beamish, 1981). Yet species differences exist. Rainbow trout increased RMR during Cu-exposure, while RMR decreased with the same exposure in common carp and gibel carp (De Boeck et al., 2006) possibly because they responded by depressing whole animal metabolism or a switch to anaerobic metabolism given the observed depression of ventilation rates (De Boeck et al., 2007). Clearly, a fish's lifestyle should be taken into account when interpreting respiratory data especially since a comparison of 30 freshwater fish species revealed those living in fast flowing rivers had a high aerobic capacity and a high SMR, while those not exposed to fast water currents had a low SMR and aerobic capacity, as well as a strong hypoxia tolerance (Pang et al., 2019).

A limitation of respirometry for measuring energetics is that it only measures aerobic metabolism. But when fish burst swim or are swum to fatigue or exhaustion they use anaerobic metabolism, as indicated by the appearance of lactate in the blood (Pagnotta and Milligan, 1991). For ecotoxicological purposes, a standardized swimming test such as critical swimming speed ($U_{crit}$) protocols that measure maximum prolonged swimming ability are ideal. $U_{crit}$ is an ecologically relevant indication of the fish's performance (Plaut, 2001) because prolonged swimming is essential for foraging, orientation against the current and fish migrations. In a $U_{crit}$ test (designed by Brett, 1964) water speeds in a swimming flume are slowly and incrementally increased until the fish fatigues and is no longer able to swim against the fastest water current, similar to a human endurance test on a treadmill or home trainer. Contaminated water typically reduces $U_{crit}$ indicating compromised performance (Corriere et al., 2020; De Boeck et al., 2006; Farrell, 2008; McGeer et al., 2000; McKenzie et al., 2007; Monteiro et al., 2021; Sahota et al., 2021; Wilson et al., 1994; Yu et al., 2015). Repeating the exercise challenge after a short recovery period can indicate the level of recovery after a first $U_{crit}$ measurement, and a contaminant effect can be detected by a lowering of the second $U_{crit}$ measurement (Farrell et al., 1998; Jain and Farrell, 2003; McKenzie et al., 2007). While the $U_{crit}$ protocol and other swimming speed measurements are standalone bioindicators, adding respirometry allows MMR to be measured. Furthermore, if oxygen uptake measurements continue after fatigue , or after a chasing to exhaustion protocol (Clark et al., 2013; Norin and Clark, 2016; Zhang et al., 2018), the excess post-exercise oxygen consumption (EPOC) can assess the anaerobic contribution to the exercise (Farrell et al., 1998; Lee et al., 2003). Full recovery from exhaustion may take more than 10h and involve different phases of recovery that occur at different rates (see Milligan, 1996; Zhang et al., 2018).

Other well-established, whole-animal measurements can be used to assess acute tolerance to progressive hypoxia or warming in contaminated waters. The Critical Thermal Maximum ($CT_{max}$) protocol steadily increases water temperature until a loss of equilibrium (LOE) occurs (Cereja, 2020; Galbreath et al., 2004). Pollutants can reduce $CT_{max}$ (Monteiro et al., 2021; Patra et al., 2007). Similarly, oxygen levels can be reduced until LOE to determine the Incipient lethal oxygen saturation (ILOS). If oxygen uptake is also measured, the critical (minimum) oxygen level ($Po_{2crit}$, $P_{crit}$ or $O_{2crit}$) to sustain SMR can be determined (for review see Claireaux and Chabot, 2016; Rogers et al., 2016). $P_{crit}$ is sensitive to pollutant exposure (De Boeck et al., 1995; Monteiro et al., 2021; Rodgers et al., 2021) and to changes in SMR (Chabot et al., 2016).

## 2.4 Oxidative stress

Free radicals, highly reactive molecules that contain unpaired electrons, are naturally formed by redox reactions in living cells (for review see Lushchak, 2016a). They play important roles in immunological responses via phagocytotic cells, and are also involved in cell signaling, e.g., the reactive nitrogen species (RNS) nitric oxide for vasodilatation. Oxidative stress typically concerns reactive oxygen species (ROS) which are primarily produced when oxygen is partly reduced in the mitochondrial electron transport chain; production of the superoxide anion radical ($O_2^{\bullet-}$) can form the hydroxyl radical ($HO^{\bullet}$) and hydrogen peroxide ($H_2O_2$) is considered a ROS having a reactivity much higher than that of molecular oxygen. ROS production can also occur in other membrane systems by different oxidases such as xanthine oxidase (Bartosz, 2009; Nanduri et al., 2013).

Oxidative stress, first defined as "the imbalance between oxidants and antioxidants in favor of the oxidants, potentially leading to damage" (Sies, 1985), was later redefined to "a transient or chronic increase in steady-state ROS levels, disturbing cellular core and signaling pathways, including ROS-based ones, leading to oxidative modification of cellular constituents which may culminate in cell death via necrosis or apoptosis" (Lushchak, 2016a). Because oxidative stress is part of natural cell functioning, and important in different cell signaling pathways, cells have also developed antioxidant systems to prevent possible damage by ROS. Naturally occurring, low molecular mass antioxidants include vitamins C (ascorbic acid) and E (α-tocopherol), carotenoids, anthocyanins, polyphenols, uric acid and glutathione (GSH), and high molecular mass antioxidant enzymes such as superoxide dismutase (SOD), catalase (CAT) and glutathione peroxidase (GPx).

A detailed review on environmentally induced oxidative stress in fish is found in Lushchak (2016a). Other reviews cover naturally fluctuating environments such as nutritional factors, temperature, and oxygen levels (Birnie-Gauvin et al., 2017; Martinez-Alvarez et al., 2005) and xenobiotics (Hellou et al., 2012; Lushchak, 2011, 2016a; van der Oost et al., 2003). Lushchak (2011, 2016a) reviewed signaling pathways that control the oxidative stress response, such as the Nrf2/Keap1 pathway which regulates the expression of antioxidant response element (ARE) driven genes including glutathione reductase (GR), GPx, or CAT, and the role of the hypoxia inducing factor HIF-1α, while Silvestre (2020) specifically reviewed the regulation of these pathways in aquatic organisms exposed to xenobiotics.

Oxidative stress is a biomarker for general stress because it is not specific to one stressor. To be useful as a biomarker for environmentally relevant toxicological studies, the oxidative stress response needs to be measured under conditions that are environmentally relevant, and in a dose- and/or time dependent manner, which has not always been the case. Lushchak (2014a,b, 2016b) proposed a time-course and intensity-based classification system in an attempt to improve the use of oxidative stress as a reliable biomarker. The most commonly used biomarker assays are presented in Birnie-Gauvin et al. (2017). Table 1 lists these and a brief discussion follows.

ROS or ROS-modified substances (ROMS) are inherently unstable substances. Steady-state levels of ROS/ROMS may exist, but through dynamic and continuous generation and elimination; therefore they should not be called "end products" (Lushchak, 2016b). The measurement of lipid peroxidation products such as malonic dialdehyde (MDA) with thiobarbituric acid (TBA) was one of the first practical and affordable techniques to measure ROMS (Draper et al., 1993). However, it also interacts with other compounds such as aldehydes, amino acids, carbohydrates, etc., therefore measuring a

**TABLE 1** Oxidative stress measures commonly used in fish biology.

| Biomarker | Method of detection | Applications |
|---|---|---|
| Protein carbonyls | Protein carbonyl formation (Levine et al., 1990; Stadman and Berlett, 1997) | Levels of protein damage, fragmentation; insight into overall oxidative stress levels |
| DNA damage | 8-hydroxy-2′-deoxyguanosine assay (Kasai, 1997) | Levels of DNA damage; insight into overall oxidative stress levels |
| Lipid peroxidation | Thiobarbituric acid reactive substances test (Draper et al., 1993) | Levels of lipid damage; insight into overall oxidative stress levels |
| Catalase (CAT) | CAT enzymatic activity assay (Sinha, 1972) | Insight into antioxidant defences; higher activities may be associated with higher $H_2O_2$levels |
| Superoxide dismutase (SOD) | SOD activity assay (Beauchamp and Fridovich, 1971; Ōyanagui, 1984) | Insight into antioxidant defences; higher activities may be associated with higher $O2^{\bullet -}$ levels |
| Glutathione peroxidase (GPX) | GPX activity assay (Flohé and Günzler, 1984; Paglia and Valentine, 1967) | Insight into antioxidant defences; higher activities may be associated with higher levels of ROS |
| Glutathione reductase (GR) | GR activity assay (Carlberg and Mannervik, 1975; Wheeler et al., 1990) | Insight into antioxidant defences; GR reduces GSSG back to GSH; higher activities of GR may be associated with higher levels of GSSG |
| Glutathione (GSH); total glutathione (TGSH); glutathione disulfide (GSSG) | Glutathione assay (Akerboom and Sies, 1981; Smith et al., 1988) | Provides insight into oxidative damage (GSH to GSSG ratio or vice versa) and antioxidant defences (GSH) |
| Vitamin C (ascorbic acid) | Ascorbic acid assay (Deutsch and Weeks, 1965; Roe and Kuether, 1943) | Insight into antioxidant defences (provides an electron to quench ROS) |
| Vitamin E (α-tocopherol) | Vitamin E assay (Prieto et al., 1999) | Insight into antioxidant defences (peroxyl radical scavenger) |
| Low molecular weight antioxidants | Oxygen radical absorbance capacity (ORAC) assay (Cao et al., 1993)<br>Ferric reducing antioxidant power assay (FRAP) (Benzie and Strain, 1996) | Insight into total low molecular weight antioxidant defences |

Adapted from Birnie-Gauvin, K., Costantini, D., Cooke, S.J., Willmore, W.G. (2017). A comparative and evolutionary approach to oxidative stress in fish: a review. Fish Fish. 18, 928–942.

range of products called TBA-reactive substances (TBARS) and should no longer be referred to as an MDA measurement (Lushchak et al., 2012). Much more accurate identification can be obtained with by liquid chromatography coupled with mass spectrometry (LC/MS). Measuring ROS-modified nucleic acids and protein is expensive and highly specialized, except for the reliable and reproducible technique of determining additional carbonyl groups in proteins (formed as a result of their ROS-induced oxidation) using 2,4-dinitrophenylhydrazine (DNP) (Levine et al., 1990; Stadman and Berlett, 1997).

Glutathione (GSH) is a reliable low molecular mass marker of oxidative stress that exists in a dynamic balance between the oxidized (GSH) and reduced (GSSG) forms. Ideally the GSH:GSSG ratio should be determined with the compounds remaining stable during extraction and the assay. HPLC and enzymatic methods are routinely used, with LC-MS being a sensitive method that needs fewer manipulations. Tandem mass spectrometry MS-MS provides the lowest limit of detection (Iwasaki et al., 2009; Monostori et al., 2009 for methodological reviews).

## 2.5 Endocrine disruption

Anthropogenic activities expose fish to endocrine active substances and endocrine disruptors. An endocrine active substance interacts with an endocrine system to cause responses that may or may not give rise to adverse effects. An endocrine disruptor (ECD) is an exogenous substance or mixture that disrupts normal endocrine function(s) and consequently causes adverse health effects directly to an individual or its progeny (WHO/IPCS, 2002). The hypothalamic–pituitary–gonadal axis and the thyroid system are considered important targets for ECDs. Modes of action include androgen and estrogen agonists and antagonists, thyroid antagonists, steroidogenesis inhibitors, aromatase inhibitors and retinoid receptor modulators (Coady et al., 2017; Hutchinson et al., 2006; Matthiessen et al., 2017; Vos et al., 2000). Biomarkers for ECD effects in laboratory and field studies include plasma steroid hormones, the egg yolk protein precursor vitellogenin (VTG), gonadosomatic indices and gonad histology, as well as secondary sexual characteristics, both. Population-relevant endpoints include effects at the individual level such as behavior, development, growth and reproduction (Ankley and Johnson, 2004; Hutchinson et al., 2006). Reproductive endpoints including survival and size, fecundity (number of eggs spawned), fertility (number of fertile eggs produced), hatch (number of fertile eggs that produce larvae), and larval viability (e.g., occurrence of malformations in hatched animals) can all be assessed (Hutchinson et al., 2006). Behavioral endpoints, which include inappropriate or ill-timed courtship or parental behavior (e.g., migration, courting, nesting), may have just as significant repercussions as disrupted ovulation or spermatogenesis (Matthiessen et al., 2017). Chronic elevation of cortisol can depress reproductive performance (Schreck, 2010) by decreasing VTG synthesis in the liver and consequently reducing egg production. For example, confinement of rainbow trout (*Oncorhynchus mykiss*) decreased VTG levels and reduced egg size, as well as offspring survival rates (Campbell et al., 1994).

Vitellogenin is a good biomarker that can be reliably measured in plasma by ELISA. It is a transport protein for yolk protein and lipids that move from the liver to the developing eggs and is a measure for estrogenic disruption. Plasma VTG can increase up to a million-fold when induced by estrogens. Increased VTG is found in male fish that are feminized due to oestrogen exposure; intersex gonads (ova–testis) of such intersex fish ( Jobling et al., 1998) are identified with histology. In female fish, early VTG synthesis can reduce the survival chances of young fish maturing too early at a life stage where energy budgets are critical (Länge et al., 2001). Androgen biomarkers have been considered for development of male gender characteristics controlled by androgens, e.g., facial tubercles and nuptial pads in fathead minnow (Ankley et al., 2001, 2003; Harries et al., 2000).

Molecular biomarkers are sensitive to ECS too. VTG mRNA is quickly induced (within a few hours of an oestrogen exposure) and could be an early-warning test. Such measurements require tissue sampling and must consider that more than one VTG mRNA can exist (e.g., two in Medaka, nine in zebrafish) (Hiramatsu et al., 2002; Trichet et al., 2000). Measurement of oestrogen receptors, enzymes involved with sex hormone biosynthesis and gonadotrophins (which control gonad development and sex steroid synthesis), and plasma levels of steroid hormones (oestradiol, testosterone, 11-ketotestosterone) concentrations are well-established techniques. Time of exposure is a critical consideration because the most sensitive stage is not necessarily reproduction per se. For example, pharmacological doses of fadrozole during the window of sexual differentiation in the zebrafish depressed aromatase mRNA expression and generated in all-male fish populations (Fenske and Segner, 2004). Additionally, effects may only occur in the next generation, especially in viviparous fish, such as the eelpout (*Zoarces viviparus*), where malformed embryos are reported after maternal exposure to 17a-ethynylestradiol (Morthorst et al., 2014). This occurrence of time-sensitive windows, the multitude of reproductive strategies in fish and delayed effects makes the risk analysis of ECDs a complex issue.

## 2.6 Immune system

Immunity is a very relevant ecologically trait of key importance for organism survival and population growth (Segner et al., 2012a,b). Fish in contaminated areas often show signs of impaired immune function and increased disease incidence (Zelikoff et al., 2000). Immunotoxicity, the capacity to induce immune dysfunction, is reported for a range of chemicals such as metals, endocrine disrupting compounds, pesticides, PCBs and PAHs, and pharmaceuticals. Typically, immunotoxicity requires a prolonged exposure to sublethal contaminant levels.

Immunotoxic effects can involve either immune suppression and increased risk for diseases, or immune stimulation which may not be adverse (unless it leads to auto-immune responses). No clear set of immune bioindicators currently exists for immunotoxicity (reviewed by Rehberger et al., 2017). Those used are mostly innate immune responses using phagocyte activity and phagocyte respiratory burst activity, which seem promising biomarkers (Bols et al., 2001; Zelikoff et al., 2000). Greater basic knowledge before immunotoxicity can be reliably added as a requirement of an ecotoxicological risk assessment (Rehberger et al., 2017; Segner et al., 2012a,b). Omics platforms including transcriptomics, proteomics, and metabolomics being used to unravel the intricate molecular mechanisms of host-pathogen interaction are found in Natnan et al. (2021a,b).

Nonetheless, the complexities of the fish immune system and its regulation by diverse signal transduction pathways, cellular components and mediators and receptors for communication and activation diffusely spread within the entire organism (Rehberger et al., 2017) are emerging. Teleosts have three lines of defence. Epithelia that are in contact with the environment (skin, gill and the gastrointestinal system) are the first line of defence. They form a mechanical, chemical (through mucus, antibodies, lysozymes, etc.) and cellular (immune cells) barrier. The innate immune system is the second (reviewed by Bols et al., 2001). It provides a fast, internal and non-specific response to many pathogens and is thought to be the most important immune component of fish (Lieschke and Trede, 2009; Magnadóttir, 2006; Tort et al., 2003). It consists of humoral factors (such as lysozyme or complement factors) and phagocytic cells (such as granulocytes, monocytes/macrophages and natural killer cells). Phagocytic cells remove tissue debris and microorganisms, secrete immune response regulating factors and bridge to innate and adaptive immune responses (Segner et al., 2012a,b). The adaptive or acquired immune system is the third line of defence and involves T- and B-lymphocytes, which mediate cellular and humoral responses, respectively. This system is pathogen-specific and generally much slower, taking weeks instead of days (as in mammals), which is why it is often not considered important in fish (Segner et al., 2012a,b). Nonetheless, fish possess antigen-presenting, T-helper cells and cytotoxic T-cells (Fischer et al., 2006). Fish B-lymphocytes produce immunoglobulins (IgM) which are primarily tetrameric IgMs (Warr, 1983).

One category of immunotoxicity testing involves pathogen challenge tests, a holistic, integrative biomarker. Survival of exposed fish then challenged with pathogens (e.g., *Aeromonas* sp., *Vibrio* sp.) is measured and such challenge tests can identify a compromised immune system post-exposure (Koellner et al., 2002). Reduced survival after toxicant exposure occurred in about 75% of the challenge studies reviewed by Rehberger et al. (2017). Fish welfare issues may limit the use of these challenge tests. Alternatively, tests with blood samples are used in lab and field studies. An increase in the absolute leukocyte count and an altered differential white blood cell count may suggest an inflammatory response probably induced by tissue damage. The other most common endpoints are phagocytic activity, respiratory burst activity and lysozyme activity, all part of the innate immune system. Granulocytes, and B cells show phagocytic activity (Li et al., 2006). Phagocytic activity is determined from the incorporation of fluorescent-labelled beads or (inactivated) bacteria into leukocytes using flow cytometry (Muller et al., 2009; Shelley et al., 2009). Respiratory burst activity determines pathogens killed by ROS within phagocytic cells either using the reduction of nitro blue tetrazolium (NBT) or the chemiluminescence of ROS (Koellner et al., 2002). Lysosome activity is determined by adding serum from the fish under examination to bacterial cell suspension and assessing the lysis of the bacterial cells (Alexander and Ingram, 1992). Even though rarely studied, methods for assessing effects on the adaptive immune system of fish are available such as B and T cell lympho-proliferation assays (Carlson et al., 2002; Koellner et al., 2002; Zelikoff, 1998).

Transcript mRNA levels of immune related genes, some of which also belong to the innate immune system, are regularly assessed by real-time quantitative polymerase chain reaction (RT-qPCR), e.g., cytokines, interleukins, tumor necrosis factor, interferons, and CXC chemokines. A lack of specific antibodies against immune proteins such as IgMs limits the use of immunochemistry and immunohistopathology although they become increasingly available.

## 2.7 Stress proteins, detoxification and metabolic biotransformation

Cellular stress responses to environmental change involve the induction of several protective proteins that are used as biomarkers for physical and chemical stressors. These are measured either at the level of gene induction or transcription (by RT-qPCR, microarrays and RNA-Seq), or at the protein level (by Western Blot, ELISA or mass spectrometry). One group of proteins that respond to general physical stressors such as temperature or osmotic stress are stress or heat shock proteins (HSP) belonging mostly to the HSP70 or HSP90 families (reviewed by Deane and Woo, 2011). HSP response can be complex, in part because cortisol can attenuate the protective HSP response (Basu et al., 2003; De Boeck et al., 2003b) and in part because laboratory and field exposures do not always concur (Köhler et al., 2001). Therefore, HSPs are not common biomarkers for pollutants.

Another group are metallothioneins (MTs), cysteine-rich, heat-stable proteins that are naturally produced and bind metal ions in all major classes of vertebrates. Their normal physiological role is to store essential metals such as copper (Cu) and zinc (Zn). However, they also bind and detoxify non-essential metals such as Cd and Hg by being induced after exposure, thereby protecting from excess metal pollution and entry into fish (Roesijadi, 1992). Several MT isoforms exist in fish organs (Vasak, 2005).

MT induction capacity is greatest in tissues actively involved in uptake, storage and excretion of metals, e.g., the small intestine, liver and gills of fish (Roesijadi and Robinson, 1994). Also other proteins such as albumin, ferritin and transferrin play a role in metal homeostasis and detoxification. Induction of these metal binding proteins is often used as a biomarker of aquatic metal exposure.

Excretion of lipophilic organic xenobiotics requires biotransformation processes to convert a more water-soluble forms (Lech and Vodicnik, 1985; Xu et al., 2005). This process takes place in two phases, in Phase I of this biotransformation is enzymatic conversion by oxidation, reduction or hydrolysis reactions that increase a water-solubility. The cytochrome P450 family (CYPs) is the cornerstone of this phase of the detoxification and excretion processes, involving mixed function oxygenase (MFO) catalyzing epoxidation, hydroxylation, dealkylation, and desulfurization (Attia et al., 2014; Tabrez and Ahmad, 2010). While most conversions detoxify, sometime harmful products are produced (bioactivation, see Section 2.8). Hepatic CYP1A protein levels are considered a sensitive biomarker of exposure. Enzyme activities also are used as biomarkers, e.g., aryl hydrocarbon hydroxylase (AHH) activity (Payne, 1976) and ethoxyresorufin O-deethylase (EROD) activity (Goksøyr and Förlin, 1992; Van der Oost et al., 2003). EROD activity is considered as one of the most reliable and widespread biomarkers for the estimation of organic pollutants (Fatima and Ahmad, 2005; Hassan et al., 2015; Stegeman and Hahn, 1994). Phase II biotransformation involves conjugation with an endogenous ligand (e.g., glutathione (GSH) and glucuronic acid (GA)) to further aid excretion (Commandeur et al., 1995).

## 2.8 DNA and tissue damage

Genotoxicity is caused by mutagenic and carcinogenic stressors that exert direct, or indirect oxidative stress effects. For example, polycyclic aromatic hydrocarbon (PAH) metabolism by CYP1A can generate activated metabolites that form DNA adducts (damaged DNA caused by the binding of a chemical to a segment of DNA) or to enhanced oxidative DNA damage caused by ROS production (Stegeman and Hahn, 1994; Van der Oost et al., 2003). Measurement of DNA adducts in liver tissue is a sensitive biomarker for genotoxicity in fish (Pampanin and Sydnes, 2013), albeit a complex one involving either a $^{32}$P-postlabeling assay or mass spectrometry analysis (Pampanin et al., 2017). Biomarkers of potential or actual genotoxic effect can be measured more easily by the alkaline Comet (Singh et al., 1988) or micronucleus (Carrasco et al., 1990; Hooftman and de Raat, 1982) assays. The Comet assay measures DNA strand breakage in single cells such as nucleated fish erythrocytes (Martins and Costa, 2015), while the micronucleus test is an index of chromosomal breaking by measuring fragments of chromatin that are separated from the nucleus (Bolognesi and Hayashi, 2011; Udroiu, 2006).

Substantial DNA damage, oxidative stress, osmotic stress and disruption of mitochondrial function can lead to either programmed apoptotic cell death or cell and tissue necrosis. Apoptosis cell death typically shows as cell shrinking, condensed chromatin, nuclear fragmentation and formation of apoptotic bodies eliminated via phagocytosis. Conversely, necrosis is characterized by swelling of the cells, nuclear dissolution and clumping of the chromatin, swelling and/or rupture of mitochondria and lysis of the plasma membrane (AnvariFar et al., 2017). Standardizing the analysis of these histopathological analysis is important because standardized control tissue pictures are not available (Wolf and Wheeler, 2018). Likewise, methods for standardizing histopathological analysis gill, liver and other tissues have been suggested (Bernet et al., 1999).

## 2.9 Neurotoxicity and behavior

Behavior is the most sensitive, most integrative and most ecologically relevant of the biomarkers of aquatic pollutants on fish populations (Scott and Sloman, 2004). Foraging, migration, predator avoidance, reproductive and social behavior are key to successful growth, reproduction and survival. Behavior, by integrating the outcomes of most of the physiological processes mentioned above (stress hormones, metabolism, gonadal or thyroid hormone levels, disease, impaired sensory input and neurotoxic effects), effectively links physiological function with ecological processes. Indeed, behavioral tests can be 10–1000 times more sensitive than lethality tests (Hellou, 2011; Hellou et al., 2008; Robinson, 2009). However, the potential exists for the experimental conditions themselves to influence behavior, especially acclimation and observation times (Melvin et al., 2017). Behavioral aquatic tests, therefore, should have a mechanistic basis, allow accurate predictions, represent what happens in the field, besides being reproducible and reliable. Promising behavioral tests include measurements of social cohesion and anxiety, such as shoaling fish and the novel tank diving test respectively (Parker, 2016) and both have potential for automation and high-throughput implementation. Automated image analysis has enabled standardization of high-throughput simple behavioral tests, such as swim speed, distance moved, activity levels, spatial distribution patterns, feeding rates, etc. (Krzykwa and Jeffries, 2020; Melvin and Wilson, 2013).

Avoidance behaviors, such as coughs, and body tremors, are useful biomarkers (for review see Tierney, 2016), but behavioral tests can also imply more complex behavioral patterns such as disruption of behaviors associated with foraging, predator avoidance, reproduction, and social hierarchies as their impact on fish populations is more important (for reviews see Kasumyan, 2001; Scott and Sloman, 2004). More complex behavior patterns can be broken down (Scott and Sloman, 2004), e.g., predator avoidance can be tested as reduced responses to chemical signals of predation threat (alarm substances), reduced responses to visual signals of predation threat, locomotory ability to escape predation and/or schooling behavior. Reproductive behaviors such as nest building, spawning, courtship behaviors, spawning site selection and natal homing can be measured too. Social behavior can be studied as agonistic acts, or the formation and maintenance of hierarchies.

Causes for changed behaviors can be implied. Sensory input such as olfaction can be impacted (reviewed by Tierney et al., 2010); pollutants may cause cell death or move along olfactory system neurons by axonal transport mechanisms. Endocrine disruption alters reproductive behavior either by acting as agonists or antagonists to endogenous hormones, or by interfering with the production or catabolism of hormones (reviewed by Söffker and Tyler, 2012), or by altering neural function. Neurotoxicants such as Hg seem to affect all these processes (Pereira et al., 2019). Choline esterase measurements, e.g., acetylcholinesterase (AChE) and butyrylcholinesterase (BChE) are well-established biomarkers to assess alterations to neural function. AChE and BChE break down the neurotransmitters acetylcholine and butyrylcholine, respectively, in neural synaptical clefts thereby ending signal transmission. Such generalized physiological processes allow comparisons among fish species. Both AChE and BChE are affected by pesticides such as organophosphates and carbamates (reviewed by Fulton and Key, 2009; Santana et al., 2021; Vieira and Nunes, 2021).

# 3 Adverse outcomes at the organismal level

## 3.1 Species sensitivity distribution (SSD) curves

Fish species differ in their likelihood of the occurrence of MIEs or KEs and so their sensitivity to environmental stress can differ even when looking at survival, growth and reproduction. Some substances such as non-polar organic chemicals that interact generally and non-specifically with lipid bilayers of cell membranes, a highly conserved structure, causing a narcotic effect that are unlikely differ considerable across species. Conversely, metals, pesticides and pharmaceuticals require binding to specific receptors or ligands to exert their toxic action and can be expected to be much more species-specific (Escher et al., 2010).

In general, two important, non-competing processes cause toxicant resistance: increased detoxification mechanisms and decreased sensitivity of target-sites, including uptake mechanisms (Doering et al., 2019; Spurgeon et al., 2020). Increased detoxification includes changes in expression and efficiency of esterases such as acetylcholinesterase, monooxygenases such as cytochrome P450, glutathione-s-transferases and metallothioneins. For example, the sevenfold difference in Cu toxicity between rainbow trout, common carp and crucian carp is best explained by the rate of activation of defence pathways and not necessarily by total accumulated metal or the total amount of defensive proteins (De Boeck et al., 2004). The more resistant crucian carp showed a strong positive correlation between newly induced MT and tissue metal levels (De Boeck et al., 2003a), a faster enhancement of antioxidant enzymes (Eyckmans et al., 2011), and stored more of the accumulated Cu in the biologically inactive pool where the metal is neutralized by MT or in metal-rich granules (Eyckmans et al., 2012). In contrast, the response of the more sensitive rainbow trout to the newly accumulated Cu was slow or absent even though it had slower gill Cu accumulation rates, and a higher baseline gill MT and GSH levels; most of the Cu in the gill was present in a biologically active pool containing heat denaturable protein such as enzymes and cell organelles (De Boeck et al., 2003a, 2004; Eyckmans et al., 2011, 2012). Decreased sensitivity of target sites can be caused by reduced numbers or point mutations in receptors and/or binding ligands.

Species sensitivity distributions (SSDs) summarize variations in sensitivity to a given toxicant across biological species (Posthuma et al., 2002) by assuming they follow a parametric statistical distribution function, i.e., a normal distribution. SSDs are formulated as the cumulative probability of toxicity as a function of toxicant concentration in the exposure medium. The toxicity data are usually obtained from laboratory tests for each species under optimal conditions and are assumed to be representative of a given target community. The concentration at which 5% of the biological species are affected ($HC_5$) is typically used for the setting of environmental quality standards (EQSs), often in combination with arbitrary "assessment factors" to account for data gaps and arbitrary acute to chronic conversion factors (Chung et al., 2021; European Commission, 2018). The outcome of the approach depends on the number of biological species included as well as the number, nature and timescale (acute vs chronic) of the toxicological endpoints considered, e.g., mortality, reproduction, olfaction, oxidative stress, behavior, etc., and the statistical model applied to fit the data. The latter point is the subject of extensive debate (Aldenberg et al., 2002; Newman et al., 2002). For example, a general assumption is that toxicity data follow a unimodal normal distribution; yet other distribution models may describe the data equally well. Indeed, a single-fit normal SSD model is often not the most appropriate (Newman et al., 2002). Furthermore, classical goodness-of-fit tests are symmetrical with respect to the data, and application of such tests to the entire SSD curve may not be appropriate especially when the most sensitive responses are of interest, i.e., the low percentile estimates (crucial for the setting of EQS) depend most strongly on the data points in the foot of the curve (Aldenberg et al., 2002). Different approaches deal with uncertainty in the data and in the parameters derived from the various statistical models used to fit the SSD curve (Aldenberg et al., 2002). Furthermore, the significance of chemical reactivity in the exposure medium and bioavailability are not considered. For example, in the case of metals, typically the dissolved or the free metal ion concentration is plotted on the *x*-axis, and factors which may exacerbate or moderate toxicity are not taken into account (temperature, pH, dissolved organic carbon (DOC), mixtures of metals and other chemicals, salinity, etc.).

## 3.2 Intraspecific variation in sensitivity

Beyond interspecific variation, intraspecific variation can exist among populations. Variations in life history or developmental life stages add further complexity in trying to arrive at a species-specific sensitivity, even for key traits such as temperature tolerance (McKenzie et al., 2021). Intraspecific variation is based on genetical heritable variation among populations or individuals, or variation in phenotypic plasticity, or differences in ontogenetic development.

In terms of ontogeny, earlier life stages are generally more sensitive to environmental stress. This sensitivity is related, in part, to their higher surface area to volume ratio which favors toxicant uptake, and in part to an underdeveloped detoxification system in larval fish (Tang et al., 2016). For this reason, ecotoxicology tends toward testing either larvae as the most sensitive life stage, e.g., in Fish Early Life Stage (FELS) toxicity testing (OECD Test No. 210, US EPA Test Guidelines OPPTS 850.1400), or Fish Embryo Tests (FET) (OECD Test No. 236) to assess effluent toxicity. Also, non-feeding developmental stages of fish are not protected stages according to European Directive 2010/63/EU on the protection of animals used for scientific purposes (European Commission, 2010), and are thus not considered as experimental animals. Therefore, they can be used as an alternative for fish acute toxicity testing (OECD Test No. 203, US EPA Test Guidelines OPPTS 850.1075) of chemicals and effluents within the framework of the Registration, Evaluation, Authorization, and Restriction of Chemicals (REACH) regulation of the European Union (European Commission, 2006) and other national and international legislations. Overall, there seems to be a good correlation between the fish embryo test and the conventional acute fish toxicity test (Braunbeck et al., 2015; Busquet et al., 2014), with FELS be more sensitive than FET because larvae tend to be more sensitive than embryos, or juveniles (Stelzer et al., 2018). The impermeability of the eggshell (chorion) prior to hatch to uptake of high molecular weight compounds is a primary reason for their reduced sensitivity. The sensitivity of FET can be increased by 30% by prolonging the exposure duration (from 48 to 96 h) and adding simple sublethal metrics such as pericardial oedema, nonhitching, and immobility (Stelzer et al., 2018). Neurotoxic substances are less toxic in the embryo than in adult fish (Klüver et al., 2015) even though their biotransformation capacity is lower (Klüver et al., 2014; Knöbel et al., 2012). Endocrine disrupting and other non-narcotic chemicals have a particularly reduced sensitivity of FET (Teixidó et al., 2020). For these reasons, FET is useful preliminary screening that can substantially reduce the number of experimental animals used for other regulatory testing.

Early in life sensitivity is clearly a bottleneck for development of fish populations in contaminated areas, even if adult fish can thrive. But early life may not be the most sensitive life-stage across species. This fact was elegantly shown when metal toxicity was compared across different life stages of white sturgeon (*Acipenser transmontanus*) and rainbow trout. Both species had similar sensitivities to Cd during the larval stage. Rainbow trout increased Cd sensitivity in later life stages, as affinity for Cd uptake and Cd-induced disruption of calcium ion homeostasis increased (Shekh et al., 2018). White sturgeon were more sensitive to Cu in the early life stages and became more tolerant in older life stages (Calfee et al., 2014; Vardy et al., 2013). Perhaps AOP and trait-based approaches looking at molecular pathways will help predict differences in sensitivity among fish species or life stages. If physiological and/or biochemical traits in a species can predict its sensitivity, phylogenetic relationships to other species of known sensitivity are a promising tool to estimate the sensitivity of a species or life stage whenever this information is lacking (Guénard et al., 2011). Unfortunately, an extensive study exposing embryonic and larval fish of six freshwater species that spanned five phylogenetic families and three phylogenetic orders to ecologically relevant concentrations of an herbicide did not reveal a direct correlation with phylogenetic proximity (Dehnert et al., 2021).

## 3.3 Trait-based approaches

Insights about the mechanisms driving sensitivity toward a given toxicant can be obtained by characterizing the traits of the more sensitive species even though SSDs make no assumptions about what they might be. Such a trait-based approach, often derived from ecotoxicological studies on model organisms, that identifies factors driving differences in sensitivity could eventually provide better predictions of unknown sensitivities for other species. A "trait" refers to a specific characteristic of an organism and encompasses physical/physiological and behavioral features. Included are body size, life cycle (timescales of growth and maturation, lifespan), salinity and temperature tolerance. For example, the decreased abundance of clupeiforms and perciforms within the Terminos lagoon, Mexico could be better predicted for a 30-year period by including maximal salinity tolerance (Sirot et al., 2015). Likewise, nitrite toxicity can be largely explained by the abundance of gill chloride channels through which $NO_2^-$ can be taken up in the more sensitive species (Spurgeon et al., 2020). Antioxidant capacity or capacity for metabolic biotransformation or detoxification are other important traits in sensitivity. To date, trait-based approaches in fish ecotoxicology have received scant attention (McKenzie et al., 2007) but are used with invertebrates (Ippolito et al., 2012; Van den Berg et al., 2020).

Among the few studies directly comparing similar toxicant exposures on different fish species, sublethal waterborne Cu exposure decreased $U_{crit}$ more so in rainbow trout than common carp (*Cyprinus carpio*) or gibel carp (*Carassius auratus gibelio*). Different tissue dependent Cu accumulation patterns was the explanation (De Boeck et al., 2006). Also, it is the rate of the sodium ion ($Na^+$) turnover and loss that largely determines the greater mortality of rainbow trout during Cu exposure compared with yellow perch or gibel carp (De Boeck et al., 2010; Pyle and Wood, 2008; Taylor et al., 2003), even though both have a maximal tolerable sodium loss of 30–40%. Rainbow trout lost a greater percentage of plasma $Na^+$ in the first 3 days than did common carp and gibel carp (De Boeck et al., 2007). Sea lamprey (*Petromyzon marinus*) sensitivity to lampricides was ascribed to high energy substrate depletion and lactate accumulation whereas bluegill (*Lepomis macrochirus*) insensitivity was ascribed to their high detoxification capacity (Lawrence et al., 2021).

Understanding the mechanistic traits that cause interspecific differences in sensitivity may help reduce the number of fish species tested which would help development of protective water quality guidelines and successful conservation measures. A good starting point would be understanding the reasons behind the 10- to 300-fold differences in acutely lethal concentrations between model species in ecotoxicology, such as rainbow trout, bluegill, guppy (*Poecilia reticulata*), carp, fathead minnow, medaka (*Oryzias latipes*), ide or orfe (*Leuciscus idus*) and zebra fish (Vittozzi and De Angeli, 1991). Knowing internal concentrations as well as epithelial surface-related toxicity, will be important because the body burden of the pollutant exerts a toxic effect within cells and organs and the internal concentration can deviate considerably from the exposure concentration. Therefore, toxicokinetics (absorption, distribution, metabolism, and excretion) will contribute to differences in species sensitivity (Escher et al., 2010) and are discussed in Section 5.

A fish's species (but not its family) is the second most important predictor of toxicity after exposure concentration (Tuulaikhuu et al., 2017). Within cyprinids, tench is a sensitive species while common carp and goldfish were more resistant (Tuulaikhuu et al., 2017). Yet different pollutants can still vary between species: rainbow trout and fathead minnow are equally sensitive to phthalates, but rainbow trout are about 10-times more sensitive to metals, while fathead minnows are more sensitive to hydrocarbons (Teather and Parrott, 2006). Likewise, the trait being studied can alter sensitivity rankings: pesticide toxicity across 21 European field populations was greatest for European minnow (*Phoxinus phoxinus*) for effects on fertility, was greatest for European brook lamprey (*Lampetra planeri*) for effects on juvenile survival and was greatest for Norther pike (*Esox lucius*) for effects on adult survival (Ibrahim et al., 2014).

# 4 Adverse outcomes from individual to population levels

Ecological situations beyond single compound exposures, e.g., mixtures of different metals or mixtures of metals and organics, as well as the effects of other stressors such as temperature and salinity, are typically not well documented or predicted by current models based on the studies described in the preceding sections. Furthermore, ecologically realistic factors such as hydrodynamic flow, light availability, food availability, etc., are rarely factored in to controlled lab studies. At the level of risk assessment, site-specific testing strategies offer a practical way to account for effects of combined stressors.

The challenge of understanding how fish will respond to the mixture of toxicants in an ecologically relevant situation will require the necessary step of linking different levels of biological organization, from molecular to population and ecosystem level. Studies in small or larger artificial ponds, micro- and mesocosms, a first step toward more complex interactions in the field, have been recommended for several decades (Crane, 1997). Such studies control over some of the exposure parameters (water quality, exposure levels) while introducing aquatic communities consisting of multiple species of different phylogeny and at different trophic levels. Thus, long-term studies of natural stressors in combination with pollutants on the functions and structures of aquatic communities under environmentally realistic conditions are possible (Brock et al., 2015; Daam et al., 2011; Van den Brink, 2006) which can then be bridged with laboratory experiments and field observations (Cardinale, 2011).

Studying changes in aquatic communities or ecosystems use two complementary assessments: structural and functional responses (Mathers et al., 2019). Structural changes describe patterns related to alterations of species composition, e.g., of biodiversity and absence or presence of species. Functional changes on the other hand describe processes related to alterations of the state of the whole community or ecosystem, e.g., primary production, decomposition and nitrification. However, upper trophic levels are often limited to different species of arthropods, not fish, in microcosms and even mesocosms due to size limitations. Yet, the importance of including fish is evident from some mesocosm experiments studying effects of climate change and nutrient loading using stickleback which predicted severe effects on fish survival (and predation) (Moran et al., 2010; Moss, 2010), but then other studies using bighead carp or crucian carp did not (Gu et al., 2018; He et al., 2018). Thus, using generalities about consequences to fish based on limited studies can be dangerous.

## 4.1 Index of biotic integrity

The "Index of biotic integrity" (IBI, reviewed by Ruaro and Gubiani, 2013) uses species-specific sensitivity to assess the ecological quality and biotic integrity of aquatic environments in field studies. Biotic integrity was defined as "the ability to support and maintain a balanced, integrated, adaptive community of organisms having a species composition, diversity, and functional organization comparable to that of the natural habitat of the region" (Frey, 1977; Karr and Dudley, 1981). IBI, first developed by Karr (Karr, 1981; Karr et al., 1986) for riverine fish assemblages in Illinois, uses a set of ten to twelve individual metrics, those related to taxonomic richness, functional groups, and community composition in the fish assemblage, to assess water quality and has since been more widely adopted to other fishes, macroinvertebrates, macrophytes and phytoplankton in rivers, lakes and transitional waters (Canning and Death, 2019). While past studies have largely focussed on freshwater systems in temperate areas, site- and ecosystem-specific IBIs are needed especially for transitional, estuarine waters and tropical areas (reviewed by Souza and Vianna, 2020). Studies are rare even for river system transitions from freshwater to estuarine in temperate areas (O'Brien et al., 2016). Extrapolating laboratory exposures from temperate to tropical areas for species-sensitivity distributions (SSDs) must be done with extreme caution, e.g., an SSD-based conclusion that the relatively small difference between temperate and tropical saltwater species' acute sensitivity to chemicals could be assigned a safe temperate-to-tropical extrapolation factor of two (Wang et al., 2014) later increased to five simply after including a more diverse dataset (Wang et al., 2019). IBIs, in contrast, are not suited to detect even large changes in abundance when different species decline to the same extent and the ratios between species do not change. Besides, when species respond differently it may take time before effects emerge against background variability (Gessner and Chauvet, 2002; Trebitz et al., 2003). Therefore, indicators at individual or population levels might provide greater sensitively for use as early warning signals (Corsi et al., 2003; O'Brien et al., 2016).

## 4.2 Passive and active biomonitoring of pollutants in the field

Field monitoring programs can assess the chemical and biological quality of water bodies and include a multi-biomarker approach for fish. Conventional monitoring relies solely on periodic collections of water and sediment samples. These are snapshots of the degree of contamination that do not accurately represent the bioavailable fraction of the pollutants (Kördel et al., 2013), nor the physiological effects potentially experienced by the biota (Delahaut et al., 2019). Passive biomonitoring, an integrative measure of bioavailable pollutants in water and sediment to these biota, provides an alternative approach. It includes evaluation of feral or native biota (bioindicators) which are collected at their natural site. Presence or absence of sensitive species and bioaccumulation of pollutants are linked to water quality evaluation. However, life-history traits, seasonal factors, and the presence and mobility of species may make it difficult to implement this type of monitoring in all sites of interest (Delahaut et al., 2019). Active biomonitoring, transplantation of a particular species from a reference site or a culture to the locations of interest, can overcome these limitations (Bervoets et al., 2004a,b; Ji et al., 2010; Oikari, 2006). In doing so, species size, age and gender are controlled, and the field experiment is easily replicated without issues of test organism availability (Besse et al., 2012). Physiological and biochemical biomarkers then can be compared with greater confidence to provide an early warning sign before higher levels of biological organization such as survival, populations and ecosystems are affected (Depledge and Fossi, 1994; Monserrat et al., 2007). Combining active and passive biomonitoring allows comparison of adaptive traits seen in relatively short-term acclimatization of transplanted organisms to possible multigenerational adaptation seen in passive monitoring (Lacroix et al., 2015) even though differentiating between genetic adaptation and phenotypic plasticity will be difficult without common garden experiments. Clearly active and passive biomonitoring with microcosms and mesocosms, as well as field studies provide a valuable link between laboratory based acute/chronic studies and landscape-level field observations, as shown in a series of probabilistic risk assessment of cotton pyrethroids (Giddings et al., 2001; Hendley et al., 2001; Maund et al., 2001; Solomon et al., 2001; Travis and Hendley, 2001).

## 5 Risk assessment and modeling: The challenge of linking exposure to effects

For a chemical to exert an adverse effect, it must contact an organism and interact with it, which initially takes place usually at the gills and skin (a water exposure) or the digestive tract (a food exposure). For hydrophilic pollutants water exposure is more important with the gill than the skin, while food exposure is more important for lipophilic compounds (Gobas et al., 2021; Mackay et al., 2013). Thus, an integrated framework linking exposure to effects must account for dynamic chemical speciation (distribution of a chemical over different physicochemical forms and their interconversion rates) both in the exposure medium, which influences bioavailability, and within an organism, which influences the nature, timescale and magnitude of adverse effects. Fig. 4 summarizes the various steps involved in the overall process from exposure to effects. The rate-limiting step in the overall process depends on the target chemical, the organism of interest, and the prevailing environmental conditions, e.g., water pH, salinity, concentration of DOC, etc.

Concepts for describing dynamic chemical speciation in aquatic media are well developed for inorganic and organic compounds, including for the case of particulate species (Duval et al., 2017, 2018; van Leeuwen et al., 2005, 2017; Zielińska et al., 2012). In brief, the potentially bioavailable concentration of a target chemical corresponds to the bioreactive form, e.g., the free metal ion, $M^{n+}$ or the neutral form of a free organic molecule, and the extent

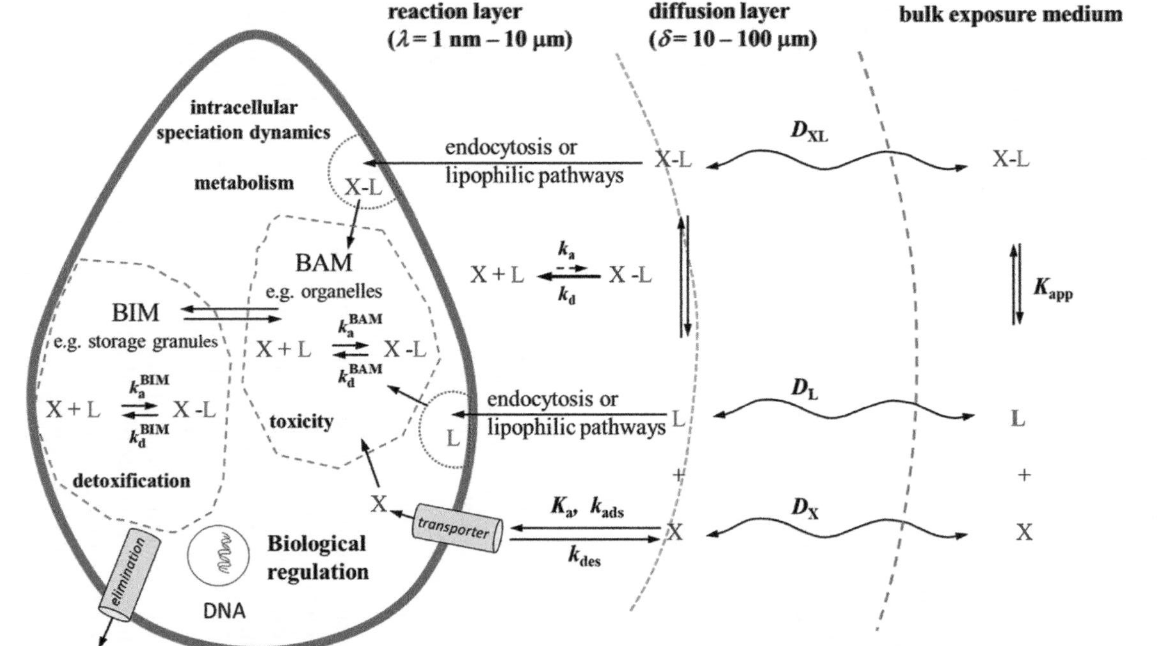


**FIG. 4** Schematic overview of the chain of processes involved in determining bioavailability, biouptake and effects. The framework applies to both aquatic exposure, in which case the bulk exposure medium is the surrounding aquatic medium, and to exposure via food, in which case the bulk exposure medium is the gut environment. X denotes a micropollutant (inorganic or organic) and L denotes an entity that associates with X to form a complex with apparent stability $K_{app}$, association rate constant, $k_a$ and dissociation rate constant, $k_d$. For illustrative purposes, L is arbitrarily shown as a macromolecular or particulate entity that may enter the cell via endocytosis or lipophilic pathways, and X is an ionic or polar species that enters the cell via transporters with affinity $K_a$ and adsorption/desorption rate constants $k_{ads}$ and $k_{des}$, respectively. $D$ represents the diffusion coefficients of the various species, $\delta$ is the diffusion layer thickness and $\lambda$ is the reaction layer thickness. Bioavailability is determined by the chemical speciation dynamics in the exposure medium ($D$, $K_{app}$, $k_a$, $k_d$) and the timescale of biouptake ($\delta$, $\lambda$, $K_a$, $k_{ads}$, $k_{des}$, and where relevant the rate constants for the endocytotic and/or lipophilic pathways). Upon biouptake, X initially enters the bioactive pool (BAM) and has the potential to exert toxic effects; subsequent transfer of X to the bioinactive pool (BIM) reduces potential toxicity. Biological species sensitivity reflects the capacity of the BIM and BAM compartments and the timescales for interconversion between these pools.

to which other physicochemical forms of the target chemical, e.g., metal complexes or organics sorbed on particles, can be converted into the bioreactive form on the timescale of the relevant biointerfacial process (van Leeuwen et al., 2005). Taking the example of metal ion biouptake at gills, determination of the potentially bioavailable concentration involves comparing the magnitude of the (unsupported) diffusive supply flux of the bioreactive free metal ion from the bulk medium toward the biointerface, $J_{diff}$, with the magnitude of the biouptake flux, $J_u$ (Jansen et al., 2002; van Leeuwen, 1999, 2000). If $J_{diff} \gg J_u$ (and in the absence of bulk depletion), then biouptake is governed by the concentration of the free metal ion; if $J_u \gg J_{dif}$, then the concentrations and kinetic features (lability) of all metal species in the exposure medium are called into play (Jansen et al., 2002; van Leeuwen, 1999, 2000). This fundamental consideration is largely ignored in aquatic (eco)toxicology and risk assessment.

Mechanistically linking chemical reactivity to biological reactivity is challenging because molecular level reactivity within organisms is largely unknown. Thus, simplifications are invoked in modeling the relationships between exposure and effects, as discussed below. Both bioavailability-based and effect-based models typically assume that chemical speciation in the exposure medium is at equilibrium and remains so throughout the exposure duration and that biological processes are overall rate-limiting. As noted above, such assumptions only hold under certain conditions, and compliance with the underlying dynamic criteria is generally not verified. A fish moving in and out of a polluted area might not reach an equilibrium state for example. Moreover, an overriding caveat is that any model could describe ecotoxicological data provided it includes a sufficiently large number of parameters. In many cases, empirical multiparametric curve fitting has provided as good an outcome, if not better, than models delineated below (Peters et al., 2021).

## 5.1 Bioavailability based models

### 5.1.1 Biotic ligand model

The biotic ligand model (BLM) describes the influence of site-specific water quality on metal ion bioavailability and toxicity (Di Toro et al., 2001; Niyogi and Wood, 2004a; Paquin et al., 2002). In brief, the BLM assumes that the bioavailable form is the *equilibrium* concentration of free metal ion in the *bulk* exposure medium. The free metal ion is assumed to bind to gill surface transporters (biotic ligands). Biotic ligands are envisaged to be ionoregulatory proteins (Di Toro et al., 2001; Paquin et al., 2002) with a fixed concentration and a single (conditional) affinity constant for the target metal ion (Playle et al., 1993). In reality, metal ion binding sites will have a distributed affinity, the average value of which will depend on the metal-to-binding site ratio, and their concentration may be up/down regulated in response to changing external and/or internal conditions. Indeed, both the conditional affinity and capacity of gill metal binding sites is modified by acclimation to water quality parameters (e.g., hardness) and to dietary composition (e.g., metal and essential ion content) (Niyogi and Wood, 2003; Niyogi et al., 2015). A BLM aims to account for the influence of chemical speciation on bioavailability via consideration of the influence of various factors (e.g., pH, DOC, concentration of major cations and anions) on the concentration of free metal ions in the bulk medium, and on the conditional affinity of biotic ligands (Mebane et al., 2020).

A BLM assumes that toxicity occurs once a certain amount of the biotic ligands on the gill surface (or another biological interface) are occupied by the target metal ion (USEPA, 2007). This is a simplification because metal accumulation does not always straightforwardly correlate with toxicity. Since binding to the biotic ligand is generally not measured, it is assumed that total metal content in gill tissue is a surrogate for the amount which is bound to the biotic ligand (Santore et al., 2001). Often acute toxicity is measured, and empirical conversion factors are used to convert the outcomes from acute to chronic effects (Chung et al., 2021).

Another assumption of a BLM is that the equilibrium concentration of the free metal ion is the relevant parameter for predictions of bioavailability and is only valid if certain dynamic criteria are met (see above; Slaveykova and Wilkinson, 2005). Such assumptions may apply for typical laboratory-based acute toxicity tests, but compliance with the underlying dynamic criteria becomes increasingly demanding under environmentally relevant conditions, i.e., low concentrations of the target bioreactive species (Zhao et al., 2016). A major shortcoming of a BLM in this regard is that the concentration of the free metal ion is computed by an equilibrium speciation code, typically WHAM, which is physicochemical unsound (Town et al., 2019). Thus, attempts to correlate free metal ion concentrations computed by WHAM with bioavailability and/or toxicity is an inherently physicochemically poor strategy which is bound to have weak predictive value. Indeed, empirical multiple linear regression has been found to perform as well as, or better than, a BLM (Peters et al., 2021). Software is available for implementation of BLMs, e.g., Bio-met (https:/bio-met.net/), and its simplistic basis is attractive to regulators. However, caution is needed because BLMs are generally biologically species-specific (Niyogi et al., 2004) and only applicable within the range of conditions and endpoints that were used to "validate" the model. For example, in the case of Cu toxicity toward fish, BLM-based criteria are not suitably protective for endpoints that were not considered in the model, e.g., behavior and chemo/mechanosensory responses (Meyer and DeForest, 2018; Morris et al., 2019). Also, several established BLMs for Cu were poorly applicable to Swedish soft freshwaters (Hoppe et al., 2015), and even when gill binding sites are known, the BLM failed to capture observed competitive effects between Cd, Zn, Cd, and Nickel (Niyogi et al., 2015).

Nevertheless, the BLM framework has highlighted the role of chemical speciation in determining metal bioavailability and, when the underlying dynamic criteria are met, the outcomes can be aligned with generally observed factors such as the role of increased major ion ($Ca^{2+}$, $Mg^{2+}$, $Na^{+}$, $K^{+}$) and DOC concentrations in reducing metal ion bioavailability (De Schamphelaere and Janssen, 2004; Esbaugh et al., 2011; Gheorghiu et al., 2010; Hollis et al., 1997; Kamunde and MacPhail, 2011; Li et al., 2016; Meinelt et al., 2001; Niyogi et al., 2008; Niyogi and Wood, 2004b; Richards and Playle, 1999; Richards et al., 1999). The BLM was originally formulated for fish gills; subsequently the approach has been applied to intestinal epithelia to model the effects of Ag on European flounder, *Platichthys flesus* (Hogstrand et al., 2002), and to olfactory epithelia to model the effect of Cu on fathead minnows (Dew et al., 2012).

### 5.1.2 Quantitative structure-activity relationships (QSARs)

QSARs are based on correlating structural features of chemicals with their effects on biota. Similar chemical structures are assumed to exhibit similar modes of action (MOA). Most QSARs are based on acute $LC_{50}$ data, which is an inherent limitation for ecological studies. Nevertheless, fish acute toxicity to some organic compounds can be predicted by the octanol-water partition coefficient ($K_{ow}$) of the chemical (Nendza et al., 2017).

Environmentally relevant chemicals can generally be classified as (non) polar baseline toxicants, oxidative phosphorylation uncouplers, photosynthesis inhibitors, acetylcholinesterase inhibitors, respiration inhibitors, thioalkylating agents, endocrine disruptors, and miscellaneous reactivity (irritants) (Nendza et al., 2017). The majority of QSARs to date have been developed for organic compounds, although the concept has also been applied to metals (Ownby and Newman, 2003) and many QSARs exist for fish toxicity (Belanger et al., 2016; Claeys et al., 2013; Huuskonen, 2003; Klüver et al., 2016; Levet et al., 2013; Netzeva et al., 2005; Pandey et al., 2020; Papa et al., 2005; Sangion and Gramatica, 2016; Sheffield and Judson, 2019; Yuan et al., 2007).

A given QSAR only holds for a given mode-of-action (MOA) and a given exposure time; multiple regression approaches have been developed to include compounds that exhibit more than one MOA (Freidig and Hermens, 2000; Sangion and Gramatica, 2016). Typically, application of a QSAR to a given set of chemicals yields many outliers (Lozano et al., 2010) and various "exclusion rules" are invoked to justify their elimination (Ellison et al., 2016; Nendza et al., 2018). A limitation of QSARs is that the MOA must be *a priori* known, and even then, the accuracy of the model may be poor (Melnikov et al., 2016). For example, the widely used OASIS (Russom et al., 1997), Cramer (Cramer et al., 1978), and Verhaar (Verhaar et al., 1992) MOA classification schemes have been reported to be unreliable for predicting the toxicity of some chemicals toward fish (Tuulaikhuu et al., 2017). Compounds with similar chemical structures and similar $K_{ow}$ can exhibit different toxicity toward a given biological species, and different biological species within the same taxonomic group can have very different sensitivity to the same compound. With most QSARs for fish being developed for a limited number of life stages and species, e.g., fathead minnow, such variability is of great concern in ecological context.

The limitations of QSARs underscore the need for mechanistic understanding of biological species sensitivity under environmentally relevant conditions and timescales.

## 5.2 Effect-based models

### *5.2.1 Toxicokinetic-toxicodynamic (TKTD) models*

Toxicokinetics (TK) refers to the uptake and distribution of chemicals within an organism (so-called ADME processes: absorption, distribution, metabolism, excretion) and toxicodynamics (TD) refers to the ensuing effects of chemicals from the cellular to the organism level. The terms "kinetic" and "dynamic" is used arbitrary and is at odds with the usual usage in a chemical reactivity context. Furthermore, chemical reactivity in the exposure medium is generally ignored. For example, a simple equilibrium model has been used to describe chemical speciation in water in a TKTD model for metal toxicity to fish (Feng et al., 2018; Gao et al., 2015, 2020) thereby constraining its validity (see Fig. 4). The timescale for both TK and TD processes can be most simply described by assuming a one-compartment model with first-order kinetics with biouptake proportional to the external concentration and elimination proportional to the internal concentration. Adverse effects are assumed to occur once a certain internal concentration is attained. When this approach does not accurately describe data, "damage" and "repair" variables which are ill-defined are included to improve data fit (Jager et al., 2011). In general, many parameters in these models cannot be directly measured; probabilistic statistics are used to implement the model, and parameterization is achieved by applying goodness-of-fit criteria (Ashauer et al., 2013; Baudrot and Charles, 2019; Jager, 2021). With survival as the endpoint, both stochastic death and individual tolerance, two very different strategies, give similar fits to the data, albeit with very different rate constants (Jager et al., 2011) and with different sensitivities to changes in parameter values (Ashauer et al., 2013). These findings underscore the empirical nature of the approach. The numerous applications of TKTD models to fish toxicity data include the effect of benzoviniflupyr on carp, and fathead minnow (Ashauer et al., 2013), the effect of pyrethroid insecticides on rainbow trout (Bradbury and Coats, 1989; Zimmer et al., 2018), the effect of Cd and Pb on zebrafish (Feng et al., 2018; Gao et al., 2015, 2020), and the effect of Ag on rainbow trout (Veltman et al., 2014).

The BIM-BAM model extends the TKTD concept by accounting for two toxicant pools within an organism (Adams et al., 2011). The biologically active (BAM) pool represents the portion of a toxicant that contributes to toxicity at the site of action and toxicity occurs once the concentration in the BAM exceeds a certain threshold value. For metals, BAM includes organelles, cell debris, and heat denaturable proteins, whereas the biologically inactive (BIM) pool includes metal-rich granules and heat-stable proteins such as metallothioneins. Adipose tissue might be considered a BIM pool for high $K_{ow}$ organics. BIM-BAM modeling has provided insights into the physiological basis for differences in sensitivity to Cu between rainbow trout, common carp, and gibel carp (Eyckmans et al., 2012). Moreover, BIM and BAM pools can be experimentally determined via subcellular fractionation protocols (Eyckmans et al., 2012).

### *5.2.2 Dynamic energy budget (DEB) models*

A dynamic energy budget (DEB) model attempts to describe rates of energy acquisition from food, and energy expenditure for somatic maintenance, growth, reproduction, and development (Jusup et al., 2017; Kearney, 2021). An organism is conceptually divided into two compartments: reserve and structure. The reserve compartment is assumed to respond to food availability and to not require physiological maintenance; the structural compartment is assumed to be involved in physiological functions and to require maintenance. Such a division is arbitrary without a straightforward means to quantify the magnitude of each compartment. Acquired energy is assumed to first enter the reserve compartment; a fixed fraction, $\kappa$, is used for somatic maintenance and growth and $1-\kappa$ is used for development and reproduction. Somatic maintenance is assumed to take priority over all other functions. Processes are described by simple differential equations (Jusup et al., 2017). DEB models invoke many assumptions including digestive efficiency, metabolic costs of food acquisition, and how the available energy is partitioned amongst physiological functions. The associated parameters—state variables, internal energy, material fluxes—are generally not directly measurable but rather derived by various means from, e.g., weight, length, fat content (Kooijman et al., 2008; Lika et al., 2014; Martin et al., 2017; Van der Meer, 2006).

A large and growing database of DEB parameters is available (Marques et al., 2018). Yet rates depend on the status of the organism (sex, growth stage, etc.) and the environmental conditions (food availability, temperature, salinity, presence of pollutants, etc.). A DEB assumes that metabolism is organized in a similar way across all organisms, albeit with species-specific parameters. Different sized organisms of the same species have different values for the state variables (reserve, structure, level of maturity). A fundamental limitation of DEB theory is that these state variables cannot be directly measured and so the equations used in DEB models are not mechanistically linked to dynamics at the cellular/molecular level (Nisbet, 2017).

The DEB framework describes toxicity in a non-specific manner at the level of changes in metabolic processes at the whole organism level. For example, under stressful conditions, energy allocations may become limited to maintenance because insufficient energy is available for growth and reproduction. While this approach rarely identifies the direct cause or mechanism of toxicity, ecologically relevant insights are possible, such as the relationship between sensitivity to chemicals and high maintenance rate (Baas and Kooijman, 2015). DEB models have described fish bioaccumulation and toxicity of chemicals, including the bioaccumulation of PCBs by European hake, *Merluccius merluccius* (Bodiguel et al., 2009), the effects of sodium pentachlorophenol, and PAHs on fathead minnows (Kooijman and Bedaux, 1996), benzo(k)-fluoranthene, phenanthrene on zebrafish (Kooijman and Bedaux, 1996), the effects of petroleum substances on cod, *Gadus morhua* (Klok et al., 2014), the effects of beta-cyfluthrin on rainbow trout (Zimmer et al., 2018), the effects of bisphenol A on three-spined sticklebacks (David et al., 2019), and the effects of flame retardants and temperature on juvenile white seabream *Diplodus sargus* (Anacleto et al., 2018). So-called DEBtox models couple the DEB framework for energy allocation with TKTD models (Jager, 2020).

## 6 Meta-analysis as a tool

Meta-analysis combines data from several sources to generate greater explanatory power, identify sources of variation across studies, and reveal effects not detected by individual studies. The primary aim of meta-analysis would be to test if pollutant effects are consistent across biological traits, taxonomic groups, habitat types or geographic regions. An overall mean estimate of treatment effects is calculated while exploring sources of variance across studies. By doing so, both general analogous responses, as well as outliers, can be identified. Most ecotoxicological studies are species-specific and it is therefore essential to synthesize effects across studies and taxonomic groups. It can generate quantitative answers to ecotoxicology challenges and inform evidence-based management practices. As such, meta-analysis is becoming increasingly prominent in ecotoxicology and is becoming a "gold-standard" tool for quantitative syntheses, preferred over narrative-style reviews.

Meta-analyses have provided insights into the effects of a range of pollutants on fishes, including heavy metals (Rodgers et al., 2019; Wang et al., 2016), endocrine disrupting chemicals (Wu and Seebacher, 2020), pesticides (Shuman-Goodier and Propper, 2016), noise pollution (Cox et al., 2018), and microplastics (Foley et al., 2018; Jacob et al., 2020). For example, a meta-analysis of traits that combined data from 179 studies, spanning 93 species, revealed that bisphenol A (BPA) from plastic pollution has consistent negative effects on aquatic organisms. BPA was found to increase abnormalities, disrupt critical behaviors, compromise the cardiovascular system and reduce both growth and survival (Wu and Seebacher, 2020). Similarly, a meta-analysis synthesizing data from 68 studies found that nitrate pollution reduces activity, growth and survival rates in fish (Gomez Isaza et al., 2020b).

Meta-analyses can identify critical pollutant thresholds for taxonomic groups or ecosystem types, such as predicted no effect concentrations (PNEC) or lowest-observed-effect concentrations (Ghose et al., 2014; Yang and Nowack, 2020). For instance, a recent meta-analysis identified PNECs for nano plastics in marine and freshwater ecosystems as 99 and 72 $\mu g\ L^{-1}$ respectively, providing tangible values for regulatory agencies (Yang and Nowack, 2020). Meta-analyses can also reveal which species or life-stages are most sensitive to pollutants. For instance, Wu and Seebacher (2020) showed that early life stages are most vulnerable to BPA, whereas Gomez Isaza et al. (2020b) showed that nitrate exposure influences all life stages except embryos. Identifying sensitive life-stages and species is critical to the development of informed management actions.

Highlighting research gaps and testing for publication biases are also key goals of many meta-analyses. For example, meta-analyses examining the effects of plastic pollution on aquatic organisms have found that research has been biased toward marine habitats (Bucci et al., 2020) and adult life-stages in fish (Jacob et al., 2020). Identifying these data gaps using quantitative methods provides valuable insight into where research efforts need to be directed. Publication bias, where studies reporting significant effects are more likely to be published than studies reporting null effects, can also be assessed. Identifying publication biases within a field can be essential to reversing this trend, so that studies with null results are equally valued.

The rising prominence of meta-analyses requires application rigor, e.g., including physiologically meaningful constraints. Ideally, meta-analyses are accompanied by a systematic literature review, and methods are sufficiently detailed to ensure reproducibility (see O'Dea et al., 2021). Meta-data are required from each study, including mean estimates, sample sizes, and variances (standard deviations, confidence intervals). Sample sizes and variances are necessary to estimate the precision of each study, so that more precise studies, with larger samples sizes and lower variance, are given greater weighting. The inclusion of explanatory (i.e., moderator) variables, such as pollutant concentration, exposure periods, and species age, sex or size, are also essential and may reveal sources of variance across studies. For instance, a meta-analysis revealed that contaminants elevate cortisol levels in teleosts, but the strength of this effect was moderated by exposure durations and contaminant class (Rohonczy et al., 2021). Studies that include a range of species must also account for phylogenetic relationships in statistical models and relative species sensitivities. Best practice recommendations for performing and reviewing biological meta-analyses are available (see Nakagawa et al., 2017; O'Dea et al., 2021), as well as guidelines for data reporting in ecotoxicological studies so they can be included in meta-analyses (see Hitchcock et al., 2018). Although there are many examples of ecotoxicological data being incorporated into biological opinion, recovery plans and conservation actions, this integration is often limited in its scope or depth. As such, there is a strong need for more meta-analyses to answer long-standing questions and reveal broad scale effects of pollutants. This includes not only a way to utilize the wealth of already published data, but also a process to identify knowledge gaps, with the overall goal in increasing the efficacy of future regulatory and research efforts.

## References

Adams, S.M., Ham, K.D., 2011. Application of biochemical and physiological indicators for assessing recovery of fish populations in a disturbed stream. Environ. Manag. 47, 1047–1063.

Adams, W.J., Blust, R., Borgmann, U., Brix, K.V., DeForest, D.K., Green, A.S., Meyer, J.S., McGeer, J.C., Paquin, P.R., Rainbow, P.S., Wood, C.M., 2011. Utility of tissue residues for predicting effects of metals on aquatic organisms. Integr. Environ. Assess. Manag. 7, 75–98.

Aerts, J., 2018. Quantification of a glucocorticoid profile in non-pooled samples is pivotal in stress research across Vertebrates. Front. Endocrinol. 9, 635.

Aerts, J., Metz, J.R., Ampe, B., Decostere, A., Flik, G., Saeger, S.D., 2015. Scales tell a story on the stress history of fish. PLoS One 10, e0123411.

Ahmed, I., Reshi, Q.M., Fazio, F., 2020. The influence of the endogenous and exogenous factors on hematological parameters in different fish species: a review. Aquac. Int. 28, 869–899.

Akerboom, T.P., Sies, H., 1981. Assay of glutathione, glutathione disulfide, and glutathione mixed disulfides in biological samples. Methods Enzymol. 77, 373–382.

Aldenberg, T., Jaworska, J.S., Traas, T.P., 2002. Normal species sensitivity distributions and probabilistic ecological risk assessment. In: Posthuma, L., Suter, G.W., Traas, T.P. (Eds.), Species Sensitivity Distributions in Ecotoxicology. CRC Press, Boca Raton, pp. 49–102.

Alexander, J., Ingram, C.A., 1992. Non-cellular non-specific defense mechanisms of fish. Ann. Rev. Fis.h Dis. 2, 249–277.

Alsop, D.H., Kieffer, J.D., Wood, C.M., 1999. The effects of temperature and swimming speed on instantaneous fuel use and nitrogenous waste excretion of the Nile tilapia. Physiol. Biochem. Zool. 72, 474–483.

Altenburger, R., Walter, H., Grote, M., 2004. What contributes to the combined effect of a complex mixture? Environ. Sci. Technol. 38, 6353–6362.

Álvarez-Muñoz, D., Farré, M., 2020. Environmental Metabolomics: Applications in Field and Laboratory Studies to Understand From Exposome to Metabolome. Elsevier Inc, ISBN: 978-0-12-818196-6, pp. 1–341.

Anacleto, P., Figueiredo, C., Baptista, M., Maulvault, A.L., Camacho, C., Pousão-Ferreira, P., Valente, L.M.P., Marques, A., Rosa, R., 2018. Fish energy budget under ocean warming and flame retardant exposure. Environ. Res. 164, 186–196.

Andrewartha, S.J., Munns, S.L., Edwards, A., 2016. Calibration of the HemoCue point-of-care analyser for determining haemoglobin concentration in a lizard and a fish. Conserv. Physiol. 4 (1), cow006.

Ankley, G.T., Johnson, R.D., 2004. Small fish models for identifying and assessing the effects of endocrine-disrupting chemicals. Inter. Lab. Anim. Res. J. 45, 467–481.

Ankley, G.T., Jensen, K.M., Kahl, M.D., Korte, J.J., Makynen, E.A., 2001. Description and evaluation of a short-term reproduction test with the fathead minnow (*Pimephales promelas*). Environ. Toxicol. Chem. 20, 1276–1290.

Ankley, G.T., Jensen, K.M., Makynen, E.A., Kahl, M.D., Korte, J.J., Hornung, M.W., et al., 2003. Effects of the androgenic growth promoter 17-β-trenbolone on fecundity and reproductive endocrinology of the fathead minnow (*Pimephales promelas*). Environ. Toxicol. Chem. 22, 1350–1360.

Ankley, G.T., Bennett, R.S., Erickson, R.J., Hoff, D.J., Hornung, M.W., Johnson, R.D., Mount, D.-R., Nichols, J.W., Russom, C.L., Schmieder, P.K., Serrano, P.K., Tietge, J.E., Villeneuve, D.-L., 2010. Adverse outcome pathways: a conceptual framework to support ecotoxicology research and risk assessment. Environ. Toxicol. Chem. 29, 730–741.

AnvariFar, H., Amirkolaie, A.K., Miandare, H.K., Ouraji, H., Jalali, M.A., Üçüncü, S.I., 2017. Apoptosis in fish: environmental factors and programmed cell death. Cell Tissue Res. 368, 425–439.

Ashauer, R., Thorbek, P., Warinton, J.S., Wheeler, J.R., Maund, S., 2013. A method to predict and understand fish survival under dynamic chemical stress using standard ecotoxicity data. Environ. Toxicol. Chem. 32, 954–965.

Attia, T.Z., Yamashita, T., Hammad, M.A., Hayasaki, A., Sato, T., Miyamoto, M., et al., 2014. Effect of cytochrome P450 2C19 and 2C9 amino acid residues 72 and 241 on metabolism of tricyclic antidepressant drugs. Chem. Pharm. Bull. 62, 176–181.

Baas, J., Kooijman, S.A.L.M., 2015. Sensitivity of animals to chemical compounds links to metabolic rate. Ecotoxicology 24, 657–663.

Bartosz, G., 2009. Reactive oxygen species: destroyers or messengers? Biochem. Pharmacol. 77, 1303–1315.

Basu, N., Kennedy, C.J., Iwama, G.K., 2003. The effects of stress on the association between hsp70 and the glucocorticoid receptor in rainbow trout. Comp. Biochem. Physiol. 134A, 655–663.

Baudrot, V., Charles, S., 2019. Recommendations to address uncertainties in environmental risk assessment using toxicokinetic-toxicodynamic models. Sci. Rep. 9, 11432.

Bautista, N.M., Crespel, A., Crossley, J., Padilla, P., Burggren, W., 2020. Parental transgenerational epigenetic inheritance related to dietary crude oil exposure in *Danio rerio*. J. Exp. Biol. 223 (16), jeb.222224.

Beauchamp, C., Fridovich, I., 1971. Superoxide dismutase: improved assays and an assay applicable to acrylamide gels. Anal. Biochem. 44, 276–287.

Belanger, S.E., Brill, J.L., Rawlings, J.M., McDonough, K.M., Zoller, A.C., Wehmeyer, K.R., 2016. Aquatic toxicity structure-activity relationships for the zwitterionic surfactant alkyl dimethyl amine oxide to several aquatic species and a resulting species sensitivity distribution. Ecotoxicol. Environ. Saf. 134, 95–105.

Bell, G., Gonzalez, A., 2009. Evolutionary rescue can prevent extinction following environmental change. Ecol. Lett. 12 (9), 942–948.

Benzie, I.F.F., Strain, J.J., 1996. The ferric reducing ability of plasma (FRAP) as a measure of "antioxidant power": the FRAP assay. Anal. Biochem. 239 (1), 70–76.

Bernet, D., Schmidt, H., Meier, W., Burkhardt-Holm, P., Wahli, T., 1999. Histopathology in fish: proposal for a protocol to assess aquatic pollution. J. Fish Dis. 22, 25–34.

Bertotto, D., Poltronieri, C., Negrato, E., Majolini, D., Radaelli, G., Simontacchi, C., 2010. Alternative matrices for cortisol measurement in fish. Aquacult. Res. 41, 1261–1267.

Bervoets, L., Blust, R., 2003. Metal concentrations in water, sediment and gudgeon (*Gobio gobio*) from a pollution gradient: relationship with fish condition factor. Environ. Pollut. 126, 9–19.
Bervoets, L., Meregalli, G., De Cooman, W., Goddeeris, B., Blust, R., 2004a. Caged midge larvae (*Chironomus riparius*) for the assessment of metal bioaccumulation from sediments in situ. *Environ. Toxicol. Chem.* 23, 443–454.
Bervoets, L., Voets, J., Chu, S., Covaci, A., Schepens, P., Blust, R., 2004b. Comparison of accumulation of micropollutants between indigenous and transplanted zebra mussels (*Dreissena polymorpha*). *Environ. Toxicol. Chem.* 23, 1973–1983.
Besse, J.-P., Geffard, O., Coquery, M., 2012. Relevance and applicability of active biomonitoring in continental waters under the Water Framework Directive. Trends Anal. Chem. 36, 113–127.
Birnie-Gauvin, K., Costantini, D., Cooke, S.J., Willmore, W.G., 2017. A comparative and evolutionary approach to oxidative stress in fish: a review. *Fish Fish.* 18, 928–942.
Bodiguel, X., Maury, O., Mellon-Duval, C., Roupsard, F., Le Guellec, A.-M., Loizeau, V., 2009. A dynamic and mechanistic model of PCB bioaccumulation in the European hake (*Merluccius merluccius*). J. Sea Res. 62, 124–134.
Bojarski, B., Witeska, M., 2020. Blood biomarkers of herbicide, insecticide, and fungicide toxicity to fish—a review. Environ. Sci. Pollut. Res. 27, 19236–19250.
Bolognesi, C., Hayashi, M., 2011. Micronucleus assay in aquatic animals. Mutagenesis 26, 205–213.
Bols, N.C., Brubacher, J.L., Ganassin, R.C., Lee, L.E.J., 2001. Ecotoxicology and innate immunity in fish. Dev. Comp. Immunol. 25, 853–873.
Bozinovic, G., Oleksiak, M.F., 2010. Embryonic gene expression among pollutant resistant and sensitive *Fundulus heteroclitus* populations. Aquat. Toxicol. 98 (3), 221–229. https://doi.org/10.1016/j.aquatox.2010.02.022.
Bozinovic, F., Pörtner, H.-O., 2015. Physiological ecology meets climate change. Ecol. Evol. 5, 1025–1030.
Bradbury, S.P., Coats, J.R., 1989. Toxicokinetics and toxicodynamics of pyrethroid insecticides in fish. Environ. Toxicol. Chem. 8, 373–380.
Brady, S.P., Richardson, J.L., Kunz, B.K., 2017. Incorporating evolutionary insights to improve ecotoxicology for freshwater species. Evol. Appl. 10 (8), 829–838. https://doi.org/10.1111/eva.12507.
Braunbeck, T., Kais, B., Lammer, E., Otte, J., Schneider, K., Stengel, D., Strecker, R., 2015. The fish embryo test (FET): origin, applications, and future. *Environ. Sci. Pollut. Res.* 22, 16247–16261.
Brett, J.R., 1964. The respiratory metabolism and swimming performance of young sockeye salmon. J. Fish. Res. Board Can. 21, 1183–1226.
Brock, T.C.M., Hammers-Wirtz, M., Hommen, U., Preuss, T.G., Ratte, H.T., Roessink, I., Strauss, T., Van den Brink, P., 2015. The minimum detectable difference (MDD) and the interpretation of treatment-related effects of pesticides in experimental ecosystems. Environ. Pollut. Sci. Res. 22, 1160–1174.
Brosset, P., Cooke, S.J., Schull, Q., Trenkel, V.M., Soudant, P., Lebigre, C., 2021. Physiological biomarkers and fisheries management. Rev. Fish Biol. Fish. 31, 797–819.
Bucci, K., Tulio, M., Rochman, C.M., 2020. What is known and unknown about the effects of plastic pollution: a meta-analysis and systematic review. Ecol. Appl. 30 (2), e02044.
Busquet, F., Strecker, R., Rawlings, J.M., Belanger, S.E., Braunbeck, T., Carr, G.J., Cenijn, P., Fochtman, P., Gourmelon, A., Hübler, N., Kleensang, A., Knöbel, M., Kussatz, C., Legler, J., Lillicrap, A., Martínez-Jerónimo, F., Polleichtner, C., Rzodeczko, H., Salinas, E.,

Schneider, K.E., Scholz, S., van den Brandhof, E.-J., van der Ven, L.T.M., Walter-Rohde, S., Weigt, S., Witters, H., Halder, M., 2014. OECD validation study to assess intra and inter-laboratory reproducibility of the zebrafish embryo toxicity test for acute aquatic toxicity testing. Regul. Toxicol. Pharmacol. 69, 496–511.

Calboli, F.C.F., Delahaut, V., Deflem, I., Hablützel, P.I., Hellemans, B., Kordas, A., Raeymaekers, J.A.M., Bervoets, L., De Boeck, G., Volckaert, F.A.M., 2021. Association between Chromosome 4 and mercury accumulation in muscle of the three-spined stickleback (*Gasterosteus aculeatus*). Evol. Appl. https://doi.org/10.1111/eva.13298.

Calfee, R.D., Little, E.E., Puglis, H.J., Scott, E., Brumbaugh, W.G., Mebane, C.A., 2014. Acute sensitivity of white sturgeon (*Acipenser transmontanus*) and rainbow trout (*Oncorhynchus mykiss*) to copper, cadmium, or zinc in water-only laboratory exposures. Environ. Toxicol. Chem. 33 (10), 2259–2272.

Callaghan, N.I., Allen, G.J.P., Robart, T.E., Dieni, C.A., MacCormack, T.J., 2016. Zinc oxide nanoparticles trigger cardiorespiratory stress and reduce aerobic scope in the white sucker, *Catostomus commersonii.* Nanoimpact 2, 29–37.

Campbell, P.M., Pottinger, T.G., Sumpter, J.P., 1994. Preliminary evidence that chronic confinement stress reduces the quality of gametes produced by brown and rainbow trout. Aquaculture 120, 151–169.

Canning, A.D., Death, R.G., 2019. Ecosystem health indicators—freshwater environments. In: Fath, B. (Ed.), Encyclopedia of Ecology, second ed. Elsevier, pp. 46–60.

Cao, G., Alessio, H.M., Culter, R., 1993. Oxygen-radical absorbance capacity assay for antioxidants. Free Radic. Biol. Med. 14, 303–311.

Cao, Y., Tveten, A.-K., Stene, A., 2017. Establishment of a non-invasive method for stress evaluation in farmed salmon based on direct fecal corticoid metabolites measurement. Fish Shellfish Immunol. 66, 317–324.

Carbajal, A., Soler, P., Tallo-Parra, O., Isasa, M., Echevarria, C., Lopez-Bejar, M., Vinyoles, D., 2019. Towards non-invasive methods in measuring fish welfare: the measurement of cortisol concentrations in fish skin mucus as a biomarker of habitat quality. Animals 9, 939.

Cardinale, B.J., 2011. Biodiversity improves water quality through niche partitioning. Nature 472, 86–U113.

Carlberg, I.N.C.E.R., Mannervik, B.E.N.G.T., 1975. Purification and characterization of the flavoenzyme glutathione reductase from rat liver. J. Biol. Chem. 250 (14), 5475–5480.

Carrasco, K.R., Tilbury, K.L., Myers, M.S., 1990. Assessment of the piscine micronucleus test as an in situ biological indicator of chemical contaminant effects. *Can. J. Fish. Aquat. Sci.* 47, 2123–2136.

Cereja, R., 2020. Critical thermal maxima in aquatic ectotherms. Ecol. Indic. 119, 106856.

Chabot, D., Steffensen, J.F., Farrell, A.P., 2016. The determination of thestandard metabolic rate in fishes. J. Fish Biol. 88, 81–121.

Chabot, D., Zhang, Y., Farrell, A.P., 2021. Valid oxygen uptake measurements: using high $r^2$ values with good intentions can bias upward the determination of standard metabolic rate. J. Fish Biol. 98, 1206–1216.

Chellapa, S., Huntingford, F.A., Strang, R.H.C., Thomson, R.Y., 1995. Condition factor and hepatosomatic index as estimates of energy status in male three-spined stickleback. J. Fish Biol. 47, 775–787.

Chung, J., Hwang, D., Park, D.-H., An, Y.-J., Yeom, D.-H., Park, T.-J., Choi, J., Lee, J.-H., 2021. Derivation of acute copper biotic ligand model-based predicted no-effect concentrations and acute-chronic ratio. Sci. Total Environ. 780, 146425.

Claeys, L., Iaccino, F., Janssen, C.J., van sprang, P., Verdonck, F., 2013. Development and validation of a quantitative structure-activity relationship for chronic narcosis to fish. Environ. Toxicol. Chem. 32, 2217–2225.
Claireaux, G., Chabot, D., 2016. Responses by fishes to environmental hypoxia: integration through Fry's concept of aerobic metabolic scope. J. Fish Biol. 88, 232–251.
Claireaux, G., Désaunay, Y., Akcha, F., Aupérin, B., Bocquené, G., Budzinski, H., Cravedi, J.-P., Davoodi, F., Galois, R., Gilliers, C., Goanvec, C., Guérault, D., Imbert, N., Mazéas, O., Nonnotte, G., Nonnotte, L., Prunet, P., Sébert, P., Vettier, A., 2004. Influence of oil exposure on the physiology and ecology of the common sole *Solea solea*: experimental and field approaches. Aquat. Living Resour. 17, 335–351.
Clark, T.D., Eliason, E.J., Sandblom, E., Hinch, S.G., Farrell, A.P., 2008. Calibration of a hand-held haemoglobin analyser for use on fish blood. J. Fish Biol. 73, 2587–2595.
Clark, T.D., Sandblom, E., Jutfelt, F., 2013. Aerobic scope measurements of fishes in an era of climate change: respirometry, relevance and recommendations. J. Exp. Biol. 216, 2771–2782.
Coady, K.K., Biever, R.C., Denslow, N.D., Gross, M., Guiney, P.D., Holbech, H., Karouna-Renier, N.K., Katsiadaki, I., Krueger, H., Levine, S.L., Maack, G., Williams, M., Wolf, J.C., Ankley, G.T., 2017. Current limitations and recommendations to improve testing for the environmental assessment of endocrine active substances. *Integr. Environ. Assess. Manag.* 13, 302–316.
Commandeur, J.N.M., Stijntjes, G.J., Vermeulen, N.P.E., 1995. Enzymes and transport systems involved in the formation and disposition of glutathione S-conjugates. Role in bioactivation and detoxication mechanisms of xenobiotics. Pharmacol. Rev. 47, 271–330.
Commission, E., 2006. Regulation (EC) 1907/2006 of the European Parliament and of the Council of 18 December 2006 concerning the registration, evaluation, authorization and restriction of chemicals (REACH), establishing a European Chemicals Agency, amending Directive 1999/45/EC and repealing Council Regulation (EEC) No. 793/93 and Commission Regulation (EC) No. 1488/94 as well as Council Directive 76/769/EEC and Commission Directives 91/155/EEC, 93/67/EEC, 93/105/EC and 2000/21/EC. Off. J. Eur. Union L136, 3–280.
Commission, E., 2010. Directive 2010/63/EU of the European Parliament and Council of 22 September 2010 on the protection of animals used for scientific purposes. Off. J. Eur. Union L276, 33–79.
Cooke, S.J., Crossin, G.T., Patterson, D.A., English, K.K., Hinch, S.G., Young, J.L., Alexander, R.F., Healey, M.C., Van Der Kraak, G., Farrell, A.P., 2005. Coupling non-invasive physiological assessments with telemetry to understand inter-individual variation in behaviour and survivorship of sockeye salmon: development and validation of a technique. J. Fish Biol. 67, 1342–1358.
Cooke, S.J., Hinch, S.G., Crossin, G.T., Patterson, D.A., English, K.K., Healey, M.C., Shrimpton, J.M., Van Der Kraak, G., Farrell, A.P., 2006. Mechanistic basis of individual mortality in pacific salmon during spawning migrations. Ecology 87 (6), 1575–1586.
Cooke, S.J., Sack, L., Franklin, C.E., Farrell, A.P., Beardall, J., Wikelski, M., Chown, S.L., 2013. What is conservation physiology? Perspectives on an increasingly integrated and essential science. Conserv. Physiol. 1, cot001.
Corriere, M., Baptista, M., Paula, J.R., Repolho, T., Rosa, R., Costa, P.R., Soliño, L., 2020. Impaired fish swimming performance following dietary exposure to the marine phycotoxin okadaic acid. Toxicon 179, 53–59.
Corsi, I., Mariottini, M., Sensini, C., Lancini, L., Focardi, S., 2003. Fish as bioindicators of brackish ecosystem health: integrating biomarker responses and target pollutant concentrations. Oceanol. Acta 26, 129–138.

Cox, K., Brennan, L.P., Gerwing, T.G., Dudas, S.E., Juanes, F., 2018. Sound the alarm: a meta-analysis on the effect of aquatic noise on fish behavior and physiology. Glob. Chang. Biol. 24 (7), 3105–3116.

Cramer, G.M., Ford, R.A., Hall, R.L., 1978. Estimation of toxic hazard—a decision tree approach *Food Cosmet.* Toxicology 16, 255–276.

Crane, M., 1997. Research needs for predictive multispecies tests in aquatic toxicology. Hydrobiologia 346, 149–155.

Crossin, G.T., Hinch, S.G., 2005. A Nonlethal, rapid method for assessing the somatic energy content of migrating adult pacific salmon. Trans. Am. Fish. Soc. 134 (1), 184–191.

Daam, M.A., Cerejeira, M.J., Van den Brink, P.J., Brock, T.C.M., 2011. Is it possible to extrapolate results of aquatic microcosm and mesocosm experiments with pesticides between climate zones in Europe? Environ. Sci. Pollut. Res. 18, 123–126.

David, V., Joachim, S., Porcher, J.-M., Beaudouin, R., 2019. Modelling BPA effects on the three-spined stickleback population dynamics in mesocosms to improve understanding of population effects. Sci. Total Environ. 692, 854–867.

De Boeck, G., De Smet, H., Blust, R., 1995. The effect of sublethal levels of copper on oxygen consumption and ammonia excretion in the common carp, *Cyprinus carpio*. Aquat. Toxicol. 32, 127–141.

De Boeck, G., Alsop, D., Wood, C.M., 2001. Cortisol effects on aerobic and anaerobic metabolism, nitrogen excretion, and whole-body composition in juvenile rainbow trout. Physiol. Biochem. Zool. 74, 858–868.

De Boeck, G., Ngo, T.T.H., Van Campenhout, K., Blust, R., 2003a. Differential metallothionein induction patterns in three freshwater fish during sublethal copper exposure. *Aquat. Toxicol.* 65, 413–424.

De Boeck, G., De Wachter, B., Vlaeminck, A., Blust, R., 2003b. Effect of cortisol treatment and/or sublethal copper exposure on copper uptake and heat shock protein levels in common carp, *Cyprinus carpio*. Environ. Toxicol. Chem. 22, 1122–1126.

De Boeck, G., Meeus, W., De Coen, W., Blust, R., 2004. Tissue-specific Cu bioaccumulation patterns and differences in sensitivity to waterborne Cu in three freshwater fish: rainbow trout (*Oncorhynchus mykiss*), common carp (*Cyprinus carpio*) and gibel carp (*Carassius auratus gibelio*). *Aquat. Toxicol.* 70, 179–188.

De Boeck, G., van der Ven, K., Hattink, J., Blust, R., 2006. Swimming performance and energy metabolism of rainbow trout, common carp and gibel carp respond differently to sublethal copper exposure. Aquat. Toxicol. 80, 92–100.

De Boeck, G., van der Ven, K., Meeus, W., Blust, R., 2007. Sublethal copper exposure induces respiratory stress in common and gibel carp but not in rainbow trout. Comp. Biochem. Physiol. C 144, 380–390.

De Boeck, G., Smolders, R., Blust, R., 2010. Copper toxicity in gibel carp *Carassius auratus gibelio*: importance of sodium and glycogen. Comp. Biochem. Physiol. C 152, 332–337.

De Jonge, M., Belpaire, C., Van Thuyne, G., Breine, J., Bervoets, L., 2015. Temporal distribution of accumulated metal mixtures in two feral fish species and the relation with condition metrics and community structure. Environ. Pollut. 197, 43–54.

De Schamphelaere, K.A.C., Janssen, C.R., 2004. Bioavailability and chronic toxicity of zinc to juvenile rainbow trout (*Oncorhynchus mykiss*): comparison with other fish species and development of a biotic ligand model. Environ. Sci. Technol. 38, 6201–6209.

Deane, E.E., Woo, N.Y.S., 2011. Advances and perspectives on the regulation and expression of piscine heat shock proteins. Rev. Fish Biol. Fish. 21, 153–185.

Decourten, B.M., Forbes, J.P., Roark, H.K., Burns, N.P., Major, K.M., White, J.W., Li, J., Mehinto, A.C., Connon, R.E., Brander, S.M., 2020. Multigenerational and transgenerational effects of environmentally relevant concentrations of endocrine disruptors in an estuarine fish model. Environ. Sci. Tech. 54 (21), 13849–13860.

Dehnert, G.K., Freitas, M.B., Sharma, P.P., Barry, T.P., Karasov, W.H., 2021. Impacts of subchronic exposure to a commercial 2,4-D herbicide on developmental stages of multiple freshwater fish species. Chemosphere 263, 127638.

Delahaut, V., Daelemans, O., Sinha, A.K., De Boeck, G., Bervoets, L., 2019. A multibiomarker approach for evaluating environmental contamination: common carp transplanted along a gradient of metal pollution. Sci. Total Environ. 669, 481–492.

Depledge, M.H., Fossi, M.C., 1994. The role of biomarkers in environmental assessment. Invertebrates. Ecotoxicology 3, 161–172.

Derocles, S.A.P., Bohan, D.A., Dumbrell, A.J., Kitson, J.J.N., Massol, F., Pauvert, C., Plantegenest, M., Vacher, C., Evans, D.M., 2018. Biomonitoring for the 21st century: integrating next-generation sequencing into ecological network analysis. Adv. Ecol. Res. 58, 1–62.

Deutsch, M.J., Weeks, C.E., 1965. Micro-fluorometric assay for vitamin C. J. Assoc. Off. Agric. Chem. 48, 1248–1256.

Dew, W.A., Wood, C.M., Pyle, G.G., 2012. Effects of continuous copper exposure and calcium on the olfactory response of fathead minnows. Environ. Sci. Technol. 46, 9019–9026.

Di Toro, D.M., Allen, H.E., Bergman, H.L., Meyer, J.S., Paquin, P.R., Santore, R.C., 2001. Biotic ligand model of the acute toxicity of metals. 1. Technical basis. Environ. Toxicol. Chem. 20, 2383–2396.

Doering, J.A., Villeneuve, D.L., Fay, K.A., Randolph, E.C., Jensen, K.M., Kahl, M.D., LaLone, C.A., Ankley, G.T., 2019. Differential sensitivity to in vitro inhibition of cytochrome P450 aromatase (cyp19) activity among 18 freshwater fishes. Toxicol. Sci. 170, 394–403.

Draper, H.H., Squires, E.J., Mahmooch, H., Wu, J., Agarwal, S., Handley, M., 1993. A comparative evaluation of thiobarbituric acid methods for the determination of malondialdehyde in biological materials. Free Radic. Biol. Med. 15, 353–363.

Duval, J.F.L., Town, R.M., van Leeuwen, H.P., 2017. Applicability of the reaction layer principle to nanoparticulate metal complexes at a macroscopic (bio)interface: a theoretical study. J. Phys. Chem. C 121, 19147–19161.

Duval, J.F.L., Town, R.M., van Leeuwen, H.P., 2018. Lability of nanoparticulate metal complexes at a macroscopic metal responsive (bio)interface: expression and asymptotic scaling laws. J. Phys. Chem. C 122, 6052–6065.

Eissa, N., Wang, H.P., 2014. Transcriptional stress responses to environmental and husbandry stressors in aquaculture species. Rev. Aquac. 8, 61–88.

Ellis, T., James, J.D., Stewart, C., Scott, A.P., 2004. A non-invasive stress assay based upon measurement of free cortisol released into the water by rainbow trout. J. Fish Biol. 65, 1233–1252.

Ellison, C.M., Piechota, P., Madden, J.C., Enoch, S.J., Cronin, M.T.D., 2016. Adverse outcome pathway (AOP) informed modelling of aquatic toxicology: QSARs, read-across, and interspecies verification of modes of action. Environ. Sci. Technol. 50, 3995–4007.

Esbaugh, A.J., Brix, K.V., Mager, E.M., Grosell, M., 2011. Multi-linear regression models predict the effects of water chemistry on acute lead toxicity to *Ceriodaphnia dubia* and *Pimephales promelas*. Comp. Biochem. Physiol. C 154, 137–145.

Escher, B.I., Ashauer, R., Dyer, S., Hermens, J.L.M., Lee, J.-H., Leslie, H.A., Mayer, P., Meador, J.P., Warne, M.S.J., 2010. Crucial role of mechanisms and modes of toxic action for understanding tissue residue toxicity and internal effect concentrations of organic chemicals. Integr. Environ. Assess. Manag. 7, 28–49.

European Commission, 2018. Technical Guidance for Deriving Environmental Quality Standards. Guidance Document No. 27.

Eyckmans, M., Celis, N., Horemans, N., Blust, R., De Boeck, G., 2011. Exposure to waterborne copper reveals differences in oxidative stress response in three freshwater fish species. *Aquat. Toxicol.* 103, 112–120.

Eyckmans, M., Blust, R., De Boeck, G., 2012. Subcellular differences in handling Cu excess in three freshwater fish species contributes greatly to their differences in sensitivity to Cu. Aquat. Toxicol. 118–119, 97–107.

Farrell, A.P., 2008. Comparisons of swimming performance in rainbow trout using constant acceleration and critical swimming speed tests. J. Fish Biol. 72, 693–710.

Farrell, A.P., 2016. Pragmatic perspective on aerobic scope: peaking, plummeting, pejus and apportioning. J. Fish Biol. 88, 322–343.

Farrell, A.P., Gamperl, A.K., Birtwell, I.K., 1998. Prolonged swimming, recovery and repeat swimming performance of mature sockeye salmon *Oncorhynchus nerka* exposed to moderate hypoxia and pentachlorophenol. J. Exp. Biol. 201, 2183–2193.

Fatima, R.A., Ahmad, M., 2005. Certain antioxidant enzymes of Allium cepa as biomarkers for the detection of toxic heavy metals in wastewater. Sci. Total Environ. 346, 256–273.

Feng, J., Gao, Y., Chen, M., Xu, X., Huang, M., Yang, T., Chen, N., Zhu, L., 2018. Predicting cadmium and lead toxicities in zebrafish (*Danio rerio*) larvae by using a toxicokinetic-toxicodynamic model that considers the effects of cations. Sci. Total Environ. 625, 1584–1595.

Fenske, M., Segner, H., 2004. Aromatase modulation alters gonadal differentiation in developing zebrafish (*Danio rerio*). Aquat. Toxicol. 67, 105–126.

Fischer, U., Utke, K., Somamoto, T., Koellner, B., Ototake, M., Nakanishi, T., 2006. Cytotoxic activities of fish leukocytes. Fish Shellfish Immunol. 20, 209–226.

Flik, G., Klaren, P.H.M., Van den Burg, E.H., Metz, J.R., Huising, M.O., 2006. CRF and stress in fish. *Gen. Comp. Endocrinol.* 146, 36–44.

Flohé, L., Günzler, W.A., 1984. Assays of glutathione peroxidase. Methods Enzymol. 105, 114–120.

Foley, C.J., Feiner, Z.S., Malinich, T.D., Höök, T.O., 2018. A meta-analysis of the effects of exposure to microplastics on fish and aquatic invertebrates. Sci. Total Environ. 631-632, 550–559.

Folmar, L.C., 1993. Effects of chemical contaminants on blood chemistry of teleost fish: a bibliography and synopsis of selected effects. Environ. Toxicol. Chem. 12, 337–375.

Freidig, A.P., Hermens, J.L.M., 2000. Narcosis and chemical reactivity QSARs for acute fish toxicity. Quant. Struct. Act. Relat. 19, 547–553.

Frey, D., 1977. Biological integrity of water: an historical approach. In: Ballentine, R.K., Guarraia, L.J. (Eds.), The Integrity of Water. Proceedings of a Symposium, March 10–12, 1975. U.S. Environmental Protection Agency, Washington, DC, pp. 127–140.

Fry, F.E.J., 1947. Effects of the environment on animal activity. *Publ. Ontario Fish. Res. Lab.* 68, 1–52.

Fry, F.E.J., Hart, J.S., 1948. The relation of temperature to oxygen consumption in the goldfish. *Biol. Bull.* 94, 66–77.

Fulton, M.H., Key, P.B., 2009. Acetylcholinesterase inhibition in estuarine fish and invertebrates as an indicator of organophosphorus insecticide exposure and effects. Environ. Toxicol. Chem. 20, 37–45.

Galbreath, P.F., Adams, N.D., Martin, T.H., 2004. Influence of heating rate on measurement of time to thermal maximum in trout. Aquaculture 241, 587–599.

Gao, Y., Feng, J., Zhu, L., 2015. Prediction of acute toxicity of cadmium and lead to zebrafish larvae by using a refined toxicokinetic-toxicodynamic model. Aquat. Toxicol. 169, 37–45.

Gao, Y., Feng, J., Zhu, J., Zhu, L., 2020. Predicting copper toxicity in zebrafish larvae under complex water chemistry conditions by using a toxicokinetic-toxicodynamic model. J. Hazard. Mater. 400, 123205.

Garcia-Reyero, N., Perkins, E.J., 2011. Systems biology: leading the revolution in ecotoxicology. *Environ. Toxicol. Chem.* 30, 265–273.

Gessner, M.O., Chauvet, E., 2002. A case for using litter breakdown to assess functional stream integrity. *Ecol. Appl.* 12, 498–510.

Gesto, M., Hernández, J., López-Patiño, M.A., Soengas, J.L., Míguez, J.M., 2015. Is gill cortisol concentration a good acute stress indicator in fish? A study in rainbow trout and zebrafish. Comp. Biochem. Physiol. A188, 65–69.

Gheorghiu, C., Smith, D.S., Al-Reasi, H.A., McGeer, J.C., Wilkie, M.P., 2010. Influence of natural organic matter (NOM) quality on Cu-gill binding in the rainbow trout (*Oncorhynchus mykiss*). Aquat. Toxicol. 97, 343–352.

Ghose, S.L., Donnelly, M.A., Kerby, J., Whitfield, S.M., 2014. Acute toxicity tests and meta-analysis identify gaps in tropical ecotoxicology for amphibians. Environ. Toxicol. Chem. 33 (9), 2114–2119.

Giddings, J.M., Solomon, K.R., Maund, S.J., 2001. Probabilistic risk assessment of cotton pyrethroids: II. Aquatic mesocosm and field studies. Environ. Toxicol. Chem. 20, 660–668.

Gobas, F.A.P.C., Lee, Y.-S., Arnot, J.A., 2021. Normalizing the biomagnification factor. Environ. Toxicol. Chem. 40, 1204–1211.

Goksøyr, A., Förlin, L., 1992. The cytochrome P450 system in fish, aquatic toxicology and environmental monitoring. Aquat. Toxicol. 22, 287–312.

Gomez Isaza, D.F., Cramp, R.L., Franklin, C.E., 2020a. Thermal acclimation offsets the negative effects of nitrate on aerobic scope and performance. J. Exp. Biol. 223. jeb224444.016.

Gomez Isaza, D.F., Cramp, R.L., Franklin, C.E., 2020b. Living in polluted waters: a meta-analysis of the effects of nitrate and interactions with other environmental stressors on freshwater taxa. Environ. Pollut. 261, 114091.

Gorissen, M., Flik, G., 2016. The endocrinology of the stress response in fish: an adaptation-physiological view. In: Schreck, C.B., Tort, L., Farrell, A.P., Brauner, C.J. (Eds.), Biology of Stress in Fish. Fish Physiology Series, vol. 35, pp. 75–111.

Groh, K.J., Carvalho, R.N., Chipman, J.K., Denslow, N.D., Halder, M., Murphy, C.A., Roelofs, D., Rolaki, A., Schirmer, K., Watanabe, K.H., 2015a. Development and application of the adverse outcome pathway framework for understanding and predicting chronic toxicity: I. Challenges and research needs in ecotoxicology. Chemosphere 120, 764–777.

Groh, K.J., Carvalho, R.N., Chipman, J.K., Denslow, N.D., Halder, M., Murphy, C.A., Roelofs, D., Rolaki, A., Schirmer, K., Watanabe, K.H., 2015b. Development and application of the adverse outcome pathway framework for understanding and predicting chronic toxicity: II. A focus on growth impairment in fish. Chemosphere 120, 778–792.

Gu, J., He, H., Jin, H., Yu, J., Jeppesen, E., Nairn, R.W., Li, K., 2018. Synergistic negative effects of small-sized benthivorous fish and nitrogen loading on the growth of submerged macrophytes—relevance for shallow lake restoration. Sci. Total Environ. 610–611, 1572–1580.

Guardiola, F.A., Cuesta, A., Esteban, M.A., 2016. Using skin mucus to evaluate stress in gilthead seabream (*Sparus aurata* L.). Fish Shellfish Immunol. 59, 323e330.

Tuulaikhuu, B.-A., Guasch, H., García-Berthou, E., 2017. Examining predictors of chemical toxicity in freshwater fish using the random forest technique. Environ. Sci. Pollut. Res. 24, 10172–10181.

Guénard, G., Von Der Ohe, P.C., De Zwart, D., Legendre, P., Lek, S., 2011. Using phylogenetic information to predict species tolerances to toxic chemicals. *Ecol. Appl.* 21 (8), 3178–3190.

Guerrera, M.C., Aragona, M., Porcino, C., Fazio, F., Laurà, R., Levanti, M., Montalbano, G., Germanà, G., Abbate, F., Germanà, A., 2021. Micro and nano plastics distribution in fish as model organisms: histopathology, blood response and bioaccumulation in different organs. *Appl. Sci.* 11, 5768.

Halsey, L.G., Killen, S.S., Clark, T.D., Norin, T., 2018. Exploring key issues of aerobic scope interpretation in ectotherms: absolute versus factorial. Rev. Fish Biol. Fish. 28, 405–415.

Harries, J.E., Runnalls, T., Hill, E., Harris, C.A., Maddix, S., Sumpter, J.P., et al., 2000. Development of a reproductive performance test for endocrine disrupting chemicals using pair-breeding fathead minnow (*Pimephales promelas*). Environ. Sci. Technol. 34, 3003–3011.

Harter, T.S., Shartau, R.B., Brauner, C.J., Farrell, A.P., 2014. Validation of the i-STAT system for the analysis of blood parameters in fish. *Conserv. Physiol.* 2, cou037.

Hassan, I., Jabir, N.R., Ahmad, S., Shah, A., Tabrez, S., 2015. Certain phase I and II enzymes as toxicity biomarker: an overview. Water Air Soil Pollut. 226, 153.

He, H., Jin, H., Jeppesen, J., Li, K., Liu, Z., Zhang, Y., 2018. Fish-mediated plankton responses to increased temperature in subtropical aquatic mesocosm ecosystems: implications for lake management. Water Res. 144, 304e311.

Hellou, J., 2011. Behavioural ecotoxicology, an "early warning" signal to assess environmental quality. Environ. Sci. Pollut. Res. 18, 1–11.

Hellou, J., Cheeseman, K., Desnoyers, E., Johnston, D., Jouvenelle, M.L., Leonard, J., Robertson, S., Walker, P., 2008. A non-lethal chemically based approach to investigate the quality of harbour sediments. Sci. Total Environ. 389, 178–187.

Hellou, J., Ross, N.W., Moon, T.W., 2012. Glutathione, glutathione S-transferase, and glutathione conjugates, complementary markers of oxidative stress in aquatic biota. Environ. Sci. Pollut. Res. 19, 2007–2023.

Hendley, P., Holmes, C., Kay, S., Maund, S.J., Travis, K.Z., Zhang, M.H., 2001. Probabilistic risk assessment of cotton pyrethroids: III. A spatial analysis of the Mississippi, USA, cotton landscape. Environ. Toxicol. Chem. 20, 669–678.

Hiramatsu, N., Matsubara, T., Weber, G.M., Sullivan, C.V., Hara, A., 2002. Vitellogenesis in aquatic animals. Fish. Sci. 68 (Suppl. I), 694–699.

Hitchcock, D.J., Andersen, T., Varpe, Ø., Borgå, K., 2018. Improving data reporting in ecotoxicological studies. Environ. Sci. Tech. 52 (15), 8061–8062.

Ho, D.H., Burggren, W.W., 2010. Epigenetics and transgenerational transfer: a physiological perspective. J. Exp. Biol. 213 (1), 3–16. https://doi.org/10.1242/jeb.019752.

Hogstrand, C., Wood, C.M., Bury, N.R., Wilson, R.W., Rankin, J.C., Grosell, M., 2002. Binding and movement of silver in the intestinal epithelium of a marine teleost fish, the European flounder (*Platichthys flesus*). Comp. Biochem. Physiol. C 133, 125–135.

Hollis, L., Muench, L., Playle, R.C., 1997. Influence of dissolved organic matter on copper binding, and calcium binding, by gills of rainbow trout. J. Fish Biol. 50, 703–720.

Hooftman, R.N., de Raat, W.K., 1982. Induction of nuclear anomalies (micronuclei) in the peripheral blood erythrocytes of the eastern mudminnow *Umbra pygmaea* by ethyl methanesulphonate. *Mutat. Res.* 104, 147–152.

Hoppe, S., Gustafsson, J.-P., Borg, H., Breitholtz, M., 2015. Evaluation of current copper bioavailability tools for soft freshwaters in Sweden. Ecotoxicol. Environ. Saf. 114, 143–149.

Hutchinson, T.H., Ankley, G.T., Segner, H., Tyler, C.R., 2006. Screening and testing for endocrine disruption in fish—biomarkers as "signposts," not "traffic lights," in risk assessment. Environmental Health Perspectives 114 (Suppl. 1), 106–114.

Huuskonen, J., 2003. QSAR modelling with the electrotopological state indices: predicting the toxicity of organic chemicals. Chemosphere 50, 949–953.

Ibrahim, L., Preuss, T.G., Schaeffer, A., Hommen, U., 2014. A contribution to the identification of representative vulnerable fish species for pesticide risk assessment in Europe—a comparison of population resilience using matrix models. Ecol. Model. 280, 65–75.

Ippolito, A., Todeschini, R., Vighi, M., 2012. Sensitivity assessment of freshwater macroinvertebrates to pesticides using biological traits. Ecotoxicology 21, 336–352.

Iwasaki, Y., Saito, Y., Nakano, Y., Mochizuki, K., Sakata, O., Ito, R., Saito, K., Nakazawa, H., 2009. Chromatographic and mass spectrometric analysis of glutathione in biological samples. J. Chromatogr. 877B, 3309–3317.

Jacob, H., Besson, M., Swarzenski, P.W., Lecchini, D., Metian, M., 2020. Effects of virgin micro- and nanoplastics on fish: trends, meta-analysis, and perspectives. Environ. Sci. Tech. 54 (8), 4733–4745.

Jager, T., 2020. Revisiting simplified DEBtox modes for analysing ecotoxicity data. Ecol. Model. 416, 108904.

Jager, T., 2021. Robust likelihood-based approach for automated optimization and uncertainty analysis of toxicokinetic-toxicodynamic models. Integr. Environ. Assess. Manag. 17, 388–397.

Jager, T., Albert, C., Preuss, T.G., Ashauer, R., 2011. General unified threshold model of survival—a toxicokinetic-toxicodynamic framework for ecotoxicology. Environ. Sci. Technol. 45, 2529–2540.

Jain, K.E., Farrell, A.P., 2003. Influence of seasonal temperature on the repeat swimming performance of rainbow trout *Oncorhynchus mykiss*. J. Exp. Biol. 206, 3569–3579.

Jansen, S., Blust, R., van Leeuwen, H.P., 2002. Metal speciation dynamics and bioavailability: Zn(II) and Cd(II) uptake by mussel (*Mytilus edulis*) and carp (*Cyprinus carpio*). Environ. Sci. Technol. 36, 2164–2170.

Jeffrey, J.D., Hasler, C.T., Chapman, J.M., Cooke, S.J., Suski, C.D., 2015. Linking landscape-scale disturbances to stress and condition of fish: implications for restoration and conservation. Integr. Comp. Biol. 55, 618–630.

Ji, Y., Lu, G., Wang, C., Song, W., Wu, H., 2010. Fish transplantation and stress-related biomarkers as useful tools for assessing water quality. *J. Environ. Sci.* 22, 1826–1832.

Jobling, S., Nolan, M., Tyler, C.R., Brighty, G., Sumpter, J.P., 1998. Widespread sexual disruption in wild fish. Environ. Sci. Technol. 32, 2498–2506.

Johnson, C., Ivanisevic, J., Siuzdak, G., 2016. Metabolomics: beyond biomarkers and towards mechanisms. Nat. Rev. Mol. Cell Biol. 17, 451–459.

Jusup, M., Sousa, T., Domingos, T., Labinac, V., Marn, N., Wang, Z., Klanjšček, T., 2017. Physics of metabolic organization. Phys. Life Rev. 20, 1–39.

Kamunde, C., MacPhail, R., 2011. Effect of humic acid during concurrent chronic waterborne exposure of rainbow trout (*Oncorhynchus mykiss*) to copper, cadmium and zinc. Ecotoxicol. Environ. Saf. 74, 259–269.

Karr, J.R., 1981. Assessment of biotic integrity using fish communities. Fisheries 6 (6), 21–27.

Karr, J.R., Dudley, D.R., 1981. Ecological perspective on water quality goals. Environ. Manag. 5, 55–68.

Karr, J.R., Fausch, K.D., Angermeier, P.L., Yant, P.R., Schlosser, I.J., 1986. Assessing Biological Integrity in Running Waters: A Method and its Rationale. Special Publication 5, Illinois Natural History Survey, Champaigne, IL.

Kasai, H., 1997. Analysis of a form of oxidative DNA damage 8-hydroxy-2′-deoxyguanosine, as a marker of cellular oxidative stress during carcinogenesis. Mutat. Res. 387, 147–163.

Kasumyan, A.O., 2001. Effects of chemical pollutants on foraging behavior and sensitivity of fish to food stimuli. J. Ichthyol. 41, 76–87.

Kearney, M.R., 2021. What is the status of metabolic theory one century after Pütter invented the von Bertalanffy growth curve? Biol. Rev. 96, 557–575.

Kearney, M.R., Jusup, M., McGeoch, M.A., Kooijman, S.A.L.M., Chown, S.L., 2021. Where do functional traits come from? The role of theory and models. Funct. Ecol. 35, 1385–1396.

Kidd, K.A., Blanchfield, P.J., Mills, K.H., Palace, V.P., Evans, R.E., Lazorchak, J.M., Flick, R.W., 2007. Collapse of a fish population after exposure to a synthetic estrogen. Proc. Natl. Acad. Sci. U. S. A. 104, 8897–8901.

Kieffer, J.D., Alsop, D., Wood, C.M., 1998. A respirometric analysis of fuel use during aerobic swimming at different temperatures in rainbow trout (*Oncorhynchus mykiss*). J. Exp. Biol. 201, 3123–3133.

Klerks, P.L., Lentz, S.A., 1998. Resistance to lead and zinc in the western mosquitofish *Gambusia affinis* inhabiting contaminated Bayou Trepagnier. Ecotoxicology 7 (1), 11–17. https://doi.org/10.1023/A:1008851516544.

Klerks, P.L., Xie, L., Levinton, J.S., 2011. Quantitative genetics approaches to study evolutionary processes in ecotoxicology, a perspective from research on the evolution of resistance. Ecotoxicology 20 (3), 513–523. https://doi.org/10.1007/s10646-011-0640-2.

Klok, C., Nordtug, T., Tamis, J.E., 2014. Estimating the impact of petroleum substances on survival in early life stages of cod (*Gadus morhua*) using the dynamic energy budget theory. Mar. Environ. Res. 101, 60–68.

Klüver, N., Ortmann, J., Paschke, H., Renner, P., Ritter, A.P., Scholz, S., 2014. Transient overexpression of adh8a increases allyl alcohol toxicity in zebrafish embryos. PLoS One 9 (3), e90619.

Klüver, N., König, M., Ortmann, J., Massei, R., Yu, H., Paschke, A., Kühne, R., Scholz, S., 2015. Fish embryo toxicity test: identification of compounds with weak toxicity and analysis of behavioral effects to improve prediction of acute toxicity for neurotoxic compounds. *Environ. Sci. Technol.* 49, 7002–7011.

Klüver, N., Vogs, C., Altenburger, R., Escher, B.I., Scholz, S., 2016. Development of a general baseline toxicity QSAR model for the fish embryo acute toxicity test. Chemosphere 164, 164–173.

Knapen, D., Vergauwen, L., Villeneuve, D.L., Ankley, G.T., 2015. The potential of AOP networks for reproductive and developmental toxicity assay development. Reprod. Toxicol. 56, 52–55.

Knöbel, M., Busser, F.J.M., Rico-Rico, A., Kramer, N.I., Hermens, J.L.M., Hafner, C., Tanneberger, K., Schirmer, K., Scholz, S., 2012. Predicting adult fish acute lethality with the zebrafish embryo: relevance of test duration, endpoints, compound properties, and exposure concentration analysis. *Environ. Sci. Technol.* 46, 9690–9700.

Koellner, B., Wasserrab, B., Kotterba, G., Fischer, U., 2002. Evaluation of immune functions of rainbow trout (*Oncorhynchus mykiss*)—how can environmental influences be detected? Toxicol. Lett. 131, 83–95.

Köhler, H.R., Bartussek, C., Eckwert, H., Farian, K., Gränzer, S., Knigge, T., Kunz, N., 2001. The hepatic stress protein (hsp70) response to interacting abiotic parameters in fish exposed to various levels of pollution. J. Aquat. Ecosyst. Stress Recovery 8, 261–279.

Kooijman, S.A.L.M., Bedaux, J.J.M., 1996. Analysis of toxicity tests on fish growth. Water Res. 30, 1633–1644.

Kooijman, S.A.L.M., Sousa, T., Pecquerie, L., van der Meer, J., Jager, T., 2008. From food-dependent statistics to metabolic parameters, a practical guide to the use of dynamic energy budget theory. Biol. Rev. 83, 533–552.

Kördel, W., Garelick, H., Gawlik, B.M., Kandile, N.G., Peijnenburg, W.J.G.M., Rüdel, H., 2013. Substance-related environmental monitoring strategies regarding soil, groundwater and surface water—an overview. *Environ. Sci. Technol.* 20, 2810–2827.

Krewski, D., Acosta Jr., D., Andersen, M., Anderson, H., Bailar 3rd, J.C., Boekelheide, K., Brent, R., Charnley, G., Cheung, V.G., Green Jr., S., Kelsey, K.T., Kerkvliet, N.I., Li, A.A., McCray, L., Meyer, O., Patterson, R.D., Pennie, W., Scala, R.A., Solomon, G.M., Stephens, M., Yager, J., Zeise, L., 2010. Toxicity testing in the 21st century: a vision and a strategy. *J. Toxicol. Environ. Health B Crit. Rev.* 13, 51–138.

Kroon, F., Streten, C., Harries, S., 2017. A protocol for identifying suitable biomarkers to assess fish health: a systematic review. PLoS One 12 (4), e0174762.

Krzykwa, J.C., Jeffries, M.K.S., 2020. Comparison of behavioral assays for assessing toxicant-induced alterations in neurological function in larval fathead minnows. Chemosphere 257, 126825.

Lacroix, C., Richard, G., Seguineau, C., Guyomarch, J., Moraga, D., Auffret, M., 2015. Active and passive biomonitoring suggest metabolic adaptation in blue mussels (*Mytilus* spp.) chronically exposed to a moderate contamination in Brest harbor (France). Aquat. Toxicol. 162, 126–137.

Länge, R., Hutchinson, T.H., Croudace, C.P., Siegmund, F., Schweinfurth, H., Hampe, P., et al., 2001. Effects of the synthetic estrogen 17α-ethinylestradiol on the life-cycle of the fathead minnow (*Pimephales promelas*). Environ. Toxicol. Chem. 20, 1216–1227.

Lardon, I., Nilsson, G.E., Steyck, J.A.W., Vu, T.N., Laukens, K., Dommisse, R., De Boeck, G., 2013. $^{1}$H-NMR study of the metabolome of an exceptionally anoxia tolerant vertebrate, the crucian carp (*Carassius carassius*). Metabolomics 9, 311–323.

Lauff, R.F., Wood, C.M., 1996. Respiratory gas exchange, nitrogenous waste excretion, and fuel usage during starvation in juvenile rainbow trout, *Oncorhynchus mykiss*. J. Comp. Physiol. B 165, 542–551.

Lawrence, M.J., Mitrovic, D., Foubister, D., Bragg, L.M., Sutherby, J., Docker, M.F., Servos, M.R., Wilkie, M.P., Jeffries, K.M., 2021. Contrasting physiological responses between invasive sea lamprey and non-target bluegill in response to acute lampricide exposure. Aquat. Toxicol. 237, 105848.

Lech, J.J., Vodicnik, M.J., 1985. Biotransformation. In: Rand, G.M., Petrocelli, S.R. (Eds.), Fundamentals of Aquatic Toxicology; Methods and Applications. Hemisphere Publishing Corporation, New York, USA, pp. 526–557.

Lee, C.G., Farrell, A.P., Lotto, A., Hinch, S.G., Healey, M.C., 2003. Excess post-exercise oxygen consumption in adult sockeye (*Oncorhynchus nerka*) and coho (*O. kisutch*) salmon following critical speed swimming. J. Exp. Biol. 206, 3253–3260.

Levet, A., Bordes, C., Clément, Y., Mignon, P., Chermette, H., Marote, P., Cren-Olivé, C., Lantéri, P., 2013. Quantitative structure-activity relationship to predict acute fish toxicity of organic solvents. Chemosphere 93, 1094–1103.

Carlson, E.A., Li, Y., Zelikoff, J.T., 2002. The Japanese medaka (Oryzias latipes) model: applicability for investigating the immunosuppressive effects of the aquatic pollutant benzo[a]pyrene (BaP). Mar. Environ. Res. 54, 565–568.

Little, A.G., Dressler, T., Kraskura, K., Hardison, E., Hendriks, B., Prystay, T., Farrell, A.P., Cooke, S.J., Patterson, D.A., Hinch, S.G., Eliason, E.J., 2020. Maxed out: optimizing accuracy, precision, and power for field measures of maximum metabolic rate in fishes. Physiol. Biochem. Zool. 93 (3), 243–254.

Levine, R.L., Garland, D., Oliver, C.N., Amici, A., Climent, I., Lenz, A.-G., Stadtman, E.R., 1990. [49] Determination of carbonyl content in oxidatively modified proteins. Methods Enzymol. 186, 464–478.

Li, D., Pi, J., Wang, J., Zhu, P., Lei, L., Zhang, T., Liu, D., 2016. Protective effects of calcium on cadmium accumulation in co-cultured silver carp (*Hypophthalmichthys molitrix*) and triangle sail mussel (*Hyriopsis cumingii*). Bull. Environ. Contam. Toxicol. 97, 826–831.

Li, J., Barreda, D.R., Zhang, Y.-A., Boshra, H., Gelman, A.E., LaPatra, S., Tort, L., Sunyer, J.O., 2006. B lymphocytes from early vertebrates have potent phagocytic and microbicidal abilities. Nat. Immunol. 7, 1116–1124.

Lieschke, G.J., Trede, N.S., 2009. Fish immunology. *Curr. Biol.* 19, R678–R682.

Liew, H.J., Pelle, A., Chiarella, D., Faggio, C., De Boeck, G., 2013. Cortisol emphasizes the metabolic strategies employed by common carp, *Cyprinus carpio* at different feeding and swimming regimes. Comp. Biochem. Physiol. A 166, 449–464.

Lika, K., Augustine, S., Pecquerie, L., Kooijman, S.A.L.M., 2014. The bijection from data to parameter space with the standard DEB model quantifies the supply-demand spectrum. J. Theor. Biol. 354, 35–47.

Loria, A., Cristescu, M.E., Gonzalez, A., 2019. Mixed evidence for adaptation to environmental pollution. Evol. Appl. 12 (7), 1259–1273. https://doi.org/10.1111/eva.12782.

Lozano, S., Lescot, E., Halm, M.-P., Lepailleur, A., Bureau, R., Rault, S., 2010. Prediction of acute toxicity in fish by using QSAR methods and chemical modes of action. J. Enzyme Inhib. Med. Chem. 25, 195–203.

Lupica, S.J., Turner Jr., J.W., 2009. Validation of enzyme-linked immunosorbent assay for measurement of fecal cortisol in fish. Aquacult. Res. 40, 437–441.

Lushchak, V.I., 2011. Environmentally induced oxidative stress in aquatic animals. Aquat. Toxicol. 101, 13–30.

Lushchak, V.I., Semchyshyn, H.M., Lushchak, O.V., 2012. Classic methods for measuring of oxidative damage: TBARS, xylenol orange, and protein carbonyls. In: Abele, D., Vazquez-Medina, J., Zenteno-Savin, T. (Eds.), Oxidative stress in aquatic ecosystems. Blackwell Publishing Ltd, pp. 420–431.

Lushchak, V.I., 2014a. Free radicals, reactive oxygen species, oxidative stress and its classification. *Chem. Biol. Interact.* 224, 164–175.

Lushchak, V.I., 2014b. Classification of oxidative stress based on its intensity. EXCLI J. 13, 922–937.

Lushchak, V.I., 2016a. Contaminant-induced oxidative stress in fish: a mechanistic approach. Fish Physiol. Biochem. 42, 711–747.

Lushchak, V.I., 2016b. Time-course and intensity-based classifications of oxidative stresses and their potential application in biomedical, comparative and environmental research. Redox Rep. 21, 262–270.

Mackay, D., Arnot, J.A., Gobase, F.A.B.C., Powell, D.E., 2013. Mathematical relationships between metrics of chemical bioaccumulation in fish. Environ. Toxicol. Chem. 32, 1459–1466.

Madliger, C.L., Cooke, S.J., Crespi, E.J., Funk, J.L., Hultine, K.R., Hunt, K.E., Rohr, J.R., Sinclair, B.J., Suski, C.D., Willis, C.K.R., Love, O.P., 2016. Success stories and emerging themes in conservation physiology. Conserv. Physiol. 4, cov057.

Mager, E.M., Grosell, M., 2011. Effects of acute and chronic waterborne lead exposure on the swimming performance and aerobic scope of fathead minnows (*Pimephales promelas*). Comp. Biochem. Physiol. C 154, 7–13.

Magnadóttir, B., 2006. Innate immunity of fish (overview). Fish Shellfish Immunol. 20, 137–151.

Martinez-Alvarez, R.M., Morales, A.E., Sanz, A., 2005. Antioxidant defenses in fish: biotic and abiotic factors. Rev. Fish Biol. Fish. 15, 75–88.

Martins, M., Costa, P.M., 2015. The comet assay in Environmental Risk Assessment of marine pollutants: applications, assets and handicaps of surveying genotoxicity in non-model organisms. Mutagenesis 30 (1), 89–106.

Mathers, K.L., Stubbington, R., Leeming, D., Westwood, C., England, J., 2019. Structural and functional responses of macroinvertebrate assemblages to long-term flow variability at perennial and nonperennial sites. Ecohydrology 12, e2112.

Maund, S.J., Travis, K.Z., Hendley, P., Giddings, J.M., Solomon, K.R., 2001. Probabilistic risk assessment of cotton pyrethroids: V. Combining landscape-level exposures and ecotoxicological effects data to characterize risks. Environ. Toxicol. Chem. 20, 687–692.

Marques, G.M., Augustine, S., Lika, K., Pecquerie, L., Domingos, T., Kooijman, S.A.L.M., 2018. The AmP project: comparing species on the basis of dynamic energy budget parameters. PLoS Comput. Biol. 14, e1006100.

Martin, B.T., Heintz, R., Danner, E.M., Nisbet, R.M., 2017. Integrating lipid storage into general representations of fish energetics. J. Anim. Ecol. 86, 812–825.

Matthiessen, P., Ankley, G.T., Biever, R.C., Bjerregaard, P., Borgert, C., Brugger, K., Blankinship, A., Chambers, J., Coady, K.K., Constantine, L., Dang, Z., Denslow, N.D., Dreier, D.A., Dungey, S., Earl Gray, L., Gross, M., Guiney, P.G., Hecker, M., Holbech, H., Iguchi, T., Kadlec, S., Karouna-Renier, N.K., Katsiadaki, I., Kawashima, Y., Kloas, W., Krueger, H., Kumar, A., Lagadic, L., Leopold, A., Levine, S.L., Maack, G., Marty, S., Meador, J., Mihaich, E., Odum, J., Ortego, L., Parrott, J., Pickford, D., Roberts, M., Schaefers, C., Schwarz, T., Solomon, K., Verslycke, T., Weltje, L., Wheeler, J.R., Williams, M., Wolf, J.C., Yamazaki, K., 2017. Recommended approaches to the scientific evaluation of ecotoxicological hazards and risks of endocrine-active substances. Integr. Environ. Assess. Manag. 13, 267–279.

McEwen, B.S., Wingfield, J.C., 2010. What is in a name? Integrating homeostasis, allostasis and stress. *Horm. Behav.* 57, 105–111.

McGeer, J.C., Szebedinszky, C., McDonald, D.G., Wood, C.M., 2000. Effects of chronic sublethal exposure to waterborne Cu, Cd or Zn in rainbow trout. 1: Iono-regulatory disturbance and metabolic costs. *Aquat. Toxicol.* 50, 231–243.

McKenzie, D.J., Garofalo, E., Winter, M.J., Ceradini, S., Verweij, F., Day, N., Hayes, R., van der Oost, R., Butler, P.J., Chipman, J.K., Taylor, E.W., 2007. Complex physiological traits as biomarkers of the sub-lethal toxicological effects of pollutant exposure in fishes. Philos. Trans. R. Soc. B 362, 2043–2059.

McKenzie, D.J., Axelsson, M., Chabot, D., Claireaux, G., Cooke, S.J., Corner, R.A., De Boeck, G., Domenici, P., Guerreiro, P.M., Hamer, B., Jørgensen, C., Killen, S.S., Lefevre, S., Marras, S., Michaelidis, B., Nilsson, G.E., Peck, M.A., Perez-Ruzafa, A., Rijnsdorp, A.D., Shiels, H.A., Steffensen, J.F., Svendsen, J.C., Svendsen, M.B.S., Teal, L.R., van der Meer, J., Wang, T., Wilson, J.M., Wilson, R.W., Metcalfe, J.D., 2016. Conservation physiology of marine fishes: state of the art and prospects for policy. Conservation. Phys. Ther. 4 (1), cow046.

McKenzie, D.J., Zhang, Y., Eliason, E.J., Schulte, P.M., Claireaux, G., Blasco, F.R., Nati, J.J.H., Farrell, A.P., 2021. Intraspecific variation in tolerance of warming in fishes. *J. Fish Biol.* 98, 1536–1555.

Mebane, C.A., Chowdhury, M.J., De Schamphelaere, K.A.C., Lofts, S., Paquin, P.R., Santore, R.C., Wood, C.M., 2020. Metal bioavailability models: current status, lessons learned, considerations for regulatory use, and the path forward. *Environ. Toxicol. Chem.* 39, 60–84.

Meinelt, T., Playle, R.C., Pietrock, M., Burnison, B.K., Wienke, A., Steinberg, C.E.W., 2001. Interaction of cadmium toxicity in embryos and larvae of zebrafish (*Danio rerio*) with calcium and humic substances. Aquat. Toxicol. 54, 205–215.

Melnikov, F., Kostal, J., Voutchkova-Kostal, A., Zimmerman, J.B., Anastas, P.T., 2016. Assessment of predictive models for estimating the acute aquatic toxicity of organic chemicals. Green Chem. 18, 4432–4445.

Melvin, S.D., Wilson, S.P., 2013. The utility of behavioral studies for aquatic toxicology testing: a meta-analysis. Chemosphere 93, 2217–2223.

Melvin, S.D., Petit, M.A., Duvignacq, M.C., Sumpter, J.P., 2017. Towards improved behavioural testing in aquatic toxicology: acclimation and observation times are important factors when designing behavioural tests with fish. Chemosphere 180, 430e436.

Metcalfe, N.B., Van Leeuwen, T.E., Killen, S.S., 2016. Does individual variation in metabolic phenotype predict fish behaviour and performance? *J. Fish Biol.* 88 (1), 298–321.

Meyer, J.N., Di Giulio, R.T., 2003. Heritable adaptation and fitness costs in killifish (*Fundulus heteroclitus*) inhabiting a polluted estuary. Ecol. Appl. 13 (2), 490–503.

Meyer, J.S., DeForest, D.K., 2018. Protectiveness of Cu water quality criteria against impairment of behavior and chemo/mechanosensory responses: an update. Environ. Toxicol. Chem. 37, 1260–1279.

Miller, D.H., Tietge, J.E., McMaster, M.E., Munkittrick, K.R., Xia, X., Griesmer, D.A., Ankley, G.T., 2015. Linking mechanistic toxicology to population models in forecasting recovery from chemical stress: a case study from Jackfish Bay, Ontario, Canada. Environ. Toxicol. Chem. 34, 1623–1633.

Milligan, C.L., 1996. Metabolic recovery from exhaustive exercise in rainbow trout. Comp. Biochem. Physiol. A 113, 51–60.

Monostori, P., Wittmann, G., Karg, E., Túri, S., 2009. Determination of glutathione and glutathione disulfide in biological samples: an in depth review. J. Chromatogr. 877B, 3331–3346.

Monserrat, J.M., Martínez, P.E., Geracitano, L.A., Lund Amado, L., Martins, M.G., C., Lopes Leães Pinho, G., Soares Chaves, I., Ferreira-Cravo, M., Ventura-Lima, J., Bianchini, A., 2007. Pollution biomarkers in estuarine animals: critical review and new perspectives. *Comp. Biochem. Physiol.* C 146, 221–234.

Monteiro, D.A., Kalinin, A.L., Rantin, F.T., McKenzie, D.J., 2021. Use of complex physiological traits as ecotoxicological biomarkers in tropical freshwater fishes. J. Exp. Zool. 335, 745–760.

Moran, R., Harvey, I., Moss, B., Feuchtmayr, H., Hatton, K., Heyes, T., Atkinson, D., 2010. Influence of simulated climate change and eutrophication on three-spined stickleback populations: a large scale mesocosm experiment. Freshw. Biol. 55, 315–325.

Morris, J.M., Brinkman, S., Takeshita, R., McFadden, A.K., Carney, M.W., Lipton, J., 2019. Copper toxicity in Bristol Bay headwaters: Part 2—olfactory inhibition in low-hardness water. Environ. Toxicol. Chem. 38, 198–209.

Morrison, N., Bearden, D., Bundy, J.G., Collette, T., Currie, F., Davey, M.P., 2007. Standard reporting requirements for biological samples in metabolomics experiments: environmental context. Metabolomics 3, 203e210.

Morthorst, J.E., Brande-Lavridsen, N., Korsgaard, B., Bjerregaard, P., 2014. 17 Beta-Estradiol causes abnormal development in embryos of the viviparous eelpout. Environ. Sci. Technol. 48, 14668–14676.

Moss, B., 2010. Climate change, nutrient pollution and the bargain of Dr Faustus. Freshw. Biol. 55, 175–187.

Muller, C., Ruby, S., Brousseau, P., Cyr, D., Fournier, M., Gagné, F., 2009. Immunotoxicological effects of an activated-sludge-treated effluent on rainbow trout (*Oncorhynchus mykiss*). Comp. Biochem. Physiol. C Toxicol. Pharmacol. 150, 390–394.

Nacci, D.E., Champlin, D., Jayaraman, S., 2010. Adaptation of the estuarine fish *fundulus heteroclitus* (Atlantic Killifish) to polychlorinated biphenyls (PCBs). Estuaries Coast 33 (4), 853–864. https:/doi.org/10.1007/s12237-009-9257-6.

Nakagawa, S., Noble, D.W.A., Senior, A.M., Lagisz, M., 2017. Meta-evaluation of meta-analysis: ten appraisal questions for biologists. BMC Biol. 15 (1), 18.

Nanduri, J., Vaddi, D.R., Khan, S.A., Wang, N., Makerenko, V., Prabhakar, N.R., 2013. Xanthine oxidase mediates hypoxia inducible factor-2α degradation by intermittent hypoxia. PLoS One 8, e75838.

Natnan, M.E., Low, C.-F., Chong, C.-M., Bunawan, H., Baharum, S.N., 2021a. Integration of omics tools for understanding the fish immune response due to microbial challenge. Front. Mar. Sci. 8, 668771.

Natnan, M.E., Mayalvanan, Y., Jazamuddin, F.M., Aizat, W.M., Low, C.-F., Goh, H.-H., Azizan, K.A., Bunawan, H., Baharum, S.N., 2021b. Omics Strategies in current advancements of infectious fish disease management. Biology 10, 1086.

Nendza, M., Müller, M., Wenzel, A., 2017. Classification of baseline toxicants for QSAR predictions to replace fish acute toxicity studies. Environ. Sci.: Processes Impacts 19, 429–437.

Nendza, M., Kühne, R., Lombardo, A., Strempel, S., Schüürmann, G., 2018. PBT assessment under REACH: screening for low aquatic bioaccumulation with QSAR classifications based on physicochemical properties to replace BCF *in vivo* testing on fish. Sci. Total Environ. 616-617, 97–106.

Netzeva, T.I., Aptula, A.O., Benfenati, E., Cronin, M.T.D., Gini, G., Lessigiarska, I., Maran, U., Vračko, M., Schüürmann, G., 2005. Description of the electronic structure of organic chemicals using semiempirical and ab initio methods for development of toxicological QSARs. J. Chem. Inf. Model. 45, 106–114.

Newman, M.C., Ownby, D.R., Mézin, L.C.A., Powell, D.C., Christensen, T.R.L., Lerberg, S.B., Andersons, B.-A., Padma, T.V., 2002. Species sensitivity distributions in ecological risk assessment: distributional assumption, alternate bootstrap techniques, and estimation of adequate number of species. In: Posthuma, L., Suter, G.W., Traas, T.P. (Eds.), Species Sensitivity Distributions in Ecotoxicology. CRC Press, Boca Raton, pp. 119–132.

Nikinmaa, M., 2013. Climate change and ocean acidification-interactions with aquatic toxicology. Aquat. Toxicol. 126, 365–372.

Nisbet, R.M., 2017. Challenges for dynamic energy budget theory. Phys. Life Rev. 20, 72–74.

Niyogi, S., Wood, C.M., 2003. Effects of chronic waterborne and dietary metal exposures on gill metal-binding: implications for the biotic ligand model. Human Ecol. Risk Assess. 9, 813–846.

Niyogi, S., Couture, P., Pyle, G., McDonald, D.G., Wood, C.M., 2004. Acute cadmium biotic ligand model characteristics of laboratory-reared and wild yellow perch (*Perca flavescens*) relative to rainbow trout (*Oncorhynchus mykiss*). Can. J. Fish. Aquat. Sci. 61, 942–953.

Niyogi, S., Wood, C.M., 2004a. Biotic ligand model, a flexible tool for developing site-specific water quality guidelines for metals. Environ. Sci. Technol. 38, 6177–6192.

Niyogi, S., Wood, C.M., 2004b. Kinetic analyses of waterborne Ca and Cd transport and their interactions in the gills of rainbow trout (*Oncorhynchus mykiss*) and yellow perch (*Perca flavescens*), two species differing greatly in acute waterborne Cd sensitivity. J. Comp. Physiol. B 174, 243–253.

Niyogi, S., Kent, R., Wood, C.M., 2008. Effects of water chemistry variables on gill binding and acute toxicity of cadmium in rainbow trout (*Oncorhynchus mykiss*): a biotic ligand model (BLM) approach. Comp. Biochem. Physiol. C 148, 305–314.

Niyogi, S., Nadella, S.R., Wood, C.M., 2015. Interactive effects of waterborne metals in binary mixtures on short-term gill-metal binding and ion uptake in rainbow trout (*Oncorhynchus mykiss*). Aquat. Toxicol. 165, 109–119.

Norin, T., Clark, T.D., 2016. Measurement and relevance of maximum metabolic rate in fishes. J. Fish Biol. 88, 122–151.

O'Brien, A., Townsend, K., Hale, R., Sharley, D., Pettigrove, V., 2016. How is ecosystem health defined and measured? A critical review of freshwater and estuarine studies. Ecol. Indic. 69, 722–729.

O'Dea, R.E., Lagisz, M., Jennions, M.D., Koricheva, J., Noble, D.W.A., Parker, T.H., Nakagawa, S., 2021. Preferred reporting items for systematic reviews and meta-analyses in ecology and evolutionary biology: a PRISMA extension. Biol. Rev. 96 (5), 1695–1722.

Oikari, A., 2006. Caging techniques for field exposures of fish to chemical contaminants. *Aquat. Toxicol.* 78, 370–381.

Oliveira, M., Tvarijonaviciute, A., Trindade, T., Soares, A.M.V.M., Tort, L., Teles, M., 2018. Can non-invasive methods be used to assess effects of nanoparticles in fish? Ecol. Indic. 95, 1118–1127.

Opinion, A.G.R., De Boeck, G., Rodgers, E.M., 2020. Synergism between elevated temperature and nitrate: impact on aerobic capacity of European grayling, *Thymallus thymallus* in warm, eutrophic waters. Aquat. Toxicol. 226, 105563.

Opinion, A.G.R., Cakir, R., De Boeck, G., 2021. Better together: cross-tolerance induced by warm acclimation and nitrate exposure improved the aerobic capacity and stress tolerance of common carp *Cyprinus carpio*. Ecotoxicol. Environ. Saf. 225, 112777.

Ownby, D.R., Newman, M.C., 2003. Advances in quantitative ion character-activity relationships (QICARs): using metal-ligand binding characteristics to predict metal toxicity. QSAR Comb. Sci. 22, 241–246.

Ōyanagui, Y., 1984. Reevaluation of assay methods and establishment of kit for superoxide dismutase activity. Anal. Biochem. 142, 290–296.

Paglia, D.E., Valentine, W.N., 1967. Studies on the quantitative and qualitative characterization of erythrocyte glutathione peroxidase. J. Lab. Clin. Med. 70, 158–169.

Pagnotta, A., Milligan, L., 1991. The role of blood glucose in the restoration of muscle glycogen during recovery from exhaustive exercise in rainbow trout (*Oncorhynchus mykiss*) and winter flounder (*Pseudopleuronectes americanus*). J. *Exp. Biol.* 161, 489–508.

Pampanin, D.M., Sydnes, M.O., 2013. Polycyclic aromatic hydrocarbons a constituent of petroleum: presence and influence in the aquatic environment. In: Kutcherov, V., Kolesnikov, A. (Eds.), Hydrocarbons. InTech, Rijeka, pp. 83–118.

Pampanin, D.M., Brooks, S.J., Grøsvik, B.E., Le Goff, J., Meier, S., Sydnes, M.O., 2017. DNA adducts in marine fish as biological marker of genotoxicity in environmental monitoring: the way forward. Mar. Environ. Res. 125, 49–62.

Pandey, S.K., Ojha, P.K., Roy, K., 2020. Exploring QSAR models for assessment of acute fish toxicity of environmental transformation products of pesticides (ETPPs). Chemosphere 252, 126508.

Pang, X., Shao, F., Ding, S.-H., Fu, S.-J., Zhang, Y.-G., 2019. Interspecific differences and ecological correlations of energy metabolism traits in freshwater fishes. Funct. Ecol. 34, 616–630.

Papa, E., Villa, F., Gramatica, P., 2005. Statistically validated QSARs, based on theoretical descriptors, for modelling aquatic toxicity of organic chemicals in *Pimephales promelas* (fathead minnow). J. Chem. Inf. Model. 45, 1256–1266.

Paquin, P.R., Gorsuch, J.W., Apte, S., Batley, G.E., Bowles, K.C., Campbell, P.G., Delos, C.G., Di Toro, D.M., Dwyer, R.L., Galvez, F., Gensemer, R.W., Goss, G.G., Hogstrand, C., Janssen, C.R., McGeer, J.C., Naddy, R.B., Playle, R.C., Santore, R.C., Schneider, U., Stubblefield, W.A., Wood, C.M., Wu, K.B., 2002. The biotic ligand model: a historical overview. Comp. Biochem. Physiol. C 133, 3–35.

Payne, J.F., 1976. Field evaluation of benzopyrene hydroxylase induction as a monitor for marine pollution. Science 191, 945–949.

Parker, M.O., 2016. Adult vertebrate behavioural aquatic toxicology: reliability and validity. Aquat. Toxicol. 170, 323–329.

Patra, R.W., Chapman, J.C., Lim, R.P., Gehrke, P.C., 2007. The effects of three organic chemicals on the upper thermal tolerances of four freshwater fishes. Environ. Toxicol. Chem. 26, 1454–1459.

Pereira, P., Korbas, M., Pereira, V., Cappello, T., Maisano, M., Canário, J., Almeida, A., Pacheco, M., 2019. A multidimensional concept for mercury neuronal and sensory toxicity in fish - From toxicokinetics and biochemistry to morphometry and behavior. Biochim. Biophys. Acta Gen. Subj. 1863, 129298.

Peters, A., Merrington, G., Stauber, J., Golding, L., Batley, G., Gissi, F., Adams, M., Binet, M., McKnight, K., Schlekat, C.E., Garman, E., Middleton, E., 2021. Empirical bioavailability corrections for nickel in freshwaters for Australia and New Zealand water quality guideline development. Environ. Toxicol. Chem. 40, 113–126.

Philippe, C., Gregoir, A.F., Thoré, E.S.J., De Boeck, G., Brendonck, L., Pinceel, T., 2018. Protocol for acute and chronic ecotoxicity testing of the turquoise Killifish *Nothobranchius Furzeri*. J. Vis. Exp. 134. https://doi.org/10.3791/57308.

Plaut, I., 2001. Critical swimming speed: its ecological relevance. Comp. Biochem. Physiol. A 131, 41–50.

Playle, R.C., Dixon, D.G., Burnison, K., 1993. Copper and cadmium binding to fish gills: estimates of metal-gill stability constants and modelling of metal accumulation. Can. J. Fish. Aquat. Sci. 50, 2678–2687.

Pörtner, H.O., 2001. Climate change and temperature-dependent biogeography: oxygen limitation of thermal tolerance in animals. Naturwissenschaften 88, 137–146.

Pörtner, H.O., 2010. Oxygen- and capacity-limitation of thermal tolerance: a matrix for integrating climate-related stressor effects in marine ecosystems. J. Exp. Biol. 213, 881–893.

Pörtner, H.O., Farrell, A.P., 2008. Physiology and climate change. Science 322, 690–692.

Pörtner, H.O., Bock, C., Mark, F.C., 2017. Oxygen- and capacity-limited thermal tolerance: bridging ecology and physiology. J. Exp. Biol. 15, 2685–2696.

Posthuma, L., Suter, G.W., Traas, T.P. (Eds.), 2002. Species Sensitivity Distributions in Ecotoxicology. CRC Press, Boca Raton.

Prieto, P., Pineda, M., Aguilar, M., 1999. Spectrophotometric quantitation of antioxidant capacity through the formation of a phosphomolybdenum complex: specific application to the determination of vitamin E. Anal. Biochem. 269, 337–341.

Pyle, G., Wood, C., 2008. Radiotracer studies on waterborne copper uptake, distribution, and toxicity in rainbow trout and yellow perch: a comparative analysis. *Hum. Ecol. Risk Assess.* 14, 243–265.

Rehberger, K., Werner, I., Hitzfeld, B., Segner, H., Baumann, L., 2017. 20 Years of fish immunotoxicology—what we know and where we are. Crit. Rev. Toxicol. 47, 516–542.

Richards, J.G., Playle, R.C., 1999. Protective effects of calcium against the physiological effects of exposure to a combination of cadmium and copper in rainbow trout (*Oncorhynchus mykiss*). Can. J. Zool. 77, 1035–1047.

Richards, J.G., Burnison, B.K., Playle, R.C., 1999. Natural and commercial dissolved organic matter protects against the physiological effects of a combined cadmium and copper exposure on rainbow trout (*Oncorhynchus mykiss*). Can. J. Fish. Aquat. Sci. 56, 407–418.

Rodgers, D.W., Beamish, F.W.H., 1981. Uptake of waterborne methylmercury by rainbow trout (*Salmo gairdneri* ) in relation to oxygen consumption and methylmercury concentration. *Can. J. Fish. Aquat. Sci.* 38, 1309–1315.

Robinson, P.D., 2009. Behavioural toxicity of organic chemical contaminants in fish: application to ecological risk assessments (ERAs). Can. J. Fish. Aquat. Sci. 66, 1179–1188.

Rodgers, E.M., Poletto, J.B., Gomez Isaza, D.F., Van Eenennaam, J.P., Connon, R.E., Todgham, A.E., Seesholtz, A., Heublein, J.C., Cech Jr., J.J., Kelly, J.T., Fangue, N.A., 2019. Integrating physiological data with the conservation and management of fishes: a meta-analytical review using the threatened green sturgeon (*Acipenser medirostris*). Conserv. Physiol. 7 (1).

Rodgers, E.M., Opinion, A.G.R., Gomez Isaza, D.F., Rašković, B., Poleksić, V., De Boeck, G., 2021. Double whammy: nitrate pollution heightens susceptibility to both hypoxia and heat in a freshwater salmonid. Sci. Total Environ. 765, 142777.

Rodríguez-Romero, A., Viguri, J.R., Calosi, P., 2021. Acquiring an evolutionary perspective in marine ecotoxicology to tackle emerging concerns in a rapidly changing ocean. Sci. Total Environ. 764, 142816. https://doi.org/10.1016/j.scitotenv.2020.142816.

Roe, J., Kuether, C., 1943. Estimation of ascorbic acid. J. Biol. Chem. 147, 3999.

Roesijadi, G., 1992. Metallothioneins in metal regulation and toxicity in aquatic animals. Aquat. Toxicol. 22, 81–114.

Roesijadi, G., Robinson, W.E., 1994. Metal regulation in aquatic animals: mechanisms of uptake, accumulation and release. In: Malins, D.C., Ostrander, G.K. (Eds.), Aquatic Toxicology; Molecular, Biochemical and Cellular Perspectives. Lewis Publishers, CRC press, pp. 387–420.

Rogers, N.J., Urbina, M.A., Reardon, E.E., McKenzie, D.J., Wilson, R.W., 2016. A new analysis of hypoxia tolerance in fishes using a database of critical oxygen level (Pcrit). *Conservation.* Phys. Ther. 4, cow012.

Rohonczy, J., O'Dwyer, K., Rochette, A., Robinson, S.A., Forbes, M.R., 2021. Meta-analysis shows environmental contaminants elevate cortisol levels in teleost fish—effect sizes depend on contaminant class and duration of experimental exposure. Sci. Total Environ. 800, 149402.

Romero, L.M., Dickens, M.J., Cyr, N.E., 2009. The reactive scope model–a new model integrating homeostasis, allostasis, and stress. Horm. Behav. 55, 375–389.

Ruaro, R., Gubiani, E.A., 2013. A scientometric assessment of 30 years of the index of biotic integrity in aquatic ecosystems: applications and main flaws. Ecol. Indic. 29, 105–110.
Russom, C.L., Bradbury, S.P., Broderius, S.J., Hammermeister, D.E., Drummond, R.A., 1997. Predicting modes of toxic action from chemical structure: acute toxicity in the fathead minnow (*Pimephales promelas*). Environ. Toxicol. Chem. 16, 948–967.
Sadoul, B., Geffroy, B., 2019. Measuring cortisol, the major stress hormone in fishes. J. Fish Biol. 94, 540–555.
Sahota, C., Hayek, K., Surbey, B., Kennedy, C.J., 2021. Lethal and sublethal effects in Pink salmon (Oncorhynchus gorbuscha) following exposure to five aquaculture chemotherapeutants. Ecotoxicology 2021. https:/doi.org/10.1007/s10646-021-02473-8.
Sangion, A, Gramatica, P, 2016. Hazard of pharmaceuticals for aquatic environment: prioritization by structural approaches and prediction of ecotoxicity. Environ. Int. 95, 131–143. https://doi.org/10.1016/j.envint.2016.08.008.
Santana, M.S., Sandrini-Neto, L., Di Domenico, M., Prodocimo, M., 2021. Pesticide effects on fish cholinesterase variability and mean activity: a meta-analytic review. Sci. Total Environ. 757, 143829.
Santore, R.C., Di Toro, D.M., Paquin, P.R., Allen, H.E., Meyer, J.S., 2001. Biotic ligand model of the acute toxicity of metals. 2. Application to acute copper toxicity in freshwater fish and *Daphnia*. Environ. Toxicol. Chem. 20, 2397–2402.
Schreck, C.B., 2010. Stress and fish reproduction: the roles of allostasis and hormesis. Gen. Comp. Endocrinol. 165, 549–556.
Schreck, C.B., Tort, L., 2016. The concept of stress in fish. In: Schreck, C.B., Tort, L., Farrell, A.P., Brauner, C.J. (Eds.), Biology of Stress in Fish. Fish Physiology Series, vol. 35, pp. 1–34.
Schwarzenbach, R.P., Escher, B.I., Fenner, K., Hofstetter, T.B., Johnson, C.A., von Gunten, U., Wehrli, B., 2006. The challenge of micropollutants in aquatic systems. Science 313, 1072–1077.
Scott, G.R., Sloman, K.A., 2004. The effects of environmental pollutants on complex fish behaviour: integrating behavioural and physiological indicators of toxicity. Aquat. Toxicol. 68, 369–392.
Segner, H., Wegner, M., Möller, A.M., Kollner, B., Casanova-Nakayama, A., 2012a. Immunotoxic effects of environmental toxicants in fish—how to assess them? Environ. Sci. Pollut. Res. 19, 2465–2476.
Segner, H., Möller, A.M., Wegner, M., Casanova-Nakayama, A., 2012b. Fish immunotoxicology: research at the crossroads of immunology, ecology and toxicology. In: Kawaguchi, M., Misaki, K., Sato, H., Yokokawa, T., Itai, T., Nguyen, T.M., Ono, J., Tanabe, S. (Eds.), Interdisciplinary Studies on Environmental Chemistry—Environmental Pollution and Ecotoxicology. Terrapub, pp. 1–12.
Sheffield, T.Y., Judson, R.S., 2019. Ensemble QSAR modeling to predict multispecies fish toxicity lethal concentrations and points of departure. Environ. Sci. Technol. 53, 12793–12802.
Shelley, L.K., Balfry, S.K., Ross, P.S., Kennedy, C.J., 2009. Immunotoxicological effects of a sub-chronic exposure to selected current-use pesticides in rainbow trout (Oncorhynchus mykiss). Aquat. Toxicol. 92, 95–103.
Shekh, K., Tang, S., Hecker, M., Niyogi, S., 2018. Investigating the role of ionoregulatory processes in the species and life-stage-specific differences in sensitivity of rainbow trout and white sturgeon to cadmium. *Environ. Sci. Technol.* 52, 12868–12876.
Shuman-Goodier, M.E., Propper, C.R., 2016. A meta-analysis synthesizing the effects of pesticides on swim speed and activity of aquatic vertebrates. Sci. Total Environ. 565, 758–766.

Sies, H., 1985. Oxidative stress: introductory remarks. In: Sies, H. (Ed.), Oxidative Stress. Academic Press, London, pp. 1–8.

Silvestre, F., 2020. Signaling pathways of oxidative stress in aquatic organisms exposed to xenobiotics. J. Exp. Zool. A 333 (6), 436–448.

Singh, N.P., McCoy, M.T., Tice, R.R., Schneider, E.L., 1988. A simple technique for quantitation of low levels of DNA damage in individual cells. Exp. Cell Res. 175, 184–191.

Sinha, A.K., 1972. Colorimetric assay of catalase. Anal. Biochem. 47, 389–394.

Sirot, C., Villéger, S., Mouillot, D., Darnaude, J., Ramos-Miranda, A.M., Flores-Hernandez, D., Panfili, J., 2015. Combinations of biological attributes predict temporal dynamics of fish species in response to environmental changes. Ecol. Indic. 48, 147–156.

Slaveykova, V.I., Wilkinson, K.J., 2005. Predicting the bioavailability of metals and metal complexes: critical review of the biotic ligand model. Environ. Chem. 2, 9–24.

Smith, I.K., Vierheller, T.L., Thorne, C.A., 1988. Assay of glutathione reductase in crude tissue homogenates using 5, 5′-dithiobis (2-nitrobenzoic acid). Anal. Biochem. 175, 408–413.

Söffker, M., Tyler, C.R., 2012. Endocrine disrupting chemicals and sexual behaviors in fish—a critical review on effects and possible consequences. Crit. Rev. Toxicol. 42, 653–668.

Solomon, K.R., Giddings, J.M., Maund, S.J., 2001. Probabilistic risk assessment of cotton pyrethroids: I. Distributional analyses of laboratory aquatic toxicity data. Environ. Toxicol. Chem. 20, 652–659.

Souza, G.B.G., Vianna, M., 2020. Fish-based indices for assessing ecological quality and biotic integrity in transitional waters: a systematic review. Ecol. Indic. 109, 105665.

Spurgeon, D., Lahive, E., Robinson, A., Short, S., Kille, P., 2020. Species sensitivity to toxic substances: evolution, ecology and applications. Front. Environ. Sci. 8, 588380.

Stadman, E.R., Berlett, B.S., 1997. Reactive oxygen mediated protein oxidation in aging and disease. Chem. Res. Toxicol. 10, 485–494.

Stegeman, J.J., Hahn, M.E., 1994. Biochemistry and molecular biology of monooxygenase: current perspective on forms, functions, and regulation of cytochrome P450 in aquatic species. In: Malins, D.C., Ostrander, G.K. (Eds.), Aquatic Toxicology; Molecular, Biochemical and Cellular Perspectives. Lewis Publishers, CRC press, Boca Raton, pp. 87–206.

Stelzer, J.A.A., Rosin, C.K., Bauer, L.H., Hartmann, M., Pulgati, F.H., Arenzon, A., 2018. Is fish embryo test (FET) according to OECD 236 sensible enough for delivering quality data for effluent risk assessment? Environ. Toxicol. Chem. 37, 2925–2932.

Stoot, L.J., Cairns, N.A., Cull, F., Taylor, J.J., Jeffrey, J.D., Morin, F., Mandelman, J.W., Clark, T.D., Cooke, S.J., 2014. Use of portable blood physiology point-of-care devices for basic and applied research on vertebrates: a review. Conserv. Physiol. 2, cou011. https://doi.org/10.1093/conphys/cou011.

Tabrez, S., Ahmad, M., 2010. Cytochrome P450 system as a toxicity biomarker of industrial wastewater in rat tissues. Food Chem. Toxicol. 48, 998–1001.

Tang, S., Doering, J.A., Sun, J., Beitel, S.C., Shekh, K., Patterson, S., Crawford, S., Giesy, J.P., Wiseman, S.B., Hecker, M., 2016. Linking oxidative stress and magnitude of compensatory responses with life-stage specific differences in sensitivity of white sturgeon (*Acipenser transmontanus*) to copper or cadmium. *Environ. Sci. Technol.* 50 (17), 9717–9726.

Taylor, L.N., Wood, C.M., McDonald, D.G., 2003. An evaluation of sodium loss and gill metal binding properties in rainbow trout and yellow perch to explain species differences in copper tolerance. *Environ. Toxicol. Chem.* 22, 2159–2166.

Teather, K., Parrott, J., 2006. Assessing the chemical sensitivity of freshwater fish commonly used in toxicological studies. Water Qual. Res. J. Canada 41, 100–105.

Teixidó, E., Leuthold, D., de Crozé, N., Léonard, M., Scholz, S., 2020. Comparative assessment of the sensitivity of fish early-life stage, daphnia, and algae tests to the chronic ecotoxicity of xenobiotics: perspectives for alternatives to animal testing. *Environ. Toxicol. Chem.* 39, 30–41.

Thoré, E.S.J., Philippe, C., Brendonck, L., Pinceel, T., 2021a. Towards improved fish tests in ecotoxicology - Efficient chronic and multi-generational testing with the killifish *Nothobranchius furzeri*. Chemosphere 273, 129697.

Thoré, E.S.J., Van Hooreweghe, F., Philippe, C., Brendonck, L., Pinceel, T., 2021b. Generation-specific and interactive effects of pesticide and antidepressant exposure in a fish model call for multi-stressor and multigenerational testing. Aquat. Toxicol. 232, 105743.

Tierney, K.B., 2016. Chemical avoidance responses of fishes. Aquat. Toxicol. 174, 228–241.

Tierney, K.B., Baldwin, D.H., Hara, T.J., Ross, P.S., Scholz, N.L., Kennedy, C.J., 2010. Olfactory toxicity in fishes. Aquat. Toxicol. 96, 2–26.

Tort, L., Balasch, J.C., Mackenzie, S., 2003. Fish immune system. A crossroads between innate and adaptive responses. Inmunologıa 22, 277–286.

Town, R.M., van Leeuwen, H.P., Duval, J.F.L., 2019. Rigorous physicochemical framework for metal ion binding by aqueous nanoparticulate humic substances: implications for speciation modelling by the NICA-Donnan and WHAM codes. Environ. Sci. Technol. 53, 8516–8532.

Travis, K.Z., Hendley, P., 2001. Probabilistic risk assessment of cotton pyrethroids: IV. Landscape-level exposure characterization. Environ. Toxicol. Chem. 20, 679–686.

Trebitz, A.S., Hill, B.H., McCormick, F.H., 2003. Sensitivity of Indices of Biotic Integrity to simulated fish assemblage changes. Environ. Manag. 32, 499–515.

Trichet, V., Buisine, N., Mouchel, N., Moran, P., Pendas, A.M., Le Pennec, J.P., et al., 2000. Genomic analysis of the vitellogenin locus in rainbow trout (*Oncorhynchus mykiss*) reveals a complex history of gene amplification and retroposon activity. Mol. Gen. Genet. 263, 828–837.

Udroiu, I., 2006. The micronucleus test in piscine erythrocytes. Aquat. Toxicol. 79, 201–204.

USEPA, 2007. Aquatic Life Ambient Freshwater Quality Criteria—Copper. Office of Water, Office of Science and Technology, Washington DC. EPA-822-R-07-001.

Van den Berg, S.J.P., Rendal, C., Focks, A., Butler, E., Peeters, E.T.H.M., De Laender, F., Van den Brink, P.J., 2020. Potential impact of chemical stress on freshwater invertebrates: a sensitivity assessment on continental and national scale based on distribution patterns, biological traits, and relatedness. Sci. Total Environ. 731, 139150.

Van den Brink, P.J., 2006. Response to recent criticism on aquatic semi field experiments: opportunities for new developments in ecological risk assessment of pesticides. Integr. Environ. Assess. Manag. 2, 202–203.

Van der Meer, J., 2006. An introduction to dynamic energy budget (DEB) models with special emphasis on parameter estimation. J. Sea Res. 56, 85–102.

Van der Oost, R., Beyer, J., Vermeulen, N.P.E., 2003. Fish bioaccumulation and biomarkers in environmental risk assessment: a review. Environ. Toxicol. Pharmacol. 13, 57–149.

Van der Oost, R., Sileno, G., Suarez Muños, M., Besselink, H., Brouwer, A., 2017a. SIMONI (Smart Integrated Monitoring) as a novel bioanalytical strategy for water quality assessment: part I. Model design and effect-based trigger values. *Environ. Toxicol. Chem.* 36, 2385–2399.

Van der Oost, R., Sileno, G., Janse, T., Nguyen, M.T., Besselink, H., Brouwer, A., 2017b. SIMONI (Smart Integrated Monitoring) as a novel bioanalytical strategy for water quality assessment: part II. Field feasibility survey. *Environ. Toxicol. Chem.* 36, 2400–2416.

Van der Oost, R., McKenzie, D.J., Verweij, F., Satumalay, S., van der Molen, N., Winter, M.J., Chipman, J.K., 2020. Identifying adverse outcome pathways (AOP) for Amsterdam city fish by integrated field monitoring. Environ. Toxicol. Pharmacol. 74, 103301.

van Leeuwen, H.P., 1999. Metal speciation dynamics and bioavailability: inert and labile complexes. Environ. Sci. Technol. 33, 3743–3748.

van Leeuwen, H.P., 2000. Speciation dynamics and bioavailability of metals. J. Radioanal. Nucl. Chem. 246, 487–492.

van Leeuwen, H.P., Duval, J.F.L., Pinheiro, J.P., Blust, R., Town, R.M., 2017. Chemodynamics and bioavailability of metal ion complexes with nanoparticles in aqueous media. Environ. Sci. Nano 4, 2108–2133.

van Leeuwen, H.P., Town, R.M., Buffle, J., Cleven, R.F.M.J., Davison, W., Puy, J., van Riemsdijk, W.H., Sigg, L., 2005. Dynamic speciation analysis and bioavailability of metals in aquatic systems. Environ. Sci. Technol. 39, 8545–8556.

Vardy, D.W., Oellers, J., Doering, J.A., Hollert, H., Giesy, J.P., Hecker, M., 2013. Sensitivity of early life stages of white sturgeon, rainbow trout, and fathead minnow to copper. Ecotoxicology 22, 139–147.

Vasak, M., 2005. Advances in metallothionein structure and functions. J. Trace Elem. Med. Biol. 19, 13–17.

Veltman, K., Hendriks, A.J., Huijbregts, M.A.J., Wannaz, C., Jolliet, O., 2014. Toxicokinetic toxicodynamic (TKTD) modeling of Ag toxicity in freshwater organisms: whole-body sodium loss predicts acute mortality across aquatic species. Environ. Sci. Technol. 48, 14481–14489.

Verhaar, H.J.M., Leeuwen, C.J.V., Hermens, J.L.M., 1992. Classifying environmental pollutants. 1: structure-activity relationships for prediction of aquatic toxicity. Chemosphere 25, 471–491.

Vieira, M., Nunes, B., 2021. Cholinesterases of marine fish: characterization and sensitivity towards specific chemicals. Environ. Sci. Pollut. Res. 28, 48595–48609.

Violle, C., Navas, M.L., Vile, D., Kazakou, E., Fortunel, C., Hummel, I., Garnier, E., 2007. Let the concept of trait be functional! Oikos 116, 882–892.

Vittozzi, L., De Angeli, G., 1991. A critical review of comparative acute toxicity data on freshwater fish. Aquat. Toxicol. 19, 167–204.

Vos, J.G., Dybing, E., Greim, H., Ladefoged, O., Lambré, C., Tarazona, J.V., et al., 2000. Health effects of endocrine disrupting chemicals on wildlife, with special reference to the European situation. Crit. Rev. Toxicol. 30, 71–133.

Wang, Z., Kwok, K.W.H., Lui, G.C.S., Zhou, G.-J., Lee, J.-S., Lam, W.H.W., Leung, K.M.Y., 2014. The difference between temperate and tropical saltwater species' acute sensitivity to chemicals is relatively small. Chemosphere 105, 31–43.

Wang, Z., Meador, J.P., Leung, K.M.Y., 2016. Metal toxicity to freshwater organisms as a function of pH: a meta-analysis. Chemosphere 144, 1544–1552.

Wang, Z., Kwok, K.W.H., Leung, K.M.Y., 2019. Comparison of temperate and tropical freshwater species' acute sensitivities to chemicals: an update. Integr. Environ. Assess. Manag. 15, 352–363.

Warr, G.W., 1983. Immunogloblin of the toadfish, Spheroides glaber. Comp. Biochem. Physiol. B 76, 507–514.

Wheeler, C.R., Salzman, J.A., Elsayed, N.M., Omaye, S.T., Korte, D.W., 1990. Automated assays for superoxide dismutase, catalase, glutathione peroxidase, and glutathione reductase activity. Anal. Biochem. 184, 193–199.

Whitehead, A., Clark, B.W., Reid, N.M., Hahn, M.E., Nacci, D., 2017. When evolution is the solution to pollution: key principles, and lessons from rapid repeated adaptation of killifish (*Fundulus heteroclitus*) populations. Evol. Appl. 10 (8), 762–783. https://doi.org/10.1111/eva.12470.

WHO/IPCS. World Health Organization/International Programme on Chemical Safety, 2002. Chapter 1: Executive summary. In: Damstra, T., Barlow, S., Bergman, A., Kavlock, R., Van Der Kraak, G. (Eds.), Global assessment of the state of the science of endocrine disruptors. WHO, Geneva (CH), pp. 1–3. Publication nr. WHO/PCS/EDC/02.2.

Wilson, R.W., Bergman, H.L., Wood, C.M., 1994. Metabolic costs and physiological consequences of acclimation to aluminum in juvenile rainbow trout (*Oncorhynchus mykiss*). 2: gill morphology, swimming performance, and aerobic Scope. Can. J. Fish. Aquat. Sci. 51, 536–544.

Wolf, J.C., Wheeler, J.R., 2018. A critical review of histopathological findings associated with endocrine and non-endocrine hepatic toxicity in fish models. Aquat. Toxicol. 197, 60–78.

Wu, N.C., Seebacher, F., 2020. Effect of the plastic pollutant bisphenol A on the biology of aquatic organisms: a meta-analysis. Glob. Chang. Biol. 26 (7), 3821–3833.

Xie, L., Klerks, P.L., 2003. Responses to selection for cadmium resistance in the least killifish, *Heterandria formosa*. Environ. Toxicol. Chem. 22 (2), 313–320.

Xu, C., Li, C.Y.-T., Kong, A.-N.T., 2005. Induction of phase I, II and III drug metabolism/transport by xenobiotics. Arch. Pharm. Res. 28, 249–268.

Yada, T., Tort, L., 2016. Stress and disease resistance: immune system and immunoendocrine interactions. In: Schreck, C.B., Tort, L., Farrell, A.P., Brauner, C.J. (Eds.), Biology of Stress in Fish. Fish Physiology Series, vol. 35, pp. 365–403.

Yang, T., Nowack, B., 2020. A meta-analysis of ecotoxicological hazard data for nanoplastics in marine and freshwater systems. Environ. Toxicol. Chem. 39 (12), 2588–2598.

Yu, X., Xu, C., Liu, H., Xing, B., Chen, L., Zhang, G., 2015. Effects of crude oil and dispersed crude oil on the critical swimming speed of puffer fish, *Takifugu rubripes*. Bull. Environ. Contam. Toxicol. 94, 549–553.

Yuan, H., Wang, Y.-Y., Cheng, Y.-Y., 2007. Mode of action-based local QSAR modelling for the prediction of acute toxicity in the fathead minnow. J. Mol. Graph. Model. 26, 327–335.

Zhang, L.-J., Qian, L., Ding, L.-Y., Wang, L., Wong, M.H., Tao, H.-C., 2021. Ecological and toxicological assessments of anthropogenic contaminants based on environmental metabolomics. Environ. Sci. Ecotechnol. 5, 100081.

Zhang, Y., Claireaux, G., Takle, H., Jørgensen, S.M., Farrell, A.P., 2018. A three-phase excess post-exercise oxygen consumption in Atlantic salmon *Salmo salar* and its response to exercise training. J. Fish Biol. 92, 1385–1403.

Zhao, C.-M., Campbell, P.G.C., Wilkinson, K.J., 2016. When are metal complexes bioavailable? Environ. Chem. 13, 425–433.

Zelikoff, J.T., 1998. Biomarkers of immunotoxicity in fish and other nonmammalian sentinel species: predictive value for mammals? Toxicology 129, 63–71.

Zelikoff, J.T., Raymond, A., Carlson, E., Li, Y., Beaman, J.R., Anderson, M., 2000. Biomarkers of immunotoxicity in fish: from the lab to the ocean. Toxicol. Lett. 112–113, 325–331.

Zielińska, K., van Leeuwen, H.P., Thibault, S., Town, R.M., 2012. Speciation analysis of aqueous nanoparticulate diclofenac complexes by solid-phase microextraction. Langmuir 28, 14672–14680.

Zimmer, E.I., Preuss, T.G., Norman, S., Minten, B., Ducrot, V., 2018. Modelling effects of time-variable exposure to the pyrethroid beta-cyfluthrin on rainbow trout early life stage. Environ. Sci. Eur. 30, 36.

Chapter 4

# Consequences for fisheries in a multi-stressor world

**Shaun S. Killen[a,*], Jack Hollins[b], Barbara Koeck[a], Robert J. Lennox[c], and Steven J. Cooke[d]**

[a]*Institute of Biodiversity, Animal Health and Comparative Medicine, College of Medical, Veterinary and Life Sciences, Graham Kerr Building, University of Glasgow, Glasgow, United Kingdom*

[b]*University of Windsor, Department of Integrative Biology, Windsor, ON, Canada*

[c]*Laboratory for Freshwater Ecology and Inland Fisheries (LFI) at NORCE, Norwegian Research Centre, Nygårdsporten 112, Bergen, Norway*

[d]*Fish Ecology and Conservation Physiology Laboratory, Department of Biology and Institute of Environmental and Interdisciplinary Science, Carleton University, Ottawa, ON, Canada*

[*]*Corresponding author: e-mail: shaun.killen@glasgow.ac.uk*

## Chapter Outline



Marine and freshwater fisheries are more important than ever for sustaining human populations but are also facing unprecedented threats from the combined effects of multiple environmental stressors. Here we review how the rapidly changing abiotic environment of fish may affect interactions between fish and fishers, at both the individual and population levels. Throughout, we highlight the role of physiological mechanisms underlying the sensitivity of fish to multiple stressors and their interactions with fishing gears. For each step in a typical capture sequence, we discuss how stressors can alter the behavioral and physiological mechanisms of capture and potential recovery after release

**Fish Physiology, Vol. 39B. https://doi.org/10.1016/bs.fp.2022.06.001**

or escape. We also consider possible feedbacks among fishing practices, environmental stressors, and physiological response of fish, including the potential for harvest-associated selection and evolutionary effects. Fisheries can also induce changes to the biotic environment, including changes in population density, species interactions, and prey density, which can in turn alter the physiology of individual fish, entire ecosystems, and the fisheries themselves. We conclude by highlighting priority research areas required to advance our understanding of the effects of multiple stressors on fish physiology and behavior within the context of global fisheries.

## 1 Introduction

With the technological advances of fishing practices and gears, fishing has transitioned from hunter-gatherer subsistence fishing to a consumptive global commodity (Pitcher and Lam, 2015). Fish are still the primary protein source for 17% of the world's population and are among the most traded food commodities (FAO, 2019). Around the globe, a diversity of fishing practices exist, spanning from recreational hook and line fishing and small-scale artisanal fisheries to commercial fishing freezer vessels targeting inland and marine species from the coast to open ocean. Together, with the industrialization of commercial fishing (Finley, 2016) and the more recent diversification of recreational fishing practices (Cooke et al., 2021), nearly all fish species are directly targeted or indirectly impacted by some form of fishing practice (FAO, 2020).

Single stressors may not have considerable deleterious effects on fitness and survival of animals, but they may when occurring in combination with other stressors. Multi-stressor effects are well known in ecotoxicological research, where co-occurring pollutants can have synergistic effects, decoupling the effects of single chemical pollutants (Kimberly and Salice, 2015). For instance, Monteiro et al. (2020) show the additive effects of hypoxic conditions and mercury contamination, which often co-occur in neotropical freshwater ecosystems, causing the impairment of cardiac output of fish. The effect of multiple stressors, in order to better capture the environments animals are experiencing in the wild, are also increasingly studied in a broader ecological context (see for, e.g., Halfwerk and Slabbekoorn, 2015 for a multi-modal approach of sensory pollution; Fu et al., 2018; Hecky et al., 2010). In the context of fisheries, research shows that the outcome of a fish-fishing gear interaction is greatly impacted when interacting with other environmental stressors. For instance, while fish may be able to adjust to fluctuating temperature regimes, including temperatures temporarily exceeding their thermal optimum (Johansen et al., 2021), increased water temperatures can reduce swim performance, possibly increasing the risk of capture by fishing trawls (Hollins et al., 2018) or reducing recovery potential and chances of survival after discard from a vessel (Gale et al., 2013).

In addition to direct fishing or harvest, fish increasingly experience a multitude of other stressors that may interact with the physiological disturbance caused during fishing, producing effects at various levels of biological organization (see other chapters in this book). To persist in a multi-stressor world, fish need to physiologically adjust and/or adapt to new habitat characteristics or shift to new habitats to meet their biological requirements and limit chronic stress that may affect fitness and survival. Despite a wide range of existing fishing gears and practices, all fishing gears exploit the natural behavior and performance of fish, such that changes in fish behavior via environmental effects on physiology, will affect the interactions between fish and fishing gears. Being predominantly ectothermic, teleost fish are particularly sensitive to changes in environmental temperature (Porter and Gates, 1969). Temperature changes near species' maximum or minimum tolerance thresholds can represent a major stressor causing impairment of cardiac and metabolic functions, increasingly the likelihood of mortality, or producing sublethal effects on physiological performance ( Jensen et al., 2017). Environmental hypoxia is an increasing global concern in freshwater and marine habitats (Breitburg et al., 2018), and in fish can have numerous effects including alteration of cardiac function, oxygen delivery (Farrell and Richards, 2009), and spontaneous swimming activity (Schurmann and Steffensen, 1994; Metcalfe and Butler, 1984). Sensory and chemical pollution are additional stressors which can affect environmental cue perception (Halfwerk and Slabbekoorn, 2015), increasing stress in fish and more generally disrupting biological rhythms (Celi et al., 2016). Waterborne pharmaceutical residues of anxiolytic drugs can affect the functioning of the central nervous system of fish by reducing overall neurotransmission, which have shown to impact the migration propensity of Atlantic salmon in riverine systems (Hellström et al., 2016). Artificial light at night in urbanized waters affects biological rythms such as sex steroids and gonadotropin production (Brüning et al., 2018), increases metabolism and disrupts natural circadian rhythms in fish (Pulgar et al., 2019). Changes to the physical environment of fish, such as water stratification, currents, and water flow will additionally affect the sensory reach of environmental cues by fish. Overall, not only may such multiple, co-occurring stressors amplify or compound the physiological stress caused in fish during fishing, but they may also affect the vulnerability of fish to various fishing practices at both the individual and population levels.

In this chapter we outline, based on examples of existing fisheries but also on knowledge of physiological processes gained from experimental work, how changes of the abiotic environment of fish may affect the interaction between fish and fishing gears interaction and ultimately shape the fisheries of our world. This includes how within-generational plasticity to environmental stressors may modulate vulnerability to fishing gears and the potential for selection by fishing practices. This chapter is constructed based on the consecutive steps of a fishing sequence, i.e., habitat selection, encounter, and interaction of fish and gear, and capture. For each step, we outline how environmental stressors modulate the corresponding behavioral and physiological mechanisms involved in the fish-gear interaction. Additionally, we discuss possible feedback loops between fishing practices, environmental stressors, and physiological response of fish. Beyond the direct impact of fishing practices on population level mortality rates of a fish population, indirect effects and sublethal effects are numerous and can affect fish populations via mechanisms of selection, possibly producing transgenerational effects (Hollins et al., 2018) and constraining adaptation to climate variability (Morrongiello et al., 2019). While we focus on how abiotic stressors can shape fish-fisheries interactions, there are likely to be numerous cascading ecosystem-level effects. To this end, we discuss how modifications of species interactions and food webs in response to environmental conditions can affect fisheries, including interactions with the expansion of non-native species.

## 2 Habitat use and availability to fisheries

To be available to a fishery, fish must overlap in space and time with the gears being used to capture them. This will largely depend on the habitat that the fish occupy, either by active choice (e.g., choosing an area with a particular prey) or by avoidance of areas outside of their abiotic environmental preferences or limits (e.g., avoiding areas above a temperature threshold). At the narrowest spatiotemporal scale, fish make decisions on a moment-to-moment basis about which microhabitat they will occupy. Conversely, range shifts can occur at the scale of hundreds or thousands of kilometres over the course of years, decades, or centuries. The exact causes of this variance in habitat use is the focus of a great deal of research, and while in most cases these remain elusive, habitat use by fish and their subsequent availability to fisheries is likely to be influenced by a range of environmental factors that interact with fish physiology.

### 2.1 Habitat selection and microhabitat use

Within and among fish species, there is wide variation in space use associated with energy requirements, performance, and behavioral traits. For example, factors such as water velocity, food abundance, predation risk, water depth, temperature and oxygen availability may all affect microhabitat selection. This has been well-studied in stream-dwelling salmonids, for example, whereby individuals can face a trade-off between occupying faster flowing waters with increased drift feeding opportunities but with greater energetic costs while holding station, and lower flow areas where less energy is spent on swimming but where there is less prey availability (e.g., Fausch et al., 1997). Intrinsic variation in risk-taking tendency, spontaneous activity, and exploration will also influence space use of fish, and these behavioral traits can show context-dependent links with various aspects of physiology including metabolic rate, locomotor performance, and hormone status (although see

Baktoft et al., 2016). Individuals that are inactive or spend more time in shelter, for example, tend to be those with a lower metabolic rate and increased stress-responsiveness (Metcalfe et al., 2016). In turn, these individuals may be less likely to encounter passive fishing methods such as traps or anglers because these depend on the fish to encounter and interact with the largely stationary gear (Hollins et al., 2018). As such, the associated fisheries-associated selection could possibly generate a "timidity syndrome" among the population of fish that remain uncaptured (Arlinghaus et al., 2017). Notably, however, the links between habitat use and vulnerability may be highly dependent on the type of gear being used. More active gear such as seines, for example, may be more likely to capture shy individuals or those in shelters, because they are less likely to escape the path of the net (Wilson et al., 2011).

Factors such as temperature and hypoxia can have independent and combined effects on the metabolic rates and aerobic capacity of fish (Claireaux and Chabot, 2016; Claireaux et al., 2000) and it is likely that, via effects on behavior, these factors will affect habitat use and the potential to encounter deployed fishing gears. Temperature, for example, has strong effects on foraging activity and choice of depth of occurrence in Arctic char *Salvelinus alpinus*, likely affecting their ability to be targeted by specific gear types in relation to the prevailing environmental conditions (Guzzo et al., 2017). Importantly, however, the exact effects of these environmental factors on habitat use will depend greatly on the magnitude of the change in conditions in relation to the "baseline" conditions, and over what time scale (Evans, 1990). For example, if an increase in temperature increases spontaneous activity in fish, there may be an increasing their space use and the likelihood of encountering a fishing gear. With acclimation, however, metabolic compensation will occur and activity will partially return to the level that occurred at the cooler temperature (Evans, 1990). At extremely high temperatures beyond a species' thermal optimum, such as that which can occur during heat waves (Mameri et al., 2020), activity may actually decrease if fish experience neuromuscular dysfunction or a diminished aerobic capacity for activity or digestion (and hence foraging). Hypoxia can have similarly variable effects on behavior: while mild hypoxia can decrease shelter use and exploration, especially for those with a high metabolic rate, severe hypoxia can suppress variation in activity among the majority of individuals and thus alter vulnerability to specific gear types at the population level (Killen et al., 2012).

Variation in depth preference among species or individuals will also influence their spatial overlap with fishing gears and can also be related to physiology. In warmer years, for example, some species will occupy greater depths with cooler water, or individuals with a general preference for cooler temperatures may consistently prefer deeper environments (Guzzo et al., 2017). In Atlantic cod *Gadus morhua*, the tendency for diel vertical migration is a repeatable trait, and individuals that make periodic migrations to shallower depths are more likely to be captured by passive fishing gears as compared to those that stay in deeper water (Olsen et al., 2012). Targeted fishing on depth-associated phenotypes has been shown to cause changes in allele frequencies in exploited Atlantic cod populations (Árnason et al., 2009). Climate-associated hypoxia is also expected to restrict the depth of many pelagic species to more well-oxygenated surface layers. In blue sharks, for example, a shallower oxygen minimum zone associated with warmer has caused a decrease in dive depths, essentially compressing their vertical distribution in the water column, and increasing their catch rates by long-line fisheries (Vedor et al., 2021).

## 2.2 Range shifts

At broader spatiotemporal scales, interactions between the environment and fish physiology are already altering the habitable ranges of economically and ecologically valuable species. Overall, there is a general trend toward more poleward distributions of species or a shift to greater depths (Dulvy et al., 2008; Gaines et al., 2018). Although models that incorporate physiological mechanisms predict a continuation of these trends (Cheung et al., 2011), there is significant controversy regarding the exact physiological mechanisms underlying such shifts. The oxygen and capacity limited thermal tolerance (OCLTT) hypothesis, for example, has posited that constrained aerobic scope at temperatures beyond a thermal optimum should limit species' capacity for activity, growth, and reproduction, and therefore limit their geographical range (Pörtner and Knust, 2007; Pörtner et al., 2017). While this basic principle has been a component of attempts to model future ranges of fish species, the empirical evidence for these effects is mixed. For example, there are very few species for which adequate physiological data have been collected to effectively model changes in aerobic scope with temperature (Nati et al., 2016), and among those species that have been studied, many show no obvious optimum temperature for aerobic scope (Lefevre, 2016). Further attempts to model the geographical distribution of fish species have included interactions between temperature and oxygen availability to derive estimates of a metabolic index that generally coincides with the current distributions of species for which sufficient physiological data is available (Deutsch et al., 2015, 2020). While this work highlights that the physiology of fish species is likely critical in determining the geographical range of species in response to stressors, there remain many unknowns in this area. For example, even if current modeling approaches predict that a species could inhabit a given range given their responses to temperature, additional stressors including food or habitat availability may render an area unsuitable.

Ongoing and future changes in the range distribution of species will strongly impact fisheries. Commercial fishers may need to relocate fishing efforts to follow the population, possibly spending more time in transit to and from fishing grounds, or shift efforts to other species. Relocations of fisheries in the northeast United States, for example, have shown a tremendous time lag relative to the range shifts of their targeted populations (Pinsky and Fogarty, 2012). It should be noted that while temperate and tropical fisheries are generally predicted to decline in response to future climate changes, fisheries in polar regions may actually experience an increase in productivity (Campana et al., 2020), though species interactions between native species in these areas and an influx of species experiencing range shifts is unknown. It is also important to note that, to date, the vast majority of research has focused on marine species while changes in the habitable shifts of freshwater species remains relatively unknown. This is a major area for future research given the large human populations across South America, Africa, and Asia, that are dependent on freshwater fisheries and are in geographical areas predicted to be strongly impacted by warming.

## 3 Gear encounter and interaction

Even if a fish and fishing gear are in the same general location at the same time, there are numerous environmental, physiological, and behavioral factors that will determine whether a fish encounters and interacts with the gear (Lennox et al., 2017). Encounter rates are modulated by fish activity such that, depending on the specific gear being used, more active individuals may be more likely to encounter a deployed gear within their home range or core activity spaces. Larger home ranges will also increase the probability of a fish overlapping with the active space of a gear across space and time. Active, compared to passive, fishing gears have different modes of function such that passive gears are more reliant on the activity rates of the fish than active gears, which actively pursue fish. In addition, passive gears may require the fish to be stimulated to interact, relying on the behavior and physiology of the individual to compel it to the gear. Here, we briefly review how stressors can enhance or reduce the rates at which fish become vulnerable followed by an assessment of how stressors affect the nature of these interactions between fish and fishers.

Catchability and availability are population-level traits relevant to fisheries that are modulated by the behavior and physiology of individuals (Arreguín-Sánchez, 1996), rendering them vulnerable or invulnerable to capture. Vulnerability is an unobservable individual trait that only becomes confirmed once a fish interacts with a gear. The internal state of the animal must be such that it is physiologically and behaviorally primed to be captured, in other words, something is motivating the individual to move or feed in an area where gear is active. Fish that are hiding or satiated are generally not vulnerable to gear and will not encounter or interact with gear. Fish that are hungry and at ease in their environment are expected to be readily vulnerable to fisheries. Within a population, individual variation in metabolism (Redpath et al., 2009), boldness (Redpath et al., 2010), sociality (Louison et al., 2018), and other traits will determine the activity levels and risk taking of fish, including relevant traits such as flight initiation distance, risk tolerance, etc., that affect the individual's response to stressors and vulnerability to capture.

When an animal's environment shifts, their physiology undergoes corresponding changes that affect the individual's likelihood of encountering or interacting with fishing gear. The effects of such stressors are threshold-dependent, such that individuals and species may have tolerance limits that enhance performance up to an optimum before decreasing performance, with consequences for their vulnerability to fisheries capture. Extreme environmental stressors such as high temperature or hypoxia, for example, shift the animal's physiology and will alter habitat selection or behavior. Van Leeuwen et al. (2021) showed how warm water temperatures in rivers affect catches of Atlantic salmon, implying that the environmental change affected the interaction of fish with the fishing gear. Migrating salmon are not feeding and are perhaps a special case, but similar effects can be anticipated when fishing with passive gears that rely on the fish's volition to be captured. Angling, longlining, fyke netting, and other traps should therefore have reduced efficacy in extreme weather. Active gears, however, may experience enhanced catch when individuals are beyond their physiological optimum and less able to escape gear such as trawls, unable to move as well to avoid capture. Killen et al. (2015a,b) demonstrated how anaerobic capacity, which may be reduced at warm temperatures for fish, affects trawl captures and Thambithurai et al. (2019) revealed similar patterns for environmental hypoxia. Taken together, fish living in rivers, lakes, and coastal zones where water temperatures are increasing with climate change should be increasingly vulnerable to active gears and decreasingly vulnerable to passive ones. Areas of the deep sea where oxygen minimum zones are rising may also yield less reactive fish that are more easily trawled. Acidification caused by acid rain in freshwater and warming on reefs may also affect swimming performance and vulnerability to fisheries in these highly exploited environments. Environmental pollutants may have variable effects on fish, with stimulants such as anxiolytic drug effluents stimulating behavior that would enhance gear encounters and vulnerability (Brodin et al., 2013). Other pollutants, such as noise, repel fish from affected sites and can be expected to reduce gear encounter rates proximate to the source, reducing gear encounter and capture probabilities (Filous et al., 2017). Koeck et al. (2020) showed that memory and learning play a key role in a population's vulnerability to capture but this probably also competes against food availability and hunger. Stressors that increase hunger or shorten memory retention in fish may have profound effects on their willingness to explore and interact with gear.

## 4 Capture and escape or release

Despite the diverse range of fishing gears across fisheries sectors, fish cannot be captured without causing some level of injury or stress which can interact

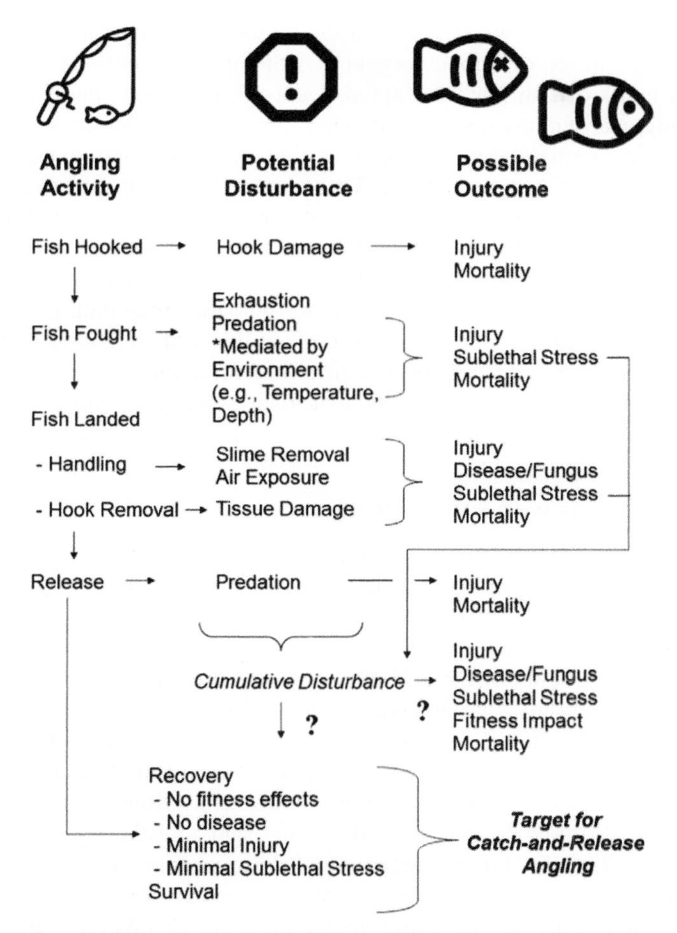


**FIG. 1** Flow chart of the potential disturbances, mediators of physiological stress, and potential fish fates occurring throughout an angling event in a recreational fishery. Question marks highlight areas where the cumulative effects of numerous stressors on the potential for recovery after release are largely unknown.

with various environmental stressors, possibly leading to immediate or delayed mortality after release or escape (Fig. 1). The number of fish that are released to comply with fisheries regulations, the conservation ethic of the fisher (e.g., as is common with voluntary catch-and-release in recreational fisheries) or because of a lack of market value (e.g., in some commercial fisheries), is by no means small. In the recreational sector alone, Cooke and Cowx (2004) estimated that as many as 30 billion fish may be released on an annual basis. Understanding the fate of fish that are released and developing strategies to reduce injury, stress, and mortality are thus of high priority to fisheries managers and fishers alike, especially in the face of numerous concurrent environmental stressors which may exacerbate the physiological stress experienced during capture or impair recovery. Although it is common to think of fishing in terms of fish that are captured and released, fish can also interact with fishing gear and escape without being landed. For example, a fish could escape from a commercial gill net or break the line when being reeled in by an angler (Chopin and Arimoto, 1995). These interactions can lead to sublethal impacts and collateral mortality but are just beginning to be explored in many fisheries (Falco et al., 2022), including interactions with various environmental factors. Here we consider the physiological consequences of different aspects of the fish capture process and the potential modulating effects of additional environmental stressors. We preface this text by noting that given the incredible diversity of fish species, fisheries techniques/gears, fisher behaviors, and environmental conditions that fish experience, our attempts to generalize will always yield exceptions.

## 4.1 Interactions with fishing gears

From the moment a fish is hooked, entrapped, or entangled, there is a neuroendocrine cascade that leads to a stress response and associated physiological adaptations (Barton, 2002). Fish will first attempt to escape and may pull on the line, struggle in a net, and/or search for an escape path from a pursuing gear. Beyond activation of the HPI axis, fish will often engage in high intensity locomotor activity that includes burst swimming. Burst swimming is fueled by anaerobic metabolism so tissue energy stores such as ATP, glycogen and PCr are rapidly depleted leading to an oxygen debt and state of physiological exhaustion (Kieffer, 2000). Warmer acclimation temperatures and exposure to hypoxia can both alter liver and muscle glycogen levels, possibly affecting the times needed for fish to become exhausted (Yang et al., 2015). At the same time metabolites accumulate in tissues and acid-base imbalances occur creating metabolic acidosis (Wood et al., 1983). Rapid depletion of tissue energy stores, and associated acidosis in fish tissues generally coincides with the onset of fish exhaustion but does not necessarily equate to fish capture. For fish caught on rod and reel, the onset of fatigue prevents fish from further resisting capture, after which point they are easily landed. However, fish caught on longlines or in gill nets may show repeated incidences of struggle and recovery throughout these gears' soak times (Guida et al., 2016), while exhausted fish pursued by a trawl often escape capture by passing under/over the net mouth (Ryer, 2008).

Increases in water temperature and reductions in oxygen availability will likely increase the rate of energetic resource depletion and the onset of acidosis in fish tissues as they attempt to escape or unhook themselves from fishing gears (Gale et al., 2013), exacerbating stress responses and subsequent physiological disruption. In the case of rod and reel fisheries, environmentally-induced reductions in swim performance (e.g., Domenici et al., 2013) may decrease the fight times of angled fish in certain instances, although fish can sometimes be landed before such severe physiological disruptions occur (Shea et al., 2022). In contrast to rod and reel fishing, the extended soak times of longlines or gillnets mean that there is greater opportunity for the extended/repeated struggling of caught fish to cause significant disruptions to fish homeostasis, with potential consequences for both recovery after release/escape (see below), and fish condition upon the retrieval of the gear. Physiological disruptions likely exacerbated by high temperatures and reduced oxygen availability, such as elevated lactate and decreased blood/tissue pH, are often associated with reductions in meat quality from harvested fish (Anders et al., 2020). Under future climate change scenarios, minimizing the stress experienced by commercially important fish species may become increasingly important. In future, fishers may need to evaluate the tradeoffs between lost income due to declining fish condition, and the relative cost of performing multiple, shorter gear sets to try and reduce the severity of physiological disruptions of target fish.

## 4.2 Handling

Once a fish is landed, how it is handled can play a large role in the ultimate outcome of the fishing interaction for the fish. In some cases a fisher may reach down into the water and slide out the hook or cut the line without bringing the fish into a vessel or onto shore. In such instances, handling is negligible but to do so may require a fish that is already exhausted from the fishing interaction (e.g., an angler using a protracted fight time to enable exhaustion). However, it is more common for a fish to be fully landed whereby they are brought into the "possession and control" of the fisher (e.g., in a landing net, on a boat deck, in the hands of the fisher, or on shore). In some cases, fish are crowded in nets (e.g., a purse or beach seine) at time of landing, which can lead to localized depletion of dissolved oxygen (i.e., hypoxia) and stress (Tenningen et al., 2012). It is common for fish to be exposed to air during handling. Air exposure is not surprisingly a rather severe stressor for fish characterized by collapse and adhesion of gill filaments, severe bradycardia, inability to respire, and a host of biochemical alterations. Beyond some threshold tissue damage arising from lack of oxygen is so severe that a fish will die. In a synthesis of air exposure studies, Cook et al. (2015) suggested 10 s as a conservative, cross-species and context value for a suggested maximum air exposure target when species- or context-specific values are unavailable. In some cases, this is easily achievable but in other cases such as challenges with removing fish from the gear, fisher inexperience, or the volume of fish that need to be sorted and handled (e.g., in non-selective gears such as seines where fish are landed *en masse*), air exposure can last 20 min or more (Raby et al., 2012). Such durations may not necessarily be injurious but require research to understand outcomes and best practices for handling such situations, and the effects of other additional stressors such as thermal history and prior oxygen availability (e.g., hypoxia during the time of capture). In trawl fisheries, air-temperature during on-deck sorting may strongly affect the degree of physiological stress experienced by fish, and so warming conditions or heat-waves may cause an increase in the mortality that occurs during this stage of the fishing process.

## 4.3 Recovery and fitness impacts

When fish are released after capture they can be in a range of physiological conditions. Some fish are able to maintain equilibrium and are vigorous while others are near death and unable to swim (Davis, 2010). Even fish that are able to swim may experience cognitive impairments that lead to them having difficulty in assessing risk and making risk averse decisions (Cooke et al., 2014). Fish with locomotor impairments may be subject to predation (Raby et al., 2014). Some researchers have evaluated different tactics for facilitating recovery of exhausted fish in an attempt to expedite physiological recovery and reduce mortality. Farrell et al. (2001) showed that use of a recovery box allowed coho salmon that were classified as lethargic to be vigorous as little as 15 min later (with accompanying recovery of tissue energy stores) while Brownscombe et al. (2013) restored locomotor activity of bonefish by temporarily holding them in flow-through recovery bags. Yet, the science is mixed and in other instances there is little evidence that recovery can be facilitated (e.g., Robinson et al., 2013). In other words, it is much better to ensure that fishing and handling practices are optimized such that fish are not exhausted at time of release than trying to facilitate recovery of exhausted fish.

The mechanisms by which fish die after release or escape are varied. Some fish that are bleeding may survive long enough to be released but die later. However, minor injuries such as scale loss or abrasion can in time (days to weeks) provide an entry route for opportunistic pathogens such that fish can die well after release. Given that stress such as that arising from capture and handling can also impair immune function (Tort, 2011), such minor injuries can turn into major infections. In some cases, fish are sufficiently exhausted that they are unable to recover, presumably due to tissue oxygen limitation and associated damage to the heart and/or brain (Farrell et al., 2009) or because of extreme acid-base imbalance (Wood et al., 1983). As noted above, if fish are exhausted and unable to escape from predators then they can be killed by predators, while exhausted fish which are obligate ram-ventilators may be unable to sufficiently oxygenate their gills through swimming, preventing recovery. While rates of recovery from exhaustive exercise show a large degree of variation among ecologically distinct species, elevated temperatures are typically associated with elevated metabolic debt after exercise, as well as increased rates and occurrence of post release mortality in released fish (Clark et al., 2017). Specifically, elevated temperature and hypoxic conditions during recovery can slow recovery of intramuscular ATP, PCr, and lactate, and plasma levels of glucose following exhaustive exercise (Suski et al., 2006). For example, Wilkie et al. (1996) revealed that angling of Atlantic salmon in warm summer water impairs restorative processes and increases the susceptibility of Atlantic salmon to delayed post-angling mortality. Under continued warming scenarios, the probability of and rate of recovery after exhaustive exercise may be lowered in fish, potentially elevating both post release mortality, and rates of depredation in future (Gale et al., 2013). In addition to the impact of capture and handling stress, environmental temperature has also been shown to directly influence the efficacy of fish immune response. While low temperatures are known to inhibit fish immune response (Butler et al., 2013), elevated temperatures can both reduce the immunocompetency of fish, as well as increase proliferation of pathogens (Shameena et al., 2021).

Although mortality is the outcome of greatest concern to fisheries managers, sublethal impacts can also be relevant to fitness. For example, stress associated with fishing that occurs prior to or during reproduction may suppress reproductive hormones (Pankhurst and Dedualj, 1994), delay reproduction (Ostrand et al., 2004), impede spawning migration (Thorstad et al., 2007), influence gamete development (Hall et al., 2009) or offspring quality (Ostrand et al., 2004), impair parental care (Kieffer et al., 1995), and even reduce reproductive success (Richard et al., 2013). Reproductive indicators related to fitness are challenging to study in wild fish but these aforementioned examples suggest that more research is needed. Other fitness impacts can occur as a result of feeding impairments (Siepker et al., 2006) or other pathways that impact growth (Meka and Margraf, 2007) although compensatory growth is common (Cline et al., 2012).

## 5 Feedbacks between fisheries and stressors

In addition to the effects on individual fish that have so far been described in this chapter, fisheries have wide ranging impacts on exploited ecosystems, and fishing is itself a "multi-stressor," potentially causing pervasive, sublethal impacts to fish populations with consequences for species of both commercial and recreational interest (Fig. 2). For example, while fisheries harvest is a direct source of mortality for many fish species, the concurrent destruction of critical habitat (Wheeler et al., 2005) and associated noise of vessel traffic (Celi et al., 2016) which also occurs, each constitute additional stressors in their own right. Furthermore, the selective nature of fisheries harvest has likely altered the life history (Heino et al., 2015), behavioral (Uusi-Heikkilä et al., 2008) and physiological (Hollins et al., 2018) traits of exploited fish populations, with potential consequences for their capacity to adapt to the other stressors discussed in this chapter (Crespel et al., 2021a).

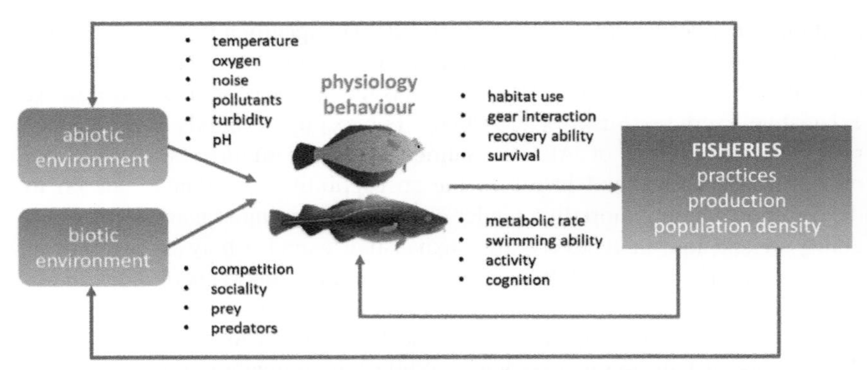


**FIG. 2** Potential feedbacks among environmental variables, fish, and fisheries practices. Various abiotic and biotic factors will have direct and indirect effects on the physiology and behavior of fish. This includes natural environmental variation but also anthropogenic stressors. The physiological state and behavior of fish will determine their vulnerability to being captured as well as the physiological stress they experience during the capture process. The fishing process itself will then feedback to affect environmental conditions, especially due to noise, pollution, habitat degradation, and changes in the population densities of targeted species. These alterations will go on to further modify fish physiology and behavior.

The following section discusses the indirect stressors fisheries activity exert on fish, and the role that fisheries harvest may play in limiting the adaptive potential of exploited fish populations to further environmental disturbances.

Fishing practices may directly alter habitats in a way that increases the stress experienced by fish living within a given region. For example, demersal trawls can clear large swathes of structured, complex benthic ecosystems which provide critical sheltered habitat for fish species (Koslow et al., 2001; Kritzer et al., 2016; Yesson et al., 2017). In addition to potentially depriving fish of habitat-structure necessary to complete critical life history functions (Caddy, 2008), lack of adequate shelter can increase the risk of predation in demersal fish species (Brooker et al., 2013; Quadros et al., 2019), potentially leading to sublethal stress effects which can impact individual fitness. For example, a lack of available shelter has been shown to increase measures of basal metabolic demand in fish (Chrétien et al., 2021; Millidine et al., 2006), potentially due to increased costs of vigilance (Killen et al., 2015b). These increased metabolic costs may reduce resources available for growth and reproductive investment, with subsequent reductions in fitness or capacity to adapt to further stress, or otherwise increase energetic demand in a resource limited environment. Fishing activity can also induce further stress responses by exposure to vessel noise, which has been shown to elevate metabolic demand and heart rate in exposed fish (Graham and Cooke, 2008; Simpson et al., 2016), as well as increasing circulating levels of biochemical stress indicators (Celi et al., 2016). Cumulatively, these indirect-fisheries stressors may contribute to reductions in fitness of wild fish populations inhabiting exploited ecosystems, with potential consequences for future fisheries yield. The potential fitness impacts of these indirect fisheries-stressors have yet to be studied in the wild, but any observed impact would be difficult to attribute to indirect effects of fisheries, as the selective nature of fisheries harvest (Heino et al., 2015), and density dependent effects (Crespel et al., 2021a,b) might also be expected to impact life history traits in similar ways.

Removal of fish from wild populations by fisheries harvest not only elevates the mortality experienced by fish stocks, but can constitute a strong selective pressure to which they must adapt (Heino et al., 2015). Where mortality extends to immature life history stages in exploited fish populations, fisheries harvest selects for individuals which can successfully reproduce at earlier ages and smaller sizes (Enberg et al., 2012; Heino et al., 2015), driving population level change in life history traits. Fisheries harvest is also selective for specific life history (Heino et al., 2015), behavioral (Uusi-Heikkilä et al., 2008), and physiological (Hollins et al., 2018) traits in fish such that individuals exhibiting certain phenotypic traits are more likely to be caught than others of the same species. If there is a heritable genetic basis for individual traits which determine capture vulnerability in fish, these phenotypic changes can constitute a true evolutionary response, in a phenomenon known as fisheries induced evolution (FIE). Fisheries harvest therefore has the capacity to drive phenotypic change in exploited fish populations, with consequences for the adaptive potential of those populations when faced with further environmental stressors.

In addition to influencing population resilience and recovery potential through impacts on life history traits, fisheries selection may also directly alter the physiological traits present in exploited fish stocks (Enberg et al., 2012; Hollins et al., 2018). Individual variation in physiological traits has been shown to correlate with risk of capture in both active (Killen et al., 2015a; Hollins et al., 2019) and passive fishing gears (Koeck et al., 2019; Redpath et al., 2010) and has also been shown to underpin a range of behaviors which influence capture vulnerability (Andersen et al., 2016; Arlinghaus et al., 2017; Diaz Pauli and Sih, 2017; Metcalfe et al., 2016). For example, angling was shown to selectively remove individual rainbow trout with low neuroendocrine stress responsiveness and high activity rates (Koeck et al., 2019), leaving an uncaptured population composed of more stress-responsive, low activity individuals. Experimental studies have shown that active gears also have the capacity to selectively remove individuals with specific physiological traits. In a simulated trawl fishery, European minnow (*Phoxinus phoxinus*) with greater anaerobic metabolic capacity were more likely to avoid trawl capture (Hollins et al., 2019; Killen et al., 2015a,b) through a mechanism of higher swim performance, suggesting physiological selection in trawl fisheries may lead to fish populations with high anaerobic capacity and associated swim performance. However, relationships between individual behavioral and physiological traits and capture vulnerability in both active and passive gears have been shown to be highly context dependent (Hollins et al., 2019, 2021), as these traits themselves show high plasticity (Killen et al., 2016). This makes the net outcome of any physiological selection difficult to predict, as the distribution of traits within a population, and thus the potential for any selection to occur, will be strongly mediated by environmental conditions. For example, in low oxygen and high temperature conditions, swim performance in targeted fish populations may be so reduced, and capture rates resultantly high, that selection is effectively obviated (Thambithurai et al., 2019). Nevertheless, many physiological traits show evidence of heritability (Ferrari et al., 2016; Volckaert et al., 2012) and repeatability (Norin and Malte, 2011), as well as correlations with other traits relevant in determining capture vulnerability in fishing gears (Hollins et al., 2018; Metcalfe et al., 2016), or influencing capture vulnerability in their own right (Hollins et al., 2019; Killen et al., 2015a,b). Therefore, where the direction of fisheries selection on physiological traits has been consistent (e.g., low activities and swim performances may be consistently selected against in trawls; Diaz Pauli et al., 2015; Hollins et al., 2019; Killen et al., 2015a,b; Thambithurai et al., 2019), resulting phenotypic change in the physiological traits of exploited fish populations seems possible.

Any fisheries-induced phenotypic change in physiological traits of wild fish populations could have consequences for ecosystems and population resilience to environmental disturbance/stress. For example, the physiological traits of fish populations which experience heavy trawl fishing may be skewed toward individuals with high anaerobic metabolic capacity, a trait which is also associated with prolonged recovery times after exhaustive exercise (Clark et al., 2017), but also resilience to low oxygen availability at high temperatures (Sørensen et al., 2014). The impacts of severe ocean weather events, such as localized extremes in temperature increase or oxygen deprivation (Bates et al., 2018) may therefore be lessened in these fish populations. However, unlike aerobic metabolism, anaerobic metabolism cannot be sustained for long periods due to continual accumulation of lactate and subsequent onset of metabolic acidosis, a process which will be accelerated under conditions of ocean warming (Clark et al., 2017). Shifts toward high performance phenotypes as a result of fisheries selectivity may therefore have further implications for fish species targeted by both trawls and catch and release/recreational fisheries, (e.g., Atlantic cod), where the stress of capture/handling/release may lead to higher rates of mortality under future climate change scenarios. The distribution of physiological traits which determine baseline energetic demand (e.g., standard metabolic rate) in fish populations may also have been altered by fisheries selectivity owing to their likely role in determining behaviors related boldness and feeding motivation (Metcalfe et al., 2016), and the relevance of those behaviors in determining capture vulnerability in passive gears (Biro and Post, 2008; Lennox et al., 2017; Redpath et al., 2010). The relative benefits/costs of high SMR are context specific (Norin and

Metcalfe, 2019; Reid et al., 2012), but fish populations comprised of low SMR individuals may be less able to translate abundant food resources or productivity pulses into enhanced growth, potentially limiting biomass available for fisheries harvest.

The role of aerobic and anaerobic metabolic traits in determining the fitness and abundance impacts of future climate change on fish populations is a contentious issue (Ejbye-Ernst et al., 2016), and so how fisheries selection on physiological traits may interact with the environmental stress of climate change is not clear. That being said, reductions in the diversity of traits present within animal populations are expected to reduce overall population resilience and limit the capacity for populations to adapt to environmental change (Schindler et al., 2010). This destabilizing effect of fisheries selection may exacerbate the impact of environmental stressors on wild fish populations, and may underlie observations of exploited fish species showing greater distributional shifts in response to climate change than unexploited populations (Hsieh et al., 2008).

The high mortality and often size-selective nature of fisheries harvest can truncate size and age structure in wild fish populations (Enberg et al., 2009; Kuparinen and Hutchings, 2012; Swain, 2011), which can lead to unpredictable recruitment success (Hsieh et al., 2006, 2010; Longhurst, 2002), and possibly increase population susceptibility to concurrent environmental stressors (Hsieh et al., 2006; Lehodey et al., 2006). Therefore, while these population-level phenotypic changes may have limited direct impact on overall economic yield in fisheries in the short term, their implications for the resilience of exploited fish stocks, and the capacity for those stocks to recover should fishing effort cease, may be more severe (Eikeset et al., 2013; Enberg et al., 2009; Kuparinen and Hutchings, 2012; (Swain, 2011)). For example, in the Atlantic cod populations off southern Labrador and eastern Newfoundland, Canada (hereafter "Northern cod"), fisheries induced reductions in size at age were apparent by 1960 (Hutchings and Rangeley, 2011; Olsen et al., 2004), but subsequent fisheries "collapse" did not occur until the 1980s and 1990s. In response to precipitous population declines, a moratorium on targeted fishing for Northern Cod was implemented in 1992. Despite this, Northern cod stocks have not recovered, and remain at approximately 2–3% of their 1960 biomass (Hutchings and Rangeley, 2011). Evolutionary reductions in size at age, and earlier maturation have likely limited the rates of population growth in Northern cod through mechanisms of reduced population fecundity (Hutchings and Rangeley, 2011; Swain, 2011), but are insufficient to explain their lack of recovery in its entirety. Indeed, despite drastically reduced fishing mortality experienced by the Northern cod population, overall mortality has actually increased since the moratorium on fishing activity was established (Swain, 2011). This elevated mortality is partially attributed to increased post reproductive mortality in smaller cod (Hutchings and Rangeley, 2011; Swain, 2011), however, increased rates of predation, reductions in egg quality, and interactive effects between earlier maturation and challenging environmental conditions (e.g., resource limitation) are likely even greater contributors to sustained low Northern cod biomass (Hutchings and Rangeley, 2011; Venturelli et al., 2009).

The evolutionary legacy of the collapsed Northern cod population highlights how fisheries harvest can alter the demographic traits of exploited fish populations, and also how these changes can cause synergistic ecosystem interactions to further hinder population recovery. While overall alterations to size at age and growth rate have clear implications for the reproductive output and recovery potential of wild fish populations (Enberg et al., 2009, 2012; Heino et al., 2015), truncated size and age structures of exploited fish stocks may also render populations more vulnerable to the continued impacts of climate change. The influence of environmental conditions on recruitment success in fish populations varies, with some showing tight linkages with variables such as temperature, while for others recruitment success is better predicted by standing biomass, or age at maturity (Longhurst, 2002; Rindorf et al., 2020). FIE impacts on these life history traits may therefore strengthen links between climate and recruitment in heavily fished species, contributing to more variable patterns of abundance over time (Hsieh et al., 2006) and exacerbating the impacts of climate change on future fisheries yield (Ottersen et al., 2013).

## 6 Environmental stressors, species interactions, and fisheries: An example with the introduction of non-native species

Ecosystem structure may be altered by many factors, including interspecific differences in response to a changing climate (Ainsworth et al., 2011; Pinsky et al., 2020; Roessig et al., 2004), the introduction of non-native species by human activities (Gozlan, 2017), and synergistic interactions between the two. The continued impacts of climate change drive increases in water temperature and acidity (Abraham et al., 2013), as well as the propagation of aquatic oxygen minimum zones (Altieri and Gedan, 2015). The combined effects of these shifting environmental conditions can constitute a significant form of physiological stress for native fauna, and biodiversity within impacted ecosystems may be altered as native fauna either leave the now disturbed environment, or struggle to compete with species more readily adaptable to changing conditions (Libralato et al., 2015). Community change in aquatic ecosystems can also occur via competitive interactions between native species and species introduced through human activity (Bando, 2006; Lovell et al., 2006; Martin et al., 2010). Establishment of invasive species can lead to population declines of native species through competitive interactions (Martin et al., 2010), but these interactions themselves will also be modulated by the continued influence of climate change (Coni et al., 2021). Fisheries may therefore be forced to continually adapt to the changing distributions/

availability of traditionally targeted species, or otherwise target newly established species of commercial value. The presence of invasive species also constitutes a biotic stressor for native fishes, possibly influencing physiological stress, habitat use, rates or energy intake, and predation, in ways that interact with various abiotic stressors.

Where non-native species have a competitive advantage over native species, or environmental conditions are otherwise no longer advantageous to native species, the availability of target fish to a given fishery may change. For example, the establishment of the invasive lionfish (*Pterois* spp.) in the Caribbean has been implicated in declines of commercially important Atlantic coral reef fish populations through both mechanisms of competition (Morris et al., 2011; O'Farrell et al., 2014), and direct predation (Green et al., 2012). While rapid somatic growth, a lack of predators, and an abundance of naïve prey have all contributed to the continued invasive success of lionfish (Côté and Smith, 2018), lionfish also exhibit physiological tolerance to a broad range of temperatures (Lower and upper critical thermal ranges of 9.5–16.5 °C and 30–40 °C, respectively) and salinities (daily fluctuations of 28‰) (Barker et al., 2018; Jud et al., 2015), while also showing high starvation tolerance (Fishelson, 1997). While these environmental tolerances are comparable to those of native reef fishes throughout its introduced range, predation pressure can limit the dispersal of native fish to habitats where abundant shelter is available. As lionfish experience very limited predation pressure (Côté and Smith, 2018), and also exhibit broad physiological tolerances, they have been able to successfully colonize a range of non-native habitats in an era of unprecedented environmental change. While reductions in populations of commercially and recreationally important Atlantic reef species have been attributed to the invasive success of the lionfish throughout its non-native range (Ballew et al., 2016; Côté and Smith, 2018), a commensurate shift in targeted fishing effort toward lionfish has not yet occurred, and management approaches including targeted removals, and incentivized harvest are often implemented as a form of population control ( Johnston et al., 2015).

In addition to reducing the availability of resources to native organisms (van Kessel et al., 2011), with subsequent reductions in fitness, the presence of invasive species can also reduce the predictability of those resources (Carpenter et al., 2011). This can influence relationships between individual physiological and behavioral traits and fitness in fish species, (Reid et al., 2012) with potential consequences for the phenotypic composition of native fish populations (Závorka et al., 2017). For example, brown trout parr (*Salmo trutta*) with higher standard metabolic rates (SMR) show more territorial behavior to secure consistent access to high quality habitat (Závorka et al., 2017). Securing territories in this way helps ensure food resources are predictably available, which can confer fitness advantages to these high metabolic rate individuals (Reid et al., 2012; Závorka et al., 2017). However, should the predictability of these resources change, this adaptive advantage may be lost, potentially leading to selection against individuals with high metabolic demands (Killen et al., 2011; Zeng et al., 2017). Indeed, competitive displacement disrupted the relationship between metabolic traits and territoriality observed in brown trout parr at sites where the invasive brook trout (*Salvelinus fontinalis*) was established (Závorka et al., 2017) and was accompanied by a reduction in space use and slower growth rates in brown trout in general. Traits related to space use (Härkönen et al., 2014; Koeck et al., 2019), energetic demand (Keiling et al., 2020; Redpath et al., 2010) and boldness/aggression (Klefoth et al., 2017; Redpath et al., 2010) may each play a role in determining capture vulnerability in fish, and so changes in the distribution of these traits amongst wild fish populations because of invasive species may in turn impact their availability to fisheries.

Although poorly understood, deleterious interactions between native and non-native species may be buffered by present-day environmental conditions. For example, high availability of shelter and food resources can mitigate competition between native and non-native species (Kernan, 2015; van Kessel et al., 2011; Stachowicz and Byrnes, 2006) while seasonal decreases in water temperature can prevent the expansion or establishment of warmwater invasive fish populations further outside of their natural range (Rahel and Olden, 2008). Therefore, further environmental change may eventually trigger a competitive imbalance and destabilizing effect in ecosystems where non-native species are already present but relatively non-disruptive, when differences in physiological tolerances and performance between native and non-native species may be revealed and translate to differences in fitness. Such a scenario is predicted to exacerbate the problem of invasive sea lamprey throughout the Laurentian Great Lakes (Lennox et al., 2020), which will benefit from enhanced growth and the expansion of thermally suitable habitat as the climate warms. This, in turn, will likely increase rates and lethality of parasitism on native fish species of both recreational (e.g., lake trout; Muir et al., 2012) and commercial (lake whitefish *Coregonus clupeaformis*; Ebener et al., 2008) fisheries importance (Lennox et al., 2020). Similarly to the lionfish example outlined above, no new commercial fishery targeting lamprey has emerged in response to their increased abundance, and lamprey are not a desirable species in recreational fisheries.

## 7 Future research and conclusions

Within the realm of comparative physiology in general, studies are only recently beginning to consider the effects of multiple stressors on animal functioning, and the potential effects of combined stressors on fisheries are mostly unknown or speculative. Most work that has been done in this has examined effects of fishing in isolation or in combination with perhaps a single additional stressor (e.g., elevated temperature). This work has been a key foundation for understanding the physiological effects of fishing, the potential for recovery after escape or release, collateral fishing mortality, and population-level effects. However, much more work is needed to understand the combined effects of the many stressors that fish regularly encounter in the wild, including hypoxia, chemical and sensory pollution, artificial light, altered pH, habitat degradation, and others, particularly in the context of fisheries. There is a broad range of potential research avenues in this field, but here we outline five general areas of especially high priority:

1. *The combined effects of multiple stressors during interactions with fishing gears.* Many studies have evaluated acute impacts of capture on fish in terms of immediate physiological stress response, behavioral impairment, and mortality, however many of these studies only consider a single environmental scenario (for example, at a single water/air temperature, or time of year). Experimental work evaluating these responses under prospective climate change scenarios, and in response to other environmental stressors, will further our knowledge on how the cumulative stress of capture and additional stressors may impact wild fish populations.
2. *Feedbacks between fisheries-induced evolution and vulnerability to environmental stressors.* While it is increasingly acknowledged that the impacts of fisheries selection alter the phenotypic composition of targeted fish species, the impacts of this phenotypic change are typically considered in terms life history and reproductive traits. How fisheries selection may have influenced the capacity for exploited fish stocks to adapt to climate change-induced stress, for example through the removal of stress-resilient phenotypes, is completely unknown. While investigating phenotypic change of cryptic traits in wild fish populations is extremely challenging, new analytical techniques such as retroactive estimation of metabolic rate via stable isotope analysis of otoliths provide new opportunities to investigate change in metabolic traits of fish stocks over time. Mesoscale fisheries simulations, where the traits of caught and uncaught fish are known and can be monitored in real time would also do much to elucidate how fishing changes the composition of fish populations, and how these altered populations may cope with future environmental stressors.
3. *Overlap between fisheries and shifting population distributions.* Poleward range expansions, and compression of vertical habitat use in response to changes in water temperature and oxygen concentrations have been observed in a range of fish taxa, but mechanistic links between physiological traits and habitat selection/use are still lacking, making future predictions of fish home ranges difficult. With the development of high-resolution telemetry devices, in addition to data loggers which can simultaneously measure fish acceleration, temperature, and heart rate, studies evaluating the metabolic costs/advantages of using specific habitats are increasingly viable. Studies

using these approaches in both fully wild and mesocosm experiments would be invaluable in predicting changes in encounter rates with fishing gears in response to a warming climate.

4. *Acute effects of multiple stressors on fish recovery after capture and subsequent release or escape.* Many studies have evaluated acute impacts of capture on fish in terms of immediate physiological stress response, behavioral impairment, and mortality, however many of these studies only consider a single environmental scenario (for example, at a single water/air temperature, or time of year). Experimental work evaluating these responses under prospective climate change scenarios, and in response to other environmental stressors, will further our knowledge on how the cumulative stress of capture and additional stressors may impact wild fish populations.
5. *Sublethal effects of fishing practices and modulating effects of multiple stressors.* Many studies consider the cumulative impact of stressors in terms of fish mortality, while more sublethal effects on performance and fitness are more rarely investigated, despite their relevance for fisheries management. The quality of harvested fish meat, reproductive investment, and growth trajectories may all be impacted by various stressors, however the degree to which this is occurring/may occur, and the economic implications for these changes in fisheries are mostly unknown.
6. *Ecosystem level impacts of fisheries in a multi-stressor world.* The myriad of stressors faced by fish populations in freshwater and marine environments will have ecosystem-wide effects, including alterations to food-webs and interactions with non-native species. This is especially true in the context of fisheries, which may exacerbate these effects by targeting specific trophic levels, altering competitive and selective landscapes, and potentially changing the phenotypic composition of targeted populations. Our knowledge of these effects is currently extremely limited and much more work is needed in this area.

There is no doubt that numerous anthropogenic stressors are having an important impact on freshwater and marine fisheries and will continue to do so well into the future. Understanding the mechanistic, physiological underpinnings of these effects is critical for developing potential solutions and effective science-based fisheries management. To date, research in this realm has largely consisted of controlled laboratory experiments and field sampling studies to isolate the physiological effects of fishing practices. Much more work is needed to understand the effects of multiple stressors at each stage in the fishing process and its relevance for not only those fish that are captured (and perhaps released or escaped) but also for those that are not captured (with potential selection or evolutionary effects). In addition, much more knowledge is needed regarding the potential feedbacks between fishing and various forms of environmental stressors encountered by fish in the wild.

The progress of various technologies for making fine-scale physiological measures in the laboratory and tracking fish movements and logging physiological measures in the wild will greatly facilitate research in these areas going forward.

## References

Abraham, J.P., Baringer, M., Bindoff, N.L., Boyer, T., Cheng, L.J., Church, J.A., Conroy, J.L., Domingues, C.M., Fasullo, J.T., Gilson, J., Goni, G., 2013. A review of global ocean temperature observations: implications for ocean heat content estimates and climate change. Rev. Geophys. 51 (3), 450–483.

Ainsworth, C.H., Samhouri, J.F., Busch, D.S., Cheung, W.W., Dunne, J., Okey, T.A., 2011. Potential impacts of climate change on Northeast Pacific marine foodwebs and fisheries. ICES J. Mar. Sci. 68 (6), 1217–1229.

Altieri, A.H., Gedan, K.B., 2015. Climate change and dead zones. Glob. Chang. Biol. 21 (4), 1395–1406.

Anders, N., Eide, I., Lerfall, J., Roth, B., Breen, M., 2020. Physiological and flesh quality consequences of pre-mortem crowding stress in Atlantic mackerel (*Scomber scombrus*). PLoS One 15 (2), e0228454.

Andersen, B.S., Jørgensen, C., Eliassen, S., Giske, J., 2016. The proximate architecture for decision-making in fish. Fish Fish. 17 (3), 680–695.

Arlinghaus, R., Laskowski, K.L., Alós, J., Klefoth, T., Monk, C.T., Nakayama, S., Schröder, A., 2017. Passive gear-induced timidity syndrome in wild fish populations and its potential ecological and managerial implications. Fish Fish. 18 (2), 360–373.

Árnason, E., Hernandez, U.B., Kristinsson, K., 2009. Intense Habitat-Specific Fisheries-Induced Selection at the Molecular Pan I Locus Predicts Imminent Collapse of a Major Cod Fishery. PLoS One 4, e5529.

Arreguín-Sánchez, F., 1996. Catchability: a key parameter for fish stock assessment. Rev. Fish Biol. Fish. 6 (2), 221–242.

Baktoft, H., Jacobsen, L., Skov, C., Koed, A., Jepsen, N., Berg, S., Boel, M., Aarestrup, K., Svendsen, J.C., 2016. Phenotypic variation in metabolism and morphology correlating with animal swimming activity in the wild: relevance for the OCLTT (oxygen- and capacity-limitation of thermal tolerance), allocation and performance models. Conserv. Physiol. 4.

Ballew, N.G., Bacheler, N.M., Kellison, G.T., Schueller, A.M., 2016. Invasive lionfish reduce native fish abundance on a regional scale. Sci. Rep. 6 (1), 1–7.

Bando, K.J., 2006. The roles of competition and disturbance in a marine invasion. Biol. Invasions 8 (4), 755–763.

Barker, B.D., Horodysky, A.Z., Kerstetter, D.W., 2018. Hot or not? Comparative behavioral thermoregulation, critical temperature regimes, and thermal tolerances of the invasive lionfish Pterois sp. versus native western North Atlantic reef fishes. Biol. Invasions 20 (1), 45–58.

Barton, B.A., 2002. Stress in fishes: a diversity of responses with particular reference to changes in circulating corticosteroids. Integr. Comp. Biol. 42 (3), 517–525.

Bates, A.E., Helmuth, B., Burrows, M.T., Duncan, M.I., Garrabou, J., Guy-Haim, T., Lima, F., Queiros, A.M., Seabra, R., Marsh, R., Belmaker, J., 2018. Biologists ignore ocean weather at their peril. Nature 560, 299–301. https://doi.org/10.1038/d41586-018-05869-5.

Biro, P.A., Post, J.R., 2008. Rapid depletion of genotypes with fast growth and bold personality traits from harvested fish populations. Proc. Natl. Acad. Sci. 105 (8), 2919–2922.

Breitburg, D., et al., 2018. Declining oxygen in the global ocean and coastal waters. Science 359, eaam7240.

Brodin, T., Fick, J., Jonsson, M., Klaminder, J., 2013. Dilute concentrations of a psychiatric drug alter behavior of fish from natural populations. Science 339 (6121), 814–815.

Brooker, R.M., Munday, P.L., Mcleod, I.M., Jones, G.P., 2013. Habitat preferences of a corallivorous reef fish: predation risk versus food quality. Coral Reefs 32 (3), 613–622.

Brownscombe, J.W., Thiem, J.D., Hatry, C., Cull, F., Haak, C.R., Danylchuk, A.J., Cooke, S.J., 2013. Recovery bags reduce post-release impairments in locomotory activity and behavior of bonefish (Albula spp.) following exposure to angling-related stressors. J. Exp. Mar. Biol. Ecol. 440, 207–215.

Brüning, A., Kloas, W., Preuer, T., Hölker, F., 2018. Influence of artificially induced light pollution on the hormone system of two common fish species, perch and roach, in a rural habitat. Conserv. Physiol. 6 (1), coy016.

Butler, M.W., Stahlschmidt, Z.R., Ardia, D.R., Davies, S., Davis, J., Guillette Jr., L.J., et al., 2013. Thermal sensitivity of immune function: evidence against a generalist-specialist trade-off among endothermic and ectothermic vertebrates. Am. Nat. 181 (6), 761–774.

Caddy, J.F., 2008. The importance of "cover" in the life histories of demersal and benthic marine resources: a neglected issue in fisheries assessment and management. Bull. Mar. Sci. 83 (1), 7–52.

Campana, S.E., Casselman, J.M., Jones, C.M., Black, G., Barker, O., Evans, M., Guzzo, M.M., Kilada, R., Muir, A.M., Perry, R., 2020. Arctic freshwater fish productivity and colonization increase with climate warming. Nat. Clim. Change 10, 428–433.

Carpenter, S.R., Stanley, E.H., Vander Zanden, M.J., 2011. State of the world's freshwater ecosystems: physical, chemical, and biological changes. Annu. Rev. Env. Resour. 36, 75–99.

Celi, M., Filiciotto, F., Maricchiolo, G., Genovese, L., Quinci, E.M., Maccarrone, V., Mazzola, S., Vazzana, M., Buscaino, G., 2016. Vessel noise pollution as a human threat to fish: assessment of the stress response in gilthead sea bream (Sparus aurata, Linnaeus 1758). Fish Physiol. Biochem. 42 (2), 631–641.

Cheung, W.W.L., Dunne, J., Sarmiento, J.L., Pauly, D., 2011. Integrating ecophysiology and plankton dynamics into projected maximum fisheries catch potential under climate change in the Northeast Atlantic. ICES J. Mar. Sci. 68, 1008–1018.

Chopin, F.S., Arimoto, T., 1995. The condition of fish escaping from fishing gears—a review. Fish. Res. 21 (3–4), 315–327.

Chrétien, E., Cooke, S.J., Boisclair, D., 2021. Does shelter influence the metabolic traits of a teleost fish? J. Fish Biol. 98, 1242–1252. https://doi.org/10.1111/jfb.14653.

Claireaux, G., Chabot, D., 2016. Responses by fishes to environmental hypoxia: integration through Fry's concept of aerobic metabolic scope. J. Fish Biol. 88, 232–251.

Claireaux, G., Webber, D.M., Lagardère, J.-P., Kerr, S.R., 2000. Influence of water temperature and oxygenation on the aerobic metabolic scope of Atlantic cod (Gadus morhua). J. Sea Res. 44, 257–265.

Clark, T.D., Messmer, V., Tobin, A.J., Hoey, A.S., Pratchett, M.S., 2017. Rising temperatures may drive fishing-induced selection of low-performance phenotypes. Sci. Rep. 7 (1), 1–11.

Cline, T.J., Weidel, B.C., Kitchell, J.F., Hodgson, J.R., 2012. Growth response of largemouth bass (Micropterus salmoides) to catch-and-release angling: a 27-year mark–recapture study. Can. J. Fish. Aquat. Sci. 69 (2), 224–230.

Coni, E.O., Booth, D.J., Nagelkerken, I., 2021. Novel species interactions and environmental conditions reduce foraging competency at the temperate range edge of a range-extending coral reef fish. Coral Reefs 40, 1–12.

Cook, K.V., Lennox, R.J., Hinch, S.G., Cooke, S.J., 2015. Fish out of water: how much air is too much? Fisheries 40 (9), 452–461.

Cooke, S.J., Cowx, I.G., 2004. The role of recreational fishing in global fish crises. BioScience 54 (9), 857–859.

Cooke, S.J., Messmer, V., Tobin, A.J., Pratchett, M.S., Clark, T.D., 2014. Refuge-seeking impairments mirror metabolic recovery following fisheries-related stressors in the Spanish flag snapper (Lutjanus carponotatus) on the Great Barrier Reef. Physiol. Biochem. Zool. 87 (1), 136–147.

Cooke, S.J., Venturelli, P., Twardek, W.M., Lennox, R.J., Brownscombe, J.W., Skov, C., Danylchuk, A.J., 2021. Technological innovations in the recreational fishing sector: implications for fisheries management and policy. Rev. Fish Biol. Fish. 31 (2), 253–288.

Côté, I.M., Smith, N.S., 2018. The lionfish Pterois sp. invasion: has the worst-case scenario come to pass? J. Fish Biol. 92 (3), 660–689.

Crespel, A., Miller, T., Rácz, A., Parsons, K., Lindström, J., Killen, S., 2021a. Density influences the heritability and genetic correlations of fish behaviour under trawling-associated selection. Evol. Appl. 14, 2527–2540. https://doi.org/10.1111/eva.13279.

Crespel, A., Schneider, K., Miller, T., Rácz, A., Jacobs, A., Lindström, J., Elmer, K.R., Killen, S.-S., 2021b. Genomic basis of fishing-associated selection varies with population density. PNAS 51 (118). https://doi.org/10.1073/pnas.2020833118.

Davis, M.W., 2010. Fish stress and mortality can be predicted using reflex impairment. Fish Fish. 11 (1), 1–11.

Deutsch, C., Ferrel, A., Seibel, B., Pörtner, H.O., Huey, R.B., 2015. Climate change tightens a metabolic constraint on marine habitats. Science 348, 1132–1135.

Deutsch, C., Penn, J.L., Seibel, B., 2020. Metabolic trait diversity shapes marine biogeography. Nature 585, 557–562.

Diaz Pauli, B., Sih, A., 2017. Behavioural responses to human-induced change: why fishing should not be ignored. Evol. Appl. 10 (3), 231–240.

Diaz Pauli, B., Wiech, M., Heino, M., Utne-Palm, A.C., 2015. Opposite selection on behavioural types by active and passive fishing gears in a simulated guppy Poecilia reticulata fishery. J. Fish Biol. 86 (3), 1030–1045.

Domenici, P., Herbert, N.A., Lefrançois, C., Steffensen, J.F., McKenzie, D.J., 2013. The effect of hypoxia on fish swimming performance and behaviour. Swimming Physiology of Fish. Springer, Berlin, Heidelberg, pp. 129–159.

Dulvy, N.K., Rogers, S.I., Jennings, S., Stelzenmüller, V., Dye, S.R., Skjoldal, H.R., 2008. Climate change and deepening of the North Sea fish assemblage: a biotic indicator of warming seas. J. Appl. Ecol. 45, 1029–1039.

Ebener, M.P., Kinnunen, R.E., Schneeberger, P.J., Mohr, L.C., Hoyle, J.A., Peeters, P., 2008. Management of commercial fisheries for lake whitefish in the Laurentian Great Lakes of North America. In: International Governance of Fisheries Ecosystems: Learning From the Past, Finding Solutions for the Future. American Fisheries Society, Bethesda, Maryland, pp. 99–143.

Eikeset, A.M., Richter, A., Dunlop, E.S., Dieckmann, U., Stenseth, N.C., 2013. Economic repercussions of fisheries-induced evolution. Proc. Natl. Acad. Sci. 110 (30), 12259–12264.

Ejbye-Ernst, R., Michaelsen, T.Y., Tirsgaard, B., Wilson, J.M., Jensen, L.F., Steffensen, J.F., Pertoldi, C., Aarestrup, K., Svendsen, J.C., 2016. Partitioning the metabolic scope: the importance of anaerobic metabolism and implications for the oxygen-and capacity-limited thermal tolerance (OCLTT) hypothesis. Conserv. Physiol. 4 (1).

Enberg, K., Jørgensen, C., Dunlop, E.S., Heino, M., Dieckmann, U., 2009. Implications of fisheries-induced evolution for stock rebuilding and recovery. Evol. Appl. 2 (3), 394–414.

Enberg, K., Jørgensen, C., Dunlop, E.S., Varpe, Ø., Boukal, D.S., Baulier, L., Eliassen, S., Heino, M., 2012. Fishing-induced evolution of growth: concepts, mechanisms and the empirical evidence. Mar. Ecol. 33 (1), 1–25.

Evans, D.O., 1990. Metabolic Thermal Compensation by Rainbow Trout: Effects on Standard Metabolic Rate and Potential Usable Power. null 119, pp. 585–600.

Falco, F., Bottari, T., Ragonese, S., Killen, S.S., 2022. Towards the integration of ecophysiology with fisheries stock assessment for conservation policy and evaluating the status of the Mediterranean Sea. Conserv. Physiol. 10 (1), coac008.

FAO, 2019. Common Oceans—A Partnership for Sustainability in the ABNJ. Food and Agriculture Organization of the United Nations, Rome.

FAO, 2020. The State of World Fisheries and Aquaculture 2020. Sustainability in Action. Food and Agriculture Organization of the United Nations, Rome.

Farrell, A., Richards, J., 2009. Defining hypoxia: an integrative synthesis of the responses of fish to hypoxia. Fish Physiol. 27, 487–503.

Farrell, A.P., Gallaugher, P.E., Fraser, J., Pike, D., Bowering, P., Hadwin, A.K., Routledge, R., 2001. Successful recovery of the physiological status of coho salmon on board a commercial gillnet vessel by means of a newly designed revival box. Can. J. Fish. Aquat. Sci. 58 (10), 1932–1946.

Farrell, A.P., Eliason, E.J., Sandblom, E., Clark, T.D., 2009. Fish cardiorespiratory physiology in an era of climate change. Can. J. Zool. 87 (10), 835–851.

Fausch, K.D., Nakano, S., Kitano, S., 1997. Experimentally induced foraging mode shift by sympatric charrs in a mountain stream. Behav. Ecol. 8, 414–420.

Ferrari, S., Horri, K., Allal, F., Vergnet, A., Benhaim, D., Vandeputte, M., Chatain, B., Bégout, M.L., 2016. Heritability of boldness and hypoxia avoidance in European seabass, Dicentrarchus labrax. PloS one 11 (12), e0168506.

Filous, A., Friedlander, A.M., Koike, H., Lammers, M., Wong, A., Stone, K., Sparks, R.T., 2017. Displacement effects of heavy human use on coral reef predators within the Molokini Marine Life Conservation District. Mar. Pollut. Bull. 121 (1–2), 274–281.

Finley, C., 2016. The industrialization of commercial fishing, 1930–2016. In: Oxford Research Encyclopedia of Environmental Science. Oxford University Press.

Fishelson, L., 1997. Experiments and observations on food consumption, growth and starvation in Dendrochirus brachypterus and Pterois volitans (Pteroinae, Scorpaenidae). Environ. Biol. Fishes 50 (4), 391–403.

Fu, C., Travers-Trolet, M., Velez, L., Grüss, A., Bundy, A., Shannon, L.J., Fulton, E.A., Akoglu, E., Houle, J.E., Coll, M., Verley, P., Heymans, J.J., John, E., Shin, Y.J., 2018. Risky business: the combined effects of fishing and changes in primary productivity on fish communities. Ecol. Model. 368, 265–276.

Gaines, S.D., Costello, C., Owashi, B., Mangin, T., Bone, J., Molinos, J.G., Burden, M., Dennis, H., Halpern, B.S., Kappel, C.V., et al., 2018. Improved fisheries management could offset many negative effects of climate change. Sci. Adv. 4, eaao1378.

Gale, M.K., Hinch, S.G., Donaldson, M.R., 2013. The role of temperature in the capture and release of fish. Fish Fish. 14 (1), 1–33.

Gozlan, R.E., 2017. Interference of non-native species with fisheries and aquaculture. In: Impact of Biological Invasions on Ecosystem Services. Springer, Cham, pp. 119–137.

Graham, A., Cooke, S., 2008. The effects of noise disturbance from various recreational boating activities common to inland waters on the cardiac physiology of a freshwater fish, the largemouth bass (*Micropterus salmoides*). Aquat. Conserv.: Mar. Freshw. Ecosyst. 18, 1315–1324.

Green, S.J., Akins, J.L., Maljković, A., Côté, I.M., 2012. Invasive lionfish drive Atlantic coral reef fish declines. PLoS One 7 (3), e32596.

Guida, L., Walker, T.I., Reina, R.D., 2016. Temperature insensitivity and behavioural reduction of the physiological stress response to longline capture by the gummy shark, Mustelus antarcticus. PLoS One 11 (2), e0148829.

Guzzo, M.M., Blanchfield, P.J., Rennie, M.D., 2017. Behavioral responses to annual temperature variation alter the dominant energy pathway, growth, and condition of a cold-water predator. Proc. Natl. Acad. Sci. U. S. A. 114, 9912.

Halfwerk, W., Slabbekoorn, H., 2015. Pollution going multimodal: the complex impact of the human-altered sensory environment on animal perception and performance. Biol. Lett. 11, 20141051.

Hall, K.C., Broadhurst, M.K., Butcher, P.A., Rowland, S.J., 2009. Effects of angling on post-release mortality, gonadal development and somatic condition of Australian bass Macquaria novemaculeata. J. Fish Biol. 75 (10), 2737–2755.

Härkönen, L., Hyvärinen, P., Paappanen, J., Vainikka, A., 2014. Explorative behavior increases vulnerability to angling in hatchery-reared brown trout (Salmo trutta). Can. J. Fish. Aquat. Sci. 71 (12), 1900–1909.

Hecky, R.E., Mugidde, R., Ramlal, P.S., Talbot, M.R., Kling, G.W., 2010. Multiple stressors cause rapid ecosystem change in Lake Victoria. Freshw. Biol. 55, 19–42.

Heino, M., Diaz Pauli, B., Dieckmann, U., 2015. Fisheries-induced evolution. Annu. Rev. Ecol. Evol. Syst. 46, 461–480.

Hellström, G., Klaminder, J., Finn, F., et al., 2016. GABAergic anxiolytic drug in water increases migration behaviour in salmon. Nat. Commun. 7, 13460.

Hollins, J., Koeck, B., Crespel, A., Bailey, D.M., Killen, S.S., 2021. Does thermal plasticity affect susceptibility to capture in fish? Insights from a simulated trap and trawl fishery. Can. J. Fish. Aquat. Sci. 78 (1), 57–67. https://doi.org/10.1139/cjfas-2020-0125.

Hollins, J., Thambithurai, D., Koeck, B., Crespel, A., Bailey, D.M., Cooke, S.J., Lindström, J., Parsons, K.J., Killen, S.S., 2018. A physiological perspective on fisheries-induced evolution. Evol. Appl. 11 (5), 561–576.

Hollins, J.P.W., Thambithurai, D., Van Leeuwen, T.E., Allan, B., Koeck, B., Bailey, D., Killen, S.S., 2019. Shoal familiarity modulates effects of individual metabolism on vulnerability to capture by trawling. Conserv. Physiol. 7 (1), coz043.

Hsieh, C.H., Reiss, C.S., Hewitt, R.P., Sugihara, G., 2008. Spatial analysis shows that fishing enhances the climatic sensitivity of marine fishes. Can. J. Fish. Aquat. Sci. 65 (5), 947–961.

Hsieh, C.-h., Reiss, C.S., Hunter, J.R., Beddington, J.R., May, R.M., Sugihara, G., 2006. Fishing elevates variability in the abundance of exploited species. Nature 443, 859–862.

Hsieh, C.-h., Yamauchi, A., Nakazawa, T., Wang, W.-F., 2010. Fishing effects on age and spatial structures undermine population stability of fishes. Aquat. Sci. 72, 165–178.

Hutchings, J.A., Rangeley, R.W., 2011. Correlates of recovery for Canadian Atlantic cod (Gadus morhua). Can. J. Zool. 89 (5), 386–400.

Jensen, D.L., Overgaard, J., Wang, T., Gesser, H., Malte, H., 2017. Temperature effects on aerobic scope and cardiac performance of European perch (Perca fluviatilis). J. Therm. Biol. 68, 162–169.

Johansen, J.L., Nadler, L.E., Habary, A., Bowden, A.J., Rummer, J., 2021. Thermal acclimation of tropical coral reef fishes to global heat waves. Elife 10, 1–30.

Johnston, M.A., Gittings, S.R., Morris, J.A., 2015. NOAA National Marine Sanctuaries Lionfish Response Plan (2015-2018): Responding, Controlling, and Adapting to an Active Marine Invasion. National Oceanic and Atmospheric Administration.

Jud, Z.R., Nichols, P.K., Layman, C.A., 2015. Broad salinity tolerance in the invasive lionfish Pterois spp. may facilitate estuarine colonization. Environ. Biol. Fishes 98 (1), 135–143.

Keiling, T.D., Louison, M.J., Suski, C.D., 2020. Big, hungry fish get the lure: size and food availability determine capture over boldness and exploratory behaviors. Fish. Res. 227, 105554.

Kernan, M., 2015. Climate change and the impact of invasive species on aquatic ecosystems. Aquat. Ecosyst. Health Manage. 18 (3), 321–333.

van Kessel, N., Dorenbosch, M., Boer, M.D., Leuven, R.S.E.W., Velde, G.V.D., 2011. Competition for shelter between four invasive gobiids and two native benthic fish species. Curr. Zool. 57 (6), 844–851.

Kieffer, J.D., 2000. Limits to exhaustive exercise in fish. Comp. Biochem. Physiol. A Mol. Integr. Physiol. 126 (2), 161–179.

Kieffer, J.D., Kubacki, M.R., Phelan, F.J.S., Philipp, D.P., Tufts, B.L., 1995. Effects of catch-and-release angling on nesting male smallmouth bass. Trans. Am. Fish. Soc. 124 (1), 70–76.

Killen, S.S., Marras, S., McKenzie, D.J., 2011. Fuel, fasting, fear: routine metabolic rate and food deprivation exert synergistic effects on risk-taking in individual juvenile European sea bass. J. Anim. Ecol. 80 (5), 1024–1033.

Killen, S.S., Marras, S., Ryan, M.R., Domenici, P., McKenzie, D.J., 2012. A relationship between metabolic rate and risk-taking behaviour is revealed during hypoxia in juvenile European sea bass. Funct. Ecol. 26, 134–143.

Killen, S.S., Nati, J.J., Suski, C.D., 2015a. Vulnerability of individual fish to capture by trawling is influenced by capacity for anaerobic metabolism. Proc. R. Soc. B: Biol. Sci. 282 (1813), 20150603.

Killen, S.S., Reid, D., Marras, S., Domenici, P., 2015b. The interplay between aerobic metabolism and antipredator performance: vigilance is related to recovery rate after exercise. Front. Physiol. 6, 111.

Killen, S.S., Adriaenssens, B., Marras, S., Claireaux, G., Cooke, S.J., 2016. Context dependency of trait repeatability and its relevance for management and conservation of fish populations. Conserv. Physiol. 4 (1).

Kimberly, D.A., Salice, C.J., 2015. Evolutionary responses to climate change and contaminants: evidence and experimental approaches. Curr. Zool. 61 (4), 690–701.

Klefoth, T., Skov, C., Kuparinen, A., Arlinghaus, R., 2017. Toward a mechanistic understanding of vulnerability to hook-and-line fishing: boldness as the basic target of angling-induced selection. Evol. Appl. 10 (10), 994–1006.

Koeck, B., Závorka, L., Aldvén, D., Näslund, J., Arlinghaus, R., Thörnqvist, P.O., Winberg, S., Björnsson, B.T., Johnsson, J.I., 2019. Angling selects against active and stress-resilient phenotypes in rainbow trout. Can. J. Fish. Aquat. Sci. 76 (2), 320–333.

Koeck, B., Lovén Wallerius, M., Arlinghaus, R., Johnsson, J.I., 2020. Behavioural adjustment of fish to temporal variation in fishing pressure affects catchability: an experiment with angled trout. Can. J. Fish. Aquat. Sci. 77 (1), 188–193.

Koslow, J.A., Gowlett-Holmes, K., Lowry, J.K., O'Hara, T., Poore, G.C.B., Williams, A., 2001. Seamount benthic macrofauna off southern Tasmania: community structure and impacts of trawling. Mar. Ecol. Prog. Ser. 213, 111–125.

Kritzer, J.P., DeLucia, M.B., Greene, E., Shumway, C., Topolski, M.F., Thomas-Blate, J., Chiarella, L.A., Davy, K.B., Smith, K., 2016. The importance of benthic habitats for coastal fisheries. BioScience 66 (4), 274–284.

Kuparinen, A., Hutchings, J.A., 2012. Consequences of fisheries-induced evolution for population productivity and recovery potential. Proc. R. Soc. B: Biol. Sci. 279 (1738), 2571–2579.

Lefevre, S., 2016. Are global warming and ocean acidification conspiring against marine ectotherms? A meta-analysis of the respiratory effects of elevated temperature, high CO2 and their interaction. Conserv. Physiol. 4 (1), cow009.

Lehodey, P., Alheit, J., Barange, M., Baumgartner, T., Beaugrand, G., Drinkwater, K., Fromentin, J.M., Hare, S.R., Ottersen, G., Perry, R.I., Roy, C.V.D.L., 2006. Climate variability, fish, and fisheries. J. Climate 19 (20), 5009–5030.

Lennox, R.J., Alós, J., Arlinghaus, R., Horodysky, A., Klefoth, T., Monk, C.T., Cooke, S.J., 2017. What makes fish vulnerable to capture by hooks? A conceptual framework and a review of key determinants. Fish Fish. 18 (5), 986–1010.

Lennox, R.J., Bravener, G.A., Lin, H.Y., Madenjian, C.P., Muir, A.M., Remucal, C.K., Robinson, K.F., Rous, A.M., Siefkes, M.J., Wilkie, M.P., Zielinski, D.P., 2020. Potential changes to the biology and challenges to the management of invasive sea lamprey Petromyzon marinus in the Laurentian Great Lakes due to climate change. Glob. Chang. Biol. 26 (3), 1118–1137.

Libralato, S., Caccin, A., Pranovi, F., 2015. Modeling species invasions using thermal and trophic niche dynamics under climate change. Front. Mar. Sci. 2, 29.

Longhurst, A., 2002. Murphy's law revisited: longevity as a factor in recruitment to fish populations. Fish. Res. 56 (2), 125–131. https://doi.org/10.1016/S0165-7836(01)00351-4.

Louison, M.J., Jeffrey, J.D., Suski, C.D., Stein, J.A., 2018. Sociable bluegill, Lepomis macrochirus, are selectively captured via recreational angling. Anim. Behav. 142, 129–137.

Lovell, S.J., Stone, S.F., Fernandez, L., 2006. The economic impacts of aquatic invasive species: a review of the literature. Agric. Resour. Econ. Rev. 35 (1), 195–208.

Mameri, D., Branco, P., Ferreira, M.T., Santos, J.M., 2020. Heatwave effects on the swimming behaviour of a Mediterranean freshwater fish, the Iberian barbel Luciobarbus bocagei. Sci. Total Environ. 730, 139152.

Martin, C.W., Valentine, M.M., Valentine, J.F., 2010. Competitive interactions between invasive Nile tilapia and native fish: the potential for altered trophic exchange and modification of food webs. PLoS One 5 (12), e14395.

Meka, J.M., Margraf, F.J., 2007. Using a bioenergetic model to assess growth reduction from catch-and-release fishing and hooking injury in rainbow trout, Oncorhynchus mykiss. Fish. Manag. Ecol. 14 (2), 131–139.

Metcalfe, J.D., Butler, P.J., 1984. Changes in activity and ventilation in response to hypoxia in unrestrained, unoperated dogfish (Scyliorhinus canicula L.). J. Exp. Biol. 108, 411–418.

Metcalfe, N.B., Van Leeuwen, T.E., Killen, S.S., 2016. Does individual variation in metabolic phenotype predict fish behaviour and performance? J. Fish Biol. 88 (1), 298–321.

Millidine, K.J., Armstrong, J.D., Metcalfe, N.B., 2006. Presence of shelter reduces maintenance metabolism of juvenile salmon. Funct. Ecol. 20 (5), 839–845.

Monteiro, D.A., Taylor, E.W., McKenzie, D.J., Ratin, F.T., Kalinin, A.L., 2020. Interactive effects of mercury exposure and hypoxia on ECG patterns in two Neotropical freshwater fish species: Matrinxã, Brycon amazonicus and traíra, Hoplias malabaricus. Ecotoxicology 29, 375–388.

Morris, J.A., Shertzer, K.W., Rice, J.A., 2011. A stage-based matrix population model of invasive lionfish with implications for control. Biol. Invasions 13 (1), 7–12.

Morrongiello, J.R., Sweetman, P.C., Thresher, R.E., 2019. Fishing constrains phenotypic responses of marine fish to climate variability. J. Anim. Ecol. 88, 1645–1656.

Muir, A.M., Krueger, C.C., Hansen, M.J., 2012. Re-establishing lake trout in the Laurentian Great Lakes: past, present, and future. In: Great Lakes Fishery Policy and Management: A Binational Perspective, second ed. Michigan State University Press, East Lansing, pp. 533–588.

Nati, J.J.H., Lindström, J., Halsey, L.G., Killen, S.S., 2016. Is there a trade-off between peak performance and performance breadth across temperatures for aerobic scope in teleost fishes? Biol. Lett. 12, 20160191.

Norin, T., Malte, H., 2011. Repeatability of standard metabolic rate, active metabolic rate and aerobic scope in young brown trout during a period of moderate food availability. J. Exp. Biol. 214 (10), 1668–1675.

Norin, T., Metcalfe, N.B., 2019. Ecological and evolutionary consequences of metabolic rate plasticity in response to environmental change. Philos. Trans. R. Soc. B 374 (1768), 20180180.

O'Farrell, S., Bearhop, S., McGill, R.A., Dahlgren, C.P., Brumbaugh, D.R., Mumby, P.J., 2014. Habitat and body size effects on the isotopic niche space of invasive lionfish and endangered Nassau grouper. Ecosphere 5 (10), 1–11.

Olsen, E.M., Heino, M., Lilly, G.R., Morgan, M.J., Brattey, J., Ernande, B., Dieckmann, U., 2004. Maturation trends indicative of rapid evolution preceded the collapse of northern cod. Nature 428 (6986), 932–935.

Olsen, E.M., Heupel, M.R., Simpfendorfer, C.A., Moland, E., 2012. Harvest selection on Atlantic cod behavioral traits: implications for spatial management. Ecol. Evol. 2, 1549–1562.

Ostrand, K.G., Cooke, S.J., Wahl, D.H., 2004. Effects of stress on largemouth bass reproduction. N. Am. J. Fish. Manag. 24 (3), 1038–1045.

Ottersen, G., Stige, L.C., Durant, J.M., Chan, K.S., Rouyer, T.A., Drinkwater, K.F., Stenseth, N.C., 2013. Temporal shifts in recruitment dynamics of North Atlantic fish stocks: effects of spawning stock and temperature. Mar. Ecol. Prog. Ser. 480, 205–225.

Pankhurst, N.W., Dedualj, M., 1994. Effects of capture and recovery on plasma levels of cortisol, lactate and gonadal steroids in a natural population of rainbow trout. J. Fish Biol. 45 (6), 1013–1025.

Pinsky, M.L., Fogarty, M., 2012. Lagged social-ecological responses to climate and range shifts in fisheries. Clim. Change 115, 883–891.

Pinsky, M.L., Selden, R.L., Kitchel, Z.J., 2020. Climate-driven shifts in marine species ranges: scaling from organisms to communities. Ann. Rev. Mar. Sci. 12, 153–179.

Pitcher, T.J., Lam, M.E., 2015. Fish commoditization and the historical origins of catching fish for profit. Marit. Stud. 14 (1), 1–19.

Porter, W., Gates, D.M., 1969. Thermodynamic equilibria of animals with environment. Ecol. Monogr. 39, 227–244.

Pörtner, H.O., Knust, R., 2007. Climate change affects marine fishes through the oxygen limitation of thermal tolerance. Science 315, 95–97.

Pörtner, H.-O., Bock, C., Mark, F.C., 2017. Oxygen- and capacity-limited thermal tolerance: bridging ecology and physiology. J. Exp. Biol. 220, 2685.

Pulgar, J., Zeballos, D., Vargas, J., Aldana, M., Manriquez, P., Manriquez, K., Quijón, P.A., Widdicombe, S., Anguita, C., Quintanilla, D., Duarte, C., 2019. Endogenous cycles, activity patterns and energy expenditure of an intertidal fish is modified by artificial light pollution at night (ALAN). Environ. Pollut. 244, 361–366.

Quadros, A.L., Barros, F., Blumstein, D.T., Meira, V.H., Nunes, J.A.C., 2019. Structural complexity but not territory sizes influences flight initiation distance in a damselfish. Mar. Biol. 166 (5), 1–6.

Raby, G.D., Donaldson, M.R., Hinch, S.G., Patterson, D.A., Lotto, A.G., Robichaud, D., Cooke, S.J., 2012. Validation of reflex indicators for measuring vitality and predicting the delayed mortality of wild coho salmon bycatch released from fishing gears. J. Appl. Ecol. 49 (1), 90–98.

Raby, G.D., Packer, J.R., Danylchuk, A.J., Cooke, S.J., 2014. The understudied and underappreciated role of predation in the mortality of fish released from fishing gears. Fish Fish. 15 (3), 489–505.

Rahel, F.J., Olden, J.D., 2008. Assessing the effects of climate change on aquatic invasive species. Conserv. Biol. 22 (3), 521–533.

Redpath, T.D., Cooke, S.J., Arlinghaus, R., Wahl, D.H., Philipp, D.P., 2009. Life-history traits and energetic status in relation to vulnerability to angling in an experimentally selected teleost fish. Evol. Appl. 2 (3), 312–323.

Redpath, T.D., Cooke, S.J., Suski, C.D., Arlinghaus, R., Couture, P., Wahl, D.H., Philipp, D.P., 2010. The metabolic and biochemical basis of vulnerability to recreational angling after three generations of angling-induced selection in a teleost fish. Can. J. Fish. Aquat. Sci. 67 (12), 1983–1992.

Reid, D., Armstrong, J.D., Metcalfe, N.B., 2012. The performance advantage of a high resting metabolic rate in juvenile salmon is habitat dependent. J. Anim. Ecol. 81 (4), 868–875.

Richard, A., Dionne, M., Wang, J., Bernatchez, L., 2013. Does catch and release affect the mating system and individual reproductive success of wild Atlantic salmon (Salmo salar L.)? Mol. Ecol. 22 (1), 187–200.

Rindorf, A., Gislason, H., Burns, F., Ellis, J.R., Reid, D., 2020. Are fish sensitive to trawling recovering in the Northeast Atlantic? J. Appl. Ecol. 57 (10), 1936–1947.

Robinson, K.A., Hinch, S.G., Gale, M.K., Clark, T.D., Wilson, S.M., Donaldson, M.R., Patterson, D.A., 2013. Effects of post-capture ventilation assistance and elevated water temperature on sockeye salmon in a simulated capture-and-release experiment. Conserv. Physiol. 1 (1), cot015.

Roessig, J.M., Woodley, C.M., Cech, J.J., Hansen, L.J., 2004. Effects of global climate change on marine and estuarine fishes and fisheries. Rev. Fish Biol. Fish. 14 (2), 251–275.

Ryer, C.H., 2008. A review of flatfish behavior relative to trawls. Fish. Res. 90, 138–146.

Schindler, D.E., Hilborn, R., Chasco, B., Boatright, C.P., Quinn, T.P., Rogers, L.A., Webster, M.S., 2010. Population diversity and the portfolio effect in an exploited species. Nature 465 (7298), 609–612.

Schurmann, H., Steffensen, J., 1994. Spontaneous swimming activity of atlantic cod Gadus morhua exposed to graded hypoxia at three temperatures. J. Exp. Biol. 197, 129–142.

Shameena, S.S., Kumar, S., Kumar, K., Raman, R.P., 2021. Role of temperature and co-infection in mediating the immune response of goldfish. Microb. Pathog. 156, 104896.

Shea, B.D., Coulter, S.K., Dooling, K.E., Isihara, H.L., Roth, J.C., Sudal, E., Donovan, H., L.A., Dove, A.D.M., Cooke, S.J., Gallagher, A.J.., 2022. Recreational fishing fight times are not correlated with physiological status of blue sharks (Prionace glauca) in the Northwestern Atlantic. Fish. Res. 248, 106220.

Siepker, M.J., Ostrand, K.G., Wahl, D.H., 2006. Effects of angling on feeding by largemouth bass. J. Fish Biol. 69 (3), 783–793.

Simpson, S.D., Radford, A.N., Nedelec, S.L., Ferrari, M.C., Chivers, D.P., McCormick, M.I., Meekan, M.G., 2016. Anthropogenic noise increases fish mortality by predation. Nat. Commun. 7 (1), 1–7.

Sørensen, C., Munday, P.L., Nilsson, G.E., 2014. Aerobic vs. anaerobic scope: sibling species of fish indicate that temperature dependence of hypoxia tolerance can predict future survival. Glob. Chang. Biol. 20 (3), 724–729.

Stachowicz, J.J., Byrnes, J.E., 2006. Species diversity, invasion success, and ecosystem functioning: disentangling the influence of resource competition, facilitation, and extrinsic factors. Mar. Ecol. Prog. Ser. 311, 251–262.

Suski, C.D., Killen, S.S., Kieffer, J.D., Tufts, B.L., 2006. The influence of environmental temperature and oxygen concentration on the recovery of largemouth bass from exercise: implications for live-release angling tournaments. J. Fish Biol. 68, 120–136.

Swain, D.P., 2011. Life-history evolution and elevated natural mortality in a population of Atlantic cod (Gadus morhua). Evol. Appl. 4 (1), 18–29.

Tenningen, M., Vold, A., Olsen, R.E., 2012. The response of herring to high crowding densities in purse-seines: survival and stress reaction. ICES J. Mar. Sci. 69 (8), 1523–1531.

Thambithurai, D., Crespel, A., Norin, T., Rácz, A., Lindström, J., Parsons, K.J., Killen, S.S., 2019. Hypoxia alters vulnerability to capture and the potential for trait-based selection in a scaled-down trawl fishery. Conserv. Physiol. 7 (1), coz082b.

Thorstad, E.B., Næsje, T.F., Leinan, I., 2007. Long-term effects of catch-and-release angling on ascending Atlantic salmon during different stages of spawning migration. Fish. Res. 85 (3), 316–320.

Tort, L., 2011. Stress and immune modulation in fish. Dev. Comp. Immunol. 35 (12), 1366–1375.

Uusi-Heikkilä, S., Wolter, C., Klefoth, T., Arlinghaus, R., 2008. A behavioral perspective on fishing-induced evolution. Trends Ecol. Evol. 23 (8), 419–421.

Van Leeuwen, T.E., Dempson, B., Cote, D., Kelly, N.I., Bates, A.E., 2021. Catchability of Atlantic salmon at high water temperatures: implications for river closure temperature thresholds to catch and release angling. Fish. Manag. Ecol. 28 (2), 147–157.

Vedor, M., Queiroz, N., Mucientes, G., Couto, A., da Costa, I., Dos Santos, A., Vandeperre, F., Afonso, P., Rosa, R., Humphries, N., et al., 2021. Climate-driven deoxygenation elevates fishing vulnerability for the ocean's widest ranging shark. Elife 10, e62508.

Venturelli, P.A., Shuter, B.J., Murphy, C.A., 2009. Evidence for harvest-induced maternal influences on the reproductive rates of fish populations. Proc. R. Soc. B: Biol. Sci. 276 (1658), 919–924.

Volckaert, F.A., Hellemans, B., Batargias, C., Louro, B., Massault, C., Van Houdt, J.K., Haley, C., De Koning, D.J., Canario, A.V., 2012. Heritability of cortisol response to confinement stress in European sea bass Dicentrarchus labrax. Genet. Sel. Evol. 44 (1), 1–5.

Wheeler, A.J., Bett, B.J., Billett, D.S.M., Masson, D.G., Mayor, D.J., 2005. The impact of demersal trawling on northeast Atlantic deepwater coral habitats: the case of the Darwin Mounds, United Kingdom. In: American Fisheries Society Symposium. Vol. 41. American Fisheries Society, pp. 807–818.

Wilkie, M.P., Davidson, K., Brobbel, M.A., Kieffer, J.D., Booth, R.K., Bielak, A.T., Tufts, B.L., 1996. Physiology and survival of wild Atlantic salmon following angling in warm summer waters. Trans. Am. Fish. Soc. 125 (4), 572–580.

Wilson, A.D.M., Binder, T.R., McGrath, K.P., Cooke, S.J., Godin, J.-G.J., 2011. Capture technique and fish personality: angling targets timid bluegill sunfish, Lepomis macrochirus. Can. J. Fish. Aquat. Sci. 68, 749–757.

Wood, C.M., Turner, J.D., Graham, M.S., 1983. Why do fish die after severe exercise? J. Fish Biol. 22 (2), 189–201.

Yang, Y., Cao, Z.D., Fu, S.J., 2015. Variations in temperature acclimation effects on glycogen storage, hypoxia tolerance and swimming performance with seasonal acclimatization in juvenile Chinese crucian carp. Comp. Biochem. Physiol. A Mol. Integr. Physiol. 185, 16–23.

Yesson, C., Fisher, J., Gorham, T., Turner, C.J., Hammeken Arboe, N., Blicher, M.E., Kemp, K.M., 2017. The impact of trawling on the epibenthic megafauna of the west Greenland shelf. ICES J. Mar. Sci. 74 (3), 866–876.

Závorka, L., Koeck, B., Cucherousset, J., Brijs, J., Näslund, J., Aldvén, D., Höjesjö, J., Fleming, I.A., Johnsson, J.I., 2017. Co-existence with non-native brook trout breaks down the integration of phenotypic traits in brown trout parr. Funct. Ecol. 31 (8), 1582–1591.

Zeng, L.Q., Zhang, A.J., Killen, S.S., Cao, Z.D., Wang, Y.X., Fu, S.J., 2017. Standard metabolic rate predicts growth trajectory of juvenile Chinese crucian carp (Carassius auratus) under changing food availability. Biol. Open 6 (9), 1305–1309.

Chapter 5

# Environmental stressors in Amazonian riverine systems

Adalberto Luis Val[a,*], Rafael Mendonça Duarte[b], Derek Campos[a], and Vera Maria Fonseca de Almeida-Val[a]

[a]*Laboratory of Ecophysiology and Molecular Evolution, Brazilian National Institute for Research of the Amazon, Manaus, Brazil*

[b]*Biosciences Institute, São Paulo State University (UNESP), Coastal Campus, São Paulo, Brazil*

[*]*Corresponding author: e-mail: dalval@inpa.gov.br*

## Chapter Outline



The Amazon has a rich history of tectonic and climatic effects that have given rise to vast, complex and dynamic interconnected landscapes. The dynamics of the system can be observed today by the oscillation of river water levels, variations in oxygen levels, pH and temperature, and the biological diversity that exists in the different systems throughout the year. This continuous environmental diversity has contributed to the emergence of a rich ichthyofauna that has developed a vast set of adaptations at all levels of biological organization to cope with the continuous environmental

**Fish Physiology, Vol. 39B. https://doi.org/10.1016/bs.fp.2022.05.001**

challenges of the biome. However, the environmental structure that was formed over some 65 million years, i.e., since the beginning of the Andes uplift, is today confronted with many challenges of a, shall we say, new era—the Anthropocene. These challenges include metal pollution, urban pollution, pesticides, oil, hydroelectric construction, and, most importantly, the effects of climate change. Many of the evolutionary adaptations incorporated by fish are not sufficient to neutralize the effects of these new challenges, many of which have synergistic effects with each other or with the natural challenges that occur in the Amazon (hypoxia, low pH, low ionic availability, naturally warmer waters). Thus, it is important that we can anticipate the responses of Amazonian fishes to the challenges imposed by their environments in order to better manage the Amazon rainforest.

## 1 The riverine systems and connecting lakes of the Amazon

The origins of the Amazon require a geological trip back in time to when the Andes cordillera began to form some 65 million years ago. The Amazon as we see it today is the product of a long geological history of tectonic and climatic effects. The product is a physiographic complex unparalleled on the planet and, therefore, a unique amphibious land. Understanding how the Amazon was formed helps us to understand how it became the land of biological diversity and how different biological groups, including fish, have adapted to the environmental variations that have occurred on an ongoing basis. More recently, these changes occur synergistically with climate change.

The Andes uplift closed the connection of the Atlantic with the Pacific and resulted in an Atlantic-oriented flow of freshwater through a colossal central channel, the Amazon River, which is given several other names (Ucayalli-Apurimac, Marañon, Solimões) before its waters meet the Atlantic Ocean. The thunderous entry of so much water entering the Atlantic has long been recognized and referred to as "pororoca," an Indigenous word meaning "great roar." The most important seasonal variation of the Amazon is the seasonal hydrological cycle, called "flood pulse" by Junk et al. (1989), that initiated in the Miocene (Vonhof and Kaandorp, 2010) where river catchments play an important role in watershed design. It is becoming increasingly clear that complex tectonic and climatic processes have given rise to biological diversity at its different levels of organization, i.e., the diversity we see today in the Amazon is not explained by a "time-limited event."

The tectonic and climatic processes that have acted, and continue to act, in the Amazon have produced a vast array of distinct environments that are colonized by a great range of species, many of which are still being described (reviewed by Tencatt et al., 2020; Val and Almeida-Val, 1995). While a great deal is known about the Amazon biome, its uniqueness may be best described by its vastness and unknowns, as follows. Out of the 7.76 million square kilometers, more than 5 million square kilometers of the biome are within the borders of Brazil. This continental territory is home to the longest river on the planet: the Amazon River, with a length of 6992 km. Four of the 10 longest rivers in the world are in the Amazon watershed: the Amazon proper, the Rio Negro, the Rio Madeira and the Rio Japurá. As if this were not enough, 20 of the 34 largest tropical rivers in the world are tributaries of the Amazon River (Latrubesse et al., 2005). It is in these dimensions that the environmental variability of this unique system arises. They consist of waters of different colors with different physical-chemical characteristics (Sioli, 1984); forests with different species of trees flooded at different times of the year; flood pulses that differ intra-regionally ( Junk et al., 1989); in short, they are environments that have evolved through distinct paths and at different rates.

Two aspects impose a dynamic characteristic to the environmental diversity in the Amazon. First, because of the region's geological formation, waters draining soils with different characteristics have different qualities. The waters of Andean origin that plunge into the Amazon valley have a muddy, cafe au lait coloration, called white waters, that carry suspended material to the mouth of the Amazon River in the Atlantic where sediments of Andean origin are deposited. Amazonian waters are generally ion poor. The white waters have higher ion content than the black and clear waters and have pH values close to neutral (Furch, 1984). Black waters, such as the Rio Negro waters, drain the Guiana's Shield, invade marginal forests seasonally, are acidic and rich in dissolved organic carbon (DOC) with levels as high as 30 mg/L in some water bodies. The levels of the main nutrients in these waters are very low, just above distilled water. The humic and fulvic acids that make up the high levels of DOC, are organic acids that contribute to the water's acidity, and in some marginal lakes, pH values as low as pH 2.5 are observed. Clear waters are the third water type frequently found in the Amazon (Fig. 1). They occur in several river systems, such as the rivers draining ancient and eroded surfaces of the Guiana and Central Brazil Shields, the headwaters of Andean rivers, and small lowland clearwater streams. The Tapajós River is the largest clearwater river in the Amazon. The waters of the Tapajós have $<1\%$ suspended material and are very similar to rainwater, with very low concentrations of the major nutrients, sodium being a predominant element. Even sodium can be at very low levels, lower than that found in rainwater. The pH of its waters is higher than the pH of the waters of the Rio Negro, ranging from 6 to 6.8. Second, because these waters differ dramatically in their characteristics, when they meet and mix it gives rise to waters with new chemical compositions, the degree to which varies seasonally, as discussed below.

## 1.1 Environmental diversity

The Amazon biome includes many ecosystems with distinct characteristics that are connected through its borders, by means of transition zones, with the other South American biomes, such as the Amazon-Cerrado and the Amazon-Pantanal. There are 53 distinct terrestrial and aquatic environments in the Amazon rainforest. Quantitatively, the Amazon biome is comprised

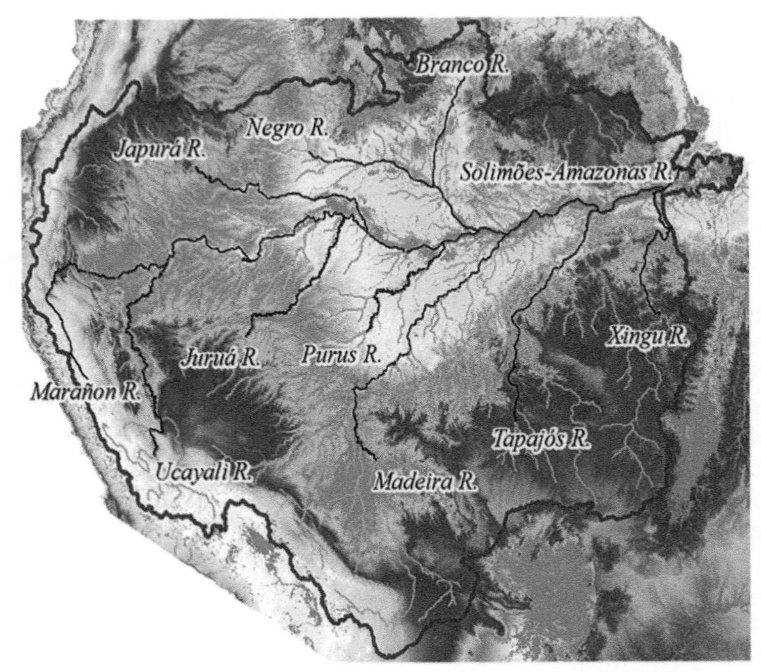

Ucayali River
Marañon River
Solimões River
Amazon River
Branco River
Japurá River

Negro River

Tapajós River
Xingu River

**FIG. 1** The major rivers of the Amazon. *Modified from Fassoni-Andrade, A.C., Fleischmann, A.S., Papa, F., Paiva, R.C.D., Sly Wongchuig, Melack, J.M., Moreira, A.A., Paris, A., Ruhoff, A., Barbosa, C., Maciel, D.A., Novo, E., Durand, F., Frappart, F., Aires, F., Abrahão, G.M., Ferreira-Ferreira, J., Espinoza, J.C., Laipelt, L., Costa, M.H., Espinoza-Villar, R., Calmant, S., Pellet, V., 2021. Amazon hydrology from space: scientific advances and future challenges. Rev. Geophys. 59, (e2020RG000728), with indication of the major types of Amazonian waters and their occurrence.*

of forests (78%), Andean areas (1.5%), flooded areas (floodplains and igapós) (5.83%), savannas (12.75%), and two tropical steppes (1.89%) (Hilty, 2012). Of the many Amazonian environments, the Andean region and Amazonian lowlands in particular are highlighted below, because their importance to the aspects discussed here.

### 1.1.1 Andean region

Throughout the geological process, the uplift of the Andes resulted in profound changes to the aquatic system, which involved the interruption of the connection of the Atlantic with the Pacific Oceans with great implications for biological diversification. There are many geological characteristics that have influenced the modern aquatic biota. Among these are rivers that flow down great elevations to the flat landscapes and then over soft rocks and complex topographic forms connecting river networks and influencing lithology (Val et al., 2021). It is in this context that divisions such as Orinoco, Essequibo, and even the Parana-Paraguay-Uruguay basin appear, suggesting that the Amazon River is still in a transitional phase (reviewed by Val et al., 2021). The characteristics of the soil where the main head waters of the Amazon River originates results in the cafe au lait appearance of the white waters, which is determined by the silty material in suspension that can be observed along the 6992km of the central channel that ultimately connects with the Atlantic. Even the Atlantic floor is impacted by large deposits that settle out shortly after the river invades the ocean. More details about this immense diversity that expands along the system can be found in the review by Val et al. (2021). Finally, it should be noted that in addition to the existing dams, back in 2012 the construction of 138 new hydroelectric dams were planned in the Andean region of countries that house the Amazon River, specifically in the six main Andean tributaries of the Amazon River (Caquetá, Madeira, Napo, Maranõn, Putumayo, Ucayalli) which has a significant impact on the water system of the biome (Finer and Jenkins, 2012). The effects of these physical interventions on the system in association with climate change are unknown and may interfere with the biology and adaptability of fish species in this part of the basin.

### 1.1.2 Amazonian lowlands

The Amazonian lowlands extend from the Andean foothills to the estuarine zone where the river splits into several branches. The Amazonian lowlands can be divided into three main parts, as proposed by Sioli (1984), the upper region, the middle Amazon basin and the lower region, all of which consist of rivers, channels (locally known as parana) and marginal lakes. Together, they constitute most of the Amazon basin and are environmentally and biologically diverse. The upper region extends to the confluence with the Negro River, a first important transition in the Amazon basin (Sioli, 1984). Up to this location, the tributaries of the Solimões-Amazonas River meander greatly resulting in many oxbow lakes. The waters of the central channel are white and non-turbulent which gives a uniform and flat character to the whole land. The rivers Juruá, Purus, Madeira and Negro are the main tributaries of this part of the lowlands. The waters of the Rio Negro are dark, as the name suggests, due to the high levels of DOC and contrast sharply with the clear waters of the Solimões-Amazonas River. The rivers that make up the Rio Negro basin drain a good part of the Guiana Shields, a very old region. In this first region of the lowlands, there are many small patches with white sand forests, that are not well studied, but thought to play an important role to diversity and ecosystem services (Adeney et al., 2016).

The second section of the lowlands, or middle Amazon basin, is compressed between the Guiana shield in the north and the Central Brazilian shield in the south. Diverse formations, with distinct stratigraphic and tectonics, show the environmental richness of this area, as is the case of Barreiras and Alter-do-Chão. This region is also home to the Tapajós River, a tributary with clear waters, the third type of Amazon water, as first described by Sioli (1984), with little DOC and a transparent appearance. The chemical characteristics of the land masses interfere with primary productivity in both the northern and southern regions of the central channel. This obviously influences the aquatic biota, including fish population diversity and size. Another tributary of this region is the Xingu River, with remarkable environmental features, including rapids, endemic species such as the cascudo zebra (*Hypancistrus zebra*), and new fish species such as the recently described *Parancistrus nudiventris* (acari) (Py-Daniel and Zuanon, 2005). The construction of the Belo Monte hydroelectric dam in this tributary has impacted the biological dynamics of aquatic life in the Xingu River (Fearnside, 2006b; Zuanon et al., 2019). River fragmentation has been an environmental issue in several places around the world, including the Amazon itself (Aragão et al., 2018; Zarfl et al., 2015).

The third section of the lowlands includes, a set of rivers, channels, and islands, including the island of Marajó. The water flow in these tributaries depends on the marine tides and is thus rhythmic. The waters advance little by little in these tributaries. It is in this region that a unique phenomenon occurs: the pororoca, when the Amazon River meets the sea, producing waves up to 4 m high that advance up to 800 km upstream. Furthermore, the chemical characteristics of the marine waters can be detected up to the city of Santarem, about 1500 km upstream, and in this stretch, there is a physical-chemical gradient determined by this difference between these two environments, the sea, and the river. There is no doubt that climate change, with rising sea levels, will significantly affect this marine introgression into the Amazon, and consequently the aquatic biota of the region. Note that several species of marine fish, such as *Pristis perotteti* (largetooth sawfish) and

*Carcharhinus leucas* (bull shark), invade the Amazon through this process, that is, through the layer of marine water that forms at the bottom of the freshwater column and invades the Amazon River (Santos and Val, 1998).

It is interesting to note that rainfall occurs at different times in different regions of the Amazon. A rainfall peak occurs during austral summer in the southern Amazon, while in the central part of the Amazon and the third sector of the lowlands, the rainfall peak occurs in austral autumn. North of the equator, rainfall peak occurs in austral winter (reviewed by Costa et al., 2021). In connection with the rainfall regime, the discharge of freshwater into the Atlantic Ocean is dynamic (Sorribas et al., 2016), and in the central part, such as Manaus, the flood peak is June. Discharge measurements made at Óbidos, where the Amazon River is narrowest, indicate a range from 100,000 $m^3/s$ during the maximum ebb period to 250,000 $m^3/s$ during the flood period (Goulding et al., 2003). This diversity of environments and dynamics has profound effects on aquatic biota, particularly on fish (Val and Almeida-Val, 1995; Val et al., 2006).

## 1.2 Environmental dynamics

### 1.2.1 Flood pulses

As mentioned, water level oscillation in Amazonian rivers began in the Miocene and became a major environmental force throughout the geological formation of the basin. Junk et al. (1989) analyzed these oscillations and called them flood pulses, which they had also described on other continents. They proposed that the forces are not just physical but include the adaptations of biota to large river-floodplain systems. In fact, these flood pulses also influence physiological variations of both the organisms of the aquatic biota (Martins et al., 2017; Val and Almeida-Val, 1995) and the organisms of the interface between the aquatic and terrestrial environments, such as *várzeas* (cyclically flooded areas along white-water rivers) and *igapós* (cyclically flooded areas along black-water rivers) (Wittmann et al., 2007).

During periods of high water, fish migrate in the morning to the interior of the flooded forest where they find food, mainly fruits, in abundance. The most typical example for this lateral movement is the tambaqui (*Colossoma macropomum*), which can feed on up to 140 different kinds of fruits in the flooded forest (Val and Oliveira, 2021). This lateral migration is also influenced by the availability of oxygen that becomes limiting in the flooded forest when light, and thus photosynthesis, is reduced with most species that enter the forest in the morning leaving by the late afternoon to avoid hypoxia. During periods of low water, in addition to the lack of food in the forest, the water quality is poor due to extensive aquatic plants degradation. In conjunction with hypoxia, high levels of carbon dioxide ($CO_2$) and hydrogen sulfide ($H_2S$) occur, which impact the behavior of the fish. In the presence of $H_2S$, the armored catfish *Hoplosternum littorale*, a facultative air-breathing fish, increases its surface breathing, avoiding aquatic gas exchange in the presence of $H_2S$ (Brauner et al., 1995).

Recent climate change brings new challenges to these dynamics that occur regularly in the flood pulses. Schöngart and Junk (2007) analyzed the flood and ebb peaks over the last 30 years and concluded that the amplitude (the difference between flood peaks and ebb minima) has progressively increased, suggesting that the region has experienced higher flood peaks and lower ebb minima. The relationship of this variation in the amplitude of Amazonian floods and ebb tides to global climate change is yet to be fully demonstrated. The causal relationship of these variations indicates that El-Niño is one of the main causal factors, with the most intense droughts generally occurring in the central-northern part of the basin, but not predictably ( Jiménez-Muñoz et al., 2016, 2018; Marengo et al., 2018).

Finally, flood pulses, with recent intensity influenced by climate change, in synergy with other environmental factors have a pronounced effect on aquatic biota, including fish. Dissolved oxygen, temperature, and pH are three other vital environmental factors that have required profound adaptive adjustments of fishes throughout the evolutionary process, and it seems that the effects of global change will affect them, as waters will become warmer, more acidic and more oxygen deprived.

### 1.2.2 Oxygen

The amount of dissolved oxygen in water is determined by many environmental factors, including temperature and altitude, and biological factors, such as photosynthesis and organic decomposition. Within a body of water, its shape, size, depth, winds, coverage by aquatic plants, the movement of water, the amount of DOC, and other factors, all influence the amount of dissolved oxygen. Furthermore, because the aquatic environments of the Amazon are interconnected, water oxygen levels in one region may directly impact another and continuous adjustments of the aquatic biota are required to ensure aerobic metabolism (Val, 1996). Oxygen levels in the Amazon can vary spatially, but also temporally, both seasonally and diurnally. In many cases, the diurnal variations in oxygen are more intense than the seasonal variations themselves; in floodplain lakes of the Amazon, conditions can often change from supersaturated (hyperoxia) to oxygen depleted (anoxia) in a matter of hours.

Throughout the evolutionary process, since the Cambrian, oxygen levels have been well below current levels (21%). Oxygen levels above present-day levels also occurred in parts of the Carboniferous and Permian (Dudley, 1998). The uplift of the Andes was yet to occur, but some groups of teleosts already existed in the region. Oxygen levels would only become similar to today's levels in the late Tertiary. Thus, it is not surprising that a myriad of adaptive strategies has evolved in fish to obtain oxygen, or even to protect them from hyperoxic or hypoxic conditions, however, this may not be sufficient to protect fish against anthropogenic changes that currently impact the environment.

Stratification in oxygen levels in the water column is very common, often with oxygen available only at the surface of the water column. Therefore, among Amazonian fish, several species have developed strategies to exploit this precious source of oxygen (Val and Almeida-Val, 1995), as will be discussed below. However, oil mining in several locations in the Amazon has put this source of oxygen at risk. Crude oil spills cover the air water interface, preventing the diffusion of gases, particularly oxygen (Kochhann et al., 2015b; Val and Almeida-Val, 1999). Other types of pollution, especially around cities, interfere with the diffusion of oxygen in the water column, such as the presence of plastics and other pollutants. Unfortunately, this has become more and more common in the hitherto pristine environments of the Amazon (Ribeiro-Brasil et al., 2020).

The increase in temperature, determined by global warming, brings additional challenges regarding the availability of oxygen in Amazonian waters, which is further compounded by the temperature induced increase of metabolic rate in aquatic organisms. The increase in temperature reduces oxygen solubility and promotes increased decomposition further depleting water oxygen levels especially in floodplains and igapós, lengthening the periods of low oxygen availability. Also, the different intensive land use processes underway in the Amazon reduce the capacity of the forests to remove $CO_2$ from the atmosphere, while contributing initially more organic matter to the aquatic ecosystem (Thurman, 1985b). Thus, global warming interferes with the complex process that affect dissolved oxygen in water in the Amazon, which directly affects the interaction of fish with their environments.

### 1.2.3 Temperature

Temperature is one of the most important environmental parameters for tropical biology, due to its physiological effects and its effects on different environmental factors, such as oxygen, water pH, among others. Analysis of thermal anomalies over the past 40 years indicates an increase in environmental temperature of about 0.6–0.7 °C, with 2016 being the warmest year (+ 0.9 °C) since 1950 for the Amazon biome (Marengo et al., 2018). An increase of up to 5 °C may occur between 2041 and 2070 in the northeastern region of the state of Amazonas in Brazil and up to 3.5 °C in the upper part of the Rio Negro (Menezes et al., 2018).

Deforestation can accentuate the environmental temperature increase in the Amazon. According to Lovejoy and Nobre (2018), when the level of deforestation reaches between 20% and 40% (17% is the current level) forest systems will be able to recover (tipping point) which will cause a significant increase in the biome temperature and therefore drive many organisms to the limit of their thermal tolerance. And of course, the current high rates of deforestation ( Jaffé et al., 2021) cause a loss of carbon sinking capacity which results in a vicious circle making the whole system even hotter.

### 1.2.4 pH

In general, Amazonian waters are ion poor and vary in pH (Sioli, 1984). The white waters are close to neutral, with a pH approaching 7, but vary spatially and seasonally (reviewed by Val and Almeida-Val, 1995). In contrast, the waters of the Negro River are acidic with pH ranging from 4 to 5 in the central part of the main rivers, but lower pH's can often be found in marginal lakes and igapós. Small streams on land (terra-firme) are also acidic, with pH <5 (Walker, 1995). The pH of clear waters is in a broader range, from pH 4.5 to pH 7.8. The characteristics of the drainage areas determine the pH of these water bodies (see Furch, 1984; Sioli, 1984). Note that these rivers connect throughout the basin and produce pH gradients and allow biota to move between waters with different pH.

Climate change may make acidic aquatic environments even more acidic, but there are no systematic studies on this effect in Amazonian water bodies. However, the study by Scanes et al. (2020) in estuarine waters of Australia over a 12-year period observed water acidification at a rate of 0.09 pH units per year. According to these authors, lakes and rivers in the region are warming and acidifying at an even greater rate. In the Amazon, warming will also reduce water pH, but given the differences in regional rates of warming, water pH values will likely differ. Note that estuarine waters would likely be well-buffered, better buffered than Amazon waters.

## 2 Fish diversity

Approximately half of the world's fishes are freshwater, 15% of those are found in the Amazon, i.e., more than 2700 species among 18 orders. Thus, the Amazon represents a hotspot for fish biodiversity. Understanding the evolutionary drivers that have led to such diversity and investigating comparative physiology and adaptive mechanisms that underly fish distributions shed insight on processes that have resulted in the Amazon basin having the highest diversity of freshwater fishes in the world. However, this diversity is vulnerable to climate change. As expounded upon by Val et al. (2021), this remarkable freshwater fish diversity is concentrated on a very small surface of the Earth (<0.5%), which when expressed as total aquatic volume is even lower (0.001%) and has led to what has been described as the "freshwater fish paradox." These authors clearly show that landscape evolution is an important driver of freshwater fish evolution in the Amazon.

The modern rainforest, as we know it, existed in the Miocene, and paleological findings in western Amazonia of that time indicate the existence of a diverse aquatic fauna, composed of mollusks, ostracods, turtles, crocodiles, and fish (Wesselingh et al., 2010) that were already "testing" adaptations to cope with the ongoing tectonic and climatic changes. The warm and humid climate and the heterogeneous edaphic substrate, amidst diverse tectonic processes with river catchments, were responsible for at least part of the biological diversification. The next period, the Quaternary, as emphasized by Wesselingh et al. (2010), was characterized by invasions into these new environments, but not as an important period for biological diversification. Even though river capture processes (river drainage diverted from its own bed)

played an important role in speciation, these same changes also caused extinctions (Albert et al., 2018). Some of these processes continue even today. Thus, species continue to evolve throughout the basin, whether as a consequence of more recent tectonic processes (Albert et al., 2018; Sá-Leitão et al., 2022), or by the isolation of populations before the falls of the rivers that enter the Amazon valley (Dutra et al., 2020) or by adaptation to the prevailing conditions in the distinct aquatic environments that exist in the region.

Complex and distinct, but not unique, processes are responsible for biological diversification in the Amazon (Bush, 1994; Wesselingh et al., 2010). Many hypotheses exist to explain the complex biogeographic history of Amazonian fishes that often contrast with one another (Albert et al., 2018; Bush, 1994; Dagosta and De Pinna, 2019; Lundberg et al., 2010). For example, recent geological studies using different methodologies have proposed two geological models: the "Old Amazon" and the "Young Amazon," which basically differ with respect to the temporal range for basin formation. The controversy is extensive as reviewed by Méndez-Camacho et al. (2021) who also produced biogeographic evidence supporting the Old Amazon hypothesis for the Amazon basin formation. There are more than 2700 species of fish living in the Amazon, a number that expands each time new environments are explored (Tencatt et al., 2020). This ichthyofauna includes animals of all sizes, from the diminutive cardinal tetra (*Paracheirodon axelrodi*) to large species such as the pirarucu (*Arapaima gigas*), which can weigh up to 250kg (Val and Almeida-Val, 1995). Fig. 2 shows the major fish groups of the Amazon, indicating their main physiological characteristics and vulnerabilities to ongoing environmental changes, including climate change.

Because speciation is a continuous process dependent on phenotypic adjustments and environmental adaptations, the Amazon is a unique place (Beheregaray et al., 2015). Several other aquatic groups have experienced the same trajectory of adaptation and diversification highlighting that the Amazon is an important source for Neotropical biodiversity in general (Antonelli et al., 2018). With domestication and globalization, it is not surprising that many Amazonian species are scattered around the world, not only in aquaria, but in breeding farms, as is the case with the tambaqui

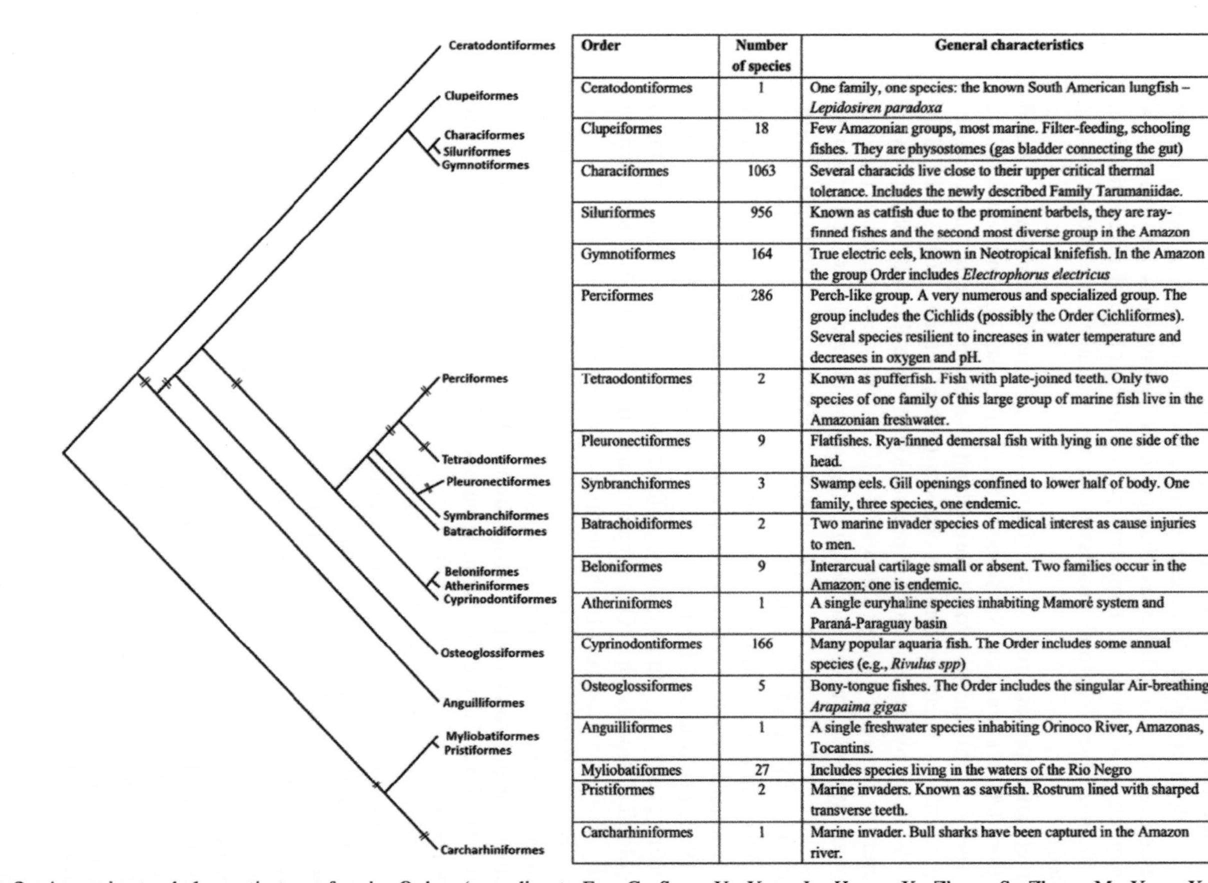


| Order | Number of species | General characteristics |
|---|---|---|
| Ceratodontiformes | 1 | One family, one species: the known South American lungfish – *Lepidosiren paradoxa* |
| Clupeiformes | 18 | Few Amazonian groups, most marine. Filter-feeding, schooling fishes. They are physostomes (gas bladder connecting the gut) |
| Characiformes | 1063 | Several characids live close to their upper critical thermal tolerance. Includes the newly described Family Tarumaniidae. |
| Siluriformes | 956 | Known as catfish due to the prominent barbels, they are ray-finned fishes and the second most diverse group in the Amazon |
| Gymnotiformes | 164 | True electric eels, known in Neotropical knifefish. In the Amazon the group Order includes *Electrophorus electricus* |
| Perciformes | 286 | Perch-like group. A very numerous and specialized group. The group includes the Cichlids (possibly the Order Cichliformes). Several species resilient to increases in water temperature and decreases in oxygen and pH. |
| Tetraodontiformes | 2 | Known as pufferfish. Fish with plate-joined teeth. Only two species of one family of this large group of marine fish live in the Amazonian freshwater. |
| Pleuronectiformes | 9 | Flatfishes. Rya-finned demersal fish with lying in one side of the head. |
| Synbranchiformes | 3 | Swamp eels. Gill openings confined to lower half of body. One family, three species, one endemic. |
| Batrachoidiformes | 2 | Two marine invader species of medical interest as cause injuries to men. |
| Beloniformes | 9 | Interarcual cartilage small or absent. Two families occur in the Amazon; one is endemic. |
| Atheriniformes | 1 | A single euryhaline species inhabiting Mamoré system and Paraná-Paraguay basin |
| Cyprinodontiformes | 166 | Many popular aquaria fish. The Order includes some annual species (e.g., *Rivulus spp*) |
| Osteoglossiformes | 5 | Bony-tongue fishes. The Order includes the singular Air-breathing *Arapaima gigas* |
| Anguilliformes | 1 | A single freshwater species inhabiting Orinoco River, Amazonas, Tocantins. |
| Myliobatiformes | 27 | Includes species living in the waters of the Rio Negro |
| Pristiformes | 2 | Marine invaders. Known as sawfish. Rostrum lined with sharped transverse teeth. |
| Carcharhiniformes | 1 | Marine invader. Bull sharks have been captured in the Amazon river. |

**FIG. 2** Approximate phylogenetic tree of major Orders (according to Fan, G., Song, Y., Yang, L., Huang, X., Zhang, S., Zhang, M., Yang, X., Chang, Y., Zhang, H., Li, Y., Liu, S., Yu, L., Chu, J., Seim, I., Feng, C., Near, T.J., Wing, R.A., Wang, W., Wang, K., Wang, J., Xu, X., Yang, H., Liu, X., Chen, N., He, S., 2020. Initial data release and announcement of the 10,000 fish genomes project (Fish10K). GigaScience 9, 1–8), and number of fish species living in Amazonian waters (according to Dagosta, F.C.P., De Pinna, M., 2019. The fishes of the Amazon: distribution and biogeographical patterns, with a comprehensive list of pecies. Bull. Am. Mus. Nat. Hist. 2019 (431), 1–163), with indication of general characteristics.

and the pirarucu. Also, in the same way, many exotic species are spreading throughout the Amazon (Doria et al., 2021).

The biodiversity of the Amazonian ichthyofauna already described has unique features. However, the number of described fish species (Dagosta and De Pinna, 2019) of the Amazon continue to increase with many more to be added. Recent work by Ximenes et al. (2021), for example, mapped the diversity hidden in the *Geophagus* sensu *strictu* group (Cichlidae: Geophagini—commonly known as acará or cará) and found six new species. The authors pointed out that of the six species, five of them live in areas under anthropogenic pressures.

The biological diversity of fish and other aquatic organisms have evolved in part in response to the slow geological and climatic dynamics of the Amazon basin as mentioned, but also in response to spatially specific environments yielding species-specific adaptations to local reigning conditions. Thus, the biodiversity of the Amazon is vulnerable to climate change, since it acts at both extremes, altering the general climatic norms of the basin and imposing physiological challenges that, in several cases, go beyond the regular adaptive abilities of organisms, particularly fish, as described later. It is also clear that it is not possible to treat biodiversity, (including fish) climate change and anthropogenic pressures separately (Pörtner et al., 2022) and an integrated approach is required.

## 3 Hypoxia driven adaptations

Intermittent aquatic hypoxia is one of the most important evolutionary and ecological drivers in the Amazon and will become more intense directly or indirectly due to global change. Amazonian fishes can cope partially or entirely with hypoxia, using adaptations developed throughout the evolutionary process and formation of the Amazon basin. These abilities include high glycolytic capacity and/or the ability to reduce metabolic rate (Almeida-Val et al., 2000; Muusze et al., 1998) and modify blood $O_2$ transport (e.g., blood $O_2$ affinities that can be rapidly increased—Val et al., 2015; Scott et al., 2017), and many have aquatic surface breathing (ASR) and/or supplemental air breathing capabilities that include accessory organs such as swim bladder, modified gills, lips, oral cavity, skin, fins, and various parts of the gastrointestinal tract (Graham, 1997; Randall et al., 1981; Val and Almeida-Val, 1995). However, there are synergistic effects of global warming in addition to oxygen availability and metabolic demand, such as a temperature induced increase in metabolic rate which leads to greater uptake of pollutants that will further negatively impact the fish.

In addition to the morphological structures mentioned above, several species have developed behavioral adaptations to hypoxia, that allow them to exploit the more oxygenated micro-environments. This is the case in tambaqui, where the expansion of their lower lip (Fig. 3) increases water oxygen

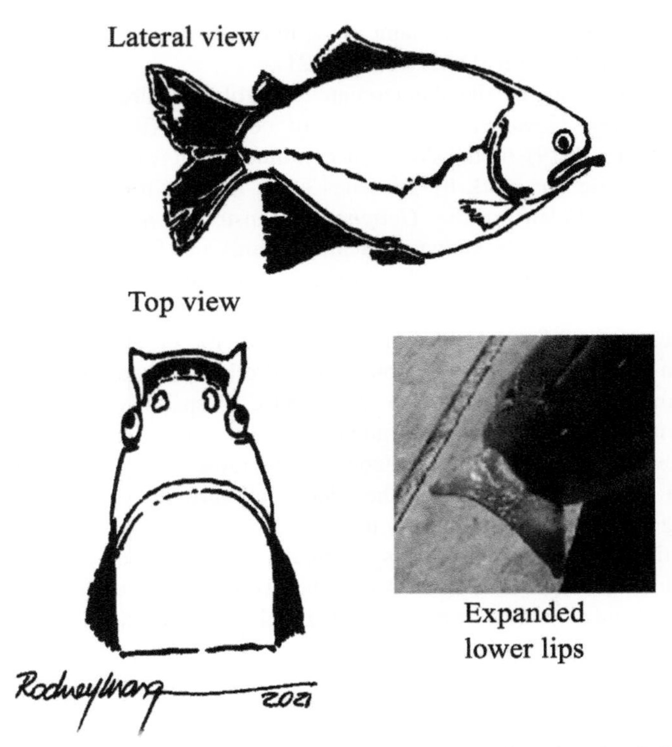


**FIG. 3** Lateral and top view illustrations of the expanded lower lips of tambaqui (*Colossoma macropomum*) when exposed to low oxygen water. *Modified from Val, A.L., Oliveira, A. M., 2021.* Colossoma macropomum*—a tropical fish model for biology and aquaculture. J. Exp. Zool. A 2021, 1–10. At the side, real image of the lower lip of the tambaqui Silva, C., Carneiro, P., 2007. Qualidade da Água na Engorda de Tambaqui em Viveiros sem Renovação da Água, EMBRAPA, Aracajú, SE.*

extraction. This rather extreme morphological adaption is present in several other species in the same group, such as the pirapitinga (*Piaractus brachypomus*) and pacus (*Myleus pacu* and *Mylossoma duriventre*), but also among distantly related groups such as the matrinchã, *Brycon amazonicus* representing a nice example of convergent evolution (Anjos et al., 2008; Rantin and Kalinin, 1996; Val and Almeida-Val, 1995; Val et al., 1998). The onset of lip expansion has been observed to occur in about 1 h with full expansion in about 2 h when the animal is exposed to environmental hypoxia (Braum, 1983; Braum and Junk, 1982; Saint-Paul, 1984; Val and Almeida-Val, 1995). The animal uses its expanded lips for skimming the surface of the water column, which is richer in oxygen. Val (1995) analyzed the contribution of the lips of the tambaqui to the oxygen content of the blood and showed that the presence of the lips associated with the behavior of exploring the surface layer of the water column (aquatic surface respiration) is responsible for up to 40% of the oxygen present in the blood in situations of extreme hypoxia. Fagundes (2012) analyzed the changes in gene expression in the expanded lips of tambaqui and observed that vascular endothelial growth factor (VEGFC), among other genes related to ion channels, increased relative to animals exposed to normoxia. It is possible that these adaptations are affected in situations where the atmospheric composition is modified, by air pollution or even climate change, but there are no definitive studies on this yet.

Another important adaptation of fish to environmental hypoxia is related to their rapid ability to regulate the levels of organic phosphates in the erythrocyte microenvironment (Val et al., 2015). These phosphates are negative allosteric modulators of hemoglobin affinity which can be rapidly reduced to increase oxygen uptake from the water. In contrast to temperate zone fish, Amazonian fishes typically have a higher concentration of GTP than ATP. Other organic phosphates are found in the erythrocytes of Amazonian fish, such as 2,3DPG (2,3 diphosphoglycerate) in tamoatá *H. littorale* and IPP (inositol pentaphosphate) in pirarucu (Bartlett, 1978; Isaacks et al., 1977; Marcon et al., 1999; Val, 2000; Weber et al., 2000). The rate and degree of change in erythrocyte organic phosphate levels are species specific and the mechanisms through which they are regulated are still being determined.

The evolution of air-breathing in fish led to a reduced dependence upon water oxygen levels, but these fish still remain dependent on water for the excretion of the majority of their metabolically produced $CO_2$. They also remain very dependent on the aquatic environment, for ammonia excretion, ionoregulation, and acid-base homeostasis (Frommel et al., 2021; Gonzalez et al., 2010). Even pirarucu, which has a high capacity for $O_2$ uptake from the air as an adult (Pelster et al., 2020) is dependent on aquatic uptake of oxygen in their early stages (Brauner et al., 2004) and thus, all air-breathing fishes are still very dependent upon the aquatic environment.

The diversity of aquatic systems in the Amazon poses challenges to air-breathers, particularly during aquatic hypoxia and elevated high hydrogen sulfide levels because of aquatic plant's degradation. These challenges include the potential loss of oxygen of aerial origin across the gills, predation, reliance on diet in acquiring ions, intoxication with $H_2S$ and ammonia, among others (Affonso et al., 2002; Brauner et al., 1995; Pelster and Wood, 2018; Pelster et al., 2016, 2018; Scott et al., 2017; Sloman et al., 2006, 2009; Wood et al., 2016). Because of these costs, it is possible that species that breathe exclusively in water will be favored in a more hypoxic future Amazon, as predicted by Val and Wood (2022).

Finally, an important aspect that has been understudied in Amazonian fish is sperm performance under different environmental conditions, including aquatic hypoxia. In tambaqui, sperm motility time decreased ($P < 0.05$) from $50.1 \pm 2.70$ s at 29°C in normoxia to $30.44 \pm 1.66$ s in hypoxia ($1\,mgO_2\,L^{-1}$) at pH 8, and this was further reduced to $27.4 \pm 1.42$ s in hypoxia, pH 4

(Castro et al., 2020). In addition, the authors showed that sperm oxygen consumption rate increased during exposure to hypoxia at both pH values. Thus, environmental hypoxia clearly affects the quality of spermatozoa of tambaqui, and likely other fish species, with great implications for the impact of eutrophication, often accentuated around cities, on the maintenance of the species' populations.

## 4 Living in ion poor and acidic waters

### 4.1 Physiological specializations to thrive in ion poor acidic waters

One of the most remarkable adaptations among Amazonian fishes is on gill physiology especially with respect to ionoregulation, acid-base balance and nitrogen excretion, which allow these animals to thrive in harsh aquatic environments. This is particularly true in the tremendous fish biodiversity found in blackwaters of the Rio Negro watershed (around 8% all freshwater fish species in the world), which are faced with acidic dilute waters (i.e., low ionic content particularly $Na^+$ and $Ca^{2+}$) with a high amount of DOC (see Chapter 7, Volume 39A: Wood, 2022). Although some non-Amazonian fish species are also considered acidophilic (Kwong et al., 2014), the exceptional tolerance and the unusual specializations in the gill physiology displayed by fish species from Rio Negro (Gonzalez et al., 2006; Morris et al., 2021) make them interesting models to understand the physiological adaptations of fishes to deal with acid stress in hypo-osmotic environments.

In freshwater environments, fish experience an osmotic gain of water and diffusive loss of ions owing to the hyper-osmotic status of their internal fluids (Evans, 2008; Evans et al., 2005). Thus, fish gill function is geared toward avoiding ionoregulatory disturbances, by the use of specialized transport mechanisms to actively uptake ions, and by limiting diffusive ion loss through a tight gill epithelium (Evans et al., 2005; McDonald et al., 1980; Milligan and Wood, 1982). During exposure to moderately low pH ($\sim$ less than 6), non-acidophilic fishes display a large stimulation of ionic passive losses (mainly $Na^+$ and $Cl^-$), which is associated with $H^+$ displacing $Ca^{2+}$ from paracellular tight junctions that reduces branchial permeability (reviewed by Kwong et al., 2014). Indeed, this increase in paracellular ionic loss mediated by low pH is influenced by water hardness, where an increase of $Ca^{2+}$ levels tend to reduce the ionic diffusive losses to most of the fish species that have been studied (Kumai et al., 2011; McDonald and Wood, 1981; McDonald et al., 1983; Wood, 1989). Another detrimental response to acid exposure is the almost complete inhibition of $Na^+$ and $Cl^-$ uptake in gills, which can impair extracellular acid-base status as these ions are normally associated with the exchange of acid ($H^+/NH_4^+$) and basic ($HCO_3^-$) products of metabolism, respectively (Freda and McDonald, 1988; Gonzalez and Dunson, 1987; Gonzalez et al., 2006; McWilliams and Potts, 1978; Wood, 1992).

The reduction in $Na^+$ and $Cl^-$ uptake in plasma, leads to a reduction in plasma ionic content leading to an osmotic loss of water from blood into the tissues. In turn, this reduces plasma volume and increases blood viscosity, which may ultimately result in mortality by cardiovascular failure (Milligan and Wood, 1982). The exposure to extreme acid conditions (pH 3.5–2.0) results in suffocation and impairment in gas transfer in fish owing to the damage of branchial epithelium and excess of mucus production and accumulation (Packer and Dunson, 1972). Note that waters of some marginal lakes of the Rio Negro in the Amazon have pH as low as 3.5, and more extreme situations due to climate change cannot be ruled out.

In contrast, in acidophilic fish species from Rio Negro ionoregulatory/acid-base disturbances arise at a much-lowered threshold of pH ($\sim$ pH 4.0–3.0), which has been associated with distinct specialization in gill function to maintain their ionic balance (Gonzalez et al., 2006). Until now, at least two different ionoregulatory strategies have previously been described in distinct phylogenetic lineages of Amazonian fishes to cope and thrive in acidic ion-poor blackwaters. Many fish species from the Characidae family (Characiformes Order), a group with relatively high abundance and diversity in blackwater systems (Beltrão et al., 2019; De Pinna, 2006; Röpke et al., 2017), display $Na^+$ transport mechanisms that are relatively insensitive to low pH and that might be unaffected by the exposure to pH 3.25, as seen in the neon tetra (*Paracheirodon innesi*) (Gonzalez and Preest, 1999). These ion uptake mechanisms have high influx rates ($J_{max}$) and a high affinity for $Na^+$ uptake (low $K_m$) allowing fish to sustain high $Na^+$ uptake rates even under extremely ion-poor acidic conditions. A branchial epithelium with low intrinsic permeability helps to reduce ionic diffusive losses, and has been reported in remarkably acid tolerant characids, such as the cardinal tetra and tambaqui (Gonzalez et al., 2021). This prevents a stimulation of $Na^+$ efflux ($J_{out}$) at moderate low pH (4.5–4.0) and results in only a slight increase in $Na^+$ $J_{out}$ at pH 3.5 at ion-poor waters (Gonzalez and Wilson, 2001; Gonzalez et al., 2021). Nevertheless, some characidae species such as *Hemigrammus* sp. (tetras) and *Carnegiella strigatta* (marbled hatchetfish) display increased $Na^+$ efflux rates (i.e., reduced control of branchial ionic permeability) during exposure to low pH (3.5) in natural DOC-rich waters from Rio Negro (Gonzalez et al., 2002). On the other hand, a non-specialized ion uptake system under acid conditions has been characterized in some cichliforms (especially of the Cichlidae family), represented by both lower $J_{max}$ and affinity for $Na^+$ (higher $K_m$), where $Na^+$ uptake rates ($J_{in}$) are almost completely inhibited following exposure to low pH (Duarte et al., 2013; Gonzalez and Wilson, 2001; Gonzalez et al., 2002, 2021). However, the ionic efflux rates are very resistant to stimulation by low pH in species like *Pterophyllum scalare* (angelfish) (Gonzalez and Wilson, 2001), *Apistogramma* sp. (dwarf cichlid) and *Geophagus* sp. (Gonzalez et al., 2002), or even unaffected as seen in *Symphysodon discus* (discus) (Duarte et al., 2013), indicating a lower intrinsic branchial permeability to ions that seems to be a fundamental specialization in this group of fish to maintain the ionic balance under ion-poor acidic waters. In these regards, a similar ionoregulatory pattern was also described to the freshwater elasmobranch *Potamotrygon* sp. (stingray) that display an acidic-unspecialized ionic transport system in the gills. This mechanism was characterized by a higher $K_m$ and a modest capacity for $Na^+$ (and $Cl^-$) uptake at circumneutral pH in dilute water, with a pronounced inhibition of both $J_{in}^{Na}$ and $J_{in}^{Cl}$ during exposure to pH 4.0 at either reference soft water or in the natural Rio Negro blackwater (DOC concentration of 8.4 mgC/L) (Wood et al., 2002, 2003). Interesting to note that the kinetic parameters of $Na^+$ and $Cl^-$ uptake (i.e., $J_{max}$ and $K_m$) in the freshwater elasmobranch sp. were virtually not affected by the acclimation for 5 days to reconstituted ion-rich waters, when compared to fish kept in natural ion-poor Rio Negro water, suggesting that the branchial $Na^+$ and $Cl^-$ uptake mechanisms in freshwater elasmobranchs were adjusted for operation in very dilute environments (Wood et al., 2002, 2003).

The extreme acidity levels found in Rio Negro blackwaters is highly correlated with DOC. For example, water quality parameters in the Rio Negro can vary seasonally (Duarte, 2013; Johannsson et al., 2017) and pH usually ranges between 4.0 and 6.0 with typical DOC levels of 5–15 mgC/L (Walker and Henderson, 1996). However, forest streams from uplands in Central Amazonia can reach pH values as low as 3.5 with DOC concentration higher than 30 mgC/L during the wet season (Duarte, 2013). DOC is the major component of natural organic matter (NOM) being characterized as a group of heterogenous molecules naturally occurring in marine and freshwater systems that are formed in water and soil by the breakdown of lignin-rich material and dead organic biomass (allochthonous or terrigenous source), and by the metabolism of aquatic microorganisms in rivers and lakes (autochthonous source) (Steinberg et al., 2006; Thurman, 1985a; Zara et al., 2006). DOC consists of humic substances (HS) with high molecular weight as humic acids (HA) and fulvic acids (FA) (e.g., the major component—50% to 90% of total component), amino acids (e.g., tryptophan and tyrosine), fatty acids, hydrocarbons, sterols, carbohydrates, urea and porphyrins (Rocha et al., 1999). Humic substances are composed by a variety of acidic functional groups such as carboxylic, phenolic and carbonyl groups that strongly contribute to the acidity in DOC-rich waters, but also have a great importance in the buffering capacity against anthropogenic acidification of surface waters (Kullberg et al., 1993). DOC plays fundamental roles in many abiotic and biotic process in aquatic ecosystems such as metabolism, nutrient transport, $CO_2$ and $O_2$ cycles and light penetration, and also affect the transport, distribution and accumulation of ions and metals in different aquatic compartments (Findlay and Parr, 2017; Steinberg et al., 2006). It is now clear that the contribution of DOC as a driving force in aquatic freshwater ecosystems is strongly dependent of both the concentration and quality of DOC, which is directly related to their molecular and optical properties. Generally, allochthonous DOC – as seen in Rio Negro watershed – are optically darker and have higher content of both HS and aromatic compounds than autochthonous molecules, which can be verified through the determination of the specific absorption coefficient at 340 nm ($SAC_{340}$) and the fluorescence index (FI) (Duarte et al., 2016; Holland et al., 2017; Johannsson et al., 2017; Rocha et al., 1999). In addition, there is a strong positive correlation between SAC and the ability of DOC molecules to bind metals and $H^+$ (formally called proton-binding index – PBI), where darker DOC has a higher binding capacity and greater ameliorative effect against metal toxicity to freshwater aquatic organisms (Al-Reasi et al., 2011, 2013).

Similarly, the same molecular properties seen in darker DOC are thought to exert physiological effects on aquatic organisms, especially making them more tolerant to ionoregulatory disturbances induced by low pH exposure (Wood et al., 2011). Previous studies have demonstrated that DOC can bind and stabilize biological membranes, as seen in algal cell membranes and isolated fish gill cells, where the adsorption of DOC molecules on the cell surface was much greater at low pH (4.0) (Campbell et al., 1997; Vigneault et al., 2000). It was suggested that at low pH values, some electronegative functional groups of DOC can bind (using hydrogen bonds) to the negative charge domains on biological membranes, or form hydrophobic bonds between cell surfaces and DOC (Campbell et al., 1997). Thus, it has been suggested that DOC may directly interact with gill membranes of freshwater fishes affecting many physiological functions, which in part may explain the acid tolerance of native fish from the Rio Negro. In four Rio Negro species - tambaqui, *Serrasalmus* cf. *holandi*, *S. rhombeus* (piranhas) and *Leporinus fasciatus* (aracu or piau), the increment of 10-fold in $Ca^{2+}$ concentration (10 to 100 mM) in water does not have a protective effect on $Na^+$ and $Cl^-$ net losses during exposure to low pH (3.0–3.5), suggesting that $Ca^{2+}$ ions are not the main component controlling the ionic permeability in gills of these Rio Negro species (Gonzalez et al., 1998). Nevertheless, $Na^+$ efflux rates in acará and *Pimelodes* sp. (piracatinga) were not stimulated during acute exposure to pH 3.75 in natural Rio Negro water, while both fish species experience significant $Na^+$ diffusive losses when exposed to low pH in de-ionized water without DOC and supplemented with NaCl, KCl, $CaNO_3$ so as to have the same major ion concentrations as Rio Negro water (Gonzalez et al., 2002). Further investigation revealed that both efflux and net fluxes rates of $Na^+$ and $Cl^-$ were only slightly stimulated in the potamotrygonid stingray (*Potamotrygon* sp.) during exposure to pH 4.0 in Rio Negro water, compared to $Na^+$ and $Cl^-$ fluxes seen in fish exposed to low pH in reference water that resembled the ionic content of Rio Negro water but without DOC (Wood et al., 2003).

These works suggest that DOC can mimic the effect of $Ca^{2+}$ ions in stabilizing paracellular tight junction against the action of $H^+$, thereby mitigating the negative effects of acid stress on ionic balance of fish. In these regards,

DOC has been shown to promote gill hyperpolarization. This effect was speculated to result from binding of DOC molecules on paracellular tight junctions altering the $Na^+$ to $Cl^-$ permeability ratio, whereby darker DOC with high $SAC_{340}$ and PBI resulted in more negative transepithelial potential (TEP) (Galvez et al., 2009; Wood et al., 2011). This effect was also reported in the Amazonian tambaqui where TEP decreased significantly after exposure to the highly colored DOC from São Gabriel da Cachoeira (SGC) village, upper Rio Negro, in ion-poor water at both pH 7.0 and pH 4.0. However, contrary to expectations, in an experimental test, DOC did not prevent but increased net losses of $Na^+$ in $Cl^-$ in tambaqui at pH 4.0, which was explained by the loss of the protective component of DOC during long-term storage (Sadauskas-Henrique et al., 2019). Nonetheless, Rio Negro DOC (SGC DOC) promoted an almost complete protection against ionoregulatory disturbances to the non-Amazonian fish *Danio rerio* (zebrafish), markedly reducing $Na^+$ efflux and $Cl^-$ net flux rates in fish acutely exposed to pH 4.0. Interestingly, this ameliorative effect was persistent during exposure to pH 4.0 in ion-poor water without DOC, after 7 days of fish acclimation at neutral pH with SGC DOC (Duarte et al., 2016), suggesting that long-term changes in gill physiology induced by DOC might provide greater tolerance to low pH in fishes.

In addition to the protective effects of DOC on ion loss, DOC also seems to be beneficial to transcellular ionic transport preventing the inhibitory action of increased $H^+$ on branchial ionic uptake. For example, $Na^+$ influx rates were unaffected by exposure to pH 4.0 in leopard corydoras (*Corydoras julii*) and marble hatchetfish at natural Rio Negro water, and only slightly reduced following exposure to pH 3.5, while a minor inhibition in $Na^+$ influx was seen in *Geophagus* sp. exposed to pH 3.75 in Rio Negro water compared to fish at low pH in de-ionized water (Gonzalez et al., 2002). Similarly, inhibition of $Na^+$ and $Cl^-$ uptake in the freshwater elasmobranch stingray in blackwater at pH 4.0 was reduced in comparison with animals kept in reference water without DOC (Wood et al., 2003). Moreover, acclimation to commercial HS (Sigma-Aldrich) at low pH (3.75) promoted a significant increase in $Na^+$ and $Ca^{2+}$ $J_{max}$ in cardinal tetra, with no effect on the transporters' affinities ($K_m$) (Matsuo and Val, 2007). In the teleost fish model, zebrafish, the SGC DOC promotes a remarkable stimulation of $Na^+$ influx during acute exposure to pH 4.0, which was also seen in fish acclimated to DOC but exposed to low pH in "clean" ion-poor water. It was also seen that SGC DOC helped fish maintain $Na^+$ uptake coupled to ammonia excretion as a compensatory response to the increase $Na^+$ influx at low pH (Duarte et al., 2016). Although the exact mechanism for the stimulation of ionic uptake by DOC remains unclear it might involve the upregulation of key transporters in fish gills, as the $Na^+/K^+$-ATPase (McGeer et al., 2002). Recently, it was demonstrated that SGC DOC from the Rio Negro significantly increased the activity of both $Na^+/K^+$-ATPase and v-type $H^+$-ATPase in gills of zebrafish acclimated to pH 7.0 in presence of SGC DOC, but this effect was absent following acute exposure (3h) to pH 4.0 (Sadauskas-Henrique et al., 2021). Given that some amphiphilic molecules of DOC can bind to gill surface and alter the physiology of ionocytes (Morris et al., 2021), the effects of Rio Negro DOC on both ionoregulatory enzymes appears to be involved in the maintenance of the ion transport capacity in order to avoid internal acid/base disturbances in fish.

Although few studies have investigated acid-base responses in Amazonian fish species under ion-poor, acidic conditions there is such evidence that Rio Negro native species are particularly tolerant against acid-base disturbances at low pH (Wilson et al., 1999; Wood et al., 1998). In the three fish species studied (tambaqui, matrinchã and tamoatá) gradual exposure to low pH (6.0 to 3.5) in "clean" ion-poor water without DOC resulted in small net acid uptake that was accompanied by only small $Na^+$ and $Cl^-$ net losses, indicating that these species can strongly control blood acidosis even in environments were $H^+$ levels are 10- to 100-fold higher than the ones where the "standard" teleost models (e.g., rainbow trout) experience severe disruptions in their internal acid-base status (Wood, 1989). In addition, ammonia excretion ($J_{Amm}$) was unaffected in tambaqui following gradual exposure to pH 4.0, and only slightly increased in matrinchã and tamoatá, but markedly stimulated after 1h exposure to pH 3.5 (Wood et al., 1998). The high tolerance of tambaqui to acid-base disturbances under acidic ion-poor conditions was further confirmed by later studies, where plasma ammonia levels were increased only after graded exposure to pH 3.0 (Wilson et al., 1999) and no stimulation of $J_{Amm}$ was seen in fish kept for 24h at pH 4.0 in ion-poor water without DOC (Wood et al., 2018). In contrast, tambaqui significantly increased $J_{Amm}$ when exposed to pH 4.0 in the presence of the highly colored SGC DOC (Sadauskas-Henrique et al., 2019), a pattern similar to that previously observed in the non-Amazonian zebrafish under comparable acidic ion-poor conditions (Duarte et al., 2016) and to the freshwater stingray *Potamotrygon* spp. exposed to pH 4.0 (2h) in natural Rio Negro water (Wood et al., 2003). In zebrafish, the increase in ammonia excretion was highly correlated and coupled to the stimulation of $Na^+$ uptake rates, whereas in stingray it was associated with lowered inhibition of $Na^+$ influx, suggesting that DOC from Rio Negro could protect fish against the inhibitory effect of high $H^+$ concentrations on $Na^+$ uptake, and facilitate $H^+/NH_4^+$ extrusion at low pH, allowing the fish to avoid severe ionoregulatory and acid-base disturbances under environmentally relevant acidity levels.

### 4.2 Environmental tolerance to stress and changes in fish distributions

Anthropogenic-induced alterations of both hydrological regime and water quality parameters (e.g., temperature, dissolved oxygen, pH, DOC and contaminants levels) may alter the relative abundance and species composition in the Amazon basin, and thereby threaten freshwater fish biodiversity. Such a change has already been reported in the floodplain lakes (Arantes et al., 2018; Röpke et al., 2017) and highland streams of the Central Amazon (Beltrão et al., 2018; Costa et al., 2020; Mendonça et al., 2005). Recently, in addition to the potential impacts of climate change on the biodiversity of aquatic ecosystems, there is a growing concern about freshwater browning, an event resulting from increasing allochthonous DOC, and iron concentrations, and their direct effects on biological systems and indirect effects on acidification of surface waters. Freshwater browning has already been reported in several freshwater bodies worldwide, particularly in northern Europe and North America. It has been caused by changes in land use that affects the net DOC production, soil DOC storage and transportation of terrestrial DOC to freshwater (Kritzberg et al., 2020; Solomon et al., 2015), as well as anthropogenic decrease of atmospheric sulfur depositions into aquatic systems (Monteith et al., 2007).

In addition to the direct effects of DOC on freshwater bodies, browning can negatively impact many ecosystems services, such as reducing the availability of potable water and increasing the cost of drinking water treatment. Browning can also modify phytoplankton communities and net primary productivity which can alter the structure of fish assemblages, and browning can alter the natural cycle of macro- and micronutrients influencing the speciation, mobility and bioavailability of cationic metals (Kritzberg et al., 2020). Moreover, the impacts of climate change on global and regional hydrological cycles, and the eutrophication by the increasing load of domestic and industrial waste effluent, might have an additive and/or synergic effect with the browning of freshwater imposing additional threats for the Amazonian ichthyofauna.

In these regards, studies have addressed the role of human land use, soil DOC transfer and autochthonous cycling of DOC and indicate that these processes are the main drivers for controlling the DOC regime in freshwater bodies (Kritzberg et al., 2020; Solomon et al., 2015; Stanley et al., 2012). The impacts of land use on soil DOC production and transportation to streams and rivers, especially the expansion of agriculture over catchment areas, remains controversial. However, there is consistent evidence that crop production can result in considerable loss of the soil pool of DOC and widely affects the organic composition of DOC. In addition, the widespread use of fertilizers can also alter the C:N ratio in the surrounding aquatic ecosystems stimulating primary production (Kritzberg et al., 2020; Stanley et al., 2012). For instance, the expansion of agriculture over catchment areas of headwaters streams might result in increased load of DOC. Consequently, DOC concentration in water column is increased and DOC composition is altered according to the crop types (Petrone et al., 2011; Stedmon and Markager, 2005). Furthermore, an expected effect of global climate warming is the alteration of regional weather patterns and hydrological cycles increasing the frequency of extreme droughts and floods, as seen in Amazon basin between 2005 and 2012 (Magrin et al., 2014). These have already had a direct impact on DOC transportation to aquatic ecosystems. According to meteorological climate models, climate warming may increase precipitation in the western Amazon basin and reduce precipitation in the eastern regions which house the majority of aquatic ecosystems that will experience more frequent and intense droughts (Duffy et al., 2015). Thus, while long-term droughts can enhance organic matter decomposition and DOC retention in soil, the subsequent further increase in precipitation, consequently enhancing the magnitude of flooding events, is expected to favor DOC transport driven by runoff (Kritzberg, 2017)—especially in areas where riparian forest was suppressed and changed by pasture and croplands.

In the Amazon basin, agricultural expansion and fragmentation of forest cover have resulted in habitat degradation and increased water runoff, resulting in a loss of soil nutrients and DOC and rise in the transport of fine sediment and DOC to water column, as documented in some areas of southeastern Amazon (Davidson et al., 2012; Leal et al., 2016). This results in an increase in DOC and a reduction in light penetration which can affect water temperature profiles with a marked effect on the metabolism and biochemical cycles in the aquatic environments (Solomon et al., 2015). In Amazonian blackwater floodplains (lakes and *igapós*) DOC composition varies seasonally (Johannsson et al., 2020) which strongly influences the composition of bacterioplankton communities and their primary productivity. During low water seasons, sediments and nutrients settle out and water temperature of the shallow waters increases. This results in an increase in phytoplankton and primary productivity which is correlated with an increase in lower molecular weight autochthonous DOC which is rapidly consumed by bacteria. In contrast, during the flooding season a predominance of humic acid-like components derived from terrestrial organic matter is observed, which was correlated with a significant reduction in chlorophyll *a* content in water (Melo et al., 2019). Therefore, DOC enrichment in Amazonian waters by browning events can enhance the dominance of DOC components from allochthonous sources and directly affect the carbon recycling process by bacterioplankton.

In addition, the autochthonous breakdown of DOC by photo-oxidation is also essential for the heterotrophic cycling of carbon in aquatic ecosystems (Wetzel, 2001); however, under high UV radiation the photo-degradation of DOC is more complete generating $CO_2$ as a final product, as well as increasing the production of reactive oxygen species (ROS) (Johannsson et al., 2017; Scully et al., 1996). Thus, under lower water and precipitation scenarios (Duffy et al., 2015), ROS production in DOC enriched waters from the Amazon is expected to markedly increase (Johannsson et al., 2020). Associated ROS results in oxidative stress in fish that can result in peroxidation of unsaturated fatty acids in biological membranes, protein and DNA damage and activation of detoxifying enzymes (Lackner, 1998). Similarly, one of main components of allochthonous DOC, the humic substances, have been shown to promote oxidative stress responses in aquatic organisms with the activation of biotransformation enzymes (Steinberg et al., 2006). This has been seen in the Amazonian tambaqui where the activities of hepatic ethoxyresorufin-*O*-deethylase (EROD) and superoxide dismutase (SOD) were strongly stimulated, with a slight inhibition of glutathione-*S*-transferase (GST), following the acclimation to natural Rio Negro water (Sadauskas-Henrique et al., 2016). Thus, there is growing evidence that DOC components might be an important predictor for the structure of aquatic communities (Steinberg et al., 2006), owing to their physiological and biochemical effects on organisms (Wood et al., 2011), but is reasonable to assume that browning of freshwater might be harmful to aquatic biota by the associated ROS production and induced oxidative stress. This is clearly an area for further research.

In small streams from the Central Amazon uplands (first to third order) differences in fish assemblage structure have been created by both structural features of water bodies (such as depth, current velocity, and discharge) and water quality parameters like the content of humic acid, pH and conductivity (Mendonça et al., 2005). In addition, the total number of species and composition are reduced during the rainy season associated with the higher conductivity and humic acid content and lower oxygen concentration (Espírito-Santo et al., 2009). Herein, the darker and acidic streams have a fish assemblage dominated by species from the Characiformes Order (around 76%) with a high abundance and frequency of species belonging to the Characidae (*Bryconops giacopinii* and *B. inpai* - inpatetra) and Crenuchidae (*Crenuchus spirulus* – sailfin tetra) families (Espírito-Santo et al., 2009; Mendonça et al., 2005). Interestingly, both Characidae species had higher gill $Na^+/K^+$-ATPase activity, but reduced activity of v-type $H^+$-ATPase activity in the rainy season, effects which were highly correlated with stream acidity (4.0–4.41) and DOC concentration (15.0–31.5 mgC/L) (Duarte, 2013). This implies that higher DOC/acidity conditions may impair ionic homeostasis and acid ($H^+$) excretion in these fish. Thus, browning might promote alterations in physicochemical properties of Amazon waters, making them more acidic and affecting the concentration and composition of DOC. This imposes changes in physical and chemical attributes of aquatic environments. The impacts of these changes on fish assemblages can be severe resulting in loss of total and functional diversity with a predominance of highly acid tolerant species, and less vulnerable to the chemical stress induced by DOC molecules.

## 5 Two sides of the same coin: Amazonian lowland fish thermal tolerance

Climate change is expected to affect biodiversity and ecosystem function, changing species range sizes, and increasing extinction probabilities (Valiente-Banuet et al., 2015). The tropics are of special concern because their organisms experience temperatures that are already close to their upper thermal limits (Campos et al., 2021; Comte and Olden, 2017; Jung et al., 2020; Tewksbury et al., 2008). According to the rise in temperature and the organism in question, species may shift their ranges to higher altitudes or latitudes, adapt to the new thermal environment, or may experience extinction due to their slow migration rates or the limited adaptation capacity (Chen et al., 2011; Lenoir and Svenning, 2015; Ohlemuller et al., 2008). The ability to adapt or migrate to a new thermal environment is related to the species' thermal limits (Valouev et al., 2008). Thus, measuring the critical thermal maxima ($CT_{max}$) will improve our predictions of species' adaptation capacity to global warming. The critical thermal minima ($CT_{min}$) is also important because it will allow us to estimate species' thermal breadth ($= CT_{max} - CT_{min}$) and to infer the species potential to respond to extreme temperature fluctuations and migrate to cold thermal habitats. Since aquatic habitats are becoming hotter in the coming decades (Maberly et al., 2020), it is imperative that we obtain thermal breadth data from a broad range of species, so we can forecast the consequences of climate change.

Many fish species, especially from lowland areas of the Amazon, appear to be extremely vulnerable to climate change. For instance, the extreme drought event of 2005 caused shifts in fish abundance and the composition of fish communities (Röpke et al., 2017). It has been hypothesized that lowland species will be able to climb the Andes mountains, and the high-altitude streams will experience an increase in biodiversity from lowland species. However, it remains unclear how and whether lowland Amazon fish species will survive at low temperatures during winter at high altitudes or will be able to cope with the predicted pace of climate warming in lowland areas. Herein, we will show the thermal niche breadth of some Amazonian fish species and predict their vulnerability to global warming.

Our knowledge of upper thermal tolerance data in fish has increased in the past few years, including in Amazonian species (Campos et al., 2017, 2018, 2019, 2021; Jung et al., 2020), but there is much less data on $CT_{min}$, which would allow us to calculate the thermal niche breadth of the species. In our survey, we found $CT_{max}$ and $CT_{min}$ for eight different Amazon species from lowland areas (Fig. 4). In Fig. 4, we show the average $CT_{max}$ and $CT_{min}$ of these species compared to two well-known eurythermal species, the goldfish (*Carassius auratus*) and zebrafish. We observed that, in general, the Amazon species has higher $CT_{max}$ compared to goldfish and zebrafish, with the exception of red-bellied piranha (*Pygocentrus nattereri*) and cardinal tetra. These findings indicate that the Amazon species can tolerate higher temperatures. These results corroborate the previous findings that shows higher thermal tolerance in tropical species compared to temperate ones (Sunday et al., 2011, 2012, 2014). Accordingly, Campos et al. (2021) reviewed the upper thermal limits of South American freshwater fish; their database showed that the $CT_{max}$ of Amazonian lowland species range from 37.07 in banded cichlid (*Heros severus*, Cichlidae) to 42.8 in longtail knifefish (*Sternopygus*

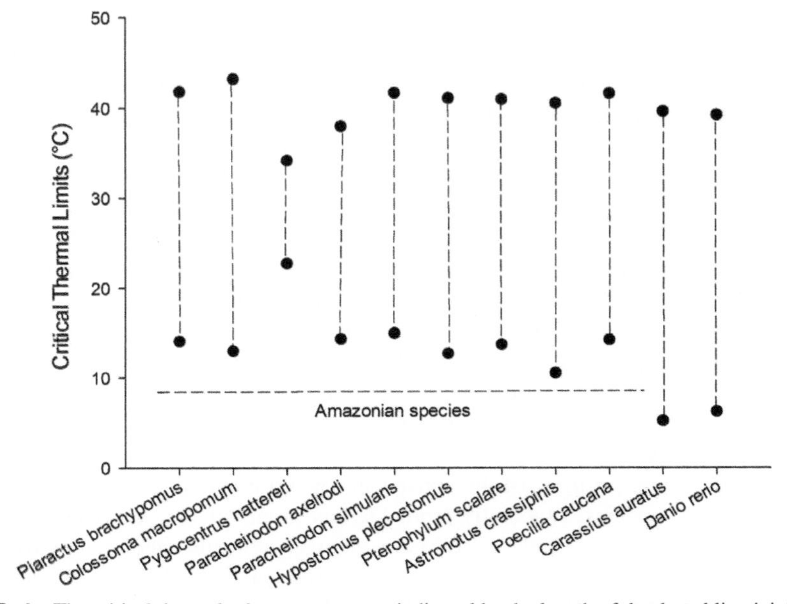


**FIG. 4** The critical thermal tolerance scope, as indicated by the length of the dotted line joining the upper thermal tolerance (upper symbols) and lower thermal tolerance (lower symbols) of eight Amazon species for comparison with two temperate eurythermal species (*C. auratus* and *Danio rerio*). Both Amazonia and temperate fish were acclimated between 29 °C and 30 °C. *Data from Di Santo V, Jordan HL, Cooper B, Currie RJ, Beitinger TL, Bennett WA. 2018. Thermal tolerance of the invasive red-bellied pacu and the risk of establishment in the United States. J. Therm. Biol. 74, 110–115; Dragan, F.G. 2014. Respostas fisiológicas e metabólicas do tambaqui (Colossoma macropomum, Cuvier 1818) às alterações climáticas previstas pelos diferentes cenários do IPCC para o ano de 2100. MSc thesis, Biologia de Água Doce e Pesca Interior, Instituto Nacional de Pesquisas da Amazônia; Divya, P.K., Ranjeet, K., 2014. Effect of temperature on oxygen consumption and ammonia excretion in red bellied piranha Pygocentrus nattereri KNER, 1858. Journal of Aquatic Biology and Fisheries 2, 97–104; Campos, D.F., Jesus, T.F., Kochhann, D., Heinrichs-Caldas, W., Coelho, M.M., Almeida-Val, V.M.F., 2017. Metabolic rate and thermal tolerance in two congeneric Amazon fishes: Paracheirodon axelrodi Schultz, 1956 and Paracheirodon simulans Géry, 1963 (Characidae). Hydrobiologia DOI 10.1007/s10750-016-2649-2; Yanar M., Erdoğan E., Kumlu, M., 2019. Thermal tolerance of thirteen popular ornamental fish species. Aquaculture, 501, 382–6; Ford, T., Beitinger, T., 2005. Temperature tolerance in the goldfish, Carassius auratus. J. Therm. Biol., 30, 147–152; López-Olmeda, J.F., Sánchez-Vázquez, F.J., 2011. Thermal biology of zebrafish (Danio rerio). J. Therm. Biol., 36: 91–104.*

*macrurus*, Sternopygidae). Despite their higher thermal limits, tropical species are speculated to be more vulnerable to global warming because they live close their thermal limits (Comte and Olden, 2017). Indeed, the average mean of maximum annual temperature in the Amazon flooded areas is ~31 °C, suggesting that the thermal safety margin (TSM; $CT_{max}$—average mean of maximum annual habitat temperature) ranges from 3.16 °C (red bellied piranha) to 12.2 °C (tambaqui). These results indicate that the Amazon lowland species are vulnerable to global warming, because they are close to their thermal limits.

In contrast to their higher upper thermal tolerance, the Amazon lowland species were relatively intolerant to cold temperatures, i.e., $CT_{min}$ in these species was higher than 10 °C. These results suggest a limited capacity to immigrate to high altitude areas and survive during the cold waves of the winter season. In the Andes, at ~2000 m above the sea, the mean winter air temperature can reach temperatures below 10 °C. At higher elevations, such as upper Andean Peru 4400 m, the average winter temperature of various wetlands can be below 8.5 °C (Salazar-Torres and Huszar, 2012). While there is an increase in the maximum and minimum temperature in the lowland areas of the Amazon basin (0.15 °C/decade), there is no trend to increase the minimum temperature in the Andes, so Andean species will have to deal with large seasonal fluctuations in temperature (Lavado Casimiro et al., 2013). These data suggest that Amazonian lowland species are unlikely to be able to immigrate to high altitude environments because their CTmin is higher than the mean temperature during winter. In accordance, the oscar (*Astronotus crassipinis*), the most cold-tolerant Amazon species known to date, has been established in the southern Florida Everglades, where the temperature ranges from 13.2 °C to 35.5 °C, but have not established in cooler regions of North America (Gutierre et al., 2016). Likewise, Bennett and colleagues (Bennett et al., 1997) showed that red-bellied piranha cannot survive temperatures below 10 °C and cannot acclimate to temperatures lower than 15 °C. Unfortunately, we do not have information about the critical thermal minima of Andean species to compare to Amazon lowland species. However, the cold tolerance of snow trout (*Schizothorax richardsonii*) living at 1200 m in the Himalayan streams is ~0 °C. These findings, corroborate the hypothesis that lowland Amazon species will be unable to immigrate and survive in high altitude areas, because they cannot survive the minimum temperature experienced during the winter season.

Animals adapt to their thermal environment, but there is a congruence between thermal tolerance and the thermal niches they exploit (Culumber and Tobler, 2018). Recently, Morgan et al. (2020) showed that both $CT_{max}$ and $CT_{min}$ evolved in an artificial selection experiment, and the selection for warm tolerance individuals caused a decrease of thermal niche breadth because $CT_{min}$ evolve faster than $CT_{max}$. These results suggest that thermal niche breadth may reflect evolutionary responses of the selection of the thermal environment. The lowland Amazon species present a narrower thermal niche breadth in comparison to temperate species (Fig. 4), supporting the hypothesis that species living in a stable thermal habitat possess a lower thermal niche breadth. The most likely explanation for the shorter thermal niche breadth in the lowland Amazon species is that climate homogeneity in the region may drive the thermal niche breadth, since temperature seasonality is stronger in the temperate region, and temperate species present larger thermal niche breadth. Therefore, temperate species having wider thermal niches seem more able to invade new climate niches. On the contrary, climatic specialization in the tropics may hinder thermal niche change as species with narrow niches might be constrained in their ability to expand their geographical ranges (Fisher-Reid et al., 2012).

The ability to survive and adapt to the changes in environmental temperature is related to acclimation capacity, where, through phenotypic plasticity, individuals increase their thermal limits ($CT_{max}$ and $CT_{min}$) with the increase in mean acclimation temperature. Therefore, reporting a single range of temperature tolerances for a species is meaningless without clearly indicating the temperature to which the fish had previously been acclimated. As the importance of acclimation became understood, researchers began to quantify thermal tolerance across a species' acclimation range, which can be graphically represented as a temperature tolerance polygon (Fry, 1947). The area of the polygon reflects the degree of eurythermicity of a species and is expressed as the temperature tolerance in units of $°C^2$, where the larger the thermal tolerance polygon area, the more eurythermic the species. Herein, we show the thermal tolerance across different temperature acclimation for seven Amazon lowland species and one eurythermal species in Fig. 5 and the polygon area in the Table 1. The thermal tolerance polygon areas are consistent with the thermal niche breadth results discussed above, where Amazon lowland species have lower thermal polygon areas compared to temperate species.

The low thermal tolerance polygon areas in the Amazon lowland species indicate that they are stenothermal, i.e., they are capable of only living within a narrow temperature range. The thermal tolerance polygon of Amazon lowland species is very similar among the species reported here and range from $210°C^2$ in angelfish *Pterophylum scalare* to $635°C^2$ in pirapitinga. Noteworthy, the red-bellied piranha shows the smallest thermal tolerance polygon area ($93.04°C^2$), which may represent the lowest for any fish species reported. However, this data should be interpreted cautiously, as the fish were measured following transport to India and a limited acclimation temperature (Divya and Ranjeet, 2014). Indeed, Bennett et al. (1997) showed lower $CT_{min}$ for red-bellied piranha introduced in the United States. The $CT_{min}$ values in Bennett et al. (1997) range from 10 °C to 19.7 °C in fish acclimated from 15 °C to 35 °C, respectively, whereas in Divya and Ranjeet (2014) $CT_{min}$ values never dropped below 21.96 °C. These results may indicate population specific differences in thermal tolerance.

The polygons reflect the species' thermal niche and, although some studies have generated the polygons over a narrow thermal acclimation range, our data very clearly show that the Amazon lowland species are stenothermal, having limited ability to acclimate and survive over a broad range of temperatures compared with temperate species. At the higher acclimation temperature, each species had a $CT_{max}$ close to 40 °C, while at the lower acclimation temperature, $CT_{min}$ was about 10 °C, this is consistent with the thermal niche breadth shown in Fig. 4. Although the $CT_{max}$ of Amazon

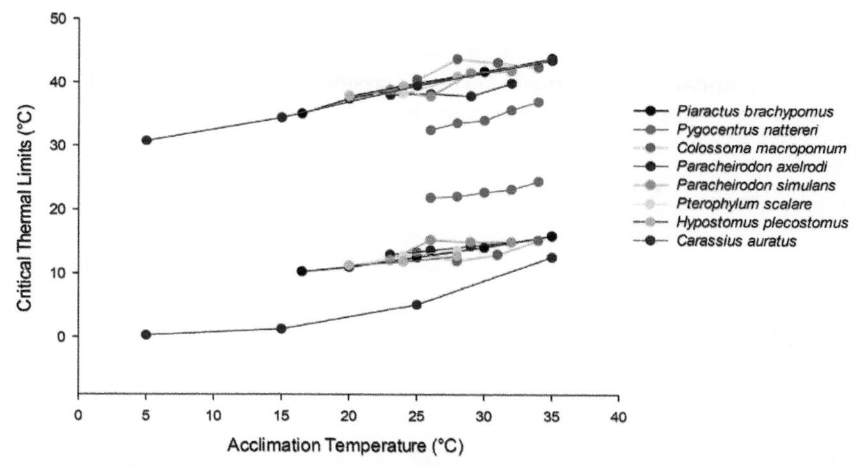


**FIG. 5** Upper and lower critical thermal tolerance values of seven species acclimated to a range of different temperatures for seven Amazonian species and one eurythermal species (*C. auratus*). *Data from Di Santo V, Jordan HL, Cooper B, Currie RJ, Beitinger TL, Bennett WA. 2018. Thermal tolerance of the invasive red-bellied pacu and the risk of establishment in the United States. J. Therm. Biol. 74, 110–115; Dragan, F.G. 2014. Respostas fisiológicas e metabólicas do tambaqui (Colossoma macropomum, Cuvier 1818) às alterações climáticas previstas pelos diferentes cenários do IPCC para o ano de 2100. MSc thesis, Biologia de Água Doce e Pesca Interior, Instituto Nacional de Pesquisas da Amazônia; Divya, P.K., Ranjeet, K., 2014. Effect of temperature on oxygen consumption and ammonia excretion in red bellied piranha Pygocentrus nattereri KNER, 1858. Journal of Aquatic Biology and Fisheries 2, 97–104; Campos, D.F., Jesus, T.F., Kochhann, D., Heinrichs-Caldas, W., Coelho, M.M., Almeida-Val, V.M.F., 2017. Metabolic rate and thermal tolerance in two congeneric Amazon fishes: Paracheirodon axelrodi Schultz, 1956 and Paracheirodon simulans Géry, 1963 (Characidae). Hydrobiologia DOI 10.1007/s10750-016-2649-2; Yanar M., Erdoğan E., Kumlu, M., 2019. Thermal tolerance of thirteen popular ornamental fish species. Aquaculture, 501, 382–6; and Ford, T., Beitinger, T., 2005. Temperature tolerance in the goldfish, Carassius auratus. J. Therm. Biol., 30, 147–152.*

lowland species is higher than temperate species, it is unlike they will survive the ongoing global warming because, as we discussed earlier, they are already living close to their thermal limits. Recently, Jung et al. (2020) showed that in a subset of 13 Amazon species, 2 species failed to acclimate at 33 °C, 9 species failed at 35 °C, and only 2 species survived up to 35 °C. These results suggest that most Amazon species are living only 4 °C below their maximum acclimation temperature and, therefore, will not be able to deal with the projected increased temperature according to climate change. In contrast to warm tolerance, the Amazon lowland species have limited cold tolerance, which may impact their emigration capacity to high altitude areas. Mean winter temperatures at or below 10 °C in the Andes would impose physiological challenges for lowland Amazon species survival.

**TABLE 1** Thermal tolerance data for seven Amazonian species compared to the temperate eurythermal species *Carassius auratus*, consisting of the critical thermal tolerance polygon areas.

| Family | Species | Critical thermal tolerance Polygon area (°$C^2$) | Source |
|---|---|---|---|
| Serrasalmidae | *Piaractus brachypomus* | 635 | Di Santo et al. (2018) |
| Serrasalmidae | *Colosoma macropomum* | 227 | Dragan (2014) |
| Serrasalmidae | *Pygocentrus nattereri* | 93 | Divya and Ranjeet (2014) |
| Characidae | *Paracheirodon axelrodi* | 218 | Campos et al. (2017) |
| Characidae | *Paracheirodon simulans* | 235 | Campos et al. (2017) |
| Loricaridae | *Hypostomus plecostomus* | 220 | Yanar et al. (2019) |
| Cichlidae | *Pterophylum scalare* | 210 | Yanar et al. (2019) |
| Cyprinidae | *C. auratus* | 1220 | Ford and Beitinger (2005) |

# 6 Anthropogenic impacts on water bodies

## 6.1 Deforestation

Deforestation rates in the Brazilian Amazonia have risen by more than 20% in the last year compared to previous years and have been rising. Deforestation has profound consequences on biodiversity, and the impact on tropical streams is well-documented. Deforestation influences stream conditions directly and indirectly by changing channel morphology, water temperature and luminosity mediated by canopy opening, and by increasing the erosion and transport of sediments, nutrients, and contaminants (Hofmann and Todgham, 2010). Streams are intimately connected to the riparian forest and, therefore, are particularly vulnerable to deforestation. In the Brazilian Amazon, deforestation has increased in the past years and 17% of the forests have already been converted into pastures and croplands, and it is expected that forest conversion or degradation will continue to occur at increasing rates (Nepstad et al., 2008).

Deforestation changes the ecological conditions in streams, and these environmental alterations determine species distribution (Córdova-Tapia et al., 2018), which can be predicted by physiological tolerance to the new environmental conditions. For instance, distributions for species that have narrow physiological tolerances may be strongly influenced by conditions such as low temperature, high oxygen, and low turbidity. As a direct result of deforestation, several studies have documented fish biodiversity loss of less tolerant species in forest streams (e.g., Casatti, 2010; Casatti et al., 2015; Leitão et al., 2018; Sutherland and Meyer, 2007). Conversely, physiologically tolerant species, such as cichlids, present an increase in occurrence and abundance in Amazonian streams with deforested margins (Ilha et al., 2018). Cichlids are known to possess physiological mechanisms that enable them to survive high temperature and hypoxic environments (Campos et al., 2018) that occur in response to deforestation, but less tolerant species can only respond in a limited way (Chown et al., 2010; Hofmann and Todgham, 2010). For the latter, some may be able to acclimatize to some degree, or respond through phenotypic plasticity to deal with changes within a specific range of conditions; while others may have the ability to remain in place because of the evolution of improved tolerance through natural selection. Such evolutionary change can occur because of changes in the limits of physiological tolerance or changes in behavior that alter the emergence time of particularly sensitive stages during development (Schaefer and Ryan, 2006).

Deforestation causes three main effects on streams: warming, hypoxia, and sedimentation. Temperature affects all biological processes, influencing many aspects of the life-history traits of fish. An increase in temperature has been shown to result in a decrease in body-size (Ilha et al., 2018), a decrease in reproduction rate (Pankhurst and Munday, 2011) and a decrease in longevity (Valenzano et al., 2006). At the cellular level, warming increases the reaction rates and energetic metabolism (Campos et al., 2019; Schulte, 2015) and increases oxidative stress (Madeira et al., 2016; Rosa et al., 2016). Species can tolerate a range of temperatures where their energetic demand is fully supplied. Campos et al. (2018) determined that the thermal tolerance of a species was related to aerobic metabolism and lifestyle. Those authors showed that slow-swimming dwarf cichlid (*Appistograma hypolitae*) and the double-sppot acara (*Aequidens pallidus*) are more tolerant than fast-swimming species of tetras (*Hyphessobrycon melazonatus*; *Hemigrammus geisleiri* and *Iguanodectes geisleri*; Characidae). Campos et al. (2019) showed that when acclimated to high-temperature *H. melazonatus* increases the levels of oxidative stress and neuronal dysfunction, while *Apistograma agassizii* (acarazinho) did not increase oxidative stress.

In contrast to the effects of warming on aerobic metabolism, hypoxia decreases oxygen demand by metabolic suppression and increases the use of anaerobic metabolism, which is one of the most effective hypoxia tolerance strategies used by some species (Almeida-Val et al., 2000). Indeed,

(Kochhann et al., 2015a) showed that acarazinho exposed 96h to hypoxia decreased aerobic metabolic rate by about fivefold. These results suggest that species with narrow physiological requirements cannot adapt to new environments in deforested areas.

The effects of suspended sediment on fish physiology have attracted less attention than temperature and hypoxia; however, some evidence points to strong negative effects. Suspended sediments have been shown to interfere with visual acuity and olfaction in some coral reef fishes, thereby affecting larval settlement (Wenger et al., 2011) and prey capture (Wenger et al., 2012, 2013). Also, an increase in suspended sediment has been related to decreased swimming performance in *Salmo trutta* (salmo trout) (Berli et al., 2014). Suspended sediment can impact other aspects of fish behavior and physiology, including growth rate (Sutherland, 2003) or feeding behavior (Olson et al., 1973). The physiological mechanisms through which suspended sediments exert their effects are not yet fully understood, however, it is suggested that affects gill morphology, resulting in an increased oxygen diffusion distance, which compromises aerobic metabolism (Crémazy et al., 2019). For instance, Hess et al. (2017) showed a reduction in aerobic scope (the difference between maximum and resting metabolic rate), in coral reef fish exposed to increase suspended sediments, indicative of a reduced energy supply capacity. In Amazonian species, we have exposed *Corydoras schwartzi* (Schwartz's catfish) to control and 45mg/L of suspended sediment (<0.6μm) for 96h (D. Campos et al., unpublished data). Our results show that Schwartz's catfish decreased routine metabolic rate, maximum metabolic rate, and aerobic scope when acclimated to 45mg/L of suspended sediments (Fig. 6). Crémazy et al. (2019) reported that background levels of TSS were around 70mg/L in Rio Solimoes and 8mg/L in Rio Negro. Schwartz's catfish is a facultative-air breathing and when exposed to high concentration of suspended sediment

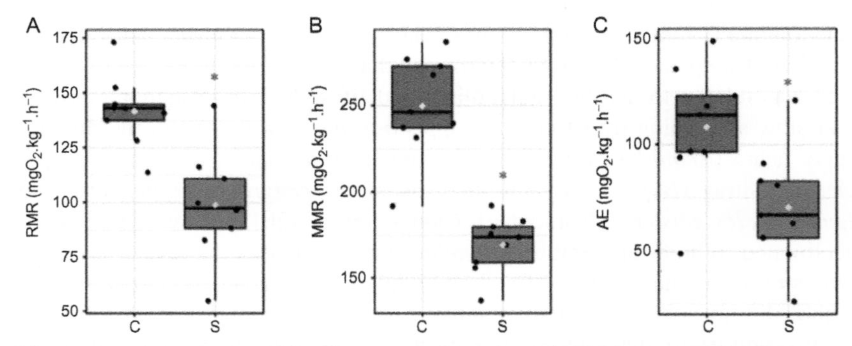


**FIG. 6** The effect of suspended sediments (C; Control and S; 45m/L suspended sediment) on (A) routine metabolic rate (RMR), (B) Maximum metabolic rate (MMR) and (C) Aerobic scope (AE) (MMR-RMR) of *Corydoras schwartzi*. The asterisk indicates significant differences between treatments by test t ($n=9$).

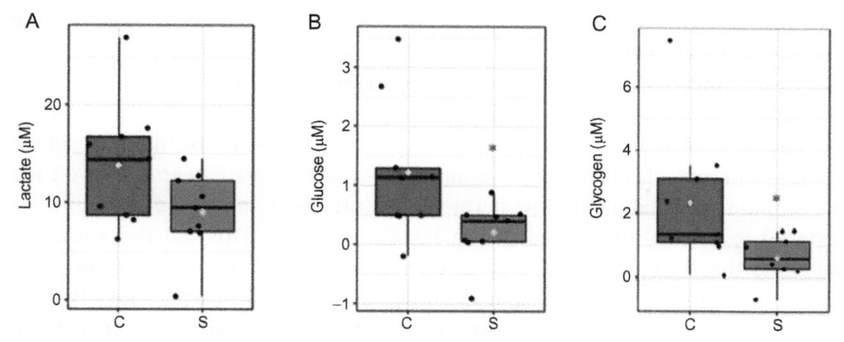


**FIG. 7** The effect of suspended sediments (C; Control and S; 45 m/L suspended sediment) on (A) Lactate, (B) Glucose and (C) Glycogen on muscle of *Corydoras schwartzi*. The asterisk indicates significant differences between treatments by test t ($n=9$).

they increase the rate of air-breathing. The increase in air breathing behavior depleted the carbohydrates reserve in muscle (glycogen and glucose) leading to a reduction in muscle glucose and glycogen (Fig. 7). Although, Schwartz's catfish can compensate for the impaired oxygen uptake from water by transitioning to air-breathing, this behavior comes at an increased predation risk from birds (Pineda et al., 2020).

Deforestation can exert strong selective pressures, driving evolutionary change and altering the organism's performance. However, studies showing evolutionary consequences and changes in phenotypic plasticity linked to deforested areas in tropical areas are still scarce (Alberti et al., 2017; Carroll et al., 2007, 2014). However, in tropical streams, deforestation-associated warming has already been shown to reduce body size of fish living in Amazonian streams (Ilha et al., 2018); nonetheless, we need to learn more about the impacts of deforestation on fish.

## 6.2 Urban pollution

Leachates from agricultural areas, discharge of domestic sewage, effluents of industries and fish farming, among others, represent additional challenges for aquatic biota in the Amazon and can cause physiological, biochemical and metabolic modifications, as found in laboratory studies. The tambaqui has been used as a model species to understand the effect of these environmental pollutions in water bodies of the Amazon ( Jacaúna et al., 2020; Santos, 2012).

Santos (2012) analyzed the effect of leachate from public landfills in Parintins (AM) on tambaqui and observed anemia and increased levels of glucose and plasma ions. In addition, the author also found serious liver and cholinergic damage. Additionally, she also found that the increase in enzyme activities was proportional to the increase in leachate concentration, indicating a dose-dependent effect. Finally, comet assay and nuclear abnormalities in erythrocytes indicated that the leachate caused increased damage to the genetic material and to the cell nucleus of tambaqui, respectively. In order to validate the findings of the laboratory experiments and, thus, to evaluate the impact in situ of the flowing leachate to Lake Macurany (Parintins/AM), the author exposed the tambaqui downstream and upstream of the leachate dump point and observed that the animals exposed downstream, showed the same physiological and biochemical disturbances observed in animals exposed to local diluted leachate in laboratory tests. This led the author to conclude that Lake Macurany is largely contaminated. Leaching from landfills is a worldwide problem and even developed countries face this type of environmental challenge. Jacaúna et al. (2020), in a similar experiment, exposed tambaqui to the water of the Igarapé do Quarenta (Manaus/AM). They observed that the animals had high bile concentrations of PAHs (polyaromatic hydrocabons), increased routine metabolic rate (RMR) and decreased activity of the electron transport system (ETS), suggesting the occurrence of metabolic impairment in the experimental fish.

These two experiments clearly indicate that aquatic environments in the Amazon near urban centers, cities and villages, show signs of environmental and ecological disturbances. These disturbances, enhanced by other anthropogenic pressures, may have profound effects regarding the conservation of the region's ichthyofauna. For example, wind-driven air circulation and water circulation on the planet brings Saharan sand and pollutants like microplastics to pristine regions of the Amazon (Cabrera et al., 2022; Yu et al., 2015). These exotic materials pose an additional challenge to fish in the region (Pegado et al., 2021; Ribeiro-Brasil et al., 2020).

## 6.3 Metals

Mining, including illegal mining, is expanding rapidly across the Amazon, from Andes to the Atlantic. According to Soares-Filho and Rajão (2018) mining is already responsible for about 10% of the deforestation in the Brazilian Amazon. The expansion of mining activities has impacted physicochemical properties of the Amazonian aquatic environments, as seen by increased metal levels in both water and sediment matrixes, and bioaccumulation of metals in tissues—especially in white muscle—of several fish species (Costa et al., 2022; Lima et al., 2015; Santana, 2016). For example, higher levels of metals in water than the limits allowed by the Resolution 357 of the Brazilian Environmental National Council (CONAMA, 2005) have been reported in the Cassiporé River/AP (particularly cadmium (Cd), lead (Pb), zinc (Zn) and Mercury (Hg)) (Lima et al., 2015), and in both the Vermelho and Sororó Rivers/PA (iron (Fe) and manganese (Mn) (Salomão et al., 2018) and in streams crossing a highly urbanized and industrialized area in Manaus/AM (cobalt (Co), Nickel (Ni), Fe, Zn, Cu, Cd and Pb) (Santana and Barroncas, 2005; Silva et al., 1999). Despite the large number of water bodies contamined with a range of different metals, toxicology studies on fish species have been limited to three main metals: Cu, Cd and Ni.

Amazonian ornamental fish species seems to be relatively sensitive to copper showing 96h $LC_{50}$ values ranging from 12.8 to 74.4μgCu/L in reference ion-poor water (Duarte et al., 2009). Interestingly, characidae species (tetras) are the most sensitive to copper (*C. strigatta*, *Hemigrammus rhodostomus* and *Hyphessobrycon socolofi*; 96h $LC_{50} < 29$ μgCu/L), while the armored catfishes Schwartz's catfish and *Dianema urostriatum* (flagtail catfish) were the most tolerant (96h $LC_{50}$ of 53.0 and 74.1μgCu/L, respectively). An exception in the highly sensitive pattern seen in characid fish was the cardinal tetra that posess a 96h $LC_{50}$ (45.0μgCu/L), almost that is almost twofold that of other tetra species (Duarte et al., 2009). Furthermore, copper nanoparticle was around 3-time less toxic than copper to both cardinal tetra and dwarf cichlid (96h $LC_{50}$ of 139.2 and116.6μg nCuO/L, respectively) (Braz-Mota et al., 2018). In addition, Cu toxicity is strongly modulated by local water chemistry as seen by the substantial decrease in copper toxicity to cardinal tetra in water from Rio Solimões (3-times higher 96h $LC_{50}$) and Rio Negro (19-times higher 96h $LC_{50}$) (Crémazy et al., 2016). Moreover, copper toxicity to the catfish *Otocinclus vitatus* (limpa vidro) is markedly reduced by more than 7-times when fish was exposed to copper in full-strength Rio Solimões water (RS) (96h $LC_{50}$ of 572.1μgCu/L), compared with toxicity tests performed with 20% of RS content (96h $LC_{50}$ of 78.7μgCu/L) (Dal Pont et al., 2017). Similarly, differences in water chemistry of reference ion-poor water, Rio Negro (blackwater), Rio Tapajós (clearwater) and Rio Solimões (whitewater) markedly influence Ni toxicity to *P. axeroldi* (Holland et al., 2017). In general, Ni was less toxic to cardinal tetra in all types of water, except in Rio Negro water at low pH (4.0) in comparison to neutral pH (7.0). In addition, Ni was remarkably less toxic to cardinal tetra than copper in Rio Negro (10- to 40-times), Rio Solimões (100- to 400-times) and reference ion-poor water (around 1000-times) (Holland et al., 2017).

As well characterized in temperate fishes, metal exposure induces mild to severe metabolic and ionoregulatory disturbances in Amazonian fish. Acute exposure to Cu (100–400μgCu/L) inhibits $Na^+$ uptake in tambaqui acclimated to extremely soft-water (10μM $Ca^{2+}$) and increased the diffusive $Na^+$ loss, resulting in a significant net $Na^+$ loss (Matsuo et al., 2005). However, the presence of commercial humic acid (HA) (5–40mgC/L) slightly attenuated $Na^+$ efflux, but stimulated $Na^+$ uptake, helping fish to control net $Na^+$ losses during copper exposure (Matsuo et al., 2005). Similarly, Cu exposure greatly stimulated $Na^+$ losses (and $K^+$ and $Cl^-$) in cardinal tetra at reference ion-poor water, and in filtered Rio Solimões and Rio Negro water (Crémazy et al., 2016, 2019). Additionally, exposure to 50% of the 96h $LC_{50}$ Cu value promoted a large stimulation of oxygen consumption rate ($MO_2$) in cardinal tetras but had no effect on $MO_2$ of dwarf cichlid (Braz-Mota et al., 2018). Regarding toxic effects of Cd, a significant reduction in whole body $Ca^{2+}$

influx rate was observed in tambaqui acclimated to extremely soft-water and exposed for 3–24h to Cd (10–80μgCd/L) (Matsuo et al., 2005).

Exposure of tambaqui to environmental transition metals causes physiological, biochemical and behavioral changes, as well as the accumulation of these metals in tissues (Giacomin et al., 2018; Lemus et al., 2013). Lemus et al. (2013) analyzed the concentrations of Cd, Cu, Ni, Fe and Pb and their associations with metallothionein and RNA/DNA ratios in tambaqui, collected in a lake in the lower Orinoco River, with greatest metal accumulation observed in liver. The increase in metallothionein in the liver was attributed to Cu, in the gills it was attributed to Cd and in the muscle to Pb and Cd. Although the authors did not find an association between the physiological condition and the concentration of metals, the RNA/DNA ratio was related to the levels of muscle and liver metallothionein.

In another study with Cu and Cd, Giacomin et al. (2018) fed tambaqui chronically (36days) with diets containing 500μg Cu/g and 500μg Cd/g food and did not observe ionoregulatory changes in the gills, intestine or plasma, nor changes in plasma cortisol and lactate, regardless of diet. However, the Cd-contaminated diet seemed to suppress feeding. Interestingly in this study the fish fed the Cd-contaminated diet appeared to have increased tolerance to hypoxia as evidenced by a decreased Pcrit (critical $O_2$ tension—partial pressure of $O_2$ below which the animal can no longer maintain a stable $MO_2$), increased time to loss of equilibrium, but blood $O_2$ transport characteristics ($P_{50}$, Bohr coefficient, hemoglobin, hematocrit) were unaffected compared to control in the same experiment.

Exposure to individual metals affected tambaqui physiology, but it has been necessary to consider mixtures of metals as they occur in the environment. The acute effects of dissolved Cu, Ni (25% of respective $CL_{50}$) and their interaction (mixture, 12.5% of their respective $CL_{50}$) has been investigated in juvenile tambaqui at pH 4 and pH 7 (Pryscilla Pavione, personal unpublished data). At pH 4, the pH effects were more pronounced than metal effects, however, at pH 7, metal effects were found to greatly affect hematological parameters ([Hb]; Ht; MCH and MCHC) and activity of gills $Na^+$/ $K^+$-ATPAse, and liver SOD, GPX, GST, LPO and AChe. Since Cu and Ni have different or even antagonistic effects, their mixture reflects these effects on the analyzed parameters.

## 6.4 Petroleum

Oil and natural gas exploration started in the Western Amazon (Peru and Ecuador) in the early 1920s, but not until 1986 in the Brazilian Amazon following the discovery of large reserves of oil and gas in Urucu province roughly 600km west of Manaus (Finer et al., 2008). Currently, the Urucu province is the third largest in oil and gas production representing 5% of Brazilian production. Since the operations began, direct environmental impacts have been reported especially regarding the contamination of water bodies from small spills during the oil operations and transportation, with moderate to large spills happening in Ecuador (Yasuní region in 2008), Peru (Camisea gas project in 2004) and Brazil (Cururu stream/Manaus in 1999) (Couceiro et al., 2006; Finer et al., 2008).

Exposure to the insoluble fraction of Urucu crude oil in suspension leads to decreased swimming capacity (Ucrit), and spontaneous swimming in tambaqui (Kochhann et al., 2015a), suggesting that exposure to this fraction may reduce aerobic capacity. In addition, exposure to this insoluble fraction resulted in a decrease in acetylcholinesterase (AChe) activity and an increase in alkaline phosphatase (FA) activity, as well as severe lamellar epithelium hypertrophy and extensive lamellar fusion in tambaqui. These results contrast with those for fish exposed to the soluble fraction of Urucu crude oil, which caused only transitory changes in these parameters. Thus, exposure to the insoluble fraction results in both behavioral and physiological changes, suggesting that the toxicity of the oil components is, in fact, enhanced by the presence of an oil phase acting as physical barrier, at least for tambaqui. In addition, tambaqui exposed for 24 h to the insoluble fraction of Urucu crude oil in reference ion-poor water displays increased levels of pyrene- and naphthalene-type metabolites in bile, genotoxic effects on red blood cells (seen as increased DNA damage) (Sadauskas-Henrique et al., 2016) and reduced levels of $K^+$ and $Ca^{2+}$ in plasma (Duarte et al., 2010).

A similar experiment in tambaqui incorporating the effect of a chemical dispersant applied to crude oil, compared to ion-poor water, was conducted by Duarte et al. (2010), who observed that crude oil that is chemically dispersed promotes a more extensive impairment of the branchial permeability to ions, particularly through stimulation of diffusive losses of $Na^+$ and net losses of $Cl^-$ and $K^+$. In addition, chemically dispersed crude oil promotes severe changes in plasma ion levels—reducing levels of $Na^+$, $K^+$, $Ca^{2+}$ and $Cl^-$—and blood parameters—increasing hematocrit and glucose and reducing hemoglobin concentration—when compared to either crude oil or chemical dispersant (Corexit 9500) exposure alone. Under these experimental conditions, chemical dispersion of Urucu crude oil enhances the exposure to hydrocarbons in tambaqui, resulting in higher levels of pyrene- and benzo[*a*]pyrene-type metabolites in bile, and induces hepatic phase II conjugation reactions and antioxidant defenses through the stimulation of GST and SOD activities, respectively (Sadauskas-Henrique et al., 2016).

It is worth noting that the toxic effects of crude oil and chemically dispersed crude oil are exacerbated in the presence of commercial humic acids (HA) and Rio Negro DOC. First, induction in expression of cytochrome P450 1A (CYP1A), and the stimulation in the activity of its catalytic form—the phase I biotransformation of EROD—were seen in liver of tambaqui exposed to crude oil. However, the levels of hepatic CYP1A protein and the EROD activity were around 2-times higher when fish was acclimated for 10 days to the commercial humic acid (Aldrich humic acid—AHA), and further exposed to crude oil+AHA (Matsuo et al., 2006). Additionally, specimens of tambaqui previously acclimated to the water of the Rio Negro and acutely exposed to either crude oil or chemically dispersed crude oil also presents an increase in the levels of hepatic lipid peroxidation (LPO) and plasma s-SDH (indicating hepatotoxicity), damage to blood cell DNA (genotoxicity) and increased bile benzo-[a]-pyrene type metabolites (Sadauskas-Henrique et al., 2016). These findings suggest that the physical-chemical characteristics of Rio Negro water, especially the presence of NOM, may make fish more susceptible to oxidative stress if facing a crude oil spill, a condition that could be particularly augmented if the oil is chemically dispersed.

Although the exact mechanism for the CYP1A induction, and the consequent activation of EROD activity, remains unclear, it is possible that some components of HA/NOM molecules might act as agonist to aryl hydrocarbon receptors (AHR) in tissues of Amazonian fish. In these regards, the studies conducted by Matsuo et al. (2006) and Sadauskas-Henrique et al. (2016) revealed that the acclimatation to a non-Amazonian NOM (Luther Marsh/Canada—40 and 80 mgC/L), and to the natural Rio Negro water (7.2 mgC/L), promotes a significant induction of EROD activity in the liver of tambaqui. Thus, the authors concluded that the use of CYP1A and EROD as a biomarker for exposure to oil in fish that live in environments rich in humic substances should be done with caution to avoid misinterpretation of toxic responses in the animals.

## 6.5 Pesticides

The agricultural expansion in the Amazon has been associated with a large indiscriminate use of pesticides, to further increase food production (Römbke et al., 2008; Waichman et al., 2002). According to the Brazilian Institute of Environment and Renewable Natural Resources the use and consumption of pesticides increased by over 380% between 2000 and 2019 in Brazil and reached more than 26 ton/year of active ingredient in Amazon region (IBAMA, 2019). The pesticides mostly used and commercialized in Amazon region are: Deltamethrin, Malathion and Methyl-parathion (Insecticides); Copper oxychloride and Mancozeb (Fungicides); and Glyphosate (Herbicides) (Römbke et al., 2008; Waichman, 2008; Waichman et al., 2002), with glyphosate accounting for more than 35% of all commercialized active ingredients (IBAMA, 2019). In addition, a substantial increase in the use of extremely toxic active ingredients (Toxicological Class I) have been reported in the Amazon region in last decade (Schiesari et al., 2013), with more than 200 new products containing active agents classified as hazardous pesticides (Coelho et al., 2019; PAN, 2019). The result has been a rapid increase in the environmental concentration of pesticides in water and soil matrices, which can result in toxic effects on non-target aquatic species such as invertebrates, amphibians, and fishes (Römbke et al., 2008; Schiesari et al., 2013).

There is recent robust evidence for pesticides contamination in Amazon with the active ingredients being detected at environmentally relevant levels in soil, water, and edible flesh of many fish species. DDT and its metabolites, for example, have been detected in muscle of many fish species—particularly in carnivorous/piscivorous species—collected at both Tapajós and Madeira Rivers (ranging from 30 to 500ng/g). Species such as *Brachyplatystoma vaillanti* (laulao catfish), *Plagioscion squamosissimus* (silver croaker) and *Pseudoplatystoma fasciatum* (barred sorubin) (D'Amato et al., 2004; Torres et al., 2002), as well as siver croaker and redtail catfish (*Phractocephalus hemioliopterus*) sampled at Iriri River (Mendes et al., 2019) have been shown to be impacted. Similarly, a relatively high concentration of Malathion, Methyl-parathion and chlorpyrifos (insecticides) were measured in eight different fish species from Tapajós and Amazon Rivers (Santarém region/Pará state) (Soumis et al., 2003). Recently, the presence of 11 pesticides in surface waters of streams in urban and peri-urban areas of Manaus (Amazonas State), Santarém and Belém (Pará State) and Macapá (Amapá State) were reported where the compounds malathion, chlorpyrifos (insecticides) and carbendazim (fungicides) had the highest environmental concentrations ($>$100ng/L) and high ecological risk to invertebrates (Rico et al., 2021).

Different pesticides have been tested and classified as moderately to highly toxic to Amazon fish. For example, based on 96h $LC_{50}$ values, the most toxic pesticide to Amazon fishes is the pyrethroid insecticide Deltamethrin (DBP) with an estimated 96h $LC_{50}$ of 2.15mg/L in the bluntnose electric knifefish (*Microsternarchus* cf. *bilineatus*) (Chaves et al., 2020), 6.69mg/L in tambaqui, 7.83mg/L in marbled hatchetfish, 22.49mg/L in cardinal tetra, 23.63mg/L in *H. rhodostomus* (rummy-nose tetra) and 183.51mg/L in Schwartz's catfish (Souza et al., 2020b). Similarly, the insecticide Malathion is highly toxic, while the fungicides Carbendazim and Parathion-methyl (insecticide) are moderately toxic, in the Chaciformes fish species tambaqui (96h $LC_{50}$ of 0.15; 4.16 and 4.98mg/L, respectively), cardinal tetra (96h $LC_{50}$ of 0.2; 1.65 and 6.09mg/L, respectively), *Nannostomus unifasciatus* (oneline pencilfish) (96h $LC_{50}$ of 0.11; 4.14 and 5.38mg/L, respectively) *Hyphessobrycon erythrostigma* (bleeding-heart tetra) (96h $LC_{50}$ of 0.25; 3.69 and 7.27mg/L, respectively) and to the catfish *Otocinclus affinis* (golden otocinclus) (96h $LC_{50}$ of 0.11; 4.24 and 6.83mg/L, respectively) (Rico et al., 2010, 2011).

Exposure to these toxic pesticides at sublethal concentrations results in histopathological damage and significant disturbances in hematological parameters and metabolism. It also affects the activity of both biotransformation and antioxidant enzymes and promotes genotoxic and neurotoxic responses in Amazonian fish species. The exposure to glyphosate-based Roundup and Malathion (75% and 50% of 96h $LC_{50}$ values, respectively) causes severe morphological damage in gills and liver of tambaqui, including hyperplasia, lamellar fusion and proliferation of both mucous and mitochondria-rich cells in gill epithelium, and by the appearance of hepatocyte deformation and cytoplasm degeneration in liver (Braz-Mota et al., 2015; Silva et al., 2019; Souza et al., 2020a).

Exposure to 75% of 96h $LC_{50}$ of Roundup (i.e., 15mg of glyphosate/L) inhibited the activity of glutathione-S-transferase (GST) in gills and liver and stimulated both the activity of the biotransformation enzyme ethoxyresorufin-O-deethylase (EROD) and the antioxidant enzyme glutathione peroxidase (GPx) in liver of tambaqui. These findings suggest that the hepatic biotransformation of glyphosate can induce the generation of ROS inducing oxidative stress in tambaqui, which was evidenced by the increases in frequency of DNA damage in red blood cells of fish (Braz-Mota et al., 2015). In barred surubim (*Pseudoplatystoma* sp.), sublethal exposure to Roundup ranging from 4.5 to 15mg/L—that represents a range from 30% to 100% of glyphosate 96h $LC_{50}$ determined for the Neotropical fish curimbatá (*Prochilodus lineatus*) (Langiano and Martinez, 2008)—resulted in an inhibition in SOD and CAT activity in brain, and CAT activity in the liver (Sinhorin et al., 2014). In tambaqui exposed to 50% of $LC_{50}$ 96h of Malathion, fish displayed a marked stimulation of GST and CAT in gills and liver, as well as in hepatic SOD and branchial GPx, with a concomitant increase in mitochondrial leak respiration and reduction in lipid peroxidation in liver (Souza et al., 2020a, 2021). Neurotoxic effects of pesticides to Amazon fish include marked inhibition in AChE activity in the brain of tambaqui exposed in vivo to 75% of 96h $LC_{50}$ values of Roundup (Braz-Mota et al., 2015), and in vitro following exposure to three organophosphates (Dichlorvos, Chlorpyrifos, TEPP) and two carbamates (Carbaryl and Carbofuran) pesticides (Assis et al., 2010). In other studies, tambaqui exposed for 96h to 7.3mg/L of Malathion (representing 50% of 96h LC50) did not show alterations in the activity of AChE in brain (Souza et al., 2021), which was also seen in bluntnose electric knifefish exposed for 96h to Deltamethrin ranging from 1 to 5μg/L (96h LC50 of 2.15μg/L) (Chaves et al., 2020). Taken together, there is growing evidence that the most common pesticides used in Brazil possess a high ecological risk to Amazonian fish, inducing several deleterious morphological, biochemical, and physiological responses.

## 6.6 Hydroelectric dams

The Amazon is the last frontier of water resources for power generation. There are 307 dams, existing and under construction, on major rivers used for power generation, with another 239 planned (Fearnside, 2006a). These dams cause profound direct and indirect impacts on aquatic and terrestrial systems, due to flooding of floodplain regions and dryland forests (Lees et al., 2016). The death of vegetation due to flooding generates significant amounts of $CO_2$ and $CH_4$ (methane) both upstream and downstream (Assahira et al., 2017; Schöngart et al., 2021). Due to the topography of the region, inundation causes the formation of many new islands with specific faunas (Berenguer et al., 2021) that must cope with a new evolutionary challenge.

The interference of migration of important commercial fish is one of the major problems caused by the construction of dams on rivers in the Amazon (Pelicice et al., 2015), and although all dams are required by law to have systems to permit fish passage, their efficiency is species specific with only a handful of species investigated to date (Hahn et al., 2022; Pelicice and Agostinho, 2008). In addition, damming of Amazonian rivers with high sediment loads, such as whitewater rivers, can have significant impacts on fish both with regard to longitudinal migration upstream (Agostinho et al., 2008), and on the physiology of the animals, particularly with regard to the maintenance of ion homeostasis (Crémazy et al., 2019). The dam induced modification of the hydrological regime has a large impact on the behavioral adaptations of fish, which have been fine-tuned over a long evolutionary timescale (Timpe and Kaplan, 2017).

Environmental fragmentation caused by successive dams along rivers is a further challenge for fish species in general, but in particular for species that depend on the riverbanks to build the great range of species specific nests using different materials (Bessa et al., 2021). The latter is a unique problem of the Amazon due to the tremendous diversity of fish that exist in this system. Although dams are perceived by the general public to provide "clean" electricity, current construction of dams and those proposed are of concern as there is clearly a large impact of hydroelectric dams on the aquatic biota of the Amazon.

## 6.7 Responses to simulations in future climate conditions

The human induced global climate changes, that are already occurring, are superimposed on the natural variations of the environment that, as mentioned above, are dynamic and complex in the Amazon, and have been since the beginning of its formation. The two main climate change relevant parameters are temperature and $CO_2$ concentration. To gain insight into how these parameters may impact biota of the Amazon, a set of four climate rooms at the National Institute for Research of the Amazon (INPA) have been built that simulate, in real time, the environmental scenarios predicted by the IPCC (Intergovernmental Panel on Climate Change). The control room reproduces current conditions based upon sensors in the INPA campus forest that measure temperature and $CO_2$, every other minute. The control room is then used as a reference to replicate temperature and $CO_2$ in the other three rooms to achieve three IPCC scenarios: mild (+200 ppm $CO_2$ and + 1.5 °C in relation to control room), intermediate (+400 ppm $CO_2$ and +2.5 °C in relation to control room) and extreme (+850 ppm $CO_2$ and +4.5 °C in relation to control room) (see details in Oliveira and Val, 2017; Prado-Lima and Val, 2016). All climate rooms are set for 12:12 h luminosity and humidity varying around 80% at the respective room temperature.

In the initial studies by Oliveira and Val (2017), tambaqui were subjected to the three scenarios, with only minor biochemical and physiological changes observed in the intermediate and drastic scenarios. However, over 90 days exposure, tambaqui exhibited higher feed consumption and lower growth in the more challenging scenarios, indicating a cost associated with these treatments to maintain physiological homeostasis ultimately resulting in reduced growth. Energy partitioning between homeostasis and growth at higher temperatures and $CO_2$ levels needs to be investigated in this species, as well as in other Amazonian species. Prado-Lima and Val (2016) investigated changes in gene transcription from these same tambaqui from the very first days of exposure of the animals to these three climate scenarios, where more than 200 genes were differentially expressed compared to controls. The genes included chaperones, energetic metabolism-related genes, translation initiation factors and ribosomal genes. How changes in the expression of these genes might affect the growth of the species, particularly in the natural environment, is unknown but clearly worthy of further study.

Following these studies, Lopes et al. (2018) analyzed the effect of the same climate scenarios on the growth and skeletal development of tambaqui. In this study, the authors observed a significant increase in mortality of larvae exposed to the intermediate and drastic scenarios, as well as up to a 40% increase in osteogenic abnormalities, such as lordosis, scoliosis, and kyphosis, in the drastic scenario. This study indicated that in the first 15 days of development, extreme climate change scenarios can compromise the development of this species which has implications for natural populations.

Our data generated in controlled climate rooms does not allow for generalizations about how Amazonian fish species may acclimate to, or be impacted by, the effects of climate change forecast for 2100. For example, the study by Fé-Gonçalves et al. (2018) showed that two congeneric species, cardinal tetra and green neon tetra (*Paracheirodon simulans*), respond differently to all three tested climate scenarios predicted for the year 2100 after 30 days. While the mortality rate of green neon tetra was zero, that of cardinal tetra was on the order of 40% after 30 days of exposure to the extreme scenario. Green neon tetra was able to upregulate the expression of lactate dehydrogenase genes, while this regulation did not occur in cardinal tetra, suggesting that the anaerobic potential of the latter species is lower. The increased anaerobic potential observed in green neon tetra was associated with better performance in warmer higher $CO_2$ environments, which are known to cause increased energy demand.

Our recent studies indicate that under controlled climate conditions the synergistic effects of temperature and $CO_2$ levels impose new challenges to Amazonian fish, which already live close to their physiological and biochemical limits. Some species such as tambaqui seem to be resilient (Val and Oliveira, 2021) while others are sensitive. Also, some physiological and biochemical processes are more vulnerable than others. In fact, a recent study by Shartau et al. (2022) showed that tambaqui did not reach LOE (loss of equilibrium) even when exposed to the maximum water $CO_2$ levels tested, which were 26.7 KPa $CO_2$. Of 17 species analyzed (5 from temperate, 2 from subtropical, and 10 from tropical regions), 3 other tropical air-breathing fish species also did not reach LOE at this $CO_2$ water tension, pirarucu, electric eel poraque (*Electrophorus electricus*), and marble swamp eel (*Synbranchus marmoratus*), indicating that these species are incredibly tolerant to $CO_2$.

Finally, increased water temperature and acidification due to atmospheric increases in $CO_2$ levels affect the toxicity of pollutants on aquatic organisms, in particular fish. The isolated effect of temperature and acidification on the toxicity of pollutants on fish is widely studied and known, both under controlled conditions and in the natural environment (Braz-Mota et al., 2017; Duarte et al., 2013; Kochhann et al., 2015a). The synergistic effects of water acidification and warming have been studied in temperate fish for some time, but restricted information is available for Amazonian fish, examples being the studies on parasitism in tambaqui exposed to the extreme environmental scenarios predicted by the IPCC (Costa and Val, 2020) and the metabolic adjustments observed in the pencilfish (*Pyrrulina aff. brevis*) (Almeida-Silva et al., 2020). Metal toxicity to aquatic biota is dependent on metal speciation and bioavailability, which in turn is mainly dependent on pH, temperature and DOC. Holland et al. (2017) highlighted this interaction by analyzing nickel toxicity in cardinal tetra in different Amazonian water types and in different seasons, showing the synergistic effect of these main factors. Thus, considering the biological diversity, the environmental diversity, dynamic nature of the Amazon environment and the many synergistic effects that can be expected to occur among climate and regional environments, we emphasize that no generalizations are possible at this point based on the current literature and clearly a great deal more research is required.

## 7 Fish conservation and the Anthropocene

Understanding the environment, we live in and its multiple dimensions is of fundamental importance in times of global challenges, including climate change. Since the seminal work of (Crutzen, 2002a,b) that proposed a new geological epoch, called the Anthropocene, the concept has been analyzed from different perspectives (López-Corona and Magallanes-Guijón, 2020; Malm, 2016; Malm and Hornborg, 2014; Moore, 2016). The world's current technological capacity has both the ability to aid in environmental conservation but also to contribute to significant environmental imbalances, even in the most isolated environments of the planet, as is the pristine Amazon. This is the case with climate change. In just over 150 years, a very short time from a geological perspective, there has been a dramatic increase of $CO_2$ in the earth's atmosphere, from about 280 ppm in 1850 during the industrial revolution in Europe, to more than 411 ppm in 2019 (de la Vega et al., 2020). As a consequence of anthropogenic emissions, and as a result of the greenhouse effect, environmental changes are occurring that include significant increases in the earth's surface temperature, acidification of water bodies, and decreases in the amount of dissolved oxygen in aquatic systems (Lloyd and Shepherd, 2020; Scanes et al., 2020). All these effects have biological implications, often involving more than one level of the biological organization. As water is the final receiver of the substances that result from land use in its most varied activities, fish are not only exposed to these substances, but also to the results of the synergistic effects between them and the physicochemical parameters altered due to climate change, such as pH, temperature, and oxygen. In the Amazon the effects of climate change are very complex given the basin formation processes, environmental diversity, biological diversity, and complex seasonal cycles (Hoorn et al., 2010). Many of the adaptations developed by the region's fish during the basin formation process play against them, either by accelerating the processes of pollutant uptake from the water or by inhibiting protective metabolic processes. We have discussed three such examples: (i) the impact on air-breathing fish of an oil spill, where going to the surface to breathe puts them in contact with oil from the surface of the water column; (ii) increased metal uptake and toxicity in hypoxic environments exacerbated with increasing temperature, where ventilator volume increases to satisfy metabolic demand; (iii) vulnerability of fish, especially those that live in the small streams of the forest, to increases in temperature, because they already live close to their upper thermal limits established throughout the evolutionary process. Here, one could list other examples related to Amazonian fish as the lateral migration for feeding, further acidification of already acidic Amazonian environments, and others. However, there is still much to describe and it is important to avoid too many generalizations in these cases in the absence of sufficient data. Indeed, these are new challenges for fish in this Anthropocene era, in which humans have had a dominant influence on the climate and the environment.

Human-caused climate change, ongoing biodiversity loss, and increasing anthropogenic pressures are intrinsically entwined, and the expansion of one affects the other. This is of utmost concern in biodiverse and pristine regions of the planet where organisms may be especially sensitive to environmental change. Campos et al. (2018) showed that species of the family Characidae have $CT_{max}$ values near the average of the highest temperatures of the Rio Negro, which puts them at risk in the face of rapid and continuous increases in global temperature. The presence of pollutants in the environment, as mentioned above, is increasingly challenging for fish in part due to increasing temperatures that increase metabolic rates. To satisfy this increased metabolic demand, gill ventilation rate of water breathing fishes is elevated, increasing gill pollutant exposure and uptake. In the case of air-breathing fishes, such as the loricariids or callichtids, pollutant uptake may be elevated across the accessory respiratory organ located at the stomach-intestine transition. One might anticipate that air-breathing fishes would be freed from aquatic pollutant exposure, but this may not be entirely true as all air-breathing species are aquatic-breathing at the earliest life stages, and in some cases, pollutants may accumulate in the air phase resulting in aerial uptake across air-breathing organs.

In addition to the climatic challenges and also those caused by the increasing presence of pollutants, one must include the physical changes imposed on the environment by humans (see Section 6.6) and overfishing. In many places, the impacts caused by overfishing have been circumvented by adopting appropriate management practices for both large fish, as is the case for pirarucu in the Mamirauá sustainable development reserve (Gonçalves et al., 2018) and the sustainable harvest of ornamental fish in the Rio Negro region that includes, among other species the cardinal tetra (Phang et al., 2019). In this case, modifications of river interconnections, flooded areas, and forest ecosystems subjected to different destructive practices, will affect the diversity and productivity of other ornamental fish species in this region which should be considered (Phang et al., 2019).

The set of environmental modifications that include human-caused modifications, physical, chemical, and biological, pose challenges often beyond the abilities of fish to maintain organic homeostasis and thus biological functions such as feeding, respiration, excretion, reproduction, among others. Even though science gathers the tools to answer challenging questions, the complexity of the aquatic environment in a region like the Amazon has hampered the possibility of readily understanding what the present and future impacts of the Anthropocene will be on fish, the main source of protein for populations of the most remote Amazonian spaces.

## 8 Concluding remarks

Tectonisms, river captures, climate change, as well as environmental and biological diversification have been part of the life of Amazonian fish to which they have adapted over millions of years. Today, extreme variations in oxygen, temperature, and pH occur in aquatic environments on a daily and seasonal basis, and fish use the adaptations they have evolved to cope, and often thrive, in these conditions. However, the extreme environmental variations may often be close to physiological limits that are buffered by the adaptations that have evolved. It is clear that natural selection continues to act and allows fish to live in Amazonian environments, and thus to evolve and diversify. This is an impressive and diverse biological world, which we know only superficially, and which reveals new species with each incursion into the region, where species continue to be unveiled by new molecular tools and more traditional methodologies. The behavioral responses of many species allow them to leave less desirable locations and invade new niches, altering their geographical distribution, but without erasing what they have "learned" throughout the evolutionary process. Biochemical and physiological responses, homogeneously spread throughout the animal and plant kingdoms, are limited in the face of the complex challenges of daily life in the Amazon, in contrast to the homogeneity and environmental regularity that exists in other regions of the planet. The resilience of diversity is lost with the disappearance or migration of species due to environmental modifications. The dispersal of seeds by fish that migrate to other regions has become increasingly evident. The set of environmental changes caused by humans is the reason for these intense modifications in recent times, times that have been considered as a new era, the Anthropocene. Even the most pristine regions, such as the Amazon, have been impacted by the anthropic actions taking place on the planet. These anthropic actions, including the massive release of $CO_2$ into the atmosphere, make the already challenging environments of the Amazon even more challenging, by superimposing warmer, more acidic, and more hypoxic conditions. These represent new pressures that require deeper biological responses that is further exacerbated by environmental degradation. Metal and oil mining in the Amazon, in combination with higher temperatures, lower pHs, and lower oxygen availability, is a reality in many parts of the region and impose a unique condition that makes it impossible for fish to exist there. The same can be said for pesticides that, with the expansion of land use for agriculture, are being leached in unprecedented amounts into the aquatic environment affecting the food chains, impacting reproduction. As if these challenges were not enough, humans have physically modified the Amazonian environment by means of deforestation, road construction, and damming rivers, which impacts, or even eliminates fish migrations. These effects all act synergistically on the fish of the region and the consequences are not fully known. The interaction of a complex region like the Amazon with the growing challenges imposed by the new era, the Anthropocene, requires that we take action to learn about the existing biological diversity and the many adaptive strategies that fish have developed to cope with their environments. Many of these strategies may play against the animals in the face of new challenges, accelerating the impacts on the fish fauna.

## Acknowledgments

Financial support by INCT ADAPTA-CNPq (465540/2014-7)/FAPEAM (062.1187/2017)/CAPES (finance code 001) is gratefully acknowledged. ALV and VMFAV are recipient of research fellowships from Brazilian National Research Council (CNPq). We thank Dr. Samara Souza for phylogeny diagram.

## References

Adeney, J.M., Christensen, N.I., Vicentini, A., Cohn-Haft, M., 2016. White-sand ecosystems in Amazonia. Biotropica 48, 7–23.

Affonso, E.G., Polez, V.L., Correa, C.F., Mazon, A.F., Araujo, M.R., Moraes, G., Rantin, F.T., 2002. Blood parameters and metabolites in the teleost fish *Colossoma macropomum* exposed to sulfide or hypoxia. Comp. Biochem. Physiol. 133C, 375–382.

Agostinho, A.A., Pelicice, F.M., Gomes, L.C., 2008. Dams and the fish fauna of the neotropical region: impacts and management related to diversity and fisheries. Braz. J. Biol. 68, 1119–1132.

Albert, J., Val, P., Hoorn, C., 2018. The changing course of the Amazon River in the Neogene: center stage for neotropical diversification. Neotrop. Ichthyol. 16.

Alberti, M., Correa, C., Marzluff, J.M., Hendry, A.P., Palkovacs, E.P., Gotanda, K.M., Hunta, V.M., Apgar, T.M., Zhou, Y.Y., 2017. Global urban signatures of phenotypic change in animal and plant populations. Proc. Natl. Acad. Sci. 114, 8951–8956.

Almeida-Silva, J., Campos, D., Almeida-Val, V., 2020. Metabolic adjustment of *Pyrrhulina aff. Brevis* exposed to different climate change scenarios. J. Therm. Biol. 92, 102657.

Almeida-Val, V.M.F., Val, A.L., Duncan, W.P., Souza, F.C.A., Paula-Silva, M.N., Land, S., 2000. Scaling effects on hypoxia tolerance in the Amazon fish *Astronotus ocellatus* (Perciformes, Cichlidae): contribution of tissue enzyme levels. Comp. Biochem. Physiol. 125B, 219–226.

Al-Reasi, H.A., Wood, C.M., Smith, D.S., 2011. Physicochemical and spectroscopic properties of natural organic matter (NOM) from various sources and implications for ameliorative effects on metal toxicity to aquatic biota. Aquat. Toxicol. 103, 179–190.

Al-Reasi, H.A., Wood, C.M., Smith, D.S., 2013. Characterization of freshwater natural dissolved organic matter (DOM): mechanistics explanations for protective effects agains metal toxicity and direct effects on organisms. Environ. Int. 59, 201–207.

Anjos, M.B., Oliveira, R.R., Zuanon, J.S., 2008. Hypoxic environments as refuge against predatory fish in the Amazonian floodplains. Braz. J. Biol. 68, 45–50.

Antonelli, A., Zizka, A., Carvalho, F., Scharn, R., Bacon, C., Silvestro, D., Dondamine, F., 2018. Amazonia is the primary source of neotropical biodiversity. Proc. Natl. Acad. Sci. 115, 6034–6039.

Aragão, L., Anderson, L., Fonseca, M., Rosan, T., Vedovato, L., Wagner, F., Silva, C., Silva Junior, C., Arai, E., Aguiar, A., Barlow, J., Berenguer, E., Deeter, M., Domingues, L., Gatti, L., Gloor, M., Malhi, Y., Marengo, J., Miller, J., Phillips, O., Saatchi, S., 2018. 21st century drought-related fires counteract the decline of Amazon deforestation carbon emissions. Nat. Commun. 9, 536.

Arantes, C.C., Winemiller, K.O., Petrere, M., Castello, L., Hess, L.L., Freitas, C.E.C., 2018. Relationships between forest cover and fish diversity in the Amazon River floodplain. J. Appl. Ecol. 55, 386–395.

Assahira, C., Piedade, M.T.F., Trumbore, S.E., Wittmann, F., Cintra, B.B.L., Batista, E.S., Schöngart, J., 2017. Tree mortality of a flood-adapted species in response of hydrographic changes caused by an Amazonian river dam. For. Ecol. Manage. 396, 113–123.

Assis, C.R.D., Castro, P.F., Amaral, I.P.G., Carvalho, E.V.M.M., Carvalho Jr., L.B., Bezerra, R.S., 2010. Characterization of acetylcholinesterase from the brain of the Amazonian tambaqui (*Colossoma macropomum*) and in vitro effects of organophosphorus and carbamate pesticides. Environ. Toxicol. Chem. 29, 2243–2248.

Bartlett, G.R., 1978. Phosphates in red cells of two south American osteoglossids: *Arapaima gigas* and *Osteoglossum bicirrhosum*. Can. J. Zool. 56, 878–881.

Beheregaray, L., Cooke, G., Chao, N., Landguth, E., 2015. Ecological speciation in the tropics: insights from comparative genetic studies in Amazonia. Front. Genet. 5, 477.

Beltrão, H., Magalhães, E.R.S., Costa, S.B., Loebens, S.C., Yamamoto, K.C., 2018. Ichthyofauna of the major urban forest fragment of the Amazon: surviving concrete and pollution. Neotrop. Biol. Conserv. 13, 124–137.

Beltrão, H., Zuanon, J.S., Ferreira, E.J., 2019. Checklist of the ichthyofauna of the Rio Negro basin in the Brazilian Amazon. ZooKeys 881, 53–89.

Bennett, W., Currie, R., Wagner, P., Beitinger, T., 1997. Cold tolerance and potential overwintering of the red-bellied piranha *Pygocentrus nattereri* in the United States. Trans. Am. Fish. Soc. 126, 841–849.

Berenguer, E., Armenteras, D., Alencar, A., Almeida, C., Aragão, L., Barlow, J., Bilbao, B., Brando, P., Bynoe, P., Fearnside, P., Finer, M., Flores, B.M., Jenkins, C.N., Silva Jr., C.H.L., Lees, A.C., Smith, C.C., Souza, C., García-Villacorta, R., 2021. Drivers and ecological impacts of deforestation and Forest degradation. In: Nobre, C., Encalada, A., Anderson, E., Alcazar, F.H.R., Bustamante, M., Mena, C., Zapata-Ríos, G. (Eds.), Amazon Assessment Report 2021. Science Panel for the Amazon (SPA). United Nations Sustainable Development Solutions Network (19.11-19.41).

Berli, B.I., Gilbert, M.J., Ralph, A.L., Tierney, K.B., Burkhardt-Holm, P., 2014. Acute exposure to a common suspended sediment affects the swimming performance and physiology of juvenile salmonids. Comp. Biochem. Physiol. 176A, 1–10.

Bessa, E., Brandão, M.L., Gonçalves-de-Freitas, E., 2021. Integrative approach on the diversity of nesting behaviour in fishes. Fish Fish., 1–20.

Braum, E., 1983. Beobachtungen uber line reversible lippenextension und ihre roler beider notatmung von *Brycon* spec. (Pisces, Characidae) und *Colossoma macropomum* (Pisces, Serrasalmidae). Amazoniana 7, 355–374.

Braum, E., Junk, W.J., 1982. Morphological adaptation of two Amazonian characoids (Pisces) for surviving in oxygen deficient waters. Int. Rev. ges. Hydrobiol. Hydrogr. 67, 869–886.

Brauner, C.J., Ballantyne, C.L., Randall, D.J., Val, A.L., 1995. Air breathing in the armoured catfish (*Hoplosternum littorale*) as an adaptation to hypoxic, acid, and hydrogen sulphide rich waters. Can. J. Zool. 73, 739–744.

Brauner, C.J., Matey, V., Wilson, J.M., Bernier, N.J., Val, A.L., 2004. Transition in organ function during the evolution of air-breathing; insights from *Arapaima gigas*, an obligate air-breathing teleost from the Amazon. J. Exp. Biol. 207, 1433–1438.

Braz-Mota, S., Sadauskas-Henrique, H., Duarte, R.M., Val, A.L., Almeida-Val, V.M.F., 2015. Roundup-R exposure promotes gills and liver impairments, DNA damage and inhibition of brain cholinergic activity in the Amazon teleost fish *Colossoma macropomum*. Chemosphere 135, 53–60.

Braz-Mota, S., Fé, L.M.L., De Lunardo, F.A.C., Sadauskas-Henrique, H., Almeida-Val, V.M.F., Val, A.L., 2017. Exposure to waterborne copper and high temperature induces the formation of reactive oxygen species and causes mortality in the Amazonian fish *Hoplosternum littorale*. Hydrobiologia 789, 157–166.

Braz-Mota, S., Campos, D.F., MacCormack, T.J., Duarte, R.M., Val, A.L., Almeida-Val, V.M.F., 2018. Mechanisms of toxic action of copper and copper nanoparticles in two Amazon fish species: dwarf cichlid (*Apistogramma agassizii*) and cardinal tetra (*Paracheirodon axelrodi*). Sci. Total Environ. 630, 1168–1180.

Bush, M.B., 1994. Amazonian speciation: a necessarily complex model. J. Biogeogr. 21, 5–17.

Cabrera, M., Moulatlet, G.M., Valencia, B.G., Maisincho, J., Rodríguez-Barroso, R., Albendín, G., Sakali, A., Lucas-Solis, O., Conicelli, B., Capparelli, M.V., 2022. Microplastics in a tropical Andean glacier: a transportation process across the Amazon basin? Sci. Total Environ. 805, 150334.

Campbell, P.G.C., Twiss, M.R., Wilkinson, K.J., 1997. Accumulation of natural organic matter on the surfaces of living cells: implications for the interaction of toxic solutes with aquatic biota. Can. J. Fish. Aquat. Sci. 54, 2543–2554.

Campos, D.F., Jesus, T.F., Kochhann, D., Heinrichs-Caldas, W., Coelho, M.M., Almeida-Val, V.M.F., 2017. Metabolic rate and thermal tolerance in two congeneric Amazon fishes:

*Paracheirodon axelrodi* Schultz, 1956 and *Paracheirodon simulans* Géry, 1963 (Characidae). Hydrobiologia. https://doi.org/10.1007/s10750-016-2649-2.

Campos, D., Val, A.L., Almeida-Val, V.M.F., 2018. The influence of lifestyle and swimming behavior on metabolic rate and thermal tolerance of twelve Amazon forest stream fish species. J. Therm. Biol. 72, 148–154.

Campos, D., Braz-Mota, S., Val, A.L., Almeida-Val, V.M.F., 2019. Predicting thermal sensitivity of three Amazon fishes exposed to climate change scenarios. Ecol. Indic. 101, 533–540.

Campos, D.F., Amanajás, R.D., Almeida-Val, V.M.F., Val, A.L., 2021. Climate vulnerability of south American freshwater fish: thermal tolerance and acclimation. J. Exp. Zool. A 2021, 1–12.

Carroll, S.P., Hendry, A.P., Reznick, D.N., Fox, C.W., 2007. Evolution on ecological time-scales. Funct. Ecol. 21, 387–393.

Carroll, S.P., Jorgensen, P.S., Kinnison, M.T., Bergstrom, C.T., Denison, R.F., Gluckman, P., Smith, T.B., Strauss, S.Y., Tabashnik, B.E., 2014. Applying evolutionary biology to address global challenges. Science 346, 1245993.

Casatti, L., 2010. Alterações no Código Florestal Brasileiro: impactos potenciais sobre a ictiofauna. Biota Neotrop. 10, 31–34.

Casatti, L., Teresa, F.B., Zeni, J.O., Ribeiro, M.D., Brejão, G.L., Ceneviva, B.M., 2015. More of the same: high functional redundancy in stream fish assemblages from tropical agroecosystems. Environ. Manag. 55, 1300–1314.

Castro, J., Braz-Mota, S., Campos, D., Souza, S., Val, A., 2020. High temperature, pH, and hypoxia cause oxidative stress and impair the spermatic performance of the Amazon fish *Colossoma macropomum*. Front. Physiol. 11, 772.

Chaves, V.S., Marcon, J.L., Duncan, W.P., Alves-Gomes, J.A., 2020. Acute toxicity of a deltamethrin based pesticide (DBP) to the neotropical electric fish *Microsternarchus* cf. *bilineatus* (Gymnotiformes). Acta Amazon. 50, 355–362.

Chen, I.-C., Hill, J.K., Ohlemüller, R., Roy, D.B., Thomas, C.D., 2011. Rapid range shifts of species associated with high levels of climate warming. Science 333, 1024–1026.

Chown, S.L., Hoffmann, A.A., Kristensen, T.N., Anguilleta, M.J., Stenseth, N.C., Pertoldi, C., 2010. Adapting to climate change: a perspective from evolutionary physiology. Climate Res. 43, 3–15.

Coelho, F.E.A., Lopes, L.C., Cavalcante, R.M.S., Corrêa, G.C., Leduc, A.O.H.C., 2019. Brazil unwisely gives pesticides a free pass. Science 365 (552), 552–553.

Comte, L., Olden, J.D., 2017. Climatic vulnerability of the world's freshwater and marine fishes. Nat. Clim. Change 7, 718–722.

CONAMA (Conselho Nacional do Meio Ambiente), 2005. Resolução no 357, de 17 de marco de 2005. Dispõe Sobre a Classificação dos Corpos de Água e Diretrizes Ambientais para seu Enquadramento, bem como Estabelece as Condições e Padrões de Lançamento de Efluentes, e da Outras Providencias. Diário Oficial da República Federativa do Brasil, Brasília, Brasil.

Córdova-Tapia, F., Hernández-Marroquín, V., Zambrano, L., 2018. The role of environmental filtering in the functional structure of fish communities in tropical wetlands. Ecol. Freshw. Fish 27, 522–532.

Costa, J.C., Val, A.L., 2020. Extreme climate scenario and parasitism affect the Amazonian fish *Colossoma macropomum*. Sci. Total Environ. 726, 138628.

Costa, F.R.C., Zuanon, J.A.S., Baccaro, F.B., Almeida, J.S., Menger, J.S., Souza, J.L.P., Borba, G.C., Esteban, E.J.L., Bertin, V.M., Gerolamo, C.S., Nogueira, A., Castilho, C.V., 2020. Effects of climate change on central Amazonian forests: a two decades synthesis of monitoring tropical biodiversity. Oecol. Aust. 24, 317–355.

Costa, M.H., Borma, L.S., Espinoza, J.C., Macedo, M., Marengo, J.A., Marra, D.M., Ometto, J.P., Gatti, L.V., 2021. The physical hydroclimate system of the Amazon. In: Nobre, C., Encalada, A., Anderson, E., Alcazar, F.H.R., Bustamante, M., Mena, C., Zapata-Ríos, G. (Eds.), Science Panel for the Amazon. Part I. the Amazon as a Regional Entity of the Earth System. Unites Nations Sustainable Development Solutions Network, New York. US, pp. 1–31.

Costa, M.S., Viana, L.F., Cardoso, C.A.L., Isacksson, E.D.G.S., Silva, J.C., Florentino, C., 2022. Landscape composition and inorganic contaminants in water and muscle tissue of *Plagioscion squamosissimus* in the Araguari River (Amazon, Brazil). Environ. Res. 208, 112691.

Couceiro, S.R.M., Forsberg, B.R., Hamada, N., Ferreira, R.L.M., 2006. Effects of an oil spill and discharge of domestic sewage on the insect fauna of Cururu stream, Manaus, AM, Brazil. Braz. J. Biol. 66 (1a), 35–44.

Crémazy, A., Wood, C.M., Scott Smith, D., Ferreira, M.S., Johannsson, O.E., Giacomin, M., Val, A.L., 2016. Investigating copper toxicity in the tropical fish cardinal tetra (*Paracheirodon axelrodi*) in natural Amazonian waters: measurements, modeling, and reality. Aquat. Toxicol. 180, 353–363.

Crémazy, A., Wood, C.M., Scott Smith, D., Val, A.L., 2019. The effects of natural suspended solids on copper toxicity to the cardinal tetra in Amazonian river waters. Environ. Toxicol. Chem. 38, 2708–2718.

Crutzen, P.J., 2002a. The Anthropocene. J. Phys., IV 12, 1–5.

Crutzen, P.J., 2002b. Geology of mankind. Nature 415, 23.

Culumber, Z.W., Tobler, M., 2018. Correlated evolution of thermal niches and functional physiology in tropical freshwater fishes. J. Evol. Biol. 31, 722–734.

Dagosta, F.C.P., De Pinna, M., 2019. The fishes of the Amazon: distribution and biogeographical patterns, with a comprehensive list of species. Bull. Am. Mus. Nat. Hist. 2019 (431), 1–163.

Dal Pont, G., Valdez-Domingos, F.X., Fernandes-Castilho, M., Val, A.L., 2017. Potential of the biotic ligand model (BLM) to predict copper toxicity in the white-water of the Solimões-Amazon River. Bull. Environ. Contam. Toxicol. 98, 27–32.

D'Amato, C., Torres, J.P., Malm, O., 2004. DDT in fishes from four different Amazon sites: exposure assessment for breast feeding infants. Organohalogen Comp. 66, 2455–2462.

Davidson, E.A., Araújo, A.C., Artaxo, P., Balch, J.K., Brown, I.F., Bustamante, M.M.C., Coe, M.T., DeFries, R.S., Keller, M., Longo, M., Munger, W., Schroeder, W., Soares-Filho, B.S., Souza Jr., C.M., Wofsy, S.C., 2012. The Amazon basin in transition. Nature 481, 321–328.

de la Vega, E., Chalk, T.B., Wilson, P.A., Bysani, R.P., Foster, G.L., 2020. Atmospheric $CO_2$ during the mid Piacenzian warm period and the M2 glaciation. Sci. Rep. 10, 11002.

De Pinna, M.C.C., 2006. Diversity of tropical fishes. In: Val, A.L., Almeida-Val, V.M.F., Randall, D.J. (Eds.), The Physiology of Tropical Fishes. Elsevier/Academic Press, San Diego, pp. 47–84.

Di Santo, V., Jordan, H.L., Cooper, B., Currie, R.J., Beitinger, T.L., Bennett, W.A., 2018. Thermal tolerance of the invasive red-bellied pacu and the risk of establishment in the United States. J. Therm. Biol. 74, 110–115.

Divya, P.K., Ranjeet, K., 2014. Effect of temperature on oxygen consumption and ammonia excretion in red bellied piranha *Pygocentrus nattereri* KNER, 1858. J. Aquat. Biol. Fish. 2, 97–104.

Doria, C.R.C., Agudelo, E., Akama, A., Barros, B., Bonfim, M., Carneiro, L., Briglia-Ferreira, S.R., Carvalho, L.N., Bonilla-Castillo, C.A., Charvet, P., Catâneo, D.T.B.S., Silva, H.P., Garcia-Dávila, C.R., Anjos, H.D.B., Duponchelle, F., Encalada, A., Fernandes, I., Florentino, A.C., Guarido, P.C.P., Guedes, T.L.O., Jimenez-Segura, L., Lasso-Alcalá, O.M.,

Macean, M.R., Marques, E.E., Mendes-Júnior, R.N.G., Miranda-Chumacero, G., Nunes, J.L.S., Occhi, T.V.T., Pereira, L.S., Castro-Pulido, W., Soares, L., Sousa, R.G.C., Torrente-Vilara, G., Van Damme, P.A., Zuanon, J.A., Vitule, J.R.S., 2021. The silent threat of non-native fish in the Amazon: ANNF database and review. Front. Ecol. Evol. 9, 646702.

Dragan, F.G., 2014. Respostas fisiológicas e metabólicas do tambaqui (*Colossoma macropomum*, Cuvier 1818) às alterações climáticas previstas pelos diferentes cenários do IPCC para o ano de 2100. MSc thesis, Biologia de Água Doce e Pesca Interior, Instituto Nacional de Pesquisas da Amazônia.

Duarte, R.M., 2013. Mecanismos de regulação de $Na^+$ nas brânquias de peixes da Amazônia: modulação por fatores ambientais e ajustes espécie-específicos. PhD thesis, Biologia de Água Doce e Pesca Interior, Instituto Nacional de Pesquisas da Amazônia.

Duarte, R.M., Menezes, A.C.L., Rodrigues, L.S., Almeida-Val, V.M.F., Val, A.L., 2009. Copper sensitivity of wild ornamental fish of the Amazon. Ecotoxicol. Environ. Saf. 72, 693–698.

Duarte, R.M., Honda, R.T., Val, A.L., 2010. Acute effects of chemically dispersed crude oil on gill ion regulation, plasma ion levels and haematological parameters in tambaqui (*Colossoma macropomum*). Aquat. Toxicol. 97, 134–141.

Duarte, R.M., Ferreira, M.S., Wood, C.M., Val, A.L., 2013. Effect of low pH exposure on $Na^+$ regulation in two cichlid fish species of the Amazon. Comp. Biochem. Physiol. Part A 166, 441–448.

Duarte, R.M., Scott Smith, D.S., Val, A.L., Wood, C.M., 2016. Dissolved organic carbon from the upper Rio Negro protects zebrafish (*Danio rerio*) against ionoregulatory disturbances caused by low pH exposure. Sci. Rep. 6, 20377–20386.

Dudley, R., 1998. Atmospheric oxygen, giant paleozoic insects and the evolution of aerial locomotor performance. J. Exp. Biol. 201, 1043–1050.

Duffy, P.B., Brando, P., Asner, G.P., Field, C.B., 2015. Projections of future meteorological drought and wet periods in the Amazon. Proc. Natl. Acad. Sci. U. S. A. 12, 13172–13177.

Dutra, G., Freitas, T., Prudente, B., Salvador, G., Leão, M., Peixoto, L., Mendonça, M., Neto-Ferreira, A., Silva, F., Montag, L., Wosiacki, W., 2020. Rapid assessment of the ichthyofauna of the southern Guiana shield tributaries of the Amazonas River in Pará, Brazil. Acta Amazon. 50, 24–36.

Espírito-Santo, H.M.V., Magnusson, W.E., Zuanon, J.A., Mendonça, F.P., Landeiro, V., 2009. Seasonal variation in the composition of fish assemblage in small Amazonian forest streams: evidence for predictable changes. Freshw. Biol. 54, 536–548.

Evans, D.H., 2008. Teleost fish osmoregulation: what have we learned since august Krogh, Homer Smith, and Ancel keys. Am. J. Physiol. Regul. Integr. Comp. Physiol. 284, 1199–1212.

Evans, D.H., Piermarini, P.M., Choe, K.P., 2005. The multifunctional fish gills: dominant site for gas exchange, osmoregulation, acid-base regulation, and excretion of nitrogen waste. Physiol. Rev. 85, 97–177.

Fagundes, D.B., 2012. Identificação de genes potencialmente envolvidos na formação do edema labial na espécie *Colossoma macropomum* (Cuvier, 1818). MSc thesis, Biologia de Água Doce e Pesca Interior, Instituto Nacional de Pesquisas da Amazônia.

Fearnside, P.M., 2006a. Dams in the Amazon - Belo Monte and Brazil s hydroelectric development of the Xingu River basin. Environ. Manag. 38, 16–27.

Fearnside, P.M., 2006b. Dams in the Amazon: Belo Monte and Brazil's hidroelectric development of the Xingu River basin. Environ. Manag. 38, 16–27.

Fé-Gonçalves, L.M., Paula-Silva, M.N., Val, A.L., Almeida-Val, V.M.F., 2018. Differential survivorship of congeneric ornamental fishes under forecasted climate changes are related to anaerobic potential. Genet. Mol. Biol. 41, 107–118.

Findlay, S.E.G., Parr, T.B., 2017. Dissolved organic matter. In: Lamberti, G.A., Hauer, F.R. (Eds.), Methods in Stream Ecology. Volume 2 - Ecosystem Function, pp. 21–36.

Finer, M., Jenkins, C.N., 2012. Proliferation of hydroelectric dams in the Andean Amazon and implications for Andes-Amazon connectivity. PLoS One 7, e35126.

Finer, M., Jenkins, C.N., Pimm, S.L., Keane, B., Ross, C., 2008. Oil and gas projects in the western Amazon: threats to wilderness, biodiversity, and indigenous peoples. PLoS One 3 (8), e2932.

Fisher-Reid, M.C., Kozak, K.H., Wiens, J.J., 2012. How is the rate of climatic-niche evolution related to climatic-niche breadth? Evolution 66, 3836–3851.

Ford, T., Beitinger, T., 2005. Temperature tolerance in the goldfish, *Carassius auratus*. J. Therm. Biol. 30, 147–152.

Freda, J., McDonald, D., 1988. Physiological correlates of interspecific variation in acid tolerance in fish. J. Exp. Biol. 136, 243–258.

Frommel, A.Y., Kwan, G.T., Prime, K.J., Tresguerres, M., Lauridsen, H., Val, A.L., Gonçalves, L.U., Brauner, C.J., 2021. Changes in gill and air-breathing organ characteristics during the transition from water- to air-breathing in juvenile *Arapaima gigas*. J. Exp. Zool. A, 1–13.

Fry, F.E.J., 1947. Effects of the environment on animal activity. Publications of the Ontario Fisheries Research Laboratory, 68 University of Toronto, Toronto, pp. 1–62.

Furch, K., 1984. Water chemistry of the Amazon basin: the distribution of chemical elements among freshwaters. In: Sioli, H. (Ed.), The Amazon. Dr W Junk Publishers, Dordrecht, Limnology and landscape ecology of a mighty tropical river and its basin, pp. 167–199.

Galvez, F., Donini, A., Playle, R.C., Smith, S., O'Donnell, M., Wood, C.M., 2009. A matter of potential concern: natural organic matter alters the electrical properties of fish gills. Environ. Sci. Tech. 42, 9385–9390.

Giacomin, M., Vilarinho, G.C., Castro, K.F., Ferreira, M.S., Duarte, R.M., Wood, C.M., Val, A.L., 2018. Physiological impacts and bioaccumulation of dietary cu and cd in a model teleost: the Amazonian tambaqui (*Colossoma macropomum*). Aquat. Toxicol. 199, 30–45.

Gonçalves, A.C.T., Cunha, J., Batista, J.S., 2018. The Amazon Giant: Sustainable Management of *Arapaima* (Pirarucu). IDSM (Instituto de Desenvolvimento Sustenntável Mamirauá), Tefé, AM.

Gonzalez, R.J., Dunson, W.A., 1987. Adaptations of sodium balance to low pH in a sunfish (*Enneacanthus obesus*) from naturally acidic waters. J. Comp. Physiol. B 157, 555–566.

Gonzalez, R.J., Preest, M.R., 1999. Ionoregulatory specializations for exceptional tolerance of ion-poor, acidic waters in the neon tetra (*Paracheirodon innesi*). Physiol. Biochem. Zool. 72, 156–163.

Gonzalez, R.J., Wilson, R.W., 2001. Patterns of ion regulation in acidophilic fish native to the ion-poor, acidic Rio Negro. J. Exp. Biol. 58, 1680–1690.

Gonzalez, R., Wood, C., Wilson, R., Patrick, M., Bergman, H., Narahara, A., Val, A., 1998. Effects of water pH and calcium concentration on ion balance in fish of the Rio Negro, Amazon. Physiol. Zool. 71, 15–22.

Gonzalez, R.J., Wilson, R.W., Wood, C.M., Patrick, M.L., Val, A.L., 2002. Diverse strategies for ion regulation in fish collected from the ion-poor, acidic Rio Negro. Physiol. Biochem. Zool. 75, 37–47.

Gonzalez, R., Wilson, R., Wood, C., 2006. Ionoregulation in tropical fishes from ion-poor, acidic blackwaters. In: Val, A.L., Almeida-Val, V.M.F., Randall, D.J. (Eds.), The Physiology of Tropical Fishes. Elsevier/Academic Press, San Diego, pp. 397–442.

Gonzalez, R.J., Brauner, C.J., Wang, Y., Richards, J., Patrick, M., Xi, W., Matey, V., Val, A.L., 2010. Impact of ontogenetic changes in branchial morphology on gill function in *Arapaima gigas*. Physiol. Biochem. Zool. 83, 322–332.

Gonzalez, R.J., Patrick, M.L., Duarte, R.M., Casciato, A., Thackeray, J., Day, N., Val, A.L., 2021. Exposure to pH 3.5 water has no effect on the gills of the Amazonian tambaqui (*Colossoma macropomum*). J. Comp. Physiol. B 191, 493–502.

Goulding, M., Barthem, R., Ferreira, E.J.G., 2003. Atlas of the Amazon. Smithsonian Books, New York, US.

Graham, J.B., 1997. Air-Breathing Fishes. Evolution, Diversity and Adaptation. Academic Press, San Diego.

Gutierre, S.M.M., Schofield, P.J., Prodocimo, V., 2016. Salinity and temperature tolerance of an emergent alien species, the Amazon fish Astronotus ocellatus. Hydrobiologia 777, 21–31.

Hahn, L., Martins, E.G., Nunes, L.D., Machado, L.S., Lopes, T.M., Câmara, L.F., 2022. Semi-natural fishway efficiency for goliath catfish (*Brachyplatystoma* spp.) in a large dam in the Amazon Basin. Hydrobiologia 849, 323–338.

Hess, S., Prescott, L.J., Hoey, A.S., McMahon, S.A., Wenger, A.S., Rummer, J.L., 2017. Species-specific impacts of suspended sediments on gill structure and function in coral reef fishes. Proc. R. Soc. B 284, 20171279.

Hilty, J.A., 2012. Climate and Conservation: Landscape and Seascape Science, Planning, and Action. Island Press (ISBN 978-1-61091-203-7).

Hofmann, G.E., Todgham, A.E., 2010. Living in the now: physiological mechanisms to tolerate a rapidly changing environment. Annu. Rev. Physiol. 72, 127–145.

Holland, A., Wood, C.M., Scott Smith, D., Correia, T.G., Val, A.L., 2017. Nickel toxicity to cardinal tetra (*Paracheirodon axelrodi*) differs seasonally and among the black, white and clear river waters of the Amazon basin. Water Res. 126, 21–29.

Hoorn, C., Wesselingh, F.P., ter Steege, H., Bermudez, M.A., Mora, A., Sevink, J., Sanmartín, I., Sanchez-Meseguer, A., Anderson, C.L., Figueiredo, J.P., Jaramillo, C., Riff, D., Negri, F.R., Hooghiemstra, H., Lundberg, J., Stadler, T., Särkinen, T., Antonelli, A., 2010. Amazonia through time: Andean uplift, climate change, landscape evolution, and biodiversity. Science 330, 927–931.

Ilha, P., Schiesari, L., Yanagawa, F., Jankowski, K., Navas, C.A., 2018. Deforestation and stream warming affect body size of Amazonian fishes. PLoS One 13, e0196560.

Instituto Brasileiro do Meio Ambiente e Recursos Renovaveis (IBAMA), 2019. Relatorio de Comercializacao de Agrotoxicos. Available at: http://www.ibama.gov.br/agrotoxicos/relatorios-de-comercializacao-de-agrotoxicos.

Isaacks, R.E., Kim, H.D., Bartlett, G.R., Harkness, D.R., 1977. Inositol pentaphosphate in erythrocytes of a freshwater fish, pirarucu (*Arapaima gigas*). Life Sci. 20, 987–990.

Jacaúna, R.P., Kochhann, D., Campos, D.F., Val, A.L., 2020. Aerobic metabolism impairment in tambaqui (*Colossoma macropomum*) juveniles exposed to urban wastewater in Manaus, Amazon. Bull. Environ. Contam. Toxicol. 105, 853–859.

Jaffé, R., Nunes, S., Santos, J.F., Gastauer, M., Giannini, T.C., Nascimento Jr., W., Sales, M., Souza Jr., C.M., Souza-Filho, P.W., Fletcher Jr., R.J., 2021. Forecasting deforestation on the Brazilian Amazon to prioritize conservation efforts. Environ. Res. Lett. 16, 084034.

Jiménez-Muñoz, J., Mattar, C., Barichivich, J., Santamaría-Artigas, A., Takahashi, K., Malhi, Y., Sobrino, J., van der Schrier, G., 2016. Record-breaking warming and extreme drought in the Amazon rainforest during the course of El Niño 2015–2016. Sci. Rep. 6, 33130.

Jiménez-Muñoz, J.C., Barichivich, J., Mattar, C., Takahashi, K., Santamaría-Artigas, A., Sobrino, J.A., Malhi, Y., 2018. Spatiotemporal patterns of thermal anomalies and drought over tropical forests driven by recent extreme climatic anomalies. Philos. Trans. R. Soc. B Biol. Sci. 373, 20170300.

Johannsson, O.E., Scott Smith, D., Sadauskas-Henrique, H., Cimprich, G., Wood, C.M., Val, A.L., 2017. Photo-oxidation processes, properties of DOC, reactive oxygen species (ROS), and their potential impacts on native biota and carbon cycling in the Rio Negro (Amazonia, Brazil). Hydrobiologia 789, 7–29.

Johannsson, O.E., Ferreira, M.S., Scott-Smith, D., Crémazy, A., Wood, C.M., Val, A.L., 2020. Effects of natural light and depth on rates of photo-oxidation of dissolved organic carbon in a major black-water river, the Rio Negro, Brazil. Sci. Total Environ. 733, 139193.

Jung, E.H., Brix, K.V., Richards, J.G., Val, A.L., Brauner, C.J., 2020. Reduced hypoxia tolerance and survival at elevated temperatures may limit the ability of Amazonian fishes to survive in a warming world. Sci. Total Environ. 748, 141349.

Junk, W.J., Bayley, P.B., Sparks, R.E., 1989. The flood pulse concept in river-floodplain systems. In: Dodge, D.P. (Ed.), Proceedings of the International Large River Symposium. Can. Spec. Publ. Fish. Aquat. Sci, Canada, pp. 110–127.

Kochhann, D., Campos, D.F., Val, A.L., 2015a. Experimentally increased temperature and hypoxia affect stability of social hierarchy and metabolism of the Amazonian cichlid *Apistogramma agassizii*. Comp. Biochem. Physiol., Part A 190, 54–60.

Kochhann, D., Meyersieck-Jardim, M., Domingos, F.X.V., Val, A.L., 2015b. Biochemical and behavioral responses of the Amazonian fish *Colossoma macropomum* to crude oil: the effect of oil layer on water surface. Ecotoxicol. Environ. Saf. 111, 32–41.

Kritzberg, E.S., 2017. Centennial-long trends of lake browning show major effect of afforestation. Limnol. Oceanogr. Lett. 2, 105–112.

Kritzberg, E., Hasselquist, E., Skerlep, M., Löfgren, S., Olsson, O., Stadmark, J., Valinia, S., Hansson, L.-A., Laudon, H., 2020. Browning of freshwaters: consequences to ecosystem services, underlying drivers, and potential mitigation measures. Ambio 49, 375–390.

Kullberg, A., Bishop, K.H., Hergeby, A., Jansson, M., Petersen, R.C., 1993. The ecological significance of dissolved organic carbon in acidified waters. Ambio 22, 331–337.

Kumai, Y., Bahubeshi, A., Steele, S., Perry, S.F., 2011. Strategies for maintaining $Na^+$ balance in zebrafish (*Danio rerio*) during prolonged exposure to acidic water. Comp. Biochem. Physiol. 160A, 52–62.

Kwong, R.W.M., Kumai, Y., Perry, S.F., 2014. The physiology of fish at low pH: the zebrafish as a model system. J. Exp. Biol. 217, 651–662.

Lackner, R., 1998. Oxidative stress in fish by environmental pollutants. In: Braunbeck, T., Hinton, D.E., Streit, B. (Eds.), Fish Ecotoxicology. Birkhäuser, Basel, p. 398.

Langiano, V.C., Martinez, C.B.R., 2008. Toxicity and effects of a glyphosate-based herbicide on the neotropical fish *Prochilodus lineatus*. Comp. Biochem. Physiol, Part C 147, 221–231.

Latrubesse, E.M., Stevaux, J.C., Sinha, R., 2005. Tropical Rivers. Geomorphology 70, 187–206.

Lavado Casimiro, W.S., Labat, D., Ronchail, J., Espinoza, J.C., Guyot, J.L., 2013. Trends in rainfall and temperature in the Peruvian Amazon–Andes basin over the last 40 years (1965–2007). Hydrol. Process. 27, 2944–2957.

Leal, C.G., Pompeu, P.S., Gardner, T.A., Leitão, R.P., Hughes, R.M., Kaufmann, P.R., Zuanon, J., Paula, F.R., Ferraz, S.F.B., Thomson, J.R., Mac Nally, R., Ferreira, J., Barlow, J., 2016. Multi-scale assessment of human-induced changes to Amazonian instream habitats. Landsc. Ecol. 31, 1725–1745.

Lees, A.C., Peres, C.A., Fearnside, P.M., Schneider, M., Zuanon, J.A.S., 2016. Hydropower and the future of Amazonian biodiversity. Biodivers. Conserv. 25, 451–466.

Leitão, R.P., Zuanon, J., Mouillot, D., Leal, C.G., Hughes, R.M., Kaufmann, P.R., Villéger, S., Pompeu, P.S., Kasper, D., Paula, F.R., Ferraz, S.F.B., Gardne, T.A., 2018. Disentangling the pathways of land use impacts on the functional structure of fish assemblages in Amazon streams. Ecography 41, 219–232.

Lemus, M., Blanco, I., Hernández, M., León, A., Centeno, L., Chung, K., Salazar-Lungo, R., 2013. Relation between metallothionein, RNA/DNA and heavy metals in juveniles of *Colossoma macropomum* (Cuvier 1818) in natural conditions. E3S Web of Conferences 1, 12001.

Lenoir, J., Svenning, J.C., 2015. Climate-related range shifts—a global multidimensional synthesis and new research directions. Ecography 38, 15+28.

Lima, D.P., Santos, C., Silva, R.D.S., Yoshioka, E.T.O., 2015. Heavy metal contamination in fish and water from Cassiporé River basin, state of Amapá, Amazonia, Brazil. Acta Amazon. 45, 405–414. https:/doi.org/10.1590/1809- 4392201403995.

Lloyd, E., Shepherd, T.G., 2020. Environmental catastrophes, climate change, and attribution. Ann. N. Y. Acad. Sci. 1469, 105–124.

Lopes, I.G., Ataújo-Daikiri, T.B., Kojima, J.T., Val, A.L., Portella, M.C., 2018. Predicted 2100 climate scenarios affects growth and skeletal development of tambaqui (*Colossoma macropomum*) larvae. Ecol. Evol. 2018, 1–10.

López-Corona, O., Magallanes-Guijón, G., 2020. It is not an Anthropocene; it is really the Technocene: names matter in decision making under planetary crisis. Front. Ecol. Evol. 8 (article 214).

Lovejoy, T., Nobre, C., 2018. Amazon tipping point. Sci. Adv. 4, (eaat2340).

Lundberg, J., Pérez, M., Dahdul, W., Aguilera, O., 2010. The Amazonian Neogene fish fauna. In: Hoorn, C., Wesselingh, F.P. (Eds.), Amazonia: Landscape and Species Evolution. A look into the past, Wiley-Blackwell, Oxford, pp. 281–310.

Maberly, S.C., O'Donnell, R.A., Woolway, R.I., Cutler, M.E.J., Gong, M., Jones, I.D., Merchant, C.J., Miller, C.A., Politi, E., Scott, E.M., Thackeray, S.J., Tyler, A.N., 2020. Global lake thermal regions shift under climate change. Nat. Commun. 11, 1232.

Madeira, C., Madeira, D., Diniz, M.S., Cabral, H.N., Vinagre, C., 2016. Thermal acclimation in clownfish: an integrated biomarker response and multi-tissue experimental approach. Ecol. Indic. 71, 280–292.

Magrin, G.O., Marengo, J.A., Boulanger, J.P., Buckeridge, M.S., Castellanos, E., Poveda, G., Scarano, F.R., Vicuña, S., 2014. Central and South America. In: Barros, V.R., Field, C.B., Dokken, D.J., Mastrandrea, M.D., Mach, K.J., Bilir, T.E., White, L.L. (Eds.), Climate Change 2014: Impacts, Adaptation, and Vulnerability. Part B: Regional Aspects. Contribution of Working Group II to the Fifth Assessment Report of the Intergovernmental Panel on Climate Change. Cambridge University Press, Cambridge, pp. 1499–1566.

Malm, A., 2016. Fossil Capital: The Rise of Steam Power and the Roots of Global Warming. Verso Books, Brooklyn, NY.

Malm, A., Hornborg, A., 2014. The geology of mankind? A critique of the anthropocene narrative. Anthr. Rev. 1, 62–69.

Marcon, J.L., Chagas, E.C., Kavassaki, J.M., Val, A.L., 1999. Intraerythrocytic phosphates in 25 fish species of the Amazon: GTP as a key factor in the regulation of Hb-$O_2$ affinity. In: Val, A.L., Alameida-Val, V.M.F. (Eds.), Biology of Tropical Fish. INPA, Manaus, pp. 229–240.

Marengo, J., Souza, C., Thonicke, K., Burton, C., Halladay, K., Betts, R., Alves, L., Soares, W., 2018. Changes in climate and land use over the Amazon region: current and future variability and trends. Front. Earth Sci. 6, 228.

Martins, R., Rezende, R., Gonçalves Jr., J., Lopes, A., Piedade, M., Cavalcante, H., Hamada, N., 2017. Effects of increasing temperature and, CO2 on quality of litter, shredders, and microorganisms in Amazonian aquatic systems. PLoS One 12, e0188791.

Matsuo, A.Y.O., Val, A.L., 2007. Acclimation to humic substances prevents whole body sodium loss and stimulates branchial calcium uptake capacity in cardinal tetras *Paracheirodon axelrodi* (Schultz) subjected to extremely low pH. J. Fish Biol. 70, 989–1000.

Matsuo, A.Y.O., Wood, C.M., Val, A.L., 2005. Effects of copper and cadmium on ion transport and gill metal binding in the Amazonian teleost tambaqui (*Colossoma macropomum*) in extremely soft water. Aquat. Toxicol. 74, 351–364.

Matsuo, A.Y.O., Woodin, B., Reddy, C.M., Val, A.L., Stegeman, J.J., 2006. Humic substances and crude oil induce cytochrome P450 1A expression in the amazonian fish species *Colossoma macropomum* (tambaqui). Environ. Sci. Tech. 40, 2851–2858.

McDonald, D.G., Wood, C.M., 1981. Branchial and renal acid and ion fluxes in the rainbow trout, *Salmo gairdneri*, at low environmental pH. J. Environ. Biol. 93, 101–118.

McDonald, D.G., Hobe, H., Wood, C.M., 1980. The influence of calcium on the physiological responses of the rainbow trout, *Salmo gairdneri*, to low environmental pH. J. Exp. Biol. 88, 109–131.

McDonald, D.G., Walker, R.L., Wilkes, P.R.H., 1983. The interaction of environmental calcium and low pH on the physiology of the rainbow trout, *Salmo gairdneri*. II. Branchial ionoregulatory mechanisms. J. Environ. Biol. 102, 141–155.

McGeer, J.C., Szebedinszky, C., McDonald, D.G., Wood, C.M., 2002. The role of dissolved organic carbon in moderating the bioavailability and toxicity of cu to rainbow trout during chronic waterborne exposure. Comp. Biochem. Physiol. 133, 147–160.

McWilliams, P.G., Potts, W.T.W., 1978. The effects of pH and calcium concentrations on gill potentials in the brown trout, *Salmo trutta*. J. Comp. Physiol. B 126, 277–286.

Melo, M., Bertilsson, S., Amaral, J., Barbosa, P., Forsberg, B., Sarmento, H., 2019. Flood pulse regulation of bacterioplankton community composition in an Amazonian floodplain lake. Freshw. Biol. 64, 108–120.

Mendes, R.A., Lima, M.O., Deus, R.J.A., Medeiros, A.C., Faial, K.C.F., Jesus, I.M., Faial, K.R.F., Santos, L.S., 2019. Assessment of DDT and mercury levels in fish and sediments in the Iriri River, Brazil: distribution and ecological risk. J. Environ. Sci. Health B 54, 915–924.

Méndez-Camacho, K., Leon-Alvarado, O., Miranda-Esquivel, M., 2021. Biogeographic evidence supports the old Amazon hypothesis for the formation of the Amazon fluvial system. PeerJ 9, e12533.

Mendonça, F.P., Magnusson, W.E., Zuanon, J., 2005. Relationships between habitat characteristics and fish assemblages in small streams of Central Amazonia. Copeia 2005, 751–764.

Menezes, J., Confalonieri, U., Madureira, A., Duval, I., Santos, R., Margonari, C., 2018. Mapping human vulnerability to climate change in the Brazilian Amazon: the construction of a municipal vulnerability index. PLoS One 13, e0190808.

Milligan, C.L., Wood, C.M., 1982. Disturbances in haematology, fluid volume distribution and circulatory function associated with low environmental pH in the rainbow trout, *Salmo gairdneri*. J. Exp. Biol. 99, 397–415.

Monteith, D.T., Stoddard, J.L., Evans, C.D., Wit, H.A., Forsius, M., Høgasen, T., Wilander, A., Skelkvåle, B.L., Jeffries, D.S., Vuorenmaa, J., Keller, B., Kopácek, J., Vesely, J., 2007. Dissolved organic carbon trends resulting from changes in atmospheric deposition chemistry. Nat. Lett. 450, 537–541.

Moore, A., 2016. Anthropocene anthropology: reconceptualizing contemporary global change. J. R. Anthropol. Inst. 22, 27–46.

Morgan, R., Finnoen, M.H., Jensen, H., Pélabon, C., Jutfelt, F., 2020. Low potential for evolutionary rescue from climate change in a tropical fish. Proc. Natl. Acad. Sci. 117, 33365–33372.

Morris, C., Val, A.L., Brauner, C.J., Wood, C.M., 2021. The physiology of fish in acidic waters rich in dissolved organic carbon, with specific reference to the Amazon basin: Ionoregulation, acid–base regulation, ammonia excretion, and metal toxicity. J. Exp. Zool. A, 1–21.

Muusze, B., Marcon, J.L., Van den Thillart, G., Almeida-Val, V.M.F., 1998. Hypoxia tolerance of Amazon fish. Respiratory and energy metabolism of the cichlid *Astronotus ocellatus*. Comp. Biochem. Physiol. 120A, 151–156.

Nepstad, D.C., Stickler, C.M., Filho, B.S., Merry, F., 2008. Interactions among Amazon land use, forests and climate: prospects for a near-term forest tipping point. Philos. Trans. R. Soc. Lond. B Biol. Sci. 363, 1737–1746.

Ohlemuller, R., Anderson, B.J., Araujo, M.B., Butchart, S.H.M., Kudrna, O., Ridgely, R.S., Thomas, C.D., 2008. The coincidence of climatic and species rarity: high risk to small-range species from climate change. Biol. Lett. 4, 567–572.

Oliveira, A.M., Val, A.L., 2017. Effects of climate scenarios on the growth and physiology of the Amazonian fish tambaqui (*Colossoma macropomum*) (Characiformes: Serrasalmidae). Hydrobiologia 789, 167–178.

Olson, W.H., Chase, D.L., Hanson, J.N., 1973. Preliminary studies using synthetic polymer to reduce turbidity in a hatchery water supply. Progress. Fish Cult. 35, 66–73.

Packer, R.K., Dunson, W.A., 1972. Anoxia and sodium loss associated with the death of brook trout at low pH. Comp. Biochem. Physiol. 41A, 17–26.

Pankhurst, N.W., Munday, P.L., 2011. Effects of climate change on fish reproduction and early life history stages. Mar. Freshw. Res. 6, 1015–1026.

Pegado, T., Brabo, L., Schmid, K., Sarti, F., Gava, T.T., Nunes, J., Chelazzi, D., Cincinelli, A., Giarrizzo, T., 2021. Ingestion of microplastics by *Hypanus guttatus* stingrays in the Western Atlantic Ocean (Brazilian Amazon coast). Mar. Pollut. Bull. 162, 111799.

Pelicice, F.M., Agostinho, A.A., 2008. Fish-passage facilities as ecological traps in large neotropical rivers. Conserv. Biol. 22, 180–188.

Pelicice, F.M., Pompeu, P.S., Agostinho, A.A., 2015. Large reservoirs as ecological barriers to downstream movements of neotropical migratory fish. Fish Fish. 16, 697–715.

Pelster, B., Wood, C.M., 2018. Ionoregulatory and oxidative stress issues associated with the evolution of air-breathing. Acta Histochem. 120, 667–679.

Pelster, B., Giacomin, M., Wood, C.M., Val, A.L., 2016. Improved ROS defense in the swimbladder of a facultative air-breathing erythrinid fish, jeju, compared to a non-air-breathing close relative, traira. J. Comp. Physiol. B 186, 615–624.

Pelster, B., Wood, C.M., Jung, E., Val, A.L., 2018. Air-breathing behavior, oxygen concentrations, and ROS defense in the swimbladders of two erythrinid fish, the facultative airbreathing jeju, and the non-air-breathing traira during normoxia, hypoxia and hyperoxia. J. Comp. Physiol. B 188, 437–449.

Pelster, B., Wood, C.M., Braz-Mota, S., Val, A.L., 2020. Gills and air-breathing organ in O2 uptake, CO2 excretion, N-waste excretion, and ionoregulation in small and large pirarucu (*Arapaima gigas*). J. Comp. Physiol. B 190, 569–583.

Pesticide Action Network International, 2019. PAN International List of Highly Hazardous Pesticides.

Petrone, K.C., Fellman, J.B., Hood, E., Donn, M.J., Grierson, P.F., 2011. The origin and function of dissolved organic matter in agro-urban coastal streams. J. Geophys. Res. 116, 1–13.

Phang, S.C., Cooperman, M., Lynch, A.J., Lynch, A.J., Steel, E.A., Elliott, V., Murchie, K.J., Cooke, S.J., Dowd, S., Cowx, I.G., 2019. Fishing for conservation of freshwater tropical fishes in the Anthropocene. Aquat. Conserv.: Mar. Freshw. Ecosys. 2019, 1–13.

Pineda, M., Aragao, I., McKenzie, D.J., Killen, S.S., 2020. Social dynamics obscure the effect of temperature on air breathing in *Corydoras* catfish. J. Exp. Biol. 223, (jeb222133).

Pörtner, H.-O., Scholes, R.J., Arneth, A., Barnes, D., Burrows, M., Diamond, S., Duarte, C.M., Kiessling, W., Leadley, P., Managi, S., McElwee, P., Midgley, G., Ngo, H.T., Obura, D., Pascual, U., Sankaran, M., Shin, Y.J., Val, A.L., 2022. Overcoming the combined climate, biodiversity and societal crises. Science. submitted.

Prado-Lima, M., Val, A.L., 2016. Transcriptomic characterization of tambaqui (*Colossoma macropomum*, Cuvier, 1818) exposed to three climate change scenarios. PLoS One.

Py-Daniel, L.H.R., Zuanon, J., 2005. Description of a new species of *Parancistrus* (Siluriformes: Loricariidae) from the rio Xingu, Brazil. Neotrop. Ichthyol. 3, 571–577.

Randall, D.J., Burggren, W.W., Farrell, A.P., Haswell, M.S., 1981. The Evolution of Air-Breathing Vertebrates. Cambridge University Press, Cambridge.

Rantin, F.T., Kalinin, A.L., 1996. Cardiorespiratory function and aquatic surface respiration in *Colossoma macropomum* exposed to graded and acute hypoxia. In: Val, A.L., Almeida-Val, V.M.F., Randall, D.J. (Eds.), Physiologt and Biochemistry of the Fishes of the Amazon. INPA, Manaus, pp. 169–180.

Ribeiro-Brasil, D.R.G., Torres, N.R., Picanço, A.B., Sousa, D.S., Ribeiro, V.S., Brasil, L.S., Montag, L.F.A., 2020. Contamination of stream fish by plastic waste in the Brazilian Amazon. Environ. Pollut. 286, 115241.

Rico, A., Geber-Corrêa, R., Campos, P.S., Garcia, M.V., Waichman, A.V., van den Brink, P.J., 2010. Effect of parathion-methyl on Amazonian fish and freshwater invertebrates: a comparison of sensitivity with temperate data. Arch. Environ. Contam. Toxicol. 58, 765–771.

Rico, A., Waichman, A.V., Geber-Corrêa, R., van den Brink, P.J., 2011. Effects of malathion and carbendazim on Amazonian freshwater organisms: comparison of tropical and temperate species sensitivity distributions. Ecotoxicology 20, 625–634.

Rico, A., de Oliveira, R., Nunes, G.S.S., Rizzi, C., Villa, S., Vizioli, B.C., Montagner, C.C., Waichman, A.V., 2021. Ecological risk assessment of pesticides in urban streams of the Brazilian Amazon. Chemosphere 2021, 132821.

Rocha, J.C., Sargentini Jr., E., Toscano, I.A.S., Rosa, A.H., Burba, P., 1999. Multi-method study on aquatic substances from the "Rio Negro"—Amazonas state/Brazil. Emphasis on molecular-size classsification of their metal contents. J. Braz. Chem. Soc. 10, 169–175.

Römbke, J., Waichman, A.V., Garcia, M.V.B., 2008. Risk assessment of pesticides for soils of the Central Amazon, Brazil: comparing outcomes with temperate and tropical data. Integr. Environ. Assess. Manag. 4, 94–104.

Röpke, C.P., Amadio, S.A., Zuanon, J.A.S., Ferreira, E.J., Deus, C.P., Pires, T.H., Winemiller, K.O., 2017. Simultaneous abrupt shifts in hydrology and fish assemblage structure in a floodplain lake in the Central Amazon. Sci. Rep. 7, 40170.

Rosa, R., Paula, J.R., Sampaio, E., Pimentel, M., Lopes, A.R., Baptista, M., Guerreiro, M., Santos, C., Campos, D., Almeida-Val, V.M.F., Calado, R., Diniz, M., Repolho, T., 2016. Neuro-oxidative damage and aerobic potential loss of sharks under elevated CO2 and warming. Mar. Biol. 163, 119.

Sadauskas-Henrique, H., Braz-Mota, S., Duarte, R.M., Almeida-Val, V.M.F., 2016. Influence of the natural Rio Negro water on the toxicological effects of a crude oil and its chemical dispersion to the Amazonian fish *Colossoma macropomum*. Environ. Sci. Pollut. Res. 23, 19764–19775.

Sadauskas-Henrique, H., Wood, C.M., Souza-Bastos, L., Duarte, R.M., Smith, D.S., Val, A.L., 2019. Does dissolved organic carbon from Amazon black water (Brazil) help a native species, the tambaqui *Colossoma macropomum* to maintain ionic homeostasis in acidic water? J. Fish Biol. 94, 595–605.

Sadauskas-Henrique, H., Scott-Smith, D., Val, A.L., Wood, C.M., 2021. Physicochemical properties of the dissolved organic carbon can lead to different physiological responses of zebrafish (*Danio rerio*) under neutral and acidic conditions. J. Exp. Zool. A 2021, 1–14.

Saint-Paul, U., 1984. Physiological adaptation to hypoxia of a neotropical characoid fish *Colossoma macropomum*, Serrasalmidae. Environ. Biol. Fishes 11, 53–62.

Salazar-Torres, G., Huszar, V.L.M., 2012. Microalgae community of the Huaytire wetland, an Andean high-altitude wetland in Peru. Acta Limnol. Bras. 24, 285–292.

Sá-Leitão, C.S., Souza, E.M.S., Santos, C.H.A., Val, P., Val, A.L., Almeida-Val, V.M.F., 2022. River reorganization affects populations of dwarf cichlid species (*Apistogramma* genus) in the lower Negro River, Brazil. Front. Ecol. Evol. 9, 760287.

Salomão, G.N., Dall'Agnol, R., Sahoo, P.K., Ferreira Júnior, J.S., Silva, M.S., Souza Filho, P.W.M., Nascimento Junior, W.R., Costa, M.F., 2018. Geochemical distribution and threshold values determination of heavy metals in stream water in the sub-basins of Vermelho and Sororó rivers, Itacaiúnas River watershed, Eastern Amazon, Brazil. Geochim. Bras. 32 (2), 180–198.

Santana, G.P., 2016. Heavy metal distribution in the sediment and *Hoplosternum littorale* from Manaus Industrial District. J. Eng. Exact Sci. 2, 70–81.

Santana, G.P., Barroncas, P.S.R., 2005. Heavy metals (Co, Cu, Fe, Cr, Ni, Mn, Pb and Zn) study in the Tarumã-Açu Basin - Manaus (AM). Acta Amazon. 37, 111–118.

Santos, I.C.C., 2012. Biomarcadores de efeito do chorume da lixeira pública de Paratins-AM em tambaqui. MSc thesis, Universidade do Estado do Amazonas, Manaus, AM, p. 110.

Santos, G.M., Val, A.L., 1998. Ocorrência do peixe-serra (*Pristis perotteti*) no rio Amazonas e comentários sobre sua história natural. Ciência Hoje 23, 66–67.

Scanes, E., Scanes, P.R., Ross, P.M., 2020. Climate change rapidly warms and acidifies Australian estuaries. Nat. Commun. 11, 1803.

Schaefer, J., Ryan, A., 2006. Developmental plasticity in the thermal tolerance of zebrafish *Danio rerio*. J. Fish Biol. 69, 722–734.

Schiesari, L., Waichman, A.V., Brock, T., Adams, C., Grillitsch, B., 2013. Pesticide use and biodiversity conservation in the Amazonian agricultural frontier. Philos. Trans. R. Soc. B Biol. Sci. 368. 20120378.

Schöngart, J., Junk, W.J., 2007. Forecasting the flood-pulse in Central Amazonia by ENSO-indices. J. Hydrol. 335, 124–132.

Schöngart, J., Wittmann, F., Resende, A.F., Assahira, C., Lobo, G.S., Neves, J.R.D., Rocha, M., Mori, G.B., Quaresma, A.C., Demarchi, L.O., Albuquerque, B.W., Feitosa, Y.O., Costa, G.S., Feitoza, G.V., Durgante, F.M., Lopes, A., Trumbore, S.E., Silva, T.S.F., Steege, H., Val, A.L., Junk, W.J., Piedade, M.T.F., 2021. The shadow of the Balbina dam: a synthesis of over 35 years of downstream impacts on floodplain forests in Central Amazonia. Aquat. Conserv.: Mar. Freshw. Ecosys. 2021, 1–19.

Schulte, P.M., 2015. The effects of temperature on aerobic metabolism: towards a mechanistic understanding of the responses of ectotherms to a changing environment. J. Exp. Biol. 218, 1856–1866.

Scott, G.R., Matey, V., Mendoza, J., Gilmour, K.M., Perry, S.F., Almeida-Val, V.M.F., Val, A.L., 2017. Air breathing and aquatic gas exchange during hypoxia in armoured catfish. J. Comp. Physiol. B 187, 117–133.

Scully, N.M., McQueen, D.J., Lean, D.R.S., Cooper, W.J., 1996. Hydrogen peroxide formation: the interaction of ultraviolet radiation and dissolved organic carbon in lake waters along a N gradient. Limnol. Oceanogr. 41, 540–548.

Shartau, R.B., Harter, T.S., Baker, D.W., Aboagye, D.L., Allen, P.J., Val, A.L., Crossley II, D.A., Kohl, Z.F., Hedrick, M.S., Damsgaard, C., Brauner, C.J., 2022. Acute CO2 tolerance in fishes is associated with air breathing but not the root effect, red cell BNHE, or habitat. Comp. Biochem. Physiol. (submitted).

Silva, M.S.R., Ramos, J.F., Pinto, A.G.N., 1999. Transition metals in the sediments of the Manaus igarapés. Acta Limnol. Bras. 11, 89–100.

Silva, G.S., Matos, L.V., Freitas, J.O.S., Campos, D.F., Almeida-Val, V.M.F., 2019. Gene expression, genotoxicity, and physiological responses in an Amazonian fish, *Colossoma macropomum* (CUVIER 1818), exposed to roundup® and subsequent acute hypoxia. Comp. Biochem. Physiol. 222C, 49–58.

Sinhorin, V.D.G., Sinhorin, A.P., Teixeira, J.M., Teixeira, J.M.S., Miléski, K.M.L., Hansen, P.C., Moreira, P.S.A., Kawashita, N.H., Baviera, A.M., Loro, V.L., 2014. Effects of the acute exposition to glyphosate-based herbicide on oxidative stress parameters and antioxidant responses in a hybrid Amazon fish Surubim (*Pseudoplatystoma* sp). Ecotoxicol. Environ. Saf. 106, 181–187.

Sioli, H., 1984. The Amazon and its main affluents: hydrogeography, morphology of the river courses and river types. In: Sioli, H. (Ed.), The Amazon. Junk Publishers, Dordrecht, Limnology and landscape ecology of a mighty tropical river and its basin. Dr. W, pp. 127–165.

Sloman, K.A., Wood, C.M., Scott, G.R., Wood, S., Kajimura, M., Johannsson, O.E., Almeida-Val, V.M.F., Val, A.L., 2006. Tribute to R. G. Boutilier: the effect of size on the physiological and behavioural responses of oscar, *Astronotus ocellatus*, to hypoxia. J. Exp. Biol. 209, 1197–1205.

Sloman, K.A., Sloman, R., De Boeck, G., Scott, G.R., Iftikar, F.I., Wood, C.M., Almeida-Val, V.M.F., Val, A.L., 2009. The role of size in synchronous air breathing of *Hoplosternum littorale*. Physiol. Biochem. Zool. 82, 625–634.

Soares-Filho, B., Rajão, R., 2018. Traditional conservation strategies still the best option. Nat. Sustain. 1, 608–610.

Solomon, C.T., Jones, S.E., Weidel, B.C., Buffam, I., Fork, M.L., Karlsson, J., Larsen, S., Lennon, J.T., Read, J.S., Sadro, S., Saros, J.E., 2015. Ecosystem consequences of changing inputs of terrestrial dissolved organic matter to lakes: current knowledge and future challenges. Ecosystems 18, 376–389.

Sorribas, M.V., Paiva, R.C.D., Melack, J.M., Bravo, J.M., Jones, C., Carvalho, L., Beighley, E., Forsberg, B., Costa, M.H., 2016. Projections of climate change effects on discharge and inundation in the Amazon basin. Clim. Change 136, 555–570.

Soumis, N., Lucotte, M., Sampaio, D., Almeida, D.C., Giroux, D., Morais, S., Pichet, P., 2003. Presence of organophosphate insecticides in fish of the Amazon River. Acta Amazon. 33, 325–338.

Souza, S.S., Machado, R.N., Costa, J.C., Campos, D.F., Silva, G.S., Almeida-Val, V.M.F., 2020a. Severe damages caused by malathion exposure in *Colossoma macropomum*. Ecotoxicol. Environ. Saf. 205, (P111340).

Souza, T.C., Silva, S.L.R., Marcon, J.L., Waichman, A.V., 2020b. Acute toxicity of deltamethrin to Amazonian freshwater fish. Toxicol. Environ. Health Sci. 12, 149–155.

Souza, S.S., Castro, J.S., Campos, D.F., Pereira, R.S., Bataglion, G.A., Silva, G.S., Almeida-Val, V.M.F., 2021. Temporal exposure to malathion: biochemical changes in the Amazonian fish tambaqui, *Colossoma macropomum*. Aquat. Toxicol. 241, 105997.

Stanley, E.M., Powers, S.M., Lottig, N.R., Buffam, I., Crawford, J.T., 2012. Contemporary changes in dissolved organic carbon (DOC) in human-dominated rivers: is there a role for DOC management? Freshw. Biol. 57, 26–42.

Stedmon, C.A., Markager, S., 2005. Resolving the variability in dissolved organic matter fluorescence in a temperate estuary and its catchment using PARAFAC analysis. Limnol. Oceanogr. 50, 686–697.

Steinberg, C.E.W., Kamara, S., Prokhotskaya, V.Y., Manusadzianas, L., Karasyova, T.A., Timofeyev, M.A., Jie, Z., Paul, A., Meinelt, T., Farjalla, V.F., Matsuo, A.Y.O., Burnison, B.K., Menzel, R., 2006. Dissolved humic substances – ecological driving forces from the individual to the ecosystem level? Freshw. Biol. 51, 1189–1210.

Sunday, J.M., Bates, A.E., Dulvy, N.K., 2011. Global analysis of thermal tolerance and latitude in ectotherms. Proc. Royal Soc. B Biol Sci. 278, 1823–1830.

Sunday, J., Bates, A., Dulvy, N., 2012. Thermal tolerance and the global redistribution of animals. Nat. Clim. Change 2, 686–690.

Sunday, J.M., Bates, A.E., Kearney, M.R., Colwell, R.K., Dulvy, N.K., Longino, J.T., Huey, R.B., 2014. Thermal-safety margins and the necessity of thermoregulatory behavior across latitude and elevation. Proc. Natl. Acad. Sci. U. S. A. 111, 5610–5615.

Sutherland, A.B., 2003. Effects of excessive sedimentation on the growth and stress response of whitetail shiner (*Cyprinella galactura*) juveniles. In: Hatcher, K.J. (Ed.), Georgia Water Resources Conference. University of Georgia, Institute of Ecology, Atlanta, Georgia, p. 4.

Sutherland, A.B., Meyer, J.L., 2007. Effects of increased suspended sediment on growth rate and gill condition of two southern Appalachian minnows. Environ. Biol. Fishes 80, 389–403.

Tencatt, L., Muriel-Cunha, J., Zuanon, J., Ferreira, M., Britto, M., 2020. A journey through the Amazon middle earth reveals *Aspidoras azaghal* (Siluriformes: Callichthyidae), a new species of armored catfish from the rio Xingu basin, Brazil. J. Fish Biol.

Tewksbury, J.J., Huey, R.B., Deutsch, C.A., 2008. Putting the heat on tropical animals. Science 320, 1296–1297.

Thurman, E., 1985a. Organic Geochemistry of Natural Waters. W Junk Publishers, Dordrecht, Boston, USA, Martinus Nijhof/Dr.

Thurman, E.M., 1985b. Aquatic humic substances. In: Thurman, E.M. (Ed.), Organic Geochemistry of Natural Waters. Dr. W Junk, Dordrecht, pp. 273–361.

Timpe, K., Kaplan, D., 2017. The changing hydrology of a dammed Amazon. Sci. Adv. 3, e1700611.

Torres, J.P.M., Pfeiffer, W.C., Markowitz, S., Pause, R., Malm, O., Japenga, J., 2002. Dichlorodiphenyltrichloroethane in soil, river sediment, and fish in the Amazon in Brazil. Environ. Res. 139, 134–139.

Val, A.L., 1995. Oxygen transfer in fish: morphological and molecular adjustments. Braz. J. Med. Biol. Res. 28, 1119–1127.

Val, A.L., 1996. Surviving low oxygen levels: lessons from fishes of the Amazon. In: Val, A.L., Almeida-Val, V.M.F., Randall, D.J. (Eds.), Physiology and Biochemistry of the Fishes of the Amazon. INPA, Manaus, pp. 59–73.

Val, A.L., 2000. Organic phosphates in the red blood cells of fish. Comp. Biochem. Physiol. 125A, 417–435.

Val, A.L., Almeida-Val, V.M.F., 1995. Fishes of the Amazon and their Environments. Springer Verlag, Heidelberg, Physiological and biochemical features.

Val, A.L., Almeida-Val, V.M.F., 1999. Effects of crude oil on respiratory aspects of some fish species of the Amazon. In: Val, A.L., Almeida-Val, V.M.F. (Eds.), Biology of Tropical Fish. INPA, Manaus, pp. 277–291.

Val, A.L., Oliveira, A.M., 2021. *Colossoma macropomum*—a tropical fish model for biology and aquaculture. J. Exp. Zool. A 2021, 1–10.

Val, A.L., Wood, C.M., 2022. Global change and physiological challenges for fish of the Amazon today and in the near future. J. Exp. Biol.

Val, A.L., Silva, M.N.P., Almeida-Val, V.M.F., 1998. Hypoxia adaptation in fish of the Amazon: a never-ending task. S. Afr. J. Zool. 33, 107–114.

Val, A., Almeida-Val, V., Randall, D., 2006. Tropical Environment. In: Val, A.L., Almeida-Val, V.M.F., Randall, D.J. (Eds.), The Physiology of Tropical Fishes. Elsevier/Academic Press, San Diego, pp. 1–45.

Val, A.L., Gomes, K.R.M., Almeida-Val, V.M.F., 2015. Rapid regulation of blood parameters under acute hypoxia in the Amazonian fish *Prochilodus nigricans*. Comp. Biochem. Physiol. 184A, 125–131.

Val, P., Figueiredo, J., Melo, G., Flantua, S.G.A., Quesada, C.A., Fan, Y., Albert, J.S., Guayasamin, J.M., Hoorn, C., 2021. Geology and geodiversity of the Amazon: three billion years of history. In: Nobre, C., Encalada, A., Anderson, E., Alcazar, F.H.R., Bustamante, M., Mena, C., Zapata-Ríos, G. (Eds.), Science Panel for the Amazon. Part I. the Amazon as a Regional Entity of the Earth System. Unites Nations Sustainable Development Solutions Network, New York, US, pp. 1–37.

Valenzano, D.R., Terzibasi, E., Cattaneo, A., Domenici, L., Cellerino, A., 2006. Temperature affects longevity and age-related locomotor and cognitive decay in the short-lived fish *Nothobranchius furzeri*. Aging Cell 5, 275–278.

Valiente-Banuet, A., Aizen, M.A., Alcántara, J.M., Arroyo, J., Cocucci, A., Galetti, M., García, M.B., García, D., Gómez, J.M., Jordano, P., Medel, R., Navarro, L., Obeso, J.R., Oviedo, R., Ramírez, N., Rey, P.J., Traveset, A., Verdú, M., Zamora, R., 2015. Beyond species loss: the extinction of ecological interactions in a changing world. Funct. Ecol. 29, 299–307.

Valouev, A., Ichikawa, J., Tonthat, T., Stuart, J., Ranade, S., Peckham, H., Zeng, K., Malek, J.A., Costa, G., McKernan, K., Sidow, A., Fire, A., Johnson, S.M., 2008. A high-resolution, nucleosome position map of *C. elegans* reveals a lack of universal sequence-dictated positioning. Genome Res. 18, 1051–1063.

Vigneault, B., Percot, A., Lafleur, M., Campbell, P.G.C., 2000. Permeability changes in model and phytoplankton membranes in the presence of aquatic humic substances. Environ. Sci. Tech. 34, 3907–3913.

Vonhof, H., Kaandorp, R., 2010. Climate variation in Amazonia during the Neogene and the quaternary. In: Hoorn, C., Wesselingh, F.P. (Eds.), Amazonia: Landscape and Species Evolution. A look into the past, Wiley-Blackwell, Oxford, p. 464p.

Waichman, A.V., 2008. A proposal for integrated risk assessment of pesticides use in Amazon State, Brazil. Acta Amazon. 38, 45–50.

Waichman, A.V., Römbke, J., Ribeiro, M.O.A., Nina, N.C.S., 2002. Use and fate of pesticides in the Amazon State, Brazil. Environ. Sci. Pollut. Res. 9, 423–428.

Walker, I., 1995. Amazonian streams and small rivers. In: Tundisi, J.G., Bicudo, C.E., Matsumura-Tundisi, T. (Eds.), Limnology in Brazil. Academia Brasileira de Ciências, Rio de Janeiro, pp. 167–193.

Walker, I., Henderson, P.A., 1996. Ecophysiological aspects of Amazonian Blackwater litterbank fish communities. In: Val, A.L., Almeida-Val, V.M.F., Randall, D.J. (Eds.), Physiology and Biochemistry of the Fishes of the Amazon. INPA, Manaus, AM, pp. 7–30.

Weber, R., Fago, A., Val, A.L., Bang, A., Van Hauwaeert, M.L., De Wilde, S., Zal, F., Moens, L., 2000. Isohemoglobin differentiation in the biomodal-breathing Amazon catfish *Hoplosternum littorale*. J. Biol. Chem. 275, 17297–17305.

Wenger, A.S., Johansen, J., Jones, G., 2011. Suspended sediment impairs habitat choice and chemosensory discrimination in two coral reef fishes. Coral Reefs 30, 879–887.

Wenger, A.S., Johansen, J.L., Jones, G.P., 2012. Increasing suspended sediment reduces foraging, growth and condition of a planktivorous damselfish. J. Exp. Mar. Biol. Ecol. 428, 43–48.

Wenger, A., McCormick, M., McLeod, I., Jones, G., 2013. Suspended sediment alters predator–prey interactions between two coral reef fishes. Coral Reefs 32, 369–374.

Wesselingh, F., Hoorn, C., Kroonenberg, S., Antonelli, A., Lundberg, J., Vonhof, H., Hooghiemstra, H., 2010. On the origin of Amazonian landscapes and biodiversity: a synthesis. In: Hoorn, C., Wesselingh, F.P. (Eds.), Amazonia: Landscape and Species Evolution. A look into the past, Wiley-Blackwell, Oxford, pp. 421–432.

Wetzel, R.G., 2001. Limnology. Lake and River Ecosystems, third ed. Academic Press, SAn Diego, CA.

Wilson, R.W., Wood, C.M., Gonzalez, R.J., Patrick, M.L., Bergman, H.L., Narahara, A., Val, A.L., 1999. Ion and acid-base balance in three species of Amazonian fish during gradual acidification of extremely soft water. Physiol. Biochem. Zool. 72, 277–285.

Wittmann, A.O., Piedade, M.T.F., Parolin, P., Wittmann, F., 2007. Germination in four low-várzea tree species of Central Amazonia. Aquat. Bot. 86, 197–203.

Wood, C.M., 1989. The physiological problems of fish in acid water. In: Taylor, E.W., Brow, D.J.A., Brow, J.A. (Eds.), Acid Toxicity and Aquatic Animals. Cambridge University Press, Cambridge, pp. 125–152.

Wood, C.M., 1992. Flux measurements as indices of $H^+$ and metal effects on freshwater fish. Aquat. Toxicol. 22, 239–264.

Wood, C.M., Wilson, R.W., Gonzalez, R.J., Patrick, M.L., Bergman, H.L., Narahara, A., Val, A.L., 1998. Responses of an Amazonian teleost, the tambaqui (*Colossoma macropomum*) to low pH in extremely soft water. Physiol. Zool. 71, 658–670.

Wood, C.M., Matsuo, A.Y., Gonzalez, R.J., Wilson, R.W., Patrick, M.L., Val, A.L., 2002. Mechanisms of ion transport in *Potamotrygon*, a stenohaline freshwater elasmobranch native to the ion-poor blackwaters of the Rio Negro. J. Exp. Biol. 205, 3039–3054.

Wood, C.M., Matsuo, A.Y., Wilson, R.W., Gonzalez, R.J., Patrick, M.L., Playle, R.C., Val, A.L., 2003. Protection by natural Blackwater against disturbances in ion fluxes caused by low pH exposure in freshwater stingrays endemic to the Rio Negro. Physiol. Biochem. Zool. 76, 12–27.

Wood, C.M., Al-Reasi, H.A., Smith, S., 2011. The two faces of DOC. Aquat. Toxicol. 105S, 3–8.

Wood, C.M., Pelster, B., Giacomin, M., Sadauskas-Henrique, H., Almeida-Val, V.M.F., Val, A.L., 2016. The transition from water-breathing to air-breathing is associated with a shift in ion uptake from gills to gut: a study of two closely related erythrinid teleosts, Hoplerythinus unitaeniatus and *Hoplias malabaricus*. J. Comp. Physiol. B 186, 431–445.

Wood, C.M., Gonzalez, R.J., Ferreira, M.S., Braz-Mota, S., Val, A.L., 2018. The physiology of the tambaqui (*Colossoma macropomum*) at pH 8.0. J. Comp. Physiol. B 188, 393–408.

Wood, C.M., 2022. Conservation aspects of osmotic, acid-base, and nitrogen homeostasis in fish. Fish Physiol. 39A.

Ximenes, A.M., Bittencourt, P.S., Machado, V.N., Hrbek, T., Farias, I.P., 2021. Mapping the hidden diversity of the *Geophagus* sensu stricto species group (Cichlidae: Geophagini) from the Amazon basin. PeerJ 9, e12443.

Yanar, M., Erdoğan, E., Kumlu, M., 2019. Thermal tolerance of thirteen popular ornamental fish species. Aquaculture 501, 382–386.

Yu, H., Chin, M., Yuan, Y., Bian, H., Remer, L.A., Prospero, J.M., Omar, A., Winker, D., Yang, Y., Zhang, Y., Zhang, Z., Zhao, C., 2015. The fertilizing role of African dust in the Amazon rainforest- a first multiyear assessment based on data from cloud-aerosol lidar and infrared path under satellite observations. Geophys. Res. Lett. 42, 1984–1991.

Zara, L.F., Rosa, A.H., Toscano, I.A.S., Rocha, J.C., 2006. A structural conformation study of aquatic humic acid. J. Braz. Chem. Soc. 17, 1014–1019.

Zarfl, C., Lumsdon, A.R., Berlekamp, J., Tydecks, L., Tockner, K., 2015. A global boom in hydropower dam construction. Aquat. Sci. 77, 161–170.

Zuanon, J., Sawakuchi, A., Camargo, M., Wahnfried, I., Sousa, L., Akama, A., Muriel-Cunha, J., Ribas, C., D'Horta, F., Pereira, T., Lopes, P., Mantovanelli, T., Lima, T.S., Garzón, B., Carneiro, C., Reis, C.P., Rocha, G., Santos, A.L., De Paula, E.M., Pennino, M., Pezzuti, J., 2019. Condições para a manutenção da dinâmica sazonal de inundação, a conservação do ecossistema aquático e manutenção dos modos de vida dos povos da volta grande do Xingu. Cadernos do Nucleo de Altos Estudos da Amazônia 28, 21–62.

Chapter 6

# Fish response to environmental stressors in the Lake Victoria Basin ecoregion

Lauren J. Chapman[a,*,†], Elizabeth A. Nyboer[b,†], and Vincent Fugère[c]
[a]*Department of Biology, McGill University, Montréal, QC, Canada*
[b]*Department of Biology, Carleton University, Carleton Technology and Training Centre, Ottawa, ON, Canada*
[c]*Département des Sciences de l'Environnement, Université du Québec à Trois-Rivières, Trois-Rivières, QC, Canada*
[*]*Corresponding author: e-mail: lauren.chapman@mcgill.ca*

## Chapter Outline



[†]Co-first authors.

Fish Physiology, Vol. 39B. https://doi.org/10.1016/bs.fp.2022.04.010

Freshwater organisms face multiple threats associated with habitat degradation, pollution, and eutrophication, in addition to overharvesting and species invasions. Furthermore, there is mounting evidence that freshwaters are highly sensitive to climate change. This chapter provides an overview of contemporary environmental changes in inland waters of the Lake Victoria Basin (LVB) ecoregion of East Africa with a focus on climate change, eutrophication, and land use. Case studies of fishes in the Lake Victoria basin and swamp-river systems of Western Uganda are used to explore potential effects of these stressors on morpho-physiological, performance, and fitness-related traits. Overall, fishes in the LVB ecoregion show acclimation capacity in upper thermal tolerance and aerobic performance, and adaptive plasticity in traits related to hypoxia tolerance (e.g., gill size). However, a trait-based climate change vulnerability assessment revealed that over 70% of LVB ecoregion fishes are vulnerable to climate change; and fish kills associated with turnover events or anoxic upwellings highlight the danger of rapid change in dissolved oxygen for some species. Plasticity may allow some fishes to persist in the face of multiple stressors in the LVB ecoregion. However, there may be consequences for fitness-related traits such as body size that could affect demographic stability and contributions of fish to food security.

## 1 Introduction

Freshwater habitats occupy only ~1% of the earth surface, yet provide critical habitat for 50.5% of known fish species (18,200 species, Fricke et al., 2021). The most diverse freshwater fish communities are found in tropical regions such as the Amazon, Mekong, and Congo Rivers, and the Great Lakes of East Africa. Freshwaters also contribute significantly to food security, with freshwater capture fisheries making up around 13% of the world's annual fish catch (FAO, 2020); and they provide the primary source of animal protein and essential micronutrients to human communities, particularly in Asia and Africa where commercial inland fishery production is highest (Funge-Smith and Bennett, 2019). Of course, freshwaters also provide many other ecosystem services. Unfortunately, both the biodiversity and services provided by freshwaters are in peril. Freshwater systems have experienced and continue to face multiple anthropogenic impacts including habitat degradation and loss, dams and water diversions, pollutants and contaminants, increase in frequency and extent of hypoxia (low dissolved oxygen), overfishing, and introductions of non-native species, which have led to imperilment of many freshwater fishes (Arthington et al., 2016; Strayer and Dudgeon, 2010; Vörösmarty et al., 2010). In addition, there is increasing evidence that freshwaters and their species are highly sensitive to climate change (Heino et al., 2009; Knouft and Ficklin, 2017; Nyboer et al., 2021; Reid et al., 2019).

The lakes of East Africa are well known for their diversity of fishes, in particular the adaptive radiations of cichlid fishes, and for their contribution to inland fisheries. Lake Victoria, the largest water body in East Africa and the largest tropical lake in the world, is Africa's most important inland fishery and once hosted 500+ species of endemic cichlids. However, this system has undergone massive and fundamental changes in its ecology over the past century that have led to dramatic changes in its fish fauna, its fisheries, and water quality. Here, we provide an overview of environmental change in the inland waters of the Lake Victoria Basin Ecoregion of East Africa with a focus on climate change, eutrophication, and land use. Case studies are used to explore potential effects of these stressors on morpho-physiological, performance, and fitness-related traits in fishes of the ecoregion.

## 2 The Lake Victoria Basin ecoregion of East Africa

The Lake Victoria Basin (LVB) ecoregion includes basins of lakes Victoria, Edward, George, Kyoga, and Kivu (Thieme et al., 2005). Lake Kivu is not included in the most recent shapefile published by the Freshwater Ecoregions of the World (FEOW) that we used in our spatial assessments and maps (https://www.feow.org/ecoregions/details/521, accessed February 2, 2022, Fig. 1). However, we retain it in our discussion of the LVB ecoregion, consistent with the original description and with evidence that Lake Kivu formerly drained into Lake Edward (Oyugi, 2019). The LVB ecoregion crosses national boundaries of six countries [Uganda (45.2% of LVB ecoregion), Kenya (13.1%), Tanzania (31.0%), Burundi (3.5%), Rwanda (5.3%), Democratic Republic of Congo (1.9%), Fig. 1, not including Lake Kivu]. It is an ecoregion with a high diversity of inland aquatic systems from small crater lakes, to enormous valley and lakeside wetlands, and large rift valley lakes that feed river systems including the Nile River. The ecoregion has an equatorial climate with two rainy seasons; April/May, and October/November.

Lake Victoria is the largest tropical lake in the world (68,800 $km^2$) with its waters shared between Tanzania (51% of the lake area), Uganda (43%), and Kenya (6%, Fig. 1) and a catchment of over 193,000 $km^2$ that reaches into Rwanda and Burundi (Hughes and Hughes, 1992). The lake (maximum depth of 79 m, mean depth of 40 m) lies in a shallow depression at 1134 m asl, between the west and east African rifts. As a result of its shallow morphometry, the edges of Lake Victoria are often fringed with papyrus (*Cyperus papyrus*) swamps. The main inflow river is the Kagera River that drains highlands of Rwanda and Burundi; however, the water balance of the lake is driven primarily by rainfall and evaporation (Cohen et al., 1996; Spigel and Coulter, 1996). Lake Victoria discharges into the Nile River passing through the Owen Falls dam near Jinja, Uganda and then flows through Lake Kyoga, over Murchison Falls, and into Lake Albert. The dendritic Lake Kyoga is part of a system of shallow lakes (satellites) and swamps, which contains 3416 $km^2$ of open water and shallow lakes and 2184 $km^2$ of permanently inundated swamp (Hughes and Hughes, 1992). Lakes Edward (2250 $km^2$ in area, maximum depth of 112 m) and George (290 $km^2$ in area, maximum depth of 3 m) are situated in the western portion of the ecoregion lying in a depression in the western rift valley. Lake George is fed from numerous rivers (e.g., Mbuku, Mulikwesi, Nonge, Mpanga, Dura) flowing off the Ruwenzori Mountains, or

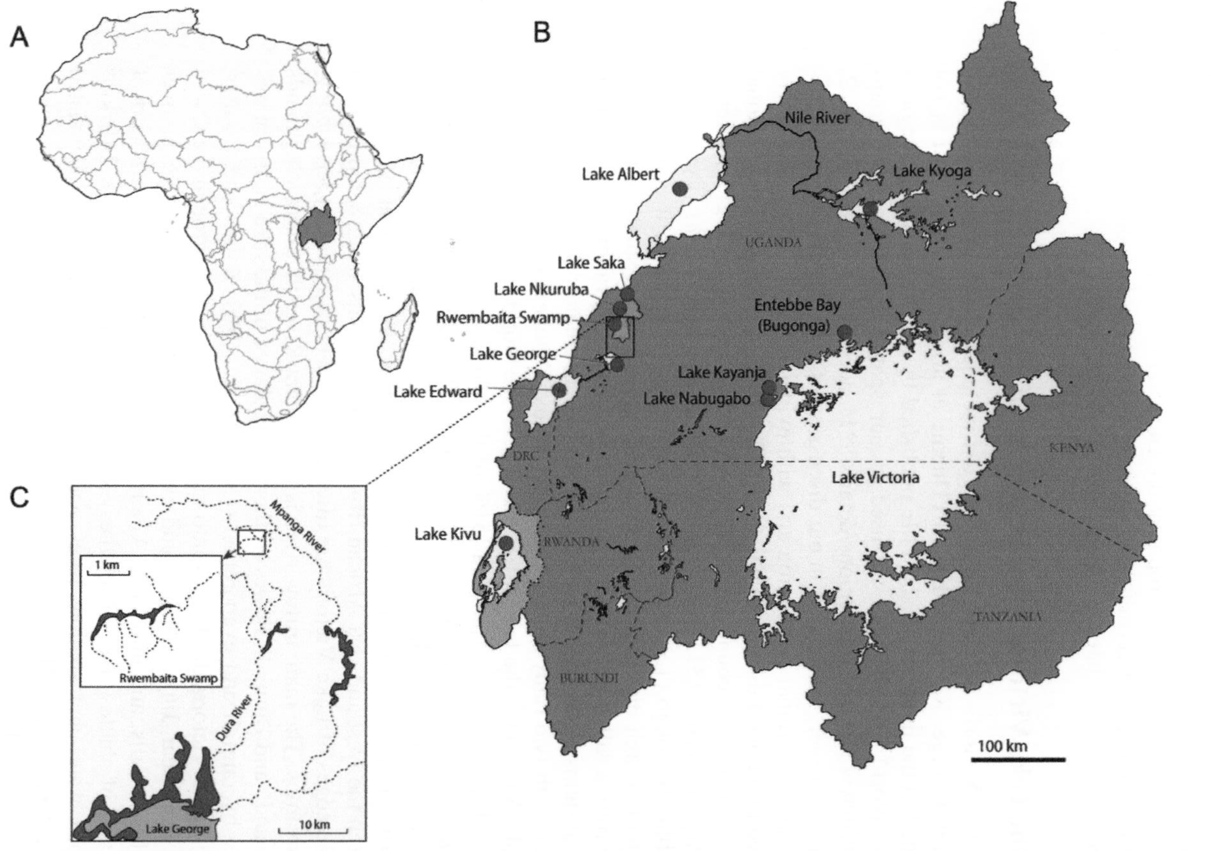


**FIG. 1** (A) Location of the Lake Victoria Basin Ecoregion within Africa; (B) a closeup of the LVB ecoregion; the darker green represents the LVB ecoregion borders as published by Freshwater Ecosystems of the World (FEOW), and the lighter green section represents the LVB ecoregion including Lake Kivu as described in the original ecoregion assessment (Thieme et al., 2005). Location of data collection sites within the LVB ecoregion are represented in red dots; and (C) A closer view of the Dura and Mpanga rivers and the Rwembaita Swamp System.

foothills thereof, filtering first through the large papyrus swamp that forms the northern part of the George watershed. Lake George flows through the 40-km long Kazinga Channel into Lake Edward (Beadle, 1981). The major inputs to Lake Edward are from rivers that drain highlands of Rwanda and the Ruwenzori Mountains, with a smaller contribution from the Kazinga Channel (Beadle, 1981). Lake Edward has one major outflow, the Semliki River, which flows north draining into Lake Albert. Lake Kivu (maximum depth > 500 m) lies on the border of Rwanda and the DRC, bordered by steep slopes. Lake Kivu drains into Lake Tanganyika via the Ruzizi River (Beadle, 1981; Worthington and Lowe-McConnell, 1994). The LVB ecoregion is also endowed with several minor lakes including satellite lakes of Victoria (e.g., Lake Kanyaboli in Kenya; Lake Nabugabo in Uganda), and more than 90 crater lakes that lie in the foothills of the Ruwenzori Mountains (Efitre et al., 2009; Melack, 1978).

Lakes Victoria, Kyoga, Edward, and George are part of the greater Nile catchment, and separated from the lower Nile catchment by Murchison Falls in Uganda. It is thought that the Lake Victoria basin began to form approximately 20 million years ago due to faulting of the Western Rift Valley and uplifting of land lying to the west of the basin (Beadle, 1981; Kaufman et al., 1996). The Katonga and Kagera Rivers previously drained the Lake Victoria region flowing towards Lake George. However, as a result of the uplift, these rivers realigned to drain to the east, backponding to become Lake Victoria. The uplifting process was slow, and Lake Victoria may have drained into Lake George well into the late Pleistocene (Beadle, 1981). Lake Victoria dried up several times during the late Pleistocene, with the last desiccation from 19,000 to 15,000 years ago (Johnson et al., 1996; Stager et al., 2009). Overall, the emerging picture is that Lake Victoria formed as a result of uplift and lake depression, but climatic fluctuations may have strongly affected the lake's fauna and flora through periods of isolation and reconnections (Kaufman et al., 1996). In the early Pleistocene (2 million years ago), there may have been connections between lakes Edward/George and Albert, but they are now separated by rapids (300 m descent). Lake Albert is not included in the LVB ecoregion because the separation has likely resulted in a fauna that is more riverine and Nilotic (Oyugi, 2019).

Africa has an estimated 3300 species of freshwater fishes (Lévêque and Paugy, 2010), which represents one of the richest ichthyofaunas on earth (Thieme et al., 2005). Africa's "living fossils" and species flocks are an important element of the continent's fish fauna. Africa has an unparalleled assemblage of archaic fish families, mostly endemic, including the African lungfishes (four species), polypterids, and mormyrids, the weakly electric fishes (Lévêque, 1997; Lundberg et al., 2000). Extraordinary species radiations of the family Cichlidae are the hallmark of the Great Lakes of East Africa including larger lakes of the LVB ecoregion (lakes Victoria, Kyoga, Edward, George), but radiations have also taken place in some of the smaller lakes, including Lake Nabugabo, a satellite of Lake Victoria (Greenwood, 1965), and some crater lakes (e.g., Lake Saka, Lemoine et al., 2018, Fig. 1). There are many examples of evolutionary convergence; cichlid species with similar morphologies and functional traits have evolved independently in various of the LVB ecoregion lakes. The adaptive radiation of cichlids in Lake Victoria is extraordinary, despite its young age, with a species flock of more than 500 endemic species of haplochromine cichlids representing many distinct ecological guilds (Meier et al., 2017; Witte et al., 2007).

The aquatic systems of the LVB ecoregion also provide critical services including drinking water, hydroelectric power, and water for irrigation, and are important to food security in the region (Chapman et al., 2008a). In Uganda alone, the fisheries sector contributes to the livelihood of nearly 5.3 million people and is a major source of critically required animal protein for about 17 million Ugandans (MAAIF, 2017). Lake Victoria is Africa's largest inland fishery; but the lake and its fauna have experienced dramatic ecological changes associated with intense harvesting of fish, eutrophication, and introduction of non-native fishes in the 1950s and 60s, including the predatory Nile perch, *Lates niloticus*, and four tilapiine species (see reviews by Balirwa et al., 2003; Chapman et al., 2008a; Kaufman, 1992; Taabu-Munyaho et al., 2016). Although populations of many fish species had declined due to harvesting prior to establishment of Nile perch, the increase in Nile perch abundance in the 1980's coincided with the further decline or loss of around 40% of the 500+ species of endemic haplochromine cichlids, and other non-cichlid fish species (Balirwa et al., 2003; Chapman et al., 2008a; Kaufman, 1992; Witte et al., 1992). In contrast, the small pelagic cyprinid *Rastrineobola argentea* increased dramatically in biomass in the 1980's (Sharpe and Chapman, 2014, 2018; Wanink, 1999). Similar changes in fish assemblages were observed in other systems in the LVB ecoregion where Nile perch were introduced, including Lake Nabugabo (Chapman et al., 1996; Ogutu-Ohwayo, 1993, 1994) and Lake Kyoga (Kaufman and Ochumba, 1993; Ogutu-Ohwayo, 1994; Schwartz et al., 2006). The rapid expansion of the Nile perch population in the Lake Victoria basin contributed to the development of a Nile perch fishery and export industry worth about 350 million USD annually (Mkumbo and Marshall, 2015). Over the last few decades, declines in Nile perch catches have coincided with a resurgence of some native fishes, in particular haplochromine cichlids in lakes Victoria and Nabugabo, which has been attributed, in part, to decreased predation pressure on native species by Nile perch (Chapman et al., 2003; Kishe-Machumu et al., 2015; Mkumbo and Marshall, 2015; Paterson and Chapman, 2009; Taabu-Munyaho et al., 2016; Witte et al., 2000).

Increasing demands on fisheries of the LVB ecoregion and cascading effects on aquatic communities need to be considered in light of other environmental changes in the ecoregion including progressive deforestation and expansion of agriculture, exponentially increasing human populations, and climate change. In the following sections, we first explore patterns of environmental change in the Lake Victoria Ecoregion with a focus on climate change, eutrophication-induced hypoxia, and agricultural expansion/deforestation. For each of these perturbations, we review a series of studies exploring the response of fishes to environmental change in the LVB ecoregion.

## 3 Effects of climate change on freshwater ecosystems of the Lake Victoria Basin ecoregion

### 3.1 Biophysical changes to freshwater ecosystems

Evidence from rivers and lakes across the African continent suggests that small climatic variations can have relatively large effects on freshwater ecosystem function (Ndebele-Murisa, 2014; Ogutu-Ohwayo et al., 2016; Olaka et al., 2010; O'Reilly et al., 2003; Saulnier-Talbot et al., 2014). In the LVB ecoregion, climate change has altered the availability and quality of freshwaters through changes to the global hydrological cycle, including changes in the timing, frequency, and intensity of precipitation events, increasing evaporation rates, and numerous drought and flood events (Harrod et al., 2018; Ogutu-Ohwayo et al., 2016).

In Africa, the rate of surface air temperature warming has increased from an average of 0.03 °C per decade from 1961–1990 to 0.25 °C per decade from 1991–2020, and average temperatures in 2020 are approximately 0.45–0.85 °C above the 1981–2010 average (World Meteorological Organization, 2021). Recent modeling has shown that global surface water temperature has increased at an average rate of 0.13 °C per decade from 1979 to 2018 (Woolway and Maberly, 2020). In Fig. 2A we present the outcome of this modeling for the LVB ecoregion, which shows an increase in surface water temperature of 0.01–0.15 °C per decade (data provided by Woolway and Maberly). The modeled expectations align well with in situ data on water temperature in several lakes within the LVB ecoregion, including Lake Edward (WWF, 2006), Lake Kivu (Katsev et al., 2014; Larke et al., 2004), and Lake Victoria (Marshall et al., 2013; Sitoki et al., 2010). Ogutu-Ohwayo et al. (2020) presented mean annual air temperature and summer lake water temperature for several lakes of Africa between 1985 and 2010, based on a global database of summer surface water temperatures (Sharma et al., 2015). In Fig. 2B, we present a subset of these data for three lakes within the LVB ecoregion (Victoria, Edward, Kivu) as well as Lake Albert, the majority of which is not included in the LVB ecoregion but which is fed by the Semliki and Nile Rivers. In general, temperatures in these lakes have been rising coincident with regional climate warming, although there is notable variation between years and no upward temperature trend for Lake Kivu. Sitoki et al. (2010) compared recent water temperature data for Lake Victoria (2000 and 2009) to data collected in 1927. Surface waters of Lake Victoria warmed by almost 1.2 °C between 1927 and 2009 and 1.57 °C in waters >50 m deep. However, the rate of increase was much faster between 2000 and 2009 (0.079 °C per year) at the surface and at >50 m (0.094 °C per year) than between 1927

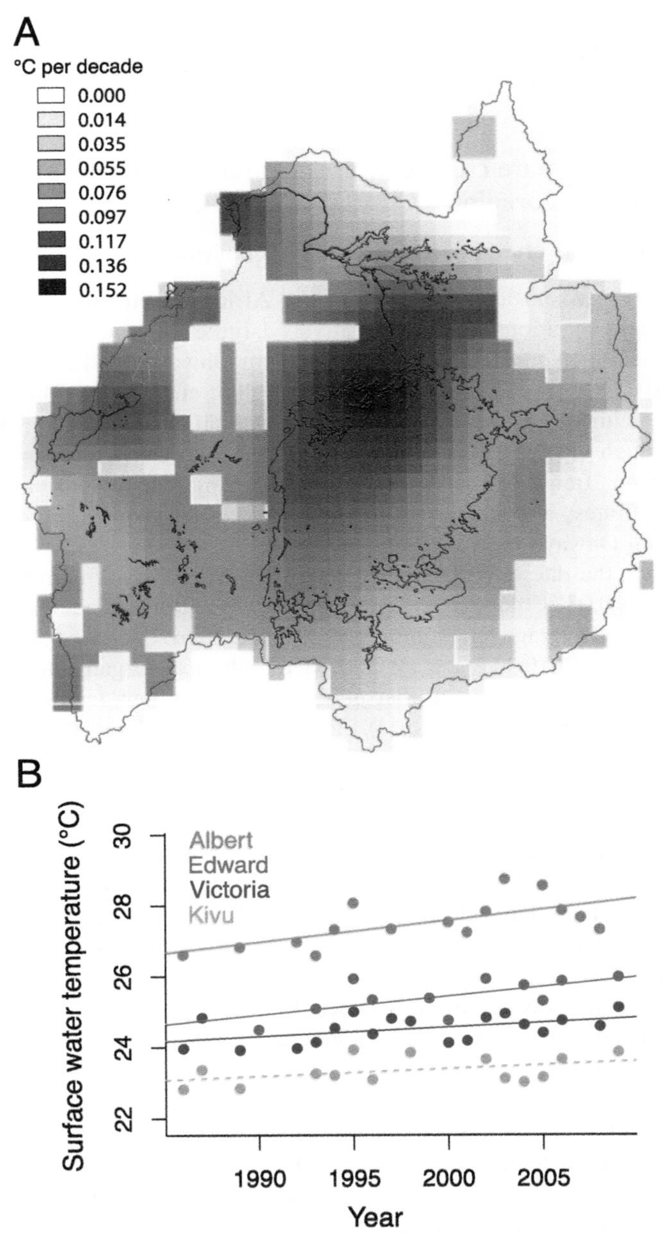


**FIG. 2** (A) Observed rate of temperature change (°C per decade) in the surface of standing waters from 1979 to 2018 as simulated by the freshwater lake (FLake) model. Annual surface water temperatures are mostly increasing across the Lake Victoria Ecoregion at a rate of 0.01–0.15 °C per decade; (B) Surface water temperature (°C) for lakes Victoria, Edward, Kivu, and Albert between 1985 and 2010, based on a global database of summer surface water temperatures (Sharma et al., 2015). Solid lines indicate significant linear trends based on linear regression; the dashed line is not significant. *Panel (A): Data from Woolway, R.I., Maberly, S.C., 2020. Climate velocity in inland standing waters. Nat. Clim. Change 10, 1124–1129. doi:10.038/s41558-020-0889-7.*

and 2000 (0.005 °C at the surface and 0.008 °C per year at >50 m, Sitoki et al., 2010). In Crater Lake Nkuruba, a small (3 ha, 38 m maximum depth) lake in western Uganda, Saulnier-Talbot et al. (2014) analyzed water temperature data between 1992 and 2012 and found that surface water temperature consistently showed a positive anomaly from 1992 to 2012, with a modest warming of air temperature (0.9 °C increase over 20 years). We extended the timeline of the Lake Nkuruba analyses from 1992 to 2019 using data from the same long-term sampling program and found no overall directional increase in water temperature (Fig. 3A). Surface water temperature showed an increasing

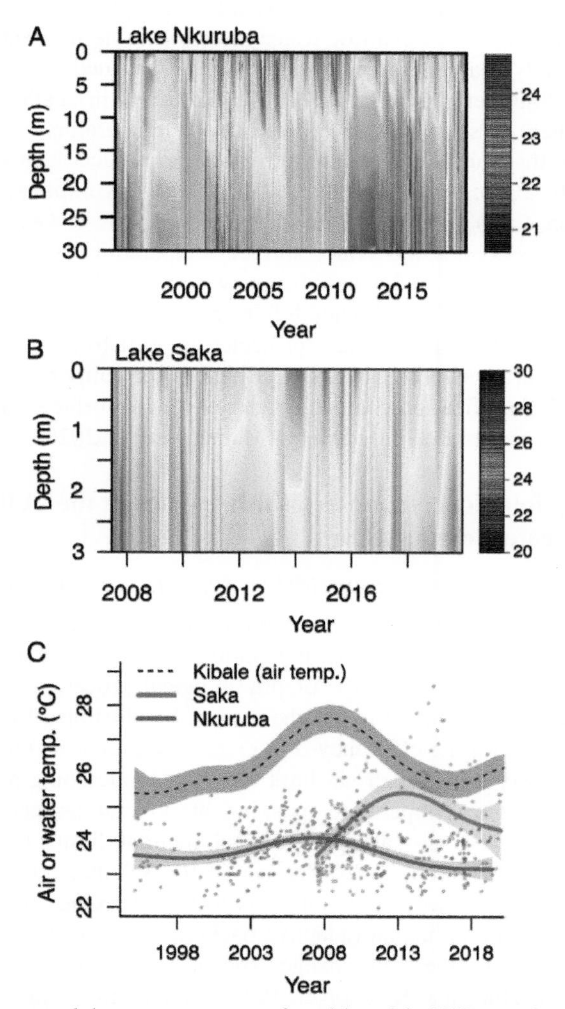


**FIG. 3** Long-term trends in water temperature of two lakes of the LVB ecoregion; (A) temperature profile over time in Lake Nkuruba and (B) Lake Saka. (C) Results of GAM models predicting surface water temperature over time. Air temperature recorded at the Makerere University Biological Field Station, in Kibale National Park (Uganda), is shown for illustrative purposes.

trend between 1992 and 2008, followed by a decreasing trend to 2017, a pattern generally reflective of maximum air temperature collected in nearby Kibale National Park (Fig. 3C). In Lake Saka, a second small (0.64 $km^2$) water body in the crater lakes region that we have monitored (mean depth of approximately 4 m), there is no consistent upward trend in water temperature between 2007 and 2020 (Fig. 3B).

While alterations to lake stratification patterns have been linked to climate change, patterns differ depending on lake position and geomorphology. In deeper lakes (e.g., Lake Nkuruba) increasing surface water temperature combined with decreasing wind speed increased the strength of lake stratification and reduced the mixing of shallow water layers with the nutrient rich deeper waters, ultimately leading to reduced biological productivity in the photo-trophic zone (Saulnier-Talbot et al., 2014). However, in shallow lakes such as Victoria, there is some evidence that bottom temperatures are increasing at a faster rate than surface temperature, leading to reduced density stratification (Marshall, 2012; Marshall et al., 2013). This has implications for dissolved oxygen (DO) availability in deep and shallow lakes in the LVB ecoregion (discussed below). Water surface temperatures are also projected to experience extended (17–250 day) heatwaves of 2–4 °C above average high lake temperatures under RCP 2.6 and 8.5, respectively (Woolway et al., 2020) with all modeled East African lakes projected to shift from a "southern hot" classification characterized by average surface water temperatures ~24 °C to a 'tropical hot' classification characterized by average surface water temperatures ~29 °C by the end of the century (Maberly et al., 2020).

## 3.2 Ecophysiological responses of fish species in the LVB ecoregion to elevated water temperature

### 3.2.1 General effects of elevated water temperature on fishes

Shifts in freshwater thermal regimes (warming and extreme temperature events) may be especially important for aquatic ectotherms, such as fishes, for which temperature controls a wide range of physiological processes. Responses of biological rates to temperature in ectotherms can be illustrated by thermal performance curves, with the efficiency of a given physiological function starting at zero at the lower critical thermal limit ($T_{crit,min}$), undergoing a rapid (exponential) increase with temperature, peaking at optimal temperatures ($T_{opt}$), and then declining to zero at upper critical thermal limits ($T_{crit,max}$) (Fry, 1947; Huey and Stevenson, 1979; Pörtner, 2010; Schulte, 2015). The shape of the thermal performance curve is thought to be determined by the temperature range of the natural habitat (adaptive background), and the ability to cope physiologically with thermal change (phenotypic plasticity; Huey and Stevenson, 1979; Huey and Kingsolver, 1989; Schulte, 2015). In terms of adaptive background, fishes that encounter a wider range of temperatures in their natural habitat (e.g., temperate species) are hypothesized to be more likely to survive thermal stress events since they have intrinsic mechanisms for coping with a wide range of temperatures (Bush and Hooghiemstra, 2005; Huey and Hertz, 1984; Tewksbury et al., 2008). Warm-adapted tropical species that inhabit relatively thermo-stable environments are expected to be more sensitive to warming. This is because they are thought to reside near their upper thermal limit and are thus likely to be driven over their upper thermal optimum by even slight temperature increases, and because a 1 °C temperature increase at higher temperatures can have a proportionately greater impact on physiological rates than a 1 °C temperature increase at lower temperatures (Deutsch et al., 2008; Nilsson et al., 2009; Tewksbury et al., 2008; Wright et al., 2009). In addition, within broad-scale latitudinal patterns, there can be variations in thermal stability depending on the environment. For example, inland water bodies in the tropics are more likely to undergo thermal fluctuations compared to their marine counterparts because of their smaller volumes and changes to surrounding landscapes (e.g., deforestation). Such considerations are important for understanding how freshwater fishes might cope with changing water temperatures in the LVB ecoregion where diverse water bodies are undergoing temperature changes at different paces and in different patterns (see above), and multiple types of land use are affecting temperature regimes. Because of these broad-scale differences among species and populations across latitudes and environments, it is highly likely that genetic adaptation to thermal regimes plays an important role in modulating metabolic rates, although this remains relatively unexplored (Munday et al., 2013). Regardless of adaptive background, species that can adjust optimal temperatures or critical thermal limits through plasticity are more likely to survive extended thermal stress events (Pörtner and Farrell, 2008; Schulte, 2015). Plastic adjustments can counteract fundamental climate-induced changes to fish populations (Nilsson et al., 2009; Visser, 2008), can allow time for genetic adaptation to occur (Chevin et al., 2010), and can happen over the lifetime of the organism (developmental plasticity) or as reversible transformations in phenotypes over time ranges of days to months (acclimation or phenotypic flexibility) (Grenchik et al., 2013; Schulte, 2015; Schulte et al., 2011). Acclimation capacity (phenotypic flexibility) has been of increasing interest as a mechanism that may allow fish to tolerate greater temperature extremes and rapid changes (Morley et al., 2019). Given their thermostable backgrounds, tropical species are generally predicted to have less capacity for acclimation; however, several meta-analyses have revealed conflicting evidence for this (Comte and Olden, 2017; Gunderson and Stillman, 2015; Seebacher et al., 2015; Sunday et al., 2011).

### 3.2.2 *Fish thermal limits, metabolism, and temperature*

The upper thermal tolerance limit in fishes is often estimated as the critical thermal maximum ($CT_{max}$), which is usually defined as the temperature at which the fish loses equilibrium in response to acute temperature increase (Becker and Genoway, 1979; Lutterschmidt and Hutchison, 1997). Responses to elevated water temperature can also be quantified by measuring alterations to fishes' metabolic scope for activity, or aerobic scope (AS), which is simply the difference between the standard (lowest) and maximum (highest) metabolic rates (SMR and MMR) of fishes (Fry, 1947; Fry and Hart, 1948). Critical thermal maximum is a widely-used estimate of a species upper thermal limit, and aerobic scope is hypothesized to represent an animal's capacity for activity over and above base survival (Farrell, 2016; Fry, 1947). As observed by Fry (1947), and quantified in several subsequent studies, SMR increases exponentially with temperature, whereas MMR is predicted to increase initially, and then plateau or decline at the highest temperatures (approaching $T_{crit,\ max}$), ultimately bringing about declines in AS at high temperatures (Farrell, 2009; Frederich and Pörtner, 2000; Fry, 1947; Lannig et al., 2004; Lee et al., 2003). Essentially, this suggests that variation in AS with temperature follows the classic thermal performance curve described above (see Schulte, 2015).

The acclimation capacity of $CT_{max}$ is often estimated as the acclimation response ratio (ARR), or the rate at which upper thermal tolerance increases relative to the acclimation temperature (Claussen, 1977). Another important measure is the thermal safety margin (TSM), or the difference between the thermal limit and maximum habitat temperature (see Campos et al., 2021). Upper thermal tolerance limits do not change linearly with latitudinal changes in temperature, so TSMs are generally lower for species living in warm tropical ecoregions (e.g., Campos et al., 2021; Comte and Olden, 2017; Dahlke et al., 2020).

Over the past decade, measurements of metabolic rate (operationalized as oxygen consumption rate) have been incorporated into predictions of how climate change could affect fishes (Comte and Grenouillet, 2013; Eliason et al., 2011; Neuheimer et al., 2011; Nilsson et al., 2009; Pörtner and Farrell, 2008; Pörtner and Knust, 2007). These predictions are based on theoretical developments that relate to Fry's aerobic scope (AS) curve to whole-animal performance and fitness through a process known as oxygen- and capacity-limited thermal tolerance (OCLTT; Pörtner, 2010; Pörtner and Farrell, 2008; Pörtner and Knust, 2007; Pörtner et al., 2017). According to the OCLTT concept, declines in AS at high temperatures are predicted because the cardiorespiratory system cannot keep pace with oxygen demands in respiring tissues as increasing temperatures elevate metabolic demands (Pörtner, 2010). Mismatches between oxygen supply and demand (manifesting in reductions in AS) are therefore hypothesized to define thermal limits in teleost fishes. OCLTT further predicts that AS is closely linked to fitness-related performance traits such as growth and reproductive success, implying that declines in AS may have negative fitness consequences for aquatic ectotherms (Pörtner, 2010; Pörtner and Farrell, 2008; Pörtner and Knust, 2007). However, numerous recent studies have shown that declines in performance at high temperatures may be driven by mechanisms other than a mismatch between oxygen supply and demand (Brijs et al., 2015; Clark et al., 2013; Healy and Schulte, 2012; Wang et al., 2014), and have indicated that the maximum oxygen uptake curves of many species lack meaningful temperature optima

(Ern et al., 2016; Gräns et al., 2014; Norin et al., 2014; Nyboer and Chapman, 2017) raising questions about the broad applicability of the OCLTT framework (Jutfelt et al., 2018). Nevertheless, measuring AS in fish provides valuable information on the energetic state of an animal under specific circumstances and insight into mechanisms that may underlie responses to elevated water temperature, even if links to overall fitness and performance are tenuous (Clark et al., 2013).

### 3.2.3 *Case studies of fish species' responses to elevated water temperature in the LVB ecoregion*

A number of studies have investigated links between water temperature and eco-physiological effects in fishes of the LVB ecoregion (Chretien and Chapman, 2016; Fugère et al., 2018; Lapointe et al., 2018; McDonnell et al., 2019; McDonnell and Chapman, 2015, 2016; Nyboer et al., 2020; Nyboer and Chapman, 2017, 2018). In this section, we review current knowledge on the eco-physiological effects of elevated water temperature on fishes of the LVB ecoregion using four species: the Nile perch (*L. niloticus*), Nile tilapia *(Oreochromis niloticus)*, Ningu *(Labeo victorianus*; a critically endangered cyprinid*)*, and a small mouthbrooding haplochromine cichlid (*Pseudocrenilabrus multicolor*). Each of these four species has a unique history in the LVB ecoregion and occupies a distinct niche within the ecosystem. The introduced Nile perch is large-bodied, piscivorous fish inhabiting a wide range of habitats but preferring open water areas and selecting zones with lower temperatures and higher dissolved oxygen (DO) concentrations (Nyboer and Chapman, 2013). Nile tilapia, also introduced into lakes of the LVB ecoregion, is commonly found in the shallow waters of lakes and slow-moving rivers across tropical and sub-tropical Africa and can tolerate a wide range of temperature and DO conditions (Lapointe et al., 2018). Ningu and *P. multicolor* are both native to the LVB ecoregion. Ningu was historically potamodromous; however, habitat loss and other stressors appear to be forcing a shift to less-migratory life history strategies (Lapointe et al., 2018). Owori-Wadunde (2009) reported optimal temperature for Ningu in riverine habitat as 25–28 °C. Ningu is also part of the LVB ecoregion fishery, but catch rates have waned since the introduction of Nile perch and Nile tilapia (Balirwa et al., 2003). *P. multicolor* is widespread haplochromine cichlid inhabiting swampy areas, rivers, and lakes throughout the Nile River basin and tolerates a wide range of DO and water temperature regimes (Chapman et al., 2000, 2002, 2008b; Reardon and Chapman, 2009).

#### 3.2.3.1 Short-term exposure: Flexibility of upper thermal tolerance in LVB ecoregion fishes

Given the diversity of life history types and temperature ranges represented by our four focal species, one might expect variability in their acclimation responses to elevated water temperature. Fig. 4 presents a comparison of the

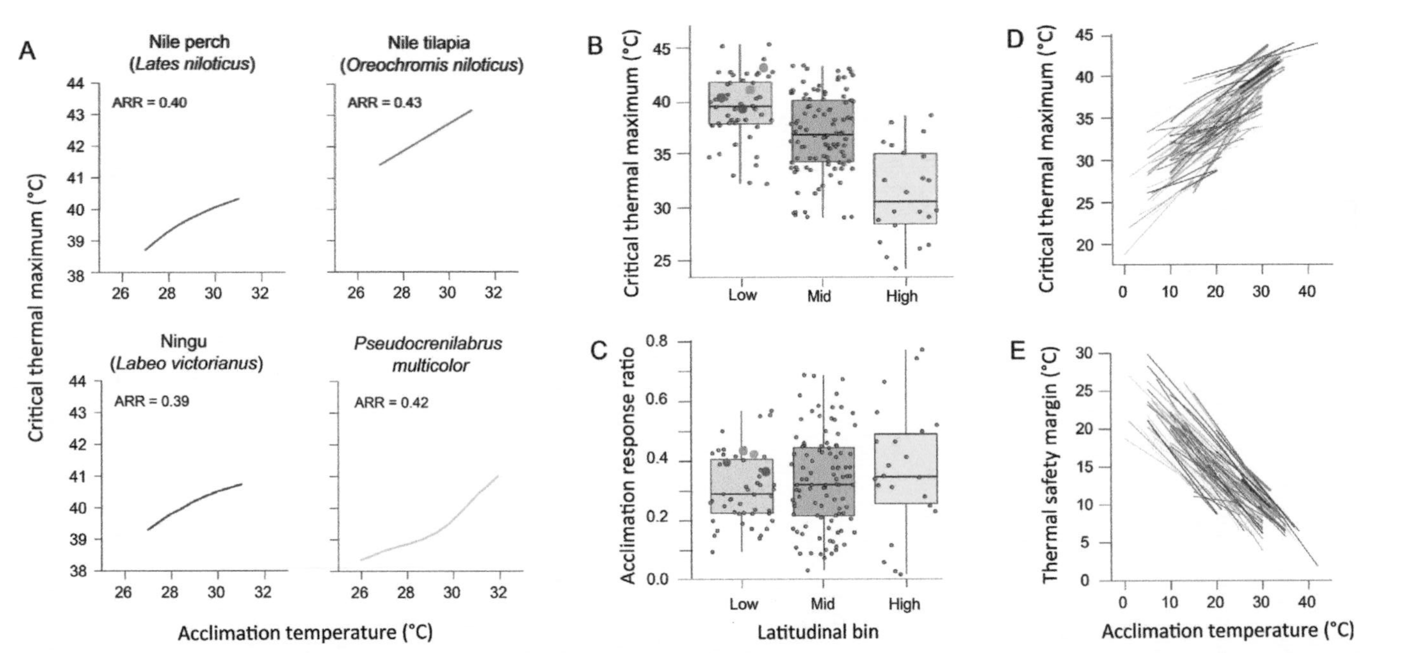


**FIG. 4** (A) Four acclimation response curves of *Lates niloticus* (Nile perch), *Oreochromis niloticus* (Nile tilapia), *Labeo victorianus* (Ningu), and *Pseudocrenilabrus multicolor* (Egyptian mouthbrooder). Critical thermal maximum increased with acclimation temperature for all species (*L. niloticus*: $F(2, 25) = 9.12$, $P = 0.022$; *O. niloticus*: $F(2, 46) = 19.01$, $P = 0.002$; *L. victorianus*: $F(2, 45) = 76.05$, $P < 0.001$; *P. multicolor*: $F(2, 15) = 158.54$, $P < 0.001$). (B) Critical thermal maximum ($CT_{max}$) plotted across latitudinal bin (low < 25° absolute latitude, mid 25°–50°, high > 50°) for freshwater fish. The four Lake Victoria species are plotted in red (*L. niloticus*), yellow (*P. multicolor*), green (*L. victorianus*), and blue (*O. niloticus*). (C) Warm acclimation capacity (i.e., acclimation response ratio, ARR) plotted against latitude for freshwater fish. (D) Acclimation response curves fitted between critical thermal maxima and the temperature of acclimation for 82 species of freshwater fish, pus four from the LVB ecoregion. Response curves for four Lake Victoria species are superimposed in color (same as above). (E) Response curves fitted between the thermal safety margin and the temperature of acclimation for 82 species of freshwater fishes, plus four from the LVB ecoregion. Response curves for four Lake Victoria species are superimposed in color (same as above). *Panel (B): Data from Comte, L., Olden, J., 2017. Climatic vulnerability of the world's freshwater and marine fishes. Nat. Clim. Change 7, 718–723; Morley, S.A., Peck, L.S., Sunday, J.M., Heiser, S., Bates, A.E., 2019. Physiological acclimation and persistence of ectothermic species under extreme heat events. Glob. Ecol. Biogeogr. 28, 1018–1037; Campos, D.R., Amanajás, R.D., Almeida-Val, V.M.F., Val, A.L., 2021. Climate vulnerability of South American freshwater fish: thermal tolerance and acclimation. J. Exp. Zool. A 335, 723–734. Panel (C): Data from Comte, L., Olden, J., 2017. Climatic vulnerability of the world's freshwater and marine fishes. Nat. Clim. Change 7, 718–723; Morley, S.A., Peck, L.S., Sunday, J.M., Heiser, S., Bates, A.E., 2019. Physiological acclimation and persistence of ectothermic species under extreme heat events. Glob. Ecol. Biogeogr. 28, 1018–1037; Campos, D.R., Amanajás, R.D., Almeida-Val, V.M.F., Val, A.L., 2021. Climate vulnerability of South American freshwater fish: thermal tolerance and acclimation. J. Exp. Zool. A 335, 723–734. Panel (D): Redrawn*

upper thermal tolerance ($CT_{max}$) and thermal plasticity (as reflected by the ARR) of the four species from experiments conducted between 2015 and 2018, where a higher value of ARR indicates higher flexibility of upper thermal tolerance (Lapointe et al., 2018; McDonnell and Chapman, 2015; Nyboer and Chapman, 2017). Fish for these studies were collected from two sites within the LVB ecoregion. Nile perch, Nile tilapia, and Ningu were collected from the Bugonga region of Lake Victoria (Fig. 1). At Bugonga, temperatures range from 24.1–30.5 °C (average 25 °C; measured between 2009 and 2015; Nyboer et al., 2020). *P. multicolor* were collected from Lake Kayanja, a small lake (1.25 $km^2$, average depth 2.6 m, Sharpe et al., 2012) in the Nabugabo region of the LVB ecoregion with water temperatures ranging from 21.6 and 31.0 °C (average 25.3 °C; McDonnell and Chapman, 2015, Fig. 1). Although experimental protocols differed slightly among studies, all experiments used short acclimation times (3–7 days) and comparable assay temperatures (26–32 °C) facilitating interspecific comparisons. All ARRs derived from $CT_{max}$ for LVB ecoregion species were slightly above the global average of $0.37 \pm 0.16$ SD, and well within the range of 0.07–0.91 (Comte and Olden, 2017). The two species in the Family Cichlidae–Nile tilapia (*O. niloticus*) and *P. multicolor*, were accustomed to the widest thermal ranges in their natural environments and showed the highest absolute $CT_{max}$ as well as the highest thermal plasticity with respective ARRs of 0.43 and 0.42 over the same approximate temperature range (i.e., 26–32 °C) (Fig. 4A). Nile perch and Ningu had lower absolute $CT_{max}$ values, and shallower respective ARRs of 0.40 and 0.39 over the same approximate temperature range (Fig. 4A). These findings show a high level of interspecific variation in thermal flexibility, which may affect persistence in the face of continued climate warming.

#### 3.2.3.2 Comparison of flexibility in upper thermal tolerance across latitudes

To determine how $CT_{max}$, ARR, and TSM of fish from the LVB ecoregion compare to (a) other tropical fish and (b) fish from other latitudes, we consulted three recent meta-analyses (i.e., Campos et al., 2021; Comte and Olden, 2017; Morley et al., 2019), each of which collated data on fish species' upper thermal limits across a range of acclimation temperatures. Although each meta-analysis had a different taxonomic focus, our final dataset included $CT_{max}$ estimates for 178 freshwater fish species (including our four species from the LVB ecoregion), with a total of 52 species from low-latitudes (0–25°), 104 species from middle latitudes (25–50°), and 22 species from high latitudes (>50°).

We split our data into three latitudinal bins to compare $CT_{max}$ and ARR of LVB ecoregion fish to fish from low, middle, and high latitudes (Fig. 4B and C). Fish species from the different latitudes were acclimated to a range of temperatures that were biologically relevant to their home range. For most species in the dataset (including the LVB species), $CT_{max}$ was determined using a standard heating rate of ~0.3 °C per minute. Previous analyses (i.e., Comte and Olden, 2017) accounted effects of heating rate on estimates of ARR and found no effect. Compared to other tropical species, LVB ecoregion fishes performed slightly above average in terms of upper critical thermal limits ($CT_{max}$) falling within the upper 50th percentile of absolute $CT_{max}$ for low latitude fish, and well above average compared to fish from all latitudes falling within the upper 35th percentile (Fig. 4B). For ARR, LVB ecoregion species were in the top 35th percentile both compared to tropical species and compared to species from all latitudes, indicating higher than average thermal plasticity (Fig. 4C). Next, following Comte and Olden (2017), we calculated thermal reaction norms of $CT_{max}$ across acclimation temperature for the LVB ecoregion species and compared them to reactions norms for 82 freshwater fish species that had a minimum of two acclimation temperatures spanning a thermal window of at least 5 °C. We followed a similar protocol to calculate reaction norms for the TSM of all species with TSM calculated as the difference between the upper thermal tolerance limit and acclimation temperature. Of all LVB ecoregion species, Nile tilapia stood out as having a high absolute $CT_{max}$ and a steep reaction norm compared to other fish species; the other three model species from the LVB ecoregion performed on par with each other (Fig. 4D). Similarly, Nile tilapia had a remarkably high TSM compared to other fish acclimated to 27–31 °C, with the other three model species having lower TSM with Nile perch being among the lowest compared to other fish acclimated to 27–31 °C (Fig. 4E). This suggests that among LVB ecoregion fishes, Nile perch will be the most sensitive to increases in water temperature associated with climate change. However, this requires both a closer analysis of the aerobic (energetic) patterns that underlie limits and flexibility in thermal tolerance and studies to determine potential fitness or performance consequences of these patterns.

#### 3.2.3.3 Flexibility in aerobic performance across exposure times: Nile perch

Here, we summarize findings of a series of studies that investigated the metabolic responses of Nile perch to elevated temperatures over multiple timescales. A physiological strategy enacted over only 1-to-2 days might have different metabolic costs and outcomes from a strategy employed over several weeks, or over a species' development. For fish from the LVB ecoregion, experiments that compare thermal responses over various timescales are rare; however, a series of experiments testing Nile perch aerobic performance over several different acclimation times provides a useful case study. In these experiments, Nile perch from the Bugonga region of Lake Victoria (Fig. 1) were exposed to temperatures at or above their current average temperature over three different timescales: 3 days (acute exposure), 3 weeks (mid-range exposure), and 3 months (developmental exposure). Acute and mid-range acclimations were carried out in a laboratory setting at 27, 29 and 31 °C after which AS was measured at the same three temperatures (Nyboer and Chapman, 2017). The developmental exposure was carried out in aquaculture ponds held at 25 and 29 °C after which AS was measured for fish from both environments at 25, 29, and 33 °C (Nyboer and Chapman, 2018). All physiological measurements were performed on fish that had been starved for 48 h prior to experimental trials to ensure comparable energetic states of fish among experiments. Body size and condition (K) were measured before and after mid-range and developmental trials.

First, we compared aerobic responses for Nile perch that were acutely exposed to those with the mid-range acclimation times. Predictions derived from OCLTT would suggest that fish with longer acclimation times would have a greater AS at higher temperatures than those with shorter acclimation times. However, Nile perch exposed for 3 weeks showed a lower overall AS than the acutely exposed fish (Fig. 5A). This pattern is especially evident at the highest experimental temperature (31 °C) where the metabolic costs are likely to be strongest. In acutely exposed fish, AS was steady at the lower two acclimation temperatures but increased at 31 °C. By contrast, there was no change in AS across the acclimation temperatures of the mid-range-acclimated fish. And, a comparison of AS across the two acclimation times (acute vs. mid-range) revealed that the mid-range acclimated fish had a lower AS than the acutely exposed group (Fig. 5A). Although patterns of lower AS in an acclimated group can be interpreted as poor aerobic compensation (Healy and Schulte, 2012), it may indicate a beneficial physiological adjustment (Norin et al., 2014; Nyboer and Chapman, 2017). This interpretation is backed by increases in condition factor (k) for fish acclimated to higher temperatures over the three-week acclimation period (Fig. 5C). Reduced AS in the mid-range acclimated fish may reflect reduced costs of metabolism after acclimation.

Next, we compared aerobic responses of Nile perch that were reared for 3 months under two temperature conditions. Nile perch reared under warmer temperatures showed a reduced AS compared to fish reared at cooler temperatures (Fig. 5B). Fish reared at cooler temperatures showed an increase in AS at the highest experimental temperature (33 °C), whereas the warm-reared fish maintain the same AS across all three experimental temperatures (Fig. 5B). These patterns are remarkably similar to the findings in the acute vs mid-range comparison. AS appears to be lower in fish that have had more time to acclimate to higher temperatures. Furthermore, comparisons of body mass and condition among fish reared at the two different temperatures revealed no differences among rearing treatments (Fig. 5D) indicating that reductions in aerobic performance are not reflected in fitness related traits even over developmental timeframes in contrast to predictions derived from OCLTT, and in line with a growing number of studies questioning the broad applicability of this hypothesis (Jutfelt et al., 2018). Taken together, findings across different acclimation times indicate that a lower AS does not induce

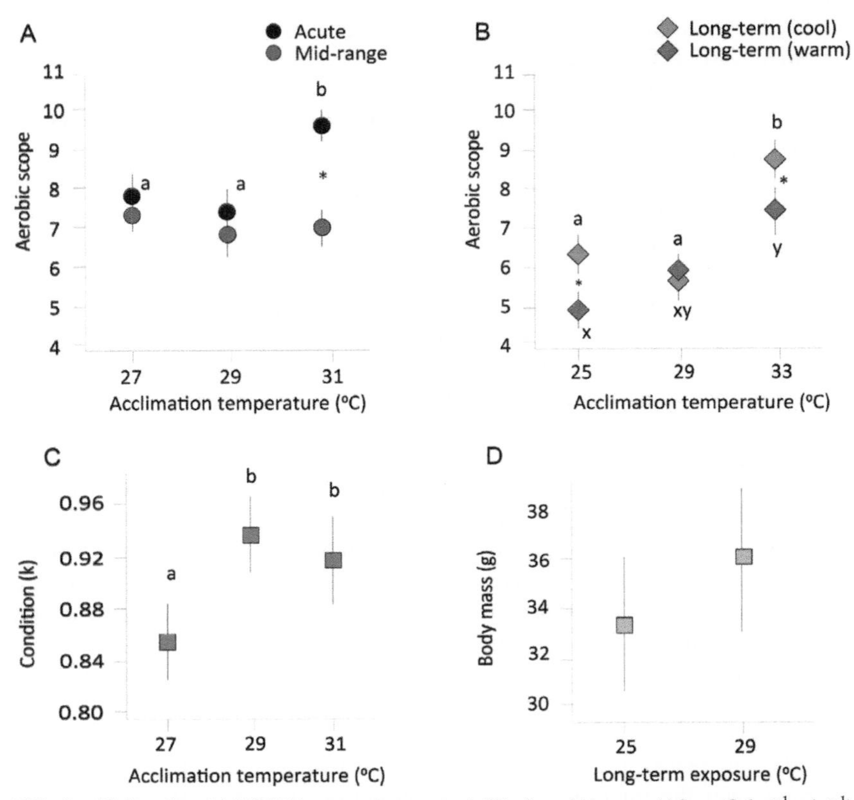


**FIG. 5** (A) Results of ANCOVA comparing means ± SE of aerobic scope (AS, $mgO_2kg^{-1}\ min^{-1}$) among acutely exposed (blue circles) and mid-range exposed (red circles) Nile perch over a range of temperatures. (B) Results of ANOVAS comparing metabolic traits and thermal tolerance limits in Nile perch among experimental temperatures and between rearing temperatures. Results of three-way ANOVA comparing means ± SE for AS between rearing temperatures (cool-reared = light blue diamonds, warm reared = pink diamonds) across a range of experimental temperatures. (C) Results of ANCOVA comparing means ± SE of condition (k) of Nile perch after a mid-range acclimation to a range of temperatures. (D) Results of ANOVA examining effects of rearing temperature on body mass. In all panels, different letters indicate significant differences among acclimation temperatures, and asterisks indicate significant differences among experimental groups. No letters or asterisks indicates no differences. *Originally published in the Journal of Experimental Biology.*

negative fitness effects. These findings also indicate that over acute, mid-range, and longer-term exposures, we see similar patterns of aerobic compensation providing evidence for both short-term and developmental thermal plasticity in Nile perch.

The degree to which these levels of plasticity will increase or buffer evolutionary change in the face of climate change is unknown. However, in introduced populations of Nile perch there is evidence for inter-populational variation in $CT_{max}$ and AS, suggesting the possibility of genetic or epigenetic variation in thermal tolerance. Nyboer et al. (2020) compared $CT_{max}$ and AS between juvenile Nile perch from lakes Victoria (Bugonga) and Nabugabo. Across acclimation temperatures, Nile perch from the cooler lake (Nabugabo) tended to have a lower $CT_{max}$ and a lower AS than Nile perch from the warmer waters of Lake Victoria. The Lake Nabugabo fish also showed less thermal plasticity in most metabolic traits. Overall, the divergence in thermal tolerance between these two introduced populations was in a direction consistent with an adaptive response to local thermal regimes. An important question is whether the levels of plasticity observed in thermal tolerance (i.e., $CT_{max}$) and AS are sufficient to facilitate persistence in the face of continued warming and extreme weather events, and/or to buy time for evolutionary change to offset any negative fitness consequences.

### 3.3 Vulnerability of African freshwater fishes to climate change—A synthesis

The ability of fishes to respond through short-term (reversible) and/or developmental plasticity to climate warming is likely to interact with many other factors to determine vulnerability to climate changes, including their ability to disperse to more favorable habitat. Range shifts have been reported for freshwater fishes (e.g., Alofs et al., 2014), but may be limited by constrained distributions of freshwater fishes within hydrographic networks. Nyboer et al. (2019) conducted a meta-analysis of the vulnerability of African freshwater fishes to climate change, collecting trait data for 2794 (84% of) freshwater fish species in Africa and pairing these with climate change projections (temperature and precipitation) across species' geographic ranges to quantify species' overall climate change vulnerability. They used two emissions scenarios or Representative Concentration Pathways (RCPs) (i.e., RCP4.5 where global carbon emissions stabilize (Thomson et al., 2011), and RCP8.5 where emissions remain high without intervention (Riahi et al., 2011)) and two future time periods (i.e., 2055 = "mid-century"; and 2085 = "end-of-century") to understand how reducing carbon emissions might affect vulnerability of Africa's freshwater fish biota. Fig. 6A–D represent these data re-analyzed to include only LVB ecoregion species, and show that LVB ecoregion fishes generally have higher than average climate change vulnerabilities compared to other freshwater fishes in Africa (NB: other vulnerable regions include the rivers of West Africa and other Rift Valley lakes). Results of their trait-based analysis for continental Africa predicted that over 40% of fishes are likely to be negatively impacted by climate change largely because of their high reliance on seasonal patterns and specific microhabitat requirements. In Lake Victoria proper, 60%–80% of fish species were shown to have high vulnerability under RCP4.5 and 80%–100% under RCP8.5 in both time periods. For the remainder of the LVB ecoregion, proportions of vulnerable species ranged from ~0% to –45% under RCP4.5 and from ~20% to 100%

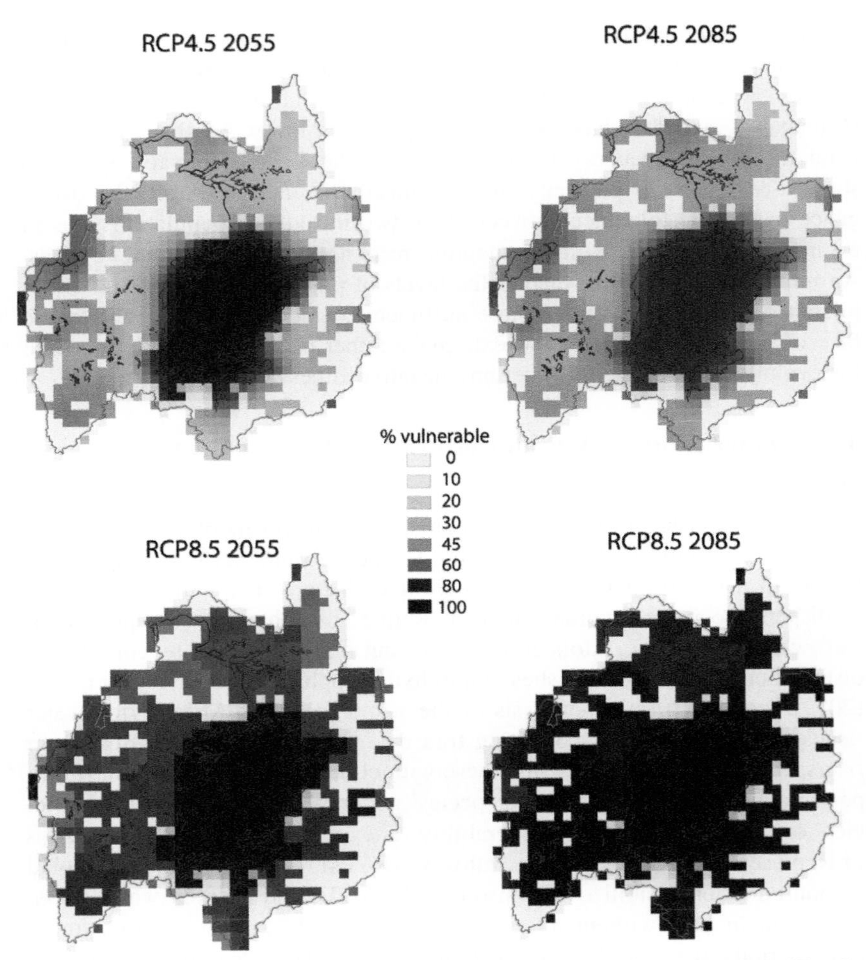


**FIG. 6** Projected effects of future climate change on freshwater fish fauna of the Lake Victoria Ecoregion. Maps show the proportion of species that are projected to be vulnerable to climate change under two emissions scenarios (RCP4.5 and RCP8.5) and for two future time periods (mid-century—2055 and late-century—2085). Fish were considered highly vulnerable if they possess traits that make them highly sensitive and of low adaptive capacity, and if they are projected to be highly exposed to the effects of climate change across their range (see Nyboer et al., 2019 for details).

under RCP8.5. These findings highlight the importance of emissions reduction and indicate that climate change mitigation policies are likely to protect substantial freshwater fish biodiversity in the LVB ecoregion (Fig. 6A–D). High rates of vulnerability in LVB ecoregion fishes were largely attributed to traits or qualities that tend to prevent species from coping with or adapting to the effects of climate change. These include being highly dependent on habitats that are sensitive to climate change, having specific dietary and/or reproductive requirements, having small, decreasing, or fragmented populations, and being exposed to other anthropogenic threats (e.g., land use change). Because of lack of data for most species, Nyboer et al. (2019) were unable to account for the potential phenotypic adjustments (either plastic or genetic) that can help species overcome climate-induced stressors. The vulnerability findings may therefore be grimmer than reality, especially considering the evidence presented here that many LVB ecoregion fishes are capable of dynamic physiological, morphological, and behavioral flexibility when faced with changing environmental conditions. Much work remains to understand the nuances of species' adaptive capacity. However, these findings highlight the potentially far-reaching negative effects of climate change on fish diversity and lake function if little is done to mitigate its pace.

## 4 Changes in aquatic oxygen regimes in the Lake Victoria Basin ecoregion

### 4.1 Aquatic hypoxia

Climate change may affect freshwater fishes through shifts in thermal regimes, but may also co-occur or interact with other stressors including hypoxia (condition of low dissolved oxygen, DO). Hypoxia occurs naturally in many freshwater systems with low mixing, high rates of organic decomposition, and/or inadequate solar radiation for photosynthetic production of oxygen, such as heavily vegetated wetlands, flooded forests, floodplains, and the deep waters of lakes and ponds (Chapman, 2015). In tropical fresh waters, hypoxic conditions are often intensified by high temperatures that promote decomposition of organic materials and reduce oxygen tension in the water (Chapman et al., 2001). In Africa, chronic hypoxia and extensive anoxia are characteristic of some deep-water meromictic lakes such as Malawi and Tanganyika, driven by strong stratification of the water column and decomposition of descending organic matter (Spigel and Coulter, 1996). In the LVB ecoregion area, deep-water anoxia is characteristic of Lake Kivu (Ogutu-Ohwayo et al., 2020), Lake Edward (Beadle, 1981), and many deep crater lakes in Western Uganda (Chapman et al., 1998; Efitre, 2007; Melack, 1978). Hypoxia is also very widespread in wetlands of the LVB ecoregion that are dominated by the emergent sedge papyrus (*C. papyrus*) or the grass *Miscanthidium violaceum* (Chapman et al., 2001). In dense papyrus swamps that average 3–4 m in height (Thompson et al., 1979), the terminal papyrus umbels form a canopy that creates cool, dark conditions, and limits wind-driven mixing producing very low DO conditions. In the Rwembaita Swamp system of Kibale National Park, Uganda, DO levels have been recorded since the early 1990s. The Rwembaita Swamp is fed by four major streams and drains into the Njuguta River, a tributary of the Mpanga River

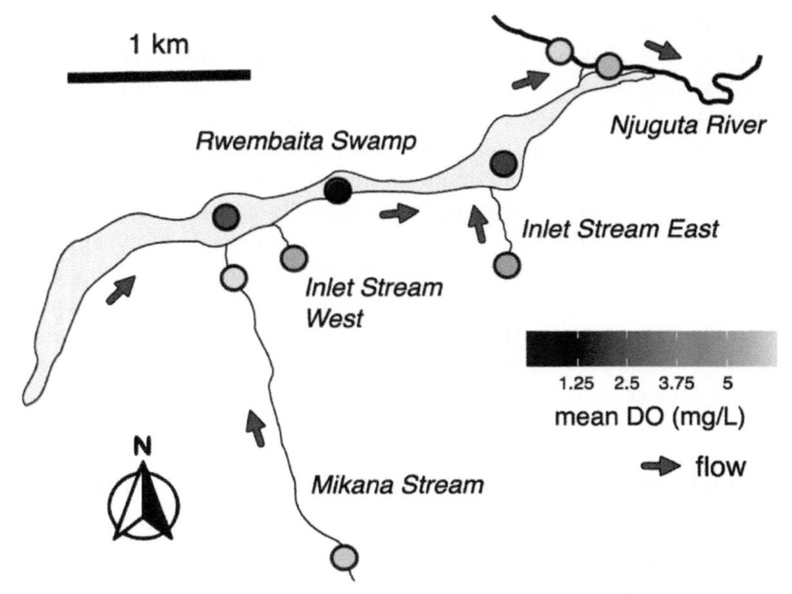


**FIG. 7** Map of the Rwembaita Swamp System of Kibale National Park, Uganda showing the average dissolved oxygen levels (mg/L) in the main papyrus swamp, inflowing streams, and upstream and downstream of where the swamp feeds into the Njuguta River. Data were collected approximately monthly between 1997 and 2019 for the Njuguta River and between 1993 and 2019 for all other sites.

system, which in turn feeds into Lake George (Figs. 1 and 7). The tributary streams form ecotonal gradients of DO and other physicochemical conditions (Chapman et al., 2001). DO is very low in the swamp interior, averaging 0.99 mg/L at a site in the middle of the swamp over a 26-year period (1993 t– 2019), much lower than the inflowing Mikana Stream, which has averaged 5.5 mg/L (upper Mikana) to 5.8 mg/L (lower Mikana) over the same time period (Fig. 7). The signature of the swamp water is apparent in the Njuguta River where DO has averaged 6.6 mg/L 20 m upstream of the swamp outflow and 5.9 mg/L downstream of the swamp outflow (1997–2019, Fig. 7). The positioning of valley papyrus swamps in river systems creates strongly divergent DO habitats: high-DO tributaries and low-DO swamp waters. Some widespread fishes persist in both low- and high-oxygen habitats in the swamp-river mosaic, while other species are restricted to either high-DO or low-DO habitats (Chapman, 2007, 2015, 2021).

Although hypoxia occurs naturally in the LVB ecoregion, hypoxic waters are becoming more widespread associated with influx of wastes and fertilizers that drive eutrophication. In addition, temperature changes associated with climate warming are expected to be accompanied by decreases in DO and longer periods of lake stratification, less predictable seasonality, and increased occurrence of extreme weather events (Niang et al., 2014; Ogutu-Ohwayo et al., 2016), with effects worsening with higher emission scenarios (Darko et al., 2019). However, trends seem to differ with lake morphometry and geological setting. In Lake Nkuruba, a crater lake, an increase in thermal stability since the early 1990s has resulted in the expansion of anoxic waters (Saulnier-Talbot et al., 2014; Fig. 8). DO concentration at 6–10 m of depth has decreased while the depth of the anoxic layer has increased, potentially limiting habitat availability for water-breathing aquatic organisms (Fig. 8). In Lake Tanganyika, an African great rift lake lying south of the LVB ecoregion, a rise in in surface-water temperature since the beginning of the 20th century has increased the stability of the water column, and a regional decrease in wind velocity has coincided with reduced mixing and a decrease in the depth of the oxycline (O'Reilly et al., 2003).

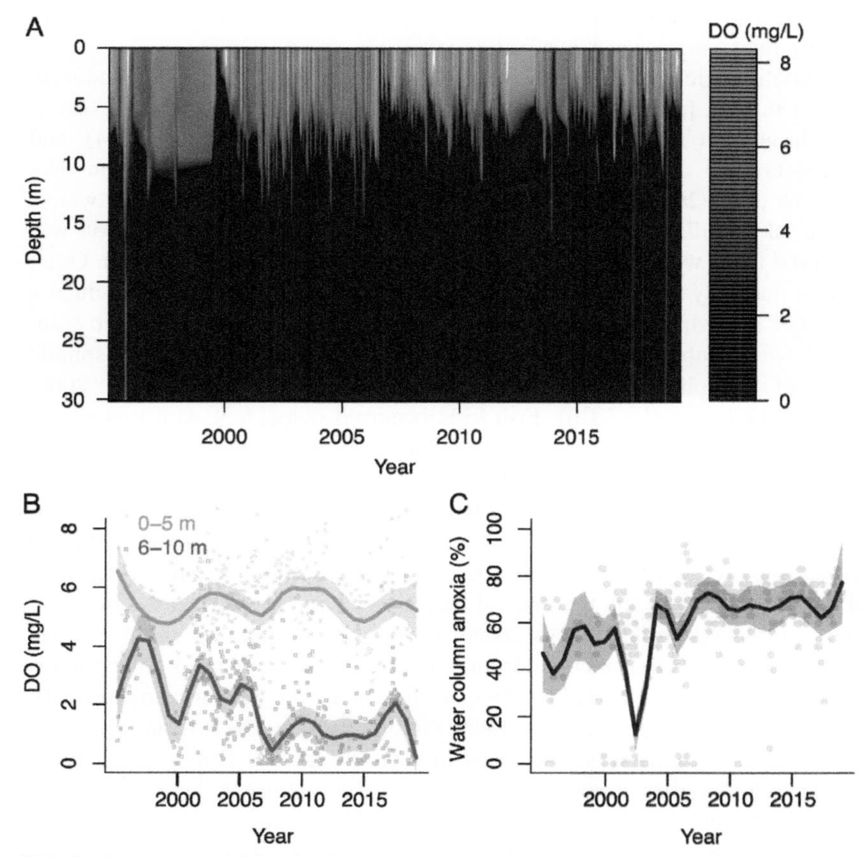


**FIG. 8** Long-term trend in dissolved oxygen (DO) concentration (mg/L) in Lake Nkuruba. (A) Water column DO profile during the period 1994–2019. DO measurements (gray symbols) and GAM model fit (lines and shaded polygons) indicating temporal variation in the mean DO is presented for (B) two depth layers and (C) % of the water column that is considered anoxic.

Changes in stratification, mixing dynamics, and DO availability characteristic of the deep rift valley lakes are not consistent with patterns observed in shallow lakes such as Lake Kariba (reservoir that filled the Kariba Gorge on the Zambezi River) and Lake Victoria. There are several studies that have explored the complex limnological dynamics of Lake Victoria (e.g., Hecky, 1993; Hecky et al., 1994, 2010; Marshall et al., 2013; Mugidde, 2001; Sitoki et al., 2010). Severe anthropogenic changes in the watershed (e.g., deforestation and agricultural expansion) coincided with a shift in Lake Victoria from a mesotrophic system in the 1930s to a eutrophic system; with severe eutrophication beginning in the 1980s (Hecky, 1993; Hecky et al., 1994; Sitoki et al., 2010; Verschuren et al., 2002). There was a dramatic increase in algal biomass (Mugidde, 1993) and a shift in algal species composition from large diatoms to blue-green algae (Hecky, 1993). The change in trophic status was also accompanied by a decrease in water transparency and spread of the invasive floating macrophyte water hyacinth (*Eichhornia crassipes*).

Hypolimnetic anoxia also developed in Lake Victoria, which was induced, at least in part, by eutrophication. The deeper part of the lake became stratified throughout much of the year (Hecky, 1993; Hecky et al., 1994), and the occurrence of hypoxia also increased in shallower areas of the lake (Wanink et al., 2001). In 1960–62, anoxia was only detected in deeper waters of the lake (Talling, 1966), while long-lasting (October–March) anoxia was observed to be widespread in deeper waters in 1990–91 (Hecky, 1993). Other factors that may have contributed to the deoxygenation of the lake include a period of low wind stress and climate warming, changes in the food web associated with establishment and expansion of the Nile perch, and atmospheric input of phosphorus (Hecky et al., 1994, 2010; Kolding et al., 2008; Ogutu-Ohwayo et al., 2020). Fish kills occurred coincident with upwellings of anoxic water (Kaufman and Ochumba, 1993; Ochumba, 1990). Interestingly, recent studies suggest that the water quality of lake has improved despite continued land use pressure in the lake basin. Since 2000, thermal gradients within the water column have decreased, weakening stratification and reducing deoxygenation. This has been attributed to the more rapid warming of the deeper waters and the general increase in wind speed over the lake that likely increased circulation (Marshall et al., 2013; Sitoki et al., 2010; van Rijssel et al., 2016). The conundrum is that the improvement to Lake Victoria's condition has occurred without a corresponding decrease in nutrients. Sitoki et al. (2010) proposed that Lake Victoria may be reaching an ecological equilibrium that may reflect, at least in part, the intense fishing pressure on Nile perch that has allowed species from lower trophic levels including *R. argentea* and some species of haplochromine cichlids to increase. However, recent fish kills in lakes Victoria and Kyoga (January 2021) coincident with acute hypoxia suggest that rapid change in dissolved oxygen availability is still likely to occur, and localized hypoxia has been noted in some deeper waters (e.g., Sharpe and Chapman, 2018).

## 4.2 Fish response to hypoxia

Fishes exhibit a diversity of strategies to cope with low environmental DO including the evolution of air-breathing organs, mechanisms to maximize oxygen uptake from the water and delivery to the tissues, reduction of metabolic rate and/or activity to reduce oxygen uptake requirements, and use of anaerobic metabolism (see reviews in Chapman, 2015, 2021; Richards et al., 2009). Behavioral responses provide additional flexibility to respond to hypoxia and include, as examples, avoidance of hypoxic habitats and/or use of aquatic surface respiration (ASR) where fish rise to the surface to skim the surface film. However, such responses may not be sufficient to offset hypoxia-induced costs on performance (e.g., swimming) and fitness-related traits.

Despite costs of hypoxia, there may also be survival benefits of life in low oxygen, such as reduced predation or competition (Chapman, 2015; Chapman and McKenzie, 2009), and such trade-offs may lead to divergent phenotypes in low- and high-DO waters (Chapman, 2007, 2021). Indeed, studies of fishes from hypoxic and normoxic sites have demonstrated that divergent DO environments (low- vs high-DO) provide a strong predictor of variation in traits related to metabolism and oxygen uptake or delivery (reviewed in Chapman, 2007, 2015, 2021). Trait variation across DO gradients may reflect genetic and/or phenotypically plastic responses, and previous studies have detected both plastic and genetic effects (reviewed in Chapman, 2015, 2021). Plasticity in respiratory traits may be important in colonizing divergent DO environments, by facilitating persistence and preserving the possibility for future evolutionary responses.

Hypoxia tolerance in fishes is often quantified by measuring critical oxygen tension ($P_{crit}$), the DO level below which a fish can no longer maintain a stable resting metabolic rate and its metabolic rate falls with further decline in DO. Although it is important to integrate $P_{crit}$ with other physiological measures (Regan et al., 2019; Wood, 2018), $P_{crit}$ serves as a valuable comparative tool in hypoxia studies. A second metric of hypoxia tolerance is the Aquatic Surface Respiration threshold often expressed as the $ASR_{10}$, $ASR_{50}$ or $ASR_{90}$ (the oxygen tension at which fish spend 10%, 50%, or 90% of their time at the surface, respectively), generally measured by exposing fish to progressive hypoxia with access to the water surface. If historical records are not available for these metrics of hypoxia tolerance, archived specimens can provide information on morphological traits may reflect changes in capacity for oxygen uptake or delivery. These can be measured on preserved specimens from different time periods or across spatial gradients in DO. Here, we focus on two gill metrics of LVB ecoregion fishes (total gill filament length, total gill surface area) in evaluating differences among populations in hypoxia tolerance and propensity for change in hypoxia tolerance associated with eutrophication-induced hypoxia. Total gill filament length is the estimated length of all the gill filaments (total number of filaments) on all hemibranchs on both sides of the branchial basket; while total gill surface area is the estimated bilateral surface area of the individual lamellae of the filaments that are the sites of oxygen exchange. In intra- and interspecific comparisons of gill size, metrics are scaled to a common body size to account for size-related variation in gill size. We also integrate other morphological, physiological, and biochemical traits to illustrate both inter-populational patterns and plasticity.

## 4.3 Response to hypoxia in LVB ecoregion fishes

To evaluate the effects of increased frequency and extent of aquatic hypoxia on LVB ecoregion fishes, we use three approaches: (1) comparisons of populations across natural DO gradients to detect traits that facilitate persistence under low DO; (2) laboratory experiments that expose organisms to stressor gradients (rearing studies) and field acclimatization experiments, and (3) comparisons of populations before and after environmental change. For (1) and (2), we focus primarily (though not exclusively) on two small-bodied widespread species that are phylogenetically distant and well-studied within the LVB ecoregion: *P. multicolor* (discussed above in *responses of LVBE fish species to elevated water temperature*) and the cyprinid *Enteromius neumayeri* Chapman et al., 1999). The widespread *P. multicolor* is a maternal mouth brooder with strong sexual dimorphism, whereas *E. neumayeri* is a shoaling fish with no post-spawning parental care. Both species eat primarily small aquatic insects and plants (Greenwood, 1966; McNeil et al., 2016; Schaack and Chapman, 2003). As with *P. multicolor*, *E. neumayeri* is found in a broad range of habitats that vary in DO content from hypoxic swamps to well-oxygenated streams and rivers (Chapman et al., 1999; Langerhans et al., 2007; Reardon and Chapman, 2009). Table 1 provides a summary of studies referred to below that examine variation in gill traits of LVB ecoregion fishes (a) across hypoxia gradients in the field, (b) in response to rearing under low- and high-DO conditions in the lab, and (c) in response (over time) to anthropogenically-induced hypoxia in the field.

### *4.3.1 Divergence between hypoxic and normoxic habitats*

Comparisons of fish populations from swamps and associated normoxic (high DO) sites in the LVB ecoregion have demonstrated that divergent DO environments provide a strong predictor of intraspecific phenotypic variation, particularly in respiratory traits, but also life-history traits, energetics, brain size, and biochemical traits (reviewed in Chapman, 2007, 2015, 2021). Such comparisons can highlight traits that may facilitate persistence under hypoxia. For example, studies comparing populations from low- and high-DO sites have detected larger gills (either total gill filament length and/or total gill surface area) in hypoxic vs normoxic populations of the LVB ecoregion cichlids *P. multicolor* (Chapman, 2021; Chapman et al., 2000, 2008b) and

**TABLE 1** Studies of LVB ecoregion fishes that report comparisons of gill metrics (total gill filament length, total gill surface area) for fish (a) across DO gradients in the field, (b) in response to rearing under low- and high-DO conditions in the lab, and (c) over time in Lake Victoria (before and after a strong increase in anthropogenically-induced eutrophication). Arrows indicate whether gill size is lower or higher under high-DO or low-DO conditions.

| Trait | Family | Species | High DO | Low DO | Study design | Data source |
|---|---|---|---|---|---|---|
| | Cichlidae | | | | | |
| Total gill filament length | | *Pseudocrenilabrus multicolor* | ↓ | ↑ | Field populations | Chapman (2021), Chapman et al. (2000, 2008b) |
| Total gill surface area | | | ↓ | ↑ | | |
| Total gill filament length | | *Astaoreochromis alluaudi* | ↓ | ↑ | Field populations | Chapman et al. (2007) |
| Total gill surface area | | | ↓ | ↑ | | |
| | Cyprinidae | | | | | |
| Total gill filament length | | *Enteriomius neumayeri* | ↓ | ↑ | Field populations | Chapman and Liem (1995), Chapman et al. (1999), Schaack and Chapman (2003) |
| Total gill filament length | | *Enteriomius apleurogramma* | ↓ | ↑ | Field populations | Hunt, Hendry, and Chapman (unpublished data) |
| | Mormyridae | | | | | |
| Total gill filament length | | *Petrocephalus degeni* | ↓ | ↑ | Field populations | Chapman and Hulen (2001) |

*Continued*

**TABLE 1** Studies of LVB ecoregion fishes that report comparisons of gill metrics (total gill filament length, total gill surface area) for fish (a) across DO gradients in the field, (b) in response to rearing under low- and high-DO conditions in the lab, and (c) over time in Lake Victoria (before and after a strong increase in anthropogenically-induced eutrophication). Arrows indicate whether gill size is lower or higher under high-DO or low-DO conditions.—Cont'd

| Trait | Family | Species | High DO | Low DO | Study design | Data source |
|---|---|---|---|---|---|---|
| Total gill filament length | | *Marcusenius victoriae* | ↓ | ↑ | Field populations | Chapman and Hulen (2001) |
| Total gill surface area | | | ↓ | ↑ | | |
| | Latidae | | | | | |
| Total gill filament length | | *Lates niloticus* | ↓ | ↑ | Field populations | Paterson and Chapman (2009) |
| | Cichlidae | | | | | |
| Total gill filament length | | *P. multicolor* | ↓ | ↑ | Lab rearing | Chapman et al. (2000), Chapman et al. (2008a,b), Crispo and Chapman (2010a,b); Wiens et al. (2014) |
| Total gill surface area | | | ↓ | ↑ | | |
| Total gill filament length | | *Astatoreochromis alluaudi* | ↓ | ↑ | Lab rearing | Chapman et al. (2007) |
| Total gill surface area | | | ↓ | ↑ | | |
| Total gill filament length | | *Astatotilapia "Mpanga blue"* | ↓ | ↑ | Lab rearing | Crispo (2010) |

| | | | | | | |
|---|---|---|---|---|---|---|
| Total gill filament length | | *Haplochromis pyrrhocephalus* | ↓ | ↑ | Lab rearing | Rutjes (2006) |
| Total gill surface area | | | ↓ | ↑ | | |
| | Cyprinidae | | | | | |
| Total gill filament length | | *Enteromius neumayeri* | ↓ | ↑ | Lab rearing | Chapman (unpublished data) |
| | Cichlidae | | | | | |
| Total gill filament length | | *A. alluaudi* | ↓ | ↑ | Historical vs Contemporary | Binning, Chapman, Witte (unpublished data) |
| Total gill filament length | | *H. pyrrhocephalus* | ↓ | ↑ | Historical vs Contemporary | Witte et al. (2008) |
| Total gill surface area | | | ↓ | ↑ | | |
| | | | | | | |
| | Cyprinidae | | | | | |
| Total gill filament length | | *Rastrineobola argentea* | ↓ | ↑ | Historical vs Contemporary | Sharpe and Chapman (2018) |
| | | | | | | |

Data from van Rijssel, J.C., Hecky, R.E., Kishe-Machumu, M.A., Meijer, S.E., Pols, J., van Tienderen, K.M., et al., 2016. Climatic variability in combination with eutrophication drives adaptive responses in the gills of Lake Victoria cichlids. Oecologia 182, 1187–1201 are not reported here, but discussed in the text.

*Astatoreochromis alluaudi* (Chapman et al., 2007); the cyprinids *E. neumayeri* (Chapman et al., 1999; Langerhans et al., 2007) and *E. apleurogramma* (Hunt, Hendry, and Chapman, unpublished data); and the mormyrids *Petrocephalus degeni* and *Marcusenius victoriae* (Chapman and Hulen, 2001) (Table 1).

It is intuitive that larger gills should be advantageous under hypoxic conditions by increasing oxygen uptake capacity; however, there may be costs associated with gill proliferation that negatively affect performance or fitness-related traits, and may affect how fish respond to anthropogenically-induced hypoxia. For example, the production of large gills may affect surrounding morphological structures (e.g., trophic muscles, gill rakers, head size, body shape), which could limit feeding performance, swimming efficiency, or other fitness-related traits. In *E. neumayeri*, large gill size in swamp-dwelling (low-DO) fish correlates with a reduction in the size of key trophic muscles and lower feeding performance than in stream-dwelling (high-DO) conspecifics (Schaack and Chapman, 2003). In a study of nine populations of *E. neumayeri*, Langerhans et al. (2007) showed that DO had an effect on body shape that was largely driven by DO-driven changes in gill size. In the cichlid *A. alluaudi*, Binning et al. (2010) found a negative correlation between pharyngeal jaw size and a suite of morphological traits related to gill size and shape across natural populations. In the cyprinid *R. argentea*, an increase in total filament length in Lake Victoria between 1966 and 2010 correlated with increased crowding of the gill rakers, suggesting a spatial trade-off between gill rakers and gill size (Sharpe and Chapman, 2018). Bouton et al. (2002) quantified ecological correlates of head shape for six species of haplochromine cichlids from Lake Victoria (one to three populations for each species). They measured the volume of three compartments of the head (oral, suspensorial, and opercular) and found that DO explained most of the variation in volume of the opercular compartment. Together these studies demonstrate links between hypoxia and divergence in both respiratory and non-respiratory traits, and suggest that an optimal trait value in one environment (large gills in hypoxia) could be indirectly associated with liability in others.

Other traits that increase oxygen uptake/delivery or increase anaerobic capacity also differ between populations and species of fish from low- and high-oxygen habitats in the LVB ecoregion. In the cyprinid *E. neumayeri*, swamp-dwelling populations are characterized by higher hematocrit (Chapman, 2007; Martinez et al., 2004) and higher liver LDH activities (important for anaerobic metabolism) than in *E. neumayeri* from connected high-DO sites (Martinez et al., 2004, 2011). In a broader interspecific study Chapman et al. (2002) compared fish species from the open waters of Lake Nabugabo and nearby Lake Kayanja to species from the dense hypoxic Lwamunda Swamp that surrounds Lake Nabugabo. The non-air breathing species from deep swamp refugia, as a group, showed higher levels of hemoglobin concentration and hematocrit than non-air-breathing lake fishes (Chapman et al., 2002).

The divergence in morpho-physiological traits between populations or species from high- and low-DO habitats should correlate with hypoxia tolerance. We have explored this idea from both intra- and inter-specific perspectives using two indicators of hypoxia tolerance (ASR thresholds and $P_{crit}$). In *E. neumayeri*, populations from high-DO stream sites show higher $ASR_{10}$, $ASR_{50}$ and $ASR_{90}$ thresholds than *E. neumayeri* from connected swamp populations (Olowo and Chapman, 1996); thus, swamp-dwelling fish can withstand a lower level of DO before rising to the surface than fish from connected stream sites. During ASR fish may be more visible to aerial predators, so a lower threshold may reduce predation risk. Reid et al. (2013a) compared ASR thresholds of *P. multicolor* from the chronically hypoxic Lwamunda Swamp surrounding Lake Nabugabo to three taxa captured in wetland ecotones of the main lake where DO is higher (juveniles of Nile perch, Nile tilapia, and endemic haplochromine cichlids) and to a literature review of ASR thresholds (81 values, Chapman and McKenzie, 2009). *P. multicolor* exhibited lower ASR thresholds than the three other taxa examined in the study and the lowest $ASR_{50}$ in the literature review indicating far-reaching adaptations to hypoxia.

Critical oxygen tension ($P_{crit}$) also varies within and among species, and seems to reflect home environments. In a comparison of *E. neumayeri* from a swamp and a nearby well-oxygenated stream site, $P_{crit}$ was lower in the swamp-dwelling fish (Chapman, 2007). In *P. multicolor* from the Lwamunda Swamp, $P_{crit}$ was lower (8.5 mmHg) than in *P. multicolor* from the well-oxygenated ecotonal waters of nearby Lake Kayanja ($P_{crit} = 13.6$ mmHg) (Chapman et al., 2002). Furthermore, in their review of $P_{crits}$ for East African cichlids, Melynchuk and Chapman (2002) found that the two groups (of 16 species/populations) that exhibited the lowest critical tensions were swamp-dwelling populations of *A. alluaudi* and *P. multicolor*.

Overall, fish from naturally occurring hypoxic sites in the LVB ecoregion are characterized by a series of morpho-physiological adjustments that correlate with higher hypoxia tolerance than populations and species from nearby well-oxygenated waters. However, it is important consider potential performance and fitness consequences of life in low oxygen. In *E. neumayeri*, Baltazar (2015) compared a suite of life-history traits between a hypoxic papyrus swamp (Rwembaita Swamp mid) and the inflowing higher-DO tributary Inlet Stream West (Figs. 1 and 7). Overall, *E. neumayeri* from the swamp exhibited lower condition, lower fecundity, lower ovary mass, and smaller size than *E. neumayeri* from the high-DO stream. In *P. multicolor*, fish from hypoxic swamps are characterized by a smaller size at maturity, a greater number of smaller eggs, and smaller juveniles than conspecifics from high-oxygen sites (Reardon and Chapman, 2009, 2012).

Despite potential costs of life in low oxygen habitats, there may be benefits such as reduced competition, predation, and/or disease. In particular, several studies have indicated that hypoxic environments can provide a refuge from predation for smaller fishes from predators less tolerant of low-oxygen conditions (Anjos et al., 2008; Chapman et al., 2002; Reid et al., 2013b; Robb and Abrahams, 2003). This has been well-studied in the LVB ecoregion where there is strong evidence that hypoxic swamps have provided important refugia for many hypoxia-tolerant species from the introduced predatory Nile perch, a fish with low tolerance to hypoxia (Chapman et al., 1996, 2002, 2003; Reid et al., 2013a,b; Schofield and Chapman, 2000). Whether these benefits will also be evident in response to anthropogenically-induced hypoxia will depend, at least in part, on the relative abilities of predator and prey species to adjust to increased frequency and intensity of hypoxic events.

### *4.3.2 Sources of phenotypic variation across DO gradients*

Phenotypic variation across DO gradients may reflect genetic and/or phenotypically plastic adaptation to local selective pressures. *P. multicolor* has been the focus of several studies looking at the role of plasticity in contributing to phenotypic divergence between low- and high-oxygen environments. Crispo and Chapman (2010a) reared young from multiple populations of *P. multicolor* in the Mpanga River (Fig. 1) under either low-DO (1.3 mg/L) or high-DO (≥7.5 mg/L) conditions. Gills were larger, and brain size was smaller in fish across multiple populations reared under hypoxia, with the former likely a strategy to increase oxygen uptake capacity and the latter likely due to the high aerobic cost of brain function (Table 1). Gills were on average 35% larger in total gill filament length, while brain size was 10% smaller in fish reared under hypoxia. Both plastic and genetic effects on brain mass were detected, as were genetic effects on brain mass plasticity, but most variation in gill size among treatments (high- vs low-DO) was due to plasticity (Crispo and Chapman, 2010a). Mitochondrial DNA and microsatellite studies provided evidence for high gene flow among *P. multicolor* populations from divergent DO environments in the Mpanga and Nabugabo River drainages of Uganda (Crispo and Chapman, 2008, 2010b). Thus, in this system, evolutionary divergence in some traits (e.g., brain mass) seems to occur despite high levels of gene flow; although plasticity in several traits is evident (Chapman et al., 2000, 2008a,b; Crispo and Chapman, 2010a). Developmental plasticity in gill size has been reported for other haplochromine cichlids in the LVB ecoregion including the Lake Victoria endemic *Haplochromis* (*Yssichromis*) *pyrrhocephalus*, which showed an increase of 80% in gill surface area in fish raised at 10% saturation when compared to full sibs reared at 80%–90% saturation (Rutjes, 2006, Table 1). Total gill filament length of *A. alluaudi* young was 27% larger in fish reared at ~1 mg/L vs full siblings reared in normoxia (~7.5 mg/L, Chapman et al., 2007, Table 1). The riverine cichlid *Haplochromis* "Mpanga blue" is found in the main river channel of the Mpanga River (Fig. 1) but is not known to penetrate hypoxic swamps. Crispo (2010) compared plasticity in gill metrics between this species to *P. multicolor* when both were reared under either low-DO or high DO, and found similar levels of plasticity in gill size (Table 1) but no plasticity in brain size in *H.* "Mpanga blue."

In *E. neumayeri*, Martinez et al. (2011) used a reciprocal transplant cage experiment in a low-DO (Rwembaita Swamp mid) and a high-DO (Inlet Stream West) site in Kibale National Park, Uganda to test for differences in tissue metabolic enzyme activities. Fish from both the low- and high-DO sites were maintained for 4 weeks in cages placed in their home and the alternative habitat, after which they were sampled for activities of phosphofructokinase (PFK), lactate dehydrogenase (LDH), citrate synthase (CS), and cytochrome *c* oxidase (CCO) in four tissues, liver, heart, brain, and skeletal muscle. PFK and LDH were chosen to reflect the tissue capacities for anaerobic metabolism, and CCO and CS were chosen to reflect aerobic capacity. Acclimatization to the low DO site resulted in lower activities of the aerobic enzyme CCO in heart, and higher activities of the glycolytic enzyme PFK in heart and skeletal muscle. Interestingly, the activity of LDH (important for anaerobic metabolism) in liver tissue was higher in fish collected from a hypoxic habitat, regardless of acclimatization treatment (Martinez et al., 2011). This pattern in LDH activity suggests the possibility of genetic differentiation between DO regimes over small spatial scales. Harniman et al. (2013) sequenced two nuclear intronic regions and a single mitochondrial region to examine whether populations of *E. neumayeri* from the Rwembaita Swamp system and a second nearby drainage (Dura River, Fig. 1) are genetically structured by DO regime and/or geographical distance. Over a large scale (between drainages), geographical distance affected the genetic structure of *E. neumayeri* populations; however, within the Rwembaita Swamp system (Figs. 1 and 7), gene flow was higher between sites that were similar in DO regimes compared to sites that differed in DO, suggesting that DO levels can act as a selective force on natural populations of *E. neumayeri*.

### *4.3.3 Phenotypic change over time*

The strong patterns of divergence in morpho-physiological observed in some LVB ecoregion fishes between low- and high-DO habitats suggest that DO can impose strong selection for traits that increase oxygen uptake and delivery and/or decrease oxygen demands. An important question is the degree to which fish can respond to changes in DO that are occurring over contemporary time scales. To address this question, we review a series of studies that quantify changes over time in gill morphology and neighboring structures. We focus on gill traits since they can be measured on archived specimens, and because they show strong patterns of divergence and a high degree of plasticity in multiple species across DO gradients.

In the LVB ecoregion, haplochromine cichlids have faced increasing exposure to hypoxia in wetland ecotones where some species have persisted in the face of Nile perch predation, but also in response to cultural eutrophication and declines in DO availability. Van Rijssel et al. (2016) explored the relationship between climate variability, environmental variables (water temperature, DO, and water transparency), and gill surface area in four haplochromine cichlids species in Mwanza Gulf, Lake Victoria between 1973 and

2011. During a period characterized by severe eutrophication and reduced DO, gill surface area in three of the four species increased; while during the 2000s when wind speed, DO, water transparency, and water temperature increased, cichlid gill surface area generally decreased (though it should be noted that they did not find a direct relationship between DO levels and gill size). Binning, Chapman, and Witte (unpublished data) found that the total gill filament length of *A. alluaudi* from Lake Victoria was smaller in specimens collected in 1970 than in specimens collected in both 1991 and 2007, when hypoxia was more widespread in the system (Table 1). In other related studies on haplochromine cichlids of Lake Victoria, morphological changes (e.g., size of the opercular compartment, trophic traits) have been detected in recovering haplochromine cichlids in Lake Victoria that correlate with environmental change (Witte et al., 2008; van Rijssel and Witte, 2013; van Rijssel et al., 2014). Although several haplochromine cichlids in Lake Victoria are presumed extinct, rapid adaptive phenotypic trait change may have contributed to the persistence of some species in the face of eutrophication and Nile perch predation.

Contemporary phenotypic change in gill traits has also been reported for *R. argentea* from Lake Victoria. Concomitant with the increase in abundance of Nile perch and other changes in the lake, *R. argentea* increased in abundance and expanded its distribution from the pelagic to the benthic zone, potentially a result of reduced competition from haplochromine cichlids, among other factors (Goldschmidt et al., 1993; Wanink, 1998; Wanink and Witte, 2000). It also broadened its trophic niche to include benthic macroinvertebrates into its previously zooplanktivorous diet. Sharpe and Chapman (2018) tested for changes in three groups of neighboring structures (respiratory, trophic and cranial) in *R. argentea* collected in 1966 and 2010. They predicted that increased use of hypoxic benthic waters should favor increased gill size; that a dietary shift to larger benthic prey should favor fewer, shorter, wider and/or more widely spaced gill rakers; and that there would be evidence of a tradeoff between these two suites of traits. Total gill filament length increased in *R. argentea* between 1966 and 2010 (Table 1), likely driving the observed increase in head size. However, gill rakers became more crowded over time, a change not consistent with a shift to larger benthic prey, but consistent with a trade-off with gill size. So, although the head size of *R. argentea* increased between 1966 and 2010, the magnitude of the change seemed insufficient to uncouple the trade-off between gill rakers and gill size. Increased gill size is likely an adaptive response to hypoxia exposure, while increased crowding of gill rakers may reflect an indirect and potentially maladaptive response to larger gills reflecting the complexity of response to environmental change.

## 5 Land use change and response of fishes

Anthropogenic land use and land cover changes represent another important threat to aquatic biodiversity (Dudgeon et al., 2006; Foley et al., 2005;

Reid et al., 2019), which may interact with warming, deoxygenation and other stressors. Inland waters are strongly linked to surrounding terrestrial ecosystems, such that land cover modifications induce biotic and abiotic changes in drainage streams, and in lakes that accumulate inputs of nutrients, contaminants, and suspended matter (Allan, 2004; Likens and Bormann, 1974; Vannote et al., 1980). Anthropogenic land cover changes in watersheds are driven by forestry, urbanization, and agriculture for food and commodity production (Curtis et al., 2018), with croplands and agricultural pollution being particularly pervasive threats to freshwater biodiversity (Vörösmarty et al., 2010). Land use changes related to agriculture include both "agricultural expansion," when natural ecosystems such as forests are cleared and converted into pastures and agricultural fields, as well as "agricultural intensification," when an increase in crop production per unit area (crop yield) is achieved due to the usage of fertilizers, pesticides, more productive cultivars, and various other technologies.

Although agricultural intensification has been globally ubiquitous and ongoing since the Green Revolution of the 1960s, deforestation and agricultural expansion are currently most pronounced in tropical areas (Ramankutty et al., 2018). For example, satellite images revealed that between the years 2000 and 2012, a total of 2.3 million $km^2$ of forest were lost worldwide, with 32% of this global forest loss in the tropics (Hansen et al., 2013). In 2001, Tilman et al. (2001) calculated rates of agricultural expansion between 1960 and 2000 as the global human population rose to 6 billion inhabitants, and then extrapolated these rates over the next 50 years to predict the amount of land required to sustain global food production in 2050. They concluded that an additional 1 billion hectares of natural ecosystems had to be converted to agriculture, which is an area roughly the size of Canada (Tilman et al., 2001). The majority of this new agricultural land is expected to come from the clearing of intact forests in South America and Sub-Saharan Africa, the only regions of the world with large expanses of land with unexploited agricultural potential (Gibbs et al., 2010; Laurance et al., 2014). In the late 1990s, when the global rate of tropical deforestation was lower than today, Benstead et al. (2003) estimated that an additional 500,000 km of stream channel were affected by tropical deforestation every year. In addition to large-scale tropical deforestation, pesticide and fertilizer usage has also increased in tropical countries (Ramankutty et al., 2018; Tang et al., 2021).

Remote sensing data indicate that rates of forest loss between 2000 and 2012 in the LVB ecoregion greatly exceeded the rates of forest gain and afforestation during the same period (Fig. 9A). Recent deforestation was particularly pronounced in western Uganda, a region with a relatively higher natural forest cover than drier (savanna) regions of the LVB ecoregion, and where deforestation was widespread during the 20th century (Aleman et al., 2018). Country-level data compiled by the Food and Agricultural Organization of the United Nations (https://www.fao.org/faostat/) also confirm that,

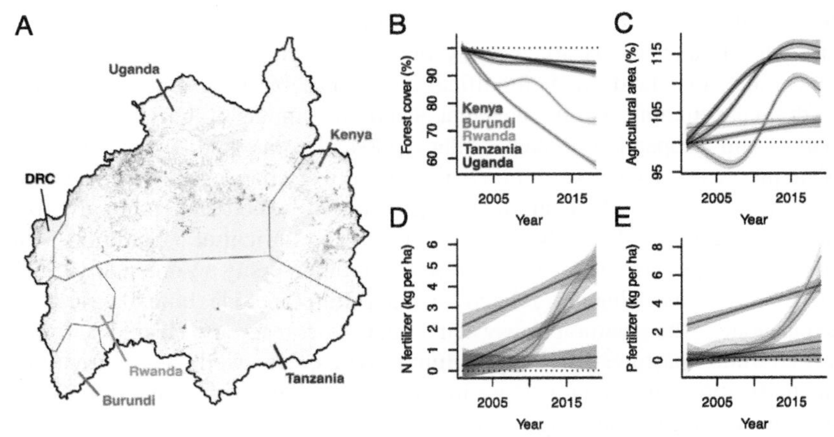


**FIG. 9** Agricultural expansion and intensification in the Lake Victoria Ecoregion between 2000 and 2019. (A) Map indicating areas with net forest loss (red pixels) and net forest gain (green pixels). Remote sensing data from Hansen et al. (2013), available at https://earthenginepartners.appspot.com/science-2013-global-forest. (B–E): Country-level change in four FAO indicators: "tree-covered areas," a measure of forest cover (B), "Agriculture," the total land area used for agriculture in any given country (C), "Nutrient nitrogen N (total)," the total amount of N fertilizer used in the country (D), and "Nutrient phosphate P2O5 (total)," the total amount of phosphate fertilizer used in the country (E). To quantify agricultural expansion (B, C) and intensification (C, D), forest (B) and agricultural (C) area was divided by total country area, while N and P fertilizer usage (D, E) were divided by agricultural land cover. For each indicator, temporal trends were visualized by dividing yearly values by the value for 2001, the reference year. The significance of temporal trends was assessed with generalized additive models (GAMs). GAMs included a single smooth term ("year") and were fitted with the function "gam" in the R package "mgcv." Lines and polygons show fitted values ±95% C.I. Significant non-linear trends ($P < 0.05$ for the "year" effect) were found for all countries and indicators except "agricultural area" in Rwanda ($P = 0.09$) and "N and P fertilizer usage" in Uganda (both with $P > 0.2$).

since 2001, relative forest cover has been declining in the five countries that comprise most of the LVB ecoregion (Fig. 9B; note that the Democratic Republic of the Congo has not been included in the analysis because most of this large country lies outside of the LVB ecoregion). Fig. 9C–E show that, in contrast to forest cover, agricultural land cover and fertilizer use per hectare of agricultural land have been increasing in these countries. Agricultural expansion and intensification are affecting water bodies in the LVB ecoregion, with reported effects that include nutrient enrichment and cultural eutrophication (Crisman et al., 2001; Stager et al., 2009), atmospheric transport and deposition of pesticides (Arinaitwe et al., 2016), increased sedimentation and water turbidity (Umwali et al., 2021), and altered flow regimes of streams and rivers (Baker and Miller, 2013). These abiotic changes could all pose challenges to freshwater fishes of the LVB ecoregion; for example, eutrophication and increased turbidity have well documented effects on the reproductive behavior of cichlids in Lake Victoria, interfering with vision and mate choice based on nuptial coloration, and relaxing barriers to hybridization among species inhabiting highly turbid areas of the lake (Seehausen et al., 1997). Turbidity differences among swamp and riverine habitats of Kibale National Park also influence the behavior of *P. multicolor*, with higher turbidity reducing activity levels and social behaviors while increasing aggressive behaviors (Gray et al., 2012).

Agricultural expansion can also influence air and water temperature and lead to 'deforestation-induced warming' (Alkama and Cescatti, 2016). The conversion of forests into grasslands or crops with a relatively lower biomass per unit area changes the amount of incident solar radiation absorbed or reflected by the land surface, which can create local 'heat islands' (Zeng et al., 2021). Moreover, forests usually have a greater evapotranspiration potential than cultivated areas and can thus dissipate more heat through the evaporation of water (Lawrence and Vandecar, 2015). In tropical regions, deforestation-induced warming can be of equal or greater magnitude than warming predicted by even the most pessimistic climate change scenarios for the 21st century (Zeppetello et al., 2020), and it occurs instantaneously as soon as the forest is cleared, thus demanding quick adjustments by aquatic organisms. A review of 25 tropical studies with in situ air temperature measurements across land use types found that agricultural sites were 1.6–13.6 °C warmer than reference (forested) sites (Senior et al., 2017). Another recent study, based on satellite imagery, found that the extent of forest loss between 2000 and 2012 in four tropical mountainous areas correlates strongly with temperature change during the same period (Zeng et al., 2021). Such microclimatic shifts can in turn influence water temperature, especially in small headwater streams. For example, in the headwaters of the Mara River, a tributary of Lake Victoria, mean diel temperature was about 4 °C warmer in agricultural than forested streams during the dry season, and about 3.5 °C warmer during the wet season (Masese et al., 2017).

Such warming should impact the energetics of ectothermic organisms living in deforested streams. For example, if we assume a median $Q_{10}$ of 2.4 for the resting metabolic rate (RMR) of fishes (Clarke and Johnston, 1999), an increase of 4 °C in mean water temperature would translate into 1.4 times greater energy requirements just to sustain the basic costs of metabolism in deforested streams. Some animal populations could adapt to the microclimatic effects of land use; for example, ants and water fleas show greater thermal tolerance in urban (warmer) water bodies (Brans et al., 2017; Diamond et al., 2018). In contrast, community-level studies of tropical reptiles, amphibians, and ants (e.g., Boyle et al., 2021; Frishkoff et al., 2015; Nowakowski et al., 2018) all found that species differences in thermal tolerance ($CT_{max}$) predict their abundance in deforested habitats, indicating that acclimation and adaptation could not prevent extirpation of cool-adapted (forest) species with low $CT_{max}$. Deforestation-induced warming will also interact with gradual climate change, potentially creating extreme thermal habitats in agricultural areas.

## 5.1 Effects of deforestation-induced warming on fishes of the LVB ecoregion

Data on the ecophysiological effects of deforestation-induced warming in the LVB ecoregion are limited, to our knowledge, to a single study of Ugandan stream-dwelling fishes (Fugère et al., 2018). In that study, the thermal environments of two rainforest streams from Kibale National Park were compared with two nearby agricultural streams situated $<2$ km away from the park boundary, in an area that was deforested roughly 50 years ago. The mean diel temperature was $\sim 2\,°C$ warmer in agricultural than forested streams, while the maximum diel temperature reached during the afternoon was $\sim 5\,°C$ warmer at agricultural sites. Comparing the energetics and growth rates of *E. neumayeri* populations found in these four streams, Fugère et al. (2018) hypothesized that fish could cope with thermal stress with two possible solutions: (1) evolve a lower RMR at farm sites to save energy, and/or (2) consume more energy (prey) at farm sites to maintain similar or even higher growth rates despite the higher costs of metabolism than in cooler, forested sites. A long-term lab acclimation experiment revealed that the temperature-dependence of RMR and growth was similar across populations, with growth (at a limited food ration) being lower at 23 °C than at 18 °C in all populations, and with growth rates correlating negatively with mass-adjusted RMR. This suggests that farm-dwelling fish did not evolve a lower RMR, and that the metabolic costs of warming should decrease growth in these populations. However, in a year-long capture-mark-recapture study of the four populations, in situ growth rates were comparable between forest and farm streams, suggesting that farm-dwelling populations could simply adjust their energy intake to maintain similar growth rates despite higher metabolic costs. Estimation of energy intake with a bioenergetics model confirmed that deforestation-induced warming was indeed likely to have increased consumption. This would then be expected to strengthen top-down predator-prey interactions at warmer sites, an effect that has been discussed widely in the context of climate change (Gilbert et al., 2014).

It should be noted that these results for one cyprinid fish, known to be tolerant of extreme environmental conditions such as hypoxic papyrus swamps (as discussed above), may or may not be representative of how other taxa in the LVB ecoregion will cope with similar thermal conditions. Some species will likely be more sensitive to warming, and food limitation might be more frequent in other environments. We therefore suggest that the ecophysiological effects of deforestation-induced warming and their potential interactions with climate change warrant further investigation, especially at tropical latitudes where 'thermal effects' of land use are substantial and well documented, and where current rates of deforestation are highest.

## 6 Implications for fish biodiversity and fisheries in the LVB ecoregion

Here we have provided an overview of the effects of elevated water temperatures, low DO, and land use change on the freshwater fishes of the LVB ecoregion. Case studies exploring how species morpho-physiological, performance, and fitness-related traits respond to these stressors have found evidence for plasticity in LVB ecoregion fishes in upper thermal tolerance, aerobic performance, and traits related to hypoxia tolerance (e.g., gill size). Several of the fishes discussed in our case studies (i.e., Nile perch, Nile tilapia, *R. argentea*, Ningu) are critical to the fisheries of the LVB ecoregion (LVFO, 2016). Although difficult to make concrete predictions about the future of the LVB ecoregion fishery based on case studies alone, recent work has suggested that climate warming will shift fisheries production to small-bodied opportunistic species that can cope with rapidly changing thermal regimes (Kolding et al., 2019; Ogutu-Ohwayo et al., 2016). The data presented here may support this idea: although the large-bodied Nile perch showed some flexibility in upper thermal tolerance, it was not as pronounced as in some smaller-bodied species (e.g., *P. multicolor, Nile tilapia*). Furthermore, while not tested in a temperature context, the small-bodied fishery species discussed here (*R. argentea*) appears to show adaptive response to other stressors in Lake Victoria, including eutrophication-induced hypoxia, indicating its robustness in the face of massive environmental change. However, shallowing of the surface stratification layer and reduced primary productivity in some lakes may affect the abundance of small fish which are highly dependent on lower trophic levels for food (Cohen et al., 2016; Descy et al., 2015; Kolding et al., 2019).

Metrics of tolerance to environmental stressors such as $CT_{max}$, $P_{crit}$, and ASR thresholds can be important in (a) identifying populations/species more tolerant than others, and thus more likely to persist in the face of increased stress and (b) offering thresholds that may be useful in evaluating threats, identifying critical habitat, and in habitat restoration efforts. For example, Nile tilapia had a remarkably high thermal safety margin compared to other fish from across the globe acclimated to 27–31 °C, whereas Nile perch was among the lowest. However, there is potential for interactive effects of the environmental stressors, which may alter thresholds. Both climate warming and hypoxia can co-occur and are likely to interact as both affect aerobic metabolism of fishes. Deforestation can increase water temperature and increase turbidity; higher temperature may increase a fish's metabolic rate, while increased turbidity may clog fish gills making it more difficult to meet increased metabolic demands. Stressors can interact in various ways. They may act synergistically or antagonistically (McBryan et al., 2013); or, acclimation to one stressor may improve tolerance to the other (i.e., cross tolerance, Steinberg, 2012). These possibilities highlight the importance of quantifying interactive effects of multiple stressors, a critical area for future studies in fishes of the LVB Ecoregion, and more broadly.

Studies of LVB Ecoregion fishes also underscore the importance of linking metrics of tolerance to environmental stressors to fitness-related traits. Both *E. neumayeri* and *P. multicolor* show higher tolerance to hypoxia in naturally hypoxic habitats, but also divergence in life-history traits (e.g., smaller egg and juvenile size in *P. multicolor* under hypoxia; lower fecundity, lower gonad mass of *E. neumayeri* under hypoxia). Whether these represent local adaptation or local maladaptation remains unclear. Some fish species in the LVB ecoregion do show apparently adaptive response to anthropogenically induced hypoxia; however, fish kills that occur coincident with mixing in some crater lakes and in lakes Victoria and Kyoga suggest that adaptive response on contemporary timescales may be limited or time lagged for some species.

## References

Aleman, J.C., Jarzyna, M.A., Staver, A.C., 2018. Forest extent and deforestation in tropical Africa since 1900. Nat. Ecol. Evol. 2, 26–33.

Alkama, R., Cescatti, A., 2016. Biophysical climate impacts of recent changes in global forest cover. Science 351, 600–604.

Allan, J.D., 2004. Landscapes and riverscapes: the influence of land use on stream ecosystems. Ann. Rev. Ecol. Evol. Syst. 35, 257–284.

Alofs, K.M., Jackson, D.A., Lester, N.P., 2014. Ontario freshwater fishes demonstrate differing range-boundary shifts in a warming climate. Divers. Distrib. 20, 123–136.

Anjos, M.B., de Oliveira, R.R., Zuanon, J., 2008. Hypoxic environments as refuge against predatory fish in the Amazonian floodplains. Braz. J. Biol. 68, 45–50.

Arinaitwe, K., Kiremire, B.T., Muir, D.C.G., Fellin, P., Li, H., Teixeira, C., et al., 2016. Legacy and currently used pesticides in the atmospheric environment of Lake Victoria, East Africa. Sci. Tot. Environ. 543, 9–18.

Arthington, A.H., Dulvy, N.K., Gladstone, W., Winfield, I.J., 2016. Fish conservation in freshwater and marine realms: status, threats and management. Aquat. Conserv. Mar. Freshwat. Ecosyst. 26, 838–857.

Baker, T.J., Miller, S.N., 2013. Using the Soil and Water Assessment Tool (SWAT) to assess land use impact on water resources in an East African watershed. J. Hydrol. 486, 100–111.

Balirwa, J.S., Chapman, C.A., Chapman, L.J., Cowx, I., Geheb, K., Kaufman, L., et al., 2003. Biodiversity and fishery sustainability in the Lake Victoria Basin: an unexpected marriage? Bioscience 53, 703–715.

Baltazar, C., 2015. Fitness Consequences of Divergent Oxygen Environments in a Widespread African Cyprinid. M.Sc. thesis, McGill University, Montreal.

Beadle, L.C., 1981. The Inland Waters of Tropical Africa. An Introduction to Tropical Limnology. Longman, London.

Becker, C.D., Genoway, R.G., 1979. Evaluation of the critical thermal maximum for determining thermal tolerance of freshwater fish. Environ. Biol. Fishes 4, 245–256.

Benstead, J.P., Douglas, M.M., Pringle, C.M., 2003. Relationships of stream invertebrate communities to deforestation in Eastern Madagascar. Ecol. Appl. 13, 1473–1490.

Binning, S.A., Chapman, L.J., Dumont, J., 2010. Feeding and breathing: trait correlations in an African cichlid fish. J. Zool. 282, 140–149.

Bouton, N., de Visser, J., Barel, C.D.N., 2002. Correlating head shape with ecological variables in rock-dwelling haplochromines (Teleostei: Cichlidae) from Lake Victoria. Biol. J. Linn. Soc. 76, 39–48.

Boyle, M.J.W., Bishop, T.R., Luke, S.H., van Breugel, M., Evans, T.A., Pfeifer, M., et al., 2021. Localised climate change defines ant communities in human-modified tropical landscapes. Funct. Ecol. 35, 1094–1108.

Brans, K.I., Jansen, M., Vanoverbeke, J., Tüzün, N., Stoks, R., Meester, L.D., 2017. The heat is on: genetic adaptation to urbanization mediated by thermal tolerance and body size. Glob. Change Biol. 23, 5218–5227.

Brijs, J., Jutfelt, F., Clark, T.D., Gräns, A., Ekstrom, A., Sandblom, E., 2015. Experimental manipulations of tissue oxygen supply do not affect warming tolerance of European perch. J. Exp. Biol. 218, 2448–2454.

Bush, M.B., Hooghiemstra, H., 2005. Tropical biotic responses to climate change. In: Lovejoy, E., Hannah, L. (Eds.), Climate Change and Biodiversity. Yale University Press, New Haven, pp. 125–137.

Campos, D.R., Amanajás, R.D., Almeida-Val, V.M.F., Val, A.L., 2021. Climate vulnerability of South American freshwater fish: thermal tolerance and acclimation. J. Exp. Zool. A 335, 723–734.

Chapman, L.J., 2007. Morpho-physiological divergence across oxygen gradients in fishes. In: Fernandes, M.N., Rantin, F.T., Glass, M.L., Kapoor, B.G. (Eds.), Fish Respiration and the Environment. Science Publishers Inc, Enfield, pp. 14–39.

Chapman, L.J., 2015. Low-oxygen lifestyles in extremophilic fishes. In: Reisch, R., Plath, M., Tobler, M. (Eds.), Extremophile Fishes—Ecology and Evolution of Teleosts in Extreme Environments. Springer, Heidelberg, pp. 9–31.

Chapman, L.J., 2021. Respiratory ecology of cichlids. In: Abate, M.E., Noakes, D.L.G. (Eds.), The Behavior, Ecology and Evolution of Cichlid Fishes: A Contemporary Modern Synthesis. Springer Academic (in press).

Chapman, L.J., Hulen, K.J., 2001. Implications of hypoxia for the brain size and gill surface area of mormyrid fishes. J. Zool. 254, 461–472.

Chapman, L.J., Liem, K.F., 1995. Papyrus swamps and the respiratory ecology of *Barbus neumayeri*. Environ. Biol. Fishes 44, 183–197.

Chapman, L.J., McKenzie, D., 2009. Behavioural responses and ecological consequences. In: Richards, J.G., Farrell, A.P., Brauner, C.J. (Eds.), Hypoxia, Fish Physiology. vol. 27. Elsevier, San Diego, pp. 26–77.

Chapman, L.J., Chapman, C.A., Ogutu-Ohwayo, R., Chandler, M., Kaufman, L., Keiter, A.E., 1996. Refugia for endangered fishes from an introduced predator in Lake Nabugabo, Uganda. Conserv. Biol. 10, 554–561.

Chapman, L.J., Chapman, C.A., Crisman, T.L., Nordlie, F.G., 1998. Dissolved oxygen regimes of a Ugandan Crater Lake. Hydrobiologia 385, 201–211.

Chapman, L.J., Chapman, C.A., Brazeau, D., McGlaughlin, B., Jordan, M., 1999. Papyrus swamps and faunal diversification: geographical variation among populations of the African cyprinid *Barbus neumayeri*. J. Fish Biol. 54, 310–327.

Chapman, L.J., Galis, F., Shinn, J., 2000. Phenotypic plasticity and the possible role of genetic assimilation: hypoxia-induced trade-offs in the morphological traits of an African cichlid. Ecol. Lett. 3, 388–393.

Chapman, L.J., Balirwa, J., Bugenyi, F.W.B., Chapman, C.A., Crisman, T.L., 2001. Wetlands of East Africa: biodiversity, exploitation, and policy perspectives. In: Gopal, B., Junk, W.J., Davis, J.A. (Eds.), Wetlands Biodiversity. Backhuys Publisher, Leiden, pp. 101–132.

Chapman, L.J., Chapman, C.A., Nordlie, F.G., Rosenberger, A.E., 2002. Physiological refugia: swamps, hypoxia tolerance, and maintenance of fish biodiversity in the Lake Victoria Region. Comp. Biochem. Physiol. A 133, 421–437.

Chapman, L.J., Chapman, C.A., Olowo, J.P., Schofield, P.J., Kaufman, L.S., Seehausen, O., Ogutu-Ohwayo, R., 2003. Fish faunal resurgence in Lake Nabugabo, East Africa. Conserv. Biol. 7, 500–511.

Chapman, L.J., DeWitt, T.J., Tzenava, V., Paterson, J., 2007. Interdemic variation in the gill morphology of a eurytopic African cichlid. In: Proceedings of the 9th International Symposium on Fish Physiology, Toxicology, and Water Quality, EPA/600/R-07/010, pp. 209–225.

Chapman, L.J., Chapman, C.A., Witte, F., Kaufman, L., Balirwa, J., 2008a. Biodiversity conservation in African inland waters: Lessons of the Lake Victoria Basin. Verh. Internat. Verein Limnol. 30 (Part I), 16–34.

Chapman, L.J., Albert, J., Galis, F., 2008b. Developmental plasticity, genetic differentiation, and hypoxia-induced trade-offs in an African cichlid fish. Open. Evol. J. 2, 75–88.

Chevin, L.-M., Lande, R., Mace, G.M., 2010. Adaptation, plasticity, and extinction in a changing environment: towards a predictive theory. PLoS Biol. 8, e1000357.

Chretien, E., Chapman, L.J., 2016. Tropical fish in a warming world: thermal tolerance of Nile perch (*Lates niloticus* L.) in Lake Nabugabo, Uganda. Conserv. Physiol. 4, co2062.

Clark, T.D., Sandblom, E., Jutfelt, F., 2013. Aerobic scope measurements of fishes in an era of climate change: respirometry, relevance and recommendations. J. Exp. Biol. 216, 2771–2782.

Clarke, A., Johnston, N.M., 1999. Scaling of metabolic rate with body mass and temperature in teleost fish. J. Anim. Ecol. 68, 893–905.

Claussen, D.L., 1977. Thermal acclimation in ambystomatid salamanders. Comp. Biochem. Physiol. 58A, 333–340.

Cohen, A., Kaufman, L.S., Ogutu-Ohwayo, R., 1996. Anthropogenic threats, impacts, and conservation strategies in the African Great Lakes: a review. In: Johnson, T.C., Odada, E.O. (Eds.), The Limnology, Climatology, and Paleoclimatology of the East African Lakes. Gordon and Breech Publishers, Amsterdam, pp. 575–624.

Cohen, A.S., Gergurich, E.L., Kraemer, B.M., McGlue, M.M., McIntyre, P.B., Russell, J.M., et al., 2016. Climate warming reduces fish production and benthic habitat in Lake Tanganyika, one of the most biodiverse freshwater ecosystems. Proc. Natl. Acad. Sci. U. S. A. 113, 9563–9568.

Comte, L., Grenouillet, G., 2013. Do stream fish track climate change? Assessing distribution shifts in recent decades. Ecography 36, 1–11.

Comte, L., Olden, J., 2017. Climatic vulnerability of the world's freshwater and marine fishes. Nat. Clim. Change 7, 718–723.

Crisman, T.L., Chapman, L.J., Chapman, C.A., Prenger, J., 2001. Cultural eutrophication of a Ugandan highland crater lake: a 25-year comparison of limnological parameters. Verh. Internat. Verein. Limnol. 27, 3574–3578.

Crispo, E., 2010. Interplay among Phenotypic Plasticity, Local Adaptation, and Gene Flow. Ph.D. dissertation, McGill University, Montreal.

Crispo, E., Chapman, L.J., 2008. Population genetic structure across dissolved oxygen regimes in an African cichlid fish. Mol. Ecol. 17, 2134–2148.

Crispo, E., Chapman, L.J., 2010a. Geographical variation in phenotypic plasticity in response to dissolved oxygen in an African cichlid fish. J. Evol. Biol. 23, 2091–2103.

Crispo, E., Chapman, L.J., 2010b. Temporal variation in the population genetic structure of a riverine African cichlid fish. J. Hered. 101, 97–106.

Curtis, P.G., Slay, C.M., Harris, N.L., Tyukavina, A., Hansen, M.C., 2018. Classifying drivers of global forest loss. Science 361, 1108–1111.

Dahlke, F.T., Wohlrab, S., Butzin, M., Pörtner, H., 2020. Thermal bottlenecks in the life cycle define climate vulnerability of fish. Science 369, 65–70.

Darko, D., Trolle, D., Asmah, R., Bolding, K., Adjei, K.A., Odai, S.N., 2019. Modeling the impacts of climate change on the thermal and oxygen dynamics of Lake Volta. J. Great Lakes Res. 45, 73–86.

Descy, J.-P., André, L., Delvaux, C., Monin, L., Bouillon, S., Morana, C., et al., 2015. East African Great Lake Ecosystem Sensitivity to Changes. Final Report. Belgian Science Policy 2015, Brussels. Research Programme Science for a Sustainable Development.

Deutsch, C.A., Tewksbury, J.J., Huey, R.B., Sheldon, K.S., Ghalambor, C.K., Haak, D.C., Martin, P.R., 2008. Impacts of climate warming on terrestrial ectotherms across latitude. Proc. Natl. Acad. Sci. U. S. A. 105, 6668–6672.

Diamond, S.E., Chick, L.D., Perez, A., Strickler, S.A., Martin, R.A., 2018. Evolution of thermal tolerance and its fitness consequences: parallel and non-parallel responses to urban heat islands across three cities. Proc. Roy. Soc. B. 285. https://doi.org/10.1098/rspb.2018.0036. 20180036.

Dudgeon, D., Arthington, A.H., Gessner, M.O., Kawabata, Z.-I., Knowler, D.J., Lévêque, C., et al., 2006. Freshwater biodiversity: importance, threats, status and conservation challenges. Biol. Rev. 81, 163–182.

Efitre, J., 2007. Life History Variation in Tilapia Populations Within the Crater Lakes of Western Uganda: The Role of Size-Selective Predation. Ph.D. dissertation, University of Florida, Gainesville.

Efitre, J., Chapman, L.J., Murie, D., 2009. Predictors of fish condition in introduced tilapias of Uganda crater lakes in relation to fishing pressure and deforestation. Environ. Biol. Fishes 85, 63–75.

Eliason, E.J., Clark, T.D., Hague, M.J., Hanson, L.M., Gallagher, Z.S., Jeffries, K.M., et al., 2011. Differences in thermal tolerance among sockeye salmon populations. Science 332, 109–112.

Ern, R., Norin, T., Gamperl, K., Esbaugh, A.J., 2016. Oxygen dependence of upper thermal limits in fishes. J. Exp. Biol. 219, 3376–3383.

FAO, 2020. The State of World Fisheries and Aquaculture. Sustainability in Action. Rome, https://doi.org/10.4060/ca9229en.

Farrell, A.P., 2009. Environment, antecedents and climate change: lessons from the study of temperature physiology and river migration of salmonids. J. Exp. Biol. 212, 3771–3780.

Farrell, A.P., 2016. Pragmatic perspective on aerobic scope: peaking, plummeting, pejus and apportioning. J. Fish Biol. 88, 322–343.

Foley, J.A., DeFries, R., Asner, G.P., Barford, C., Bonan, G., Carpenter, S.R., et al., 2005. Global consequences of land use. Science 309, 570–574.

Frederich, M., Pörtner, H.O., 2000. Oxygen limitation of thermal tolerance defined by cardiac and ventilatory performance in spider crab, *Maja squinado*. Am. J. Physiol. Regul. Integr. Comp. Physiol. 279, R1531–R1538.

Fricke, R., Eschmeyer, W.N., Van der Laan, R., 2021. Eschmeyer's Catalog of Fishes: Genera, Species, References. http://researcharchive.calacademy.org/research/ichthyology/catalog/fishcatmain.asp. Electronic version accessed 20 February, 2021.

Frishkoff, L.O., Hadly, E.A., Daily, G.C., 2015. Thermal niche predicts tolerance to habitat conversion in tropical amphibians and reptiles. Glob. Change Biol. 21, 3901–3916.

Fry, F.E.J., 1947. Effects of the environment on animal activity. Ont. Fish. Res. Lab. 68, 1–52.

Fry, F.E.J., Hart, J.S., 1948. The relation of oxygen consumption in the goldfish. Biol. Bull. 94, 66–77.

Fugère, V., Mehner, T., Chapman, L.J., 2018. Impacts of deforestation-induced warming on the metabolism, growth and trophic interactions of an afrotropical stream fish. Funct. Ecol. 32, 1343–1357.

Funge-Smith, S., Bennett, A., 2019. A fresh look at inland fisheries and their role in food security and livelihoods. Fish. Fish. 20, 1176–1195.

Gibbs, H.K., Ruesch, A.S., Achard, F., Clayton, M.K., Holmgren, P., Ramankutty, N., Foley, J.A., 2010. Tropical forests were the primary sources of new agricultural land in the 1980s and 1990s. Proc. Natl. Acad. Sci. U. S. A. 107, 16732–16737.

Gilbert, B., Tunney, T.D., McCann, K.S., DeLong, J.P., Vasseur, D.A., Savage, V., et al., 2014. A bioenergetic framework for the temperature dependence of trophic interactions. Ecol. Lett. 17, 902–914.

Goldschmidt, T., Witte, F., Wanink, J.H., 1993. Cascading effects of the introduced Nile perch on the detritivorous/phytoplanktivorous species in the sublittoral areas of Lake Victoria. Conserv. Biol. 7, 686–700.

Gräns, A., Jutfelt, F., Sandblom, E., Jönsson, E., Wiklander, K., Seth, H., et al., 2014. Aerobic scope fails to explain the detrimental effects on growth resulting from warming and elevated $CO_2$ in Atlantic halibut. J. Exp. Biol. 217, 711–717.

Gray, S.M., McDonnell, L.H., Cinquemani, F.G., Chapman, L.J., 2012. As clear as mud: turbidity induces behavioral changes in the African cichlid *Pseudocrenilabrus multicolor*. Curr. Zool. 58, 146–157.

Greenwood, P.H., 1965. The cichlid fishes of Lake Nabugabo, Uganda. Bull. British Mus. Nat. Hist. (Zool.) 12, 315–357.

Greenwood, P.H., 1966. The Fishes of Uganda. The Uganda Society, Kampala.

Grenchik, M.K., Donelson, J.M., Munday, P.L., 2013. Evidence for developmental thermal acclimation in the damselfish *Pomacentrus moluccensis*. Coral Reefs 32, 85–90.

Gunderson, A.R., Stillman, J.H., 2015. Plasticity in thermal tolerance has limited potential to buffer ectotherms from global warming. Proc. Roy. Soc. B 282. 20150401.

Hansen, M.C., Potapov, P.V., Moore, R., Hancher, M., Turubanova, S.A., Tyukavina, A., et al., 2013. High-resolution global maps of 21st-century forest cover change. Science 342, 850–853.

Harniman, R., Merritt, T.J.S., Chapman, L.J., Lesbarreres, D., Martinez, M.L., 2013. Population differentiation of the African cyprinid *Barbus neumayeri* across dissolved oxygen regimes. Ecol. Evol. 3, 1495–1506.

Harrod, C., Ramirez, A., Valbo-Jorgensen, J., Funge-Smith, S., 2018. Current anthropogenic stress and projected effect of climate change on global inland fisheries. In: Barange, M., Bahri, T., Beveridge, M., Cochrane, K., Funge-Smith, S. (Eds.), Impacts of Climate Change on Fisheries and Aquaculture: Synthesis of Current Knowledge, Adaptation and Mitigation Options. Food and Agriculture Organization of the UN, Rome, pp. 393–448.

Healy, T.M., Schulte, P.M., 2012. Thermal acclimation is not necessary to maintain a wide thermal breadth of aerobic scope in the common killisfish (*Fundulus heteroclitus*). Physiol. Biochem. Zool. 85, 107–119.

Hecky, R.E., 1993. The eutrophication of Lake Victoria. Verh. Internat. Verein. Limnol. 25, 39–48.

Hecky, R.E., Bugenyi, F.W.B., Ochumba, P., Talling, J.F., Mugidde, R., Gophen, M., Kaufman, L., 1994. Deoxygenation of the deep-water of Lake Victoria, East-Africa. Limnol. Oceanogr. 39, 1476–1481.

Hecky, R.E., Mugidde, R., Ramlal, P.S., Talbot, M.R., Kling, G.W., 2010. Multiple stressors cause rapid ecosystem change in Lake Victoria. Freshwat. Biol. 55, 19–42.

Heino, J., Virkkala, R., Toivonen, H., 2009. Climate change and freshwater biodiversity: detected patterns, future trends and adaptations in northern regions. Biol. Rev. 84, 39–54.

Huey, R.B., Hertz, P.E., 1984. Is a jack-of-all temperatures a master of none? Evolution 38, 441–444.

Huey, R.B., Kingsolver, J.G., 1989. Evolution of thermal sensitivity of ectotherm performance. Trends Ecol. Evol. 4, 131–135.

Huey, R.B., Stevenson, R.D., 1979. Integrating thermal physiology and ecology of ectotherms: discussion of approaches. Am. Zool. 19, 357–366.

Hughes, R.H., Hughes, J.S., 1992. A Directory of African Wetlands, World Conservation Union (IUCN), Gland/United National Environment Program (UNEP). Nairobi/World Conservation Monitoring Centre (WCMC), Cambridge.

Johnson, T.C., Scholz, C.A., Talbot, M.R., Kelts, K., Ricketts, R.D., Ngobi, G., et al., 1996. Late Pleistocene desiccation of Lake Victoria and rapid evolution of cichlid fishes. Science 273, 1091–1093.

Jutfelt, R., Norin, T., Ern, R., Overgaard, J., Wang, T., McKenzie, D.J., et al., 2018. Oxygen- and capacity-limited thermal tolerance: blurring ecology and physiology. J. Exp. Biol. 211, 1–4.

Katsev, S., Aaberg, A.A., Crowe, S.A., Hecky, R.E., 2014. Recent warming of Lake Kivu. PLoS One 10, e109084.

Kaufman, L.S., 1992. The lessons of Lake Victoria: catastrophic change in species rich freshwater ecosystems. Bioscience 42, 846–858.

Kaufman, L., Ochumba, P., 1993. Evolutionary and conservation biology of cichlid fishes as revealed by faunal remnants in northern Lake Victoria. Conserv. Biol. 7, 719–730.

Kaufman, L.S., Chapman, L.J., Chapman, C.A., 1996. The great lakes. In: McClanahan, T.R., Young, T.P. (Eds.), Ecosystems and Their Conservation in East Africa. Oxford University Press, New York, pp. 178–204.

Kishe-Machumu, M.A., van Rijssel, J.C., Wanink, J.H., Witte, F., 2015. Differential recovery and spatial distribution pattern of haplochromine cichlids in the Mwanza Gulf of Lake Victoria. J. Great Lakes Res. 41, 454–462.

Knouft, J.H., Ficklin, D.L., 2017. The potential impacts of climate change on biodiversity in flowing freshwater systems. Ann. Rev. Ecol. Evol. Syst. 48, 111–133.

Kolding, J., Haug, L., Stefansson, S., 2008. Effect of ambient oxygen on growth and reproduction in Nile tilapia (*Oreochromis niloticus*.). Can. J. Fish. Aquat. Sci. 65, 1413–1424.

Kolding, J., van Zwieten, P., Marttin, F., Funge-Smith, S., Poulain, F., 2019. Freshwater small pelagic fish and their fisheries in major African lakes and reservoirs in relation to food security and nutrition. In: FAO, Fisheries and Aquaculture Technical Paper 642, Rome.

Lake Victoria Fisheries Organization (LVFO), 2016. Fisheries Management Plan III (FMP III) for Lake Victoria Fisheries, 2016–2020. Jinja, Uganda.

Langerhans, R.B., Chapman, L.J., DeWitt, T.J., 2007. Complex phenotype-environment associations revealed in an East African cyprinid. J. Evol. Biol. 20, 1171–1181.

Lannig, G., Bock, C., Sartoris, F.J., Pörtner, H.O., 2004. Oxygen limitation of thermal tolerance in cod, *Gadus morhua* L. studied by non-invasive NMR techniques and on-line venous oxygen monitoring. Am. J. Physiol. 287, R902–R910.

Lapointe, D., Cooperman, M., Chapman, L., Clark, T., Aldalberto, V., Ferreira, M., et al., 2018. Predicted impacts of climate warming on aerobic performance and upper thermal tolerance of six tropical freshwater fishes spanning three continents. Conserv. Physiol. 6. coy056.

Larke, A., Tietze, K., Halbwachs, M., Wüest, A., 2004. Response of Lake Kivu stratification to lava inflow and climate warming. Limnol. Oceanogr. 49, 778–783.

Laurance, W.F., Sayer, J., Cassman, K.G., 2014. Agricultural expansion and its impacts on tropical nature. Trends Ecol. Evol. 29, 107–116.

Lawrence, D., Vandecar, K., 2015. Effects of tropical deforestation on climate and agriculture. Nat. Clim. Change 5, 27–36.

Lee, C.G., Farrell, A.P., Lotto, A., MacNutt, M.J., Hinch, S.G., Healey, M.C., 2003. The effect of temperature on swimming performance and oxygen consumption in adult sockeye (*Oncorhynchus nerka*) and coho (*O. kisutch*) salmon stocks. J. Exp. Biol. 206, 3239–3251.

Lemoine, M., Barluenga, M., Lucek, K., Haesler, M., Chapman, L.J., Chapman, C.A., Seehausen, O., 2018. Recent sympatric speciation involving habitat-associated nuptial colour polymorphism in a crater lake cichlid. Hydrobiologia 832, 297–315.

Lévêque, C., 1997. Biodiversity Dynamics and Conservation: The Freshwater Fish of Tropical Africa. Cambridge University Press, Cambridge.

Lévêque, C., Paugy, D., 2010. Animal resources and diversity in Africa, freshwater fishes in Africa. In: Encyclopedia of Life Support Systems (EOLSS). EOLSS Publishers, Paris. Developed under the auspices of the UNESCO.

Likens, G.E., Bormann, F.H., 1974. Linkages between terrestrial and aquatic ecosystems. BioScience 24, 447–456.

Lundberg, J.G., Kottelat, M., Smith, G.R., Stiassny, M.L.J., Gill, A.C., 2000. So many fishes, so little time: an overview of recent ichthyological discovery in continental waters. Ann. Miss. Bot. Gard. 87, 26–62.

Lutterschmidt, W.I., Hutchison, V.H., 1997. The critical thermal maximum: history and critique. Can. J. Zool. 75, 1561–1574.

MAAIF, 2017. National Fisheries and Aquaculture Policy: Optimising Benefits from Fisheries and Aquaculture Resources for Socio-Economic Transformation. Retrieved from https://www.worldfishcenter.org/sites/default/files/u24/Annex-35-National-Aquaculture-Policy_Uganda.pdf.

Maberly, S.C., O'Donnell, R.A., Woolway, R.I., Cutler, M.E.J., Gong, M., Jones, I.D., et al., 2020. Global lake thermal regions shift under climate change. Nat. Commun. 11, 1232.

Marshall, B.E., 2012. Does climate change really explain changes in the fisheries productivity of Lake Kariba (Zambia-Zimbabwe)? Trans. R. Soc. S. Afr. 67, 45–51.

Marshall, B.E., Ezekiel, C.N., Gichuki, J., Mkumbo, O.C., Sitoki, L., Wanda, F., 2013. Has climate change disrupted stratification patterns in Lake Victoria, East Africa? Afr. J. Aquat. Sci. 38, 249–253.

Martinez, M.S., Chapman, L.J., Grady, J.M., Rees, B.B., 2004. Interdemic variation in hematocrit and lactate dehydrogenase in the African cyprinid *Barbus neumayeri*. J. Fish Biol. 65, 1056–1069.

Martinez, M.L., Raynard, E.L., Rees, B.B., Chapman, L.J., 2011. Oxygen limitation and tissue metabolic potential of the African fish *Barbus neumayeri*: roles of native habitat and acclimatization. BMC Ecol. 11, 2. www.biomedcentral.com/1472-6785/11/2.

Masese, F.O., Salcedo-Borda, J.S., Gettel, G.M., Irvine, K., McClain, M.E., 2017. Influence of catchment land use and seasonality on dissolved organic matter composition and ecosystem metabolism in headwater streams of a Kenyan river. Biogeochemistry 132, 1–22.

McBryan, T.L., Anttila, K., Healy, T.M., Schulte, P.M., 2013. Responses to temperature and hypoxia as interacting stressors in fish: implications for adaptation to environmental change. Integr. Comp. Biol. 53, 648–659.

McDonnell, L.J., Chapman, L.J., 2015. At the edge of the thermal window: effects of elevated temperature on the resting metabolism, hypoxia tolerance and upper critical thermal limit of a widespread African cichlid. Conserv. Physiol. 3, 1–13.

McDonnell, L.H., Chapman, L.J., 2016. Effects of thermal increase on aerobic capacity and swim performance in a tropical inland fish. Comp. Biochem. Physiol. Part A. 199, 62–70.

McDonnell, L.H., Reemeyer, J., Chapman, L.J., 2019. Independent and interactive effects of long-erm exposure to hypoxia and elevated water temperature on behavior and thermal tolerance of an equatorial cichlid. Physiol. Biochem. Zool. 92, 253–265.

McNeil, G., Friesen, C., Gray, S.M., Aldredge, A., Chapman, L.J., 2016. Male colour variation in a eurytopic African cichlid: the role of diet and hypoxia. Biol. J. Linn. Soc. 118, 551–568.

Meier, J.I., Marques, D.A., Mwaiko, S., Wagner, C.E., Excoffier, L., Seehausen, O., 2017. Ancient hybridization fuels rapid cichlid fish adaptive radiations. Nat. Commun. 8, 14363.

Melack, J.M., 1978. Morphometric, physical and chemical features of volcanic crater lakes of western Uganda. Arch. Hydrobiol. 84, 430–453.

Melnychuk, M.C., Chapman, L.J., 2002. Hypoxia tolerance of two haplochromine cichlids: swamp leakage and potential for interlacustrine dispersal. Environ. Biol. Fishes 65, 99–110.

Mkumbo, O.C., Marshall, B.E., 2015. The Nile perch fishery of Lake Victoria: current status and management challenges. Fish. Manage. Ecol. 22, 56–63.

Morley, S.A., Peck, L.S., Sunday, J.M., Heiser, S., Bates, A.E., 2019. Physiological acclimation and persistence of ectothermic species under extreme heat events. Glob. Ecol. Biogeogr. 28, 1018–1037.

Mugidde, R., 1993. The increase in phytoplankton primary productivity and biomass in Lake Victoria (Uganda). Verh. Internat. Verein. Limnol. 25, 846–849.

Mugidde, R., 2001. Nutrient Status and Planktonic Nitrogen Fixation in Lake Victoria, Africa. Ph. D. dissertation, University of Waterloo, Waterloo.

Munday, P.L., Warner, R.R., Monro, K., Pandolfi, J.M., Marshall, D.J., 2013. Predicting evolutionary responses to climate change in the sea. Ecol. Lett. 16, 1488–1500.

Ndebele-Murisa, M.R., 2014. Associations between climate, water environment, and phytoplankton production in African lakes. In: Sebastia, M.T. (Ed.), Phytoplankton Biology, Classification and Environmental Impacts. Nova Science Publishers Inc, New York, pp. 37–62.

Neuheimer, A.B., Thresher, R.W., Lyle, J.M., Semmens, J.M., 2011. Tolerance limit for fish growth exceeded by warming waters. Nat. Clim. Change. https://doi.org/10.1038/NCLIMATE1084.

Niang, I., Ruppel, O.C., Abdrabo, M.A., Essel, A., Lennard, C., Padgham, J., Urquhart, P., 2014. Africa. In: Barros, V.R., Field, C.B., Dokken, D.J., Mastrandrea, K.J., Mach, K.J., Bilir, T.E., et al. (Eds.), Climate Change 2014: Impacts, Adaptation, and Vulnerability. Part B: Regional Aspects. Contribution of Working Group II to the Fifth Assessment Report of the Intergovernmental Panel on Climate Change. Cambridge University Press, Cambridge, New York, NY, pp. 1199–1265.

Nilsson, G.E., Crawley, N., Lunde, I.G., Munday, P.L., 2009. Elevated temperature reduces the respiratory scope of coral reef fishes. Glob. Change Biol. 15, 1405–1412.

Norin, T., Malte, H., Clark, T.D., 2014. Aerobic scope does not predict the performance of a tropical eurythermal fish at elevated temperatures. J. Exp. Biol. 217, 244–251.

Nowakowski, A.J., Watling, J.I., Thompson, M.E., Brusch IV, G.A., Catenazzi, A., Whitfield, S.-M., et al., 2018. Thermal biology mediates responses of amphibians and reptiles to habitat modification. Ecol. Lett. 21, 345–355.

Nyboer, E.A., Chapman, L.J., 2013. Movement and home range of introduced Nile perch (*Lates niloticus*) in Lake Nabugabo, Uganda: implications for ecological divergence and fisheries management. Fish. Res. 13, 18–29.

Nyboer, E.A., Chapman, L.J., 2017. Elevated temperature and acclimation time affect metabolic performance in the heavily exploited Nile perch of Lake Victoria. J. Exp. Biol. 220, 3782–3793.

Nyboer, E.A., Chapman, L.J., 2018. Cardiac plasticity influences aerobic performance and thermal tolerance in a tropical, freshwater fish at elevated temperatures. J. Exp. Biol. 221. jeb178087.

Nyboer, E.A., Liang, C., Chapman, L.J., 2019. Assessing the vulnerability of Africa's freshwater fishes to climate change: a continent-wide trait-based analysis. Conserv. Biol. 236, 505–520.

Nyboer, E.A., Chretien, E., Chapman, L.J., 2020. Divergence in aerobic scope and thermal tolerance of an introduced piscivore is related to local thermal regime. J. Fish Biol. 97, 231–245.

Nyboer, E.A., Lin, H.Y., Bennett, J.R., Gabriel, J., Twardek, W., Chhor, A., et al., 2021. Global assessment of marine and freshwater recreational fish reveals mismatch in climate change vulnerability and conservation. Glob. Change Biol., 1–26.

Ochumba, P.B.O., 1990. Massive fish kills within the Nyanza Gulf of Lake Victoria, Kenya. Hydrobiologia 208, 93–99.

Ogutu-Ohwayo, R., 1993. The effects of predation by Nile perch, *Lates niloticus* L., on the fish of Lake Nabugabo, with suggestions for conservation of endangered endemic cichlids. Conserv. Biol. 7, 701–711.

Ogutu-Ohwayo, R., 1994. Adjustments in Fish Stocks and in Life History Characteristics of the Nile perch, *Lates niloticus* L. in Lakes Victoria, Kyoga, and Nabugabo. Ph.D. dissertation, University of Manitoba, Winnipeg.

Ogutu-Ohwayo, R., Natugonza, V., Musinguzi, L., Olokitum, M., Naigaga, S., 2016. Implications of climate variability and change for African lake ecosystems, fisheries productivity, and livelihoods. J. Great Lakes Res. 42, 498–510.

Ogutu-Ohwayo, R., Natugonza, V., Olokotum, M., Rwezawula, P., Lugya, J., Musinguzi, L., 2020. Biogeography: Lakes—African Great Lake. In: Goldstein, M.I., DellaSala, D. (Eds.), Encyclopedia of the World's Biomes, 4. vol. 4. Elsevier Inc, pp. 243–260.

Olaka, L.A., Odada, E.O., Trauth, M.H., Olago, D.O., 2010. The sensitivity of East African rift lakes to climate fluctuations. J. Paleolimnol. 44, 629–644.

Olowo, J.P., Chapman, L.J., 1996. Papyrus swamps and variation in the respiratory behaviour of the African fish *Barbus neumayeri*. Afr. J. Eco. 34, 211–222.

O'Reilly, C.M., Alin, S.R., Plisnier, P.-D., Cohen, A.S., McKee, B.A., 2003. Climate change decreases aquatic ecosystem productivity of Lake Tanganyika, Africa. Nature 424, 766–768.

Owori-Wadunde, A., 2009. The Feeding Ecology, Ontogeny and Larval Feeding in *Labeo victorianus* Boulenger 1901 (Pisces: Cyprinidae). Ph.D. Dissertation, Makerere University, Kampala.

Oyugi, D., 2019. 521. Lake Victoria Basin. In: Freshwater Ecoregions of the World. WWF/TNC. https://www.feow.org/.

Paterson, J.A., Chapman, L.J., 2009. Fishing down and fishing hard: ecological change in the Nile perch of Lake Nabugabo, Uganda. Ecol. Freshwat. Fish 18, 380–394.

Pörtner, H.O., 2010. Oxygen- and capacity-limitation of thermal tolerance: a matrix for integrating climate-related stressor effects in marine ecosystems. J. Exp. Biol. 213, 881–893.

Pörtner, H.O., Farrell, A.P., 2008. Physiology and climate change. Ecology 322, 690–692.

Pörtner, H.O., Knust, R., 2007. Climate change affects marine fishes through the oxygen limitation of thermal tolerance. Science 315, 95–97.

Pörtner, H.O., Bock, C., Mark, F.C., 2017. Oxygen- and capacity-limited thermal tolerance: bridging ecology and physiology. J. Exp. Biol. 220, 2685–2696.

Ramankutty, N., Mehrabi, Z., Waha, K., Jarvis, L., Kremen, C., Herrero, M., et al., 2018. Trends in global agricultural land use: implications for environmental health and food security. Ann. Rev. Plant Biol. 69, 789–815.

Reardon, E.E., Chapman, L.J., 2009. Hypoxia and life-history traits in a eurytopic African cichlid. J. Fish Biol. 75, 1683–1699.

Reardon, E.E., Chapman, L.J., 2012. Fish embryo and juvenile size under hypoxia in the mouth-brooding African cichlid *Pseudocrenilabrus multicolor*. Curr. Zool. 58, 410–412.

Regan, M.D., Mandic, M., Dhillon, R.S., Lau, G.Y., Farrell, A.P., Schulte, P.M., et al., 2019. Don't throw the fish out with the respirometry water. J. Exp. Biol. 222. jeb.200253.

Reid, A.J., Farrell, M.J., Luke, M.N., Chapman, L.J., 2013a. Implications of hypoxia tolerance for wetland refugia use in Lake Nabugabo, Uganda. Ecol. Freshwat. Fish 22, 421–429.

Reid, A.J., Chapman, L.J., Ricciardi, A., 2013b. Wetland edges as peak refugia from an introduced piscivore. Aquat. Conserv. 23, 646–655.

Reid, A.J., Carlson, A.K., Creed, I.F., Eliason, E.J., Gell, P.A., Johnson, P.T.J., et al., 2019. Emerging threats and persistent conservation challenges for freshwater biodiversity. Biol. Rev. 94, 849–873.

Riahi, K., Rao, S., Krey, V., Cho, C., Chirkov, V., Fischer, G., Kindermann, G., Nakicenovic, N., Rafaj, P., 2011. RCP 8.5—a scenario of comparatively high greenhouse gas emissions. Clim. Change 109 (33).

Richards, J.G., Farrell, A.P., Brauner, C.J. (Eds.), 2009. Hypoxia. Fish Physiology. vol. 27. Elsevier Inc, Amsterdam.

Robb, T., Abrahams, M.V., 2003. Variation in tolerance to hypoxia in a predatory and prey species: an ecological advantage of being small? J. Fish Biol. 62, 1067–1081.

Rutjes, H.A., 2006. Phenotypic Responses to Lifelong Hypoxia in Cichlids. Ph.D. dissertation, Leiden University, Leiden.

Saulnier-Talbot, E., Gregory-Eaves, I., Simpson, K.G., Efitre, J., Nowlan, T.E., Taranu, Z., Chapman, L.J., 2014. Small changes in climate can profoundly alter the dynamics and ecosystem services of tropical crater Lakes. PLoS One 9, 1–8.

Schaack, S.R., Chapman, L.J., 2003. Interdemic variation in the African cyprinid *Barbus neumayeri*: correlations among hypoxia, morphology, and feeding performance. Can. J. Zool. 81, 430–440.

Schofield, P.J., Chapman, L.J., 2000. Hypoxia tolerance of introduced Nile perch: implications for survival of indigenous fishes in the Lake Victoria basin. Afr. Zool. 35, 35–42.

Schulte, P.M., 2015. The effects of temperature on aerobic metabolism: towards a mechanistic understanding of the responses of ectotherms to a changing environment. J. Exp. Biol. 218, 1856–1866.

Schulte, P.M., Healy, T.M., Fangue, N.A., 2011. Thermal performance curves, phenotypic plasticity, and the time scales of temperature exposure. Integr. Comp. Biol. 51, 691–702.

Schwartz, J.D.M., Pallin, M.J., Michener, R.H., Mbabazi, D., Kaufman, L., 2006. Effect of Nile perch, *Lates niloticus*, on functional and specific fish diversity in Uganda's Lake Kyoga system. Afr. J. Ecol. 44, 145–156.

Seebacher, F., White, C.R., Franklin, C.E., 2015. Physiological plasticity increases resilience of ectothermic animals to climate change. Nat. Clim. Change 5, 61–66.

Seehausen, O., van Alphen, J.J.M., Witte, F., 1997. Cichlid fish diversity threatened by eutrophication that curbs sexual selection. Science 277, 1808–1811.

Senior, R.A., Hill, J.K., González del Pliego, P., Goode, L.K., Edwards, D.P., 2017. A pantropical analysis of the impacts of forest degradation and conversion on local temperature. Ecol. Evol. 7, 7897–7908.

Sharma, G., Gray, D.K., Read, J.S., O'Reilly, C.M., Schneider, P., Qudrat, A., et al., 2015. A global database of lake surface temperatures, collected by in situ and satellite methods from 1985 to 2009. Sci. Data 2. https://doi.org/10.1038/sdata.2015.8, 150008.

Sharpe, D.M.T., Chapman, L.J., 2014. Niche expansion in a resilient endemic species following introduction of a novel top predator. Freshwat. Biol. 59, 2539–2554.

Sharpe, D.M.T., Chapman, L.J., 2018. Contemporary phenotypic change in correlated characters in the African cyprinid, *Rastrineobola argentea*. Biol. J. Linn. Soc. 124, 85–98.

Sharpe, D., Wandera, S.B., Chapman, L.J., 2012. Life history change in response to fishing and an introduced predator in the East African cyprinid *Rastrineobola argentea*. Evol. Appl. 5, 677–693.

Sitoki, L, Gichuki, J., Ezekiel, C., Wanda, F., Mkumbo, O.C., Marshall, B., 2010. The environment of Lake Victoria (East Africa): current status and historical changes. Internat. Rev. Hydrobiol. 95, 209–223.

Spigel, R.H., Coulter, G.W., 1996. Comparison of hydrology and physical limnology of the East African Great Lakes: Tanganyika, Malawi, Victoria, Kivu and Turkana (with references to some North American Great Lakes). In: Johnson, C., Odada, E.O. (Eds.), The Limnology, Climatology, and Paleoclimatology of the East African Lakes. Gordon and Breach Publishers, Amsterdam, pp. 103–135.

Stager, J.C., Hecky, R.E., Grzesik, D., Cumming, B.F., Kling, H., 2009. Diatom evidence for the timing and causes of eutrophication in Lake Victoria, East Africa. Hydrobiologia 636, 463–478.

Steinberg, C.E.W., 2012. One stressor prepares for the next one to come: cross-tolerance. In: Steinberg, C.E.W. (Ed.), Stress Ecology: Environmental Stress as Ecological Driving Force and Key Player in Evolution. Springer, Netherlands, Dordrecht, pp. 311–325.

Strayer, D.L., Dudgeon, D., 2010. Freshwater biodiversity conservation: recent progress and future challenges. J. N. Am. Benth. Soc. 29, 344–358.

Sunday, J.M., Bates, A.E., Dulvy, N.K., 2011. Global analysis of thermal tolerance and latitude in ecotherms. Proc. Roy. Soc. B. 278, 1823–1830.

Taabu-Munyaho, A., Marshall, B.E., Tomasson, T., Marteinsdottir, G., 2016. Nile perch and the transformation of Lake Victoria. Afr. J. Aquat. Sci. 41, 127–142.

Talling, J.F., 1966. The annual cycle of stratification and phytoplankton growth in Lake Victoria (East Africa). Int. Revue Ges. Hydrobiol. 51 (545), 621.

Tang, F.H.M., Lenzen, M., McBratney, A., Maggi, F., 2021. Risk of pesticide pollution at the global scale. Nat. Geosci. 14, 206–210.

Tewksbury, J.J., Huey, R.B., Deutsch, C.A., 2008. Putting the heat on tropical animals. Science 320, 1296.

Thieme, M.L., Teugels, G., Abell, R., Burgess, N., Dinerstein, E., Kamdem-Toham, A., et al., 2005. Freshwater Ecoregions of Africa and Madagascar: A Conservation Assessment. World Wildlife Fund. Island Press, Washington, pp. 361–372. Ch 2: Approach pp, 22–34. Appendix F: Ecoregion descriptions.

Thompson, K., Shewry, P.R., Woolhouse, H.W., 1979. Papyrus swamp development in the Upemba Basin, Zaire: studies of population structure in *Cyperus papyrus* stands. Bot. J. Linn. Soc. 78, 299–316.

Thomson, A.M., Calvin, K.V., Smith, S.J., Kyle, G.P., Volke, A., Patel, P., Delgado-Arias, S., Bond-Lamberty, B., Wise, M.A., Clarke, L.E., Edmonds, J.A., 2011. RCP4. 5: a pathway for stabilization of radiative forcing by 2100. Clim. Change 109 (1–2), 77.

Tilman, D., Fargione, J., Wolff, B., D'Antonio, C., Dobson, A., Howarth, R., et al., 2001. Forecasting agriculturally driven global environmental change. Science 292, 281–284.

Umwali, E.D., Kurban, A., Isabwe, A., Mind'je, R., Azadi, H., Guo, Z., et al., 2021. Spatio-seasonal variation of water quality influenced by land use and land cover in Lake Muhazi. Sci. Rep. 11, 17376.

van Rijssel, J.C., Witte, F., 2013. Adaptive responses in resurgent Lake Victoria cichlids over the past 30 years. Evol. Ecol. 27, 253–267.

van Rijssel, J.C., Hoogwater, E.S., Kishe-Machumu, M.A., van Reenen, E., Spits, K.V., van der Stelt, R.C., et al., 2014. Fast adaptive responses in the oral jaw of Lake Victoria cichlids. Evolution 1, 179–189.

van Rijssel, J.C., Hecky, R.E., Kishe-Machumu, M.A., Meijer, S.E., Pols, J., van Tienderen, K.M., et al., 2016. Climatic variability in combination with eutrophication drives adaptive responses in the gills of Lake Victoria cichlids. Oecologia 182, 1187–1201.

Vannote, R.L., Minshall, G.W., Cummins, K.W., Sedell, J.R., Cushing, C.E., 1980. The river continuum concept. Can. J. Fish. Aquat. Sci. 37, 130–137.

Verschuren, D., Johnson, T.C., Kling, H.J., Edgington, D.N., Leavitt, P.R., Brown, E.T., et al., 2002. History and timing of human impact on Lake Victoria, East Africa. Proc. R. Soc. Lond. Ser. B 269, 289–294.

Visser, M.E., 2008. Keeping up with a warming world: assessing the rate of adaptation to climate change. Proc. R. Soc. Lond. B 275, 649–659.

Vörösmarty, C.J., McIntyre, P.B., Gessner, M.O., Dudgeon, D., Prusevich, A., Green, P., et al., 2010. Global threats to human water security and river biodiversity. Nature 467, 555–561.

Wang, T., Lefevre, S., Iverson, N.K., Findorf, I., Buchanan, R., McKenzie, D.J., 2014. Anemia only causes a small reduction in the upper critical temperature of sea bass: is oxygen delivery the limiting factor of acute warming in fishes? J. Exp. Biol. 217, 4275–4278.

Wanink, J.H., 1998. The Pelagic Cyprinid *Rastrineobola argentea* as a Crucial Link in the Disrupted Ecosystem of Lake Victoria. Ph.D. Dissertation, University of Leiden, Leiden.

Wanink, J.H., 1999. Prospects for the fishery on the small pelagic *Rastrineobola argentea* in Lake Victoria. Hydrobiologia 407, 183–189.

Wanink, J.H., Witte, F., 2000. Rapid morphological changes following niche shift in the zooplanktivorous cyprinid *Rastrineobola argentea* from Lake Victoria. Neth. J. Zool. 50, 365–372.

Wanink, J.H., Kashindye, J.J., Goudswaard, K.P.C., Witte, F., 2001. Dwelling at the oxycline: does increased stratification provide a predation refugium for the Lake Victoria sardine *Rastrineobola argentea*? Freshwat. Biol. 46, 75–85.

Wiens, K., Crispo, E., Chapman, L.J., 2014. Phenotypic plasticity is maintained despite geographical isolation in an African cichlid fish, *Pseudocrenilabrus multicolor*. Integr. Zool. 9, 85–96.

Witte, F., Goldschmidt, T., Wanink, J.H., van Oijen, M.J.P., Goudswaard, P.C., Witte-Maas, E.L.M., Bouton, N., 1992. The destruction of an endemic species flock: quantitative data on the decline of the haplochromine cichlids of Lake Victoria. Environ. Biol. Fishes 34, 1–28.

Witte, F., Msuku, B.S., Wanink, J.H., Seehausen, O., Katunzi, E.F.B., Goudswaard, K.P.C., Goldschmidt, T., 2000. Recovery of cichlid species in Lake Victoria: an examination of factors leading to differential extinction. Rev. Fish. Biol. Fish. 10, 233–241.

Witte, F., Wanink, J.H., Kishe-Machumu, M.A., 2007. Species distinction and the biodiversity crisis in Lake Victoria. Trans. Am. Fish. Soc. 136, 1146–1159.

Witte, F., Welten, M., Heemskerk, M., van der Stap, I., Ham, L., Rutjes, H., Wanink, J., 2008. Major morphological changes in a Lake Victoria cichlid fish within two decades. Biol. J. Linn. Soc. 94, 41–52.

Wood, C.M., 2018. The fallacy of the Pcrit—are there more useful alternatives? J. Exp. Biol. 221. jeb163717.

Woolway, R.I., Maberly, S.C., 2020. Climate velocity in inland standing waters. Nat. Clim. Change 10, 1124–1129.

Woolway, R.I., Kraemer, B.M., Lenters, J.D., Merchant, C.J., O'Reilly, C.M., Sharma, S., 2020. Nat. Rev. Earth Environ. 1, 388–403.
World Meteorological Organization (WMO), 2021. The State of the Climate in Africa 2020. WMO-No. 1275. World Meteorological Organization, Geneva.
Worthington, E.B., Lowe-McConnell, R.H., 1994. African lakes reviewed: creation and destruction of biodiversity. Environ. Conserv. 21, 199–213.
Wright, S.J., Muller-Landau, H.C., Schipper, J., 2009. The future of tropical species on a warmer planet. Conserv. Biol. 23, 1418–1426.
WWF, 2006. Climate Change Impacts on East Africa. WWF, Gland, Switzerland.
Zeng, Z., Wang, D., Yang, L., Wu, J., Ziegler, A.D., Liu, M., et al., 2021. Deforestation-induced warming over tropical mountain regions regulated by elevation. Nat. Geosci. 14, 23–29.
Zeppetello, L.R.V., Parsons, L.A., Spector, J.T., Naylor, R.L., Battisti, D.S., Masuda, Y.J., et al., 2020. Large scale tropical deforestation drives extreme warming. Environ. Res. Lett. 15, 084012.

Chapter 7

# Coral reef fishes in a multi-stressor world

Jodie L. Rummer[a,b,*] and Björn Illing[a,c]
[a]*ARC Centre of Excellence for Coral Reef Studies, James Cook University, Townsville, QLD, Australia*
[b]*College of Science and Engineering, James Cook University, Townsville, QLD, Australia*
[c]*Thünen Institute of Fisheries Ecology, Bremerhaven, Germany*
[*]*Corresponding author: e-mail: jodie.rummer@jcu.edu.au*

## Chapter Outline



Coral reef fishes and the ecosystems they support represent some of the most biodiverse and productive ecosystems on the planet yet are under threat as they face dramatic increases in multiple, interacting stressors that are largely intensified by anthropogenic influences, such as climate change. Coral reef fishes have been the topic of 875 studies between 1979 and 2020 examining physiological responses to various abiotic and biotic stressors. Here, we highlight the current state of knowledge regarding coral reef fishes' responses to eight key abiotic stressors (i.e., pollutants, temperature, hypoxia and ocean deoxygenation, pH/$CO_2$, noise, salinity, pressure/depth, and turbidity) and four key biotic stressors (i.e., prey abundance, predator threats, parasites, and disease) and discuss stressors that have been examined in combination. We conclude with a horizon scan to discuss acclimation and adaptation, technological advances, knowledge gaps, and the future of physiological research on coral reef fishes. As we proceed through this new epoch, the Anthropocene, it is critical that the scientific and general communities work

**Fish Physiology, Vol. 39B. https://doi.org/10.1016/bs.fp.2022.04.011**

to recognize the issues that various habitats and ecosystems, such as coral reefs and the fishes that depend on and support them, are facing so that mitigation strategies can be implemented to protect biodiversity and ecosystem health.

## 1 Introduction

Coral reef fishes represent the most speciose assemblage of vertebrates on the planet today (Hixon and Randall, 2019) and continue to be the focus of myriad research programs, where studies are becoming even more essential as anthropogenic activities negatively affect coral reef ecosystems worldwide. Coral reef fish species numbers are estimated to be between 5000 and 8000, making up anywhere from 16% to 25% of all named, extant fishes (Victor, 2015). Coral reef fishes exhibit an array of body morphologies, fin arrangements, locomotory types, feeding strategies, physiological adaptations, and reproductive modes. They range in size from less than 50mm (e.g., the cryptobenthic species; Gobiidae and Blenniiformes; Brandl et al., 2018), up to 18m in the largest fish in today's oceans, the reef-associated whale shark (*Rhincodon typus*). Moreover, coral reef fishes have long been investigated across a multitude of—morphology, systematics, evolution, ecology, and conservation, to name a few—that integrate naturally with physiological research. Such integrative studies have been of particular, recent (i.e., 21st century) importance as well and key to addressing the effects of multiple environmental and anthropogenic stressors, not only on individual coral reef fish species and within particular taxa, but also on whole ecosystems. Undeniably, given the current epoch, the Anthropocene, where the dominant influences on the climate and environment come from human-based (i.e., anthropogenic) activities, there has never been a more important time to be researching coral reef fishes and how they respond to multiple, simultaneous, and often interacting stressors. We begin this chapter with an overview of the state of knowledge, for which we use a systematic literature search and bibliometric analysis to illustrate. We then provide case studies and discuss how the individual and combined stressors, ordered by their frequency of occurrence in the meta-analysis, affect coral reef fish physiology. We conclude with a horizon scan to discuss potential for acclimation and adaptation, technological advances, knowledge gaps, and the future of physiological research on coral reef fishes with implications toward conservation and protecting biodiversity and ecosystem health.

## 2 Current knowledge and trends over time

We conducted a systematic literature search to find studies investigating effects of (multiple) environmental stressors on the physiology of (sub-) tropical coral reef fishes. We collated studies examining physiological effects of abiotic stressors, including (1) temperature, (2) pH/$CO_2$, (3) hypoxia/deoxygenation, (4) salinity, (5) turbidity, (6) pollution, (7) noise, and (8) pressure and biotic factors, such as (9) predator threats, (10) prey abundance, (11) parasites, and (12) disease on coral reef fishes.

The optimal search strategy included separate search terms for each of the 12 factors and was librarian-verified at James Cook University (Townsville, Australia) (see Appendix for more details and the search terms used (Supporting Information in the online version at https://doi.org/10.1016/bs.fp.2022.04.011)). Results of the different searches were pooled, duplicates were removed, and then the remaining articles were manually checked for their relevance, and subsequently retrieved for bibliometric analyses. We included (i) original research (i.e., no reviews or meta-analyses) on subtropical or tropical coral reef fishes (i.e., based on Fishbase's environment classification), and (ii) studies that investigated a physiological metric. Inter-disciplinary studies were included if they helped gain mechanistic insight into physiological processes (e.g., molecular studies looking at gene expression patterns of enzymes or stable isotope analyses). Studies investigating highly mobile pelagic species that can occur on (sub-) tropical coral reefs (e.g., whale sharks, etc.) and studies examining the effects of fisheries and angling related stressors were excluded. Purely ecological field and laboratory studies (e.g., those reporting biting rates on benthos and plant material) were excluded; however, studies that experimentally examined feeding rates were included if the authors investigated the mechanistic relationship between a stressor and feeding frequency. Studies examining swimming performance and kinetics (e.g., acceleration, velocity) were included, but those examining purely behavioral metrics (e.g., boldness, habitat preference, and risk assessment trials) were excluded. In line with this, biotelemetry studies were only included if they also examined physiological parameters (e.g., body temperature, acceleration, etc.). Methodological studies were largely excluded. Furthermore, an overview of all existing multi-stressor studies was also created by checking each of the studies resulting from the search for the number and type of stressor investigated. The resulting articles were analyzed in R (R Core Development Team, 2018) using the R package "bibliometrix" (Aria and Cuccarullo, 2017).

In total, the search resulted in 875 scientific articles, of which 862 could be retrieved for further bibliometric analysis. All articles were generated over a 42-year period (i.e., 1979–2020, inclusive). Yet, only 67 of these studies were published in the first 20 years of this analysis (i.e., 1979–1998), meaning that approximately 92% of the studies have been published between 1999 and 2020. In fact, there were only four studies published before the 1990s after which, until the year 2000, an average of eight articles were published annually. Research outputs for studies examining these 12 stressors on coral reef fishes accelerated at the turn of the century (i.e., from 2000 onward), with an average of 38 related studies being published annually through 2020. There were a few global events related to coral reef health that occurred around this time, which may have catalyzed some of these studies. For example, the Great Barrier Reef (GBR, Australia) underwent mass coral bleaching (Fig. 1) due to marine heatwaves (MHWs) associated with ocean warming in 1998, 2002, 2006, 2016, 2017, and 2020, which had never before been documented in

**FIG. 1** Blue-green and black axil chromis (i.e., *Chromis viridis* and *C. atripectoralis*; Pomacentridae) swimming among fully bleached coral (i.e., *Acropora* sp.) near Lizard Island, Australia, in the northern part of the Great Barrier Reef 1 week into the marine heatwave, February–March 2016. *Photo credit: J.L. Rummer.*

human history (Hughes et al., 2021). It is noteworthy that 93% of the studies assessing the effects of temperature stress (e.g., simulated ocean warming conditions or heatwave events) on coral reef fishes, including studies investigating more than one stressor, were published between 2001 and 2020.

However, only 29 studies investigated two of the aforementioned stressors in combination and no studies examined three or more stressors together. Therefore, multiple stressor studies represented only 3% of the literature (Fig. 2). The most common combination for these dual-stressor studies included temperature and $CO_2$/pH, further highlighting the emphasis on climate change stressors (e.g., ocean warming and acidification) throughout the analysis (see Section 4). Of all the studies assessed, 52% examined one of the eight aforementioned abiotic stressors, and 45% examined one of the four biotic stressors (Fig. 2). Studies examining the effects of pollutants on coral reef fishes dominated the literature (i.e., 29% of all studies), which also represented the most examined of the abiotic stressors, followed by temperature (i.e., 13% of all studies) (Fig. 2). Of the biotic stressors, prey abundance was the most examined, representing 26% of all studies (Fig. 2) (see Appendix for further details (Supporting Information in the online version at https://doi.org/10.1016/bs.fp.2022.04.011)).

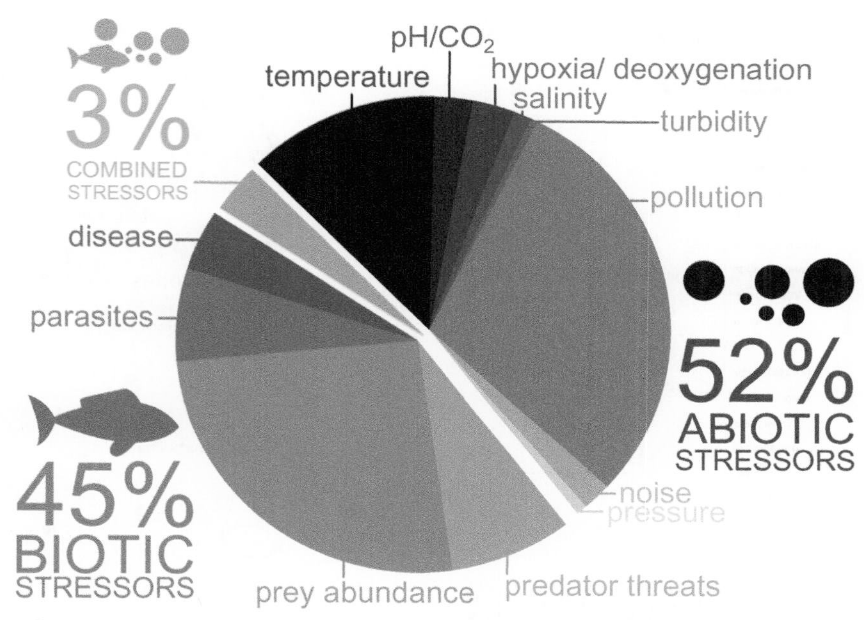


**FIG. 2** Visual representation of the proportion of studies investigating the physiological effects of abiotic (i.e., temperature, pH/$CO_2$, hypoxia/deoxygenation, salinity, turbidity, pollution, noise, and pressure), biotic (i.e., predator threats, prey abundance, parasites, and disease) and combined stressors on coral reef fishes from 1979 to 2020, inclusive. Data were derived from 875 studies obtained from a systematic literature analysis (see Section 2).

Trends in research topics, assessed through the studies' author keywords, shifted over time (Fig. 3A). Several terms, such as stable isotopes, mercury, bioaccumulation, and foodwebs, occurred more frequently from 2016 onward, suggesting physiological metrics being used for examining feeding ecology and dietary shifts (see Section 3.2.1) as well as bioaccumulation and biomagnification of pollutants (see Section 3.1.1). A network analysis generated three main clusters of author keywords and visualized how the terms were connected (Fig. 3B). The most dominant keywords in the first cluster were fish, mercury, bioaccumulation, and biomagnification, with nine other minor keywords in this cluster, again, suggesting considerable emphasis on abiotic stressors associated with pollutants (see Section 3.1.1). The second cluster was dominated by keywords including stable isotopes, coral reefs, food web, and elasmobranchs, with seven other minor keywords, suggesting that biotic stressors associated with predator-prey relationships and food availability were heavily emphasized and perhaps most so on elasmobranch species, as shark was also one of the minor keywords. The third cluster was dominated only by two phrases: coral reef fish and climate change, with the other 18 keywords weighted similarly and most associated to climate change stressors like ocean warming and the various approaches that are used in physiological

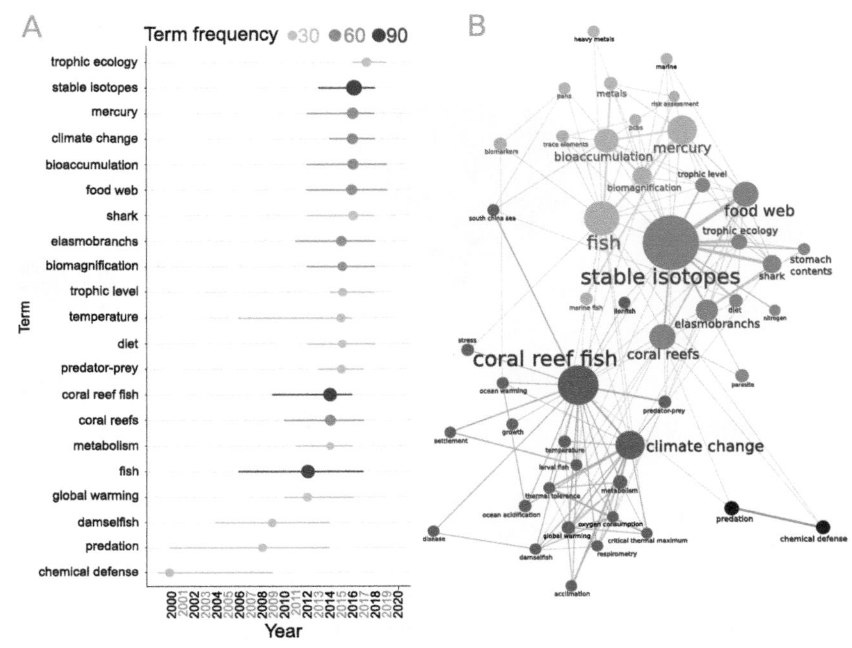


**FIG. 3** The trend over time (A) and network analysis (B) of keywords (author generated) that were most frequently used to describe studies examining the physiological effects of various abiotic and biotic stressors on coral reef fishes. In panel A, the size of the circle represents the frequency (i.e., 30, 60, 90 times, and corresponding spectrum from yellow, to gold, to rust colors) by which the term was used as a keyword. While bibliometric results were derived from 875 studies spanning 1979 to 2020, due to the low number of studies prior to 2000, only keywords from 2000 to 2020 (x-axis) are represented here. In panel B, each circle represents a node, with the size of the node emphasizing the frequency by which the term is used. The nodes are connected by lines that represent edges, with the weight of the lines emphasizing the strength of the connection. The color map represents clusters or aggregations of nodes and edges. Together, the network of nodes and edges and where they aggregate depicts the relationships between terms. See Section 2 for more details.

studies to assess the effects of those stressors (e.g., respirometry, critical thermal maximum, etc.) on coral reef fishes. There was a fourth, minor cluster containing only two keywords: predation and chemical defense. Of these, predation was most frequently utilized as a keyword in 2008; whereas, chemical defense peaked as a keyword in 2000 (Fig. 3). The trends in keyword use have also seemingly shifted over time, with those most commonly using terms related to ocean warming, climate change, and elasmobranchs peaking in the last decade, which also reflects the trends described above (Fig. 3).

## 3 Stress in coral reef fishes (primary, secondary, and tertiary responses)

Classically defined, stress is "the nonspecific response of the body to any demand made upon it" (Selye, 1973); yet, the word "stress" invokes an array of definitions depending on the context and discipline. Here, stress is defined as the collective responses at primary, secondary, and tertiary levels underpinned by physiological mechanisms but often supported by behavioral changes that are key to re-establishing and maintaining homeostasis and ultimately surviving and thriving under an altered condition (Barton, 2002). A fish will respond to a threatening situation, the stressor, whether the threat originates abiotically or biotically, and whether physical, chemical, or perceived, via initiating a stress response. Therefore, it is the stressor that evokes the stress response. Primary responses are largely neuro-endocrine in nature via the hypothalamic–pituitary–adrenal (HPA) axis and involve a suite of catecholamines and corticosteroids (glucocorticoids) that are released into circulation (Wendelaar Bonga, 1997; see Chapter 3, Volume 39A: Castro-Santos et al., 2022). Secondary responses involve the initiation of heat shock proteins (HSPs) and changes in hematological parameters that instigate changes at the metabolic, cardio-respiratory, immune, and ion-balance levels (Mommsen et al., 1999). Primary responses can initiate secondary responses, and therefore, the relationships can be difficult to separate. Whereas, tertiary responses are often more behavioral and whole-organism level, resulting in changes in growth, movement, and decreased disease resistance, to name a few, and often stem from primary and secondary responses (Wedemeyer et al., 1990). The stress response can be immediately beneficial (e.g., fight or flight) or adaptive over the longer term, which is important to note given the classic definitions and misconceptions that stress is always negative (Chrousos, 1998). However, some stress responses may also be maladaptive, for example, considering responses that alter growth, feeding, digestion, immune function, and/or reproduction (Barton and Iwama, 1991). Despite the considerable variation across taxa and life history stages, the hormonal underpinnings of the stress response and resulting mechanistic responses are relatively well-understood and have been examined extensively across the teleost fishes, albeit less so in the elasmobranchs. The effects of multiple stressors, however, are not well-understood but are more relevant to the fishes—coral reef fishes or otherwise—today, living in the multi-stressor world of the Anthropocene.

### 3.1 Abiotic stressors (natural and anthropogenic)

Abiotic stressors arise from non-living influences on living organisms and can originate from chemical or physical sources and can be both natural and anthropogenic in origin, as well as originate both locally and globally. The order of the following sections is based on the frequency of studies investigating the respective stressors in the bibliometric analysis. The most pervasive, individual abiotic stressors include pollutants (i.e., including heavy metals and other toxicants), temperature, low oxygen (i.e., hypoxia, deoxygenation), changes in pH or $CO_2$ (often communicated together), noise, salinity, pressure or depth, and turbidity.

### 3.1.1 Pollutants

Toxicology studies on coral reef fishes have a long and extensive history, with some of the first physiological studies commencing at least by the 1970s (reviewed in Wood, 2012; also see Chapter 3, Volume 39B: De Boeck et al., 2022). While much is known regarding the effects of various metals and their bioaccumulation capacities, emerging pollutants (e.g., poly-aromatic hydrocarbons (PAHs), polychlorinated biphenyls (PCBs), flame retardants (e.g., polybrominated diphenyl ethers, PBDEs, and polybrominated biphenyls, PBBs), surfactants, and pesticides and interacting stressors are a product of the Anthropocene. The foundational work in these early toxicology studies, however, has been key in integrating and assessing the effects of emerging pollutants such as micro- and nano-plastics (reviewed in John et al., 2021), antibiotics and other pharmaceuticals, and although not unique to the Anthropocene, the increased prevalence of ciguatoxins (CTXs). Yet, it is important to note that, even though the physiological effects of heavy metal exposure have been thoroughly investigated over the years, with technological advancements (e.g., computers, smartphones, smartwatches) and their rapid turnover rates and associated waste issues, metal pollution has remained a pervasive issue.

Heavy metals largely come from agricultural, technological, medical, and industrial sectors are categorized by whether they are biologically essential or nonessential to the organism. Biologically essential metals include copper (Cu), zinc (Zn), chromium (Cr), nickel (Ni), cobalt (Co), molybdenum (Mo), and iron (Fe), and while these metals have known biological roles, toxicity can occur if concentrations are too low or too high. In contrast, nonessential metals, such as aluminum (Al), cadmium (Cd), mercury (Hg), tin (Sn), and lead (Pb) have no demonstrated biological function in fish, and so toxicity tends to commensurate with concentration. Due to increased heavy metal demand in the Anthropocene and issues associated with runoff, effluent discharge, and atmospheric fallout, heavy metal distribution in coastal waters has become widespread. Sea ports, having concentrated economic and recreational activities, have long been recognized as environments that are susceptible to heavy metal pollution. In Queensland, Australia, for example, there are 21 ports ranging from small community, multi-cargo, and multi-national coal export terminals, all of which are in close proximity to the Great Barrier Reef, meaning that coral reef fishes are highly susceptible to heavy metal exposure, and as such have been heavily investigated.

Heavy metals tend to accumulate in fish as they feed and respire, with consequences throughout other physiological systems, such as the liver, due to its metabolizing and detoxifying properties. Metals are also well-known inducers of oxidative stress as well, and so many studies have examined oxidative damage (reactive oxygen species, ROS, production) and antioxidant defenses (e.g., superoxide dismutase; SOD, catalase; CAT, glutathione peroxidase; GPx, and glutathione-s-transferase; GST) in fish species exposed to metals (reviewed in Sevcikova et al., 2011). Yet, the earliest studies were likely executed because vertebral deformities were observed in fish exposed to cadmium. The different pathways of accumulation lead to differential distribution of heavy metals across the various tissues of the fish but can also be influenced by the presence of metallothioneins (i.e., primary metal-binding proteins), the metabolic activity of the particular organ or tissue, the rate of blood flow, and metal-specific binding properties. While many studies on heavy metal exposure/accumulation in coral reef fishes have taken a mechanistic perspective and spanning developmental stages, much focus has been for applied outcomes. This is likely given the proximity of heavy metal sources and recreational and commercial fishing sectors and the potential for bioaccumulation in economically prized species (e.g., coral trout, *Plectropomus* spp.). As such, many coral reef fish species—and specific tissues, such as the liver—have become bioindicator species for harmful levels of heavy metals (e.g., Great Barrier Reef, Rayment and Barry, 2000), and this area of research will be important in addressing the effects of interacting stressors.

The demand for petroleum products has increased dramatically since the mid-20th century, with oil extraction and transport activities growing. More than 6 million metric tons of petroleum products have entered the oceans, largely due to industrial discharge, urban run-off, and shipping operations, and more than 300 major marine oil spills since the 1970s (summarized in Johansen et al., 2017). The heavy crude oils that are a key component of these processes include PAHs that are toxic, carcinogenic, mutagenic, and teratogenic to marine life (Negri et al., 2016). Because PAHs come from many sources—pyrogenic (combustion-derived) and petrogenic (petroleum-derived)—are often lipophilic, and break down slowly, they are pervasive in the marine environments, especially near urban areas, industrial or shipping operations and oil drilling/extraction sites. Yet their effects on coral reef fish species, trophic transfer, and potential for bioaccumulation (over ontogeny) and biomagnification (across trophic levels) are still not well understood.

While many studies on PAH exposure in coral reef fishes come from an applied, human health angle (i.e., consumer driven), further researchers have started investigating potential adverse health effects for the fish as well. Liver and bile metabolites, enzymes, and muscle tissue are typically analyzed across species, ontogeny, trophic levels, and source proximity to determine bioaccumulation, biomagnification, and if/when species would be good bioindicators (Juma et al., 2017; De Albergaria-Barbosa et al., 2017). Indeed, such analytical approaches determined that PAH contamination in coral reef fishes from the South China Sea originated from biomass combustion, petroleum sources, and vehicular emissions (Li et al., 2019). Similar to those associated with antioxidant roles with metal exposure, enzymes that have been specifically identified for tracking PAH exposure because of their elimination pathways include ethoxyresorufin-O-deethylase (EROD) and GST (Cullen et al., 2019; King et al., 2005). It has also been determined that liver burdens are indicative of acute exposure to PAHs because the tissue is highly dynamic when compared to muscle. However, muscle burdens are useful to assess chronic exposure because of the slower turnover rate and lower likelihood of concentrating such contaminants due to lipid mobilization. Invasive species (e.g., lionfish, *Pterois volitans*) have been used as biomonitoring species with PAH accumulation via biomarkers including bile fluorescence and liver enzyme activities (Van Den Hurk et al., 2020). Two economically valued Australian reef-associated species, the gold-spotted trevally (*Carangoides fulvoguttatus*) and bar-cheeked coral trout (*Plectropomus maculatus*), have also been used as bioindicators (King et al., 2005). Yet, such studies continue to highlight that, factors such as the fish's lipid content, length, and weight can affect PAH accumulation ( Jafarabadi et al., 2019, 2020; Sun et al., 2019). Trophic ecology is key as well, with sharks prone to bioaccumulation and biomagnification since they occupy high trophic levels in their ecosystems. Although Cullen et al. (2019) measured higher than expected PAH accumulation in the liver and muscle samples from several shark species, they also noted that diet, niche partitioning, and life history characteristics, including those spanning ontogeny, affect PAH accumulation.

It is known from studies on PCBs that PAH accumulation results in similar adverse health effects in fishes, such as, but not limited to decreased vitamin A concentrations, thyroid hormone deficiency, and immunosuppression. PAH exposure also alters growth and has a suite of other physiological effects as demonstrated in salmonids (e.g., Meador et al., 2006), cod (Sørensen et al., 2019), haddock (Meier et al., 2010; Sørhus et al., 2016), herring (Incardona et al., 2012) and mahi mahi (Mager et al., 2017). Other studies have indicated that PAH exposure and accumulation in fishes can cause genotoxicity (and associated carcinogenesis), as well as endocrine and metabolic disruption (reviewed in Cullen et al., 2019) and enhanced photo-toxicity upon exposure to ultra-violet (UV) radiation (Aranguren-Abadía et al., 2022). While the interaction with PAH and UV exposure has been thoroughly demonstrated in cod (Aranguren-Abadía et al., 2022) and mahi mahi (Alloy et al., 2016), given the proximity of coral reefs to UV radiation and the surge of studies documenting adverse effects of PAH and UV exposure on corals (Overmans et al., 2018), it follows that this combination of stressors will continue to increase in relevance in the Anthropocene. Early life history stages of coral reef fishes may be most vulnerable as well, as determined by Johansen et al. (2017) in six pre-settlement stages of coral reef fishes, where PAH exposure resulted in greater mortality, stunted growth rates, altered habitat settlement, and changes in anti-predator behaviors (reduced sheltering and shoaling and increased risk-taking). Such results suggest a novel path of PAH injury whereby higher-order cognitive processing and behaviors necessary for successful settlement and recruitment of larval coral reef fishes are impaired ( Johansen et al., 2017). More work is needed, and this will be an area of research on coral reef fishes that will continue to increase in importance and necessity.

Environmental levels of PCBs and various organochlorin-based pesticides (e.g., dichlorodiphenyl trichloroethane; DDT) have been on the decline (e.g., Rachel Carson's *Silent Spring* was published in 1962, and both were banned in the United States in the 1970s and other countries soon after), but many fish species still exhibit significant levels. This is likely because their overall toxicity, capacity for bioaccumulation and biomagnification (Kobayashi et al., 2019), low elimination (Wang and Wang, 2005) in fishes, long half-lives, continued use in some areas (e.g., DDT for malaria control in Kenya), and prevalence in landfills, sediments, and rivers mean that they remain in marine ecosystems. Studies like those examining the effects of PAH exposure on fishes have been executed to determine the effects of PCB and DDT exposure, but perhaps for longer, since the advent of modern agriculture and industrial pesticides. Moreover, because there is a relationship between, for example, PCB exposure time and age, some coral reef fish species (e.g., *Abudefduf sordidus*, blackspot sergeant; Kerr et al., 1997) have been used as bioindicators. Much work has incorporated trophic positions using various organisms (e.g., phytoplankton, copepods, and fish; Wang and Wang, 2005) to determine trophic magnification factors of such environmental pollutants (e.g., benthic food webs compared to pelagic food webs; Kobayashi et al., 2019). While much of this research is executed because of the implications these pollutants have on human health when persisting in marine ecosystems, these compounds also directly affect the physiological health of coral reef fishes (e.g., including sharks and rays, Storelli et al., 2011; Cullen et al., 2019). Most effects are at the level of endocrine pathways (e.g., steroid biosynthesis, oogenesis, spermatogenesis, etc.), and liver is the tissue most often analyzed (reviewed in Reijnders and Brasseur, 1997; Storelli et al., 2011). Climatic and ecological factors, temperatures, rainy versus dry seasons, etc. can affect how such compounds degrade or persist in marine environments (Wandiga et al., 2002), and given that they persist despite bans, this area of research will likely continue to be important.

Emerging pollutants such as, for example, micro- and nano-plastics will continue to pose health concerns for coral reef fishes and their ecosystems, given their global abundance and persistent use, even as single-use, by humans. The sources of these plastics and why they end up in the oceans (i.e., and more specifically why there are more micro- and nano-plastics in shallow, productive areas, such as coral reefs) are clear (Jambeck et al., 2015). Yet, exposure is seemingly unavoidable, especially under the current global waste management and production trends. Models predict that, by 2050, 12,000 million tonnes (MT) of plastic waste will be incinerated, 9000 MT will be recycled, and 12,000 MT of waste will be disposed of in landfills or in the natural environment (reviewed in John et al., 2021). Moreover, it is not just the chemical composition (e.g., polypropylene, polyethylene, polypropylene ether, polyethylene terephthalate, polyester, etc., all of which have been well-studied in terms of their toxicity), but also the chemical additives, which are often used to improve the structural properties of the plastic, that also contribute toxic effects (Galloway et al., 2017). While the size of micro- and nano-plastics is key, as it is within optimal prey ranges for many organisms in the marine environment, the shape and color also influence ingestion rates and toxicity (Galloway et al., 2017). Like with other pollutants, bioaccumulation via ingestion and transfer across trophic levels is the starting point, and studies commencing in the early 2000s through to the present emphasize that microplastics can now be found at every trophic level (Kroon et al., 2018). However, the mechanisms that underpin transfer from digestive tissues across to other organ systems are not yet clear. From freshwater fish models, studies have determined that nano-plastic particles can cross the blood-brain-barrier, which may underpin behavioral issues observed upon exposure (Mattsson et al., 2017). And, when coupled with the stress of degraded habitat, coral reef fishes (e.g., Ambon damselfish, *P. amboinensis*) that ingest micro-plastics, exhibit bold, risk-taking behaviors that decrease survival, which may relate to their nutritional status (McCormick et al., 2020). Given that another study found no effect of micro-plastic ingestion on settlement stage surgeonfish (*Acanthurus triostegus*) when faced with the threat of predatory lionfish (*Pterois radiata*) (Jacob et al., 2019), it is clear that more work is needed in this area, especially when considering that coral reef fishes exposed to increasing concentrations of micro- and nano-plastics will also be facing other stressors in concert. Although plastic toxicity has been studied for decades, this is an emerging stressor coral reef fishes are experiencing and one they will continue to face if waste management protocols remain unabated.

### 3.1.2 Temperature

Temperature is one of the most well-studied abiotic stressors in coral reef fishes, second only to pollutants. Whether warm or cold temperatures, over acute or chronic timescales, the physiological effects of thermal stress on most ectotherms, including the coral reef fishes, are generally understood. Moreover, with increasingly pervasive MHWs and widespread warming due to climate change, contemporary studies focus heavily on temperature. However, the temporal scale and the magnitude over which temperature changes occur, the order in which certain physiological systems respond to thermal stressors, and how those physiological responses affect behavior, movement, distribution, and other fitness-relevant traits will depend on species, life history stage, prior thermal history, and the presence of other stressors.

Physiological sensitivity to changing temperatures can commence at the most basic level but the mechanistic underpinnings are not always easy to interpret. Ectotherms in general, coral reef fishes included, typically exhibit an increase in biochemical rate functions with an increase in temperature. This temperature quotient or $Q_{10}$ relationship, on average, suggests that for every 10 °C increase in temperature, rates double or triple (i.e., $Q_{10}=2–3$; Clarke and Johnston, 1999; Schmidt-Nielsen, 1997). Classic examples of these relationships come from studies on enzyme activities and organ tissue preparations (e.g., muscle contractions) under controlled laboratory conditions (Fields and Somero, 1997; Gelman et al., 1992; Johns and Somero, 2004; Johnson and Johnston, 1991; Lin and Somero, 1995). In contemporary studies, this relationship is also often reflected in estimates of metabolic rates (e.g., standard, resting, active, and maximum), where species compensating for a temperature increase can maintain $Q_{10}$ at or around 2 (Eme and Bennett, 2009a), while those unable to compensate exhibit $Q_{10}$ relationships well exceeding 2 (Rummer et al., 2014). Trade-offs may occur as well, where compensation occurs for functions at rest but not under maximal performance, or energy is saved in one trait at the cost of another, which may depend on the behavior and functional role of the species ( Johansen and Jones, 2011; Rummer et al., 2014).

Several theories have been proposed to predict the causes of the various physiological responses to compensate (or not) for temperature increases. The Gill-Oxygen Limitation Theory (GOLT) proposes that body size and function in fish is limited by the gills' inability to adjust and supply sufficient oxygen to satisfy increasing metabolic costs under elevated temperatures (Pauly, 2019). The Oxygen and Capacity Limited Thermal Tolerance (OCLTT) hypothesis proposes cardio-respiratory transport and tissue demand as the main determinants of an organism's performance under ocean warming (Pörtner, 2014; Pörtner et al., 2017). However, mixed empirical evidence has led to a controversy about the exact mechanisms affecting species' performance under elevated temperatures, as none of the current theories can explain all observed responses (reviewed in Audzijonyte et al., 2019; Ern et al., 2017; Jutfelt et al., 2018). More broadly unifying principles are currently lacking (but see Audzijonyte et al., 2019; Clark et al., 2013; Ern, 2019). Reviews on this topic have, therefore, emphasized the urgent need for cross-disciplinary, mechanistic studies that explore the timescales over which thermal responses occur to assess the molecular and physiological mechanisms underpinning temperature compensation, especially in thermally sensitive species (Audzijonyte et al., 2019; Jutfelt et al., 2018).

Coral reef fishes, given their latitudinal distribution, especially populations living in particularly low latitudes (i.e., closer to the equator), are likely adapted to narrow temperature ranges (i.e., stenothermal, as opposed to eurythermal). This suggests that, despite living in warmer climes than their temperate latitude counterparts, coral reef fishes will exhibit greater sensitivity to temperature increases associated with ocean warming as they are already living close to their upper thermal limits (Eme and Bennett, 2009b; Rummer et al., 2014; Tewksbury et al., 2008). For example, five species of coral reef fishes from three latitudinally distinct populations spanning more than 2300 km from the southern Great Barrier Reef to near-equatorial locations of Papua New Guinea exhibited no differences in optimal temperatures for aerobic performance (Rummer et al., 2014). This was despite the equatorial populations residing in a very narrow range (e.g., approximately 2 °C) and the southern Great Barrier Reef populations residing in a much wider range (e.g., approximately 10 °C) of annual, seasonal temperatures (Rummer et al., 2014), suggesting a wide thermal buffer zone for high latitude populations and high temperature sensitivity in the near-equatorial populations. This is not unprecedented either with loss of hypoxia tolerance (Nilsson et al., 2010), increased metabolic costs (Gardiner et al., 2010; Rodgers et al., 2018, 2019), reductions in swimming performance (Johansen and Jones, 2011), and worsened predator escape responses (i.e., slower reaction times, slower escape speeds, and shorter escape distances; Allan et al., 2015; Motson and Donelson, 2017; Warren et al., 2017) when compared to their control temperature counterparts. Although these trends in coral reef fishes have only been revealed since the early 21st century, this topic has been historically well-researched in terrestrial species (Deutsch et al., 2008; Tewksbury et al., 2008).

Many examples of temperature sensitivity in coral reef fishes exist, and factors such as latitude/biogeography, activity level, and size (Di Santo and Lobel, 2016, 2017; Messmer et al., 2017; Ospina and Mora, 2004), help determine temperature sensitivity. Interestingly, in equatorial populations of two damselfish species, calculated $Q_{10}$ values for resting metabolic rate estimates at 29 and 34 °C well exceeded $Q_{10}=2$ (i.e., 4.8 and 7.2), suggesting that, at 34 °C, over twice as much energy is required for these species to maintain routine metabolic processes than at their summer average temperatures of 29 °C (Rummer et al., 2014). It could be that the high $Q_{10}$ values are a product of the stable, narrow thermal range experienced by these small, tropical, coral reef fishes residing near the equator. Likewise, two closely related coral reef fish species (*Abudefduf vaigiensis* and *A. whitleyi*) exhibited different energetic costs upon exposure to cooler temperatures, which resulted in reduced growth rates, feeding rates, burst escape speed and metabolic rates that were more pronounced in the species with the narrower latitudinal range (Djurichkovic et al., 2019). Species with wide thermal ranges (e.g., *A. vaigiensis* in the former example), however, have the potential to expatriate into new habitats, suggesting that climate-driven range shifts could result in species introductions and alter trophic interactions and predator-prey dynamics (Barker et al., 2018; Figueira et al., 2019; Rowe et al., 2018). Life stage may play a role as well. Given that the majority of coral reef fishes have a bipartite life cycle, consisting of a pelagic larval stage, many species could be venturing several hundred kilometers from their natal reefs and for weeks to months at a time in vastly different thermal regimes, thereby necessitating thermal compensation. However, if coral reef fishes are locally adapted to their thermal environment, do not regularly experience seasonal temperature fluctuations, and do not move far from these microhabitats, there would be no drive to possess such metabolic compensation for changes in temperature (e.g., acclimation or acclimatization, also see Section 5), further underpinning their temperature sensitivity as warming continues.

Indeed, understanding such temperature compensation strategies is also important when disentangling the responses of an organism upon acute changes (e.g., MHWs, Fig. 1; Allan et al., 2015; Bernal et al., 2020; Johansen et al., 2021) from how an organism responds to slowly increasing but chronically elevated temperatures (e.g., as with ocean warming). By definition, MHWs, which are when temperatures are warmer than the 90th percentile based on a 30-year historical baseline period and last for five or more days (Hobday et al., 2016), have been of particular concern on coral reefs, as heatwaves have been increasing in severity and frequency since the beginning of the 21st century. The rate of temperature increase may be important as well, as found in laboratory studies on coral reef fishes (Illing et al., 2020), but aside from a few studies (e.g., Allan et al., 2015), MHW scenarios have only been applied to coral reef fishes since coral reefs started bleaching (Fig. 1), globally, due to climate change mediated ocean warming (Hughes et al., 2021). Upon simulation of MHW conditions (e.g., 3 °C above ambient), two coral reef fish species (e.g., *Caesio cuning* and *Cheilodipterus quinquelineatus*) elicited coordinated responses in 13 tissue and organ systems over 5 weeks (Johansen et al., 2021). The onset and duration of biomarker responses (e.g., red muscle citrate synthase and lactate dehydrogenase activities, blood glucose and hemoglobin concentrations, spleen somatic index, and gill lamellar perimeter and width; Johansen et al., 2021), differed between species as well. The more active, mobile species (*C. cuning*) initiated responses to the simulated heatwave within the first week of exposure (Johansen et al., 2021). However, the more sessile, territorial cardinalfish species (*C. quinquelineatus*) exhibited a comparatively reduced response that was delayed over time. Perhaps the more mobile species, that would normally move to more favorable thermal microhabitats in the face of a MHW, once unable to do so instigated responses right away. In contrast, the more sessile species may be more prone to "wait it out" before initiating physiological responses. The study identified seven biomarkers, including red muscle citrate synthase and lactate dehydrogenase activities, blood glucose and hemoglobin concentrations, spleen somatic index, and gill lamellar perimeter and width, that proved critical in evaluating the progression as to how fish responded over the course of the simulated heatwave (Johansen et al., 2021). Some work has emphasized the role of gill biomarkers (Madeira et al., 2017a; Rodgers et al., 2019) or antioxidant chaperones (e.g., catalase, ubiquitin, lipid peroxidase; Madeira et al., 2017a), but ultimately, the degree of thermal sensitivity depends on the trait examined.

At the molecular level, gene expression patterns can help to rapidly and thoroughly survey the physiological processes that are key to maintaining homeostasis during a thermal event, with a plastic transcriptional response indicating varying degrees of thermal tolerance (Bernal et al., 2020).

Such changes in gene expression patterns have been found to be species and population specific (Veilleux et al., 2018a), which may factor into activity, behavior, and home range (e.g., cardinalfishes). Populations from higher latitudes with wider thermal ranges have exhibited greater plasticity in gene expression patterns in response to simulated heatwave events when compared to their lower latitude counterparts (Veilleux et al., 2018a), which supports findings for whole organism physiological traits as well. Gene expression patterns may also depend on the duration of exposure, as was found in wild coral reef fishes collected before, during, and after the MHW (Fig. 1) that was pervasive throughout the northern Great Barrier Reef in the austral summer of 2015–2016 (Bernal et al., 2020). Although it was the first time that gene expression patterns had been directly evaluated in wild fish populations during a MHW, it was perhaps not surprising that differentially expressed genes associated with immune function and cellular stress responses, including HSPs, mitochondrial activity, and toxin metabolism in the liver in response to the changing temperatures were uncovered (Bernal et al., 2020), given the physiological traits that have been assessed in previous studies. Other changes that were species-specific and temporally evident were related to fatty acid and cholesterol metabolism and glucose levels that may also be associated with secondary stressors these coral reef fish populations experienced during the MHW (e.g., changes in food availability and trophic interactions, increased algal cover) (Bernal et al., 2020). This is not unprecedented, however, as Norin et al. (2018) determined that, for orange-fin anemonefish (*Amphiprion chrysopterus*), the combination of warming conditions and resulting habitat damage (i.e., bleaching of their symbiont sea anemones; *Heteractis magnifica*) resulted in a significant increase in metabolic costs, which, over the longer term can lead to stress-induced changes to reproductive hormones and decreased fecundity (Beldade et al., 2017). Indeed, findings so far have emphasized the complex role of multiple stressors and cascading effects they will have on coral reef fishes as we move through the Anthropocene.

Certainly, many studies examining temperature stressors in coral reef fishes have done so within a life stage and population. Species or populations with narrow thermal ranges and those that do not experience dramatic changes in seasonal temperatures may not possess considerable capacity for thermal acclimation. In contrast, species or populations residing in more subtropical or temperate latitudes with greater seasonal variation in temperatures may acquire more tolerance for warm temperatures during the summer months and more cold tolerance (i.e., while losing high temperature tolerance) during the winter months (Fangue and Bennett, 2003). For example, low latitude populations of *Acanthochromis polyacanthus* exhibit increases in gill pathologies not found in mid or high latitude populations of the same species upon acclimation to elevated temperatures emphasizing thermal specialization in low latitude populations (Rodgers et al., 2019). Yet, in the pan-tropical clownfish species (*Amphiprion ocellaris*), thermal preference, tolerance, and aerobic metabolic scope were all found to depend heavily on acclimation (Velasco-Blanco et al., 2019), and similar findings were reported for subtropical *Hippocampus erectus* (Mascaro et al., 2019)—noting that both are important species in the aquarium trade. Acclimation or acclimatization may play a substantial role in how species respond to changing temperatures, which varies by species, population, and/or may depend on demography (Eme and Bennett, 2009b).

While acclimation can confer plasticity in some performance traits, it may not be the most reliable indicator of the ultimate survival and distribution of stenothermal coral reef fishes, especially the more mobile species, under ocean warming scenarios. Thermal preference and behavioral (e.g., movement) thermoregulation (Barker et al., 2018; Gervais et al., 2018; Habary et al., 2017; Hight and Lowe, 2007; Reyes et al., 2011; Speed et al., 2012b) can often supersede acclimation in some instances. Indeed, physiological thermal sensitivity and thermoregulatory behavior are likely coadapted (Angilletta et al., 2002, 2006; Huey and Bennett, 1987) as the thermal history that defines a species' optimal temperatures for performance (e.g., metabolic traits, swimming, etc., as discussed above) often determines its preferred temperature range as well (Bryan et al., 1990; Johnson and Kelsch, 1998). Therefore, in the wild, most species will likely pursue temperatures that coincide with their optimal performance temperatures during a given life stage (Beitinger and Fitzpatrick, 1979; Brett, 1971; Payne et al., 2016; Pörtner and Farrell, 2008; Pörtner and Knust, 2007). Critically, for behavioral thermoregulation to help mitigate the effects of rapid climate change, evolutionary changes in optimal temperatures for certain traits should also provide a strong selective pressure for changes in preferred temperatures (Angilletta et al., 2002; Bryan et al., 1990). This inherent relationship has been examined more frequently throughout contemporary studies in coral reef fish species with the aim of predicting phenotypic shifts in temperature sensitivity of various physiological performance traits (e.g., see Donelson et al., 2011, 2012) that may also lead to changes in preferred temperatures—such relationships may be the primary driver escalating the poleward migration of species. Indeed, current evidence suggests that 365 different species across 55 families of tropical fishes are either on the move or have already undergone bio-geographical redistributions or range shifts as a result of climate change and more specifically ocean warming (Feary et al., 2014; Figueira and Booth, 2010; Gervais et al., 2021).

### 3.1.3 Hypoxia and ocean deoxygenation

Oxygen is the greatest factor limiting physiological performance and survival of marine life, including the coral reef fishes (Sampaio et al., 2021). Nearly all vertebrate life requires oxygen ($O_2$) to support and sustain aerobic activities, and while anerobic (i.e., without $O_2$) metabolism is possible, it is time-limited and species- and context-dependent. Because $O_2$ is the final electron acceptor in the electron transport chain, $O_2$ partial pressures ($pO_2$) set the rates of aerobic metabolism. Aerobic metabolic rates are both dependent and independent of environmental $pO_2$ along a continuum; species that regulate metabolic rate independent of environmental $pO_2$ are referred to as "oxyregulators," and species whose metabolic rates conform to environmental $pO_2$ are referred to as "oxyconformers" (Mueller and Seymour, 2011). While most species are thought to be oxyregulators (Svendsen et al., 2019), and maximum metabolic rates in fishes (and ectotherms in general) are reduced with decreasing environmental $pO_2$, such rates do not increase when environmental $pO_2$ exceeds saturation (Seibel and Deutsch, 2020). Instead, standard metabolic rate (i.e., the energetic costs required to sustain basic metabolic functions) is regulated until $pO_2$ is too low, and fishes transition from oxyregulating to oxyconforming (see Fig. 1 in Heinrich et al., 2014).

Hypoxia is usually defined when dissolved $O_2$ concentrations fall below 2.8 mg $O_2$ $L^{-1}$ (Breitburg et al., 2018), but this is an arbitrary threshold, because hypoxia tolerance is species- and context-specific, at the very least. It is also important to note that oxygen uptake is driven by the partial pressure gradient of the gas between the water and the blood, and therefore the associated parameters that determine oxygen solubility (e.g., temperature, salinity, pressure, etc.) are key. Species may also respond differently depending on life stage, energetic demands, and habitat conditions.

Most coral reef fishes, due to a bipartite life cycle, transition from a pelagic larval stage where they may spend weeks to months in the pelagic environment to settling onto the reef. In the pelagic, they exhibit record aerobic swimming and $O_2$ uptake capacities (Downie et al., 2021). Upon settling onto the reef and on into adulthood, however, reef fishes acquire notable hypoxia tolerance (Nilsson et al., 2007). Other species that mouth brood also exhibit unparalleled levels of hypoxia tolerance (Ostlund-Nilsson and Nilsson, 2004; Takegaki and Nakazono, 1999). Indeed, hypoxia tolerance may be a necessity for coral reef fishes to benefit from sheltering within the reef matrix at night because nighttime $O_2$ levels decrease dramatically (i.e., below 20% air saturation) when coral and other benthic organisms transition from photosynthesis to respiration (Nilsson et al., 2007; Nilsson and Ostlund-Nilsson, 2004). Moreover, small reef flats and tidepools that become isolated during low tide will also become hypoxic at night, necessitating hypoxia tolerance (Rummer et al., 2009). Several studies have assessed the capacity for coral reef fishes, from settlement to adulthood, to tolerate varying levels of hypoxia and the potential underlying physiological mechanisms (Nilsson et al., 2007; Nilsson and Ostlund-Nilsson, 2004). This is particularly interesting from a purely physiological perspective, such as hemoglobin (Hb) $O_2$ binding affinity. Usually, if an organism has a high capacity for $O_2$ delivery (low Hb-$O_2$ affinity) that would come with elite aerobic performance, they will not also have a high capacity for enhanced $O_2$ uptake (high Hb-$O_2$ affinity), which is usually observed in hypoxia tolerant species. This could change with life history stage, however, as was demonstrated with the salmonids in the late 1970s (Giles and Vanstone, 1978), or with hypoxia exposure (Bianchini and Wright, 2013). Indeed, several larval coral reef fishes have the highest mass-specific maximum $O_2$ uptake rates of any larval teleost (i.e., also any other ectothermic vertebrate measured) as well as the fastest swimming speeds for their body sizes (reviewed in Downie et al., 2021), suggesting high aerobic capacity during their larval life history stage. However, when coral reef fishes transition to the reef, they exhibit a dramatic decrease in critical oxygen tensions, suggesting they shift to being hypoxia tolerant (Nilsson et al., 2007). The reductions in aerobic metabolism and swimming performance that occur in coordination with settlement and hypoxia tolerance are seemingly unequivocal, but the exact timeline and underlying mechanisms are likely species-specific and not yet well-understood, but an area certainly warranting further investigation.

Beyond the teleost fishes, several elasmobranch species have been investigated for their typically uncharacteristic capacity to tolerate hypoxia and even anoxia (zero oxygen), which may also be related to the shallow, benthic, reef flat microhabitats that some shark species inhabit. In sharks, hypoxia initiates a suite of physiological responses, including ventilatory depression (Chapman et al., 2010) and expression of hypoxia inducible factor and heat-shock proteins (Renshaw et al., 2012). While several studies have investigated hypoxia and even anoxia tolerance in sharks and their relatives (reviewed in Pereira Santos et al., 2021; Rummer et al., 2022), only a few coral reef or reef-associated species have been investigated (e.g., Bouyoucos et al., 2020; Carlson and Parsons, 2001, 2003; Crear et al., 2019; Dabruzzi and Bennett, 2013; Hickey et al., 2012; Musa et al., 2020; Routley et al., 2002; Speers-Roesch et al., 2012; Wise et al., 1998). One such species is the epaulet shark (*Hemiscyllium ocellatum*), which has been the focus of the majority (i.e., at least 13 since the 1990s) of the studies on hypoxia and anoxia tolerance (in addition to studies on other environmental stressors) on sharks and their relatives. Unlike in other hypoxia- and anoxia-tolerant species, the epaulet shark exhibits no adenosine-mediated increase in cerebral blood flow and likely activates adenosine receptors that initiate metabolic depression and aid in maintaining brain adenosine triphosphate (ATP) levels—which would normally deplete—during an unprecedented 4h of anoxia (Renshaw et al., 2002; Söderstrom et al., 1999). Moreover, unlike other vertebrates, the epaulet shark preserves mitochondrial function—which would otherwise lead to cell damage and cell death—upon re-oxygenation post-anoxia exposure (Devaux et al., 2019). Given the small size of the epaulet shark and vulnerability to predation, it may make sense that this species exploits shallow, tidally influenced reef flats for shelter, even if such habitats exhibit dramatic declines in $O_2$. Therefore, this species must possess the physiological mechanisms that allow them to do so. Yet, it is interesting that these mechanisms used by the epaulet shark are different than what is understood for hypoxia tolerant teleosts and unprecedented in other elasmobranchs.

Despite the array of studies that have investigated low oxygen stress on teleost and elasmobranch fishes, no study published before mid-2021 has done so within a climate change context (i.e., ocean deoxygenation). It is important to clarify that the term hypoxia is not interchangeable with ocean deoxygenation (Klein et al., 2020), but quantifying hypoxia tolerance strategies and determining species' thresholds is important in defining the effects of ocean deoxygenation. This is especially important, given that hypoxia events elicit a stronger effect than ocean warming, ocean acidification, or the combination, and across biological traits (e.g., survival, abundance, development, metabolism, growth, and reproduction), taxonomic groups, ontogenetic stages, and climatic regions (Sampaio et al., 2021). Still, within the context of anthropogenic stressors, issues related to the effects of low oxygen on coral reef fishes have attracted far less attention in the scientific community when compared to other stressors (Sampaio et al., 2021).

Ocean deoxygenation is noted as the third global ocean syndrome but one that operates on different spatial and temporal scales than warming and acidification; yet, the term "ocean deoxygenation" was only first defined in 2009 (reviewed in Klein et al., 2020). Since the middle of the 20th century, the $O_2$ content of the oceans has decreased by more than 0.5–3%, low oxygen events (1–3.5 $O_2$ $mg L^{-1}$) are becoming more frequent and severe, and oxygen minimum zones (OMZs) are expanding (reviewed in Gregoire et al., 2021). While the causes are not fully understood, it is recognized that this process involves decreased $O_2$ and heightened biological consumption, which is worsened by enhanced stratification and induced by ocean warming. These changes, paired with rising ocean temperatures throughout the 21st century will further accelerate reductions in ocean $O_2$ content (reviewed in Gregoire et al., 2021; Klein et al., 2020).

### 3.1.4 pH/$CO_2$

Coral reef fishes are experiencing changes in water pH and $CO_2$ that stem from elevated atmospheric $CO_2$ ($pCO_2$; partial pressure) due to human-related emissions (ocean acidification; see above). Increased biological activity in shallow water habitats associated with coastal development and agricultural/industrial runoff can also contribute to elevated $pCO_2$, as can the ongoing global expansion of high intensity aquaculture (reviewed in Munday et al., 2019). While evolution suggests adaptations are in place for fishes to cope with such changes to maintain acid-base, ionoregulatory, and osmotic balance (reviewed in Hannan and Rummer, 2018; also see Chapter 5, Volume 39A: Eliason et al., 2022, and some extant fish species already live in elevated

$pCO_2$ environments or experience them on a diel basis (e.g., coral reef fishes at night when corals shift from photosynthesis to respiration; Hannan et al., 2020a), laboratory experiments show different responses across species, life history stage, and exposure duration. Such differences in fitness-related traits, differential capacity for acclimation and/or adaptation, and the influence of multiple stressors will factor into predicting these impacts over the timescales at which $CO_2$ levels are rising.

Changes in environmental $pCO_2$ and/or pH can dramatically affect the most basic most yet critical physiological processes in fishes—$O_2$ uptake, transport, delivery, and $CO_2$ removal, let alone ion and osmoregulatory processes (see Chapter 5, Volume 39A: Eliason et al., 2022). There is an intimate interaction between $O_2$ and $CO_2$ transport at the gills, and in other tissues, due to their interactions with Hb within the red blood cells (RBCs), which can vary by species (reviewed in Hannan and Rummer, 2018). Generally speaking, most cartilaginous fishes, such as sharks, skates, and rays, can efficiently compensate an acid-base disturbance due to the buffering capacity of their blood and plasma (Berenbrink et al., 2005). Moreover, most sharks, skates, and rays possess relatively pH-insensitive Hbs (Berenbrink et al., 2005), meaning that a pH disturbance associated with an acidosis, such as elevated $pCO_2$, may not compromise $O_2$ transport in the way we understand for teleost fishes. Modern teleost fishes, such as the coral reef fishes, have a different physiological response to an acidosis than the cartilaginous fishes. Teleost fishes possess extremely pH-sensitive Hb—probably evolving nearly 400 million years ago (MYA) in basal Actinopterygians (reviewed in Randall et al., 2014)—and low buffering capacity in the blood and plasma. Adrenergically-activated transporters on the RBCs help to regulate pH, and plasma-accessible carbonic anhydrase in select locations enhances $O_2$ release from the tissues during an acidosis, such as elevated $pCO_2$ (Randall et al., 2014; Rummer et al., 2013; see Fig. 1 in Hannan and Rummer, 2018). These traits result in an enhanced capacity for $O_2$ transport, especially during conditions that would normally preclude efficient $O_2$ uptake (Randall et al., 2014; Rummer et al., 2013). While these physiological traits may have facilitated the successful radiation of the fishes throughout geological history (Randall et al., 2014) and may be imperative in coping with ongoing and future changes in ocean $pCO_2$, it is important to note that other morphological adjustments (e.g., gill remodeling) and physiological compensation mechanisms, such as bicarbonate-mediated ion exchange from the environment to correct extracellular pH (Deigweiher et al., 2008; Heuer and Grosell, 2014) are energetically expensive and may not be sustainable over the long term (Lefevre, 2016). Because extracellular pH compensation is limited by the amount of bicarbonate that can be exchanged (Brauner and Baker, 2009), teleost fishes can only tolerate and function in extremely high $CO_2$ conditions for a finite period. Indeed, work has been done in naturally high $pCO_2$ conditions to understand the various mechanisms that underpin the physiological response to elevated $pCO_2$ conditions. For example, within the coral reefs near natural $CO_2$ seeps $pCO_2$ is similar to that predicted for the end of the century (i.e., ~1000 µatm; Munday et al., 2014). These seeps support diverse communities of coral reef fishes, yet fewer species occur near intense vents where $pCO_2$ ranges between 5000 and 10,000 µatm (reviewed in Munday et al., 2019). Therefore, these fish likely experience very high $pCO_2$ for short periods of time. Regardless of the exact mechanism, understanding how coral reef fishes perform at the physiological level under elevated $pCO_2$ may help in predicting ecosystem-level responses now and into the future.

Many of the adaptations potentially in place to maintain $O_2$ transport, $CO_2$ removal, and acid-base, ionoregulatory, and osmotic balance have been investigated in experimentally high $pCO_2$ settings and include hundreds of studies (reviewed in Hannan and Rummer, 2018). While a lot of these studies have focussed on more primitive and basal fishes and/or were for purely mechanistic understandings, there has been a surge of studies since the beginning of the 21st century investigating the physiological responses, acclimation processes, and adaptive capacity of coral reef fishes under ocean acidification relevant conditions. Such studies on coral reef fishes have spanned life history stages, species, activity levels, and habitats. Some of the first studies investigated larvae of the orange clownfish, *Amphiprion percula*, and the spiny chromis, *A. polyacanthus*. While elevated $pCO_2$ had no detectable effects on embryonic development, hatching time, or survival, let alone swimming performance, in orange clownfish, there were substantial increases in growth rates noted (Munday et al., 2009b). A study on spiny chromis also detected no effects of elevated $pCO_2$ in terms of growth or skeletal (i.e., otolith, ear bone) development (Munday et al., 2011a), which was surprising, as the basic chemistry associated with elevated $pCO_2$ and reduced carbonate saturation states would suggest an impact on bone calcification. However, exposure to much higher $pCO_2$ levels did result in larger otoliths in orange clownfish, which may be associated with increased acid-base regulation and increased precipitation of $CaCO_3$ (Munday et al., 2011b). It may be that coral reef fish species physiologically tolerate elevated $pCO_2$ levels because of the daily cyclic changes they already experience on the reef (Hannan et al., 2020a; Jarrold and Munday, 2018a,b). However, similar findings have been documented for large, highly mobile, pelagic, and widely distributed fish species as well, such as cobia, *Rachycentron canadum*, mahi-mahi, *Coryphaena hippurus*, and kingfish, *Seriola lalandi* (Bignami et al., 2013, 2014; Frommel et al., 2019; Laubenstein et al., 2018; Pan et al., 2020). Yet, these studies examined larval and juvenile stages of the pelagic species that likely also use shallow, nearshore habitats that would experience natural cycles of elevated $pCO_2$; however, when juvenile kingfish are reared under much higher, recirculating aquaculture system relevant $pCO_2$ levels, negative impacts on growth, swimming, and metabolism can be detected (Pan et al., 2020).

Studies on adults of reef-associated pelagic species are needed, as they are expected to be much more heavily impacted by elevated $pCO_2$ (Munday et al., 2016), but such studies are logistically challenging.

Other work on coral reef fishes has focussed on adult physiological performance and fitness-related behavioral traits in response to climate change relevant $pCO_2$ levels. Some of these results support the notion that teleost fishes can maintain aerobic performance under a mild pH disturbance, such as after short term exposure (e.g., weeks) to elevated $pCO_2$ conditions, possibly due to their unique capacity for maintaining $O_2$ transport (Rummer et al., 2013). These findings for maintained or enhanced performance (Couturier et al., 2013; Rummer et al., 2013) differ from the 47% decrease in aerobic scope observed in coral reef cardinalfishes exposed to similar $CO_2$ levels (Munday et al., 2009a). Yet, there may, indeed, be species specific differences (Couturier et al., 2013). Such differences may be related to when those species might be most active (diurnal vs nocturnal) and experiencing the highest $pCO_2$ levels in their natural reef habitats (exposure to constant elevated vs. fluctuating elevated $pCO_2$ levels) (Hannan et al., 2020a, 2020b, 2020c). For example, nocturnal cardinalfishes (e.g., *C. quinquelineatus*) have been found more sensitive to elevated, fluctuating $pCO_2$ levels (e.g., in terms of swimming and aerobic performance) than their diurnal counterparts (Hannan et al., 2021). Shallow water, benthic elasmobranchs such as the epaulet (*H. ocellatum*) and white spotted bamboo (*Chiloscyllium plagiosum*) sharks, as well as reef sharks that use shallow, lagoonal habitats as newborns (e.g., *Carcharhinus melanopterus*) exhibit minimal effects upon exposure to elevated $pCO_2$ conditions, even though elasmobranchs in general exhibit a slight yet significant negative response to ocean acidification relevant elevated $pCO_2$ conditions (reviewed in Rosa et al., 2017; Pereira Santos et al., 2021; Rummer et al., 2022). Some of these findings may be related to differences in physiological adaptations to maintaining performance under a mild acidosis, but various behavioral alterations as a result of exposure to elevated $pCO_2$ have been identified as well.

The physiological mechanisms underpinning altered behaviors (e.g., responses to alarm cues, behavioral lateralization, anti-predator responses (Allan et al., 2013), including fast-starts and reactions to chemical alarm cues, and sheltering), are likely related to acid–base regulatory processes interfering with γ-aminobutyric acid (GABA) receptor function (see fig. 4 in Schunter et al., 2018). The GABA-A receptor is the primary inhibitory neurotransmitter receptor in the vertebrate brain (Hamilton et al., 2014; Nilsson et al., 2012). Normally, ion gradients over the neuronal membrane result in an inflow of chloride ($Cl^-$) and bicarbonate ($HCO_3^-$) upon binding of the GABA-A receptor, which then leads to hyperpolarization and neuron inhibition. However, when fish are exposed to elevated $pCO_2$, pH compensation will change ion concentrations (see Fig. 1 in Hannan and Rummer, 2018) that could alter the receptor function and therefore could explain the behavioral changes that have been noted in coral reef fishes upon exposure to elevated $pCO_2$ (Nilsson and Lefevre, 2016). Depending on the magnitude of changes in these ions during acid–base regulation, resultant alterations in ion gradients could either potentiate GABA-A receptor function or reverse its action, making it excitatory rather than inhibitory (Heuer and Grosell, 2014). Indeed, Heuer et al. (2016) were the first to pair some of the key behavioral assays with measurements of relevant intracellular and extracellular acid-base parameters in spiny chromis (*A. polyacanthus*) exposed to elevated $pCO_2$. Even vision is related to GABA-A receptor function, as the negative effects of elevated $pCO_2$ exposure on *A. polyacanthus* retinal function—where impairments could preclude a fish's capacity to quickly respond to threatening events—can be counteracted upon exposure to a GABA antagonist (Chung et al., 2014). Also, the effects of elevated $pCO_2$ on fish behavior and sensory abilities occur when fish are exposed to levels >600 μatm $pCO_2$, which is well within climate change relevant ocean acidification levels for the 21st century (Munday et al., 2010, 2012). Interestingly, few behavioral effects have been detected upon exposure to levels below 600 μatm and none below 500 μatm $pCO_2$, which are levels that would resemble summer night time hours on most coral reefs (Hannan et al., 2020a). Therefore, behaviors (i.e., via acid-base regulation) may also be adapted to the daily fluctuations in $pCO_2$ fishes experience within coral reefs, however, those behaviors could be sensitive to continuously elevated (as opposed to fluctuating) $pCO_2$ conditions. Studies that link the timing of behavioral and acid–base regulatory responses to $pCO_2$ fluctuations in coral reef environments have since been important in determining thresholds for ocean acidification relevant $pCO_2$ levels and daily $pCO_2$ fluctuations.

Understanding species-, context-, and temporally specific effects of elevated $pCO_2$ on certain fitness-related traits is critical. However, molecular responses that underpin developmental, parental, and transgenerational effects of elevated $pCO_2$ will be key in determining the long-term implications for coral reef ecosystem health. Predicting the potential for acclimation and adaptation cannot be done by acutely exposing animals to elevated $pCO_2$ for days to weeks alone. Moreover, conditions experienced early in life can affect—via developmental plasticity—how an organism responds to those conditions later in life, which can also be mediated epigenetically (Schunter et al., 2018). The environment experienced by the parents can also influence how offspring respond (Munday, 2014; Schunter et al., 2016). Indeed, studies investigating transgenerational effects of elevated $pCO_2$ exposure demonstrate that metabolic performance and growth rates are recovered in juvenile fish when both parents and offspring are exposed to elevated $pCO_2$ (Miller et al., 2012). Heritability can underpin variations in the responses that offspring exhibit in response to elevated $pCO_2$, which may be based on parental environments and responses (Welch and Munday, 2017). In *A. polyacanthus*, altered gene expression for the majority of within-generation responses return to baseline levels following parental exposure to elevated $pCO_2$ conditions, suggesting that both parental variation in tolerance and transgenerational exposure to elevated $pCO_2$ are crucial factors supporting the response of coral reef fishes to ocean acidification relevant conditions (Schunter et al., 2018). Indeed, long-term developmental and generational studies will be important in understanding the role of individual variation in how coral reef fish species respond to elevated $pCO_2$, which will collectively be key in understanding and predicting the effects of elevated $pCO_2$ on populations and their capacity to adapt (Vargas et al., 2017).

### *3.1.5 Noise*

Sound pollution or anthropogenic noise in the marine environment may originate from boat noise (commercial shipping, fishing, cruise, and recreational motorboats), seismic testing, and pile driving, activities that have all increased dramatically since the Industrial Revolution due to urbanization, resource extraction, tourism, and transportation. Anthropogenic noise is changing natural soundscapes worldwide (Duarte et al., 2021). Near coral reefs, anthropogenic noise predominantly comes from boat noise and, as such, is becoming recognized in international legislation as a prevalent and increasing anthropogenic pollutant (International Maritime Organization, 2014). However, it is important to distinguish between frequencies of sounds that naturally occur on the reefs (i.e., and their importance to fish health) and in the marine environment (e.g., 20 Hz—15 kHz, the sounds of the reefs, snapping shrimp, fish calls ranging from popping, trumpeting, to banging sounds, and the crushing of coral by parrotfishes as they feed), from artificial sounds associated with noise pollution (Gordon et al., 2018; Simpson et al., 2005a).

The Great Barrier Reef, Australia, has been the setting for work investigating the effects of anthropogenic noise on coral reef fishes, spanning species, life stages, experimental approaches, and exposure simulations (Gordon et al., 2018). This site is highly impacted by anthropogenic noise, and increasingly so, with predictions that 0.5 million recreational motorboats will be using the GBR by 2040 (GBR Marine Park Authority outlook report 2014). Motorboat noise affects physiological processes in coral reef fishes that can impact parental care (Nedelec et al., 2017a), navigation (Holles et al., 2013), foraging (Voellmy et al., 2014), surviving a predator threat or predator avoidance (Simpson et al., 2016), and various aspects of morphological development (Fakan and Mccormick, 2019). The physiological mechanisms underpinning these impacts are likely via metabolic (Simpson et al., 2016) and endocrine pathways (e.g., androgen/glucocorticoid pathways) and may also interact with how fish species respond to additional anthropogenic stressors (Mills et al., 2020).

Some species may be more sensitive to detecting sound pressure and frequency than others (Colleye et al., 2016), and both preconditioning (Staaterman et al., 2020), exposure duration, and the timeline over which sound impacts a fish (Egner and Mann, 2005; Parmentier et al., 2009; Wright et al., 2005, 2011) are also important. While closely related damselfish species can respond differently to anthropogenic noise (Fakan and Mccormick, 2019), populations of the same species may also respond differently. For example, populations of *Halichoeres bivittatus* living in noisy areas had differing levels of baseline stress (measured as whole-body cortisol) than populations living in quiet areas (Staaterman et al., 2020). Indeed, the period of noise exposure matters as well, being brief or acute, or chronic (Holmes et al., 2017; Mills et al., 2020; Nedelec et al., 2016, 2017b; Staaterman et al., 2020). Moreover, some species (e.g., Ambon damselfish, *P. amboinensis*) may habituate or desensitize to boat noise over extended periods as well (Holmes et al., 2017). Timing of exposure and associated effects in coral reef fishes may also have a lot to do with development. Of the 100 families of coral reef fishes, 36 families are brooders. They lay their eggs within the reef matrix, and parents often guard these eggs, during which time boat traffic and other anthropogenic noise could influence developmental milestones. This is all assuming that coral reef fishes have developed auditory sensory organs sufficiently while still developing in the egg and upon hatching, which is likely species specific. However, it has been determined for two species so far (i.e., *Amphiprion melanopus* and *A. polyacanthus*) that the effects of noise pollution begin during embryogenesis (Fakan and Mccormick, 2019). Regardless, the developing embryos, upon hatch either stay on the reef (i.e., where they may continue to experience anthropogenic noise) or leave for the pelagic where they spend weeks to months. For those species with a pelagic larval stage, navigating back to the coral reefs (i.e., whether natal or new) to settle is a crucial component of life history and may involve cues, such as the sounds of the reef (Simpson et al., 2004). These critical life history stages, if affected, could not only impact proper growth, development, and settlement of coral reef fish species, but could also impact demography and distribution patterns and therefore ecosystem health (Fakan and Mccormick, 2019).

Various methods have been used to physiologically assess the effects of anthropogenic noise on coral reef fishes and under different simulated soundscapes. Approaches assessed various levels of the stress response and have used heart rate monitoring, stress hormone analyses, the Auditory Brainstem Response (Egner and Mann, 2005), and Auditory Evoked Potential (AEP) audiometry (Colleye et al., 2016; Parmentier et al., 2009). Indeed, via the primary stress response, exposure to anthropogenic noise can result in increased glucocorticoid levels in fishes (Mills et al., 2020; Staaterman et al., 2020). Via the secondary stress responses, exposure to anthropogenic noise can result in increases in blood glucose and hematocrit (Filiciotto et al., 2013), and metaboloic rates can be altered as well (Simpson et al., 2005b; Staaterman et al., 2020). Indeed, Simpson et al. (2005b) and Jain-Schlaepfer et al. (2018) used changes in the secondary stress response (i.e., heart rate) to assess the acoustic sensitivity of early life stages of coral reef fishes. They found that clownfish embryos (*Amphiprion ephippium* and *A. rubrocinctus*) exhibit increased heartrate in response to noise, and their sensitivity increases across development from fertilization to near hatching (Simpson et al., 2005b). Via the tertiary stress response, exposure to anthropogenic noise can affect various aspects of movement (Holmes et al., 2017; Picciulin et al., 2010) that may be important for schooling and foraging as well as anti-predator behaviors (Simpson et al., 2015), parental care (Nedelec et al., 2017a; Picciulin et al., 2010), interactions between species (Nedelec et al., 2017b) and survival (Simpson et al., 2016). Studies often use playbacks of recorded sounds or real noise, which has been helpful in translating findings to management solutions. For example, exposure to the sound profile of a two- versus a four-stroke engine resulted in twice the stress response in *Amblyglyphidodon curacao* damselfish embryos, as measured by changes in heart rate ( Jain-Schlaepfer et al., 2018) and twice the response at the level of escape performance in whitetail damselfish, *Pomacentrus chrysurus* (McCormick et al., 2019a). As a result, boat noise is starting to be included in environmental management plans as a stressor, and recommendations can be made regarding boat engine type based on empirical evidence. Anthropogenic noise is an area of research that will continue to grow into the Anthropocene.

### 3.1.6 Salinity

Salinity stress or iono- or osmo-regulatory stress (see Chapter 5, Volume 39A: Eliason et al., 2022) in relation to changes in environmental salinity have not been heavily investigated in coral reef fish species. Given that coral reef fish species are not as prone to such scenarios compared to estuarine and mangrove-dwelling species, this is expected. However, extreme scenarios do exist for coral reef fishes. For example, in areas such as the environmentally challenging Arabian/Persian Gulf, coral reef fish species (e.g., black-spot snapper, *Lutjanus ehrenbergii* and yellowbar angelfish, *Pomacanthus maculosus*) can incur the life-history and metabolic costs of osmoregulation in hypersaline environments, which are reflected in growth various parameters (D'Agostino et al., 2021). On a more temporal basis, reefs in some areas (e.g., Kaneohe Bay, Hawai'i) can succumb to storm flooding and freshwater inundation, making for hyposaline challenges and even mortality for resident fish species ( Jokiel et al., 1993). Mechanistic studies (e.g., in the economically valuable coral trout, *P. maculatus* and *Plectropomus leopardus*, and also associated with capture stress; Frisch and Anderson, 2005) have been done including determining salinity preference (Serrano et al., 2010) and understanding the role of salinity in modulating reproductive hormones (Hung et al., 2010) and stress (e.g., heat shock proteins and cytoprotection; Tang et al., 2014a). In many cases, changes in gill morphology are observed, sodium potassium pump ($Na^+$, $K^+$, ATPase) activities and $Na^+$, $K^+$, $2Cl^-$ cotransporter proteins are measured and/or isoforms assessed to determine a coral reef fish's status in hypo- or hyper-saline conditions (Tang et al., 2014b). Studies have also been designed for applied outcomes, such as for streamlining aquaculture practices, understanding range expansions for invasive species (e.g., lionfishes; Jud et al., 2014; Schofield et al., 2015), and interpreting interactions with other stressors. It is important to note, however, that some coral reef associated species may use estuarine and mangrove habitats as nursery areas and may therefore experience salinity fluctuations (Prodocimo and Freire, 2001; Shirai et al., 2018). However, still, few studies have investigated these issues specifically in coral reef fish species.

### 3.1.7 Pressure/depth

Studies assessing pressure and depth relationships in coral reef fishes have been largely from a mechanistic perspective or an ecological perspective, with few if any evaluating pressure or depth as a pervasive anthropogenic stressor. Studies have assessed the role of pressure—and therefore depth—on various components of vision in four coral reef fish species, yellowstripe goatfish (*Mulloidichthys flavolineatus*), manybar goatfish (*Parupeneus multifasciatus*), convict surgeonfish (*Acanthurus triostegas*), and (orangespine unicornfish) *Naso lituratus*, and concluded a minor role of habitat depth with ocular transmission (Nelson et al., 2003). Other studies have assessed the role of depth or pressure in modulating the release of brain hormones important for circadian rhythms (e.g., including dopamine, serotonin, etc.), reproductive hormones (e.g., follicle stimulating and luteinizing hormones) and environmental cues such as time of day with respect to important ecological processes (e.g., spawning synchrony in threespot wrasse, *Halichoeres trimaculatus*; Takemura et al., 2010, 2012)). Indeed, reproduction seems to be the most sensitive to pressure and depth, but it is important to note that it may be challenging to disentangle the effects associated with depth from those related to light. Depth has a negative impact on ovarian development in the sapphire devil, *Chrysiptera cyanea* (Fukuoka et al., 2017). In contrast, an opposite relationship was found upon comparing bicolor damselfish (*Stegastes partitus*) between shallow ($<10$m), deep shelf (20–30m), and mesophotic (60–70m) reefs. Populations were less dense, but individuals were older and larger on the deeper reefs with potentially longer life spans, a broader diet niche, and higher reproductive investments producing high condition larvae, when compared to shallow reefs (Goldstein et al., 2016, 2017). Indeed, there has been an eco-physiological component to these studies that, while not directly addressing pressure and depth as anthropogenic stressors, have investigated shifts and requirements for species distribution as habitat suitability declines with the continuing global loss of shallow water reefs.

### 3.1.8 Turbidity

Since the middle of the 20th century, many nearshore coral reefs have experienced decreasing water quality, in particular increasing turbidity and increasing concentrations of suspended sediments (often referred to as total suspended solids, TSS), due to coastal development, land conversion, mining, shipping, and dredging (Foley et al., 2005; Syvitski et al., 2005). Although often used interchangeably, often complementary, and both indicate the clarity of the water, turbidity and TSS measure different things. Turbidity examines at how well a light passes through liquid, while TSS is a quantitative expression of suspended particles. Indeed, increasing TSS concentrations are leading to changes in the composition of fish assemblages on coastal reefs (Bejarano and Appeldoorn, 2013; Cheal et al., 2013; Moustaka et al., 2018), either indirectly through changes in benthic composition and/or directly through impacts on the fishes themselves (Fabricius et al., 2005; Hamilton et al., 2017). Yet, the direct effects of TSS exposure and other sources of turbidity on the physiology of coral reef fishes have been seldom investigated, until recently.

Some of the most profound physiological effects of TSS exposure in coral reef fishes have been observed at the gills. Orange-spotted grouper (*Epinephelus coioides*) exhibit a decrease in gas diffusion distance at the gill in response to suspended sediment exposure (Wong et al., 2013). However, suspended sediments can also directly damage the gill epithelium. Studies found shortened gill lamellae (Hess et al., 2017; Lake and Hinch, 1999; Sutherland and Meyer, 2007), increased growth of protective cell layers, which increases gas diffusion distances (Cumming and Herbert, 2016; Hess et al., 2015; Lowe et al., 2015), and an increase in mucous secretion (Hess et al., 2015; Humborstad et al., 2006). Moreover, coral reef fishes exposed to elevated suspended sediment levels exhibit gill microbiome shift from "healthy" to pathogenic bacterial communities, which can further compromise immune function (Hess et al., 2015). All of these changes can reduce gas exchange efficiency and interfere with oxygen uptake across the gills (Evans et al., 2005; Lappivaara et al., 1995). That said, it may not be surprising that exposure to suspended sediments can decrease maximum oxygen uptake rates ($\dot{M}O_{2max}$) in juvenile anemone fish, *A. melanopus*, and increase resting oxygen uptake rates ($\dot{M}O_{2rest}$, a proxy for metabolic costs required to sustain basic metabolic functions) (Hess et al., 2017). These changes can reduce aerobic scope (i.e., difference between $\dot{M}O_{2max}$ and $\dot{M}O_{2rest}$) and thus overall capacity for aerobic activity. A reduction in aerobic scope can affect aerobic activities such as growth and development, with negative consequences for the survival and fitness of fishes (Norin and Clark, 2016). In contrast, however, two confamilial species, *A. percula* and *A. melanopus,* have been found to maintain aerobic performance, despite changes in gill morphology following suspended sediment exposure (Hess et al., 2017).

While the consequences of gill alterations to aerobic metabolism and performance may seem straightforward, the effects of suspended sediment exposure on anerobic metabolic pathways and performance traits are not as clear. Sediment-exposed coral reef fish species, such as juvenile anemonefish (*A. melanopus*) can respond faster to a mechanical stimulus, achieve higher escape speeds and acceleration, and escape further distances than their control counterparts (Hess et al., 2019). This kind of response counters what is expected when fish are exposed to other stressors (e.g., elevated temperatures). However, visual acuity is reduced for fishes living in turbid waters, which could also impact predator detection times. Therefore, the effects of suspended sediments on escape performance may be more related to the behavioral changes that are required as fish increase vigilance and perceived predation risk in turbid waters than direct effects of TSS on physiological mechanisms. Indeed, newly settled *Chromis atripectoralis* are preyed upon by an ambush predator, *P. fuscus*, much more heavily under medium suspended sediment concentrations (Wenger et al., 2013). It is important to note that the effects of suspended sediment exposure on aerobic metabolic pathways could impact a coral reef fish's capacity to recover from these anerobically driven responses. Increased vigilance can also increase metabolic costs due to trade-offs associated with activity and foraging (Killen et al., 2015). Over the longer-term, this could result in non-consumptive costs (e.g., compromised immune function, or reductions in growth and condition) (Hawlena and Schmitz, 2010). Regardless, predator escape performance plays an important role in predator-prey interactions, and any changes whether physiological, behavioral, or a combination of both, are likely to directly influence survival of juvenile and adult fishes on coral reefs (McCormick et al., 2018). Finally, it is important to note, like with many of the stressors affecting coral reef fishes today, seldom does one stressor occur in isolation (e.g., in the wild, TSS may contain pollutants; increased turbidity may co-occur with the warmest seasons and therefore elevated temperatures, flood plumes and storms; dredging and coastal maintenance and thus increased noise, etc.); therefore, their effects cannot be assumed independent from other abiotic or biotic influences—this is an area requiring much more investigation.

## 3.2 Biotic stressors

Biotic stressors arise from living organisms and can originate from natural and anthropogenic sources. The most pervasive, individual biotic stressors include prey abundance, predator threats, parasites, and disease.

### *3.2.1 Prey abundance*

Changes in prey abundance—in some cases related to starvation, dietary shifts over temporal (e.g., ontogeny) and spatial scales, trophic interactions, resource partitioning, and food webs represent the largest biotic stressors studied in coral reef fishes. One of the most common approaches used to discern these relationships is via stable isotope analysis of bones and tissues. Stable isotopes of nitrogen, carbon, sulfur, and oxygen can be useful in determining trophic position of consumers and nutrient sources (i.e., linking consumers to their food sources; Speed et al., 2012a), influential factors in the environment (e.g., flow velocities, substrate type, amount of rainfall, and possibly temperature), oxidation/reduction conditions of certain habitats, and overall habitat use (Bouillon et al., 2008). Changes in certain stable isotopes over time can be useful in determining diet switching or habitat shifts. Moreover, because different tissues respond at different rates, they can provide a time history of diet and habitat use. The widely used stable isotope approach does not directly assess physiological stress in coral reef fishes. The approach can, however, help infer when there are resource-use overlaps and trophic redundancies in certain ecosystems, ecologically distinct regions, and can provide baseline information to help identify potential for fisheries pressure or other anthropogenic stress (e.g., in elasmobranchs; Morgan et al., 2020; Peterson et al., 2020).

Metabolic responses to different feeding regimes, starvation, digestion and assimilation rates (e.g., specific dynamic action; SDA) have been examined in many coral reef fish species. Such approaches may be especially important during the larval stage, with the extremely high mortality rates (see above) and rapid growth rates that require a large supply of planktonic prey (reviewed in McLeod and Clark, 2016). With increasing temperatures resulting from climate change, tropical marine plankton communities, which are the primary food sources for larval coral reef fishes, are predicted to shift in composition and distribution. This could very well result in a mismatch between coral reef fish larvae and their prey. Studies assessing metabolic costs to different food rations as well as determining the energetic budgets of different feeding regimes (e.g., *A. percula*, McLeod et al., 2013; McLeod and Clark, 2016) provide important information for assessing individual performance and predicting recruitment success and ecosystem health under various stressors. In controlled laboratory studies, this is an interesting avenue, given that many protocols feed experimental animals ad libitum, which could be masking the effects of other treatments (e.g., McMahon et al., 2018, 2019). Such food ration and feeding regime information are useful for applied outcomes as well, such as in ensuring efficient aquaculture practices and in understanding adverse effects of eco-tourism provisioning and other non-consumptive wildlife activities (Birnie-Gauvin et al., 2017). For example, in the Cook Islands, tourists often feed bread to threadfin butterfly fish (*Chaetodon auriga*) and striated surgeonfish (*Ctenochaetus striatus*), which results in reductions in their foraging on natural food sources (Prinz et al., 2020). Some reef sharks may shift their behaviors in response to long-term provisioning. For example, at a long-term provisioning site in French Polynesia, blacktip reef sharks (*C. melanopterus*) now exhibit smaller home ranges and have changed how they use their habitat (Mourier et al., 2021). Interestingly, after tourism activities ceased for six consecutive weeks (i.e., due to COVID-19 pandemic lockdowns in 2020), all animals left the area, and pre-lockdown abundances were not restored for at least 1 month after tourism resumed (Séguigne, 2022). However, some species exhibit no effect of provisioning at all, like Caribbean reef sharks (*Carcharhinus perezi*; Maljkovic and Cote, 2011) and juvenile lemon sharks (*Negaprion brevirostris*; Heinrich et al., 2020). In the latter species, provisioning neither affected spatial distribution nor mean daily activity or energy requirements, but lemon sharks did start exhibiting anticipatory behaviors as soon as 11 days following regular provisioning (Heinrich et al., 2020).

Dietary stress and the effects of food availability on trophic interactions can also be assessed via the microbiome, which, with the advent of high throughput sequencing, has become much cheaper, faster, and informative. Given the important role the gut microbiome has in nutrient acquisition and pathogen resilience, establishing baseline data and having a rapid means for determining dietary stress (e.g., as in surgeonfishes, Family Acanthuridae; Miyake et al., 2015) will increase in importance, especially with interacting anthropogenic stressors. Coral reef fish species inhabiting disturbed reefs have changes in gut bacterial composition and fermentative bacteria ratios, suggesting that this disturbance not only affects the gut microbiome community but also impacts ecosystem function through microbial processes (Cheutin et al., 2021). But moreover, coral reef fishes living in reefs characterized by high suspended sediment concentrations exhibit changes in gill microbiome as well, with implications toward immune function, metabolic performance, growth, and development (Hess et al., 2015). Indeed, this approach can also elucidate feeding stress in post-disturbance reefs (e.g., following coral bleaching, Fig. 1, or crown-of-thorns starfish outbreaks). Finally, if gastrointestinal microbial fauna, which is fundamental to fish health, is not flexible enough to accommodate new habitats and food sources (e.g., rabbitfish, *Siganus fuscescens*; Jones et al., 2018), this may be problematic as coral reef fish distribution patterns and ranges change due to ocean warming.

### 3.2.2 Predator threats

The primary ecological interaction between organisms that most significantly shapes selection and fitness is predation (i.e., eat, but do not get eaten); coral reef fishes are no exception. Because most coral reef fishes produce larvae that spend some portion of early life in the pelagic before settling onto new or natal reefs, predation is a serious threat. Settling juveniles are small and naïve (Almany and Webster, 2006); thus, the time around settlement can be a life-history bottleneck for coral reef fish populations due to predation mortality. Moreover, many coral reef fishes occupy low to mid trophic levels, meaning they are subject to piscivory through all life stages. While predator-prey dynamics represent naturally occurring stressors within coral reef ecosystems, continued human exploitation of reef predators, habitat degradation (i.e., nurseries, shelter; McCormick et al., 2019b), and additional stressors into the Anthropocene will potentially further alter coral reef fish population dynamics and ecosystem health.

While much work on predator threats has been from an ecological perspective, most predator-prey interactions have physiological underpinnings (Palacios et al., 2016). This includes vast array of body morphologies, color changes (e.g., mimicry, Cortesi et al., 2015), and chemical defenses (e.g., toxic gobies, Gratzer et al., 2015, but also toxins from other non-fish species, like polychaetes, sponges, soft corals, etc.). Predator-prey interactions, perhaps especially on coral reefs, also represents an area where non-consumptive effects (NCEs) have been well-investigated (Mitchell and Harborne, 2020). Indeed, NCEs due to the risk, fear, and other non-lethal interactions associated with predator-prey dynamics can change prey behavior, physiological performance, such as metabolism and swimming, morphology, and development, which may result in changes in distribution patterns and overall fitness (Arvizu et al., 2021; Hess et al., 2019; Mitchell and Harborne, 2020). Ultimately, most physiological underpinnings of predator-prey dynamics are associated with the primary (e.g., catecholamine release) and secondary (e.g., glucocorticoid release) stress responses and metabolic performance (e.g., escape responses). Glucocorticoids can cross the blood-brain barrier (e.g., unlike most catecholamines) and therefore interact with receptors in several brain regions meaning their role in the stress response is quite important in modulating behaviors (reviewed in Soares et al., 2012). Cleaning gobies (*Elacatinus evelynae*) release cortisol when encountering a "client" that could be a potential predator, but of the "fight," "flight," or "freeze" responses typically associated with stress, *E. evelynae* exhibits more of a "fight" response by cleaning more thoroughly and being more proactive, which may be an effort to reduce predatory danger or conflict (Soares et al., 2012). The metabolic and neurophysiology that underpins fast-start, escape responses and are triggered by an approaching predator include anerobically powered myotomal blocks of fast glycolytic muscle and Mauthner neurons in the brain (reviewed in Allan et al., 2013; Ramasamy et al., 2015). However, behavioral lateralization—the preferential use of one side of the body or another (Domenici et al., 2012)—is another trait that is key to predator escape responses (Ferrari et al., 2017). Escape performance in coral reef fishes, overall, has been well investigated in many species, and whether fast starts or lateralized movements, both reveal a high degree of inter, intra-specific and within-individual plasticity (Allan et al., 2013, 2014, 2015, 2017, 2020). Such capacity to modulate these responses may be beneficial given the "high cost of repaying $O_2$ debt" upon an exhaustive challenge and the metabolic (e.g., stress) and behavioral disruption associated with escape responses (Allan et al., 2015). Additional energy saving occurs, as coral reef fishes can discriminate between predator and non-predator, which can be detected in certain metabolic traits and honed with a history of predator exposure (Hall and Clark, 2016; Ramasamy et al., 2015). When juvenile spiny chromis (*A. polyacanthus*) are pre-exposed to visual or olfactory predator cues, they exhibit stronger escape response (i.e., reduced latency, increased escape distance, mean response speed, maximum response speed and maximum acceleration) than if they had no prior experience (Ramasamy et al., 2015). Although morphological and performance variables are most often measured in response to predator interactions, the behaviors associated with these traits ultimately have the strongest effects on survival (McCormick et al., 2017).

The larval life stage of coral reef fishes is challenging to investigate because of its narrow window of time; yet, this stage has been the focus of the majority of predation studies on reef fishes (Almany and Webster, 2006). Yet, traits developed during this phase—such as, but not limited to, growth rates, body condition, lipid content, swimming, boldness, and escape performance—influence overall survival (e.g., in Pomacentridae, Booth and Beretta, 2004; Hoey and McCormick, 2004; Figueira et al., 2008; Fuiman et al., 2010). Even while parents are guarding eggs, predator threats can alter maternal cortisol levels (e.g., in *Pomacentrus amboinensis*), which can influence larval morphology via stress responses (McCormick, 1998). Embryonic cinnamon clownfish, *A. melanopus*, while *in ovo*, that are exposed to conspecific chemical alarm cues or conspecific alarm cues combined with predator cues respond by modulating their heart rate. This response indicates that they can detect and react to cues suggesting a conspecific has been injured. Moreover, they can also use such information to learn about predation, all of which may influence development and behavior (Atherton and McCormick, 2015). Evidence also suggests a strong transgenerational (i.e., parental) effect of predator recognition exists in coral reef fishes (Atherton and McCormick, 2020). If coral reef fishes are continuously exposed to high-risk conditions, they exhibit behavioral and physiological antipredator traits that are important when faced with an actual predator interaction, suggesting a strong ontogenetic pre-conditioning role (Ferrari et al., 2015a). Indeed, plasticity in traits and "as needed" responses may be key for some species, as predation is not always strong enough, consistent in space or time, and not always unidirectional to result in genetic adaptations and may also promote greater resilience in species as habitats continue to change (i.e., habitat degradation) into the Anthropocene (McCormick et al., 2019b).

### *3.2.3 Parasites*

Parasites remain poorly studied due to sheer numbers and their cryptic nature; yet, parasites constitute the majority of biodiversity found on coral reefs (Sikkel et al., 2018). Parasitic interactions in coral reef fishes are exacerbated in the Anthropocene due to habitat loss, water quality reductions, and top-level predator removal (Artim et al., 2020). Among coral reef fishes, species that are commonly cultured as ornamentals and key for the aquarium industry are subject to parasites, often due to high density holding and immune suppression. However, parasites also occur in the wild naturally. Classic examples include grooming and cleaning stations (Grutter et al., 2003). While the diversity of parasites that interact with coral reef fishes is largely unknown, estimates suggest that there could be as many as 20 different parasite species for every coral reef fish species on the Great Barrier Reef and New Caledonia (reviewed in Justine et al., 2012). With this high diversity of parasites ranging from various worms, (e.g., flatworms such as Turbellaria, Monopisthocotylea, Polyopisthocotylea, Digenea, and Cestoda; roundworms, such as Nematoda; thorny-headed worms such as Acanthocephala; segmented worms, such as Hirudinea), protozoans, copepods, and isopods, and the array of associated infection and removal strategies, it follows that there would also be a vast suite of physiological effects assessed in coral reef fishes.

The most common stress associated with parasitic interactions in coral reef fishes is documented as changes in growth and development, but immune suppression, reductions in reproduction, and other metabolic costs are common Indeed, when ectoparasitic isopods infect the five-lined cardinalfish, *Cheliodipterus quinquelineatus*, they experience reductions in growth (e.g., 20% decrease in body mass) and reproductive output (e.g., 42% fewer ova) (Fogelman et al., 2009). In other fish species, these ectoparasitic isopods cause reductions in aerobic scope and swimming performance, including reductions in fast starts (Allan et al., 2020; Binning et al., 2013; Östlund-Nilsson et al., 2005). While it has also been found that interactions between cleaner fishes and their clients may elicit pathogen transmission and disease transfer (reviewed in Narvaez et al., 2021), it is also important to note that cleaner fish interactions with their clients can also elicit benefits beyond parasite removal (e.g., tactile stimulation; Paula et al., 2015, and stress reduction; Paula et al., 2019).

Blood samples can be used to determine the degree of the immune response (e.g., leukocytes, granulocytes, etc.), blood loss (e.g., hematocrit), and hormone levels (e.g., cortisol, reproductive hormones; Allan et al., 2020), all of which can affect growth, aerobic performance, and reproduction (Demaire et al., 2020). Not only can these physiological effects be detected in the presence of a parasite, but such parameters may also represent trade-offs if the fish is unable to rid itself of parasites, for example, via the use of mutualistic cleaning stations (Demaire et al., 2020). Many ectoparasites that externally attach to their host fish can increase the drag of the host fish and therefore impact swimming and escape kinematics and behaviors, while increasing overall metabolic costs (e.g., bridled monocle bream, *Scolopsis bilineata*, Binning et al., 2013, 2014; Ambon damselfish, *P. amboinensis,* Allan et al., 2020). Effects of ectoparasites can vary by species and the size of the fish, among other factors, including diurnal as opposed to nocturnal species (Cook et al., 2015). For example, French grunt (*Haemulon flavolineatum*) and brown chromis (*Chromis multilineata*) are closely related species and both commonly infected by an isopod (*Anilocra* spp.), but the energetic effects (i.e., as per condition factor, percent moisture in the muscle tissue, total muscle tissue calories, and gut content volume) were different between species (Welicky et al., 2018), highlighting that such generalizations cannot always be made. Other infections can occur via consumption of an intermediate host. This is the case with blood flukes, which often enter fish hosts via polychaete worms. Adult blood flukes have been found in 26 species of butterflyfishes (Chaetodontidae) on the Great Barrier Reef, with fluke eggs observed in hearts and gills (Yong et al., 2013). Another example is in the ecologically and commercially important dusky grouper, *Epinephelus marginatus*, where histopathology revealed parasitic flatworms on the gills, resulting in inflammation and reduced immune function (Polinas et al., 2018). Indeed, much research has also focussed on the recreationally and commercially important coral reef barramundi (*Lates calcarifer*) as well, given its high value in aquaculture industry and the propensity for parasitic infections to interact with other environmental stressors (e.g., temperature and salinity; Brazenor and Hutson, 2015). Technological advancements in, for example environmental deoxyribonucleic acid (eDNA), are now making these outbreaks more detectable and predictable (Gomes et al., 2017). Undoubtedly, such energetic costs associated with parasite loads can be problematic on an array of levels, given that some ectoparasites infect their hosts for months to years. That said, parasites can greatly impact coral reef fish energetics and therefore population dynamics and ecosystem health. This biotic stressor will be continually examined, especially as it interacts with other biotic (e.g., disease) and abiotic (e.g., temperature, warming) stressors that are increasing in severity in the Anthropocene.

### *3.2.4 Disease*

Novel pathogen exposure in coral reef fishes is increasing. However, not all diseases that are increasing in prevalence in coral reef fishes in the Anthropocene (e.g., carcinomas, melanomas from increased UV exposure, potentially due to ozone depletion) are from microbial exposure (i.e., from viruses, bacteria, and fungi). Indeed, disease prevalence in coral reef fishes is not a new topic. Rather, it has been a topic of intense investigation since the early 20th century, as many of the traditionally investigated diseases were common in coral reef fishes used in aquarium trades and human consumption (e.g., especially live trade). Today, stress, poor water quality (e.g., coastal development, sewage treatment issues, including an increased prevalence of pharmaceuticals in sewage, etc.), overcrowding (e.g., for cultured fish species), and runoff (e.g., industrial and agricultural) likely contribute to most of the diseases in coral reef fishes.

In cultured fish species (e.g., for food fish, ornamentals, aquarium trade, etc.), disease, whether viral, bacterial, or otherwise, represents a serious economic loss, but this could be the case for disease in wild fish species as well (e.g., fisheries, recreation, eco-tourism, etc.). Neurofibromatosis tumors or nerve sheath tumors (e.g., in bicolor damselfishes, *S. partitus*) may stem from a virus and can be fatal and is often associated with density issues (Fieber and Schmale, 1994; Schmale, 1995; Schmale et al., 2002, 2004). Some viruses (e.g., iridovirus) result in anemia and eventually spleen and kidney necrosis as well as general immunosuppression, which makes fish susceptible to other diseases (Mahardika et al., 2004). The hematopoietic necrosis virus, which originated in freshwaters species (e.g., the salmonids) but can now affect coral reef fishes, enters the fish from the base of the fins with the fish first exhibiting a bulging abdomen and eyes, and then external hemorrhaging commences, ultimately leading to mortality (Harmache et al., 2006). Another issue originating from a virus is hemorrhagic septicemia; the virus enters the fish via the gills, moves to the internal organs, and weakens the blood vessels, which eventually collapse, causing the hemorrhaging. However, septicemia can also come from bacterial infections and has been found to impact various reef-associated cartilaginous and teleost fish species (Briones et al., 1998; Camus et al., 2013; Keirstead et al., 2014). Natural outbreaks of *Streptococcus agalactiae*, which can be transmitted between humans and aquatic species, in wild groupers and various shark species have also resulted in many cases of septicemia (Bowater et al., 2012). As many pathogens can be transmitted between waterways via boats, trailers, and nets, some of these pathogens that originated in freshwater systems, probably in North America in the Pacific Northwest (i.e., likely via salmonids) or the Great Lakes, now impact coral reef fishes.

Density issues, eutrophic bodies of water, and pollution can underpin these bacterial infections and even result in tumor formation. Mycobacteriosis, for example, enters through open wounds or the gastro-intestinal system, spreads through the whole body of the fish via the circulatory and lymphatic systems, and can be rampant in many coral reef fishes, resulting in high mortality rates (Diamant et al., 2000). Evidence of melanoma in coral trout *P. leopardus* likely stems from increased ultraviolet radiation exposure (Sweet et al., 2012), which will continue to increase into the Anthropocene, especially in shallow water coral reef environments. Tumor formation can also come from chronic trauma, viral infections, or pollution as well as inbreeding and low genetic diversity. The latter can be most prevalent in the ornamental and aquarium fish trade while attempting to breed for certain traits (e.g., clown anemonefish, *A. ocellaris*). Odontomas are benign tumors of dental tissue and/or abnormal tooth formation that can lead to difficulties eating. Such tumors can also occur in the lips, gill arches, or esophagus as neoplasms or in the fat tissue as liposarcomas (Vorbach et al., 2018). Ultimately, the stressors that result in disease in coral reef fishes are becoming more prevalent and severe moving into the Anthropocene. Determining host ranges, timing, and environmental cues will be important in identifying the cause, source, or origin of the pathogen, potential pathogen habitat, how it might interact with other stressors, and how it can be best managed.

## 4 Interacting stressors

To date, the scientific community understands the least about the effects of combined (i.e., two or more) stressors on the physiology of coral reef fishes. Moreover, it is not always straightforward whether effects will be additive, synergistic, or antagonistic, or whether the combined effect of the stressors will equal, be greater than, or be less than the sum of their individual effects, respectively (see Gunderson et al., 2016 for review). The most common combination of interacting stressors that has been addressed has been ocean warming and acidification. Such studies found a decline in offspring quality in anemonefish, *A. melanopus* (Miller et al., 2015). In juvenile spiny chromis, *A. polyacanthus*, results suggest a negative correlation between behavioral (i.e., responding to an alarm odor) and physiological (i.e., aerobic metabolic scope) traits (Laubenstein et al., 2019). Moreover, such findings suggest the potential for trade-offs that limit fish performance and even the capacity for populations to adapt (Laubenstein et al., 2019). Combined warming and acidification conditions also negatively impact metabolic traits such as aerobic performance in two nocturnal cardinalfish species (*Ostorhinchus doedeleini* and *O.* cyanosoma; Munday et al., 2009a) and predator-prey kinematics and swimming performance in Ward's damselfish (*P. wardi*; Allan et al., 2017) from the Great Barrier Reef, Australia. Indeed, such combined stressors can affect behavioral traits that are key for predator-prey dynamics, but perhaps not unidirectionally (i.e., synergistically on overall predation rate, but antagonistically on predator selectivity; Ferrari et al., 2015b). Elevated temperatures in combination with elevated $pCO_2$ results in opposing effects in some traits, such as swimming performance (e.g., yellowtail kingfish, *S. lalandi*; Watson et al., 2018). Other studies have found neurobiological changes in response to the combined stressors in the form of dopaminergic and serotoninergic systems are important for cleaner fish species (e.g., *Labroides dimidiatus*) and their clients (e.g., *Naso elegans*); such cascading effects could impact community structures (Paula et al., 2019). While *in ovo*, embryonic bamboo sharks (*Chiloscyllium punctatum*) exhibit increased routine metabolic rates under the combined stressors of warming and acidification (Rosa et al., 2014). Upon hatching, juveniles exhibited decreased body condition (Rosa et al., 2014) as well as neuro-oxidative damage and loss of aerobic potential under such combined conditions (Rosa et al., 2016a). However, at 30 days post hatch, juveniles exhibited an antagonistic response to ocean warming and acidification conditions in terms of digestive enzyme activities (Rosa et al., 2016b), while metabolic costs increased and survival decreased (Rosa et al., 2014). Collectively, these studies demonstrate the importance of life stage, physiology, species, and habitat/life history strategies.

While these studies examining the combined effects of ocean warming and acidification relevant scenarios represent an important step in assessing and predicting the effects of climate change on coral reef fishes, they still lack the third primary abiotic stressor, being hypoxia and ocean deoxygenation. As discussed throughout this chapter, an array of experimental approaches simulating these scenarios have suggested potential sensitivities of coral reef fishes to any of these stressors in isolation, but we can only surmise the combination (Poloczanska et al., 2016; Richardson et al., 2012). Yet, this so-called "deadly trio" of ocean warming, acidification, and deoxygenation has been previously involved in some of Earth's mass extinction events (Bijma et al., 2013). Therefore, understanding how the physiological performance of coral reef fishes will be affected by climate change is of paramount importance. Despite the wealth of empirical data collected in the past decade, more advanced experimental approaches are needed to address existing knowledge gaps: the intraspecific variability in vulnerability to climate change stressors, the nature of the stressor interaction (e.g., additive, synergistic, or antagonistic), and the adaptative potential. The predicted co-occurrence of climate change stressors highlights the need for multi-stressor experimentation to realistically assess effects of climate change on coral reef fishes.

Ocean warming and prey abundance as interacting stressors has also been a topic of research priority, as increasing temperatures resulting from climate change are causing tropical marine plankton communities—the primary food sources for larval coral reef fishes—to shift in composition and distribution. Studies assessing metabolic costs to different food rations as well as determining the energetic budgets of different feeding regimes under different warming scenarios have been important steps to addressing this issue (McLeod et al., 2013). Newly hatched anemonefish, *A. percula*, exhibited decreased survival rates, and those that made it to settlement stage took longer to do so, longer to metamorphose, and exhibited lower body condition if they had been exposed to warm temperatures and low quantity diets (McLeod et al., 2013). These fish also exhibited higher energetic costs (i.e., via routine $O_2$ uptake rates) than their control counterparts, all of which suggests severe impacts of warming waters and declining food availability on larval coral reef fishes (McLeod et al., 2013). In adult breeding pairs of spiny chromis (*A. polyacanthus*) maintained under one of three temperatures and one of two feeding regimes, not only did elevated temperatures in isolation negatively impact reproduction and egg size, but none of the pairs reproduced under elevated temperatures and low quantity diets (Donelson et al., 2010). Trade-offs could occur to maintain one process over another, and therefore fitness, which could be seen in reproduction success and egg size/quality, as in Donelson et al. (2010), but also growth and immune responses. Moreover, in the wild, declining food availability may also result in fish spending more time and energy foraging, making them more vulnerable to predation. Indeed, this combination of stressors—elevated temperatures and decreased food availability—has already been examined within the context of predator (i.e., chemical alarm cue) recognition. Lemon damselfish, *Pomacentrus moluccensis*, while maintained under elevated temperatures and low food availability, not only depleted energy reserves and reduced growth, as expected, but were also unable to effectively elicit an anti-predatory response (Lienart et al., 2016). Only one study to date on a coral reef associated fish species has addressed the effects of elevated temperatures, reductions in food supply, and ocean acidification conditions and concluded that such conditions may increase starvation risk in larval cobia, *R. canadum* (Bignami et al., 2017).

While each of these 12 stressors (i.e., and others not discussed here) must be first understood in isolation, conclusions as to how coral reef fish species, populations, and ecosystems will be affected in the Anthropocene can only be drawn if multiple stressors are investigated together. Indeed, experiments that address combinations of global and local stressors (i.e., warming and turbidity/TSS, parasites and habitat degradation, noise and elevated $pCO_2$, etc.) will be crucial. While it seems logical to emphasize the "deadly trio," local stressors are as damaging particularly when paired with global stressors. Undeniably, this is also further complicated when investigating an array of species with diverse life history strategies, across life history stages, and traits spanning various levels of biological organization. Effects of multiple stressors can be additive, synergistic, or antagonistic, and therefore if extrapolations are made based on the effects of just one stressor, whole organismal, species, population, and even ecosystem-level responses can be grossly underestimated or mis-represented. Indeed, multi-stressor experiments are the next step in predicting the impact of the Anthropocene on coral reef fishes and the ecosystems they support.

## 5 Acclimation and adaptation potential

Acclimation (or acclimatization), as nominally discussed earlier in this chapter as relevant to aforementioned stressors, and adaptation are the primary ways that organisms, such as the coral reef fishes, can survive and thrive under new environmental conditions. Addressing the potential for and limitations to acclimation and adaptation will be key research priorities for coral reef fishes in the Anthropocene.

Acclimation represents a type of phenotypic plasticity where one genotype has the capacity to express varying phenotypes when exposed to different environmental conditions (Angilletta, 2009). Phenotypic plasticity represents a rapid response mechanism (i.e., when compared to adaptation via selection for certain phenotypes) acting at the level of the individual (i.e., as opposed to populations) and may be key for organisms to respond to and survive changing environmental temperatures without genetic selection (Munday, 2014). Such "rapid" strategies may play a substantial role in compensation to new conditions, that is, if the new phenotype is beneficial. Yet, the capacity for acclimation also depends on the species, population, and/or its demography (Donelson and Munday, 2012; Eme and Bennett, 2009b) as well as prior history regarding environmental conditions (e.g., temperature; Angilletta, 2009). Developmental and transgenerational acclimation as well as epigenetic effects also play a role in how species respond to changes in their environments.

Early life-history traits and development are notoriously susceptible to altered environmental conditions, but in some cases, developmental acclimation can prime later stages for altered environmental conditions. It has been found that the thermal environment of the eggs (Gagliano et al., 2007), embryos, and juvenile early life stages—rearing temperatures—are important (Illing et al., 2020) and will affect the thermal tolerance traits of the adults

(Donelson et al., 2011). Increased rearing temperatures in cobia (*R. canadum*) results in decreased digestion efficiency (Yufera et al., 2019). When spiny chromis (*A. polyacanthus*) were exposed to elevated temperatures upon hatch and following hatch for set periods of time, those that were exposed to at least 30 days of elevated temperatures exhibited enhanced escape performance traits at this early life stage (Spinks et al., 2019). However, those with extended exposure (i.e., 108 days) to elevated temperatures exhibited reductions in body size, suggesting some trade-offs to long-term exposure and/or developmental windows that are more or less receptive to thermal history (Spinks et al., 2019). Developmental acclimation at the level of resting metabolic rates is also evident in newly settled lemon damselfish (*P. moluccensis*; Grenchik et al., 2012). Additionally, Donelson et al. (2011) determined that, while spiny chromis, *A. polyacanthus*, reared for their entire lives at temperatures that were 3 °C above ambient reduced their metabolic costs, they were also smaller and had poorer body condition than their control counterparts. In many species, early life stages exploit more thermally challenging habitats than adults, which may be for protection and/or resources, suggesting there is a perceived advantage of developmental plasticity (Dabruzzi et al., 2013; Bouyoucos et al., 2022); yet, some of these habitats and therefore species—especially at the edges of their thermal safety margins—might be most at risk under future climate scenarios (Madeira et al., 2017b).

The environment of the parents can also dramatically influence tolerance traits of the offspring (Donelson et al., 2012), and the molecular processes that underpin both developmental and transgenerational acclimation to various climate change stressors (e.g., warming) are becoming more and more clear (Veilleux et al., 2015). In some species (e.g., *A. polyacanthus*), paternal influence (e.g., fewer and poorer quality offspring) is more profound than maternal influence with regards to elevated temperatures (Spinks et al., 2021). In this example, such alterations could stem from the role stress has in regulating sex hormones. Non-genetic inheritance (i.e., epigenetic changes) can underpin various aspects of within-generation phenotypic plasticity in coral reef fishes (Ryu et al., 2020). Environmental history can also carry over for many generations, as has been demonstrated in spiny chromis (*A. polyacanthus*) where stepwise increases in temperature over generations can increase reproductive performance (Donelson et al., 2016; Veilleux et al., 2018b) and restore aerobic metabolic traits (Bernal et al., 2018). When parent cinnamon anemone fish, *A. melanopus*, were exposed to elevated $pCO_2$, their offspring (i.e., via non-genetic inheritance) exhibited improved escape performance, metabolic rate estimates, and growth, but benefits were not observed in all traits examined (Allan et al., 2014; Miller et al., 2012). Limitations to transgenerational acclimation were also noted in juvenile spiny damselfish, *A. polyacanthus*, from parents exposed to elevated $pCO_2$, in terms of olfactory preferences and behavioral lateralization (Welch et al., 2014). While the presence of rapid, developmental and transgenerational acclimation still necessitates environmental preferences and behaviors (Donelson et al., 2012), it will buffer coral reef fish populations against the challenges of climate change and other anthropogenic stressors over the short term. However, it is important to note that, although transgenerational acclimation may provide offspring with increased tolerance to challenging environmental conditions, very few of the studies that have been executed to date have been truly transgenerational; studies have been generally conducted with one generation of progeny, despite the primordial germ cells of that generation being exposed when the parental generation is exposed to such conditions, which is a topic worthy of future investigation. Non-existent or even incomplete developmental or transgenerational acclimation for coral reef fishes living in the changing waters of the Anthropocene necessitates genetic adaptation.

Where acclimation can occur over relatively short time scales, adaptation typically requires many generations and therefore, depending on the species, much longer time frames. That said, the way in which various forms of acclimation interact with adaptation are still not understood and will be the focus of research in the coming decades. Research questions centering around how acclimation can shift mean phenotypes, and thus the strength or direction of selection, without genetic change as well as how selection acts on traits expressed by phenotypic plasticity, and whether changes in phenotypes can become decoupled from changes in genotypes are at the forefront (reviewed in Munday, 2014). The role of epigenetic variation is also factored as well, as non-genetic inheritance, although considered for decades, has only just been detailed under this umbrella. Indeed, when compared to genetic inheritance, epigenetics may provide a faster route of informational transmission across generations and is becoming better understood via the molecular processes that are implicated in this phenomenon (e.g., methylation, histone modification, and non-coding ribonucleic acid (RNA) gene silencing; Jablonka and Raz, 2009). The idea that we are no longer investigating these issues of how organisms respond to change under the Darwinian "all or nothing" concept of adaptation, is exciting. Future studies assessing the interactions between developmental and transgenerational plasticity with epigenetic signatures and adaptation will be imperative for understanding how coral reef fishes and the ecosystems they support fare in the Anthropocene.

## 6 Knowledge gaps, technological advancements, and future directions

While experimental studies assessing how coral reef fishes respond to the stressors discussed here have rapidly expanded and produced a wealth of empirical data, especially since the start of the 21st century, knowledge gaps remain. Most notably, and because of the dynamic and multi-stressor nature of both local and global climate change, interactions between stressors are still challenging to anticipate. Such challenges preclude our capacity to predict species, population, and ecosystem health in the Anthropocene. However, technological advances (e.g., the National Sea Simulator at the Australian Institute of Marine Sciences, which allows for an array of stressors to be simulated in static or cycling conditions (e.g., Jarrold and Munday, 2018a, 2018b) over relevant time courses, and the array of innovations in genomics) will allow for multi-stressor, multi-level experimentation to not only be possible, but also time- and cost-effective. It is also important to recognize short-term and long-term variabilities in many of the discussed stressors. Different species may be sensitive to stressors over different timescales. Cycling of various stressors, especially in combination, has not always been possible under laboratory experimentation, but is an important next step. For example, studies revealed differential responses in the exercise physiology of coral reef fishes exposed to elevated, yet cycling $pCO_2$ conditions (i.e., to more closely represent natural reef conditions; Hannan et al., 2020a) when compared to stable, yet elevated $pCO_2$ conditions (Hannan et al., 2020b, 2020c). Moreover, individual experiments do not precisely predict the fate of future populations of coral reef fishes, but as we accumulate more empirical evidence, we can more robustly estimate species reaction norms and thus have better information for trait-based modeling. However, as demonstrated, studies addressing physiological effects of multiple stressors on coral reef fishes tend to take mechanistic approaches; yet experimental designs will need to grow in size and complexity to be more effective. Moreover, molecular underpinnings will need to be integrated more frequently into these studies, and such approaches are far more tenable today than before with ongoing advances in genomics (e.g., Bernal et al., 2020; Kang et al., 2022; Schunter et al., 2021). Data—and long-term datasets that *can* and *should be* accessed—can then be used to parameterize, test, and refine models to predict how combined stressors will affect coral reef ecosystems and biodiversity.

Mesocosm experiments and other approaches where laboratory studies are paired with field behaviors and responses (Cortese, 2021; Norin, 2018) can expedite understanding the consequences of anthropogenic stressors across the levels of biological organization, spanning from species to community structure and ecosystem function (Fordham, 2015). When physiological (e.g., as outlined and discussed in the above sections) and traditional ecological (e.g., collections, surveys, monitoring, and translocation) approaches are combined, studies can effectively start to bridge findings from the sometimes single-stressor conditions in the laboratory to the natural world. A mesocosm experiment, for example, is beneficial because it allows for replicated experimental designs (e.g., statistical power) and standardized physiochemical conditions, thus revealing elements observable in the wild (e.g., multi-species interactions, community structure, diversity, trophic complexity, nutrient cycling, etc.). Such combined eco-physiological studies can help identify where plasticity in certain physiological traits could aid resilience to climate change, for example (Seebacher et al., 2015). Mesocosm approaches can also reflect the consequences of sublethal and delayed effects on abundance and community structure, and identify feedback loops, species interactions, conditions that might drive species to endangerment, and the potential for restoration programs. Ultimately, maintaining the fine-scale resolution typically associated with physiological approaches while incorporating larger scale, more bio-complex aspects of an ecosystem via mesocosm experimentation will also help improve the accuracy of predictive models.

The variation in responses between individuals is also often overlooked in studies investigating the effects of various abiotic and biotic stressors on multiple traits and/or levels of organization (Sunday et al., 2014). Many overarching predictions are based on the average response of a population, which can often mask the—sometimes extreme—variability in how individuals respond. Identifying "winners and losers" and determining species' capacity for adaptation in the face of global change is crucial, but, especially in the case of "winners," can be detrimental, however, to abatement strategies (i.e., giving a false sense that species will thrive despite adverse conditions). Basic information on acclimation capacity as well as genetic (and epigenetic) variation in fitness-associated traits also aids in making more informed decisions about the impacts of various stressors on coral reef fishes over the timeframes that their habitats are changing. However, modern molecular methods—high throughput sequencing, the -omics (transcriptomics, genomics, metabolomics, proteomics), bioinformatics—paired with physiological approaches, elucidates the mechanistic basis for this crucial within and between generation plasticity.

It is also critical to understand how selection acts on variations in traits and heritability to better predict the fate of coral reef fishes in a multi-stressor world. This has been well studied with respect to fitness-related traits that are influenced by elevated temperatures (Rummer and Munday, 2017) and $CO_2$ (Munday et al., 2019), but only in a few select species and not in combination or with any other stressor. While the importance of investigating multiple traits simultaneously was highlighted earlier in this chapter, here within the context of selection, it is especially important. For example, if two traits are positively correlated, then selection can act unimpeded on the population by acting in the same direction as the most variation in the population (Munday et al., 2013). However, if two traits are negatively correlated, then selection may act orthogonally in the direction of the most variation in the population, which may result in limited influence on a population. Therefore, selection for one trait will decrease performance in the other and vice versa (Sunday et al., 2014). Describing this relationship between traits can therefore aid in predicting whether selection can act freely on a population, or will be constrained, thus limiting species' ability to adapt to future conditions. Identifying correlations among key traits is, indeed, an important step in predicting species persistence (Munday et al., 2019; Sunday et al., 2014). Incorporating evolutionary and environmental parameters, such as sensitivity analyses and evolutionary rescue models that have perhaps been more traditionally associated with terrestrial species and populations, may aid in predicting how coral reef fish populations will respond to an array of stressors in the Anthropocene.

## 7 Conservation and the future of coral reef fishes in the Anthropocene

Unequivocally, coral reef fishes, the ecosystems they support, and the services they provide are facing unprecedented threats from a wide range of multiple stressors, including global climate change. Coral reefs have the highest biodiversity in the marine realm and accommodate ~5000 fish species (Bellwood et al., 2012). Yet, the capacity of coral reefs to provide ecosystem services such as food and jobs, relied upon by millions of people worldwide, has declined by half since the 1950s (Eddy et al., 2021). Indeed, the rate at which conditions on coral reefs are changing today, since pre-Industrial times, as well as the increased frequency and severity of extreme environmental events (e.g., MHWs) is staggering. Since 2016, only 2% of the 2300 km-long GBR has not succumb to ocean warming induced coral bleaching (Hughes et al., 2021). These conditions keep coral reef fishes at a heightened risk when it comes to climate change and other anthropogenic influences, especially when other stressors are considered. Even worse, due to such high habitat specialization, coral reef fishes in biodiversity hotspots (e.g., Coral Triangle) are at an even greater risk of local extinction than regions with lower species richness (Holbrook et al., 2015).

In this chapter, we highlight that a mechanistic understanding of physiological processes governing individual organismal performance is the first step for identifying drivers of coral reef fish health and population dynamics in a multi-stressor world (Illing and Rummer, 2017). The array of physiological approaches we outlined in this chapter, in conjunction with new and emerging technologies, will help to reveal potential cause-and-effect relationships and enable scientists to advise conservation managers by scaling results from molecular, cellular, and individual organismal up to population levels (Illing and Rummer, 2017). In a perfect world, most of the anthropogenic stressors that coral reef fishes face today can and will be mitigated and hopefully quickly. However, it is more realistic that the coral reef fishes of the Anthropocene will continue to face new and emerging stressors. Now, more than ever, the future of coral reef fishes and the ecosystems they support depend on a diverse, passionate, and engaged interdisciplinary scientific community, knowledge co-production, evidence-based decision making, and the most innovative management and conservation strategies. In the spirit of conservation physiology and the quest for a "good Anthropocene" (Madliger et al., 2017), the coral reef fishes represent a great flagship for public engagement in the climate change crisis and an umbrella for the conservation of marine biodiversity in a rapidly changing future (Jepson and Barua, 2015).

## Acknowledgments

The authors thank Erin Walsh for support with figure illustrations and Lisa Louth and Luen Warneke for assistance with references and formatting. Funding to J.L.R. is from the Australian Research Council Centre of Excellence for Coral Reef Studies and College of Science and Engineering at James Cook University. B.I. was funded by a postdoctoral fellowship from the German Research Foundation (BI 220/3-1). The authors would like to thank Bridie J.M. Allan, Jacey van Wert, and two anonymous reviewers for their constructive comments and suggestions on the chapter.

## References

Allan, B.J.M., et al., 2013. Elevated $CO_2$ affects predator-prey interactions through altered performance. PLoS One 8 (3), e58520. https://doi.org/10.1371/journal.pone.0058520.

Allan, B.J.M., et al., 2014. Parental effects improve escape performance of juvenile reef fish in a high- $CO_2$ world. Proc. R. Soc. B Biol. Sci. 281, 20132179. https://doi.org/10.1098/rspb.2013.2179.

Allan, B.J.M., et al., 2015. Feeling the heat: the effect of acute temperature changes on predator-prey interactions in coral reef fish. Conserv. Physiol. 3 (1), cov011. https://doi.org/10.1093/conphys/cov011.

Allan, B.J.M., et al., 2017. Warming has a greater effect than elevated $CO_2$ on predator-prey interactions in coral reef fish. Proc. R. Soc. B Biol. Sci. 284, 20170784. https://doi.org/10.1098/rspb.2017.0784.

Allan, B.J.M., et al., 2020. Parasite infection directly impacts escape response and stress levels in fish. J. Exp. Biol. 223 (16), jeb230904. https://doi.org/10.1242/jeb.230904.

Alloy, M., et al., 2016. Ultraviolet radiation enhances the toxicity of Deepwater horizon oil to Mahi-mahi (*Coryphaena hippurus*) embryos. Environ. Sci. Tech. 50 (4), 2011–2017. https://doi.org/10.1021/acs.est.5b05356.

Almany, G.R., Webster, M.S., 2006. The predation gauntlet: early post-settlement mortality in reef fishes. Coral Reefs 25, 19–22.

Angilletta, M.J., 2009. Thermal Adaptation: A Theoretical and Empirical Synthesis. Oxford University Press, New York.

Angilletta, M.J., Bennett, A.F., Guderley, H., Navas, C.A., Seebacher, F., Wilson, R.S., 2006. Coadaptation: a unifying principle in evolutionary thermal biology. Physiol. Biochem. Zool. 79 (2), 282–294. https://www.jstor.org/stable/10.1086/499990.

Angilletta, M.J., Niewiarowski, P.H., Navas, C.A., 2002. The evolution of thermal physiology in ectotherms. J. Therm. Biol. 27 (4), 249–268. https://doi.org/10.1016/S0306-4565(01)00094-8.

Aranguren-Abadía, L., et al., 2022. Photo-enhanced toxicity of crude oil on early developmental stages of Atlantic cod (*Gadus morhua*). Sci. Total Environ. 807, 150697. https://doi.org/10.1016/j.scitotenv.2021.150697.

Aria, M., Cuccarullo, C., 2017. Bibliometrix: an R-tool for comprehensive science mapping analysis. J. Informet. 11, 959–975.

Artim, J.M., Nicholson, M.D., Hendrick, G.C., Brandt, M., Smith, T.B., Sikkel, P.C., 2020. Abundance of a cryptic generalist parasite reflects degradation of an ecosystem. Ecosphere 11, e03268. https://doi.org/10.1002/ecs2.3268.

Arvizu, B., Allan, B.J., Rizzari, J.R., 2021. Indirect predator effects influence behaviour but not morphology of juvenile coral reef Ambon damselfish *Pomacentrus amboinensis*. J. Fish Biol. 99 (2), 679–683.

Atherton, J.A., McCormick, M.I., 2015. Active in the sac: damselfish embryos use innate recognition of odours to learn predation risk before hatching. Anim. Behav. 103, 1–6. https://doi.org/10.1016/j.anbehav.2015.01.033.

Atherton, J.A., McCormick, M.I., 2020. Parents know best: transgenerational predator recognition through parental effects. PeerJ. 8, e9340. https://doi.org/10.7717/peerj.9340.

Audzijonyte, A., et al., 2019. Is oxygen limitation in warming waters a valid mechanism to explain decreased body sizes in aquatic ectotherms? Glob. Ecol. Biogeogr. 28 (2), 64–77. https://doi.org/10.1111/geb.12847.

Barker, B.D., Horodysky, A.Z., Kerstetter, D.W., 2018. Hot or not? Comparative behavioral thermoregulation, critical temperature regimes, and thermal tolerances of the invasive lionfish *Pterois* sp. versus native western North Atlantic reef fishes. Biol. Invasions 20, 45–58. https://doi.org/10.1007/s10530-017-1511-4.

Barton, B.A., 2002. Stress in fishes: a diversity of responses with particular reference to changes in circulating corticosteroids. Integr. Comp. Biol. 42, 517–525. https://doi.org/10.1093/icb/42.3.517.

Barton, B.A., Iwama, G.K., 1991. Physiological changes in fish from stress in aquaculture with emphasis on the response and effects of corticosteroids. Annu. Rev. Fish Dis. 1, 3–26.

Beitinger, T.L., Fitzpatrick, L.C., 1979. Physiological and ecological correlates of preferred temperature. Am. Zool. 19, 319–329.

Bejarano, I., Appeldoorn, R.S., 2013. Seawater turbidity and fish communities on coral reefs of Puerto Rico. Mar. Ecol. Prog. Ser. 474, 217–226. https://doi.org/10.3354/meps10051.

Beldade, R., et al., 2017. Cascading effects of thermally-induced anemone bleaching on associated anemonefish hormonal stress response and reproduction. Nat. Commun. 8, 716. https://doi.org/10.1038/s41467-017-00565-w.

Bellwood, D.R., Renema, W., Rosen, B.R., 2012. Biodiversity hotspots, evolution and coral reef biogeography. In: Gower, D., Johnson, K., Richardson, J., Rosen, B., Ruber, L., Williams, S. (Eds.), Biotic Evolution and Environmental Change in Southeast Asia. Cambridge Univ Press, pp. 2–32.

Berenbrink, M., Koldkjær, P., Kepp, O., Cossins, A.R., 2005. Evolution of oxygen secretion in fishes and the emergence of a complex physiological system. Science 307 (5716), 1752–1757. http://www.jstor.org/stable/3841821.

Bernal, M.A., et al., 2018. Phenotypic and molecular consequences of stepwise temperature increase across generations in a coral reef fish. Mol. Ecol. 27, 4516–4528. https://doi.org/10.1111/mec.14884.

Bernal, M.A., et al., 2020. Species-specific molecular responses of wild coral reef fishes during a marine heatwave. Sci. Adv. 6, eaay3423. https://doi.org/10.1126/sciadv.aay3423.

Bianchini, K., Wright, P.A., 2013. Hypoxia delays hematopoiesis: retention of embryonic hemoglobin and erythrocytes in larval rainbow trout, *Oncorhynchus mykiss*, during chronic hypoxia exposure. J. Exp. Biol. 216, 4415–4425. https://doi.org/10.1242/jeb.083337.

Bignami, S., Sponaugle, S., Cowen, R.K., 2013. Response to ocean acidification in larvae of a large tropical marine fish, *Rachycentron canadum*. Glob. Chang. Biol. 19, 996–1006. https://doi.org/10.1111/gcb.12133.

Bignami, S., Sponaugle, S., Cowen, R.K., 2014. Effects of ocean acidification on the larvae of a high-value pelagic fisheries species, mahi-mahi *C. hippurus*. Aquat. Biol. 21, 249–260.

Bignami, S., et al., 2017. Combined effects of elevated $pCO_2$, temperature, and starvation stress on larvae of a large tropical marine fish. ICES J. Mar. Sci. 74, 1220–1229. https://doi.org/10.1093/icesjms/fsw216.

Bijma, J., et al., 2013. Climate change and the oceans—what does the future hold? Mar. Pollut. Bull. 74, 495–505. https://doi.org/10.1016/j.marpolbul.2013.07.022.

Binning, S.A., Roche, D.G., Layton, C., 2013. Ectoparasites increase swimming costs in a coral reef fish. Biol. Lett. 9, 20120927. https://doi.org/10.1098/rsbl.2012.0927.

Binning, S.A., et al., 2014. Ectoparasites modify escape behaviour, but not performance, in a coral reef fish. Anim. Behav. 93, 1–7. https://doi.org/10.1016/j.anbehav.2014.04.010.

Birnie-Gauvin, K., Peiman, K.S., Raubenheimer, D., Cooke, S.J., 2017. Nutritional physiology and ecology of wildlife in a changing world. Conserv. Physiol. 5, cox030. https://doi.org/10.1093/conphys/cox030.

Booth, D.J., Beretta, G.A., 2004. Influence of recruit condition on food competition and predation risk in a coral reef fish. Oecologia 140, 289–294. https://doi.org/10.1007/s00442-004-1608-1.

Bouillon, S., Connolly, R.M., Lee, S.Y., 2008. Organic matter exchange and cycling in mangrove ecosystems: recent insights from stable isotope studies. J. Sea Res. 59, 44–58.

Bouyoucos, I.A., Morrison, P., Weideli, O., Jacquesson, E., Planes, S., Simpfendorer, C., Brauner, C.J., Rummer, J.L., 2020. Thermal tolerance and hypoxia tolerance are associated in blacktip reef shark (*Carcharhinus melanopterus*) neonates. J. Exp. Biol. 223 (14), jeb221937. https://doi.org/10.1242/jeb.221937.

Bouyoucos, I.A., Simpfendorfer, C.A., Planes, S., Schwieterman, G.D., Weideli, O.C., Rummer, J.L., 2022. Thermally insensitive physiological performance allows neonatal sharks to use coastal habitats as nursery areas. Mar. Ecol. Prog. Ser. 682, 137–152. https://doi.org/10.3354/meps13941.

Bowater, R.O., et al., 2012. Natural outbreak of *Streptococcus agalactiae* (GBS) infection in wild giant Queensland grouper, *Epinephelus lanceolatus* (Bloch), and other wild fish in northern Queensland, Australia. J. Fish Dis. 35, 173–186. https://doi.org/10.1111/j.1365-2761.2011.01332.x.

Brandl, S.J., Goatley, C.H.R., Bellwood, D.R., Tornabene, L., 2018. The hidden half: ecology and evolution of cryptobenthic fishes on coral reefs. Biol. Rev. 93, 1846–1873. https://doi.org/10.1111/brv.12423.

Brauner, C., Baker, D., 2009. Patterns of acid–base regulation during exposure to hypercarbia in fishes. In: Glass, M., Wood, S. (Eds.), Cardio-Respiratory Control in Vertebrates. Springer, Berlin, Heidelberg. https://doi.org/10.1007/978-3-540-93985-6_3.

Brazenor, A.K., Hutson, K.S., 2015. Effects of temperature and salinity on the life cycle of *Neobenedenia* sp. (Monogenea: Capsalidae) infecting farmed barramundi (*Lates calcarifer*). Parasitol. Res. 114, 1875–1886. https://doi.org/10.1007/s00436-015-4375-5.

Breitburg, D., et al., 2018. Declining oxygen in the global ocean and coastal waters. Science 359 (6371), eaam7240. PMID: 29301986. https://doi.org/10.1126/science.aam7240

Brett, J., 1971. Energetic responses of salmon to temperature. A study of some thermal relations in the physiology and freshwater ecology of sockeye salmon (*Oncorhynchus nerka*). Integr. Comp. Biol. 11 (1), 99.

Briones, V., et al., 1998. Haemorrhagic septicaemia by *Aeromonas salmonicida* subsp. *salmonicida* in a black-tip reef shark (*C. melanopterus*). J. Vet. Med. B. 45, 443–445. https://doi.org/10.1111/j.1439-0450.1998.tb00814.x.

Bryan, J.D., Kelsch, S.W., Neill, W.H., 1990. The maximum power principle in behavioral thermoregulation by fishes. Trans. Am. Fish. Soc. 119, 611–621.

Camus, A.C., et al., 2013. *Serratia marcescens* associated ampullary system infection and septicaemia in a bonnethead shark, *Sphyrna tiburo* (L.). J. Fish Dis. 36, 891–895. https://doi.org/10.1111/jfd.12107.

Carlson, J.K., Parsons, G.R., 2001. The effects of hypoxia on three sympatric shark species: physiological and behavioral responses. Environ. Biol. Fishes 61, 427–433.

Carlson, J.K., Parsons, G.R., 2003. Respiratory and hematological responses of the bonnethead shark, *Sphyrna tiburo*, to acute changes in dissolved oxygen. J. Exp. Mar. Biol. Ecol. 294, 15–26.

Castro-Santos, T., Goerig, E., He, P., Lauder, G.V., 2022. Applied aspects of locomotion and biomechanics. Fish Physiol. 39A, 91–140.

Chapman, C.A., Harahush, B.K., Renshaw, G.M., 2010. The physiological tolerance of the grey carpet shark (*Chiloscyllium punctatum*) and the epaulette shark (*Hemiscyllium ocellatum*) to anoxic exposure at three seasonal temperatures. Fish Physiol. Biochem. 37, 387–399.

Cheal, A.J., Emslie, M., MacNeil, M.A., Miller, I., Sweatman, H., 2013. Spatial variation in the functional characteristics of herbivorous fish communities and the resilience of coral reefs. Ecol. Appl. 23 (1), 174–188.

Cheutin, M.C., et al., 2021. Microbial shift in the enteric bacteriome of coral reef fish following climate-driven regime shifts. Microorganisms 9, 1711. https://doi.org/10.3390/microorganisms9081711.

Chrousos, G.P., 1998. Stressors, stress, and neuroendocrine integration of the adaptive response. Ann. N. Y. Acad. Sci., 851311–851335.

Chung, W.S., et al., 2014. Ocean acidification slows retinal function in a damselfish through interference with GABA(a) receptors. J. Exp. Biol. 217, 323–326. https://doi.org/10.1242/jeb.092478.

Clark, T.D., Sandblom, E., Jutfelt, F., 2013. Aerobic scope measurements of fishes in an era of climate change: respirometry, relevance and recommendations. J. Exp. Biol. 216 (15), 2771–2782.

Clarke, A., Johnston, N.M., 1999. Scaling of metabolic rate with body mass and temperature in teleost fish. J. Anim. Ecol. 68, 893–905.

Colleye, O., et al., 2016. Auditory evoked potential audiograms in post-settlement stage individuals of coral reef fishes. J. Exp. Mar. Biol. Ecol. 483, 1–9. https://doi.org/10.1016/j.jembe.2016.05.007.

Cook, C.A., et al., 2015. Blood parasite biodiversity of reef-associated fishes of the eastern Caribbean. Mar. Ecol. Prog. Ser. 533, 1–13. https://doi.org/10.3354/meps11430.

Cortese, D., et al., 2021. Physiological and behavioural effects of anemone bleaching on symbiont anemonefish in the wild. Funct. Ecol. 35 (3), 663–674. https://doi.org/10.1111/1365-2435.13729.

Cortesi, F., et al., 2015. Phenotypic plasticity confers multiple fitness benefits to a mimic. Curr. Biol. 25, 949–954. https://doi.org/10.1016/j.cub.2015.02.013.

Couturier, C.S., et al., 2013. Species-specific effects of near-future $CO_2$ on the respiratory performance of two tropical prey fish and their predator. Comp. Biochem. Physiol. A Mol. Integr. Physiol. 166, 482–489. https://doi.org/10.1016/j.cbpa.2013.07.025.

Crear, D.P., et al., 2019. The impacts of warming and hypoxia on the performance of an obligate ram ventilator. Conserv. Physiol. 7 (1), coz026. https://doi.org/10.1093/conphys/coz026.

Cullen, J.A., Marshall, C.D., Hala, D., 2019. Integration of multi-tissue PAH and PCB burdens with biomarker activity in three coastal shark species from the northwestern Gulf of Mexico. Sci. Total Environ. 650, 1158–1172. https://doi.org/10.1016/j.scitotenv.2018.09.128.

Cumming, H., Herbert, N.A., 2016. Gill structural change in response to turbidity has noeffect on the oxygen uptake of a juvenile sparid fish. Conserv. Physiol. 4 (1), cow033.

D'Agostino, D., et al., 2021. Growth impacts in a changing ocean: insights from two coral reef fishes in an extreme environment. Coral Reefs 40 (2), 433–446.

Dabruzzi, T.F., Bennett, W., 2013. Hypoxia effects on gill surface area and blood oxygen-carrying capacity of the Atlantic stingray, *Dasyatis sabina*. Fish Physiol. Biochem. 40, 1011–1020.

Dabruzzi, T.F., et al., 2013. Juvenile ribbontail stingray, *Taeniura lymma* (Forsskayenl, 1775) (Chondrichthyes: Dasyatidae), demonstrate a unique suite of physiological adaptations to survive hyperthermic nursery conditions. Hydrobiologia 701, 37–49. https://doi.org/10.1007/s10750-012-1249-z.

De Albergaria-Barbosa, R., et al., 2017. *Mugil curema* as a PAH bioavailability monitor for Atlantic west sub-tropical estuaries. Mar. Pollut. Bull. 114, 609–614. https://doi.org/10.1016/j.marpolbul.2016.09.039.

De Boeck, G., Rodgers, E., Town, R.M., 2022. Using ecotoxicology for conservation: From biomarkers to modeling. Fish Physiol. 39B.

Deigweiher, K., Koschnick, N., Pörtner, H.O., Lucassen, M., 2008. Acclimation of ion regulatory capacities in gills of marine fish under environmental hypercapnia. Am. J. Physiol. Regul. Integr. Comp. Physiol. 295, R1660–R1670. https://doi.org/10.1152/ajpregu.90403.2008.

Demaire, C., et al., 2020. Reduced access to cleaner fish negatively impacts the physiological state of two resident reef fishes. Mar. Biol. 167, 48. https://doi.org/10.1007/s00227-020-3658-2.

Deutsch, C.A., Tewksbury, J.J., Huey, R.B., Martin, P.R., 2008. Impacts of climate warming on terrestrial ectotherms across latitude. Proc. Natl. Acad. Sci. U. S. A. 105 (18), 6668–6672. https://doi.org/10.1073/pnas.0709472105.

Devaux, J.B.L., Hickey, A.J.R., Renshaw, G.M.C., 2019. Mitochondrial plasticity in the cerebellum of two anoxia-tolerant sharks: contrasting responses to anoxia/re-oxygenation. J. Exp. Biol. 222, jeb191353. UNSP https://doi.org/10.1242/jeb.191353.

Di Santo, V., Lobel, P.S., 2016. Size affects digestive responses to increasing temperature in fishes: physiological implications of being small under climate change. Mar. Ecol. 37, 813–820. https://doi.org/10.1111/maec.12358.

Di Santo, V., Lobel, P.S., 2017. Body size and thermal tolerance in tropical gobies. J. Exp. Mar. Biol. Ecol. 487, 11–17. https://doi.org/10.1016/j.jembe.2016.11.007.

Diamant, A., Banet, A., Ucko, M., Colorni, A., Knibb, W., Kvitt, H., 2000. Mycobacteriosis in wild rabbitfish *Siganus rivulatus* associated with cage farming in the Gulf of Eilat, Red Sea. Dis. Aquat. Organ. 39, 211–239.

Djurichkovic, L.D., et al., 2019. The effects of water temperature on the juvenile performance of two tropical damselfishes expatriating to temperate reefs. Sci. Rep. 9, 13937. https://doi.org/10.1038/s41598-019-50303-z.

Domenici, P., Allan, B., McCormick, M.I., Munday, P.L., 2012. Elevated carbon dioxide affects behavioural lateralization in a coral reef fish. Biol. Lett. 8 (1). https://doi.org/10.1098/rsbl.2011.0591.

Donelson, J.M., Munday, P.L., 2012. Thermal sensitivity does not determine acclimation capacity for a tropical reef fish. J. Anim. Ecol. 81, 1126–1131. https://doi.org/10.1111/j.1365-2656.2012.01982.x.

Donelson, J.M., et al., 2010. Effects of elevated water temperature and food availability on the reproductive performance of a coral reef fish. Mar. Ecol. Prog. Ser. 401, 233–243. https://doi.org/10.3354/meps08366.

Donelson, J.M., et al., 2011. Acclimation to predicted ocean warming through developmental plasticity in a tropical reef fish. Glob. Chang. Biol. 17, 1712–1719. https://doi.org/10.1111/j.1365-2486.2010.02339.x.

Donelson, J.M., et al., 2012. Rapid transgenerational acclimation of a tropical reef fish to climate change. Nat. Clim. Chang. 2, 30–32. https://doi.org/10.1038/NCLIMATE1323.

Donelson, J.M., et al., 2016. Transgenerational plasticity of reproduction depends on rate of warming across generations. Evol. Appl. 9, 1072–1081. https://doi.org/10.1111/eva.12386.

Downie, A.T., et al., 2021. Habitat association may influence swimming performance in marine teleost larvae. Fish Fish. 22 (6), 1187–1212. https://doi.org/10.1111/faf.12580.

Duarte, C.M., et al., 2021. The soundscape of the Anthropocene ocean. Science 371, eaba4658. https://doi.org/10.1126/science.aba4658.

Eddy, T.D., et al., 2021. Global decline in capacity of coral reefs to provide ecosystem services. One Earth 4, 1278. https://doi.org/10.1016/j.oneear.2021.08.016.

Egner, S.A., Mann, D.A., 2005. Auditory sensitivity of sergeant major damselfish *Abudefduf saxatilis* from post-settlement juvenile to adult. Mar. Ecol. Prog. Ser. 285, 213–222. https://doi.org/10.3354/meps285213.

Eliason, E.J., Van Wert, J.C., Schwieterman, G.D., 2022. Applied aspects of the cardiorespiratory system. Fish Physiol. 39A, 189–252.

Eme, J., Bennett, W.A., 2009a. Acute temperature quotient responses of fishes reflect their divergent thermal habitats in the Banda Sea, Sulawesi, Indonesia. Aust. J. Zool. 57, 357–362. https://doi.org/10.1071/ZO09081.

Eme, J., Bennett, W.A., 2009b. Critical thermal tolerance polygons of tropical marine fishes from Sulawesi, Indonesia. J. Theor. Biol. 34, 220–225. https://doi.org/10.1016/j.jtherbio.2009.02.005.

Ern, R., 2019. A mechanistic oxygen- and temperature-limited metabolic niche framework. Philos. Trans. R. Soc. B 374, 20180540. https://doi.org/10.1098/rstb.2018.0540.

Ern, R., et al., 2017. Effects of hypoxia and ocean acidification on the upper thermal niche boundaries of coral reef fishes. Biol. Lett. 13, 20170135. https://doi.org/10.1098/rsbl.2017.0135.

Evans, D.H., Piermarini, P.M., Choe, K.P., 2005. The multifunctional fish gill: dominant site of gas exchange, osmoregulation, acid-base regulation, and excretion of nitrogenous waste. Physiol. Rev. 85 (1), 97–177.

Fabricius, K., De'ath, G., McCook, L., Turak, E., Williams, D.M., 2005. Changes in algal, coral and fish assemblages along water quality gradients on the inshore Great Barrier Reef. Mar. Pollut. Bull. 51 (1–4), 384–398.

Fakan, E.P., Mccormick, M.I., 2019. Boat noise affects the early life history of two damselfishes. Mar. Pollut. Bull. 141, 493–500. https://doi.org/10.1016/j.marpolbul.2019.02.054.

Fangue, N.A., Bennett, W.A., 2003. Thermal tolerance responses of laboratory-acclimated and seasonally acclimatized Atlantic stingray, *D. sabina*. Copeia 2003, 315–325.

Feary, D.A., et al., 2014. Latitudinal shifts in coral reef fishes: why some species do and others do not shift. Fish Fish. 15 (4), 593–615. https://doi.org/10.1111/faf.12036.

Ferrari, M.C.O., et al., 2015a. Living in a risky world: the onset and ontogeny of an integrated antipredator phenotype in a coral reef fish. Sci. Rep. 5, 15537. https://doi.org/10.1038/srep15537.

Ferrari, M.C.O., et al., 2015b. Interactive effects of ocean acidification and rising sea temperatures alter predation rate and predator selectivity in reef fish communities. Glob. Chang. Biol. 21, 1848–1855. https://doi.org/10.1111/gcb.12818.

Ferrari, M.C.O., et al., 2017. Daily variation in behavioural lateralization is linked to predation stress in a coral reef fish. Anim. Behav. 133, 189–193. https://doi.org/10.1016/j.anbehav.2017.09.020.

Fieber, L.A., Schmale, M.C., 1994. Differences in a K-current in Schwann-cells from normal and neurofibromatosis-infected damselfish. Glia 11, 64–72. https://doi.org/10.1002/glia.440110109.

Fields, P.A., Somero, G.N., 1997. Amino acid sequence differences cannot fully explain interspecific variation in thermal sensitivities of gobiid fish A4-lactate dehydrogenases (A4-LDHs). J. Exp. Biol. 200, 1839–1850.

Figueira, W.F., Booth, D.J., 2010. Increasing ocean temperatures allow tropical fishes to survive over winter in temperate waters. Glob. Chang. Biol. 16, 506–516. https://doi.org/10.1111/j.1365-2486.2009.01934.x.

Figueira, W.F., Booth, D.J., Gregson, M.A., 2008. Selective mortality of a coral reef damselfish: role of predator-competitor synergisms. Oecologia 156, 215–226. https://doi.org/10.1007/s00442-008-0985-2.

Figueira, W.F., Curley, B., Booth, D.J., 2019. Can temperature-dependent predation rates regulate range expansion potential of tropical vagrant fishes? Mar. Biol. 166, 73. https://doi.org/10.1007/s00227-019-3521-5.

Filiciotto, F., et al., 2013. Effect of acoustic environment on gilthead sea bream (Sparus aurata): sea and onshore aquaculture background noise. Aquaculture 414-415, 36–45. https://doi.org/10.1016/j.aquaculture.2013.07.042.

Fogelman, R.M., Kuris, A.M., Grutter, A.S., 2009. Parasitic castration of a vertebrate: effect of the cymothoid isopod, *Anilocra apogonae*, on the five-lined cardinalfish, *Cheilodipterus quinquelineatus*. Int. J. Parasitol. 39, 577–583. https://doi.org/10.1016/j.ijpara.2008.10.013.

Foley, J.A., DeFries, R., Asner, G.P., Barford, C., Bonan, G., Carpenter, S.R., Chapin, F.S., Coe, M.T., Daily, G.C., Gibbs, H.K., Helkowski, J.H., 2005. Global consequences of land use. Science 309 (5734), 570–574.

Fordham, D.A., 2015. Mesocosms reveal ecological surprises from climate change. PLoS Biol. 13 (12), e1002323. https://doi.org/10.1371/journal.pbio.1002323.

Frisch, A., Anderson, T., 2005. Physiological stress responses of two species of coral trout (*Plectropomus leopardus* and *Plectropomus maculatus*). Comp. Biochem. Physiol. A Mol. Integr. Physiol. 140, 317–327. https://doi.org/10.1016/j.cbpb.2005.01.014.

Frommel, A.Y., et al., 2019. Organ health and development in larval kingfish are unaffected by ocean acidification and warming. PeerJ 7, e8266. https://doi.org/10.7717/peerj.8266.

Fuiman, L.A., Meekan, M.G., McCormick, M.I., 2010. Maladaptive behavior reinforces a recruitment bottleneck in newly settled fishes. Oecologia 164, 99–108. https://doi.org/10.1007/s00442-010-1712-3.

Fukuoka, T., et al., 2017. Does depth become a permissive factor for reproductive habitat selection in the reef-associate damselfish *Chrysiptera cyanea*? Cybium 41, 343–350.

Gagliano, M., McCormick, M.I., Meekan, M.G., 2007. Temperature-induced shifts in selective pressure at a critical developmental transition. Oecologia 152, 219–225. https://doi.org/10.1007/s00442-006-0647-1.

Galloway, T.S., Cole, M., Lewis, C., 2017. Interactions of microplastic debris throughout the marine ecosystem. Nat. Ecol. Evol. 1, 0116. https://doi.org/10.1038/s41559-017-0116.

Gardiner, N.M., Munday, P.L., Nilsson, G.E., 2010. Counter-gradient variation in respiratory performance of coral reef fishes at elevated temperatures. PLoS One 5, e13299. https://doi.org/10.1371/journal.pone.0013299.

Gelman, A., Cogan, U., Mokady, S., 1992. The thermal-properties of fish enzymes as a possible indicator of the temperature adaptation potential of the fish. Comp. Biochem. Physiol. B Biochem. Mol. Biol. 101, 205–208. https://doi.org/10.1016/0305-0491(92)90180-Y.

Gervais, C.R., et al., 2018. Too hot to handle? Using movement to alleviate effects of elevated temperatures in a benthic elasmobranch, *H. ocellatum*. Mar. Biol. 165, 162. https://doi.org/10.1007/s00227-018-3427-7.

Gervais, C.R., Champion, C., Pecl, G.T., 2021. Species on the move around the Australian coastline: a continental-scale review of climate-driven species redistribution in marine systems. Glob. Chang. Biol. 27 (14), 3200–3217. https://doi.org/10.1111/gcb.15634.

Giles, M.A., Vanstone, W.E., 1978. Ontogenetic variation in the multiple hemoglobins of Coho salmon (*Oncorhynchus kisutch*) and effect of environmental factors on their expression. J. Fish. Res. Board Can. 33, 1144–1149.

Goldstein, E.D., D'Alessandro, E.K., Sponaugle, S., 2016. Demographic and reproductive plasticity across the depth distribution of a coral reef fish. Sci. Rep. 6, 34077. https://doi.org/10.1038/srep34077.

Goldstein, E.D., D'Alessandro, E.K., Sponaugle, S., 2017. Fitness consequences of habitat variability, trophic position, and energy allocation across the depth distribution of a coral-reef fish. Coral Reefs. 36, 957–968. https://doi.org/10.1007/s00338-017-1587-4.

Gomes, B., et al., 2017. Use of environmental DNA (eDNA) and water quality data to predict protozoan parasites outbreaks in fish farms. Aquaculture 479, 467–473. https://doi.org/10.1016/j.aquaculture.2017.06.021.

Gordon, T.A.C., Hardine, H.R., Wong, K.E., Merchant, N.D., Meekan, M.G., McCormick, M.I., Radford, A.N., Simpson, S.D., 2018. Habitat degradation negatively affects auditory settlement behavior of coral reef fishes. Proc. Natl. Acad. Sci. U. S. A. 115, 5193–5198.

Gratzer, B., et al., 2015. Skin toxins in coral-associated *Gobiodon* species (Teleostei: Gobiidae) affect predator preference and prey survival. Mar. Ecol. 36, 67–76. https://doi.org/10.1111/maec.12117.

Gregoire, M., et al., 2021. A global ocean oxygen database and atlas for assessing and predicting deoxygenation and ocean health in the open and coastal ocean. Front. Mar. Sci. 8, 724913. https://doi.org/10.3389/fmars.2021.724913.

Grenchik, M.K., Donelson, J.M., Munday, P.L., 2012. Evidence for developmental thermal acclimation in the damselfish, *Pomacentrus moluccensis*. Coral Reefs. 32 (1), 85–90. https://doi.org/10.1007/s00338-012-0949-1.

Grutter, A.S., Murphy, J.M., Choat, J.H., 2003. Cleaner fish drives local fish diversity on coral reefs. Curr. Biol. 13, 64–67. https://doi.org/10.1016/S0960-9822(02)01393-3.

Gunderson, A.R., Armstrong, E.J., Stillman, J.H., 2016. Multiple stressors in a changing world: the need for an improved perspective on physiological responses to the dynamic marine environment. Ann. Rev. Mar. Sci. 8, 357–378.

Habary, A., et al., 2017. Adapt, move or die—how will tropical coral reef fishes cope with ocean warming? Glob. Chang. Biol. 23, 566–577. https://doi.org/10.1111/gcb.13488.

Hall, A.E., Clark, T.D., 2016. Seeing is believing: metabolism provides insight into threat perception for a prey species of coral reef fish. Anim. Behav. 115, 117–126. https://doi.org/10.1016/j.anbehav.2016.03.008.

Hamilton, T.J., Holcombe, A., Tresguerres, M., 2014. $CO_2$-induced ocean acidification increases anxiety in rockfish via alteration of GABAA receptor functioning. Proc. R. Soc. B Biol. Sci. 281 (1775). https://doi.org/10.1098/rspb.2013.2509.

Hamilton, R.J., Almany, G.R., Brown, C.J., Pita, J., Peterson, N.A., Choat, J.H., 2017. Logging degrades nursery habitat for an iconic coral reef fish. Biol. Conserv. 210, 273–280.

Hannan, K.D., Rummer, J.L., 2018. Aquatic acidification: a mechanism underpinning maintained oxygen transport and performance in fish experiencing elevated carbon dioxide conditions. J. Exp. Biol. 221 (5), jeb154559. https://doi.org/10.1242/jeb.154559.

Hannan, K.D., et al., 2020a. Diel $pCO_2$ variation among coral reef sites and microhabitats at Lizard Island, great barrier reef. Coral Reefs 39, 1391–1406. https://doi.org/10.1007/s00338-020-01973-z.

Hannan, K.D., et al., 2020b. Contrasting effects of constant and fluctuating $pCO_2$ conditions on the exercise physiology of coral reef fishes. Mar. Environ. Res. 163, 105224. https://doi.org/10.1016/j.marenvres.2020.105224.

Hannan, K.D., et al., 2020c. The effects of constant and fluctuating elevated $pCO_2$ levels on oxygen uptake rates of coral reef fishes. Sci. Total Environ. 741, 140334. https://doi.org/10.1016/j.scitotenv.2020.140334.

Hannan, K.D., McMahon, S.J., Munday, P.L., Rummer, J.L., 2021. Contrasting effects of constant and fluctuating pCO2 conditions on the exercise physiology of coral reef fishes. Mar. Environ. Res. 163, 105224. https://doi.org/10.1016/j.marenvres.2020.105224.

Harmache, A., et al., 2006. Bioluminescence imaging of live infected salmonids reveals that the fin bases are the major portal of entry for *Novirhabdovirus*. J. Virol. 80, 3655–3659. https://doi.org/10.1128/JVI.80.7.3655-3659.2006.

Hawlena, D., Schmitz, O.J., 2010. Physiological stress as a fundamental mechanism linking predation to ecosystem functioning. Am. Nat. 176 (5), 537–556.

Heinrich, D.D.U., et al., 2014. A product of its environment: the epaulette shark (*H. ocellatum*) exhibits physiological tolerance to elevated environmental $CO_2$. Conserv. Physiol. 2, cou047. https://doi.org/10.1093/conphys/cou047.

Heinrich, D.D.U., et al., 2020. Short-term impacts of daily feeding on the residency, distribution, and energy expenditure of sharks. Anim. Behav. 172, 55–71. https://doi.org/10.1016/j.anbehav.2020.12.002.

Hess, S., et al., 2015. Exposure of clownfish larvae to suspended sediment levels found on the great barrier reef: impacts on gill structure and microbiome. Sci. Rep. 5, 10561. https://doi.org/10.1038/srep10561.

Hess, S., et al., 2017. Species-specific impacts of suspended sediments on gill structure and function in coral reef fishes. Proc. R. Soc. B Biol. Sci. 284, 20171279. https://doi.org/10.1098/rspb.2017.1279.

Hess, S., et al., 2019. Enhanced fast-start performance and anti-predator behaviour in a coral reef fish in response to suspended sediment exposure. Coral Reefs. 38, 103–108. https://doi.org/10.1007/s00338-018-01757-6.

Heuer, R.M., Grosell, M., 2014. Physiological impacts of elevated carbon dioxide and ocean acidification on fish. Am. J. Physiol. Regul. Integr. Comp. Physiol. 307 (9), R1061–R1084. https://doi.org/10.1152/ajpregu.00064.2014.

Heuer, R.M., et al., 2016. Altered brain ion gradients following compensation for elevated $CO_2$ are linked to behavioural alterations in a coral reef fish. Sci. Rep. 6, 33216. https://doi.org/10.1038/srep33216.

Hickey, A.J.R., et al., 2012. A radical approach to beating hypoxia: depressed free radical release from heart fibres of the hypoxia-tolerant epaulette shark (*Hemiscyllum ocellatum*). J. Comp. Physiol. B 182, 91–100. https://doi.org/10.1007/s00360-011-0599-6.

Hight, B.V., Lowe, C.G., 2007. Elevated body temperatures of adult female leopard sharks, *Triakis semifasciata*, while aggregating in shallow nearshore embayments: evidence for behavioral thermoregulation? J. Exp. Mar. Biol. Ecol. 352, 114–128. https://doi.org/10.1016/j.jembe.2007.07.021.

Hixon, M.A., Randall, J.E., 2019. Coral reef fishes. In: Cochran, J.K., Bokuniewicz, J.H., Yager, L.P. (Eds.), Encyclopedia of ocean sciences, third ed. vol. 2. Elsevier, pp. 142–150. ISBN: 978-0-12-813081-0.

Hobday, A.J., et al., 2016. A hierarchical approach to defining marine heatwaves. Prog. Oceanogr. 141, 227–238. https://doi.org/10.1016/j.pocean.2015.12.014.

Hoey, A.S., McCormick, M.I., 2004. Selective predation for low body condition at the larval-juvenile transition of a coral reef fish. Oecologia 139, 23–29. https://doi.org/10.1007/s00442-004-1489-3.

Holbrook, S.J., et al., 2015. Reef fishes in biodiversity hotspots are at greatest risk from loss of coral species. PLoS One 10 (5), e0124054. https://doi.org/10.1371/journal.pone.0124054.

Holles, S., Simpson, S.D., Radford, A.N., Berten, L., Lecchini, D., 2013. Boat noise disrupts orientation behaviour in a coral reef fish. Mar. Ecol. Prog. Ser. 485, 295–300.

Holmes, L.J., McWilliam, J., Ferrari, M.C.O., McCormick, M.I., 2017. Juvenile damselfish are affected but desensitize to small motor boat noise. J. Exp. Mar. Biol. Ecol. 494, 63–68. https://doi.org/10.1016/j.jembe.2017.05.009.

Huey, R.B., Bennett, A.F., 1987. Phylogenetic studies of coadaptation: preferred temperatures versus optimal performancetemperatures of lizards. Evolution 41, 1098–1115.

Hughes, T.P., et al., 2021. Emergent properties in the responses of tropical corals to recurrent climate extremes. Curr. Biol. 31, 5393–5399.

Humborstad, O.B., Jørgensen, T., Grotmol, S., 2006. Exposure of cod *G. morhua* to resuspended sediment: an experimental study of the impact of bottom trawling. Mar. Ecol. Prog. Ser. 309, 247–254.

Hung, Q.P., et al., 2010. Reproductive cycle in female Waigieu seaperch (*Psammoperca waigiensis*) reared under different salinity levels and the effects of dopamine antagonist on steroid hormone levels. J. Exp. Mar. Biol. Ecol. 383, 137–145. https://doi.org/10.1016/j.jembe.2009.12.010.

Illing, B., Rummer, J.L., 2017. Physiology can contribute to better understanding, management, and conservation of coral reef fishes. Conserv. Physiol. 5 (1), cox005. https://doi.org/10.1093/conphys/cox005.

Illing, B., et al., 2020. Critical thermal maxima of early life stages of three tropical fishes: effects of rearing temperature and experimental heating rate. J. Therm. Biol. 90, 102582. https://doi.org/10.1016/j.jtherbio.2020.102582.

Incardona, J.P., et al., 2012. Potent phototoxicity of marine bunker oil to translucent herring embryos after prolonged weathering. PLoS One 7 (2), e30116. https://doi.org/10.1371/journal.pone.0030116.

International Maritime Organization (IMO), 2014. Guidelines for the reduction of underwater noise from commercial shipping to address adverse impacts on marine life. Circulation No. MEPC.1/Circ.833. IMO, London.

Jablonka, E., Raz, G., 2009. Transgenerational epigenetic inheritance: prevalence, mechanisms, and implications for the study of heredity and evolution. Q. Rev. Biol. 84 (2), 131–176. https://doi.org/10.1086/598822.

Jacob, H., Gilson, A., Lanctôt, C., Besson, M., Metian, M., Lecchini, D., 2019. No effect of polystyrene microplastics on foraging activity and survival in a post-larvae coral-reef fish, *Acanthurus triostegus*. Bull. Environ. Contam. Toxicol. 102, 457–461.

Jafarabadi, A.R., et al., 2019. Distributions and compositional patterns of polycyclic aromatic hydrocarbons (PAHs) and their derivatives in three edible fishes from Kharg Coral Island, Persian Gulf, Iran. Chemosphere 215, 835–845. https://doi.org/10.1016/j.chemosphere.2018.10.092.

Jafarabadi, A.R., et al., 2020. Dietary intake of polycyclic aromatic hydrocarbons (PAHs) from coral reef fish in the Persian Gulf: human health risk assessment. Food Chem. 329, 127035. https://doi.org/10.1016/j.foodchem.2020.127035.

Jain-Schlaepfer, S., et al., 2018. Impact of motorboats on fish embryos depends on engine type. Conserv. Physiol. 6, coy014. https://doi.org/10.1093/conphys/coy014.

Jambeck, J.R., et al., 2015. Plastic waste inputs from land into the ocean. Science 347, 768–771. https://doi.org/10.1126/science.1260352.

Jarrold, M.D., Munday, P.L., 2018a. Elevated temperature does not substantially modify the interactive effects between elevated $CO_2$ and diel $CO_2$ cycles on the survival, growth and behavior of a coral reef fish. Front. Mar. Sci. 5, 458. UNSP https://doi.org/10.3389/fmars.2018.00458.

Jarrold, M.D., Munday, P.L., 2018b. Diel $CO_2$ cycles do not modify juvenile growth, survival and otolith development in two coral reef fish under ocean acidification. Mar. Biol. 165, 49. https://doi.org/10.1007/s00227-018-3311-5.

Jepson, P., Barua, M., 2015. A theory of flagship species action. Conserv. Soc. 13, 95–104. https://doi.org/10.4103/0972-4923.161228.

Johansen, J.L., Jones, G.P., 2011. Increasing ocean temperature reduces the metabolic performance and swimming ability of coral reef damselfishes. Glob. Chang. Biol. 17, 2971–2979. https://doi.org/10.1111/j.1365-2486.2011.02436.x.

Johansen, J.L., et al., 2017. Publisher correction: oil exposure disrupts early life-history stages of coral reef fishes via behavioural impairments (vol 1, pg 1146, 2017). Nat. Ecol. Evol. 1, 1412. https://doi.org/10.1038/s41559-017-0292-6.

Johansen, J.L., Nadler, L.E., Habary, A., Bowden, A.J., Rummer, J.L., 2021. Thermal acclimation of tropical reef fishes to global heat waves. Elife 10, e59162. https://doi.org/10.7554/elife.59162.

John, J., Nandhini, A.R., Velayudhaperumal Chellam, P., Sillanpää, M., 2021. Microplastics in mangroves and coral reef ecosystems: a review. Environ. Chem. Lett. 20, 397–416. https://doi.org/10.1007/s10311-021-01326-4.

Johns, G.C., Somero, G.N., 2004. Evolutionary convergence in adaptation of proteins to temperature: $A_4$-lactate dehydrogenases of pacific damselfishes (*Chromis* spp.). Mol. Biol. Evol. 21, 314–320. https://doi.org/10.1093/molbev/msh021.

Johnson, T.P., Johnston, I.A., 1991. Temperature adaptation and the contractile properties of live muscle-fibers from teleost fish. J. Comp. Physiol. B 161, 27–36. https://doi.org/10.1007/BF00258743.

Johnson, J.A., Kelsch, S.W., 1998. Effects of evolutionary thermal environment on temperature-preference relationships in fishes. Environ. Biol. Fishes 53, 447–458. https://doi.org/10.1023/A:1007425215669.

Jokiel, P.L., Hunter, C.L., Taguchi, S., Watarai, L., 1993. Ecological impact of a fresh-water "reef kill" in Kaneohe Bay, Oahu, Hawaii. Coral Reefs 12 (3), 177–184.

Jones, J., et al., 2018. The microbiome of the gastrointestinal tract of a range-shifting marine herbivorous fish. Front. Microbiol. 9, 2000. https://doi.org/10.3389/fmicb.2018.02000.

Jud, Z.R., Nichols, P.K., Layman, C.A., 2014. Broad salinity tolerance in the invasive lionfish *Pterois* spp. may facilitate estuarine colonization. Environ. Biol. Fishes 98, 135–143.

Juma, R.R., et al., 2017. Potential of *Periophthalmus sobrinus* and *Siganus sutor* as bioindicator fish species for PAH pollution in tropical waters. Reg. Stud. Mar. Sci. 18, 170–176. https://doi.org/10.1016/j.rsma.2017.09.016.

Justine, J.L., et al., 2012. An annotated list of fish parasites (isopoda, Copepoda, Monogenea, Digenea, Cestoda, Nematoda) collected from snappers and bream (Lutjanidae, Nemipteridae, Caesionidae) in New Caledonia confirms high parasite biodiversity on coral reef fish. Aquat. Biosyst. 8, 22. https://doi.org/10.1186/2046-9063-8-22.

Jutfelt, F., et al., 2018. Oxygen- and capacity-limited thermal tolerance: blurring ecology and physiology. J. Exp. Biol. 221 (1), jeb169615. https://doi.org/10.1242/jeb.169615.

Kang, J., Nagelkerken, I., Rummer, J.L., Rodolfo-Metalpa, R., Munday, P.L., Ravasi, T., Schunter, C., 2022. Rapid evolution fuels transcriptional plasticity to ocean acidification. Global Change Biol. https://doi.org/10.1111/gcb.16119.

Keirstead, N.D., et al., 2014. Fatal septicemia caused by the zoonotic bacterium *streptococcus iniae* during an outbreak in Caribbean reef fish. Vet. Pathol. 51, 1035–1041. https://doi.org/10.1177/0300985813505876.

Kerr, L.M., Lang, K.L., Lobel, P.S., 1997. PCB contamination relative to age for a Pacific damselfish, *Abudefduf sordidus* (Pomacentridae). Biol. Bull. 193, 279–281. https://doi.org/10.1086/BBLv193n2p279.

Killen, S.S., Reid, D., Marras, S., Domenici, P., 2015. The interplay between aerobic metabolism and antipredator performance: vigilance is related to recovery rate after exercise. Front. Physiol. 6, 111.

King, S.C., et al., 2005. Summary results from a pilot study conducted around an oil production platform on the northwest shelf of Australia. Mar. Pollut. Bull. 50, 1163–1172. https://doi.org/10.1016/j.marpolbul.2005.04.027.

Klein, S.G., Steckbauer, A., Duarte, C.M., 2020. Defining $CO_2$ and $O_2$ syndromes of marine biomes in the Anthropocene. Glob. Chang. Biol. 26 (2), 355–363. https://doi.org/10.1111/gcb.14879.

Kobayashi, J., et al., 2019. Comparison of trophic magnification factors of PCBs and PBDEs in Tokyo Bay based on nitrogen isotope ratios in bulk nitrogen and amino acids. Chemosphere 226, 220–228. https://doi.org/10.1016/j.chemosphere.2019.03.133.

Kroon, F.J., Motti, C.E., Jensen, L.H., Berry, K.L.E., 2018. Classification of marine microdebris: a review and case study on fish from the great barrier reef, Australia. Sci. Rep. 8, 15. https://doi.org/10.1038/s41598-018-34590-6.

Lake, R.G., Hinch, S.G., 1999. Acute effects of suspended sediment angularity on juvenile coho salmon (*Oncorhynchus kisutch*). Can. J. Fish. Aquat. Sci. 56 (5), 862–867.

Lappivaara, J., Nikinmaa, M., Tuurala, H., 1995. Arterial oxygen tension and the structure of the secondary lamellae of the gills in rainbow trout (Oncorhynchus mykiss) after acute exposure to zinc and during recovery. Aquat. Toxicol. 32 (4), 321–331.

Laubenstein, T.D., Rummer, J.L., Nicol, S., Parsons, D.M., Pether, S.M.J., Pope, S., Smith, N., Munday, P.L., 2018. Correlated effects of ocean acidification and warming on behavioural and metabolic traits of a large pelagic fish. Diversity 10 (2), 35. https://doi.org/10.3390/d10020035.

Laubenstein, T.D., et al., 2019. A negative correlation between behavioural and physiological performance under ocean acidification and warming. Sci. Rep. 9, 4265. https://doi.org/10.1038/s41598-018-36747-9.

Lefevre, S., 2016. Are global warming and ocean acidification conspiring against marine ectotherms? A meta-analysis of the respiratory effects of elevated temperature, high CO2 and their interaction. Conserv. Physiol. 4 (1), cow009. https://doi.org/10.1093/conphys/cow009.

Li, Y., et al., 2019. Occurrence of polycyclic aromatic hydrocarbons (PAHs) in coral reef fish from the South China Sea. Mar. Pollut. Bull. 139, 339–345. https://doi.org/10.1016/j.marpolbul.2019.01.001.

Lienart, G.D.H., Ferrari, M.C.O., McCormick, M.I., 2016. Thermal environment and nutritional condition affect the efficacy of chemical alarm cues produced by prey fish. Environ. Biol. Fishes 99, 729–739. https://doi.org/10.1007/s10641-016-0516-7.

Lin, J.J., Somero, G.N., 1995. Thermal adaptation of cytoplasmic malate-dehydrogenases of eastern Pacific barracuda (*Sphyraena* spp.): the role of differential isoenzyme expression. J. Exp. Biol. 198, 551–560.

Lowe, M.L., Morrison, M.A., Taylor, R.B., 2015. Harmful effects of sediment-induced turbidity on juvenile fish in estuaries. Mar. Ecol. Prog. Ser. 539, 241–254.

Madeira, C., et al., 2017a. Comparing biomarker responses during thermal acclimation: a lethal vs non-lethal approach in a tropical reef clownfish. Comp. Biochem. Physiol. A Mol. Integr. Physiol. 204, 104–112. https://doi.org/10.1016/j.cbpa.2016.11.018.

Madeira, C., et al., 2017b. Thermal stress, thermal safety margins and acclimation capacity in tropical shallow waters-an experimental approach testing multiple end-points in two common fish. Ecol. Indic. 81, 146–158. https://doi.org/10.1016/j.ecolind.2017.05.050.

Madliger, C.L., et al., 2017. Conservation physiology and the quest for a "good" Anthropocene. Conserv. Physiol. 5, cox003. https://doi.org/10.1093/conphys/cox003.

Mager, E.M., et al., 2017. Assessment of early life stage mahi-mahi windows of sensitivity during acute exposures to Deepwater horizon crude oil. Environ. Toxicol. Chem. 36 (7), 1887–1895. https://doi.org/10.1002/etc.3713.

Mahardika, K., et al., 2004. Susceptibility of juvenile humpback grouper *Cromileptes altivelis* to grouper sleepy disease iridovirus (GSDIV). Dis. Aquat. Organ. 59, 1–9. https://doi.org/10.3354/dao059001.

Maljkovic, A., Cote, I.M., 2011. Effects of tourism-related provisioning on the trophic signatures and movement patterns of an apex predator, the Caribbean reef shark. Biol. Conserv. 144, 859–865. https://doi.org/10.1016/j.biocon.2010.11.019.

Mascaro, M., et al., 2019. Effect of a gradually increasing temperature on the behavioural and physiological response of juvenile hippocampus erectus: thermal preference, tolerance, energy balance and growth. J. Therm. Biol. 85, 102406. UNSP https://doi.org/10.1016/j.jtherbio.2019.102406.

Mattsson, K., et al., 2017. Brain damage and behavioural disorders in fish induced by plastic nanoparticles delivered through the food chain. Sci. Rep. 7, 11452. https://doi.org/10.1038/s41598-017-10813-0.

McCormick, M.I., 1998. Behaviorally induced maternal stress in a fish influences progeny quality by a hormonal mechanism. Ecology 79 (6), 1873–1883. https://doi.org/10.1890/0012-9658(1998)079[1873:BIMSIA]2.0.CO;2.

McCormick, M.I., Chivers, D.P., Allan, B.J.M., Ferrari, M.C.O., 2017. Habitat degradation disrupts neophobia in juvenile coral reef fish. Glob. Chang. Biol. 23 (2), 719–727. https://doi.org/10.1111/gcb.13393.

McCormick, M.I., et al., 2018. Effect of elevated $CO_2$ and small boat noise on the kinematics of predator—prey interactions. Proc. R. Soc. B Biol. Sci. 285, 20172650. https://doi.org/10.1098/rspb.2017.2650.

McCormick, M.I., et al., 2019a. Effects of boat noise on fish fast-start escape response depend on engine type. Sci. Rep. 9, 6554. https://doi.org/10.1038/s41598-019-43099-5.

McCormick, M.I., Fakan, E.P., Palacios, M.M., 2019b. Habitat degradation and predators have independent trait-mediated effects on prey. Sci. Rep. 9, 15705. https://doi.org/10.1038/s41598-019-51798-2.

McCormick, M.I., et al., 2020. Microplastic exposure interacts with habitat degradation to affect behaviour and survival of juvenile fish in the field. Proc. Royal Soc. Biol. 287 (1937), 20201947. https://doi.org/10.1098/rspb.2020.1947.

McLeod, I.M., Clark, T.D., 2016. Limited capacity for faster digestion in larval coral reef fish at an elevated temperature. PLoS One 11, e0155360. https://doi.org/10.1371/journal.pone.0155360.

McLeod, I.M., et al., 2013. Climate change and the performance of larval coral reef fishes: the interaction between temperature and food availability. Conserv. Physiol. 1, cot024. https://doi.org/10.1093/conphys/cot024.

McMahon, S.J., Donelson, J.M., Munday, P.L., 2018. Food ration does not influence the effect of elevated $CO_2$ on antipredator behaviour of a reef fish. Mar. Ecol. Prog. Ser. 586, 155–165.

McMahon, S.J., Munday, P.L., Wong, M.Y., Donelson, J.M., 2019. Elevated CO2 and food ration affect growth but not the size-based hierarchy of a reef fish. Sci. Rep. 9 (1), 1–10.

Meador, J., et al., 2006. Altered growth and related physiological responses in juvenile Chinook salmon (*Oncorhynchus tshawytscha*) from dietary exposure to polycyclic aromatic hydrocarbons (PAHs). Can. J. Fish. Aquat. Sci. 63, 2364–2376. https://doi.org/10.1139/f06-127.

Meier, S., et al., 2010. Development of Atlantic cod (*Gadus morhua*) exposed to produced water during early life stages: effects on embryos, larvae, and juvenile fish. Mar. Environ. Res. 70 (5), 383–394. https://doi.org/10.1016/j.marenvres.2010.08.002.

Messmer, V., et al., 2017. Global warming may disproportionately affect larger adults in a predatory coral reef fish. Glob. Chang. Biol. 23, 2230–2240. https://doi.org/10.1111/gcb.13552.

Miller, G.M., Watson, S.A., Donelson, J.M., McCormick, M.I., Munday, P.L., 2012. Parental environment mediates impacts of increased carbon dioxide on a coral reef fish. Nat. Clim. Chang. 2 (12), 858–861.

Miller, G.M., et al., 2015. Temperature is the evil twin: effects of increased temperature and ocean acidification on reproduction in a reef fish. Ecol. Appl. 25, 603–620. https://doi.org/10.1890/14-0559.1.

Mills, S.C., et al., 2020. Hormonal and behavioural effects of motorboat noise on wild coral reef fish. Environ. Pollut. 262, 114250. https://doi.org/10.1016/j.envpol.2020.114250.

Mitchell, M.D., Harborne, A.R., 2020. Non-consumptive effects in fish predator–prey interactions on coral reefs. Coral Reefs 39 (4), 867–884.

Miyake, S., Ngugi, D.K., Stingl, U., 2015. Diet strongly influences the gut microbiota of surgeonfishes. Mol. Ecol. 24, 656–672. https://doi.org/10.1111/mec.13050.

Mommsen, T.P., Vijayan, M.M., Moon, T.W., 1999. Cortisol in teleosts: dynamics, mechanisms of action, and metabolic regulation. Rev. Fish Biol. Fish. 9, 211–268.

Morgan, C., Shipley, O.N., Gelsleichter, J., 2020. Resource-use dynamics of co-occurring chondrichthyans from the first coast, North Florida, USA. J. Fish Biol. 96, 570–579. https://doi.org/10.1111/jfb.14238.

Motson, K., Donelson, J.M., 2017. Limited capacity for developmental thermal acclimation in three tropical wrasses. Coral Reefs 36 (2), 609–621.

Mourier, J., Claudet, J., Planes, S., 2021. Human-induced shifts in habitat use and behaviour of a marine predator: the effects of bait provisioning in the blacktip reef shark. Anim. Conserv. 24, 230–238. https://doi.org/10.1111/acv.12630.

Moustaka, M., et al., 2018. The effects of suspended sediment on coral reef fish assemblages and feeding guilds of north-West Australia. Coral Reefs 37 (3), 659–673.

Mueller, C.A., Seymour, R.S., 2011. The regulation index: a new method for assessing the relationship between oxygen consumption and environmental oxygen. Physiol. Biochem. Zool. 84 (5), 522–532. https://doi.org/10.1086/661953.

Munday, P.L., 2014. Transgenerational acclimation of fishes to climate change and ocean acidification. F1000prime Rep. 6, 99.

Munday, P.L., Crawley, N.E., Nilsson, G.E., 2009a. Interacting effects of elevated temperature and ocean acidification on the aerobic performance of coral reef fishes. Mar. Ecol. Prog. Ser. 388, 235–242. https://doi.org/10.3354/meps08137.

Munday, P.L., et al., 2009b. Effects of ocean acidification on the early life history of a tropical marine fish. Proc. R. Soc. B Biol. Sci. 276, 3275–3283. https://doi.org/10.1098/rspb.2009.0784.

Munday, P.L., Dixson, D.L., McCormick, M.I., Meekan, M., Ferrari, M.C.O., Chivers, D.P., 2010. Replenishment of fish populations is threatened by ocean acidification. Proc. Natl. Acad. Sci. U. S. A. 107 (29), 12930–12934. https://doi.org/10.1073/pnas.1004519107.

Munday, P.L., et al., 2011a. Effect of ocean acidification on otolith development in larvae of a tropical marine fish. Biogeosciences 8, 1631–1641. https://doi.org/10.5194/bg-8-1631-2011.
Munday, P.L., et al., 2011b. Ocean acidification does not affect the early life history development of a tropical marine fish. Mar. Ecol. Prog. Ser. 423, 211–221. https://doi.org/10.3354/meps08990.
Munday, P.L., McCormick, M.I., Nilsson, G.E., 2012. Impact of global warming and rising $CO_2$ levels on coral reef fishes: what hope for the future? J. Exp. Biol. 215 (22), 3865–3873.
Munday, P.L., et al., 2013. Elevated $CO_2$ affects the behavior of an ecologically and economically important coral reef fish. Mar. Biol. 160, 2137–2144. https://doi.org/10.1007/s00227-012-2111-6.
Munday, P.L., et al., 2014. Behavioural impairment in reef fishes caused by ocean acidification at $CO_2$ seeps. Nat. Clim. Change. 4, 487–492. https://doi.org/10.1038/NCLIMATE2195.
Munday, P.L., et al., 2016. Effects of elevated CO2 on early life history development of the yellowtail kingfish, *Seriola lalandi*, a large pelagic fish. ICES J. Mar. Sci. 73 (3), 641–649. https://doi.org/10.1093/icesjms/fsv210.
Munday, P.L., et al., 2019. Testing the adaptive potential of yellowtail kingfish to ocean warming and acidification. Front. Ecol. Evol. 7, 253.
Musa, S.M., Ripley, D.M., Moritz, T., Shiels, H.A., 2020. Ocean warming and hypoxia affect embryonic growth, fitness and survival of small-spotted catsharks, *Scyliorhinus canicular*. J. Fish Biol. 97 (1), 257–264. https://doi.org/10.1111/jfb.14370.
Narvaez, P., et al., 2021. New perspectives on the role of cleaning symbiosis in the possible transmission of fish diseases. Rev. Fish Biol. Fish. 31, 233–251. https://doi.org/10.1007/s11160-021-09642-2.
Nedelec, S.L., et al., 2016. Repeated exposure to noise increases tolerance in a coral reef fish. Environ. Pollut. 216, 428–436. https://doi.org/10.1016/j.envpol.2016.05.058.
Nedelec, S.L., et al., 2017a. Motorboat noise impacts parental behaviour and offspring survival in a reef fish. Proc. R. Soc. B Biol. Sci. 284, 20170143. https://doi.org/10.1098/rspb.2017.0143.
Nedelec, S.L., et al., 2017b. Motorboat noise disrupts co-operative interspecific interactions. Sci. Rep. 7, 6987. https://doi.org/10.1038/s41598-017-06515-2.
Negri, A.P., et al., 2016. Acute ecotoxicology of natural oil and gas condensate to coral reef larvae. Sci. Rep. 6, 21153.
Nelson, P.A., et al., 2003. Ontogenetic changes and environmental effects on ocular transmission in four species of coral reef fishes. J. Comp. Physiol. A Neuroethol. Sens. Neural Behav. Physiol. 189, 391–399. https://doi.org/10.1007/s00359-003-0418-y.
Nilsson, G.E., Lefevre, S., 2016. Physiological challenges to fishes in a warmer and acidified future. Physiology 31 (6), 409–417.
Nilsson, G.E., Ostlund-Nilsson, S., 2004. Hypoxia in paradise: widespread hypoxia tolerance in coral reef fishes. Proc. R. Soc. B Biol. Sci. 271, S30–S33. https://doi.org/10.1098/rsbl.2003.0087.
Nilsson, G.E., et al., 2007. From record performance to hypoxia tolerance: respiratory transition in damselfish larvae settling on a coral reef. Proc. R. Soc. B Biol. Sci. 274, 79–85. https://doi.org/10.1098/rspb.2006.3706.
Nilsson, G.E., Ostlund-Nilsson, S., Munday, P.L., 2010. Effects of elevated temperature on coral reef fishes: loss of hypoxia tolerance and inability to acclimate. Comp. Biochem. Physiol. A Mol. Integr. Physiol. 156, 389–393. https://doi.org/10.1016/j.cbpa.2010.03.009.
Nilsson, G.E., et al., 2012. Near-future carbon dioxide levels alter fish behaviour by interfering with neurotransmitter function. Nat. Clim. Chang. 2, 201–204. https://doi.org/10.1038/NCLIMATE1352.

Norin, T., Clark, T.D., 2016. Measurement and relevance of maximum metabolic rate in fishes. J. Fish Biol. 88 (1), 122–151.

Norin, T., et al., 2018. Anemone bleaching increases the metabolic demands of symbiont anemonefish. Proc. R. Soc. B Biol. Sci. 285, 20180282. https://doi.org/10.1098/rspb.2018.0282.

Ospina, A.F., Mora, C., 2004. Effect of body size on reef fish tolerance to extreme low and high temperatures. Environ. Biol. Fishes 70, 339–343. https://doi.org/10.1023/B:EBFI.0000035429.39129.34.

Ostlund-Nilsson, S., Nilsson, G.E., 2004. Breathing with a mouth full of eggs: respiratory consequences of mouthbrooding in cardinalfish. Proc. R. Soc. B Biol. Sci. 271, 1015–1022. https://doi.org/10.1098/rspb.2004.2700.

Östlund-Nilsson, S., et al., 2005. Parasitic isopod *a. apogonae*, a drag for the cardinal fish *C. quinquelineatus*. Mar. Ecol. Prog. Ser. 287, 209–216. https://doi.org/10.3354/meps287209.

Overmans, S., et al., 2018. Phototoxic effects of PAH and UVA exposure on molecular responses and developmental success in coral larvae. Aquat. Toxicol. 198, 165–174. https://doi.org/10.1016/j.aquatox.2018.03.008.

Palacios, M.M., Killen, S.S., Nadler, L.E., White, J.R., McCormick, M.I., 2016. Top predators negate the effect of mesopredators on prey physiology. J. Anim. Ecol. 85 (4), 1078–1086.

Pan, H.-H., et al., 2020. Elevated CO2 concentrations impacts growth and swimming metabolism in yellowtail kingfish, *Seriola lalandi*. Aquaculture 523, 735157. https://doi.org/10.1016/j.aquaculture.2020.735157.

Parmentier, E., Colleye, O., Mann, D., 2009. Hearing ability in three clownfish species. J. Exp. Biol. 212 (13), 2023–2026. https://doi.org/10.1242/jeb.030270.

Paula, J.R., et al., 2015. The role of serotonin in the modulation of cooperative behavior. Behav. Ecol. 26, 1005–1012. https://doi.org/10.1093/beheco/arv039.

Paula, J.R., et al., 2019. Neurobiological and behavioural responses of cleaning mutualisms to ocean warming and acidification. Sci. Rep. 9, 12728. https://doi.org/10.1038/s41598-019-49086-0.

Pauly, D., 2019. A précis of gill-oxygen limitation theory (GOLT), with some emphasis on the eastern Mediterranean. Mediterr. Mar. Sci. 20 (4), 660–668. https://doi.org/10.12681/mms.19285.

Payne, N.L., et al., 2016. Temperature dependence of fish performance in the wild: links with species biogeography and physiological thermal tolerance. Funct. Ecol. 30 (6), 903–912. https://doi.org/10.1111/1365-2435.12618.

Pereira Santos, C., Sampaio, E., Pereira, B., Pegado, M.R., Borges, F.O., Wheeler, C., Bouyoucos, I., Rummer, J., Frazão Santos, C., Rosa, R., 2021. Elasmobranch responses to experimental warming, acidification, and oxygen loss—a meta-analysis. Front. Mar. Sci. 8, 735377. https://doi.org/10.3389/fmars.2021.735377.

Peterson, C.T., Grubbs, R.D., Mickle, A., 2020. Trophic ecology of elasmobranch and teleost fishes in a large subtropical seagrass ecosystem (Florida big bend) determined by stable isotope analysis. Environ. Biol. Fishes 103, 683–701. https://doi.org/10.1007/s10641-020-00976-7.

Picciulin, M., Sebastianutto, L., Codarin, A., Farina, A., Ferrero, E.A., 2010. In situ behavioural responses to boat noise exposure of *Gobius cruentatus* (Gmelin, 1789; fam. Gobiidae) and *Chromis chromis* (Linnaeus, 1758; fam. Pomacentridae) living in a marine protected area. J. Exp. Mar. Biol. Ecol. 386 (1–2), 125–132. https://doi.org/10.1016/j.jembe.2010.02.012.

Polinas, M., et al., 2018. Ecological and histopathological aspects of *Didymodiclinus* sp. (Trematoda: Didymozoidae) parasite of the dusky grouper, I (osteichthyes: serranidae), from the Western Mediterranean Sea. J. Fish Dis. 41, 1385–1393. https://doi.org/10.1111/jfd.12836.

Poloczanska, E.S., et al., 2016. Responses of marine organisms to climate change across oceans. Front. Mar. Sci. 3, 62. https://doi.org/10.3389/fmars.2016.00062.

Pörtner, H.-O., 2014. How to and how not to investigate the oxygen and capacity limitation of thermal tolerance (OCLTT) and aerobic scope. J. Exp. Biol. 217, 4432–4435. https://doi.org/10.1242/jeb.114181.

Pörtner, H.O., Farrell, A.P., 2008. Physiology and climate change. Science 322, 690–692.

Pörtner, H.-O., Knust, R., 2007. Climate change affects marine fishes through the oxygen limitation of thermal tolerance. Science 315, 95–97. https://doi.org/10.1126/science.1135471.

Pörtner, H.O., Bock, C., Mark, F.C., 2017. Oxygen- and capacity-limited thermal tolerance: bridging ecology and physiology. J. Exp. Biol. 220 (15), 2685–2696. https://doi.org/10.1242/jeb.134585.

Prinz, N., et al., 2020. To feed or not to feed? Coral reef fish responses to artificial feeding and stakeholder perceptions in the Aitutaki lagoon, Cook Islands. Front. Mar. Sci. 7, 145. https://doi.org/10.3389/fmars.2020.00145.

Prodocimo, V., Freire, C.A., 2001. Ionic regulation in aglomerular tropical estuarine pufferfishes submitted to sea water dilution. J. Exp. Mar. Biol. Ecol. 262, 243–253. https://doi.org/10.1016/S0022-0981(01)00293-3.

R Core Development Team, 2018. R: A Language and Environment for Statistical Computing. R Foundation for Statistical Computing, Vienna, Austria. http://www.R-project.org.

Ramasamy, R.A., Allan, B.J.M., McCormick, M.I., 2015. Plasticity of escape responses: prior predator experience enhances escape performance in a coral reef fish. PLoS One 10, e0132790. https://doi.org/10.1371/journal.pone.0132790.

Randall, D.J., Rummer, J.L., Wilson, J.M., Wang, S., Brauner, C.J., 2014. Review: a unique mode of tissue oxygenation and the success of teleost fish. J. Exp. Biol. 217, 1205–1214. https://doi.org/10.1242/jeb.093526.

Rayment, G.E., Barry, G.A., 2000. Indicator tissues for heavy metal monitoring: additional attributes. Mar. Pollut. Bull. 41, 353–358. https://doi.org/10.1016/S0025-326X(00)00128-4.

Reijnders, P.J.H., Brasseur, S.M.J.M., 1997. Xenobiotic induced hormonal and associated developmental disorders in marine organisms and related effects in humans; an overview. J. Clean Technol. Environ. Toxicol. Occup. Med. 6 (4), 367–380.

Renshaw, G.M.C., Kerrisk, C.B., Nilsson, G.E., 2002. The role of adenosine in the anoxic survival of the epaulette shark, *H. ocellatum*. Comp. Biochem. Physiol. A Mol. Integr. Physiol. 131 (2), 133–141. https://doi.org/10.1016/S1096-4959(01)00484-5.

Renshaw, G.M.C., Kutek, A.K., Grant, G.D., Anoopkumar-Dukie, S., 2012. Forecasting elasmobranch survival following exposure to severe stressors. Comp. Biochem. Physiol. A Mol. Integr. Physiol. 162 (2), 101–112. https://doi.org/10.1016/j.cbpa.2011.08.001.

Reyes, I., et al., 2011. Behavioral thermoregulation, temperature tolerance and oxygen consumption in the Mexican Bullseye puffer fish, *Sphoeroides annulatus* Jenyns (1842), acclimated to different temperatures. J. Therm. Biol. 36, 200–205. https://doi.org/10.1016/j.jtherbio.2011.03.003.

Richardson, A.J., et al., 2012. Climate change and marine life. Biol. Lett. 8, 907–909. https://doi.org/10.1098/rsbl.2012.0530.

Rodgers, G.G., et al., 2018. In hot water: Sustained Ocean warming reduces survival of a low-latitude coral reef fish. Mar. Biol. 165, 73. https://doi.org/10.1007/s00227-018-3333-z.

Rodgers, G.G., et al., 2019. Impacts of increased ocean temperatures on a low-latitude coral reef fish: processes related to oxygen uptake and delivery. J. Therm. Biol. 79, 95–102. https://doi.org/10.1016/j.jtherbio.2018.12.008.

Rosa, R., et al., 2014. Early-life exposure to climate change impairs tropical shark survival. Proc. R. Soc. B Biol. Sci. 281, 20141738. UNSP https://doi.org/10.1098/rspb.2014.1738.

Rosa, R., et al., 2016a. Deficit in digestive capabilities of bamboo shark early stages under climate change. Mar. Biol. 163. https://doi.org/10.1007/s00227-016-2840-z.

Rosa, R., et al., 2016b. Neuro-oxidative damage and aerobic potential loss of sharks under elevated $CO_2$ and warming. Mar. Biol. 163, 119. https://doi.org/10.1007/s00227-016-2898-7.

Rosa, R., Rummer, J.L., Munday, P.L., 2017. Biological responses of sharks to ocean acidification. Biol. Lett. 13 (3), 20160796. https://doi.org/10.1098/rsbl.2016.0796.

Routley, M.H., Nilsson, G.E., Renshaw, G.M.C., 2002. Exposure to hypoxia primes the respiratory and metabolic responses of the epaulette shark to progressive hypoxia. Comp. Biochem. Physiol. A Mol. Integr. Physiol. 131, 313–321. https://doi.org/10.1016/S1095-6433(01)00484-6.

Rowe, C.E., et al., 2018. Effects of temperature on macronutrient selection, metabolic and swimming performance of the indo-Pacific damselfish (*Abudefduf vaigiensis*). Mar. Biol. 165, 178. https://doi.org/10.1007/s00227-018-3435-7.

Rummer, J.L., Munday, P.L., 2017. Climate change and the evolution of reef fishes: past and future. Fish Fish. 18 (1), 22–39.

Rummer, J.L., et al., 2009. Physiological tolerance to hyperthermia and hypoxia and effects on species richness and distribution of rockpool fishes of loggerhead key, dry Tortugas National Park. J. Exp. Mar. Biol. Ecol. 371, 155–162. https://doi.org/10.1016/j.jembe.2009.01.015.

Rummer, J.L., et al., 2013. Elevated $CO_2$ enhances aerobic scope of a coral reef fish. Conserv. Physiol. 1, cot023. https://doi.org/10.1093/conphys/cot023.

Rummer, J.L., et al., 2014. Life on the edge: thermal optima for aerobic scope of equatorial reef fishes are close to current day temperatures. Glob. Chang. Biol. 20, 1055–1066. https://doi.org/10.1111/gcb.12455.

Rummer, J.L., et al., 2022. Climate change and sharks, chapter 25. In: Carrier, J.C., Simpfendorfer, C.A., Heithaus, M.R., Yopak, K.E. (Eds.), Biology of Sharks and their Relatives, third ed. CRC Press (in press).

Ryu, T., et al., 2020. An epigenetic signature for within-generational plasticity of a reef fish to ocean warming. Front. Mar. Sci. 7, 284. https://doi.org/10.3389/fmars.2020.00284.

Sampaio, E., Santos, C., Rosa, I.C., et al., 2021. Impacts of hypoxic events surpass those of future ocean warming and acidification. Nat. Ecol. Evol. 5, 311–321. https://doi.org/10.1038/s41559-020-01370-3.

Schmale, M.C., 1995. Experimental induction of neurofibromatosis in bicolor damselfish. Dis. Aquat. Organ. 23, 201–212. https://doi.org/10.3354/dao023201.

Schmale, M.C., Gibbs, P.D.L., Campbell, C.E., 2002. A virus-like agent associated with neurofibromatosis in damselfish. Dis. Aquat. Organ. 49, 107–115. https://doi.org/10.3354/dao049107.

Schmale, M.C., Vicha, D., Cacal, S.M., 2004. Degranulation of eosinophilic granule cells in neurofibromas and gastrointestinal tract in the bicolor damselfish. Fish Shellfish Immunol. 17, 53–63. https://doi.org/10.1016/j.fsi.2003.12.002.

Schmidt-Nielsen, K., 1997. Animal Physiology: Adaptation and Environment. Cambridge University Press, Cambridge.

Schofield, P.J., et al., 2015. Survival and growth of invasive indo-Pacific lionfish at low salinities. Aquat. Invasions 10, 333–337. https://doi.org/10.3391/ai.2015.10.3.08.

Schunter, C., et al., 2016. Molecular signatures of transgenerational response to ocean acidification in a species of reef fish. Nat. Clim. Chang. 6, 1014–1018. https://doi.org/10.1038/NCLIMATE3087.

Schunter, C., Welch, M.J., Nilsson, G.E., Rummer, J.L., Munday, P.L., Ravasi, T., 2018. An interplay between plasticity, epigenetics, and parental phenotype determines impacts of ocean acidification on a reef fish. Nat. Ecol. Evol. 2, 334–342. https://doi.org/10.1038/s41559-017-0428-8.

Schunter, C., Jarrold, M.D., Munday, P.L., Ravasi, T., 2021. Diel $pCO_2$ fluctuations alter the molecular response of coral reef fishes to ocean acidification conditions. Mol. Ecol. 30, 5105–5118. https://doi.org/10.1111/mec.16124.

Seebacher, F., White, C.R., Franklin, C.E., 2015. Physiological plasticity increases resilience of ectothermic animals to climate change. Nat. Clim. Chang. 5, 61–66.

Séguigne, C., et al., 2022. Effects of a COVID-19 lockdown-induced pause and resumption of artificial provisioning on blacktip reef sharks (*C. melanopterus*) and pink whiprays (*Pateobatis fai*) in French Polynesia (East-Pacific). Ethology 128 (2), 119–130. https://doi.org/10.1111/eth.13246.

Seibel, B.A., Deutsch, C., 2020. Oxygen supply capacity in animals evolves to meet maximum demand at the current oxygen partial pressure regardless of size or temperature. J. Exp. Biol. 223 (12), jeb210492. https://doi.org/10.1242/jeb.210492.

Selye, H., 1973. The evolution of the stress concept. Am. Sci. 61, 692–699.

Serrano, X., Grosell, M., Serafy, J.E., 2010. Salinity selection and preference of the grey snapper *Lutjanus griseus*: field and laboratory observations. J. Fish Biol. 76, 1592–1608. https://doi.org/10.1111/j.1095-8649.2010.02585.x.

Sevcikova, M., Modra, H., Slaninova, A., Svobodova, Z., 2011. Metals as a cause of oxidative stress in fish: a review. Vet. Med. 56 (11), 537–546.

Shirai, K., et al., 2018. Reconstruction of the salinity history associated with movements of mangrove fishes using otolith oxygen isotopic analysis. Mar. Ecol. Prog. Ser. 593, 127–139. https://doi.org/10.3354/meps12514.

Sikkel, P.C., Cook, C.A., Renoux, L.P., Bennett, C.L., Tuttle, L.J., Smit, N.J., 2018. The distribution and host-association of a haemoparasite of damselfishes (Pomacentridae) from the eastern Caribbean based on a combination of morphology and 18S rDNA sequences. Int. J. Parasitol. 7 (2), 213–220. https://doi.org/10.1016/j.ijppaw.2018.05.004. PMID: 29988386; PMCID: PMC6024192.

Simpson, S.D., Meekan, M.G., McCauley, R.D., Jeffs, A., 2004. Attraction of settlement-stage coral reef fishes to reef noise. Mar. Ecol. Prog. Ser. 276, 263–268. https://doi.org/10.3354/meps276263.

Simpson, S.D., et al., 2005a. Homeward sound. Science 308, 221. https://doi.org/10.1126/science.1107406.

Simpson, S.D., et al., 2005b. Response of embryonic coral reef fishes (Pomacentridae: *Amphiprion* spp.) to noise. Mar. Ecol. Prog. Ser. 287, 201–208. https://doi.org/10.3354/meps287201.

Simpson, S.D., Purser, J., Radford, A.N., 2015. Anthropogenic noise compromises antipredator behaviour in European eels. Glob. Chang. Biol. 21 (2), 586–593. https://doi.org/10.1111/gcb.12685.

Simpson, S.D., et al., 2016. Anthropogenic noise increases fish mortality by predation. Nat. Commun. 7, 10544. https://doi.org/10.1038/ncomms10544.

Soares, M.C., et al., 2012. Face your fears: cleaning gobies inspect predators despite being stressed by them. PLoS One 7, e39781. https://doi.org/10.1371/journal.pone.0039781.

Söderstrom, V., Renshaw, G.M.C., Nilsson, G.E., 1999. Responses of cerebral blood flow and blood pressure to hypoxia in the epaulette shark, an hypoxia tolerant tropical elasmobranch. FASEB J. 13, A383.

Sørensen, L., et al., 2019. Accumulation and toxicity of monoaromatic petroleum hydrocarbons in early life stages of cod and haddock. Environ. Pollut. 251, 212–220. https://doi.org/10.1016/j.envpol.2019.04.126.

Sørhus, E., et al., 2016. Crude oil exposures reveal roles for intracellular calcium cycling in haddock craniofacial and cardiac development. Sci. Rep. 6 (31058), 1–21. https://doi.org/10.1038/srep31058.

Speed, C.W., et al., 2012a. Trophic ecology of reef sharks determined using stable isotopes and telemetry. Coral Reefs 31, 357–367. https://doi.org/10.1007/s00338-011-0850-3.

Speed, C.W., et al., 2012b. Heat-seeking sharks: support for behavioural thermoregulation in reef sharks. Mar. Ecol. Prog. Ser. 463, 231–244. https://doi.org/10.3354/meps09864.

Speers-Roesch, B., et al., 2012. Hypoxia tolerance in elasmobranchs. I. Critical oxygen tension as a measure of blood oxygen transport during hypoxia exposure. J. Exp. Biol. 215, 93–102.

Spinks, R.K., Munday, P.L., Donelson, J.M., 2019. Developmental effects of heatwave conditions on the early life stages of a coral reef fish. J. Exp. Biol. 222, jeb202713. UNSP https://doi.org/10.1242/jeb.202713.

Spinks, R.K., Bonzi, L.C., Ravasi, T., Munday, P.L., Donelson, J.M., 2021. Sex- and time-specific parental effects of warming on reproduction and offspring quality in a coral reef fish. Evol. Appl. 14 (4), 1145–1158. https://doi.org/10.1111/eva.13187.

Staaterman, E., et al., 2020. Exposure to boat noise in the field yields minimal stress response in wild reef fish. Aquat. Biol. 29, 93–103. https://doi.org/10.3354/ab00728.

Storelli, M.M., Perrone, V.G., Barone, G., 2011. Organochlorine residues (PCBs and DDTs) in two torpedinid species liver from the Southeastern Mediterranean Sea. Environ. Sci. Pollut. Res. 18, 1160–1165. https://doi.org/10.1007/s11356-011-0463-y.

Sun, R.X., et al., 2019. Polycyclic aromatic hydrocarbons in marine organisms from mischief reef in the South China Sea: implications for sources and human exposure. Mar. Pollut. Bull. 149, 110623. UNSP https://doi.org/10.1016/j.marpolbul.2019.110623.

Sunday, J.M., et al., 2014. Evolution in an acidifying ocean. Trends Ecol. Evol. 29, 117–125. https://doi.org/10.1016/j.tree.2013.11.001.

Sutherland, A.B., Meyer, J.L., 2007. Effects of increased suspended sediment on growth rate and gill condition of two southern Appalachian minnows. Environ. Biol. Fishes 80 (4), 389–403.

Svendsen, M.B.S., et al., 2019. Are all bony fishes oxygen regulators? Evidence for oxygen regulation in a putative oxygen conformer, the swamp eel *Synbranchus marmoratus*. J. Fish Biol. 94 (1), 178–182. https://doi.org/10.1111/jfb.13861.

Sweet, M., et al., 2012. Evidence of melanoma in wild marine fish populations. PLoS One 7, e41989. https://doi.org/10.1371/journal.pone.0041989.

Syvitski, J.P., Vörösmarty, C.J., Kettner, A.J., Green, P., 2005. Impact of humans on the flux of terrestrial sediment to the global coastal ocean. Science 308 (5720), 376–380.

Takemura, A., Uchimura, M., Shibata, Y., 2010. Dopaminergic activity in the brain of a tropical wrasse in response to changes in light and hydrostatic pressure. Gen. Comp. Endocrinol. 166, 513–519. https://doi.org/10.1016/j.ygcen.2010.01.001.

Takegaki, T., Nakazono, A., 1999. Responses of the egg-tending gobiid fish *Valenciennea longipinnis* to the fluctuation of dissolved oxygen in the burrow. Bull. Mar. Sci. 65 (3), 815–823.

Takemura, A., et al., 2012. Effects of hydrostatic pressure on monoaminergic activity in the brain of a tropical wrasse, *Halicoeres trimaculatus*: possible implication for controlling tidal-related reproductive activity. Gen. Comp. Endocrinol. 175, 173–179. https://doi.org/10.1016/j.ygcen.2011.11.019.

Tang, C.H., et al., 2014a. Short-term effects of thermal stress on the responses of branchial protein quality control and osmoregulation in a reef-associated fish, *Chromis viridis*. Zool. Stud. 53, 21. https://doi.org/10.1186/s40555-014-0021-7.

Tang, C.H., et al., 2014b. Exploration of the mechanisms of protein quality control and osmoregulation in gills of *C. viridis* in response to reduced salinity. Fish Physiol. Biochem. 40, 1533–1546. https://doi.org/10.1007/s10695-014-9946-3.

Tewksbury, J.J., Huey, R.B., Deutsch, C., 2008. Putting the heat on tropical animals. Science 320 (5881), 1296–1297. https://doi.org/10.1126/science.1159328.

Van Den Hurk, P., et al., 2020. Lionfish (*Pterois volitans*) as biomonitoring species for oil pollution effects in coral reef ecosystems. Mar. Environ. Res. 156, 104915. UNSP https://doi.org/10.1016/j.marenvres.2020.104915.

Vargas, C.A., et al., 2017. Species-specific responses to ocean acidification should account for local adaptation and adaptive plasticity. Nat. Ecol. Evol. 1, 1–7. https://doi.org/10.1038/s41559-017-0084.

Veilleux, H.D., et al., 2015. Molecular processes of transgenerational acclimation to a warming ocean. Nat. Clim. Chang. 5 (12), 1074–1078.

Veilleux, H.D., et al., 2018a. Molecular response to extreme summer temperatures differs between two genetically differentiated populations of a coral reef fish. Front. Mar. Sci. 5, 349. https://doi.org/10.3389/fmars.2018.00349. UNSP.

Veilleux, H.D., Donelson, J.M., Munday, P.L., 2018b. Reproductive gene expression in a coral reef fish exposed to increasing temperature across generations. Conserv. Physiol. 6, cox077. https://doi.org/10.1093/conphys/cox077.

Velasco-Blanco, G., et al., 2019. Thermal preference, tolerance, and thermal aerobic scope in clownfish *Amphiprion ocellaris* (Cuvier, 1830) predict its aquaculture potential across tropical regions. Int. Aquat. Res 11, 187–197. https://doi.org/10.1007/s40071-019-0228-7.

Victor, B., 2015. How many coral reef fish species are there? Cryptic diversity and the new molecular taxonomy. In: Mora, C. (Ed.), Ecology of Fishes on Coral Reefs. Cambridge University Press, Cambridge, pp. 76–87, https://doi.org/10.1017/CBO9781316105412.010.

Voellmy, I.K., et al., 2014. Acoustic noise reduces foraging success in two sympatric fish species via different mechanisms. Anim. Behav. 89, 191–198. https://doi.org/10.1016/j.anbehav.2013.12.029.

Vorbach, B.S., Wolf, J.C., Yanong, R.P., 2018. Odontomas in two long-finned ocellaris clownfish (*a. Ocellaris*). J. Vet. Diagn. Invest. 30, 136–139. https://doi.org/10.1177/1040638717729726.

Wandiga, S.O., et al., 2002. Accumulation, distribution and metabolism of $^{14}$C-1,1,1-Trichloro-2,2-, Bis- (P-Chlorophenyl) Ethane (p,p '-DDT) residues in a model tropical marine ecosystem. Environ. Technol. 23, 1285–1292. https://doi.org/10.1080/09593332308618327.

Wang, X.H., Wang, W.X., 2005. Uptake, absorption efficiency and elimination of DDT in marine phytoplankton, copepods and fish. Environ. Pollut. 136, 453–464. https://doi.org/10.1016/j.envpol.2005.01.004.

Warren, D.T., Donelson, J.M., McCormick, M.I., 2017. Extended exposure to elevated temperature affects escape response behaviour in coral reef fishes. PeerJ 5, e3652.

Watson, S.A., et al., 2018. Ocean warming has a greater effect than acidification on the early life history development and swimming performance of a large circumglobal pelagic fish. Glob. Chang. Biol. 24 (9), 4368–4385. https://doi.org/10.1111/gcb.14290.

Wedemeyer, G.A., Barton, B.A., McLeay, D.J., 1990. Stress and acclimation. In: Schreck, C.B., Moyle, P.B. (Eds.), Methods for Fish Biology. American Fisheries Society, Bethesda, Maryland, pp. 451–489.

Welch, M.J., Munday, P.L., 2017. Heritability of behavioural tolerance to high $CO_2$ in a coral reef fish is masked by nonadaptive phenotypic plasticity. Evol. Appl. 10 (7), 682–693.

Welch, M.J., Watson, S.A., Welsh, J.Q., McCormick, M.I., Munday, P.L., 2014. Effects of elevated $CO_2$ on fish behaviour undiminished by transgenerational acclimation. Nat. Clim. Chang. 4 (12), 1086–1089.

Welicky, R.L., Parkyn, D.C., Sikkel, P.C., 2018. Host-dependent differences in measures of condition associated with *Anilocra* spp. parasitism in two coral reef fishes. Environ. Biol. Fishes 101, 1223–1234. https://doi.org/10.1007/s10641-018-0770-y.

Wendelaar Bonga, S.E., 1997. The stress response in fish. Physiol. Rev. 77, 591–625.

Wenger, A.S., McCormick, M.I., McLeod, I.M., Jones, G.P., 2013. Suspended sediment alters predator–prey interactions between two coral reef fishes. Coral Reefs 32 (2), 369–374.

Wise, G., Mulvey, J.M., Renshaw, G.M.C., 1998. Hypoxia tolerance in the epaulette shark (*H. ocellatum*). J. Exp. Zool. 281, 1–5. https://doi.org/10.1002/(SICI)1097-010X(19980501)281:1<1::AID-JEZ1>3.0.CO;2-S.

Wong, C.K., Pak, I.A.P., Jiang Liu, X., 2013. Gill damage to juvenile orange-spotted grouper *Epinephelus coioides* following exposure to suspended sediments. Aquacult. Res. 44, 1685–1695.

Wood, C.M., 2012. An introduction to metals in fish physiology and toxicology: basic principles. In: Wood, C.M., Farrell, A.P., Brauner, C.J. (Eds.), Fish Physiology: Homeostasis and Toxicology of Essential Metals. Academic Press/Elsevier, San Diego, CA, pp. 1–51. https://doi.org/10.1016/S1546-5098(11)31001-1.

Wright, K.J., et al., 2005. Auditory and olfactory abilities of pre-settlement larvae and post-settlement juveniles of a coral reef damselfish (Pisces: Pomacentridae). Mar. Biol. 147, 1425–1434. https://doi.org/10.1007/s00227-005-0028-z.

Wright, K.J., Higgs, D.M., Leis, J.M., 2011. Ontogenetic and interspecific variation in hearing ability in marine fish larvae. Mar. Ecol. Prog. Ser. 424, 1–13. https://doi.org/10.3354/meps09004.

Yong, R.Q.Y., et al., 2013. The ghost of parasites past: eggs of the blood fluke *Cardicola chaetodontis* (Aporocotylidae) trapped in the heart and gills of butterflyfishes (Perciformes: Chaetodontidae) of the great barrier reef. Parasitology 140, 1186–1194. https://doi.org/10.1017/S0031182013000681.

Yufera, M., et al., 2019. Effect of increased rearing temperature on digestive function in cobia early juvenile. Comp. Biochem. Physiol. A Mol. Integr. Physiol. 230, 71–80. https://doi.org/10.1016/j.cbpa.2019.01.007.

Chapter 8

# Restoration physiology of fishes: Frontiers old and new for aquatic restoration

**Katherine K. Strailey**[a] **and Cory D. Suski**[b,*]
[a]*Program in Ecology, Evolution, and Conservation Biology, University of Illinois at Urbana-Champaign, Champaign, IL, United States*
[b]*Department of Natural Resources and Environmental Sciences, University of Illinois at Urbana-Champaign, Champaign, IL, United States*
[*]*Corresponding author: e-mail: suski@illinois.edu*

## Chapter Outline



Aquatic ecosystems have been extensively modified by human activity. The resulting degradation of aquatic habitats has led to devastating declines in fish biodiversity, with freshwater-dependent species suffering disproportionately. Restoration has become an increasingly popular strategy aimed at remedying the degradation of aquatic systems and reversing declines in fish biodiversity. Billions of dollars are spent on aquatic restoration each year within the United States alone, and yet the majority of restoration projects do not achieve success in reaching the goals of mitigating degradation and halting fish biodiversity loss. Restoration monitoring efforts typically rely on population- and community-level metrics that cannot respond to restoration on the short temporal

Fish Physiology, Vol. 39B. https://doi.org/10.1016/bs.fp.2022.04.014

scales that projects are usually monitored (<2 years), making it difficult to evaluate success or garner lessons that can be applied to future projects. Physiology is able to "fill in the gaps" that traditional monitoring cannot as it is mechanistic and can respond to environmental changes across multiple scales, particularly within short timeframes. Physiology can be thought of as a "filter," whereby any change within the environment must first affect individual fish at the physiological level before it can affect fish at the population or community level. We discuss this link between physiological responses and population-level effects and ways in that physiology might be integrated into the restoration process. We propose that the integration of physiology into restoration monitoring is essential to improve the success of restoration projects for fish.

# 1 The "Anthropocene"

The term *"Anthropocene"* refers to the most recent epoch in the Earth's history, starting around the late 18th century, denoting a time when humans began to exert a pronounced impact on the planet (Crutzen, 2006; Crutzen and Stoermer 2000). The Anthropocene has been characterized by environmental challenges that include the release of greenhouse gasses into the atmosphere, a decrease in forested areas, the proliferation of dams, pollution, and an escalation in the consumption of freshwater (Bruns and Frick, 2014; Crutzen, 2006). These environmental alterations have been responsible for an alarming decline in biodiversity, defined as the variation of life ranging from genes to species to functional traits (Cardinale et al., 2012), across several biomes (Díaz et al., 2019; Reid et al., 2019). This decline is much more pronounced in freshwater environments than in marine or terrestrial environments (Albert et al., 2021; Reid et al., 2019; Strayer and Dudgeon, 2010). While the loss of individual species can translate into changes at the ecosystem and/or community level, it can also result in declines in functional interactions within an ecosystem (Díaz et al., 2019; Geist, 2011; Naeem, 2016) that can lead to negative consequences for human health and well-being through the loss of critical ecosystem services (Reid et al., 2005). Thus, the concerning decline in biodiversity that has occurred during the Anthropocene has the potential to cause negative consequences for human well-being due to the loss of ecosystem services.

## 1.1 Fish in the Anthropocene

### *1.1.1 Freshwater systems*

The alarming recent decline in freshwater fish diversity is among the most concerning impacts of the Anthropocene. Despite containing only 0.02% of habitable volume and around 0.01% of the world's water (Dawson, 2012; Gleick, 2011), freshwater habitats contain more than 40% of fish species on Earth (Lundberg et al., 2000). Freshwater fishes provide a number of critical ecosystem services that include regulating food webs, providing nutrients across ecosystem boundaries, serving as bioindicators, along with food and cultural services (Aylward et al., 2005; Cowx and Portocarrero Aya, 2011). Likely due in part to their important role as a natural resource, freshwater fishes are the second most threatened vertebrate group on the planet after amphibians (Bruton, 1995). The IUCN Red List considers approximately 31% of freshwater and freshwater-dependent fish to be either critically endangered, endangered or vulnerable (after accounting for extinct species and those that are data deficient) (IUCN, 2021), a number higher than many other taxa (Darwall and Freyhof, 2016). During the period from 1989 to 2008, the number of imperiled fish species in North America grew by 92% with only 6% improving in their conservation status ( Jelks et al., 2008), and, over roughly this same time period, the number of freshwater fish extinctions grew by 25% (Burkhead, 2012). The loss of freshwater fish biodiversity stems primarily from large-scale anthropogenic alterations of aquatic systems and surrounding landscapes. Expanding human populations have altered aquatic landscapes for the purposes of obtaining food and generating hydropower, and freshwaters near or within developed catchments receive inputs of pollutants from the surrounding landscape, and such problems can easily spread to due to the high degree of connectivity within aquatic landscapes (Darwall and Freyhof, 2016; Dudgeon, 2019; Vörösmarty et al., 2010). Other causes of biodiversity loss in freshwater include invasion by non-native species, disease, climate change, pollution, overexploitation, and diversion and withdrawal of water. Overall, freshwater fishes around the globe have suffered tremendous declines, which may translate to negative consequences for human health and well-being, and these declines are largely occurring because freshwater ecosystems have been highly influenced and modified by human activity.

Habitat loss and degradation are among the most impactful of factors responsible for the loss of biodiversity in freshwaters. The term "habitat" as it relates to fisheries and conservation is highly variable, with both legal and biological considerations (Naiman and Latterell, 2005; Orth and White, 1993; Rosenfeld and Hatfield, 2006), as well as consumable and non-consumable components (Hayes et al., 1996). For this chapter, we will use a definition of habitat that draws from Magnuson–Stevens Fishery Conservation and Management Act (Public Law 94–265, 16 U.S.C. 1801–1891(d)), which is, "waters and substrate necessary to fish for spawning, breeding, feeding or growth to maturity." By extension, the term "habitat loss" is therefore meant to encompass a number of different "types" of degradation that can result in impairment to individuals or declines in fish populations, including (1) the inability of an organism to *access* quality habitat (e.g., dams that prevent movement of organisms or gametes), (2) the *destruction* of habitat (e.g., siltation from agriculture that destroys critical spawning areas), or

(3) the *degradation or modification* of habitat to the point that it is no longer able to facilitate an organism carrying out essential life functions (e.g., alteration of riparian habitat that increases water temperature beyond the thermal limits of a fish species). When this definition of habitat loss is considered, the impacts of this stressor on freshwater biodiversity are staggering (Darwall and Freyhof, 2016; Dudgeon, 2019; Jelks et al., 2008). For example, Jelks et al. (2008) indicate that 92% of threatened fishes in North America have been impacted by habitat loss, and 30–50% of the freshwater habitat assessed in the United States in 2000 failed to fully support water quality standards evaluated by the Environmental Protection Agency, with habitat-related factors such as runoff and siltation listed as key drivers of impairment (USEPA, 2000). In Europe, where more than one-third of all freshwater fish species are considered threatened, 56% of freshwater habitats were considered to be "unfavorable-inadequate" and 16% were considered "unfavorable-bad" by the European Union Agency (European Union Agency, 2015; Gozlan et al., 2019). These issues are not specific to any one country or region. Dudgeon et al. (2006) listed habitat-related factors (e.g., flow modification, habitat degradation, pollution) as a major threats driving global loss of freshwater biodiversity, Arthington et al. (2016) identified habitat loss and pollution as key factors threatening freshwater fishes globally, and Cowx (2002) showed that habitat-related challenges are among the most important threats facing freshwater fishes. Though a number of factors are responsible for declines in freshwater fish populations, habitat loss is among the most important.

### *1.1.2 Marine systems*

Within this chapter, we will focus largely on freshwater ecosystems (particularly flowing water ecosystems) as freshwater fish as a whole are more imperiled than marine fishes (Arthington et al., 2016). However, fish declines due to human activity are not unique to freshwater systems as marine ecosystems face challenges that are both different from, and similar to, those facing freshwater ecosystems. As in freshwater systems, habitat loss is a major concern in marine systems, where nearshore and shallow water areas are most sensitive to and most at risk of habitat loss (Arthington et al., 2016). Issues in nearshore zones often tend to originate from the surrounding landscapes, and threats to fish and their habitats often arise with coastal development, whether it be commercial, residential, or industrial (Arthington et al., 2016). For example, nearly one-third of the historical extent of mangroves were lost in less than a century (Alongi, 2002), a loss that is predicted to have serious consequences for fish biodiversity and fishery catches, due in part to the essential nursery habitat they provide (Blaber, 2007). Between 1879 and 2006, seagrass beds declined in area by nearly 30%, which is similarly expected to have consequences for fish biodiversity and fishery catch rates (Blandon and Zu Ermgassen, 2014; Orth et al., 2006). In the Caribbean, reefs have lost an estimated 80% of coral cover and complexity, resulting in decreases in fish abundance, and up to a third of reef-building corals species have declined to the point of being considered "Threatened" by the IUCN Red List (Alvarez-Filip et al., 2019; Carpenter et al., 2008; Gardner et al., 2005; Paddack et al., 2009). Coupled with these habitat-related challenges, marine fisheries are also experiencing overfishing (Arthington et al., 2016; Essington et al., 2006; Pauly and Palomares, 2005) and ocean warming (Ullah et al., 2018). Much like freshwater ecosystems, human activity threatens the biodiversity and abundance of marine fishes, and habitat loss is a particularly strong driver of this.

#### *1.1.3 Chapter overview*

The restoration of aquatic habitats following declines has an unfortunate history of being unsuccessful despite extensive efforts, and this lack of success been driven, in part, by shortcomings in traditional restoration science and monitoring practices. As such, the goal of this chapter is to demonstrate how physiological tools can be integrated into habitat restoration practices, and how these tools can bolster the success of future restoration projects. The United Nations General Assembly has declared 2021–30 to be the "UN Decade on Ecosystem Restoration," and we believe that novel tools and approaches to restoration will be imperative in truly embracing restoration this decade (United Nations General Assembly, 2019; Waltham et al., 2020). In this chapter, we aim to (1) provide an overview of aquatic restoration and explain why aquatic restoration often fails to achieve its goals, (2) detail why physiological tools can fill in some shortcomings of traditional monitoring, (3) describe the theory underlying the use of physiology in restoration, (4) explore example scenarios in which physiological tools can be integrated into the restoration process, and (5) share examples of successful uses of physiological tools in aquatic restoration. Together, these sections will provide an improved understanding and a path forward to integrate physiological tools into aquatic restoration to mitigate and prevent further declines in aquatic biodiversity from human impacts.

## 2 Restoration: The remedy for habitat degradation?

### 2.1 Theories, processes, and practices of restoration in the aquatic world

Restoration has emerged as a critical tool to reverse both habitat loss and degradation, as well as the associated declines in biodiversity and ecosystem services. The term "restoration" has been widely applied and many definitions exist (Table 1), but, in general, restoration encompasses a broad range of techniques intended to remedy environmental degradation. It is worth noting that the term is often coupled with other modifiers, including "ecological restoration," "habitat restoration," or simply "restoration." We distinguish between these terms because the aims of the three are not necessarily the same. Ecological restoration emphasizes the recovery of whole ecosystems and often ecosystem processes (Higgs et al., 2018; Hobbs and Harris, 2001; McDonald et al., 2016). Habitat restoration, not necessarily in contrast, tends

**TABLE 1 Definitions of the term "restoration" as drawn from several sources.**

| Definition | Source |
|---|---|
| The process of assisting the recovery of an ecosystem that has been degraded, damaged, or destroyed | SER (Society for Ecological Restoration) International Science and Policy Working Group (2004) |
| The act of assisting in the recovery of ecosystems that have been degraded or destroyed, as well as conserving the ecosystems that are still intact | United Nations General Assembly (2019) |
| The remediation and recovery of key habitats and recovery of natural processes | Cowx and Portocarrero Aya (2011) |
| The compensation for environmental damage or loss of habitat through replacement of functions, values, and/or acreage | Race and Fonseca (1996) |
| The process of repairing damage caused by humans to the diversity and dynamics of indigenous ecosystems | Jackson et al. (1995) |
| The act of attempting to bring an anthropogenically-degraded system (or part of a system) to a less-degraded state through human intervention | Current document |

to emphasize the physical presence and qualities of habitat, and is often focused on recovering habitat for specific biota (Beechie et al., 2013; Miller and Hobbs, 2016; Roni et al., 2008). Restoration, used on its own, may refer to one of these, but can also refer to any efforts to remediate anthropogenic degradation, regardless of its intention or methodology (Miller and Kochel, 2010; Shields et al., 2003). The common thread among these terms is that restoration specifically targets systems that have been damaged by human activity (Allison, 2012). In this chapter, we will simply use the term "restoration" and employ the following definition: restoration is the act of attempting to bring an anthropogenically-degraded system (or parts of a system) to a less-degraded state through human intervention. In using this definition, we seek to encompass both restoration activities conducted to benefit specific taxa, those that target physical habitat qualities, as well as activities aimed at any level of the recovery of ecosystem and habitat function.

Just as the definitions of restoration can vary, the goals of restoration activities are also quite diverse. The field of restoration ecology, for example, focuses on achieving ecological and biological goals such as the improvement of previously-impaired ecosystem functions (Hobbs and Harris, 2001).

Outside of ecology, restoration often focuses on physical goals, commonly with the assumption that physical restoration of a system will drive the restoration of biota and ecological function (Manners and Doyle, 2008; Miller and Kochel, 2010); this is often referred to as the "Field of Dreams" hypothesis: if you build it, they will come (Palmer et al., 1997). The underlying motivations of restoration are often socially or economically driven (for example, the restoration of spawning habitat for a fish species of economic or recreational importance), but ultimately, restoration is most commonly aimed at evoking changes within the system of interest itself. Regardless of the specific focus, the desired outcome typically is to return a disturbed or degraded system to a condition similar to some pre-disturbance/pre-degradation state (Swetnam et al., 1999; White and Walker, 1997). In North and South America, restoration often targets conditions pre-European settlement (Rhoads, 2020). Alternatively, restoration efforts may opt to replicate a similar, minimally-impacted location nearby that exemplifies how the restored system is expected to look and function (i.e., a reference site). Not all stakeholders agree that restoration should aim to replicate a pre-disturbance state, particularly considering that a reference site may not be available, and that restoring a location to its previous state may be impossible, especially in situations where human activity is likely to continue impacting a site even after it has been restored (Beechie et al., 2010; Dufour and Piegay, 2009; Palmer et al., 2014). In such cases, it has been suggested that projects should instead restore lost ecosystem processes and services that benefit humans or focus on achieving a "historical range of variability" (Morgan et al., 1994), acknowledging that the physical, biological, and ecological characteristics of a system vary over time (Landres et al., 1999; Morgan et al., 1994; Palmer et al., 2003). This is of key importance to freshwater ecosystems, particularly streams and rivers, systems that are strongly influenced by temporal variations in temperature and precipitation (Schumm, 2005; Shorthouse and Arnell, 1999). Clearly, there are many opinions regarding what restoration activities should aim to achieve and regarding what a restored system should look like.

Restoration has grown to become a major global industry, with billions of dollars invested annually into remedying habitat degradation, and this number is expected to continue growing (Malakoff, 2012; Woodworth, 2006). A cursory Web of Science search using the following search string TS=(habitat OR ecological AND river OR stream OR aquatic OR coral OR marine OR coastal AND restor* OR rehab* OR mitigat*) shows that close to 25,000 papers related to aquatic restoration were published in the year 2000; just over two decades later, in 2021, over 156,000 such papers were published (Fig. 1). In the United States alone, it is estimated that practitioners spend upwards of $25 billion USD per year on restoration, although costs are often not published making a definitive assessment challenging (BenDor et al., 2015; Kimball et al., 2015). Restoration costs are often much higher for aquatic systems than for terrestrial systems, likely due in part to difficulties posed by the necessity of working in and around water. Freshwater restoration has been

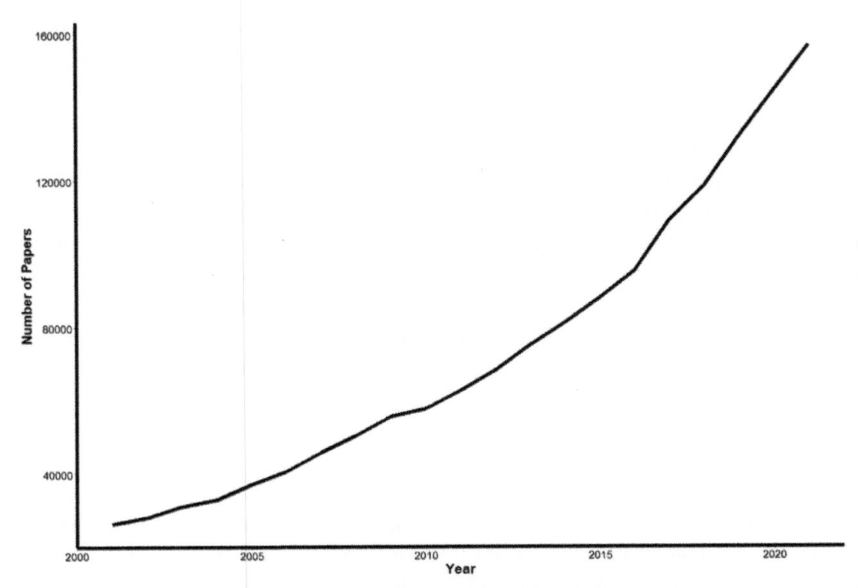


**FIG. 1** The results of a Web of Science search using the string TS = (habitat OR ecological AND river OR stream OR aquatic OR coral OR marine OR coastal AND restor* OR rehab* OR mitigat*) show that, over the past two decades, the restoration of aquatic systems has become increasingly common.

reported to have a median cost of $16,000/ha, and costs for coastal marine systems are even higher at $80,000/ha, while costs for terrestrial ecosystems range from $1000/ha to $7000/ha (De Groot et al., 2013). A review by Bayraktarov et al. (2019) found $400,000 USD was the median cost of a coral reef restoration project, but that project costs could soar as high as $4,000,000 USD. Bernhardt et al. (2007) suggested an "idealized restoration process" that should be followed, which simply required that a project have a clearly defined goal, objective criteria for determining success, and analytical evaluations of success based on these criteria. The median cost of river restoration projects that had followed this process and that had been ecologically successful was $580,000 USD, while the median project cost of all river restoration projects surveyed was calculated to less than half that, at $150,000 USD (Bernhardt et al., 2007). These findings suggest that, not only is aquatic restoration a large financial investment, but also that greater success often requires greater financial investment.

## 2.2 Challenges with aquatic restoration

Though a vast quantity of money and resources are invested into aquatic restoration, a concerning proportion of restoration projects fail to achieve their biological goals and ultimately have little to no positive impact on fish or other aquatic biota. This is especially true in the field of river and stream restoration, in which startlingly low levels of restoration effectiveness have been reported. Many restoration techniques may induce desired changes in the physical and chemical quality of a restored area (e.g., bank stabilization, riparian enhancement) while having little effect on fish populations and communities (Baldigo et al., 2010; Lepori et al., 2005; Palmer et al., 2010; Rosi-Marshall et al., 2006). For example, Alexander and Allan (2007), found that only 11% of aquatic restoration projects reached their desired status for a particular ecological indicator, while Palmer et al. (2010) synthesized the results of 78 separate aquatic restoration activities and found that 84% of projects had no significant effect on taxa richness (the target for research projects). When project managers do report improvements in biological metrics following restoration, they are often slight or short-lived (Palmer et al., 2010; Roni et al., 2008; Thomas et al., 2015). Perhaps more troubling than the lack of success in restoration is the way in which success is determined. There is no universal means of evaluating success for restoration activities, even within an ecosystem type or across similar projects, and it is not uncommon for success to be evaluated based on practitioner perception rather than on a quantifiable metric (Zedler, 2007). This problem is not unique to aquatic restoration or to a specific region (Bernhardt and Palmer, 2011; Nilsson et al., 2016; Suding, 2011). Though empirical evaluations are increasing in frequency, qualitative assessments of success remain all too common (Palmer et al., 2010; Wortley et al., 2013). For example, while investigating the effectiveness of river restoration projects in Germany, Jähnig et al. (2011) indicated that nearly 40% of respondents considered their projects to be successful based on a "gut feeling" alone, and a similar evaluation of French river restoration projects discovered that projects with the least rigorous methods of evaluating success tended to report the highest levels of success (Morandi et al., 2014). Clearly, there are significant challenges underlying the implementation and evaluation of many restoration efforts, and these issues in turn likely contribute to the low success rates of many projects.

We believe there are two major reasons, both related to how success is evaluated, that contribute to why restorative actions commonly do not reach their intended biological outcomes. First, monitoring following the completion of a restoration project is often insufficient, making it impossible to ascertain whether a project has reached its goals. To determine if a project has been successful, long-term, targeted monitoring of biota and habitats is necessary. A large-scale assessment of over 37,000 river restoration efforts in the United States revealed that only 10% of projects indicated monitoring ever occurred, and that, even in these cases, the monitoring performed was not designed to effectively evaluate success. Similarly, in Germany, a survey of 26 projects showed that only 45% received any level of monitoring ( Jähnig et al., 2011; Palmer et al., 2005) These widespread inadequacies in project monitoring stem primarily from funding challenges. Though billions are invested into restoration projects, the costs of traditional monitoring are often too great, and thus monitoring is often not prioritized (Holl and Howarth, 2000; Rohr et al., 2018). In Washington State, USA, a survey of 94 stream restoration project managers found that 34% considered a lack of funding to be the primary barrier to project monitoring (Bash and Ryan, 2002). A similar synthesis by Follstad Shah et al. (2007) reported that over half of project managers in the US Southwest indicated that a lack of funding constrained monitoring efforts, an unsurprising result given that they also found that monitoring costs could account for as much as 61% of total project cost. When monitoring does occur, it rarely is long enough to detect changes within communities. Boström-Einarsson et al. (2020) found that 60% of surveyed coral reef projects were monitored for less than 18 months post completion, an amount of time that is likely insufficient to project long-term success (Bayraktarov et al., 2019). Vehanen et al. (2010) showed that a monitoring period of 3 years, relatively long in comparison to field activities for most projects, showed no recovery of brown trout (*Salmo trutta*) in a restored stream; a follow-up 12 years post-project detected positive responses from the target species (Louhi et al., 2016). However, even long-term monitoring may not reveal changes in metrics of interest (e.g., species abundance, community composition, species diversity), and improvement with increased restoration project age is not guaranteed (Leps et al., 2016). Together, it may be difficult or even impossible to assess whether or not a project has succeeded, as relevant data may not be collected or may be missed due to infrequent monitoring, and lack of post-project data also makes it challenging to identify successful or unsuccessful restoration activities that can guide future work and increase success. Unfortunately, given that the financial needs of restoration projects often exceed available funding (Rohr et al., 2018), it is unlikely that traditional monitoring can consistently be funded for the length of time and at the scale needed to guide and assess restoration practices, and so under monitoring is likely to continue.

A second reason that restoration projects fail to achieve success is that the metrics selected to evaluate restoration are often not well-suited to demonstrate positive changes within the ecosystem being restored. Common metrics of interest to restoration practitioners that could be quantified in post-project monitoring, such as the abundance of a target species or species assemblage, often take years or even decades to respond following restoration (Bayraktarov et al., 2019; Geist and Hawkins, 2016; Louhi et al., 2016), and may be entirely incapable of reflecting alterations in habitat quality or changes within the biotic community (Geist and Hawkins, 2016; Palmer et al., 2010; Rose, 2000; Wohl et al., 2005). A meta-analysis of 26 studies examining stream restoration found that, while all were interested in biotic responses to increased habitat heterogeneity, half measured the alteration of the habitat itself, not biotic responses (Rubin et al., 2017). In marine habitats, where ecosystems are often reliant on a key foundational organism (i.e., coral, mangroves, seagrasses), the most common metric of success was the survival rate of a foundational organism following reintroduction or replanting, though such a metric says little regarding the status of the community as a whole (Hein et al., 2017; Powers et al., 2009; Ruiz-Jaen and Aide, 2005). For example, assessments of coral reef restoration success overwhelmingly focused on the survival and growth of transplanted corals, with over 80% of such projects using these metrics (Hein et al., 2017), and restoration of oyster reefs in the Gulf of Mexico similarly emphasized metrics such as presence and recruitment of oysters (Powers et al., 2009). In the freshwater realm, species- and community-level metrics such as abundance, richness, and community assemblage are commonly employed to measure restoration success (Fausch et al., 1990), but may not show clear responses to restoration efforts or reflect the status of an area as "degraded" or "restored" (Moerke and Lamberti, 2003; Zampella and Bunnell, 1998). Community metrics are also influenced by factors that may have nothing to do with changes in habitat quality. For example, community composition data may be reported incorrectly if a rare or cryptic species is missed during surveys (MacKenzie, 2005). Following restoration, improvements in community-level metrics are dependent on the availability of nearby sources of organisms that are available to recolonize the restored area, and improvements in habitat quality may not be reflected in community metrics (Stoll et al., 2014; Sundermann et al., 2011; Tonkin et al., 2014). Links between population-level metrics and habitat quality are similarly tenuous, and it is often unclear as to whether restoration increases fish abundance or simply concentrates fish (Kail et al., 2015; Lavelle et al., 2021). Population- and community-level metrics are also not mechanistic and cannot directly explain why biotic changes have occurred; comparisons at this level can only be made before and after restoration has been implemented, a difficult task given the lack of both pre- and post-project monitoring conducted for most restoration projects. Given the inconsistencies in restoration success, it is evident that traditional methods of restoration implementation and assessment are not working as well as needed, and that novel strategies will be essential to help increase restoration success and complement traditional techniques as we move forward into the Decade on Restoration.

## 3 Physiology, environmental stressors, and restoration

Physiological techniques provide a suite of tools to guide restoration and supplement some of the challenges associated with traditional, field-based assessment methods. Unlike the population and community metrics typically targeted in traditional monitoring, many physiological metrics respond quickly and are sensitive to environmental stressors, and thus can be used for adaptive management throughout the restoration process (Cooke and Suski, 2008). Physiological measures have high specificity to specific threats, and thus can mechanistically link fish responses to specific changes within their environment as a result of restoration, and do so across both short and long time scales, and at both small and large spatial scales (Adams and Ham, 2011; Cooke and Suski, 2008). As such, we believe that it is vital that physiological tools be employed to help guide the restoration process as well as complement more traditional monitoring practices.

## 3.1 Linking restoration and physiology

Like all living organisms, individual fish are complex organisms that contain a hierarchical network of components that range from molecules, to genes to cells to individuals; individuals can then be grouped into populations, and several populations can be integrated into communities and ecosystems (Dobzhansky, 1964; Nielsen, 2000; Simon, 1962). Within this hierarchical framework, alterations or perturbations at one level result in changes at the next higher level, along with a feedback mechanism to control conditions at a lower level, such that there is vertical integration and coordination of responses across levels of organization (Jørgensen and Nielsen, 2013) (Fig. 2). Physiology therefore acts as a link between an individual and its environment, which, in turn, functions as a conduit through which environmental stressors such as habitat loss translate to changes in population-level parameters such as abundance and distribution (Ricklefs and Wikelski, 2002; Weissburg et al., 2005). More specifically, the physiological status of

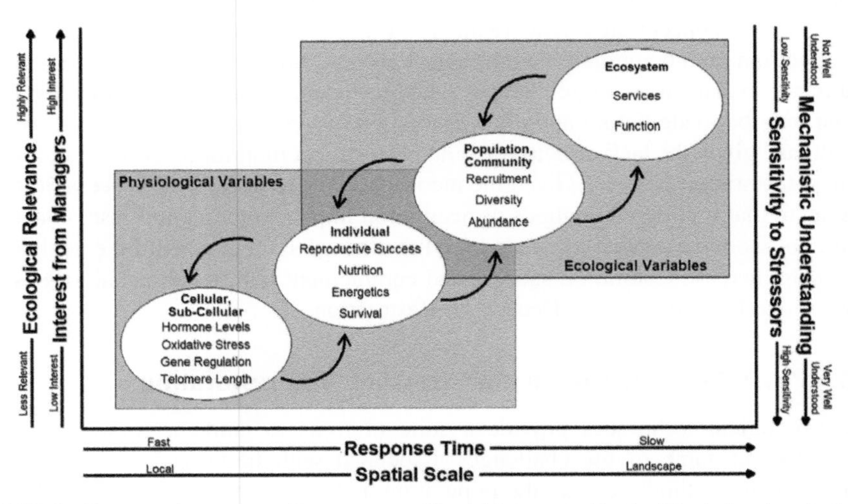


**FIG. 2** Conceptual overview of how stressors (X axis) intersect with physiological and ecological variables. Stressors can vary in space and time, and the response to those stressors can respond over small to large spatial scales. Traditionally, restoration practitioners have been interested in response variables that change over long time periods, across large spatial scales, and that are not sensitive to stressors (ecological variables). Physiological responses, while traditionally having less ecological relevance and interest from managers, are more sensitive to environmental stressors, provide a mechanistic understanding of stressors and respond at smaller temporal and spatial scales.

an individual (e.g., nutritional condition, reproductive potential, disease state, level of oxidative damage, etc.) is an integration of responses to environmental factors such as temperature, oxygen levels, food resources, and other factors (Ricklefs and Wikelski, 2002). Environmental factors that alter molecular/cellular responses, in turn, can result in changes manifested at the level of the individual that have the potential to be reflected at the population or community level, including emigration, impaired performance, reduced reproduction, or even death (Calow, 1989; Ricklefs and Wikelski, 2002) (Fig. 3). Seminal work by Fry (1947), for example, showed that a range of environmental factors (e.g., temperature, dissolved oxygen) can influence the metabolism and behavioral responses of fishes, largely through physiological mechanisms, while Huey (1991) argued that habitat selection of animals can influence physiological properties, behavior and short-term performance, that, in turn, influence fitness. Similarly, Ricklefs and Wikelski (2002) proposed the "physiology/life-history nexus" where the life history responses of species are influenced by physiological factors (stress hormone concentrations, tolerance limits) that are heavily influenced by environmental conditions such as food resources, diseases, and predation, that can alter behavior, performance, and fitness. These relationships were further illustrated by Hayes et al. (1996) who showed that habitat features can be linked to changes in fish growth, survival and ultimately to reproduction, highlighting links between levels of biological organization. As such, due to the hierarchical relationship between levels of biological organization (Fig. 2), quantifying the physiological responses of individuals is a key component of understanding the responses of populations to environmental stressors such as habitat loss or restoration (Fig. 3).

One of the central tenets linking physiology to restoration is that physiological tools can be used to define "health" and condition of a fish captured in the wild. There are a host of indices, metrics and biomarkers that can be used to define the health and condition of fish, many of which can be obtained

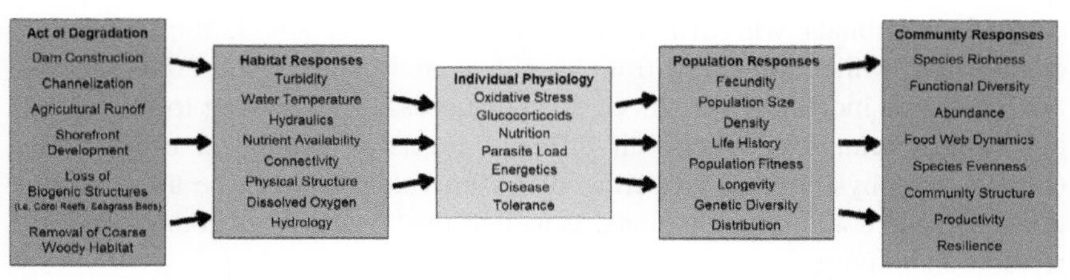


**FIG. 3** Conceptual model showing how characteristics of an individual's physiology link environmental degradation and community-level responses. Physiology can be thought of as a filter. Environmental degradation (or restoration) generates environmental stressors that first impact the physiology of individual fish. Then, due to the hierarchical organization of life, these physiological impacts accumulate to manifest in changes to fish populations or communities. This gives physiological responses the potential to predict community-level outcomes.

through rapid, non-lethal sampling in the field (e.g., Madliger et al., 2018; Table 1). For many years, managers, conservation practitioners and ecologists have relied on biological criteria (e.g., data on the presence/absence/abundance of fishes, presence of sensitive vs. tolerant species, often compared to a reference site) to infer the health and condition of an ecosystem. The use of such biological criteria has been common for several reasons: the known relationship between habitat characteristics and species abundances described above; the development of standardized collection techniques; and the perceived links between fish communities and environmental perturbations (Canning and Death, 2019; Karr, 1999; Simon, 2000; Simon and Evans, 2017). The classic example of using fish communities to infer the health of an ecosystem is the Index of Biotic Integrity (IBI), developed by Karr (1981) to define the health of watersheds in wadeable streams of the Midwestern United States. Owing to the ease of implementation and its ability to define the health of aquatic resources (Fausch et al., 1984; Simon and Evans, 2017), the IBI developed for fishes in the Midwestern USA has been expanded and modified to quantify the health of lentic and lotic waterbodies around the world, and modified to use on populations of invertebrates or macrophytes as part of the assessment (Canning and Death, 2019).

Just as an IBI can use community composition data to infer "good" vs. "poor" habitat for a watershed, it is possible to extend the impacts of landscape-level perturbations to the level of physiological responses of fishes and use physiology to infer the health and condition of an individual fish and quantify habitat quality. The environment in which an animal resides is known to result in physiological responses (e.g., Ricklefs and Wikelski, 2002), and Jeffrey et al. (2015) suggested that habitat-related disturbances for fishes result from a reduction in habitat quality, decreased habitat connectivity, or reduction in resources. Habitat-related impairments have the potential to induce a host of different physiological responses in fishes that can be quantified with different response variables (Schreck and Tort, 2016). Building upon this concept, Lennox et al. (2018) suggested that fish physiology spans a continuum from "healthy" to "sick," such that individuals from "poor" (degraded) habitats will display altered homeostasis, stress, impairment or degradation, while individuals from "healthy" environments (e.g., quality habitat) will show indices of health, vigor, and success. Thus, owing to the nested, hierarchical relationship between different levels of biological organization (Fig. 2), the physiological properties of an individual fish can be used to infer the health and condition of habitats, as well as the "quality" of habitat following restoration activity.

The basis for the relationship between habitat quality and physiology resides within the framework of how fishes respond to environmental stressors (Schreck and Tort, 2016). Physiology can be thought of as a filter through which all stressors must pass before they are able to affect higher levels of organization: fish first respond to stressors at the physiological level, the cumulative responses of many individuals manifests in population-level responses, and then community-level responses (Fig. 3). Over the past few decades there have been a number of influential papers that have shaped our understanding of the stress response in fishes (e.g., Adams, 2002; Barton, 2002; Mazeaud et al., 1977; Pickering, 1981; Selye, 1976; Wendelaar Bonga, 1997) culminating in the recent edited volume by Schreck and Tort (2016) dedicated exclusively to this topic. A comprehensive overview of the physiological responses of fishes to environmental perturbations exceeds the scope of this chapter, and we would refer the reader to the above resources for specifics and details on this topic, and our coverage of this topic will be cursory. Briefly, the term "stress" as used in this chapter refers to the physiological cascade of events that result when a fish attempts to reestablish homeostasis and resist death in the face of external challenges (Schreck and Tort, 2016). When confronted with an environmental challenge, fish experience a suite of physiological changes including (1) the secretion of corticosteroid hormones to liberate energy intended to overcome, escape or resist the challenge (the primary stress response), and (2) alterations to internal physiology to liberate energy and increase oxygen distribution (the secondary stress response). Prolonged activation of the stress response can lead to negative, maladaptive consequences from the tertiary stress response, such as reduced fecundity, impaired growth, and compromised immune function, that can ultimately lead to death (Schreck and Tort, 2016). Almost all environmental factors have the potential to induce a stress response (e.g., temperature, salinity, oxygen, light, food availability) (Barton, 2002), and the magnitude and duration of a stress response depends on stressor severity and length, coupled with the genetic composition of an individual fish, age, and epigenetic effects, along with prior exposure to a particular stressor (Giesing et al., 2010; Schreck and Tort, 2016). Overall, the physiological responses of fish to stressors can be linked with changes in habitat quality, and thus physiological tools have the potential to provide a broad range of insights into how fish and fish populations are interacting with their environment.

A number of synthesis articles have detailed the variety of responses of fishes to external stressors, highlighting the countless physiological tools and techniques that can be used to quantify the responses of fishes to environmental perturbations. Barton (2002), for example, listed a range of physiological responses that are part of the primary, secondary, and tertiary stress responses, as well as their expected ranges, strengths, and weaknesses, that could be used to quantify stress in fish, while Sopinka et al. (2016) compiled a suite of tools to quantify stress in fish, ranging from molecular, to hormonal to individual-based metrics. Rice and Arkoosh (2002), as well as Greeley (2002), showed how immunological and reproductive tools (respectively) can be used to define environmental impacts on the tertiary stress response in fishes, while Madliger et al. (2018) developed a "conservation physiology toolbox" showcasing a range of tools used to assist with conservation of wild populations. Colin et al. (2016) highlighted a range of biomarkers that can be integrated into monitoring studies for fishes (and invertebrates), Bernos et al. (2020) showed how genomic tools can be utilized to aid with the conservation of fishes, while Chapman et al. (2021) described current tools and techniques for incorporating epidemiological metrics into field ecology for fishes. Owing to the already expansive literature cataloging the physiological tools that can quantify how fish respond to environmental or restorative conditions, we have elected to organize tools and techniques around specific restoration problems. We highlight where, when, and how physiological tools can be used to distinguish this section from previous studies that provide a more comprehensive overview of tools and techniques.

# 4 Integrating physiology into the restoration process

## 4.1 Stream restoration: A hypothetical case study

A scenario that can help demonstrate how physiological tools can be integrated into restoration relates to the remediation of riparian and in-stream habitat for a stream segment, with the hopes of improving fish populations (Martens et al., 2019; Opperman and Merenlender, 2004). This example was chosen as stream restoration is common around the world and is of interest to a wide range of disciplines (e.g., biologists, geomorphologists, restoration agencies, etc.) (Federal Interagency Stream Restoration Working Group, 1998; Roni et al., 2002; Wohl et al., 2015). Restoration practice varies across agency, habitat, location, and project goals, but a typical project will often fit within the framework of "adaptive management," which involves a cyclical process of (I) defining a problem, (II) identifying objectives, (III) evaluating a range of options to address the problem, (IV) implementing a solution, (V) monitoring, and (VI) re-visiting the initial problem/objectives (Allen et al., 2011; Kondolf and Micheli, 1995; Skidmore et al., 2012; Walters, 1986); this process may be paired with structured decision making to develop solutions and improve project outcomes (Lyons et al., 2008). The adaptive management approach has been praised as a way that practitioners can define a path to success despite the uncertainties related to management options, and as a way to learn more about the system being manipulated (Allen et al., 2011; Lyons et al., 2008). Following this framework, there are multiple points at which physiology could be integrated into the restoration process (Fig. 4).

I. Defining the problem

In our restoration scenario, established stream monitoring techniques (e.g., on-site electrofishing) could document declines in population- and community-level metrics such as biomass/abundance of fishes within the site, or the disappearance of sensitive/important fish, such as those important for the ecosystem or that are targeted by recreational anglers (however, fish may not always be the

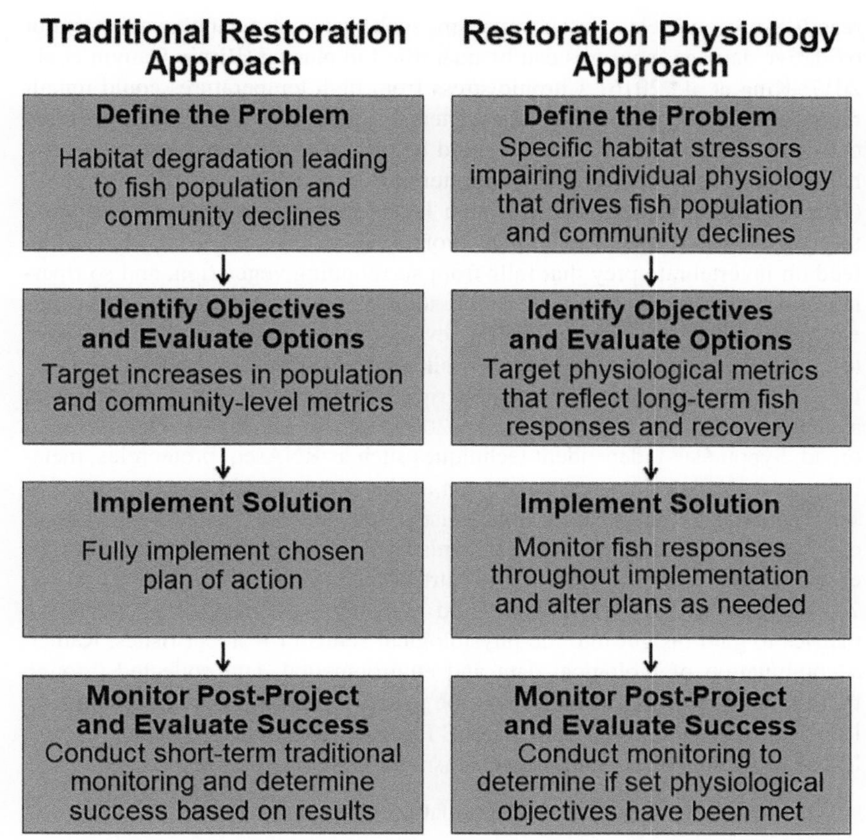


**FIG. 4** Adaptive management is a framework in which the restoration process is broken down into a series of steps. Here, we illustrate how the adaptive management framework might be applied taking either a traditional restoration approach or a restoration physiology approach. Physiology can provide new insight throughout all steps of the restoration process, from pre-project monitoring to implementation to post-project monitoring.

motivation or target for restoration activities). This level of monitoring may be sufficient to broadly define the problem at hand, but may be unable to identify the underlying cause of the problem due to the lack of direct mechanistic links between such metrics and environmental stressors. Physiological monitoring could be used to *refine* the problem, and identify specific stressors within the area of interest that could be targeted for restoration.

The degradation of instream and riparian habitat in this scenario could manifest in a host of physiological responses. For example, the loss of shading from riparian vegetation may increase water temperature, which can lead to chronic "stress" if these elevated temperatures exceed the thermal niche of resident species. Elevated temperatures can also influence the presence of reactive oxygen species in an organism, such that antioxidant capacity and/or oxidative damage measures can be quantified in plasma (Birnie-Gauvin et al., 2017; King et al., 2016). Chronic stress from high temperatures could impair reproductive outputs due to limited energy supply and/or extended elevation of hormones, and this hypothesis could be tested by quantifying reproductive hormones (e.g., testosterone, progesterone, estradiol as in Tucker et al. (2020)). Similarly, heat shock protein levels may increase as animals work to preserve the structure of critical proteins (Currie et al., 1999). Fish often feed on invertebrate prey that falls from surrounding vegetation, and so riparian destruction may remove this food source. A number of plasma constituents can reflect short-term feeding events or long-term nutritional status (e.g., total protein, total cholesterol, alkaline phosphatase), and thus can be used to test hypotheses related to how riparian degradation can influence food availability (Congleton and Wagner, 2006; Liss et al., 2014). Finally, more broad, hypothesis-independent techniques such as RNAseq, proteomics, metabolomics, or other data generated from analyses generated by next-generation sequencing can query many mechanisms relevant for defining individual-environment interactions and mechanisms of population declines (Bernos et al., 2020; Connon et al., 2018; Curtis-Quick et al., 2021; Jeffrey et al., 2019). Restoration practitioners would not need to collect all of the above metrics to gain insight into the physiological status of resident fishes. Rather, a combination of biological data and environmental data collected through traditional monitoring would ideally be paired to broadly identify what physiological metrics may be most useful. Physiological data could then be collected to identify the specific mechanism driving fish declines.

II. Identifying objectives and evaluating options to reach objectives

Once a problem has been determined, restoration practitioners can then identify objectives. Declines in metrics such as species abundance or biodiversity tend to be the "warning flag" that indicates a need for restoration, and, in turn, increasing metrics like fish abundance or biodiversity is a common objective for restoration projects (Simon, 2000; Simon and Evans, 2017). However, these targeted increases often do not come to fruition following restoration projects, and thus it can be difficult to know what kind of impacts, if any, a project may have had (Palmer et al., 2010; Rosi-Marshall et al., 2006; Wohl et al., 2005). Instead of solely identifying population- and community-level responses for their biological objectives, restoration practitioners could additionally identify physiological objectives such as increased nutritional status, elevated levels of reproductive hormones, or reduced stress levels in resident fish. Such objectives could be targeted more directly through restoration action, and, given the hierarchical manner in which organisms are linked (Fig. 3), achieving these physiological objectives would contribute to achieving objectives for fish populations and communities as well (e.g., Fig. 1 in Jeffrey et al., 2015) (see Chapter 1, Volume 39B: Yanagitsuru et al., 2022).

Without physiological data, the exact cause of a problem that restoration seeks to address may be unknown. As a result, practitioners must often resort to using restoration methods that rely heavily on value judgments and pre-conceived notions of what a particular area *should* be like. In the case of stream restoration, one commonly used method is "natural channel design" in which streams are sorted into an array of classifications based upon their geomorphic characteristics and restored to match the conditions of a reference site with similar classifications (Rosgen, 2011). Projects using such systems often do not achieve the biological objectives they have set (Ernst et al., 2012). Thus, without the use of physiological data to help determine biological objectives and guide the decision-making process, restoration practitioners may select project designs that do not address the underlying mechanism driving fish declines.

Following our hypothetical example of stream restoration, a lack of recruitment within a fish population of interest, such as Pacific salmon (*Oncorhynchus* spp.), could be attributed to a wide number of factors. Monitoring of just the environmental conditions within the area might lead a project manager to determine that the source of the issue is a lack of spawning gravel leading to reduced reproductive success, and the restorative action chosen may be to supplement spawning grounds with gravel. However, were physiological monitoring conducted in concert with population assessments, the collected data may have indicated that mature females had poor nutritional status and were less fecund. In such a case, replanting riparian vegetation to provide a source of invertebrate prey could have been a preferable action to take. While fairly simple, this is just one example of how physiological monitoring could benefit and complement the restoration decision-making process and ensure that the restoration action taken is most well-suited to addressing the problem at hand.

III. Implementing the solution

In our example scenario, restorative actions could include alterations to in-stream and riparian habitats (e.g., placement of woody structures, addition/removal of substrates, development of buffer strips, planting of riparian vegetation) (Kondolf and Micheli, 1995; Roni et al., 2002), with the intent to enhance or re-establish fish populations. The implementation of such actions may take place over a relatively short period of time, on the scale of weeks to months (e.g., installation of in-stream habitat structures), but it may take years for a project to reach completion (e.g., maturation of riparian plantings). Any biological assessment is typically withheld until projects are fully implemented because the population- and community-level metrics favored for such assessments cannot respond in any meaningful way within the timeframe of a project's implementation.

Physiological monitoring could offer a unique opportunity to assess restoration, not only after it has been completed, but also while it is in the process of being implemented. This, in turn, would allow restoration practitioners to adapt and adjust their design as needed based upon the physiological responses of local fish. In-stream restoration structures significantly alter flow conditions when they are installed; oxygen consumption, a measurement of metabolism, could reflect if these altered flow conditions are beneficial or deleterious for fish. If an element of an in-stream structure was shown to result in elevated oxygen consumption, modifications could then be made to structures to maximize the energetic benefits for fish (Strailey et al., 2021). Biomarkers that reflect impaired water quality and heavy contaminant load could be sampled throughout the process of constructing riparian buffer strips, and, in turn, guide decisions regarding what plant species should be planted where and in what amount (see Chapter 3, Volume 39B: De Boeck et al., 2022). The use of physiology in this manner would help to ensure that the benefits of a restoration project could be maximized.

IV. Post-project monitoring and evaluating success

Following implementation, restoration activities would then ideally be paired with a monitoring period to quantify the success or failure of this work to restore fish populations. This would allow practitioners to adjust their approach for future projects. As discussed earlier within this chapter, stream restoration projects often lack the funding for extended monitoring periods, and so projects may only be monitored for a few years following completion. However, this typically entails monitoring at the population- and community-levels, which take far longer than a few years to respond to environmental changes, and so such monitoring may show no change at all. Given that traditional monitoring is generally unable to quantify fish responses to restoration within a short time frame, it may be ineffective to attempt traditional monitoring at all if funding is only available for such a timeframe. For certain projects, such as those targeting long-lived species like paddlefish (family Polyodontidae) and gar (family Lepisosteidae) whose populations are likely to respond slowly to any restoration, the use of traditional monitoring for a short period may be challenging. In such cases, restoration practitioners could instead utilize physiological monitoring, as the physiology of individual fish would be able to respond more quickly to the restoration actions taken and indicate if fish responses are moving in or likely to move in the direction of meeting project objectives.

For example, if stream temperatures had previously exceeded the temperature thresholds of resident species, as in the described in Section 1, fish stress levels could be sampled to evaluate if a riparian planting had succeeded in shading a stream sufficiently to reduce water temperatures below those thresholds. Because chronic stress can increase fish energy consumption, reduction in these levels would not merely indicate that fish were less stressed, but would also suggest that fish then had more energy available for reproduction. Blood-based tools such as leukocrit, lysozymes, and serum immunoglobulin levels can all be used as immunological indices of stress in fish that can provide information on disease or pathogen levels. Because these metrics can result from chronic stress due to habitat degradation, they could additionally be used to reflect increased health following habitat restoration (Rice and Arkoosh, 2002).

This is but one example scenario that demonstrates how physiological tools can be employed throughout the restoration process; similar examples are shown in Table 2. Physiological data can be incorporated throughout the restoration process and can be applied before or following restoration activities. Habitat improvements through restoration should hypothetically yield improvements in physiological condition that can suggest if population trends will be reversed. Thus, there are a number of physiological tools and techniques that can be incorporated into "traditional," field-based monitoring that can identify mechanisms for population declines and document subsequent recovery following restoration activities.

## 4.2 Integrating physiology into the restoration process: Examples to date

The integration of physiological monitoring with restoration activities of aquatic environments has not been common to date, but successful examples are emerging, particularly in the context of pollution abatement and remediation (Cooke and Suski, 2008). Facey et al. (2005), for example, found increased incidence of macrophage aggregates within fish splenic tissue, a biomarker correlated with contaminant exposure, in rock bass (*Ambloplites rupestris*) collected from an area receiving sewage effluent compared to fish from reference sites not receiving effluent. Following relocation of the effluent outlet, sampling revealed a decline in this biomarker, indicating that contaminant levels and environmental conditions had improved. Adams and Ham (2011) investigated the effect of pollution abatement strategies on the physiological status of redbreast sunfish (*Lepomis auritus*), a keystone and sentinel species, in a historically polluted stream in Eastern Tennessee. With a suite of biochemical, bioenergetic, and nutritional bioindicators, Adams and Ham (2011) compared fish from several sites within the polluted stream to an undisturbed reference stream nearby. The activity of liver detoxification enzymes, a metric indicating contaminant exposure, decreased toward levels seen in reference stream fish, and overall health and condition of fish within the disturbed stream improved. Combined, the authors were able to determine that not only was the physicochemical recovery of the stream underway, but biological recovery was underway as well. McLennan et al. (2021) found that young Atlantic salmon (*Salmo salar*) had longer telomeres, associated with a reduced rate of cellular aging, when reared in streams supplemented with adult salmon carcasses, while young salmon in streams that were not supplemented had shorter telomeres. This demonstrated the positive impact of nutrient restoration on the physiological state of these fish. While there is clearly

**TABLE 2** Examples of how reductions in habitat quality (ultimate stressor) can lead to the rise of various proximate stressors that may act mechanistically to impact the physiology of individual fish.

| Example of ultimate stressor | Example of proximate stressor | Mechanism of action | Individual-level (physiological) response | Example reference | Population-level responses |
|---|---|---|---|---|---|
| Loss of tidal connectivity | Increased cover by invasive plant species | Exclusion from prime feeding grounds | Lipid reserves | Weinstein et al. (2009) | Impaired population growth |
| Effluent discharge from industrial facility | Increased concentration of contaminants | Alterations to gill structure and function | Organ dysfunction | Adams and Ham (2011) | Mortality |
| Installation of physical structure | Altered hydraulics | Changes in swimming behavior | Energetics | Strailey et al. (2021) | Reduced reproductive output |
| Shoreline development | Increased dissolved solids | Respiration difficulties | Stress response | Kjelland et al. (2015) | Emigration |

The accumulation of responses from individual fish, in turn, can lead to a response at the population level.

still a need for increased integration of physiology into restoration practice, these cases demonstrate that there are successful examples of physiological tools being coupled with traditional in-field assessments of populations.

## 4.3 Challenges and opportunities

A number of challenges have slowed the integration of physiological monitoring with "traditional" monitoring of restoration activities. Conservation and management activities tend to monitor and emphasize metrics at higher levels of organization (e.g., species abundance, species richness, biodiversity), often because these types of metrics are believed to integrate the effects of stressors operating at a wide range of spatial and temporal scales, allowing the quantification of stressors even if the perturbation is not ongoing at the time of sampling (Friberg et al., 2011; Niemi and McDonald, 2004). Consequently, many restoration practitioners primarily consider such metrics when carrying out restoration projects. However, links between environmental disturbances, individual physiology, and populations are not simple linear, correlative relationships. In contrast, they are complex, owing to the fact that multiple stressors can be synergistic, additive or antagonistic (Côté et al., 2016; Folt et al., 1999). The variation observed at a particular level of organization is typically smaller than the sum of variation across lower levels, the impacts of environmental stressors can take long periods of time to be observed as levels of organization increase ( Jørgensen and Nielsen, 2013), animals can make decisions and move between different environmental compartments in an effort to minimize disturbance and maximize performance (Claireaux and Lefrançois, 2007), and links between physiological responses and Darwinian fitness have not been well-defined (Bonier et al., 2009). Thus, a key challenge for integrating physiology and restoration lies in the fact that, when combined, even large changes at the cellular or subcellular level can translate to small changes to organs or individuals, and even smaller changes to populations and biodiversity, and the quantification of perturbations using organismal physiology has the potential to over-represent environmental changes and under-represent alterations to biodiversity. In addition to these conceptual challenges, a number of technical and social challenges have contributed to the historical separation of physiology from aquatic restoration. Practitioners and fisheries managers may be unfamiliar with physiological tools, and consequentially may not see a use for them; even if interested in utilizing physiological tools, practitioners are unlikely to have training in their use (Friberg et al., 2011). Given that many projects already lack the funding for monitoring of any kind, practitioners may perceive physiological monitoring to be more expensive than traditional monitoring and thus be averse to its use. For small-scale projects with limited budgets, it simply may not be feasible to use physiological tools (Cooke and Suski, 2008). Although governments may commit to improving water quality and restoring aquatic habitat, there is often administrative inertia to mandate the inclusion of novel techniques into existing restoration programs (Moog and Chovanec, 2000). Given this wide array of challenges, it is not surprising that physiology has not been developed as a component of traditional restoration monitoring.

Despite these challenges, there has been a call to incorporate new and novel techniques into restoration activities, with physiological tools being a priority for development. For example, Wortley et al. (2013) highlighted the need to move beyond species abundance as a metric for restoration and emphasized the value of restoration data that can define ecosystem processes to help measure the progress or success. Friberg et al. (2011) called for restoration practitioners to adapt a more *a priori*, predictive approach to monitoring restoration activities to better understand how ecosystems respond to stressors, specifically calling for the integration of physiological tools into restoration monitoring, and Niemi and McDonald (2004) called for additional integration of physiological biomarkers into monitoring programs to define environmental perturbations. Kimball et al. (2016) highlighted the value in incorporating physiological tools into restoration activities and monitoring to assess restoration success and ecosystem services, while Norris and Hawkins (2000) advocated for a multimetric approach to monitoring river health, incorporating multiple levels of organization to obtain a holistic overview of biological integrity.

While facilitating the integration of physiological metrics into traditional restoration activities may face a number of challenges, there are just as many possible opportunities that can overcome these challenges. In particular, this potential integration presents an excellent opportunity for increased collaboration and communication between academic scientists and the practitioners and managers directly carrying out restoration actions. If small budgets limit the monitoring potential for a restoration project, such collaborations may be able to pool financial resources to fund physiological monitoring. In other cases, the development of affordable, kit-style assays may be able to improve both the ease and accessibility of physiology for restoration projects. These may require a number of actions and efforts, but should be achievable and have value for restoration. Much of the onus in facilitating this integration may be on physiologists, but there are also steps that restoration practitioners and other stakeholders can take. For example, at the outset, it is important for restoration practitioners to define the goals of their restoration project to ensure that physiological tools and techniques would be appropriate for helping address project objectives. Assuming that physiological tools are an appropriate addition to a monitoring activity, the choice of physiological biomarker should be easy to collect in the field, and should have a clear, mechanistic cause-and-effect relationship between the physiological response and perceived environmental perturbations, ideally with ranges of low-stress values defined (e.g., Barton, 2002), and preferably linking to population-level outcomes if possible. Field crews need to be trained in the collection, storage, analysis, and interpretation of physiological metrics (Iwama et al., 1995), and funders need to commit to the analysis of physiological samples, which can take weeks or months to complete following the completion of field monitoring activities. Finally, the relevance of physiological tools need to be emphasized to policy makers to help promote their use (Cooke and O'Connor, 2010), which can be facilitated by integrating stakeholders into the generation of restoration data, incorporating citizen science into data collection, and communicating research findings in traditional (e.g., conference presentations, peer-reviewed journal articles) and non-traditional (e.g., social media, popular press articles, stakeholder engagements) venues (Laubenstein and Rummer, 2020).

The actual tools, techniques, and sample design required for the successful integration of physiological metrics into a restoration study will vary based on budget, time, project goals, and other constraints. However, a number of papers exist that can aid in developing a framework for how to choose relevant biomarkers, integrate activities within the framework of field-based population monitoring, and achieve sound, robust results that can complement existing activities (Adams et al., 1993; Iwama et al., 1995; Morgan and Iwama, 1991; Norris and Hawkins, 2000; Power, 2002; Smith, 2002). Thus, despite inherent challenges and difficulties, a number of strategies exist to facilitate the incorporation of physiological tools into traditional monitoring activities to better understand how organisms interact with their environment, and improve outcomes for restoration activities.

## 5 Conclusions

Fish populations worldwide have been severely impacted by anthropogenic habitat degradation and loss. This problem is likely to become amplified in the future, especially for freshwater species, as nearly 80% of the world's human population is threatened with water insecurity (Vörösmarty et al., 2010). As human populations continue to grow, the demands placed on our environment will intensify, in turn placing increased stress on fish. Accordingly, the UN has declared 2021–2030 to be the "Decade on Ecosystem Restoration" to encourage the remediation of habitat degradation and loss through restoration (Waltham et al., 2020). However, while widescale restoration efforts have been underway, these efforts, though financially costly, have often been partially or entirely unsuccessful. Many factors contribute to this lack of success, but a dependence on unreliable and unsuitable monitoring methods have made it difficult to identify what restoration techniques are most effective or what avenues we should take moving forward. We believe that physiological tools can offer new insights that can guide future restoration efforts and increase success. While we do not propose that physiological tools should replace the use of population and community metrics, we suggest that the integration of physiological tools with field-based restoration monitoring can complement traditional metrics, and "fill in the gaps" at shorter time scales before traditional metrics are able to respond, helping increase the likelihood of project success and sharing data on effective restoration techniques. The successful integration of physiology into the restoration process will require new lines of collaboration between restoration practitioners and fish physiologists, and much of the impetus may fall upon physiologists to demonstrate the value of physiological tools. However, this integration will drive measurable innovations in the restoration process, leading to a more successful Decade on Ecosystem Restoration and a brighter future for our fish.

## References

Adams, S.M., Ham, K.D., 2011. Application of biochemical and physiological indicators for assessing recovery of fish populations in a disturbed stream. Environ. Manag. 47 (6), 1047–1063. https://doi.org/10.1007/s00267-010-9599-7.

Adams, S.M., 2002. Biological Indicators of Aquatic Ecosystem Stress. American Fisheries Society, Bethesda.

Adams, S.M., Brown, A.M., Goede, R.W., 1993. A quantitative health assessment index for rapid evaluation of fish condition in the field. Trans. Am. Fish. Soc. 122 (1), 63–73. https://doi.org/10.1577/1548-8659(1993)122<0063:aqhaif>2.3.co;2.

Albert, J.S., Destouni, G., Duke-Sylvester, S.M., Magurran, A.E., Oberdorff, T., Reis, R.E., Winemiller, K.O., Ripple, W.J., 2021. Scientists' warning to humanity on the freshwater biodiversity crisis. Ambio 50 (1), 85–94. Springer Netherlands. https://doi.org/10.1007/s13280-020-01318-8.

Alexander, G.G., Allan, J.D., 2007. Ecological success in stream restoration: case studies from the Midwestern United States. Environ. Manage. 40, 245–255. https://doi.org/10.1007/s00267-006-0064-6.

Allen, C.R., Fontaine, J.J., Pope, K.L., Garmestani, A.S., 2011. Adaptive management for a turbulent future. J. Environ. Manage. 92 (5), 1339–1345. Elsevier Ltd. https://doi.org/10.1016/j.jenvman.2010.11.019.

Allison, S.K., 2012. How did we get here? A brief history of ecological restoration. In: Ecological Restoration and Environmental Change: Renewing Damaged Ecosystems. Routledge, Abingdon, pp. 20–46.

Alongi, D.M., 2002. Present state and future of the world's mangrove forests. Environ. Conserv. 29 (3), 331–349. https://doi.org/10.1017/S0376892902000231.

Alvarez-Filip, L., Estrada-Saldívar, N., Pérez-Cervantes, E., Molina-Hernández, A., González-Barrios, F.J., 2019. A rapid spread of the stony coral tissue loss disease outbreak in the Mexican Caribbean. PeerJ 7, e8069. https://doi.org/10.7717/peerj.8069.

Arthington, A.H., Dulvy, N.K., Gladstone, W., Winfield, I.J., 2016. Fish conservation in freshwater and marine realms: status, threats and management. Aquat. Conserv. Mar. Freshwat. Ecosyst. 26 (5), 838–857. https://doi.org/10.1002/aqc.2712.

Aylward, B., et al., 2005. Freshwater ecosystem services. In: Chopra, K., Leemans, R., Kumar, P., Simons, H. (Eds.), Ecosystems and Human Well-being: Policy Responses. vol. 3. Island Press, Washington, DC, pp. 213–255.

Baldigo, B.P., Ernst, A.G., Warren, D.R., Miller, S.J., 2010. Variable responses of fish assemblages, habitat, and stability to natural-channel-design restoration in Catskill Mountain streams. Trans. Am. Fish. Soc. 139 (2), 449–467. https://doi.org/10.1577/T08-152.1.

Barton, B.A., 2002. Stress in fishes: a diversity of responses with particular reference to changes in circulating corticosteroids. Integr. Comp. Biol. 42 (3), 517–525. https://doi.org/10.1093/icb/42.3.517.

Bash, J.S., Ryan, C.M., 2002. Stream restoration and enhancement projects: is anyone monitoring? Environ. Manag. 29 (6), 877–885. https://doi.org/10.1007/s00267-001-0066-3.

Bayraktarov, E., et al., 2019. Motivations, success, and cost of coral reef restoration. Restor. Ecol. 27 (5), 981–991. https://doi.org/10.1111/rec.12977.

Beechie, T.J., et al., 2010. Process-based principles for restoring river ecosystems. Bioscience 60 (3), 209–222. https://doi.org/10.1525/bio.2010.60.3.7.

Beechie, T., et al., 2013. Restoring salmon habitat for a changing climate. River Res. Appl. 29, 939–960. https://doi.org/10.1002/rra.

BenDor, T., Lester, T.W., Livengood, A., Davis, A., Yonavjak, L., 2015. Estimating the size and impact of the ecological restoration economy. PLoS One 10 (6), 1–15. https://doi.org/10.1371/journal.pone.0128339.

Bernhardt, E.S., Palmer, M.A., 2011. River restoration: the fuzzy logic of repairing reaches to reverse catchment scale degradation. Ecol. Appl. 21 (6), 1926–1931. https://doi.org/10.1890/10-1574.1.

Bernhardt, E.S., et al., 2007. Restoring rivers one reach at a time: results from a survey of U.S. river restoration practitioners. Restor. Ecol. 15 (3), 482–493. https://doi.org/10.1111/j.1526-100X.2007.00244.x.

Bernos, T.A., Jeffries, K.M., Mandrak, N.E., 2020. Linking genomics and fish conservation decision making: a review. Rev. Fish Biol. Fish. 30 (4), 587–604. https://doi.org/10.1007/s11160-020-09618-8.

Birnie-Gauvin, K., Costantini, D., Cooke, S.J., Willmore, W.G., 2017. A comparative and evolutionary approach to oxidative stress in fish: a review. Fish Fish. 18 (5), 928–942. https://doi.org/10.1111/faf.12215.

Blaber, S.J.M., 2007. Mangroves and fishes: issues of diversity, dependence, and dogma. Bull. Mar. Sci. 80, 457–472.

Blandon, A., Zu Ermgassen, P.S.E., 2014. Quantitative estimate of commercial fish enhancement by seagrass habitat in southern Australia. Estuar. Coast. Shelf Sci. 141, 1–8. Elsevier Ltd. https://doi.org/10.1016/j.ecss.2014.01.009.

Bonier, F., Martin, P.R., Moore, I.T., Wingfield, J.C., 2009. Do baseline glucocorticoids predict fitness? Trends Ecol. Evol. 24 (11), 634–642. https://doi.org/10.1016/j.tree.2009.04.013.

Boström-Einarsson, L., Babcock, R.C., Bayraktarov, E., Ceccarelli, D., Cook, N., Ferse, S.C.A., Hancock, B., Harrison, P., Hein, M., Shaver, E., Smith, A., Suggett, D., Stewart-Sinclair, P.J., Vardi, T., McLeod, I.M., 2020. Coral restoration – a systematic review of current methods, successes, failures and future directions. PLoS One 15 (1), e0226631. https://doi.org/10.1371/journal.pone.0226631.

Bruns, A., Frick, F., 2014. The notion of the global water crisis and urban water realities. In: Bhaduri, A., et al. (Eds.), The Global Water System in the Anthropocene. Springer International Publishing Switzerland, Heidelberg, pp. 415–426, https://doi.org/10.1007/978-3-319-07548-8_27.

Bruton, M.N., 1995. Have fishes had their chips? The dilemma of threatened fishes. Environ. Biol. Fishes 43 (1), 1–27. https://doi.org/10.1007/BF00001812.

Burkhead, N.M., 2012. Extinction rates in north American freshwater fishes, 1900-2010. Bioscience 62 (9), 798–808. https://doi.org/10.1525/bio.2012.62.9.5.

Calow, P., 1989. Proximate and ultimate responses to stress in biological systems. Biol. J. Linn. Soc. 37 (1–2), 173–181. https://doi.org/10.1111/j.1095-8312.1989.tb02101.x.

Canning, A.D., Death, R.G., 2019. Ecosystem health indicators—freshwater environments. In: Fath, B. (Ed.), Encyclopedia of Ecology, second ed. vol. 1. Elsevier, Amsterdam, pp. 46–60, https://doi.org/10.1016/B978-0-12-409548-9.10617-7.

Cardinale, B.J., et al., 2012. Biodiversity loss and its impact on humanity. Nature 486 (7401), 59–67. https://doi.org/10.1038/nature11148.

Carpenter, K.E., et al., 2008. One-third of reef-building corals face elevated extinction risk from climate change and local impacts. Science 321 (5888), 560–563. https://doi.org/10.1126/science.1159196.

Chapman, J.M., Kelly, L.A., Teffer, A.K., Miller, K.M., Cooke, S.J., 2021. Disease ecology of wild fish: opportunities and challenges for linking infection metrics with behaviour, condition, and survival. Can. J. Fish. Aquat. Sci. 78 (8), 995–1007. https://doi.org/10.1139/cjfas-2020-0315.

Claireaux, G., Lefrançois, C., 2007. Linking environmental variability and fish performance: integration through the concept of scope for activity. Philos. Trans. R. Soc. B 362 (1487), 2031–2041. https://doi.org/10.1098/rstb.2007.2099.

Colin, N., et al., 2016. Ecological relevance of biomarkers in monitoring studies of macro-invertebrates and fish in Mediterranean rivers. Sci. Total Environ. 540, 307–323. Elsevier B.V. https://doi.org/10.1016/j.scitotenv.2015.06.099.

Congleton, J.L., Wagner, T., 2006. Blood-chemistry indicators of nutritional status in juvenile salmonids. J. Fish Biol. 69 (2), 473–490. https://doi.org/10.1111/j.1095-8649.2006.01114.x.

Connon, R.E., Jeffries, K.M., Komoroske, L.M., Todgham, A.E., Fangue, N.A., 2018. The utility of transcriptomics in fish conservation. J. Exp. Biol. 221 (2), jeb148833. https://doi.org/10.1242/jeb.148833.

Cooke, S.J., O'Connor, C.M., 2010. Making conservation physiology relevant to policy makers and conservation practitioners. Conserv. Lett. 3 (3), 159–166. https://doi.org/10.1111/j.1755-263X.2010.00109.x.

Cooke, S.J., Suski, C.D., 2008. Ecological restoration and physiology: an overdue integration. Bioscience 58 (10), 957–968. https://doi.org/10.1641/b581009.

Côté, I.M., Darling, E.S., Brown, C.J., 2016. Interactions among ecosystem stressors and their importance in conservation. Proc. R. Soc. B Biol. Sci. 283 (1824), 20152592. https://doi.org/10.1098/rspb.2015.2592.

Cowx, I.G., 2002. Analysis of threats to freshwater fish conservation: past and present challenges. In: Conservation of Freshwater Fishes: Options for the Future. Blackwell Science, Oxford, pp. 201–220.

Cowx, I.G., Portocarrero Aya, M., 2011. Paradigm shifts in fish conservation: moving to the ecosystem services concept. J. Fish Biol. 79 (6), 1663–1680. https://doi.org/10.1111/j.1095-8649.2011.03144.x.

Crutzen, P.J., 2006. The "Anthropocene". In: Ehlers, E., Krafft, T. (Eds.), Earth System Science in the Anthropocene. Springer, Berlin, Heidelberg, pp. 13–18.

Crutzen, P., Stoermer, E., 2000. Sustaining Earth's Life Support Systems—the Challenge for the Next Decade and Beyond. IGBP newsletter, pp. 17–18 (41).

Currie, S., Tufts, B.L., Moyes, C.D., 1999. Influence of bioenergetic stress on heat shock protein gene expression in nucleated red blood cells of fish. Am. J. Physiol. Regul. Integr. Comp. Physiol. 276 (4), R990–R996. https://doi.org/10.1152/ajpregu.1999.276.4.r990.

Curtis-Quick, J.A., Ulanov, A.V., Li, Z., Bieber, J.F., Tucker-Retter, E.K., Suski, C.D., 2021. Why the stall? Using metabolomics to define the lack of upstream movement of invasive bigheaded carp in the Illinois River. PLoS One 16 (10), e0258150. https://doi.org/10.1371/journal.pone.0258150.

Darwall, W.R.T., Freyhof, J., 2016. Lost fishes, who is counting? The extent of the threat to freshwater fish biodiversity. In: Gloss, G.P., Krkosek, M., Olden, J.D. (Eds.), Conservation of Freshwater Fishes. Cambridge University Press, Cambridge.

Dawson, M.N., 2012. Species richness, habitable volume, and species densities in freshwater, the sea, and on land. Front. Biogeogr. 4 (3), 105–116. https://doi.org/10.21425/f5fbg12675.

De Boeck, G., Rodgers, E., Town, R.M., 2022. Using ecotoxicology for conservation: From biomarkers to modeling. Fish Physiol. 39B, 111–174.

De Groot, R.S., Blignaut, J., van der Ploeg, S., Aronson, J., Elmqvist, T., Farley, J., 2013. Benefits of investing in ecosystem restoration. Conserv. Biol. 27 (6), 1286–1293. https://doi.org/10.1111/cobi.12158.

Díaz, S., et al., 2019. Summary for policymakers of the global assessment report on biodiversity and ecosystem services of the intergovernmental science-policy platform on biodiversity and ecosystem services.

Dobzhansky, T., 1964. Biology, molecular and organismic. Am. Zool. 4 (4), 443–452.

Dudgeon, D., 2019. Multiple threats imperil freshwater biodiversity in the Anthropocene. Curr. Biol. 29 (19), R960–R967. Elsevier https://doi.org/10.1016/j.cub.2019.08.002.

Dudgeon, D., et al., 2006. Freshwater biodiversity: importance, threats, status and conservation challenges. Biol. Rev. 81, 163–182. https://doi.org/10.1017/S1464793105006950.

Dufour, S., Piegay, H., 2009. From the myth of a lost paradise to targeted river restoration: forget natural references and focus on human benefits. River Res. Appl. 25, 568–581. https://doi.org/10.1002/rra.

Ernst, A.G., Warren, D.R., Baldigo, B.P., 2012. Natural-channel-design restorations that changed geomorphology have little effect on macroinvertebrate on macroinvertebrate communities in headwater streams. Restor. Ecol. 20 (4), 532–540.

Essington, T.E., Beaudreau, A.H., Wiedenmann, J., 2006. Fishing through marine food webs. Proc. Natl. Acad. Sci. U. S. A. 103 (9), 3171–3175. https://doi.org/10.1073/pnas.0510964103.

European Union Agency, 2015. SOER. The European Environment—State and Outlook. A Comprehensive Assessment of the European Environment's State, Trends, and Prospects, in a Global Context. EEA, Copenhagen, Denmark.

Facey, D.E., Blazer, V.S., Gasper, M.M., Turcotte, C.L., 2005. Using fish biomarkers to monitor improvements in environmental quality. J. Aquat. Anim. Health 17 (3), 263–266. https://doi.org/10.1577/H04-055.1.

Fausch, K.D., Karr, J.R., Yant, P.R., 1984. Regional application of an index of biotic integrity based on stream fish communities. Trans. Am. Fish. Soc. 113 (1), 39–55. https://doi.org/10.1577/1548-8659(1984)113<39:raoaio>2.0.co;2.

Fausch, K.D., Lyons, J., Karr, J.R., Angermeier, P.L., 1990. Fish communities as indicators of environmental degradation. Am. Fish. Soc. Symp. 8, 123–144.

Federal Interagency Stream Restoration Working Group, 1998. Stream Corridor Restoration: Principles, Process, and Practices. National Technical Information Service (NTIS), Springfield, VA.

Follstad Shah, J.J., Dahm, C.N., Gloss, S.P., Bernhardt, E.S., 2007. River and riparian restoration in the southwest: results of the National River Restoration Science Synthesis project. Restor. Ecol. 15 (3), 550–562. https://doi.org/10.1111/j.1526-100X.2007.00250.x.

Folt, C.L., Chen, C.Y., Moore, M.V., Burnaford, J., 1999. Synergism and antagonism among multiple stressors. Limnol. Oceanogr. 44 (3), 864–877. https://doi.org/10.1002/ece3.1465.

Friberg, N., et al., 2011. Biomonitoring of human impacts in freshwater ecosystems. The good, the bad and the ugly. In: Woodward, G. (Ed.), Advances in Ecological Research. Academic Press, London, pp. 1–68, https://doi.org/10.1016/B978-0-12-374794-5.00001-8.

Fry, F.E.J., 1947. Effects of the Environment on Animal Activity. Publications of the Ontario Fisheries Research Laboratory No. 68. University of Toronto Studies, Biological Series, pp. 1–52. 55.

Gardner, T.A., Côté, I.M., Gill, J.A., Grant, A., Watkinson, A.R., 2005. Hurricanes and Caribbean coral reefs: impacts, recovery patterns, and role in long-term decline. Ecology 86 (1), 174–184.

Geist, J., 2011. Integrative freshwater ecology and biodiversity conservation. Ecol. Indic. 11 (6), 1507–1516. https://doi.org/10.1016/j.ecolind.2011.04.002.

Geist, J., Hawkins, S.J., 2016. Habitat recovery and restoration in aquatic ecosystems: current progress and future challenges. Aquat. Conserv. Mar. Freshwat. Ecosyst. 26 (5), 942–962. https://doi.org/10.1002/aqc.2702.

Giesing, E.R., Suski, C.D., Warner, R.E., Bell, A.M., 2010. Female sticklebacks transfer information via eggs: effects of maternal experience with predators on offspring. Proc. R. Soc. B Biol. Sci. 278, 1753–1759. https://doi.org/10.1098/rspb.2010.1819.

Gleick, P.H., 2011. Water Resources. Encyclopedia of Climate and Weather, second ed. Oxford University Press, https://doi.org/10.5860/choice.49-2412.

Gozlan, R.E., Karimov, B.K., Zadereev, E., Kuznetsova, D., Brucet, S., 2019. Status, trends, and future dynamics of freshwater ecosystems in Europe and Central Asia. Inland Waters 9 (1), 78–94. https://doi.org/10.1080/20442041.2018.1510271.

Greeley Jr., M.S., 2002. Reproductive indicators of environmental stress in fish. In: Biological Indicators of Aquatic Ecosystem Stress. American Fisheries Society, Bethesda, pp. 321–377.

Hayes, D.B., Ferreri, C.P., Taylor, W.W., 1996. Linking fish habitat to their population dynamics. Can. J. Fish. Aquat. Sci. 53 (Suppl. 1), 383–390. https://doi.org/10.1139/cjfas-53-s1-383.

Hein, M.Y., Willis, B.L., Beeden, R., Birtles, A., 2017. The need for broader ecological and socioeconomic tools to evaluate the effectiveness of coral restoration programs. Restor. Ecol. 25 (6), 873–883. https://doi.org/10.1111/rec.12580.

Higgs, E., et al., 2018. On principles and standards in ecological restoration. Restor. Ecol. 26 (3), 399–403. https://doi.org/10.1111/rec.12691.

Hobbs, R.J., Harris, J.A., 2001. Restoration ecology: repairing the earth's ecosystems in the new millennium. Restor. Ecol. 9 (2), 239–246. https://doi.org/10.1046/j.1526-100X.2001.009002239.x.

Holl, K.D., Howarth, R.B., 2000. Paying for restoration. Restor. Ecol. 8 (3), 260–267. https://doi.org/10.1046/j.1526-100X.2000.80037.x.

Huey, R.B., 1991. Physiological consequences of habitat selection. Am. Nat. 137, S91–S115.

IUCN, 2021. The IUCN Red List of Threatened Species. Version 2021-2. https://www.iucnredlist.org. Downloaded on [6 December 2021].

Iwama, G.K., Morgan, J.D., Barton, B.A., 1995. Simple field methods for monitoring stress and general condition of fish. Aquacult. Res. 26 (4), 273–282. https://doi.org/10.1111/j.1365-2109.1995.tb00912.x.

Jackson, L.L., Lopoukhine, N., Hillyard, D., 1995. Ecological restoration- a definition and comments. Restor. Ecol. 3 (2), 71–75.

Jähnig, S.C., Lorenz, A.W., Hering, D., Antons, C., Sundermann, A., Jedicke, E., Haase, P., 2011. River restoration success: a question of perception. Ecol. Appl. 21 (6), 2007–2015. https://doi.org/10.1890/10-0618.1.

Jeffrey, J.D., Hasler, C.T., Chapman, J.M., Cooke, S.J., Suski, C.D., 2015. Linking landscape-scale disturbances to stress and condition of fish: implications for restoration and conservation. Integr. Comp. Biol. 55 (4), 618–630. https://doi.org/10.1093/icb/icv022.

Jeffrey, J.D., Jeffries, K.M., Suski, C.D., 2019. Physiological status of silver carp (*Hypophthalmichthys molitrix*) in the Illinois River: an assessment of fish at the leading edge of the invasion front. Comp. Biochem. Physiol. Part D Genomics Proteomics 32 (July), 100614. Elsevier. https://doi.org/10.1016/j.cbd.2019.100614.

Jelks, H.L., et al., 2008. Conservation status of imperiled north American freshwater and diadromous fishes. Fisheries 33 (8), 372–407.

Jørgensen, S.E., Nielsen, S.N., 2013. The properties of the ecological hierarchy and their application as ecological indicators. Ecol. Indic. 28, 48–53. Elsevier Ltd. https://doi.org/10.1016/j.ecolind.2012.04.010.

Kail, J., Brabec, K., Poppe, M., Januschke, K., 2015. The effect of river restoration on fish, macroinvertebrates and aquatic macrophytes: a meta-analysis. Ecol. Indic. 58, 311–321. Elsevier Ltd. https://doi.org/10.1016/j.ecolind.2015.06.011.

Karr, J.R., 1981. Assessment of biotic integrity using fish communities. Fisheries 6 (6), 21–27. https://doi.org/10.1577/1548-8446(1981)006<0021:aobiuf>2.0.co;2.

Karr, J.R., 1999. Defining and measuring river health. Freshw. Biol. 41 (2), 221–234. https://doi.org/10.1046/j.1365-2427.1999.00427.x.

Kimball, S., et al., 2015. Cost-effective ecological restoration. Restor. Ecol. 23 (6), 800–810. https://doi.org/10.1111/rec.12261.

Kimball, S., Funk, J.L., Sandquist, D.R., Ehleringer, J.R., 2016. Ecophysiological Considerations for Restoration. In: Palmer, M.A., Zedler, J.B., Falk, D.A. (Eds.), Foundations of Restoration Ecology, second ed. Island Press, Washington, DC, pp. 153–181.

King, G.D., Chapman, J.M., Cooke, S.J., Suski, C.D., 2016. Stress in the neighborhood: tissue glucocorticoids relative to stream quality for five species of fish. Sci. Total Environ. 547, 87–94. Elsevier B.V. https://doi.org/10.1016/j.scitotenv.2015.12.116.

Kjelland, M.E., Woodley, C.M., Swannack, T.M., Smith, D.L., 2015. A review of the potential effects of suspended sediment on fishes: potential dredging-related physiological, behavioral, and transgenerational implications. Environ. Syst. Decis. 35, 334–350. https://doi.org/10.1007/s10669-015-9557-2.

Kondolf, G.M., Micheli, E.R., 1995. Evaluating stream restoration projects. Environ. Manag. 19 (1), 1–15. https://doi.org/10.1007/BF02471999.

Landres, P.B., Morgan, P., Swanson, F.J., 1999. Overview of the use of natural variability concepts in managing ecological systems. Ecol. Appl. 9 (4), 1179–1188. https://doi.org/10.1890/1051-0761(1999)009[1179:OOTUON]2.0.CO;2.

Laubenstein, T.D., Rummer, J.L., 2020. Communication in conservation physiology. In: Madliger, C.L., Franklin, C.E., Love, O.P., Cooke, S.J. (Eds.), Conservation Physiology: Applications for Wildlife Conservation and Management. Oxford University Press, Oxford, pp. 303–317.

Lavelle, A.M., Chadwick, M.A., Chadwick, D.D.A., Pritchard, E.G., Bury, N.R., 2021. Effects of habitat restoration on fish communities in urban streams. Water (Switzerland) 13 (16), 2170. https://doi.org/10.3390/w13162170.

Lennox, R.J., Suski, C.D., Cooke, S.J., 2018. A macrophysiology approach to watershed science and management. Sci. Total Environ. 626, 434–440. Elsevier B.V. https://doi.org/10.1016/j.scitotenv.2018.01.069.

Lepori, F., Palm, D., Brannas, E., Malmqvist, B., 2005. Does restoration of structural heterogeneity in streams enahnce fish and macroinvertebrate diversity? Ecol. Appl. 15 (6), 2060–2071.

Leps, M., Sundermann, A., Tonkin, J.D., Lorenz, A.W., Haase, P., 2016. Time is no healer: increasing restoration age does not lead to improved benthic invertebrate communities in restored river reaches. Sci. Total Environ. 557–558, 722–732. Elsevier B.V. https://doi.org/10.1016/j.scitotenv.2016.03.120.

Liss, S.A., Sass, G.G., Suski, C.D., 2014. Influence of local-scale abiotic and biotic factors on stress and nutrition in invasive silver carp. Hydrobiologia 736 (1), 1–15. https://doi.org/10.1007/s10750-014-1880-y.

Louhi, P., Vehanen, T., Huusko, A., Maki-Petays, A., Muotka, T., 2016. Long-term monitoring reveals the success of salmonid habitat restoration. Can. J. Fish. Aquat. Sci. 73 (12), 1733–1741. https://doi.org/10.1139/cjfas-2015-0546.

Lundberg, J.G., Kottelat, M., Smith, G.R., Stiassny, M.L.J., Gill, A.C., 2000. So many fishes, so little time: an overview of recent ichthyological discovery in continental waters. Ann. Mo. Bot. Gard. 87 (1), 26–62.

Lyons, J.E., Runge, M.C., Laskowski, H.P., Kendall, W.L., 2008. Monitoring in the context of structured decision-making and adaptive management. J. Wildl. Manag. 72 (8), 1683–1692. https://doi.org/10.2193/2008-141.

MacKenzie, D.I., 2005. What are the issues with presence-absence data for wildlife managers? J. Wildl. Manag. 69 (3), 849–860. https://doi.org/10.2193/0022-541x(2005)069[0849:watiwp]2.0.co;2.

Madliger, C.L., Love, O.P., Hultine, K.R., Cooke, S.J., 2018. The conservation physiology toolbox: status and opportunities. Conserv. Physiol. 6 (1), 1–16. https://doi.org/10.1093/conphys/coy029.

Malakoff, D., 2012. Researchers hail new restoration program funds. Science 337 (6090), 22. https://doi.org/10.1126/science.337.6090.22.

Manners, R.B., Doyle, M.W., 2008. A mechanistic model of woody debris jam evolution and its application to wood-based restoration and management. River Res. Appl. 24, 1104–1123. https://doi.org/10.1002/rra.

Martens, K.D., Devine, W.D., Minkova, T.V., Foster, A.D., 2019. Stream conditions after 18 years of passive riparian restoration in small fish-bearing watersheds. Environ. Manag. 63 (5), 673–690. Springer US. https://doi.org/10.1007/s00267-019-01146-x.

Mazeaud, M.M., Mazeaud, F., Donaldson, E.M., 1977. Primary and secondary effects of stress in fish: some new data with a general review. Trans. Am. Fish. Soc. 106 (3), 201–212.

McDonald, T., Gann, G.D., Jonson, J., Dixon, K.W., 2016. International Standards for the Practice of Ecological Restoration—Including Principles and Key Concepts. Society for Ecological Restoration, Washington, D.C.

McLennan, D., Auer, S.K., McKelvey, S., McKelvey, L., Anderson, G., Boner, W., Duprez, J.S., Metcalfe, N.B., 2021. Habitat restoration weakens negative environmental effects on telomere dynamics. Mol. Ecol. (April), 1–14. https://doi.org/10.1111/mec.15980.

Miller, J.R., Hobbs, R.J., 2016. Habitat restoration—do we know what we're doing? Restor. Ecol. 15 (2), 382–390.

Miller, J.R., Kochel, R.C., 2010. Assessment of channel dynamics, in-stream structures and post-project channel adjustments in North Carolina and its implications to effective stream restoration. Environ. Earth Sci. 59, 1681–1692. https://doi.org/10.1007/s12665-009-0150-1.

Moerke, A.H., Lamberti, G.A., 2003. Responses in fish community structure to restoration of two Indiana streams. N. Am. J. Fish Manag. 23 (3), 748–759. https://doi.org/10.1577/m02-012.

Moog, O., Chovanec, A., 2000. Assessing the ecological integrity of rivers: walking the line among ecological, political and administrative interests. Hydrobiologia 422–423, 99–109. https://doi.org/10.1007/978-94-011-4164-2_8.

Morandi, B., Piegay, H., Lamouroux, N., Vaudor, L., 2014. How is success or failure in river restoration projects evaluated? Feedback from French restoration projects. J. Environ. Manage. 137, 178–188. Elsevier Ltd https://doi.org/10.1016/j.jenvman.2014.02.010.

Morgan, J.D., Iwama, G.K., 1991. Effects of salinity on growth, metabolism, and ion regulation in juvenile rainbow and steelhead trout (*Oncorhynchus mykiss)* and fall Chinook salmon (*Oncorhynchus tshawytscha*). Can. J. Fish. Aquat. Sci. 48, 2083–2094.

Morgan, P., Aplet, G.H., Haufler, J.B., Humphries, H.C., Moore, M.M., Wilson, W.D., 1994. Historical range of variability: a useful tool for evaluating ecosystem change. J. Sustain. For. 2 (1–2), 87–111. https://doi.org/10.1300/J091v02n01_04.

Naeem, S., 2016. Biodiversity as a goal and driver of restoration. In: Palmer, M.A., Zedler, J.B., Falk, D.A. (Eds.), Foundations of Restoration Ecology. Island Press/Center for Resource Economics, Washington, DC, pp. 57–89, https://doi.org/10.5822/978-1-61091-698-1_3.

Naiman, R.J., Latterell, J.J., 2005. Principles for linking fish habitat to fisheries management and conservation. J. Fish Biol. 67 (Suppl. B), 166–185. https://doi.org/10.1111/j.0022-1112.2005.00921.x.

Nielsen, S.N., 2000. Thermodynamics of an ecosystem interpreted as a hierarchy of embedded systems. Ecol. Model. 135 (2–3), 279–289. https://doi.org/10.1016/S0304-3800(00)00379-3.

Niemi, G.J., McDonald, M.E., 2004. Application of ecological indicators. Annu. Rev. Ecol. Evol. Syst. 35 (Rapport 1992), 89–111. https://doi.org/10.1146/annurev.ecolsys.35.112202.130132.

Nilsson, C., et al., 2016. Evaluating the process of ecological restoration. Ecol. Soc. 21 (1), 41. https://doi.org/10.5751/ES-08289-210141.

Norris, R.H., Hawkins, C.P., 2000. Monitoring river health. Hydrobiologia 435, 5–17. https://doi.org/10.1023/A:1004176507184.

Opperman, J.J., Merenlender, A.M., 2004. The effectiveness of riparian restoration for improving instream fish habitat in four hardwood-dominated California streams. N. Am. J. Fish Manag. 24 (3), 822–834. https://doi.org/10.1577/m03-147.1.

Orth, D.J., White, R.J., 1993. Stream habitat management. In: Kohler, C., Hubert, W. (Eds.), Inland Fisheries Management in North America. American Fisheries Society, Bethesda, pp. 205–230.

Orth, R.J., et al., 2006. A global crisis for seagrass ecosystems. Bioscience 56 (12), 987–996. https://doi.org/10.1641/0006-3568(2006)56[987:AGCFSE]2.0.CO;2.

Paddack, M.J., et al., 2009. Recent region-wide declines in Caribbean reef fish abundacne. Curr. Biol. 19 (7), 590–595. https://doi.org/10.1016/j.cub.2009.02.041.

Palmer, M.A., Ambrose, R.F., Poff, N.L., 1997. Ecological theory and community restoration ecology. Restor. Ecol. 5 (4), 291–300. https://doi.org/10.1046/j.1526-100X.1997.00543.x.

Palmer, M.A., et al., 2003. Bridging engineering, ecological, and geomorphic science to enhance riverine restoration: local and national efforts. In: Proceedings of A National Symposium on Urban and Rural Stream Protection and Restoration. EWRI World Water and Environmental Congress, https://doi.org/10.1061/40695(2004)3.

Palmer, M.A., et al., 2005. Standards for ecologically successful river restoration. J. Appl. Ecol. 42 (2), 208–217. https://doi.org/10.1111/j.1365-2664.2005.01004.x.

Palmer, M.A., Menninger, H.L., Bernhardt, E.S., 2010. River restoration, habitat heterogeneity and biodiversity: a failure of theory or practice? Freshw. Biol. 55 (Suppl. 1), 205–222. https://doi.org/10.1111/j.1365-2427.2009.02372.x.

Palmer, M.A., Hondula, K.L., Koch, B.J., 2014. Ecological restoration of streams and rivers: shifting strategies and shifting goals. Annu. Rev. Ecol. Evol. Syst. 45, 247–269. https://doi.org/10.1146/annurev-ecolsys-120213-091935.

Pauly, D., Palomares, M.L., 2005. Fishing down marine food web: it is far more pervasive than we thought. Bull. Mar. Sci. 76 (2), 197–211.

Pickering, A.D. (Ed.), 1981. Stress and Fish. Academic Press, New York.

Power, M., 2002. Assessing fish population responses to stress. In: Adams, S.M. (Ed.), Biological Indicators of Aquatic Ecosystem Stress. American Fisheries Society, Bethesda, pp. 379–429.

Powers, S.P., Peterson, C.H., Grabowski, J.H., Lenihan, H.S., 2009. Success of constructed oyster reefs in no-harvest sanctuaries: implications for restoration. Mar. Ecol. Prog. Ser. 389, 159–170. https://doi.org/10.3354/meps08164.

Race, M.S., Fonseca, M.S., 1996. Fixing compensatory mitigation: what will it take? Ecol. Appl. 6 (1), 94–101. https://doi.org/10.2307/2269556.

Reid, A.J., et al., 2019. Emerging threats and persistent conservation challenges for freshwater biodiversity. Biol. Rev. 94 (3), 849–873. https://doi.org/10.1111/brv.12480.

Reid, W.V., Mooney, H.A., Cropper, A., Capistrano, D., Carpenter, S.R., Chopra, K., Dasgupta, P., Dietz, T., Duraiappah, A.K., Hassan, R., Kasperson, R., 2005. Ecosystems and Human Well-Being-Synthesis: A Report of the Millennium Ecosystem Assessment. Island Press, ISBN: 1-59726-040-1.

Rhoads, B.L., 2020. Flow dynamics in rivers. In: River Dynamics. Cambridge University Press, Cambridge, pp. 72–96.

Rice, C.D., Arkoosh, M.R., 2002. Immunological indicators of environmental stress and disease susceptibility in fishes. In: Biological Indicators of Aquatic Ecosystem Stress. American Fisheries Society, Bethesda, pp. 187–220.

Ricklefs, R.E., Wikelski, M., 2002. The physiology/life-history nexus. Trends Ecol. Evol. 17 (10), 462–468. https://doi.org/10.1016/S0169-5347(02)02578-8.

Rohr, J.R., Bernhardt, E.S., Cadotte, M.W., Clements, W.H., 2018. The ecology and economics of restoration: when, what, where, and how to restore ecosystems. Ecol. Soc. 23 (2), 15. https://doi.org/10.5751/ES-09876-230215.

Roni, P., Beechie, T.J., Bilby, R.E., Leonetti, F.E., Pollock, M.M., Pess, G.R., 2002. A review of stream restoration techniques and a hierarchical strategy for prioritizing restoration in Pacific northwest watersheds. N. Am. J. Fish Manag. 22 (1), 1–20. https://doi.org/10.1577/1548-8675(2002)022<0001:arosrt>2.0.co;2.

Roni, P., Hanson, K., Beechie, T.J., 2008. Global review of the physical and biological effectiveness of stream habitat rehabilitation techniques. N. Am. J. Fish Manag. 28 (3), 856–890. https://doi.org/10.1577/M06-169.1.

Rose, K.A., 2000. Why are quantitative relationships between environmental quality and fish populations so elusive? Ecol. Appl. 10 (2), 367–385. https://doi.org/10.1890/1051-0761(2000)010[0367:WAQRBE]2.0.CO;2.

Rosenfeld, J.S., Hatfield, T., 2006. Information needs for assessing critical habitat of freshwater fish. Can. J. Fish. Aquat. Sci. 63 (3), 683–698. https://doi.org/10.1139/f05-242.

Rosgen, D.L., 2011. Natural channel design: fundamental concepts, assumptions, and methods. Geophys. Monogr. Ser. 194, 69–93. https://doi.org/10.1029/2010GM000990.

Rosi-Marshall, E.J., Moerke, A.H., Lamberti, G.A., 2006. Ecological responses to trout habitat rehabilitation in a northern Michigan stream. Environ. Manag. 38 (1), 99–107. https://doi.org/10.1007/s00267-005-0177-3.

Rubin, Z., Kondolf, G.M., Rios-Touma, B., 2017. Evaluating stream restoration projects: what do we learn from monitoring? Water (Switzerland) 9 (3), 1–16. https://doi.org/10.3390/w9030174.

Ruiz-Jaen, M.C., Aide, T.M., 2005. Restoration success: how is it being measured? Restor. Ecol. 13 (3), 569–577. https://doi.org/10.1111/j.1526-100X.2005.00072.x.

Schreck, C.B., Tort, L., 2016. The concept of stress in fish. In: Schreck, C.B., Tort, L., Farrell, A.P., Brauner, C.J. (Eds.), Biology of Stress in Fish. Academic Press, London, pp. 1–34.

Schumm, S.A., 2005. River Variability and Complexity. Cambridge University Press, Cambridge.

Selye, H., 1976. The stress concept. Can. Med. Assoc. J. 115, 78.

SER (Society for Ecological Restoration) International Science & Policy Working Group, 2004. The SER International Primer on Ecological Restoration. Accessed in July 2005. Society for Ecological Restoration International, Tucson, Arizona. Accessed in December 2021.

Shields, F.D., Copeland, R.R., Klingeman, P.C., Doyle, M.W., Simon, A., 2003. Design for stream restoration. J. Hydraul. Eng. 129 (8), 575–584. https://doi.org/10.1061/(asce)0733-9429(2003)129:8(575).

Shorthouse, C., Arnell, N., 1999. The effects of climatic variability on spatial characteristics of European river flows. Phys. Chem. Earth Part B 24 (1–2), 7–13. https://doi.org/10.1016/S1464-1909(98)00003-3.

Simon, H.A., 1962. The architecture of complexity. Proc. Am. Philos. Soc. 106 (6), 467–482. https://doi.org/10.1007/978-3-642-27922-5_23.

Simon, T.P., 2000. The use of biological criteria as a tool for water resource management. Environ. Sci. Policy 3 (Suppl. 1), 43–49. https://doi.org/10.1016/s1462-9011(00)00026-5.

Simon, T.P., Evans, N.T., 2017. Environmental Quality Assessment Using Stream Fishes, Methods in Stream Ecology. Elsevier Inc., https://doi.org/10.1016/B978-0-12-813047-6.00017-6.

Skidmore, P., Beechie, T., Pess, G., Castro, J., Cluer, B., Thorne, C., Shea, C., Chen, R., 2012. Developing, designing, and implementing restoration projects. In: Stream and Watershed Restoration: A Guide to Restoring Riverine Processes and Habitats. Wiley-Blackwell, Chichester, pp. 215–253.

Smith, E.P., 2002. Statistical considerations in the development, evaluation, and use of biomarkers in environmental studies. In: Adams, S.M. (Ed.), Biological Indicators of Aquatic Ecosystem Stress. American Fisheries Society, Bethesda, pp. 565–590.

Sopinka, N.M., Donaldson, M.R., O'Connor, C.M., Suski, C.D., Cooke, S.J., 2016. Stress indicators in fish. In: Fish Physiology. Academic Press, pp. 405–462, https://doi.org/10.1016/B978-0-12-802728-8.00011-4.

Stoll, S., Jochem, K., Lorenz, A.W., Sundermann, A., Haase, P., 2014. The importance of the regional species pool, ecological species traits and local habitat conditions for the colonization of restored river reaches by fish. PLoS One 9 (1), 1–10. https://doi.org/10.1371/journal.pone.0084741.

Strailey, K.K., Osborn, R.T., Tinoco, R.O., Cienciala, P., Rhoads, B.L., Suski, C.D., 2021. Simulated instream restoration structures offer smallmouth bass (*Micropterus dolomieu*) swimming and energetic advantages at high flow velocities. Can. J. Fish. Aquat. Sci. 78 (1), 40–56. https://doi.org/10.1139/cjfas-2020-0032.

Strayer, D.L., Dudgeon, D., 2010. Freshwater biodiversity conservation: recent progress and future challenges. J. N. Am. Benthol. Soc. 29 (1), 344–358. https://doi.org/10.1899/08-171.1.

Suding, K., 2011. Understanding successes and failures in restoration ecology. Annu. Rev. Ecol. Evol. Syst. 42, 465–487. https://doi.org/10.1146/annurev-ecolsys-102710-145115.

Sundermann, A., Stoll, S., Haase, P., 2011. River restoration success depends on the species pool of the immediate surroundings. Ecol. Appl. 21 (6), 1962–1971. https://doi.org/10.1890/10-0607.1.

Swetnam, T.W., Allen, C.D., Betancourt, J., 1999. Applied historical ecology: using the past to manage for the future. Ecol. Appl. 9 (4), 1189–1206.

Thomas, G., Lorenz, A.W., Sundermann, A., Haase, P., Peter, A., Stoll, S., 2015. Fish community responses and the temporal dynamics of recovery following river habitat restorations in Europe. Freshwater Sci. 34 (3), 975–990. https://doi.org/10.1086/681820.

Tonkin, J.D., Stoll, S., Sundermann, A., Haase, P., 2014. Dispersal distance and the pool of taxa, but not barriers, determine the colonisation of restored river reaches by benthic invertebrates. Freshwater Biol. 59 (9), 1843–1855. https://doi.org/10.1111/fwb.12387.

Tucker, E.K., Zurliene, M.E., Suski, C.D., Novak, R.A., 2020. Gonad development and reproductive hormones of invasive silver carp (*Hypophthalmichthys molitrix*) in the Illinois River. Biol. Reprod. 102 (3), 647–659. https://doi.org/10.1093/biolre/ioz207.

Ullah, H., Nagelkerken, I., Goldenberg, S.U., Fordham, D.A., 2018. Climate change could drive marine food web collapse through altered trophic flows and cyanobacterial proliferation. PLoS Biol. 16 (1), e2003446. https://doi.org/10.1371/journal.pbio.2003446.

United Nations General Assembly, 2019. Resolution 73/284: United Nations Decade on Ecosystem Restoration (2021 – 2030) (1 March 2019). [Online]. A/RES/73/284. Accessed 6 December 2021.

US-Environmental Protection Agency, 2000. The Quality of Our Nation's Waters. EPA-841-R-02–001. US EPA.

Vehanen, T., Huusko, A., Maki-Petays, A., Louhi, P., Mykdra, H., Muotka, T., 2010. Effects of habitat rehabilitation on brown trout (Salmo trutta) in boreal forest streams. Freshwater Biol. 55 (10), 2200–2214. https://doi.org/10.1111/j.1365-2427.2010.02467.x.

Vörösmarty, C.J., et al., 2010. Global threats to human water security and river biodiversity. Nature 467, 555–561. https://doi.org/10.1038/nature09440.

Walters, C., 1986. In: Getz, W.M. (Ed.), Adaptive Mangement of Renewable Resources. Macmillan Publishing Company, New York.

Waltham, N.J., et al., 2020. UN decade on ecosystem restoration 2021–2030—what chance for success in restoring coastal ecosystems? Front. Mar. Sci. 7 (February 2020), 1–5. https://doi.org/10.3389/fmars.2020.00071.

Weinstein, M.P., Litvin, S.Y., Guida, V.G., 2009. Essential fish habitat and wetland restoration success: a tier III approach to the biochemical condition of common mummichog *Fundulus heteroclitus* in common reed *Phragmites australis*- and smooth cordgrass *Spartina alterniflora*-dominated salt marshes. Estuar. Coasts 32, 1011–1022. https://doi.org/10.1007/s12237-009-9185-5.

Weissburg, M.J., et al., 2005. Sensory biology: linking the internal and external ecologies of marine organisms. Mar. Ecol. Prog. Ser. 287, 263–307. https://doi.org/10.3354/meps287263.

Wendelaar Bonga, S.E., 1997. The stress response in fish. Physiol. Rev. 77 (3), 591–625. https://doi.org/10.1152/physrev.1997.77.3.591.

White, P.S., Walker, J.L., 1997. Approximating nature's variation: selecting and using reference information in restoration ecology. Restor. Ecol. 5 (4), 338–349.

Wohl, E., et al., 2005. River restoration. Water Resour. Res. 41, W10301. https://doi.org/10.1029/2005WR003985.

Wohl, E., Lane, S.N., Wilcox, A.C., 2015. The science and practice of river restoration. Water Resour. Res. 51, 5974–5997. https://doi.org/10.1002/2014WR016874.Received.

Woodworth, P., 2006. What price ecological restoration? In putting a price tag on endangered species and degraded ecosystems, ecologists and economists have joined forces to formulate a new rationale for environmental issues: restoring natural capital. Scientist 20 (4), 39–45.

Wortley, L., Hero, J.M., Howes, M., 2013. Evaluating ecological restoration success: a review of the literature. Restor. Ecol. 21 (5), 537–543. https://doi.org/10.1111/rec.12028.

Yanagitsuru, Y.R., Davis, B.E., Baerwald, M.R., Sommer, T.R., Fangue, N.A., 2022. Using physiology to recover imperiled smelt species. Fish Physiol. 39B, 1–37.

Zampella, R.A., Bunnell, J.F., 1998. Use of reference-site fish assemblages to assess aquatic degradation in pinelands streams. Ecol. Appl. 8 (3), 645–658.

Zedler, J.B., 2007. Success: an unclear, subjective descriptor of restoration outcomes. Ecol. Restor. 25 (3), 162–168. https://doi.org/10.3368/er.25.3.162.

Chapter 9

# A conservation physiological perspective on dam passage by fishes

**Scott G. Hinch[a,*], Nolan N. Bett[a], and Anthony P. Farrell[b]**
[a]*Department of Forest and Conservation Sciences, University of British Columbia, Vancouver, BC, Canada*
[b]*Department of Zoology, University of British Columbia, Vancouver, BC, Canada*
[*]*Corresponding author: e-mail: scott.hinch@ubc.ca*

## Chapter Outline



For many freshwater and diadromous fishes, dams create a significant conservation challenge by fragmenting migratory routes, modifying habitats and altering patterns of water movement. Despite advances in fish passage strategies and infrastructure, passage can still be delayed or prevented. Fish can experience a variety of physiological issues as a consequence of both attempted and successful passage with upstream migrants affected by disorientation, energetics and anaerobic processes associated with movements through dam tailraces. Whereas fishway ascension has generally been found to be neither seriously physiologically stressful nor energetically taxing, disorientation

**Fish Physiology, Vol. 39B. https://doi.org/10.1016/bs.fp.2022.04.015**

and stress from dam passage can be severe in downstream migrants, compromising energy reserves, immune systems, and smoltification (in salmonids). There are also many ways that downstream migrants can be seriously injured or delayed during dam passage. All of these negative physiological responses can be compounded by carryover effects that continue to threaten fish survival and fitness even after successful passage. Physiological assessments have played a significant role in helping those who manage dams, water flow, and passage infrastructure gain insight into passage problems. A few physiological studies have guided modifications to passage infrastructure or led to changes in operational approaches, including addressing physiologically mediated carry-over effects. These efforts demonstrate the importance of an interdisciplinary and integrative research and monitoring approach, including conservation physiology, for improving dam passage.

# 1 General introduction

The objective of this chapter is to review the conservation physiology associated with dam passage in fish. This is an important issue because dams are so abundant on the landscape and they are crucial for the management of water resources. They can reduce flooding, provide and protect potable water supplies, generate hydropower and facilitate recreational and other commercial activities. More than 58,700 dams that exceed 15 m in height exist worldwide (International Commission on Large Dams, 2018), with an additional 3700 hydroelectric dams of at least 1 MW capacity planned or under construction (Zarfl et al., 2015). The highest density of large dams exists in North America, Europe, Southern Africa, East Asia, and South Asia, while China, India and African countries currently lead in the construction and planning of large dams (Wang et al., 2014). Lower head dams (<15 m high), meanwhile, exist in substantially larger numbers although they are difficult to detect and enumerate. In Europe, for example, they are believed to account for >99% of all barriers (Belletti et al., 2020).

Unfortunately, damming of rivers has a major impact on freshwater habitat and threatens the sustainability of freshwater and diadromous fishes (Lehner et al., 2011). Dams fragment natural watersheds, which can obstruct fish dispersal and the migration routes crucial for spawning or feeding (Barbarossa et al., 2020), often leading to extirpation (Nehlsen et al., 1991; Slaney et al., 1996). Dams impede the downstream transport of sediments that replenishes eroding beaches (Wang et al., 2016). Dams also alter natural river flow regimes, either by holding back water and reducing seasonal peak flows, or by drawing so much water that the downstream habitat becomes marginalized or even fragmented as unconnected pools. Downstream water temperatures, which have significant effects on fish physiology and survival, can be profoundly affected in one of three ways. Foremost, the impounded surface water behind the dam (aka the reservoir) can become warmer than normal during summer months (Barbarossa et al., 2020; Wang et al., 2014) and become thermally stratified if sufficiently deep. Second, downstream water temperatures can become much warmer if water is released from the surface epilimnion or much cooler if released from the deeper hypolimnion (Thornton et al., 1990). Lastly, if very little water discharge is flowing downstream, summer months can see unusually high diel thermal fluctuations resulting from direct insolation of the water, as well as radiation, reflection and conduction from surrounding rocks.

The most common infrastructure used to assist upstream migrations is a fishway (aka fish ladders) that facilitates volitional upstream swimming past the dam. Fishway designs vary tremendously in style, size and permanency (reviewed in Clay, 1995; Odeh, 1999). The use of "nature-like" fishways, which are intended to mimic natural side channels, has become popular (Parasiewicz et al., 1998). Also, where fishways were not or could not be built, fish can be transported upstream over a dam using operator-controlled elevators (lifts) or locks (Travade and Larinier, 2002), and even using barges or truck operations (Bond et al., 2017; Muir et al., 2006). It is troubling, however, that only a small number of large dams, particularly older ones, have upstream passage facilities. Indeed, only 34 of the 280 non-federal hydroelectric projects in the United States (a mere 12%) had measures in place to allow upstream passage (Čada, 1998). Furthermore, low head dams rarely have fish passage assists. Equally troubling is that fishway designs have largely focused on assisting economically important, anadromous fish species (Roscoe and Hinch, 2010) and, as a result, many non-target species are not able to adequately utilize them (Office of Technology Assessment (OTA), 1995). This issue is compounded by the fact that studies indicate salmonids (*Oncorhynchus* spp.) can be nearly three-times more successful than non-salmonids in passing upstream (61.7% vs 21.1%) (and nearly twice as successful in passing downstream [74.6% vs 39.6%] [Noonan et al., 2012]).

Downstream migrants can also use fishways, and other passage routes (Nyqvist et al., 2017a,b,c,d, 2016). Water spillways, whose original purpose may have been to regulate or lower water levels behind a dam, can often allow fish passage for downstream migrations. If electricity is being generated at the dam, fish can also pass through the turbine system. However, neither of these two downstream pathways is ideal because fish can suffer physical injury or be killed especially during passage through turbines (Ferguson, 2005; Ferguson et al., 2006; Muir et al., 2001). Infrastructure can be added to assist downstream migrations: screens, angled bar racks and enhanced surface bypasses have been used to divert migrating fish from passing through turbines (Larinier, 2001). Consequently, fishways have become increasingly sophisticated with designs emerging as a product of collaboration between hydraulic engineers and biologists (Castro-Santos et al., 2009), although most designs are focused on enabling upstream movement, with relatively little concern for moving fish downstream.

Noonan et al. (2012) provided the most recent review of fish passage efficiency when they examined 65 such studies. Many factors were implicated in the success of fish to successfully utilize fish passage facilities. They found, as might be expected, that upstream passage efficiency (41.7%) was lower than downstream passage efficiency (68.5%), which they attributed to a more directed and energetically demanding upstream swimming by fish vs more "passive" downstream movements. Facility characteristics were also important. Pool-and-weir, pool-and-slot and nature-like fishways had the highest efficiencies, whereas Denil fishways, which use a series of symmetrical, close-spaced baffles to slow flowing water, and fish locks/elevators had the lowest efficiencies. Noonan et al. (2012) also found that upstream passage efficiency was reduced with fishway slope, but increased with fishway length, and water velocity.

Many fish passage facilities still prevent or delay passage of both target (Aarestrup et al., 2003; Bunt et al., 2000; Caudill et al., 2007; Naughton et al., 2005) and non-target species (Haro and Kynard, 1997; Mallen-Cooper and Brand, 2007; Parsley et al., 2007). In their review of European passage facilities, Fjeldstad et al. (2018) found few well-functioning facilities for downstream migration of fish. Thus, the presence of passage infrastructure alone does not guarantee that fish are able to surmount barriers to their movement (Noonan et al., 2012; Nyqvist et al., 2017a,b,c,d, 2016; Roscoe and Hinch, 2010). After all, successful dam passage should not be ascribed to a semelparous or iteroparous fish that has only successfully ascended a fishway, but is in such a bad physiological state afterwards that it might die, be more easily captured by predators or otherwise fail to reproduce (i.e., carryover effects and fitness consequences).

Fish morphology, origin and behavior can contribute to passage success or failure. For example, larger bodied and stronger swimming species are often more successful (Bunt et al., 2012), such as salmonids, which are often more successful than non-salmonids (Noonan et al., 2012). Wild fish may be more successful than hatchery fish, perhaps because hatchery fish are less motivated to migrate to natal areas (Cooke and Hinch, 2013). Inter-individual differences in swimming behaviors have also been linked to passage success (Castro-Santos, 2005). For example, adult sockeye salmon (*O. nerka*) that exhibited fast and erratic swim speeds were less likely to pass the Hells Gate fishway in Canada than those exhibiting slower, steadier speeds (Hinch and Bratty, 2000). Ultimately physiological processes underlie swimming abilities and the motivation to pass through dam infrastructure, as well as the level of stress imposed by fish passage. Thus, to more holistically understand fish passage success and failure, a physiological perspective is needed. Yet, Roscoe and Hinch (2010) identified only 7 studies out of 96 studies (<10%) which had examined relationships between physiology and fish passage success in their review (Fig. 1). Only a few additional studies that directly incorporate

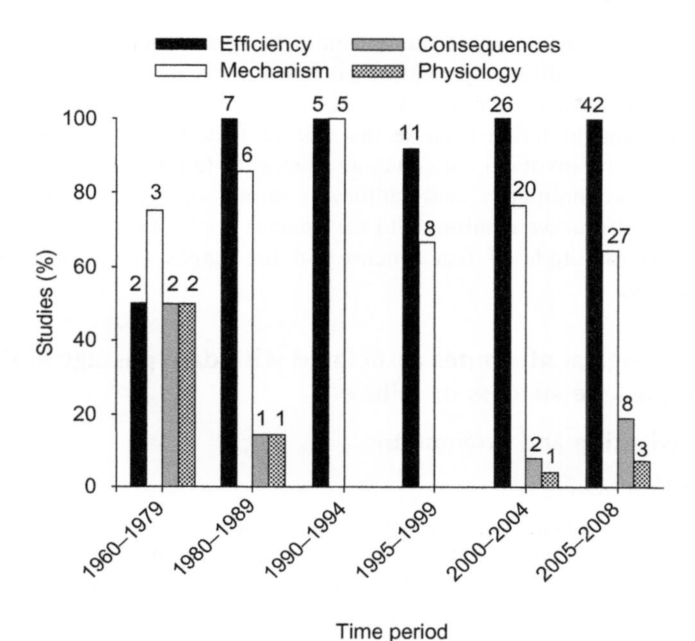


**FIG. 1** Percentage of passage evaluation articles published in different time periods that included four types of research questions. "Efficiency" questions quantified the proportion of individuals able to pass a structure, or qualitatively assessed which species in a community were able to pass. "Mechanism" questions examined environmental, biological or structural factors that affected passage. "Consequences" questions quantified post-passage effects on individual fish. "Physiology" questions examined the relationship between passage and fish physiology. Numerals above bars are the number of studies represented by percentages. *Used with permission from Roscoe, D.W., Hinch, S.G., 2010. Effectiveness monitoring of fish passage facilities: historical trends, geographic patterns and future directions. Fish Fish 11, 12–33.*

physiological measures into fish passage research have been published since that review (e.g., Ammar et al., 2020; Bido et al., 2018; Burnett et al., 2014a,b; Cocherell et al., 2011; Roscoe et al., 2011), so this is still an emerging and evolving field.

Our chapter focusses on the efficacy of these various infrastructure passage options not just by considering the percentage of fish that successfully pass, but by also considering the physiological state of the fish before, during and after passage. In particular, we examine the specific physiological attributes and mechanisms associated with successful fish passage, and consider what physiological issues are emerging for failed passage at dams. We will also examine the strong potential for carryover effects and fitness consequences because attraction into, and successful passage through dam infrastructure can have large negative impacts on fish physiology.

Lastly, we will consider the fisheries and habitat management applications, and potential utilization of physiological information for improving passage. Unfortunately, most of our review draws primarily, though not exclusively, on the salmonid literature because the science underlying the physiological attributes that are involved with passage success or failure has been primarily focussed on economically and culturally important species such as salmonids. Therefore, we caution as to the general applicability of our conclusions to the multitude of fish species and life stages that utilize passage infrastructure.

# 2 Physiological attributes associated with dam passage and their roles in passage success or failure

## 2.1 Navigation and orientation

### 2.1.1 Olfaction

Many fish behaviors are mediated by complex sensory systems (see Chapter 2, Volume 39A: Horodysky et al., 2022). With respect to fish migrations, the use of olfactory cues by migrating salmonids has been well documented, but the effects that dams have on such olfactory-mediated movement have only recently begun to emerge. Juvenile salmonids "imprint" on the unique odor of their natal water, and returning adults use their sense of smell to detect these imprinted odors. Olfactory imprinting is believed to occur during the parr-smolt transformation as juveniles prepare for their downstream migration to the ocean (Cooper and Hasler, 1974; Cooper et al., 1976; Dittman et al., 1997; Yamamoto et al., 2010). Concurrent with this imprinting is the above-mentioned surge in circulating thyroxine (Morin et al., 1994) which may even regulate the imprinting process (Dittman and Quinn, 1996; Lema and Nevitt, 2004). All the same, recent research has confirmed that odor imprinting can occur even earlier at the alevin stage in some species like pink salmon (*O. gorbuscha*), which migrate to the ocean immediately following emergence (Bett et al., 2016a). Likewise, and perhaps as might be expected, pink salmon also emerge with their gill $Na^+/K^+$-ATPase activity partially prepared for seawater entry (Sackville et al., 2012). This odor imprinting process, however, can be disrupted by activities that are designed to facilitate dam passage. One such activity is the collection at dams and trucking or barging downstream of juvenile salmon in the Columbia River basin, USA, which is sometimes used to increase dam passage survival (the alternative is passage through turbines) and to reduce interactions with predators (Chapman et al., 1997). However, when hatchery-reared, fall-run Chinook salmon *(O. tshawytscha)* were PIT-tagged, the returning adults were 10–19 times more likely to stray (i.e., they did not arrive to their natal streams) if they had been barged as juveniles (Bond et al., 2017). The authors speculated that the barging process, which pumps water from the centre of the large navigational channels, may not have provided the appropriate natal chemical cues for imprinting, or the fish were not yet physiologically prepared to imprint when they were barged. The adult straying was further exacerbated by high river temperatures (Bond et al., 2017) because upriver migrants sought thermal refuges in cool water tributaries that were not on their homeward routes (Goniea et al., 2006).

Even with successful imprinting of out-migrating juvenile salmonids, dams and their hydro-power systems can negatively affect the detection of natal cues for returning adults if olfactory cues are diluted in some manner. Indeed, the multiple dam impoundments that characterize the Columbia River, USA, have led to "odor diffusion" from natal tributaries across river channels (Keefer and Caudill, 2014; Quinn, 2018), which might contribute to explaining the wandering behaviors exhibited by adult salmon in this system (Keefer et al., 2008). Also, diverting natal water sources through discharge channels can inadvertently attract homing adult salmon to these power station outlets along migration routes, presumably through an olfactory attraction to these sites (Andrew and Geen, 1958; Middleton et al., 2018; Thorstad et al., 2003, 2008). Such false olfactory cues can then increase search time for natal streams and delay migration (Caudill et al., 2007; Fretwell, 1989).

Middleton et al. (2018) sampled blood from over 300 adult sockeye salmon from two populations that were then radio-tracked as they migrated upstream past a power generating station on a tributary of the Fraser River, Canada, which emanated daily varying concentrations of their natal water. Migrants must pass this facility (Fig. 2) and enter the Seton River 1.5 km further along to reach their natal spawning grounds. The levels of natal water concentration in the powerhouse effluent and the Seton River are managed by the local power company to minimize "olfactory confusion" and delay by homeward migrating adult salmon (Bett et al., 2020). Individuals were nonetheless attracted to the power house effluent and delayed there on average 5–25 h (summer-run vs fall-run populations, respectively). The delay of the summer-run salmon was further shortened if they had an elevated blood glucose level. Unlike the fall-run, the summer-run salmon experienced relatively high migration temperatures during their 312 km migration before reaching the Seton River and so their elevated glucose could reflect enhanced stress and reduced aerobic capacity. Entering the Seton River can alleviate the thermal stress because it is several degrees cooler than the Fraser River in the summer. This finding further suggests that migratory stressors like high temperatures can override certain olfactory cues to assist physiological recovery of exhausted or stressed individuals.

In the first whole watershed experiment of its kind, Drenner et al. (2018) manipulated olfactory cues in the Seton River (Fig. 2) by adjusting

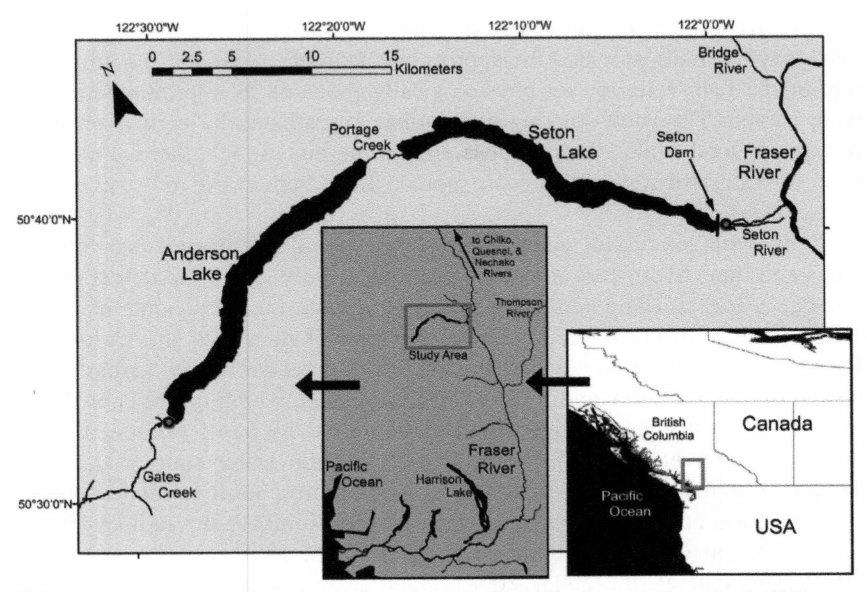


**FIG. 2** Each summer, adult sockeye salmon from the Gates Creek population migrate 312km up the Fraser River from the Pacific Ocean to the Seton River, past the Seton Generating Station, then westward in the Seton River and through the Seton Dam fishway, and then migrate an additional 40km through two lakes before arriving at their spawning stream. Inset maps show the study location in western Canada, and in the lower Fraser River. The white dot on the right indicates the location of numerous hydrological and fish tracking assessments discussed in this chapter. The white dot on the left indicates the entrance to spawning areas and further tracking assessments. *Modified and used with permission from Bass, A.L., Hinch, S.G., Casselman, M.T., Bett, N.N., Burnett, N.J., Middleton, C.T., Patterson, D.A., 2018. Visible gill-net injuries predict migration and spawning failure in adult sockeye salmon. Trans. Am. Fish. Soc. 147, 1085–1099.*

hydropower operations and followed the upstream migration behavior of adult sockeye salmon fitted with radio tags. This was the same summer-run population and study system used by Middleton et al. (2018) above. An experimental reduction to 72% of the natal water olfactory cue in the Seton River was associated with 80% reduced odds of a radio tagged fish entering the Seton River from the Fraser River mainstem (Fig. 2), a behavioral decision that would have negative carryover effects for spawning success.

Pheromones are another cue used for navigation. For example, a Y-maze experiment exposed adult sockeye salmon to conspecific odors with and without an imprinted natal cue (Bett and Hinch, 2015). Sockeye salmon were attracted to conspecifics only when imprinting was absent, suggesting that pheromones provided a directional cue secondary to a primary odor imprint.

Such a "hierarchical approach to navigation" (Bett and Hinch, 2016) could increase the reproductive success of stray migrants, promoting dispersal and genetic diversity. At fishways, high conspecific presence and abundance could attract stray individuals from nearby populations, as was witnessed anecdotally in the Seton River (Bett et al., 2017).

Conspecifics can alert each other using disturbance cues, chemicals released by a fish in the presence of a threat (Bett et al., 2016b; Byford et al., 2016). In fact, a Y-maze experiment (Bett et al., 2016b) revealed adult migrating sockeye salmon avoided the odor of a disturbed conspecific (an air-exposed fish held in confinement) independent of sex. Yet, females elevated their plasma cortisol, signaling a clear physiological stress (see Section 2.2) in response to the pheromone. Such signaling could be important at fishways. While avoidance of disturbed conspecifics at passage structures could limit exposure to risks, it could also delay the migration or even limit conspecific assistance in attraction to fishways. Disturbance cues, however, could be used to guide migrants towards structures if they could be strategically deployed (Byford et al., 2016).

### 2.1.2 *Rheotaxis and response to flow fields*

Diadromous fish typically undertake rheotactically directed migrations, using flow fields and changes in flow velocities to orient and locate, or avoid, entrances to passage structures. The lateral line, which extends from the head to the tail along both sides of the body, is a sensory structure that consists of a dense array of neuromasts functioning as water pressure sensors (Bleckmann, 2008; Chagnaud et al., 2008). This organ enables fish to very rapidly detect changes in water velocity due to specialized synaptic connections (Bleckmann, 1994) and greatly aids in rheotaxis, especially in turbulent water (see Chapter 3, Volume 39A: Castro-Santos et al., 2022). Sufficient water flow is an important facility characteristic for enhancing passage success (Larinier, 2001). Upstream migrating fish, which are positively rheotactic, are often drawn to areas of higher flow, so higher-velocity water emanating near the entrance of passage infrastructure may improve attraction efficiency particularly with large structure fishways (Bunt et al., 2012). Downstream migrating fish, which are negatively rheotactic, also benefit from higher flows. For example, when spill flows were enhanced, over 70% of the Atlantic salmon (*Salmo salar*) kelts (post-spawn adults that are migrating back to the ocean) used the spillway at a dam on the River Klarälven in Sweden compared to only 16% that passed through an alternative turbine route (Nyqvist et al., 2016). Similarly, downstream-migrating Atlantic salmon smolts were more attracted to fish by-passes and spillways with higher flows in the Winooski River, Vermont (Nyqvist et al., 2017a). Physical structures, such as fine-mesh trash racks with alternative escape routes and bypass arrangements, have improved downstream passage efficiency to 90% at some facilities for brown trout (*Salmo trutta*), Atlantic salmon and European eels (*Anguilla anguilla*) (Fjeldstad et al., 2018).

Swimming behaviors largely determine the success of fish as they navigate complex flow fields around dams, and much of the work on this topic has focused on salmonids. Downstream migrating juvenile salmonids use surface-oriented, active swimming that can generate ground speeds often greater than that of the river's bulk water flow in hydraulically simple reaches (Achord et al., 1996; Hockersmith et al., 2003; Smith et al., 2002). In hydraulically more complex reaches, ground speeds may be similar to, or slower than, riverine speeds (Clark et al., 2016). A controlled flume experiment with Chinook salmon (Enders et al., 2012) discovered that while smolts actively swam into the current whenever they drifted into areas with a reduced water velocity, they avoided rapidly accelerating flows. Fish screens or other structures have been installed at many dams to guide fish away from turbines or attract them to bypasses, spillways or fishways (Algera et al., 2020; Crew et al., 2017). However, these guidance structures rapidly decelerate and accelerate water flows which could result in less effective directional cues for the desirable passage pathway (Enders et al., 2012). Goodwin et al. (2014) combined a computational fluid dynamics model of dam flow-fields with a behavioral model that allowed the model fish to adjust swim orientation and speed in relation to encountered water acceleration and pressure. They then fit the combined model to data on the passage of juvenile Pacific salmonids at 7 dams in the Columbia/Snake River system and were able to reproduce observed fish movement and passage patterns across 47 flow-field conditions. Their results confirmed that juvenile salmonids can perceive and react to a vast range of flow fields and conditions (water velocities and accelerations) near to dams, and they concluded that fish behavioral choices were a key component of passage performance.

For upriver migrating fish, numerous tracking studies have confirmed that swimming patterns are strongly affected by hydrodynamics in dam or fishway tailraces (see review, Bunt et al., 2012). By orientating into the current, salmon derive an energetic benefit from ram gill ventilation, which is estimated to save up to 10% of metabolic rate (Farrell and Steffensen, 1987). Complex flow patterns can delay upstream progress for hours to days, suggesting a difficulty in locating clear flow cues (Brown and Geist, 2002; Hinch and Rand, 1998; Standen et al., 2002). Delays of days will deplete finite energy stores at a rate dependent on the fish's standard metabolic rate, which increases exponentially with temperature (Clark et al., 2008; Eliason et al., 2011; Farrell, 2007; Steinhausen et al., 2008). Even if the appropriate directional flow cues are located, extremely fast water flows may surpass burst swimming capabilities and consequently prevent passage by some migrants. Dam tailraces or entrances to fishways can create situations like this (e.g., Burnett et al., 2014a,b; Hinch and Bratty, 2000; Li et al., 2021).

Path selection through complex bottom topography can play a significant role in reducing energetic costs of swimming upstream when flow conditions are slow and salmon can receive forward-assists during their upstream migration. For example, underwater stereovideography (Standen et al., 2004) revealed that adult sockeye salmon selected low speed pathways to migrate upstream and the current speeds they encountered were lower than the mean current speeds measured in the immediate vicinity of the fish. Electromyogram (EMG) telemetry studies have further revealed that sockeye salmon can move faster through water than their tails are propelling them (Hinch and Rand, 1998, 2000). These fish are likely locating and exploiting very small-scale reverse-flow vortices created by rough substrates or banks (Vogel, 1994). However, these naturally occurring "energetically-advantageous" reverse-flow fields may be less common or absent in dam tailraces owing to the artificial nature of the structures.

Directly assessing swimming speeds and behaviors requires sophisticated tracking technologies that are not commonly used in most dam passage studies. Nevertheless, Hinch and Bratty (2000) did use EMG radio-transmitters to measure swim speeds and passage success for adult sockeye salmon migrating through the Hell's Gate Fishway, BC, Canada, a reach in the Fraser River canyon that is notorious for creating hydraulic conditions that impede salmon migrations. For fish that successfully entered the fishway, swim speeds in the approach zone were less than half those of unsuccessful migrants ($1.85\,BL \cdot s^{-1}$ vs $4.23\,BL \cdot s^{-1}$, respectively). Visual observations of the flow dynamics suggested that the unsuccessful migrants swam faster because they used relatively turbulent and fast-flowing migration pathways. Whether this was a bad behavioral choice or forced upon these fish for some reason is unknown. In addition, Pon et al. (2009a) used EMG telemetry to assess swim speeds and behaviors adopted by adult sockeye salmon to ascend a vertical-slot fishway on the Seton River, British Columbia, Canada (Fig. 2). Tagged fish placed in a lower section of the fishway were tracked to the exit. Unlike Hinch and Bratty (2000), no difference in mean swim speeds was found between successful and unsuccessful migrants. However, and importantly, the maximum water velocities in the Seton River fishway were substantially lower ($\sim 2\,m\,s^{-1}$) compared with average surface velocities of $6\,m\,s^{-1}$ in the approach to the Hell's Gate fishway (Hinch and Bratty, 2000).

A complete understanding of how fish respond to hydrodynamics at passage structures for management purposes will need direct measures of the flow fields that fish encounter, as well as the fish's physiological and behavioral responses. To this end, Li et al. (2021) helped bridge this knowledge gap by examining the flow field downstream of the Seton Dam Fishway (British Columbia, Canada; Fig. 2) using both acoustic Doppler current profilers (ADCPs) and computational fluid dynamic (CFD) modeling to generate a dynamic picture of the flow environment through which fish must pass.

These were then linked to swim speeds and passage trajectories from telemetry of individual fish (see Energetics case study).

### 2.1.3 Phototaxis and responses to light

Many freshwater and anadromous fish, such as sturgeon or juvenile salmonids, are often negatively phototactic and migrate nocturnally, presumably reducing predation risk (Beauchamp, 1995; Elvidge et al., 2019; Ginetz and Larkin, 1976; Hartman et al., 1962). Sockeye salmon smolts, for example, only migrate at night when water is slow moving and clear, but will migrate throughout the day and night when water is turbid or turbulent and predation risk is reduced (Clark et al., 2016; Furey et al., 2015). Sockeye fry delay downstream migration if they encounter nocturnal light pollution from urban areas but, in doing so, suffer increased predation (Tabor et al., 2004). These empirical findings highlight the importance of an evolved phototactic response to natural (and artificial) light regimes in controlling downstream migrations and protecting fish from predation, and indicate that light conditions and their management will be an extremely important issue at downstream passage infrastructure. For example, recent studies on lake sturgeon (*Acipenser fulvescens*) and white sturgeon (*A. transmontanus*) show that strobing lights can be effective in deterring downstream migrants away from hazardous and potentially lethal hydropower infrastructure (Elvidge et al., 2019; Ford et al., 2017). In addition, adequate light levels may be needed for fish to avoid screens which are installed to divert them away from turbines and toward by-pass structures. Indeed, a laboratory flume experiment found that shiner surfperch (*Cymatogaster aggregata)* and staghorn sculpin (*Leptocottus armatus)* contacted such screens less frequently during the day than at night (Mussen and Cech, 2019). When strobe lights and industrial vibrators were used in association with the screens, surfperch contacted the vibrating screens less frequently whereas sculpins contacted the strobe-illuminated screens less frequently. A laboratory flume experiment on white sturgeon and green sturgeon (*A. medirostris*), meanwhile, found that white sturgeon contacted screens more frequently during the night than during the day, whereas green sturgeon did not (Poletto et al., 2014). Clearly, sensory utilization of light and vibration for increased visual and mechano-receptor guidance at night is a species-specific consideration for fishway infrastructure.

Adult salmonids can certainly migrate upstream during both daylight and nighttime under certain environmental conditions but species-specific considerations are still needed. Upstream migrating adult pink salmon in the Fraser River, British Columbia Canada, will not ascend rapids in the dark, whereas comigrating adult sockeye salmon will (Hinch et al., 2002; Standen et al., 2002). Several species have been observed to migrate through long and dark tunnels associated with passage infrastructure (Fjeldstad et al., 2018), and passage can be enhanced by "darkening" the artificial channel (Lindmark and Gustavsson, 2008). Yet, at some locales, artificial light is required for successful passage (Fjeldstad et al., 2013). The upstream migratory behavior of adult Atlantic salmon was investigated using acoustic telemetry at a large natural waterfall in a small Irish river (Kennedy et al., 2013). They discovered that the complex flows created an obstacle that could only be passed during daylight. Dam passage in the Columbia River hydro-power system is strongly affected by light levels for several species of adult salmon (Naughton et al., 2005; Quinn and Adams, 1996; Robards and Quinn, 2002). Low visibility seems to make navigation through complex flows associated with dams and fishways too difficult for nocturnal passage. Ultimately, the role of light in facilitating upstream migrations is complex and can be both species- and flow-specific (Fjeldstad et al., 2018).

## 2.2 Physiological stress

### *2.2.1 Background*

Maintaining internal homeostasis when faced with a stressor represents a unique intersection of physiology, behavior and life history (Ricklefs and Wikelski, 2002). When faced with challenging environmental conditions, fish such as salmonids mount an evolved physiological stress response that rapidly mobilizes energy stores (Bonga, 1997; Schreck and Tort, 2016; see Chapter 6, Volume 39A: Bernier and Alderman, 2022). The primary humoral response is cortisol release from interrenal cells located in the head kidney, which promotes a secondary response, the conversion of hepatic glycogen into glucose, which can fuel rapid swimming activities (Barton and Iwama, 1991) that do not rely immediately on increasing oxygen supply to tissues (substrate-phosphorylation fuelled metabolism as opposed to oxidative-phosphorylation fuelled metabolism) (Zhang et al., 2018). Besides mobilizing energy reserves, cortisol triggers the release of catecholamines into the circulation, which contributes to increase in oxygen supply to the tissues, and helps maintain ionic balance (Pickering and Pottinger, 1995). Collectively, this stress response helps restore homeostasis after an acute stressful event, but depending on the severity of the stress, this may require a recovery period of 12–20 h, even after a brief bout (just several minutes) of chasing to exhaustion (Milligan, 1996; Zhang et al., 2018). Of course, if the stressor persists or if the magnitude of the stressor pushes the organism beyond its physiological limits, homeostasis may not be restored and sublethal impairments accrue, such as reductions in swim performance by sequestering energy away from anabolic processes and maintenance functions (Bonga, 1997), declines in immune function (Bonga, 1997; Pickering and Pottinger, 1989), or impaired ionoregulation (Brauner et al., 2000).

In extreme cases, delayed mortality is possible (Black, 1958; Selye, 1936; Wood et al., 1983), sometimes several days after the stressor was applied. More recently, it has been acknowledged that all stress is not bad, i.e., eustress (Schreck and Tort, 2016). Indeed, any fish that acclimates in a beneficial way to a new environmental condition does so by establishing a new homeostatic state. The migration into seawater following smoltification would be considered a eustress, as would the physiological changes associated with a prolonged change in water temperature and water oxygenation, or with exercise training. In this chapter, we concern ourselves primarily with the negative impacts of stress as these are invariably signaled by a marked elevation in plasma cortisol levels in a fish (Zhang et al., 2018).

Numerous factors can cause a stress response in association with migration through fish passage infrastructure, including swimming through excessive water velocities (Woodhead, 1975), experiencing acute elevations in water temperature (Teffer et al., 2018), and becoming physically injured (Nguyen et al., 2014). During both smoltification and adult maturation, the interpretation of the stress response can be confounded by the naturally elevated cortisol levels often associated with these two life stages (Hinch et al., 2006). In fact, additive effects of elevated cortisol could increase the vulnerability to the negative consequences of passage stressors.

Perhaps the most pervasive yet least understood of the passage stressors is burst swimming. Short duration burst swimming is required by fishes whenever rapid and/or exceedingly fast movements are required (predator avoidance, prey capture, high water velocity, etc.). Burst swimming is fuelled by the immediate consumption of phosphocreatine (PCr) and adenosine triphosphate (ATP) stores contained in white skeletal muscle, and by the anaerobic breakdown glycogen to lactate (Dobson and Hochachka, 1987) (more correctly termed substrate phosphorylation) and the associated production of $H^+$. As a result, PCr, ATP and glycogen levels characteristically fall in white muscle, while in both white muscle and the plasma, the pH falls and the lactate ion concentration increase (Milligan, 1996; Wood, 1991). The decrease in blood pH is a combined metabolic acidosis (i.e., increased $H^+$) and respiratory (i.e., increased $PCO_2$) acidosis (Wang et al., 1994). Furthermore, this acidification can disrupt ion-osmoregulatory balance as water moves passively from blood to muscle tissue, resulting in temporary increases in concentrations of some plasma ions in freshwater, followed by depressed ion concentrations over the longer term (Wood, 1991). The time-course of each of these various physiological changes differs (Zhang et al., 2018).

While burst swimming is employed when the capacity for aerobic activity is exceeded, burst swimming can only be maintained on the order of minutes and, if prolonged, results in exhaustion. If the burst swimming period is brief, other bursts may follow (Mckenzie et al., 2007) perhaps at a reduced velocity (the effect of muscle fatigue), but prolonged burst swimming will ultimately result in the fish becoming refractory to further stimulation and unable to perform further burst swimming in the short-term (Jones, 1982). Complete recovery, however, is on the order of hours (Zhang et al., 2018, 2019). It is this prolonged, post-exercise recovery period that could impact salmon swimming behavior for many hours and results in a higher energy use even though the fish may not be moving. This phenomenon could have profound effects on fish at passage infrastructure.

A surge of catecholamines and corticosteroids mediate much of this exhaustive exercise stress response (Milligan, 1996) although the cortisol surge plays an important role in delaying metabolic recovery (Eros and Milligan, 1996; Milligan et al., 2000). Moreover, the extent of these surges to some degree can indicate the level of stress encountered, i.e., a higher peak cortisol is associated with a greater stress and a longer recovery period. Consequently, the levels of plasma cortisol, lactate and several ions, which can be sampled rapidly and non-lethally from fish (Cook et al., 2011; Cooke et al., 2005), have been used regularly used as reliable indices of physiological and metabolic stress, even for studies of fishway infrastructure, several of which are reviewed below.

### 2.2.2 *Stress indices and upstream passage studies*

Pioneering work by Connor et al. (1964) used large experimental fishways on the Columbia River, USA, to investigate the design of pool and weir fishways (length and slope) on swimming performance and tissue physiology. Blood and muscle lactate, as well as muscle and liver glycogen levels were compared in adult Chinook, sockeye and steelhead trout before and after ascents of various durations in the different fishways to assess stress response among the fishway designs. The limited changes in blood lactate and muscle glycogen concentrations suggested that burst exercise was used only moderately even for prolonged ascents. There was no evidence of any physiological impairments associated with passage with all species ascending even the steepest fishway (1:8 slope), indicating that relatively steep and longer fishways could potentially be used in this river system. Yet, some individuals of each species ceased swimming during their ascent without a clear physiological explanation; rather individual motivation was important for passage success, a now commonly observed phenomenon.

Subsequent studies have confirmed that passage is often only modestly stressful. Dominy (1971) measured blood lactate concentration of anadromous alewives (*Alosa pseudoharengus*) during their upstream migration through a pool and weir fishway in Nova Scotia, Canada. Again, lactate values were low in fish that ascended the fishway (less than half compared with fish subjected to a severe experimental exercise challenge that elevated lactate levels five times higher than in resting fish). Moreover, the low lactate concentrations in fish collected at the first fishway pool suggested that fish used behavior to avoid the turbulent water in the main portion of the tailrace where captured fish had relatively high lactate concentrations: instead they had entered the fishway from slow water along the stream bank—an observation also confirmed using scuba. Thus, stress metabolites were reliably used to infer a fish passage route, i.e., avoidance of a more demanding region of turbulent water. Similarly, Amaral et al. (2021) studied the movement of Iberian barbel (*Luciobarbus bocagei*) across a low-headed ramped weir in a laboratory and found that lactate concentration was lower under high discharge conditions than low discharge conditions, which was likely due the higher amount of movement recorded during low discharge conditions. Schwalme et al. (1985) examined the passage of post-spawned, actively feeding northern pike (*Esox lucius*) at a newly constructed passage facility on Lesser Slave River, Alberta, Canada. Pike that successfully ascended the fishway elevated their plasma glucose and lactate above resting levels, but to a much lesser degree than those seen with angling stress.

Hatry et al. (2014) evaluated the physiological capacity of upstream migrating adults from three potadromous redhorse species (*Moxostoma anisurum, M. carinatum, M. macrolepidotum*) to pass the Vianney-Legendre vertical slot fishway on the Richelieu River, Quebec, Canada. Blood lactate, glucose, and pH levels measured from fish sampled at the upstream end of the fishway were similar to those of control fish that had been held for 24 h in sensory deprivation chambers but differed from another group of fish that had been manually exercised to exhaustion in tanks, triggering extremely high blood lactate, glucose and low pH levels. These results confirm that fishway ascension was neither physiologically stressful nor energetically taxing, and that physiological capacity was likely not a limiting factor in relation to passage success.

Cocherell et al. (2011) measured upstream passage performance of adult, wild-caught white sturgeon (123–225 cm TL) in a simulated mid-section of a 24.4-m-long experimental fishway incorporating vertical barriers, a 4% bed slope, and a series of five, paired vertical baffles (with 0.61-m slot widths) to dissipate flowing-water kinetic energy as well as to provide guidance for upstream migration. In general, faster velocities (0.76–1.07 m/s) cued fish to swim upstream sooner (<100 s, Webber et al., 2007). Fish, in good physical condition, reaching the upstream end of the flume in both the low (50% of the fish) and high (48%) tailwater treatments exceeded that of fish in poor condition (5%, Cocherell et al., 2011). The physiological stress responses of cannulated fish showed post-swimming peaks in hematocrit and plasma cortisol concentrations, compared with pre-swimming and (24-h) post-experiment levels. In addition, plasma pH showed decreases and mean plasma lactate showed increases post-swimming, indicating white-muscle recruitment (Cocherell et al., 2011). Overall, a relatively high percentage of adult white sturgeon in good physical condition, successfully ascended the sturgeon-compatible fishway and displayed burst-swimming-associated stress responses typical of many teleostean fishes, followed by complete recovery after 24 h (Cocherell et al., 2011).

Bido et al. (2018) examined migration of the streaked prochilod (*Prochilodus lineatus*), a tropical species, over a dam on the Upper Paraná River, Brazil. Compared to fish sampled downstream of the fishway, those that successfully ascended the dam did not increase plasma cortisol but doubled plasma glucose, increased plasma lactate by just 25% and lowered hematocrit by 30%. Taken together, these results suggest an ascent with low to modest physiological stress.

Adult sockeye salmon passage through a small pool and weir fishway at the Seton River Diversion Dam, British Columbia, Canada (Figs. 2 and 3) was assessed over several years (2012–2018) by sampling salmon from the top pool of the fishway (e.g., Burnett et al., 2014a,b; Pon et al., 2009a,b, 2012; Roscoe et al., 2011) and from a full span fence 200m downstream of

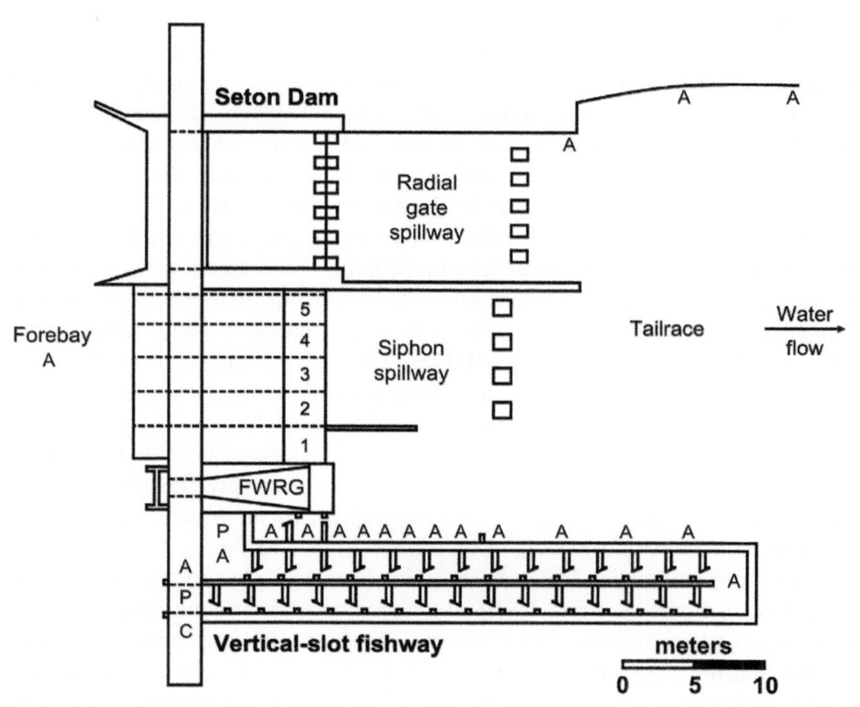


**FIG. 3** Structural layout of the Seton River Dam and fishway (see Fig. 2 for geographic location details). Under routine operations during sockeye salmon migrations, the discharge from the dam is primarily through the fish water release gate (FWRG) and Siphon Spillway 1, with small but consistent flows from the fishway entrance located at the dam face adjacent to the FWRG and minor seepage from the radial spillway gate. Acoustic receivers (A) were situated throughout the tailrace, adjacent to and inside the fishway, and in the forebay. PIT antennas (P) were located at the entrance and exit of the fishway. A resistivity "fish" counter (C) was situated at the fishway exit. *Used with permission from Burnett, N.J., Hinch, S.G., Bett, N.N., Braun, D.C., Casselman, M.T., Cooke, S.J., Gelchu, A., Lingard, S., Middleton, C.T., Minke-Martin, V., White, C.F.H., 2017. Reducing carryover effects on the migration and spawning success of sockeye salmon through a management experiment of dam flows. River Res. Appl. 33, 3–15.*

the fishway (e.g., Bett et al., 2020; Middleton et al., 2018). Travel times through the fishway (based on radio-tracking individuals) was short (<50 min for most fish; Pon et al., 2009b), and well within the recovery period needed for modestly stressed salmon. Every year produced little evidence that fish were severely stressed or they had extensively used burst swimming while ascending the fishway (Pon et al., 2009a,b; Roscoe et al., 2011). For example, plasma lactate (1.7–2.9 $mmol L^{-1}$) and glucose (4.4–4.8 $mmol L^{-1}$) were similar to or lower than levels previously reported for adult sockeye sampled in a similar late stage of migration not migrating through a fishway (e.g., Young et al., 2006), as were plasma ion concentrations ($Na^+$ 144–164 $mmol L^{-1}$; $Cl^-$ 130-140 $mmol L^{-1}$, $K^+$ 2.54–3.11 $mmol L^{-1}$) and total osmolality (310–306 $mOsm kg^{-1}$) (Crossin et al., 2008; Young et al., 2006). Likewise, plasma cortisol levels, which naturally differ widely between the sexes (males 198–238 $ng mL^{-1}$, females 323–349 $ng mL^{-1}$), were within the range reported by other studies of sockeye (e.g., Cooke et al., 2005; Hinch et al., 2006), noting that plasma cortisol titres can naturally vary considerably during migration (Hinch et al., 2006). Tracking of salmon placed into the pool in front of the fishway entrance (Pon et al., 2009b) revealed little evidence that successful entry and ascent of the fishway was negatively affected by changing the magnitude of attraction flows, or the fish's physiological state at arrival. Neither failed nor successful migrants employed burst swimming during passage (Pon et al., 2009a). The salmon with surgically implanted EMG transmitters that failed to ascend the fishway (6 of 13) did not differ from successful migrants in their plasma physiology, mean swim speeds and energy use (Pon et al., 2009a). Although plasma $Na^+$ concentration was lower in unsuccessful fish, suggesting these fish were exhibiting modest physiological stress, all other plasma metrics, including $Na^+$, were within the expected range for typical migrating adult sockeye. The failed migrants remained in the dam tailrace for more than a day actively seeking a means to enter the fishway before dropping further downstream with no clear physiological explanation for passage failure (Pon et al., 2009a).

In sum, most studies that have examined passage physiology have found only low to moderate stress levels associated with successful ascents within fishways, and even some failed ascents. Several review papers have concluded that the attraction to, and motivation to enter into a passage structure entrance may be larger issues in terms of successful upstream migrations through fishways (e.g., Bunt et al., 2012; Noonan et al., 2012). Therefore, it may be more important to assess the physiology associated with tailrace movements and fishway attraction to fully understand the limitations of upstream passage.

### 2.2.3 Stress indices and downstream passage studies

As with upstream migrating fish, stress-related physiological indices have been used to investigate downstream passage of juvenile salmonids. Stress resulting from dam passage by juvenile salmonids can be acute or chronic and can compromise energy reserves, immune systems, ability to smolt, and propensity to migrate (reviewed in Schreck et al., 2006). For example, Ferguson et al. (2007) not only found that plasma cortisol and lactate concentrations in yearling and subyearling Chinook salmon were twice as high in fish that used the fish bypass system at the Bonneville Dam on the Columbia River, USA, to move downstream, but many lost scales and had visible injuries. After the bypass system was extensively modified, fish survival increased, injury levels dramatically decreased, and only modest stress responses were observed in salmon using the new bypass system. The use of injury and stress assessments provided mechanistic explanations for poor migration survival and altered timing patterns, therefore conferred empirical support for the beneficial infrastructure modifications (Ferguson et al., 2007). To help in by-passing dams, some juvenile salmonids are transported by barge or truck. Schreck et al. (1989) found that plasma cortisol was increased by at least 60-fold in yearling coho salmon smolts (*O. kisutch*) that were trucked for up to 2h from a hatchery prior to release. Recovery to resting levels occurred within 3h after transport if it occurred in the winter but was elevated for over a day if done in the spring (Schreck et al., 1989) revealing the importance of transport timing of smolts around dams to minimize stress-related consequences.

Ammar et al. (2020) experimentally introduced Atlantic salmon smolts into a hydroelectric turbine on the Meuse River, Belgium and used plasma and gene expression parameters to evaluate the physiological response. While over 96% of the downstream migrants survived turbine passage, they had a large decline in plasma glucose levels, enhanced complement and peroxidase activities, and expressed immunity-related genes (lysg, igm and mpo) compared with control fish. Thus, turbine passage increased energy expenditure and disrupted stimulated innate immunity. This stimulation may be viewed as a benefit of turbine passage, but additional stresses further downstream could produce chronic stress, which could lead to immune system depression and increase the susceptibility to pathogens (Ammar et al., 2020).

While much of our understanding regarding stress responses during downstream passage comes from salmonids, changes in stress-related indices during passage have been found in other fish species as well. Ammar et al. (2021) assessed downstream passage of European eels, also on the Meuse River in Belgium. Unlike Atlantic salmon smolts but similar to the Pacific salmon studied on the Columbia River, plasma cortisol and glucose levels in the migrating eels were elevated following passage, with glucose remaining higher relative to a control group even 144h after passage. The effect of capture and handling during the experiment seemed to overshadow the impact of passage on cortisol levels, which also returned to levels similar to those found in non-stressed eels (Teles et al., 2004) after 144h. Even an absence of elevated cortisol does not necessarily mean an absence of stressors (Bonga, 1997), however, and the lasting rise in circulating glucose suggests the involvement of mechanisms preventing hypoglycaemia and exhaustion. Stressors can affect antibody production in fish (Tort, 2011), and immunoglobulin was higher in eels that underwent passage relative to those in the control group after 144 h, suggesting a relatively prolonged immune suppressive action on antibody production due to the stress of downstream passage, as has been observed elsewhere (reviewed in Bonga, 1997).

## 2.3 Energetics and anaerobic metabolism

### *2.3.1 Background*

Fish use different swimming gaits to vary swimming speed and acceleration when moving through water (Videler, 1993; Weihs, 1974). The difference between standard metabolism and that observed with maximum aerobic swimming is termed aerobic scope for activity (ASA). Temperature which influences several aspects of swimming performance (e.g., $MO_{2max}$, $U_{crit}$ [critical swimming speed, above which the fish uses burst swimming], ASA), all of which have a unimodal relationship with temperature, displaying a maximum at an optimal temperature, $T_{opt}$ (reviewed in Brett, 1995). $T_{opt}$ values are species-specific and population-specific (Eliason et al., 2011, 2013). The critical temperature ($T_{crit}$) indicates when aerobic swimming is no longer possible, i.e., the temperature when standard and maximum metabolic rates are equal at exceptionally cold or warm temperatures.

Saving energy is important to fish during migrations and especially so for adult migratory salmon that do not feed during their up-river migration and rely on endogenous energy reserves to complete migration and spawn. Thus, where possible, fish appear to choose pathways where they can encounter optimal temperatures (Minke-Martin et al., 2018). Indeed, unusually warm temperatures are known to alter the timing or even prevent upstream adult salmon migrations ( Juanes et al., 2004; Quinn and Adams, 1996; Quinn et al., 1997). Transport costs also can be minimized by invoking the swimming speed that imparts the lowest net energy expenditure to move a unit mass through a unit distance (Bernatchez and Dodson, 1987; Hinch and Rand, 2000; Lee et al., 2003a,b). The relationship between swimming speed and oxygen cost has a nadir in fishes (e.g., Brett, 1995; Lee et al., 2003a,b) with minimum energy expenditure in the absence of water currents occurring near the optimal cruising speed (i.e., minimum cost of transport) of $\sim 1\,BL \cdot s^{-1}$ (Trump and Leggett, 1980). Hinch and Rand (2000) used underwater stereo-videography to monitor migrations of sockeye swimming up tributaries of the Fraser River finding that fish maintained swimming speeds just above $1\,BL \cdot s^{-1}$ under low to moderate encountered water speeds (up to $20\,cm\,s^{-1}$) but accelerated their swimming speeds ($>2$–$3\,BL \cdot s^{-1}$) when currents exceeded $20\,cm\,s^{-1}$. In their review of >200 papers involving marine tagging and tracking of salmonids, Drenner et al. (2012) found that salmonids at all life stages were consistently found to travel at an average speed of approximately $1\,BL\,s^{-1}$.

At speeds approaching their critical swimming speed (typically around 60–80% of $U_{crit}$), Pacific salmon and rainbow trout (*O. mykiss*) begin to utilize burst-and-coast swimming (Burgetz et al., 1998; Lee et al., 2003a; Zhang et al., 2020). Similarly, smallmouth bass (*Micropterus dolomieu*) voluntarily ascending a 50 m raceway against different water currents maintained a constant ground speed independent of water velocity until they reached swimming speeds that required mixed gait swimming (steady aerobic swimming and burst-and-coast swimming gaits) which increased ground speeds and decreased passage time (Peake and Farrell, 2004). During burst-and-coast swimming, the oxygen cost of swimming may increase or remain high, but a proportion of the actual cost of locomotion is deferred until after swimming has ceased (or can be paid back at a lower swimming speed, e.g., Wagner et al., 2006). By measuring excess post-exercise oxygen consumption (EPOC) following $U_{crit}$, the deferred aerobic costs in adult sockeye and coho were estimated to add 20–50% to the actual oxygen consumed measured at $U_{crit}$ (Lee et al., 2003a). Thus, EPOC must be considered in the net cost of transportation (see Lee et al., 2003b).

Fish often encounter fast or turbulent water as they approach or ascend passage facilities (e.g., Li et al., 2021). Using burst swimming behaviors that generate swimming velocities greater than those possible with aerobic swimming are critically important for negotiating hydraulically challenging areas. While burst-and-coast swimming (Brown and Geist, 2002; Hinch and Rand, 2000; Hinch et al., 2002; Kramer and McLaughlin, 2001) has been suggested to reduce the net transport cost by 60% in non-salmonid fish (Weihs, 1974), more recent studies of the delayed oxygen costs of burst swimming (Zhang et al., 2018, 2020) warrant a re-evaluation of this benefit. As shown by the studies described above, burst swimming seems to be used sparingly because either they are more energetically costly than previously thought, or they have carryover costs associated with depletion of high energy stores and/or toxic waste accumulation. Moreover, extended bouts of bursting can lead to exhaustion and accumulation of wastes such as lactic acid, which in both laboratory and field situations can cause metabolic acidosis and delayed mortality (Black, 1958; Farrell et al., 2001a; Hinch et al., 2006). Indeed, hyperactive swimming by Fraser River sockeye, recorded by EMG telemetry at Hell's Gate fishway, was associated with passage failure and in-river mortality (Hinch and Bratty, 2000). Furthermore, if migrants experience temperatures above $T_{opt}$ while swimming at $U_{crit}$, cardiorespiratory collapse can occur due to an insufficient scope for heart rate (Eliason et al., 2013).

These challenges are likely even greater for catadromous fish such as the European eel, which must ascend passage facilities during their juvenile phase. As with other salmon and other fish species (e.g., *Cyprinus carpio*; Wu et al., 2007), eel use burst-and-coast swimming when they encounter high velocities (Vezza et al., 2020), but struggle when swimming through structures that lack adequate resting places (Barbin and Krueger, 1994). Consequently, fishway velocities considered appropriate for eels (e.g., Langdon and Collins, 2000) are much lower than those often encountered by stronger swimmers such as salmonids. For anguilliforms that comigrate with large populations of salmonids, such as Pacific lamprey (*Lampetra tridentata*), the swimming speeds required to negotiate fishways could result in energetic exhaustion, even though upstream passage for this species of lamprey occurs in adults. For example, Pacific lamprey exercised to exhaustion exhibited $U_{crit}$ at speeds surpassed by water velocities found in fishways on the Columbia River (Mesa et al., 2003), with plasma lactate concentrations increasing up to 100-fold relative to resting levels, alongside a 40% decrease in muscle glycogen levels and a 22% increase in hematocrit. While these measures usually returned to resting levels within 4h after fatigue, whether this metabolic dysfunction following exhaustive exercise could have a carryover effect on post-passage fitness (see Section 3 of this chapter) is unknown. Furthermore, a recent experiment by Vezza et al. (2020) on European eels found that the fish consistently swam near the bottom of the flume, and CFD modeling revealed that velocities along the bottom of the flume are lower than the bulk average water velocity. As such, the authors suggest that swimming trials that fail to account for this could overestimate $U_{crit}$ for bottom-dwelling fish.

Anadromous fish can increase oxygen uptake and cardiac output during the migration, which enables them to provide more oxygen to locomotory muscles (Cooke et al., 2012). This could explain why sockeye salmon tested during the summer spawning season show a higher swimming capacity than those tested during the winter (Hammer, 1995). Such a seasonal effect may occur in potadromous fish as well, although information is limited. In one of the few studies on potadromous fish, Romão et al. (2018) compared passage of Iberian barbel during their reproductive season in the spring to the non-reproductive season in the fall. While differences in passage performance between the two seasons were limited, barbel tested during the reproductive season had a lower concentration of plasma lactate, suggesting they might have a higher aerobic swimming capacity during the reproductive migration. The physiological adjustments seen in anadromous fish during their migrations may therefore also occur in potadromous fish, but more studies are needed.

Ultimately, fish that undergo arduous freshwater migrations and are likely to encounter barriers have experienced selective pressures that favor energy conservation and physiological adaptations that improve the likelihood of migration success. Whether this be through the utilization of optimal thermal pathways, the adoption of energetically efficient swimming strategies, the development of advantageous morphological traits (Crossin et al., 2003, 2004), or a combination of these and other adaptations, natural selection clearly influences fish swimming energetics. Given that passage success is associated with these types of characteristics and traits (Bunt et al., 2012), fish passage structures may act as an additional selective pressure for these traits. While perhaps too early to see any population-level effects of passage structures on the physiology of migratory fish, it is possible that these barriers might further shape fish swimming energetics in the coming decades.

### 2.3.2 A case-study of upstream salmon passage: Seton Dam fishway

Few investigations have directly examined the role of energetics and burst swimming behaviors associated with passage infrastructure. However, extensive work focussed on adult sockeye salmon attempting to use a vertical slot fishway to pass the Seton Dam in British Columbia, Canada (Figs. 2 and 3; Burnett et al., 2014a,b, 2017; Pon et al., 2009a,b) provides insight into this issue. We examine these energetics papers as a case study, while acknowledging the on-going need to discover the physiological mechanisms responsible for sockeye salmon having difficulty locating and entering the Seton Dam fishway (see Section 2.2; Pon et al., 2009a; Roscoe et al., 2011).

Burnett et al. (2014b) tracked >60 adult Gates Creek sockeye salmon that had been gastrically implanted with triaxial accelerometer transmitters. About half were captured downstream, tagged and released at a fish fence fully spanning the dam entrance 200m downstream. The other half were captured upstream at the top fishway pool (as in Pon et al. (2009a,b)) then tagged and returned for release at the fence for a volitional upstream passage through the fishway. As salmon approached the dam, they swam at about $1\,BL \cdot s^{-1}$ with just a few individuals exhibiting burst swimming (Fig. 4). But to reach the fishway entrance fish had to burst swim (most individuals traveled at >80% $U_{crit}$; Fig. 4) while traversing high flows released from Siphon 1 near the fishway entrance (Fig. 3; Burnett et al., 2014b). This was especially evident in individuals that traversed the high flows multiple times. Within the fishway, however, individual swim speed was typically less than critical ($<2.1\,BL\,s^{-1}$) and <20% of individuals used burst swimming. A similar slower swimming pattern was evident for sockeye salmon passing through the forebay after ascending the fishway.

Water temperature below Seton Dam also influenced fish passage (Fig. 5). Seton River daily temperatures in 2013 were generally above the $T_{opt}$ for aerobic scope in Gates Creek sockeye (17.5°C; Lee et al., 2003b). Of the fish that failed to pass the dam, 90% experienced water temperatures >21°C on some days which were near acutely lethal (Martins et al., 2011) and would have resulted in cardiorespiratory collapse in the dam tailrace arising from exhaustive exercise and insufficient scope for heart rate (Eliason et al., 2013). Exposure to supra-optimal water temperatures while swimming at critical speeds can decrease aerobic scope and cardiac scope and increase EPOC which prolongs recovery time in migrating sockeye salmon (Eliason et al., 2013; Farrell et al., 2008; Lee et al., 2003a,b).

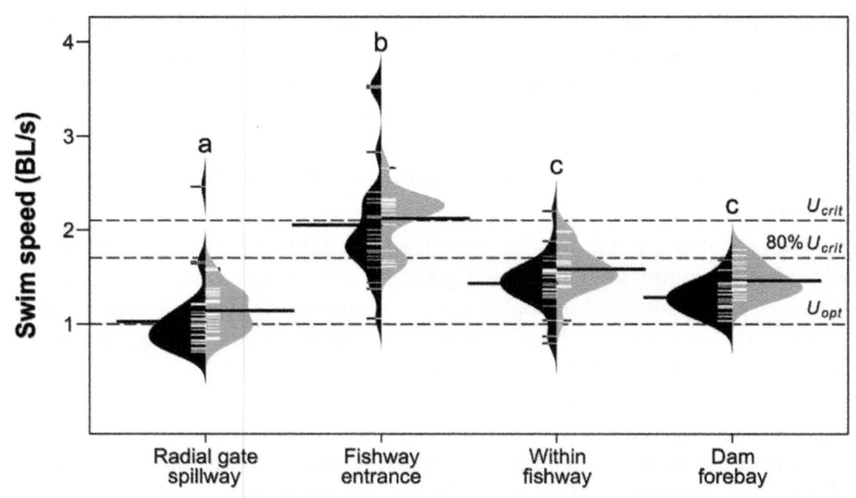


**FIG. 4** Beanplot (black horizontal lines = means) comparing the swimming speed (body lengths per second [$BL\,s^{-1}$]) of male (black) and female (gray) Gates Creek sockeye salmon in the radial gate spillway, surrounding the fishway entrance, within the fishway, and in the dam forebay. Shaded curved polygons (beans) depict the estimated density of the distribution of individual swimming speed values (white horizontal lines). Optimal ($U_{opt}$; $1.0\,BL\,s^{-1}$) and critical ($U_{crit}$; $2.1\,BL\,s^{-1}$) swimming speeds are shown as dashed horizontal lines; the swimming speed at which anaerobic muscle fibers start to be recruited (80% $U_{crit}$) is also shown. Sample sizes (n) are presented below each bean, and lowercase letters represent significant differences ($P \leq 0.05$) from one-way ANOVA and Tukey post hoc tests. *Used with permission from Burnett, N.J., Hinch, S.G., Braun, D.C., Casselman, M.T., Middleton, C.T., Wilson, S.M., Cooke, S.J., 2014b. Burst swimming in areas of high flow: delayed consequences of anaerobiosis in wild adult sockeye salmon. Physiol. Biochem. Zool. 87, 587–598.*

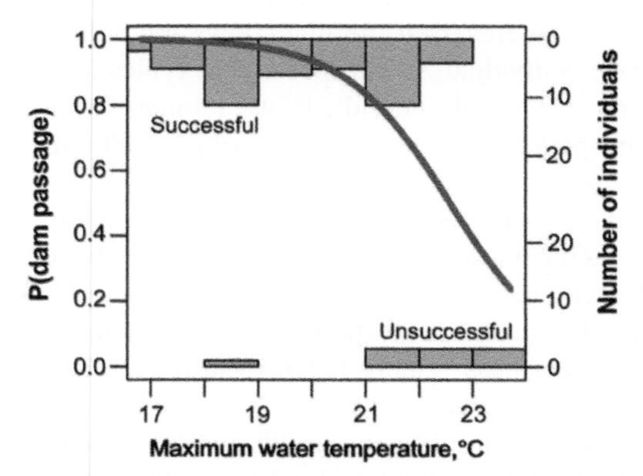


**FIG. 5** Predicted probability of Gates Creek sockeye salmon passing Seton Dam, visualized by fitting a logistic regression to the maximum Seton River temperature successful and unsuccessful fish experienced while directly below the dam. *Used with permission from Burnett, N.J., Hinch, S.G., Braun, D.C., Casselman, M.T., Middleton, C.T., Wilson, S.M., Cooke, S.J., 2014b. Burst swimming in areas of high flow: delayed consequences of anaerobiosis in wild adult sockeye salmon. Physiol. Biochem. Zool. 87, 587–598.*

Fish captured at the top of the fishway and transported downriver for release, i.e., they had previously passed Seton Dam, were 15% less likely to enter the fishway entrance and 16% less likely to pass Seton Dam fishway than "naïve" fish (Burnett et al., 2014b). The duration of holding behavior in the low flows of the radial gate spillway (Fig. 3) increased and they used less burst swimming at the fishway entrance. A parsimonious explanation for this result is that these fish had not fully recovered their EPOC before the second attempt at ascent. This explanation would also help explain why nearly half of EMG-tagged fish in an earlier study did not re-ascend the fishway (Pon et al., 2009a).

Using ADCPs and CFD modeling, Li et al. (2021) directly measured the flow fields at the Seton Dam that adult sockeye encounter as they migrate upstream, and used triaxial accelerometry telemetry (Burnett et al., 2014a,b) to link physiological and behavioral responses to flow field dynamics (Fig. 6). Dam discharge flows were experimentally altered to vary encountered water velocities and hydrodynamics. *En route* to the fishway entrance in the dam tailrace, tagged fish avoided areas of high velocity ($>2.4\,\mathrm{m\,s^{-1}}$ or $4\,\mathrm{BL\,s^{-1}}$, which is twice the critical swim speed for this population (Burnett et al., 2014b)), high Reynolds shear stress ($>21\,\mathrm{Pa}$), and high turbulence kinetic energy ($>0.12\,\mathrm{m^2\,s^{-2}}$) (Li et al., 2021). Telemetry accelerometers revealed that most fish swam at speeds below their maximum prolonged swimming speed ($U_{crit,}$ $2.1\,\mathrm{BL\,s^{-1}}$; Burnett et al., 2017) as they approached the fishway entrance despite other available pathways representing a large range of flow velocities (Li et al., 2021). These findings highlight the importance of migratory path selection, as well as the fish's ability to limit energy expenditures and stress exposure during this phase of the upstream migration. These results mirror those of Baktoft et al. (2020), who similarly used a combination of CFD modeling and telemetry to track Atlantic salmon kelts as they navigated a hydropower facility in Central Norway. They found a level of energy depletion that, according to the authors, could lead to reduced post-spawning survival and even negate the benefits of repeat migration and iteroparous breeding.

Sex also played an important role in fish passage (Burnett et al., 2014a,b, 2017). Females were 17% less likely to locate and enter the fishway and 9% less likely to pass the dam. Sex-specific differences in swimming behavior were also observed, with males apparently searching more for a lower velocity route, but in doing so making 50% more tailrace crossings towards the fishway entrance compared to females. Females, on the other hand, swam more directly towards the fishway entrance and the unsuccessful female migrants used greater burst swimming effort in areas of high flow (Burnett et al., 2014b). Sex-specific differences in migration physiology, and its role in dam passage infrastructure, is now discussed more generally below.

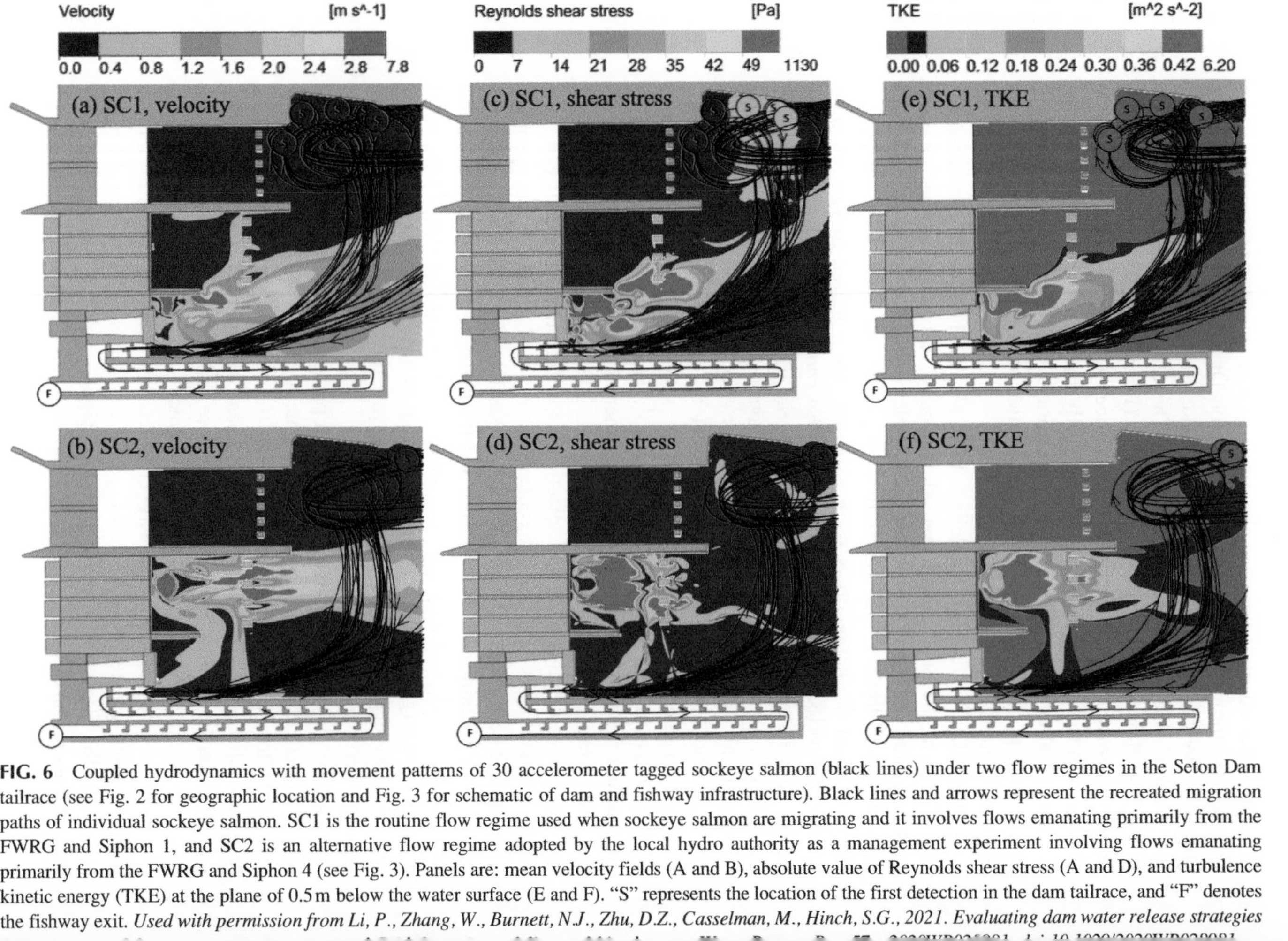


**FIG. 6** Coupled hydrodynamics with movement patterns of 30 accelerometer tagged sockeye salmon (black lines) under two flow regimes in the Seton Dam tailrace (see Fig. 2 for geographic location and Fig. 3 for schematic of dam and fishway infrastructure). Black lines and arrows represent the recreated migration paths of individual sockeye salmon. SC1 is the routine flow regime used when sockeye salmon are migrating and it involves flows emanating primarily from the FWRG and Siphon 1, and SC2 is an alternative flow regime adopted by the local hydro authority as a management experiment involving flows emanating primarily from the FWRG and Siphon 4 (see Fig. 3). Panels are: mean velocity fields (A and B), absolute value of Reynolds shear stress (A and D), and turbulence kinetic energy (TKE) at the plane of 0.5 m below the water surface (E and F). "S" represents the location of the first detection in the dam tailrace, and "F" denotes the fishway exit. *Used with permission from Li, P., Zhang, W., Burnett, N.J., Zhu, D.Z., Casselman, M., Hinch, S.G., 2021. Evaluating dam water release strategies*

## 2.4 Sex effects in adult passage studies

It is becoming increasingly clear that differences in physiology and swimming behavior can result in higher levels of female mortality in areas of complex hydraulics and passage infrastructure. For example, female adult Atlantic salmon had a much lower passage success compared to males through a Denil fishway on the River Ätran, Sweden (Nyqvist et al., 2017c). Similarly, female Atlantic salmon passing the fishway on the Vindelalven River, Sweden were nearly twice as slow as males (Lundqvist et al., 2008). But when a more natural fishway was built on the River Ätran to improve the attraction flows and slow the fishway water velocity, female passage success improved considerably, with both sexes achieving a 96% passage success (Nyqvist et al., 2017c). Furthermore, a review directly comparing adult salmon mortality in 22 migration studies found that female mortality averaged twice that of males, and could be up to eightfold higher. Indeed, female mortality was highest when migration conditions were challenging (e.g., high/turbulent flows associated with passage infrastructure, high temperatures, confinement), as well as when females were reaching the end of their river migration (reviewed in Hinch et al., 2021). Unfortunately, studies regularly do not account for differences between sexes, and much of the knowledge we have is restricted to salmonids.

There are several potential explanations as to why females might have more difficulty with passage than males. One could be energy depletion. For fish that are migrating to reproductive grounds, the energetic requirements of gonad development are often much greater in females than in males (for example, gonad mass relative to body mass differ by sixfold in mature sockeye salmon; Crossin et al., 2004), and the extreme high flows that can be experienced in dam tailraces could pose a significant challenge for females in terms of energy depletion. For example, the more directed movements of female Gates Creek sockeye at the Seton Dam (Burnett et al., 2014a) may reflect their need to minimize migration time and the number of burst swimming attempts compared with males. Alternatively, differential stress responses between the sexes might affect passage. Females appear to respond more strongly to stress than males, as evidenced by higher levels of plasma cortisol following exposure to various stressors (e.g., exhaustive exercise—Gale et al., 2011; high temperature—Jeffries et al., 2012). With their elevated baseline cortisol levels, they may have an impaired response to additional stressors, such that they cannot effectively mobilize additional energy stores or clear waste products. Alternatively, a prolonged post-stress cortisol recovery period could impact salmon swimming behavior, and cause an undesirable higher use of energy stores. A third potential explanation is reduced cardiac performance. Mature female sockeye salmon have a routine heart rate that is 21% higher than in males (Sandblom et al., 2009), which will lead to a reduced scope to increase heart rate during activity and therefore decrease the ability to mobilize cardiac capacity. Females also likely divert more blood to develop their relatively larger gonads than males during maturation. Since the fish heart does not have the capacity to fully perfuse all organs at the same time (Farrell, 2016; Farrell and Jones, 1992), it may not be possible to partition blood flow according to the needs of all organs. A prioritization of blood flow to gonads rather than swimming muscles could force females to disproportionately utilize burst swimming, with negative carryover effects. If females exert a greater anaerobic effort, as seen at the Seton Dam (Burnett et al., 2014a,b), they could incur a greater EPOC and take longer to fully recover. Delayed recovery could result in an inability of females to ascend fishways (e.g., Burnett et al., 2014b), or lead to delayed mortality of females days after fishway passage (e.g., Burnett et al., 2017). Finally, immunosuppression might contribute to differential effects of passage on the two sexes. Stress in general and elevated cortisol levels in particular can contribute to immunosuppression and increased susceptibility to infectious disease (Pickering and Pottinger, 1989), potentially enhancing migration mortality. It is possible that difficulties encountered with hydraulic and/or thermal conditions associated with passage infrastructure could lead to disease which has either immediate or delayed negative effects on passage. Whether any or a combination of these factors contribute to reduced passage success in females, or whether differences in sexes exist at all for non-salmonid species, remains to be seen.

## 2.5 Physical injury

### *2.5.1 Upstream migrations*

Physical injury arises from contact with predators and solid objects, such as those encountered when fish move either upstream or downstream past dams. Skin damage can be an entryway or site for infections, while gill damage can lead to blood loss. The loss of the skin's barrier function could lead to osmoregulatory challenges, which if large could be a proximate cause of mortality for fish (Mateus et al., 2017; Olsen et al., 2012). All the same, physically abrading a small portion of the skin of tiny juvenile pink salmon did not compromise their whole-body ion balance (Sackville et al., 2011). Ultimately, the greater the injury the more likely a fish's functioning becomes impaired.

Threatened delta smelt (*Hypomesus transpacificus*) are vulnerable to >2000 water diversions in the Sacramento–San Joaquin Delta system of California. To increase successful smelt passage (i.e., minimize fish entrainment) past these diversions, some are equipped with fish screens, where fish are exposed to (resultant) two-vector flows, from approach (through the screen) and sweeping (past the screen) velocities. Although fish screens should decrease direct losses from the system, the physiological stress associated with exposure to two-vector flows near a fish screen may increase delta smelt mortality (Swanson et al., 2005; Young et al., 2010). Using a large laboratory-based flume, delta smelt experienced frequent temporary contacts with the screen, and contact rates were influenced by flow and time of day (i.e., light level). Contact was injurious, and post-experiment mortality rates were directly related to both contact frequency and severity as well as temperature (Swanson et al., 2005; White et al., 2007). Combinations of high approach velocity (10 or 15 cm/s) and high sweeping velocity (31 or 62 cm/s) increased the primary stress response (plasma cortisol concentrations) and the number of moribund adult delta smelt (Young et al., 2010). Fish screens designed to protect the delta smelt population could minimize physiological stress by decreasing approach velocities in areas with high sweeping velocities (Young et al., 2010). Quantitative models from these laboratory experiments showed that both behavioral responses such as swimming velocity and physiological responses to fish screen contact, as modified by environmental conditions, controlled the smelts' performance and their risk from the diversion and the screen. The results illustrated that ecologically effective protection strategies and regulatory criteria developed on the basis of multiple integrated responses of the organism to the stressor offer greater benefits and certainty to both the organism and the regulated activities (Swanson et al., 2005).

Examples of the ways that fish can get injured while using passage infrastructure to move upstream are numerous. Olfactory attraction to the outflow of power stations and turbines can lead to lesions and abrasions from encountering grates and screens (Bett et al., 2020; Fretwell, 1989; Middleton et al., 2018). Migrants have also incurred scale loss and skin lesions during fishway entrance and passage in some pool and weir and Denil fishways, caused by striking concrete or steel baffles. Predatory fish, birds and mammals can accumulate at fishway entrances and also in fishways which can lead to high levels of immediate predation, but also result in significant injury to fish that successfully pass (Agostinho et al., 2012). In neotropical regions where numerous predator and prey fish species exist in high density, the restricted environment of a fishway can transform this environment to a "hotspot" for predation (Agostinho et al., 2012).

Injuries incurred prior to reaching a fishway may also influence post-passage migration success at fishways. For example, Bass et al. (2018) collected moribund adult sockeye salmon immediately below the tailrace of the Seton Dam Fishway (Figs. 2 and 3) and found that over 50% had physical wounds (variously resulting from escapement from a downstream gill-net fishery, marine sea louse scars, and powerhouse screen encounters). Moreover, wounds and scars occurred in a greater percentage of females than males, possibly because their small bodies made them more able to escape from gill-nets. Compared to those with no wounds, PIT-tagged sockeye salmon with wounds had a 62% lower odds of successfully reaching the spawning area 40 km upstream of the fishway. Clearly, the association between visible injuries and migration failure indicates that damage to the integument can have significant consequences for upstream migrating adult fish at passage facilities, although the exact mechanism by which it does so is still unclear.

### 2.5.2 Downstream migrations

Similarly, anadromous and resident fishes can incur injuries as they migrate downstream through passage infrastructure (Čada, 2001). A review of >260 studies on consequences to fish associated with downstream passage at hydroelectric dams noted injury in nearly 50% of studies (Algera et al., 2020). Some of the greatest injuries were caused by turbine entrainment ((Čada, 1997; EPRI, 2011), which exposes fish to rapid pressure changes and barotrauma (Brown et al., 2014, 2012a; Crew et al., 2017) as well as injuries from mechanical contacts. Turbine-associated mortality can be induced in several ways, including blade strikes, shear forces, cavitation and pressure changes (Čada, 2001), and the prevalence of mortality or injury can be strongly influenced by both turbine type and the size of the fish (Pracheil et al., 2016).

Barotrauma, which is particularly common during turbine entrainment, can be caused in two ways. One is through the expansion of gases present within a fish, particularly within the swim bladder (Brown et al., 2014, 2012b). A rapid reduction in external pressure when passing through turbines causes the expansion of internal gases, leading to exophthalmia (pop eye), internal hemorrhaging, and a bloated or ruptured swim bladder (Brown et al., 2009, 2015; Stephenson et al., 2010). The other barotrauma injury arises when gases in blood and other body fluids become supersaturated, causing emboli in blood, organs, gills or fins as well as internal hemorrhaging (Brown et al., 2014; Colotelo et al., 2012). In their review, Brown et al. (2012c) suggested that the majority of barotrauma-related injuries are due to the expansion of existing gases in the fish, particularly the expansion and rupture of the swim bladder, and that the force of swim bladder rupture may push gases into the surrounding musculature, resulting in exopthalmia, emboli and hemorrhaging in the fins and surrounding tissues. Injury can vary among species owing to different types of swim bladders (Crew et al., 2017). Species with open swim bladders (physostomes such as salmonids) can swallow or expel air to rapidly adjust swim bladder volume and lower the risk of barotrauma compared to species with closed swim bladders (physoclists such as percids) which cannot rapidly regulate swim bladder volume (Brown et al., 2014, 2012c; Čada and Schweizer, 2012). Nonetheless, barotrauma can be severe in physostomes when passing through turbines (Colotelo et al., 2012). Species with no swim bladders (e.g., petromyzontids, anguillids) are at the lowest risk of barotrauma (Colotelo et al., 2012).

Turbine-entrained fish can also become trapped between the boundaries of water moving in different directions (Algera et al., 2020). Exposure to this sheer stress can cause a range of injuries of different severities including bruising, bleeding, descaling, swimming impairment, disorientation, and loss of equilibrium (Deng et al., 2010). Reductions in turbine speeds can reduce the level of major injuries caused by sheer stress (Deng et al., 2005). Entrained fish may also come in contact with the moving turbine blades or other components of the turbine resulting in mechanical injury ranging in severity from minor bruising and lacerations to hemorrhaging, amputation and decapitation (Pracheil et al., 2016). Injuries are thought to be more severe in larger fish, and with slower flow rates and more turbine blades (Crew et al., 2017). Blade strikes may be a more important cause of injuries than sheer stress given that different turbine types have similar pressure profiles but different survival outcomes for migrants (Crew et al., 2017; Pracheil et al., 2016). It is difficult, however, to quantify the injuries attributed solely to shear stress (Crew et al., 2017).

Intake louvers, trash racks, screens and grates placed in the dam forebay are used to discourage fish from becoming entrained into turbines and to guide them towards alternative routes such as spillways or bypasses (Crew et al., 2017). Yet fish are often impinged on these structures and injured (Barnthouse, 2013). Structural elements like water velocity, the size of screen openings and screen construction can contribute to impingement (Crew et al., 2017), as can fish swimming ability and morphology. Fish that exhibit subcarangiform locomotion (e.g., salmonids) can avoid rapid changes in velocity, making them less susceptible to impingement, whereas fish that have elongated bodies and exhibit anguilliform locomotion (e.g., anguillids) are less adept at avoiding rapid changes (reviewed in Crew et al., 2017). Freefalling over a spillway and the use of fish bypasses or fishways can also cause abrasions, injury and mortality, although generally less than for those that get entrained and/or encounter fish screens (Algera et al., 2020). The most common injury-type reported across all studies in the review of Algera et al. (2020) was descaling, which is an important finding to help understand latent mortality, particularly in diadromous fish given the role of the integument in ion balance and the osmoregulatory challenges smolts are about to encounter when they reach the ocean.

Another type of barotrauma is possible whenever water spills from a dam and plunges to depth, supersaturating dissolved atmospheric gases in the process (Weitkamp and Katz, 1980). These supersaturated gases can freely diffuse into fishes found in the vicinity. Total dissolved gas supersaturation (TDGS) is a significant injury and mortality risk for fish in the tailraces of dams and it has been researched for over 50 years (reviewed in Crew et al., 2017). TDGS causes gas bubbles to accumulate in the lateral line, under skin, in gills, in blood, behind eyes and in other tissues (reviewed in Pleizier et al., 2020). These physical effects can cause hemorrhaging, tissue necrosis, impaired development, and infection (Pleizier et al., 2020). Together, these physiological consequences are termed gas bubble trauma (GBT) which is considered by some as a non-infectious disease state (Bouck, 1980). The effects of

GBT can be lethal if bubble formation affects the cardiovascular system (Weitkamp and Katz, 1980), especially the gill lamellae of the respiratory system (Fidler, 1988). Suffocation as a result of bubbles in the gills is one of the primary putative causes of mortality from GBT (Pleizier et al., 2020). Gas bubbles in the heart impair cardiac compression of blood during systole too. If blockage of blood vessels of the gills by bubbles is causing rapid suffocation by TDGS, then species or life stage differences in hypoxia tolerance and gill morphology should influence susceptibility to TDGS. However, species differences in vulnerability to disease might be an important factor during long exposures to low levels of TDGS (Pleizier et al., 2020). GBT can have sublethal effects by reducing growth, causing blindness, increasing stress, and decreasing lateral line sensitivity, which can all affect navigation and migration (Schiewe, 1974; Weitkamp and Katz, 1980). GBT can also cause the over-inflation or rupturing of swim bladders affecting buoyancy, swimming and predator avoidance (Pleizier et al., 2020; Shrimpton et al., 1990). Severity of GBT seems to differ between species and life stages (reviewed in Crew et al., 2017), but the mechanisms responsible for these differences have not been determined (Pleizier et al., 2021).

One of the most compelling new technologies being used to assess potential injuries incurred by downstream migrants passing through hydroelectric generating stations is the "*Sensor Fish*" (Deng et al., 2014; Hou et al., 2018; Martinez et al., 2019). The device, which has the length and mass of a typical out-migrating juvenile Chinook salmon and is neutrally buoyant in freshwater, contains several sensors that assess 3-D linear acceleration, 3-D rotational velocity, 3-D orientation/compass, pressure and temperature. A low-power microcontroller collects data from the sensors at a sampling frequency of 2048 Hz and stores up to 5 min of data. After a set time, the unit floats to the water surface and emits a radio-frequency to help with device retrieval. With a rechargeable battery, it can be re-used. *Sensor Fish* can be deployed to get entrained into turbines or spillways to help characterize hydraulic passage infrastructure and determine the potential causes of actual fish injury (Deng et al., 2014; Hou et al., 2018; Martinez et al., 2019). *Sensor Fish* used at three different turbine-type facilities in Bavaria, Germany (Boys et al., 2018) revealed a low likelihood of barotrauma injury with a very low head turbine despite the potential for mechanical injury resulting from blade strikes. Similarly, blade strike was likely at the screw turbine, and physical injury was likely in its tailrace. The Kaplan turbine, however, generated rapid decompression that likely resulted in barotrauma injury. When paired with studies that utilize live-fish passage trials at turbines or in a laboratory, *Sensor Fish* data provide a means of assessing the mechanisms of passage injury and mortality, which can inform infrastructure or operational modifications that reduce specific types of fish injury (Boys et al., 2018).

Despite these promising advances through new technologies, however, several challenges still exist when considering management of dam passage to minimize injury. The exact mechanisms responsible for injury and mortality when passing through turbines are not fully understood, and this knowledge gap has hampered the development of fish-friendly turbines (Schwevers and Adam, 2020). Furthermore, the types of turbines receiving the most research attention do not necessarily align with those that pose the greatest risk of injury, nor are the species most commonly studied necessarily those that are most threatened. For example, while Francis turbines generate 56% of hydropower in the United States and are associated with more fish mortality than any other turbine type, these are relatively understudied in comparison to the less-common and less-injurious Kaplan turbines (Pracheil et al., 2016). Similarly, most of the research into hydropower infrastructure and injury have focussed on juvenile salmonids (Crew et al., 2017), despite evidence that other types of fish such as percids may experience higher mortality rates (Pracheil et al., 2016). Future studies on a broader variety of turbine types and fish species (such as potadromous fish) would help identify which issues relating to passage-associated injuries are most pressing.

## 2.6 Summary: Contrasting upstream vs downstream physiological effects

There are clear differences in the physiological issues depending on whether fish are passing upstream or downstream. Across several species, fishway ascension is generally neither physiologically stressful nor energetically taxing, and physiological capacity is not a limiting factor (Bido et al., 2018; Dominy, 1971; Hatry et al., 2014; Pon et al., 2009a,b; Roscoe et al., 2011; Schwalme et al., 1985). Physiological stress may pose a challenge to upstream passage from some species, but establishing a causal relationship between even extreme changes in flow (for example, during hydropeaking) and physiological stress has proven difficult (Costa et al., 2017). In addition, injuries incurred during passage are not generally viewed as a major conservation concern for upstream migrating fish. Energetics and anaerobic processes, however, can influence pathway selection when approaching the fishway, and have carryover effects on post-passage migration (reviewed in the next section of this chapter). For fish migrating downstream, on the other hand, stress from dam passage can be severe, compromising energy reserves, immune systems, and smoltification (Ammar et al., 2020; Ferguson et al., 2007; Schreck et al., 2006). The many ways that downstream migrants can be injured during dam passage can also have serious negative consequences (Algera et al., 2020; Pleizier et al., 2020). Energetic issues, meanwhile, have generally not been studied in downstream migrants, possibly because these migrations are not viewed as being energetically demanding (swimming can be assisted by prevailing flows). Further, and specific to salmonids, downstream smolt migrations tend to occur in the spring when temperatures are not contributing to energetically demanding swimming, unlike many upstream migrations that occur during peak summer temperatures.

## 3 Carryover effects

In an ecological context, carryover effects occur when an individual's previous history and experience explains their current performance in a given situation (O'Connor et al., 2014). From the perspective of infrastructure passage, carryover effects could be reflected in alterations to fish behavior, survival or reproduction following successful, yet physiologically challenging, passage experiences. EPOC could be considered a short-term carryover effect of exercise.

In their review of fish passage at dam infrastructure, Roscoe and Hinch (2010) identified 13 studies out of 96 (14%), (Fig. 1) that examined "post-passage consequences" of dam passage on fish. Only a few of these studies examined physiological causes of the consequences. Since that review, there have been additional studies examining physiological causes of carryover effects on adult upstream migrants. Nevertheless, the research is limited. This section will examine how some aspects of fish physiology, in particular energetics, stress, injury and navigation, can result in significant carryover effects to fish that move through passage infrastructure.

### 3.1 Upstream passage

Upstream passage by adult sockeye salmon at the Seton Dam in British Columbia, Canada (Figs. 2 and 3) is one of the few situations where physiological carryover effects have been thoroughly investigated. Earlier in this chapter, we reviewed the roles that physiological stress, olfaction, swimming behavior and energetics had on passage success or failure to locate and ascend this fishway (Burnett et al., 2014a; Pon et al., 2009a,b; Roscoe et al., 2011). After sockeye salmon pass the dam, they travel through two lakes, an additional 40km, to reach their spawning stream (Fig. 2). Using radio tracking and blood sampling of tagged fish, Roscoe et al. (2011) observed that post-dam passage mortality in 2007 was higher (27%) for fish that were captured in the fishway and released downstream of the dam than for fish released upstream of the dam (7%), indicating that dam passage had post-passage consequences on survival. Glucose concentrations were higher in fish that died following successful dam passage but did not reach the spawning grounds when compared to those that successfully ascended the fishway and reached the spawning grounds ($\sim$7.6mmol$\,L^{-1}$ and $\sim$4.6mmol$\,L^{-1}$, respectively). Fish mobilize glucose as an energy substrate for aerobic fuel, and high glucose levels have been used as indicators of exhaustive exercise (Farrell et al., 2001b; Wood, 1991; Zhang et al., 2018). All the same, glucose levels were not critically high and no other blood parameter (i.e., cortisol, lactate, or osmolality) indicated that mortality would be expected based on acute physiological stress (Roscoe et al., 2011). Gross somatic energy levels at the time of release were not lower for the fish that subsequently died en route and in-lake compared to successful migrants, suggesting that body energy reserves were not a limiting factor.

Burnett et al. (2014b) helped to resolve this mortality and glucose mechanism dilemma. In 2014, they used accelerometry telemetry to track sockeye through the Seton Dam tailrace, fishway, and to spawning grounds (Figs. 2 and 3), and found that fish which exhibited high levels of anaerobiosis in the tailrace were able to more readily pass the dam (Fig. 7A). However, extremely high levels of anaerobiosis led to post-passage mortality in the lakes (Fig. 7B). Nearly half of the mortality observed in their study occurred in Seton Lake within 1–3 days of dam passage, likely owing to the greater EPOC and long recovery times associated with prolonged periods of swimming at critical speeds (Lee et al., 2003a). Seton River water temperatures

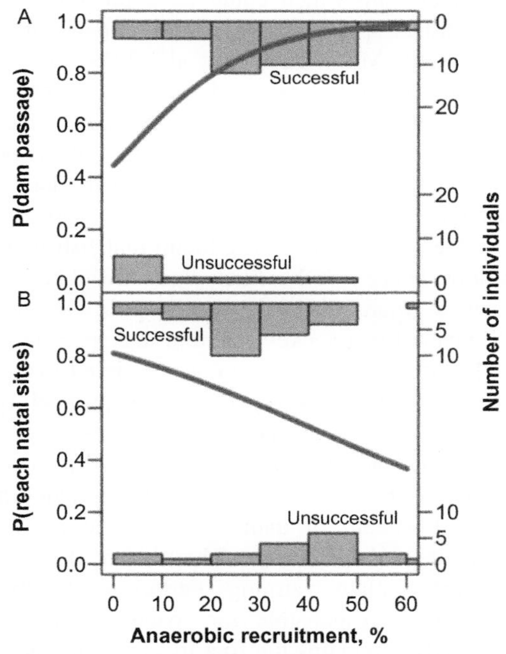


**FIG. 7** Predicted probability of Gates Creek sockeye salmon passing Seton Dam (A; $n = 54$) and reaching natal spawning streams (B; $n = 44$), visualized by fitting a logistic regression to the anaerobic recruitment of successful and unsuccessful individual migrants. *Used with permission from Burnett, N.J., Hinch, S.G., Braun, D.C., Casselman, M.T., Middleton, C.T., Wilson, S.M., Cooke, S.J., 2014b. Burst swimming in areas of high flow: delayed consequences of anaerobiosis in wild adult sockeye salmon. Physiol. Biochem. Zool. 87, 587–598.*

were often above the $T_{opt}$ for the study population (i.e., 17.5 °C; Lee et al., 2003b), which would decrease aerobic scope and cardiac scope, and hence critical swimming speed, perhaps further increasing and prolonging EPOC (Farrell et al., 2008; Lee et al., 2003a,b) if burst swimming was used to maintain groundspeed at these elevated temperatures and ultimately causing cardiovascular collapse (Eliason et al., 2013). Fish that were captured and tagged downstream of the dam, and not initially captured at the top of the fishway, and forced to re-ascend it, were over 50% more likely to reach spawning grounds after passing Seton Dam, which could also be related to lower levels of EPOC they incurred (Burnett et al., 2014b).

Roscoe et al. (2010) sampled blood and tagged sockeye salmon with acoustic transmitters and archival temperature loggers at the Seton Dam, tracking fish to their spawning grounds (Figs. 2 and 3). More reproductively advanced females (i.e., those with a low blood estradiol level and a low gross somatic energy, as assessed with microwave energy meter) transited through cooler portions of the lakes en route to spawning grounds compared to less mature females with higher energy stores. This behavior reduced metabolic costs and slowed maturation, which could have prevented "over-ripening" of gametes and thus failed spawning (Flett et al., 1996), although spawning success was not assessed. Minke-Martin et al. (2018) conducted a similar study to Roscoe et al. (2010), but assessed spawning success and examined the thermal selection of sockeye post-dam passage. Specifically, they examined the proportion of time fish spent within their optimal aerobic scope temperature window ($T_{optAS}$) while traversing the lakes. This window (13.4–19.5 °C) provides ≥90% of maximum aerobic scope (Eliason et al., 2011). As with Roscoe et al. (2011), they found that fish with lower plasma glucose concentrations, hence fish that likely had utilized burst swimming less prior to capture, were more likely to reach spawning grounds. Equally important, they discovered that sockeye salmon that spent more of their lake migration time within the $T_{optAS}$ window not only lived longer on spawning grounds before natural death, but also had a higher probability of spawning. Exposure to higher dam discharge, however, was associated with a lower probability of successful spawning. In sum, these findings demonstrate the importance of variable thermal habitats that can be behaviourally selected in lakes and reservoirs upstream of dams as they provide thermal options for dealing with strenuous dam passage. They could also be of importance for coping with future climate warming (Ferrari et al., 2007). While the physiological mechanisms responsible for spawning failures are unclear, Minke-Martin et al. (2018) remains the first to assess the direct reproductive fitness consequences of successful fishway passage.

Bass et al. (2018) captured and tagged migrating sockeye below the Seton Dam (Fig. 2) that had a range of visible injuries, including numerous open wounds caused by escape from downstream fisheries, ocean sea lice and other causes. While there have been no studies examining the fitness consequences of injuries obtained by fish while passing upstream through fishways, their study is a good analog. When compared with the salmon that had no visible injuries, they found that injured fish had a 16% lower probability of completing their 40km migration to spawning grounds, and injured females had an 18% lower probability of successfully spawning (no gametes left in the body). The annual proportion of effective female spawners that died in the final 40km of their migration due to injuries ranged up to 10% of the total female population. (Note that spawning success could not be assessed in males by visual assessment of gametes.) Although these injuries to the females were not caused by dam passage, it is possible that the cumulative effects of injury and dam passage could have a significant fitness consequence. Further studies are needed to determine whether such an interaction exists.

The studies examining upstream migration of sockeye salmon at the Seton Dam demonstrate the importance of monitoring fish after they pass dams to incorporate potential carryover effects and post-passage consequences in evaluations of fishway performance. Other more recent studies have also measured these carryover effects, such as Gosselin et al. (2021) who found a negative effect of percent of water spilled at the Bonneville Dam in the lower Columbia River, USA, on post-passage survival of Chinook salmon. Specific consequences of dam passage such as physiological stress, energy use or physical injury are likely to be associated with fitness costs (Castro-Santos et al., 2009) and should be more readily incorporated into research and monitoring. Studies on heavily-dammed rivers such as the Columbia River, and elsewhere in the world, need to examine cumulative effects arising from multiple dam passages. It is worth noting that carryover effects relating to passage conditions (such as river temperature) in anadromous fish, during both upstream and downstream migrations, can be strongly overshadowed by the effect of ocean conditions (Gosselin et al., 2021). Nevertheless, high cumulative mortality associated with multiple dam passage is common in freshwater systems with Pacific salmon (Keefer et al., 2008; Naughton et al., 2005) and Atlantic salmon (Gowans et al., 2003; Lundqvist et al., 2008; Nyqvist et al., 2017b). Physiological assessments are rarely made at consecutive passage facilities, so the physiological mechanisms for cumulative carryover effects are not known.

## 3.2 Downstream passage

Carryover effects on downstream migrants have been more broadly examined in terms of study locations, with a focus on the cumulative effects of sequential infrastructure passage than for upstream passage. Most research has focused on "delayed mortality" associated with infrastructure passage and has primarily involved out-migrating salmonid smolts.

In the upper Columbia River, USA, salmonid smolts must pass through eight large hydroelectric dams to reach the ocean which they can do via direct dam passage or barging. Budy et al. (2002), using PIT-tag data from 1994 to 1996, reported that both in-river dam passage and transportation around the dam lowered the smolt-to-adult survival rate for Snake River Chinook salmon and steelhead trout, although the relative degree of this delayed mortality caused by the two passage routes was controversial. The work by Gosselin et al. (2021) expanded on this research, as they assessed potential carryover effects relating to downstream and upstream passage on the survival of Chinook salmon across multiple life stages in the upper Columbia River. They found that survival during adult upstream migration was negatively affected by being transported around a dam as juveniles. Evans et al. (2014) explored the individual fish characteristics associated with the probability of returning as an adult to the Snake River and upper Columbia River. They tagged >47,000 smolts in 2007–2010 after they had passed at least one dam facility, though there were still numerous additional dams to pass to reach the ocean. Before release, they used photography to examine body size and external condition. The probability of a juvenile surviving to adulthood was lowest for smolts with body injuries, fin damage, and external signs of disease confirming the strong carryover effects to adult hood of these juvenile conditions. Efforts to reduce fish injury rates by modifying dam and dam operational strategies (see Section 2.2; Ferguson et al., 2007), along with efforts to reduce disease and disease transmission are important ways to reduce these carryover consequences.

Schreck et al. (2006) examined the survival of radio-tagged Chinook salmon smolts after passing the Bonneville Dam on the lower Columbia River. USA. The salmon moved downstream, either in-river past dams (through spillways, turbines or bypasses), or by being transported around the dam. As reviewed above, all of these passage approaches can result in immediate physiological stress, injury, and navigational disorientation, which could result in carryover effects, albeit to differing degrees. While mortality was very low between the dam and estuary, a distance of 180km, delayed mortality in the estuary was high and caused by avian predators such as Caspian terns (*Sterna caspia*) and double-crested cormorants (*Phalacrocorax auratus*). Predation appeared to be preferential on smolts that were infected with *Renibacterium salmoninarum*, the causative agent of Bacterial Kidney Disease which can affect osmoregulation, and for fish possessing low smoltification levels (relatively low gill $Na^+,K^+$ -ATPase activity) (Schreck et al., 2006). Many outmigrating Chinook salmon are carriers of the pathogen, its virulence made worse by physiological stress. The authors suggested stress caused by dam passage led to higher predation on smolts that were physiologically and behaviorally compromized by a pathogen which slowed the smoltification process (Schreck et al., 2006). Thus, juvenile salmon in advanced states of smoltification would prefer and entered seawater (McInerney, 1964) more quickly (Seals-Price and Schreck, 2003), potentially leaving less time for avian predation in the estuary. Furey et al. (2021) found direct support for the predator-pathogen association in out-migrating sockeye salmon smolts by demonstrating a 15- to 26-fold greater chance of being found in the stomach of downstream-residing predatory bull trout (*Salvelinus confluentus*) if smolts carried the naturally occurring infectious hematopoietic necrosis virus. How predators like sea-birds or piscivores identify physiologically compromized prey is still uncertain. Nonetheless, Schreck et al. (2006) suggest that mitigating delayed mortality will require improvements to upstream fish passage and transportation facilities so that juvenile salmon encounter minimal "stress deficits" before they reach estuaries and their associated salinity and predator challenges (Schreck and Tort, 2016).

Predation by piscivores (e.g., northern pike minnow; *Ptychocheilus oregonensis*) on smolts can also be extensive immediately downstream of fish bypasses in the Pacific northwest, USA, when outfalls are shallow and slow-moving (Petersen, 1994; Rieman et al., 1991) but significantly reduced if outfalls have deeper and faster water where predators are less efficient (Shively et al., 1996). It is not clear if dam passage stress, injury or prior pathogen loads (e.g., Furey et al., 2021; Schreck et al., 2006) affects this immediate downstream carryover effect, but this is an area of needed research.

Barrier passage studies in Europe have also identified that stress, smoltification timing and immune responses are important carryover effects. During their downstream migration in the Albert Canal, which connects the Meuse River to the Scheldt Estuary in Belgium, European eel adults and Atlantic salmon smolts must pass five intermediate-head navigation locks. Migration success through the entire system is generally poor. Vergeynst et al. (2019) tracked fish with acoustic tags and found that 20% and 44% of gate-passing eels and smolts, respectively, stopped their migration after passing just the first lock system. The authors suspect that injury or physiological stress at this first lock system may have been responsible for the cessation of migration. While very few tagged smolts passed all locks to reach the sea, eels were more successful. Repeated delays experienced by smolts were thought to have moved them out of their physiological smolt window making them more susceptible to ionoregulatory imbalances, which could lead to higher predator risk (McCormick et al., 1998). Ammar et al. (2020) noted that injuries and stress caused by multiple turbine encounters leads to immunity and osmoregulatory changes compromising Atlantic salmon smolt's ability to escape through consecutive barriers and successfully reach the ocean.

## 4 Conservation physiology and fish passage

### 4.1 Using physiology to understand and solve passage problems

Conservation physiology, the study of physiological responses of organisms to human alteration of the environment (Wikelski and Cooke, 2006), is a relatively new subdiscipline of conservation science. Its premise is that physiological knowledge can be used to document problems, and generate appropriate management responses (Cooke et al., 2012). Dams have created a major conservation issue for fish yet practical examples of how physiology has helped to understand *and* solve conservation problems associated with dams remain relatively scarce. The performance of fish passage infrastructure has been assessed mostly through empirical observations of survival and behavior, rather than physiology (Roscoe and Hinch, 2010). Without a physiological perspective, a mechanistic understanding of survival and behavior is difficult to achieve.

As reviewed in this chapter, physiological assessments have played a significant role in helping those that manage dams, water flow, and passage infrastructure to gain insight into passage problems by examining specific physiological aspects of fish: navigation and orientation in particular the alteration of olfactory, visual or flow cues, as well as stress, and injury. Most of the studies cited in this chapter that examined physiological aspects of dam passage stated that their results would have implications to passage infrastructure or water management, yet it was not clear how results in many of these contributed to solving passage problems. However, there are some studies whose results played important roles in the modification of passage infrastructure or led to changes in operational approaches. Most of these involved examinations of injury incurred during downstream migrations, usually in conjunction with survival information, to modify turbine designs and develop approaches to direct fish away from turbines and spillways (e.g., reviewed in: Čada, 2001; Crew et al., 2017; Office of Technology Assessment (OTA), 1995; Martinez et al., 2019). There are two hydro systems where physiological information was used to support large scale infrastructure modifications (Columbia River, USA—Ferguson et al., 2007) and flow operational changes (Seton River, Canada—Bett et al., 2020), and also to assess the consequences of physiologically mediated carry-over effects which were incorporated into management decision-making (Columbia River, USA—Budy et al., 2002; Schreck et al., 2006; Seton River, Canada—Bett et al., 2020; Burnett et al., 2017).

## 4.2 Knowledge gaps and the need for integrative research

While there have been several conservation physiology studies conducted on dam passage since the review of Roscoe and Hinch (2010), there remain numerous gaps in our understanding. In particular, few have investigated the potential role of injuries associated with upstream passage structures and its environments (e.g., predators), on passage failure or latent mortality. In terms of downstream passage, few studies have examined the energetics and aerobic capacity of fish moving downstream (e.g., Baktoft et al., 2020), yet this will become an exceedingly important issue particularly for juvenile salmonids as climate change continues to warm rivers. Also, conservation physiological studies are generally lacking in dam passage systems outside of North America, and on non-salmonid species (Roscoe and Hinch, 2010). Consequently, fish passage facilities have been predominantly designed for salmonids (Clay, 1995) and may not be suitable to many other species of migratory fish (Lucas and Baras, 2001). Furthermore, decades of research bias has created a substantial imbalance in our understanding of the basic life history traits of migratory fish in North America compared to those in the tropics and subtropics (Baras and Lucas, 2001), let alone our understanding of their physiological responses to barriers. Research on the movement patterns of fish in biodiversity hotspots such as the neotropics are on the rise (e.g., Silva et al., 2018), and there is a strong need for this trend to continue given that the construction of current and future dams is largely concentrated in South America, Africa, and Asia. As more becomes known about the biology of native species in these areas, physiological studies could provide important contributions to the development and management of fish passage structures. Another current knowledge gap is the physiological response of fish to lower head dams or barriers, which make up the vast majority of barriers around the world. While most passage studies focus on larger dams, there is evidence that smaller barriers can also have significant impacts on fish movement and behavior (e.g., Alexandre and Almeida, 2010; Carpenter-Bundhoo et al., 2020; Tiemann et al., 2004). How these smaller barriers affect fish physiology, however, remains largely unknown. While we know the effects of habitat modification stemming from hydropower facilities (such as the creation of upstream reservoirs) were beyond the scope of our review, these could interact with passage carryover effects and are another area that warrants more investigation. Lastly, while we were able to identify some physiological generalities in terms of causes of passage success or failure, each passage facility offers unique hydrological challenges to fish so specific physiological results from one facility might not be applicable to another, thus requiring physiological assessments to be more widely included in effectiveness monitoring programs (Noonan et al., 2012; Roscoe and Hinch, 2010).

Passage facilities can create environments with complex hydraulics and thermal regimes, and also alter olfactory cues, light and water pressure in ways that cause disorientation and injury. Few natural movement corridors have this array of factors affecting fish within such a relatively small spatial footprint, which is why an integrative research approach is so important to adopt when examining infrastructure passage. Bett et al. (2020) overviewed a case study that incorporated this approach in a 7-year examination of returning adult salmon passage through a hydropower system in British Columbia, Canada (Figs. 2 and 3). Studies on this system integrated biotelemetry (radio, acoustic, PIT, EMG, accelerometers), biologging (ibuttons), riverside-based behavioral tests, physiological (blood) and molecular analyses (genomic), and hydraulic monitoring (ADCP and computational fluid hydrodynamic models) (Bett and Hinch, 2015; Bett et al., 2016b, 2018; Burnett et al., 2014a,b, 2017; Drenner et al., 2018; Li et al., 2021; Minke-Martin et al., 2018; Middleton et al., 2018; Pon et al., 2009a,b, 2012; Roscoe et al., 2010, 2011). The investigators worked in cooperation with the local power company to characterize the effects of hydrosystem manipulations on upstream navigation, fishway passage success, post-passage survival, and spawning. These studies confirmed that established operational guidelines for natal water dilution were appropriate for reducing negative physiological and survival outcomes, and informed the adoption of new operational conditions for dam flow releases (Harrower et al., 2020). Specifically, the latter involved an adaptive management experiment whereby dominant fishway attraction flows were moved away from the fishway entrance (termed "alternative flows") and compared to the historical approach (termed "routine flows") (Fig. 8). Individuals exposed to routine flows spent two times longer recovering from dam passage (some fish had excessively high EPOC) and over all they exhibited 14% higher mortality following passage than those exposed to alternative flows. Release of alternative flows for 10 days, which was only one-third of the run timing, assisted approximately 550 fish (or 3% of total spawners) in reaching spawning grounds. Given this population of salmon is in decline (Rand, 2011), the alternative flows can play a significant role in its sustainability. The local power company (BCHydro) has adopted the alternative flows as the new flow conditions during the timing for the summer sockeye run.

To realize the full benefits that conservation physiology has to offer to the study of fish passage, researchers need to engage hydro and dam facility managers (and vice versa) with open dialogue regarding the capabilities of physiological tools and how they can be used to address conservation and management problems, as was done in Bett et al. (2020). Recent advances in biologging (e.g., Minke-Martin et al., 2018, temperature loggers), biotelemetry (e.g., Burnett et al., 2014b, triaxial accelerometers) and other sensor technologies (e.g., Martinez et al., 2019, Sensor Fish) are enabling conservation physiologists to study free-roaming fish, and their analogues, at dam infrastructure to obtain powerful physiological insights into passage problems. This information can be paired with current profilers and computational fluid dynamic modeling (e.g., Baktoft et al., 2020; Li et al., 2021) to link physiology to hydrodynamics. For conservation physiology to be truly effective, it must represent a meaningful collaboration of researchers and practitioners (Cooke et al., 2012), and we encourage more researchers to integrate these tools into collaborative dam passage monitoring programs. Using this approach to analyze the relationship between fish physiology and dam passage will lead to the modification, operation and development of fish passage facilities that minimize their effects on naturally occurring fish species.

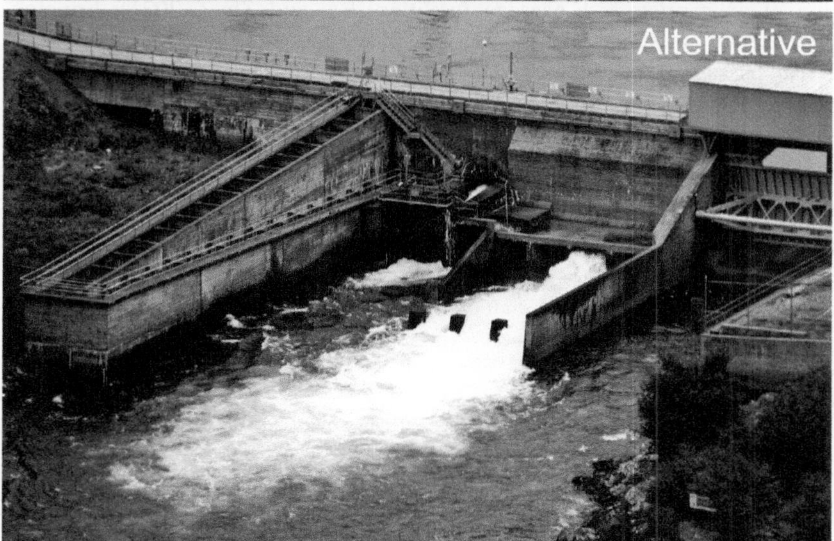


**FIG. 8** Images of the routine and alternative flows in 2014 in the tailrace of the Seton River Dam. Routine flows largely emanate from the fish water release gate situated next to the fishway entrance and alternative flows largely from siphon 4 situated away from the fishway entrance. Alternative flows reduced latent mortality of dam passage by sockeye salmon. See Fig. 3 for more details on the dam and fishway infrastructure. *Used with permission from Bett, N.B., Hinch, S.G., Bass, A.L., Braun, D.C., Burnett, N.J., Casselman, M.T., Cooke, S.J., Drenner, S.M., Gelchu, A., Harrower, W.L., Ledoux, R., Lotto, A.G., Middleton, C.T., Minke-Martin, V., Patterson, D.A., Zhang, W., Zhu, D.Z., 2020. Using an integrative research approach to improve fish migrations in regulated rivers: a case study on Pacific salmon in the Fraser River, Canada. Hydrobiologia, 849, 385–405.*

## 4.3 Conclusions

There is a large history of research on fish passage, reflecting the value (economic, ecological and cultural) that is placed on mitigating the effects of barriers on migratory fish. The majority of this research, however, has focused on the measurement of basic variables such as passage duration and rate of passage success. The impact of barriers on migration success and survival is undoubtedly influenced by a fish's capacity to meet the physiological demands of passage, and this field of work remains severely understudied. Integrating physiological analyses with other means of study will provide a much more thorough and nuanced understanding of the challenges associated with fish passage, and can be accomplished through the use of novel technologies. As our knowledge of passage issues continues to grow on a global scale, it is also becoming increasingly apparent that generalized approaches to dam operations and fish management are not sufficient. Instead, each region's unique hydrological characteristics and assemblage of fish species needs to be accounted for. There is growing concern that the rapid increase in hydropower development in emerging nations may proceed without the creation of appropriate fish-passage facilities (Bai and Zhou, 2021), bringing some level of urgency to this issue. Incorporation of physiological analyses in fish passage research and management programs will contribute to the conservation of the many species and populations of fish that must contend with the risks associated with dam passage.

## Acknowledgments

We appreciate the library and reference assistance of Anthony Pica. We are also thankful to Derrick Alcott, Nann Fangue, Kim Birnie-Gauvin and an anonymous reviewer for their insightful comments on an earlier draft of this chapter.

## References

Aarestrup, K., Lucas, M.C., Hansen, J.A., 2003. Efficiency of a nature-like bypass channel for sea trout (*Salmo trutta*) ascending a small Danish stream studied by PIT telemetry. Ecol. Freshw. Fish 12, 160–168.

Achord, S., Matthews, G.M., Johnson, O.W., Marsh, D.M., 1996. Use of passive integrated transponder (PIT) tags to monitor migration timing of Snake River chinook salmon smolts. N. Am. J. Fish. Manag. 16, 302–313.

Agostinho, A.A., Agostinho, C.S., Pelicice, F.A., Marques, E.E., 2012. Fish ladders: safe fish passage or hotspot for predation? Neotrop. Ichthyol. 10, 687–696.

Alexandre, C.M., Almeida, P.R., 2010. The impact of small physical obstacles on the structure of freshwater fish assemblages. River Res. Appl. 26, 977–994.

Algera, D.A., Rytwinski, T., Taylor, J.J., Bennett, J.R., Smokorowski, K.E., Harrison, P.M., Clarke, K.D., Enders, E.C., Power, M., Bevelhimer, M.S., Cooke, S.J., 2020. What are the relative risks of mortality and injury for fish during downstream passage at hydroelectric dams in temperate regions? A systematic review. Environ. Evid. 9, 3. https://doi.org/10.1186/s13750-020-0184-0.

Amaral, S.D., Branco, P., Romão, F., Ferreira, M.T., Pinheiro, A.N., Santos, J.M., 2021. Evaluation of low-headed ramped weirs for a potadromous cyprinid: effects of substrate addition and discharge on fish passage performance, stress and fatigue. Water 13, 765.

Ammar, I.B., Baekleandt, S., Cornet, V., Antipine, S., Sonny, D., Mandiki, S.N.M., Kestemont, P., 2020. Passage through a hydropower plant affects physiological and health status of Atlantic salmon smolts. Comp. Biochem. Phys. A 247, 110745.

Ammar, I.B., Cornet, V., Houndji, A., Baekelandt, S., Antipine, S., Sonny, D., Mandiki, S.N.M., Kestemont, P., 2021. Impact of downstream passage through hydropower plants on the physiological and health status of a critically endangered species: the European eel *Anguilla anguilla*. Comp. Biochem. Phys. A 254, 110876.

Andrew, F.J., Geen, G.H., 1958. Sockeye and pink salmon investigations at the Seton Creek hydroelectric installation. In: International Pacific Salmon Fisheries Commission Progress Report No. 4. New Westminster, BC. 74 pp.

Bai, L., Zhou, L., 2021. Aiming for fish-friendly hydropower plants. Science 374, 1062–1063.

Baktoft, H., Gjelland, K.Ø., Szabo-Meszaros, M., Silva, A.T., Riha, M., Økland, F., Alfredsen, K., Forseth, T., 2020. Can energy depletion of wild Atlantic salmon kelts negotiating hydropower facilities lead to reduced survival? Sustainability 12, 7341.

Baras, E., Lucas, M.C., 2001. Impacts of man's modifications of river hydrology on the migration of freshwater fishes: a mechanistic perspective. Ecohydrol. Hydriobiol. 1, 291–304.

Barbarossa, V., Schmitt, R.J.P., Huijbregts, M.A.J., Zarfl, C., King, H., Schipper, A.M., 2020. Impacts of current and future large dams on the geographic range connectivity of freshwater fish worldwide. Proc. Natl. Acad. Sci. 117, 3648–3655.

Barbin, G.P., Krueger, W.H., 1994. Behaviour and swimming performance of elvers of the American eel, *Anguilla rostrata*, in an experimental flume. J. Fish Biol. 45, 111–121.

Barnthouse, L.W., 2013. Impacts of entrainment and impingement on fish populations: a review of the scientific evidence. Environ. Sci. Policy 31, 149–156.

Barton, B.A., Iwama, G.K., 1991. Physiological changes in fish from stress in aquaculture with emphasis on the response and effects of corticosteroids. Annu. Rev. Fish Dis. 1, 3–26.

Bass, A.L., Hinch, S.G., Casselman, M.T., Bett, N.N., Burnett, N.J., Middleton, C.T., Patterson, D.A., 2018. Visible gill-net injuries predict migration and spawning failure in adult sockeye salmon. Trans. Am. Fish. Soc. 147, 1085–1099.

Beauchamp, D.A., 1995. Riverine predation on sockeye salmon fry migrating to Lake Washington. N. Am. J. Fish. Manag. 15, 358–365.

Belletti, B., de Leaniz, C.G., Jones, J., Bizzi, S., Börger, L., Segura, G., Castelletti, A., et al., 2020. More than one million barriers fragment Europe's rivers. Nature 558, 436–441.

Bernatchez, L., Dodson, J.J., 1987. Relationship between bioenergetics and behaviour in anadromous fish migration. Can. J. Fish. Aquat. Sci. 44, 399–407.

Bernier, N.J., Alderman, S.L., 2022. Applied aspects of fish endocrinology. Fish Physiol. 39A, 253–320.

Bett, N.N., Hinch, S.G., 2015. Attraction of migrating adult sockeye salmon to conspecifics in the absence of natal chemical cues. Behav. Ecol. 26, 1180–1187.

Bett, N.N., Hinch, S.G., 2016. Olfactory navigation during spawning migrations: a review and introduction of the hierarchical navigation hypothesis. Biol. Rev. 91, 728–759.

Bett, N.N., Hinch, S.G., Dittman, A.H., Yun, S.S., 2016a. Evidence of olfactory imprinting at an early life stage in pink salmon (*Oncorhynchus gorbuscha*). Sci. Rep. 6, 36393.

Bett, N.N., Hinch, S.G., Yun, S.-S., 2016b. Behavioural responses of Pacific salmon to chemical disturbance cues during the spawning migration. Behav. Processes 132, 76–84.

Bett, N.N., Hinch, S.G., Burnett, N.J., Donaldson, M.R., Naman, S.M., 2017. Causes and consequences of straying into small populations of Pacific salmon. Fisheries 42, 220–230.

Bett, N.N., Hinch, S.G., Casselman, M.T., 2018. Effects of natal water dilution in a regulated river on the migration of Pacific salmon. River Res. Appl. 34, 1151–1157.

Bett, N.B., Hinch, S.G., Bass, A.L., Braun, D.C., Burnett, N.J., Casselman, M.T., Cooke, S.J., Drenner, S.M., Gelchu, A., Harrower, W.L., Ledoux, R., Lotto, A.G., Middleton, C.T., Minke-Martin, V., Patterson, D.A., Zhang, W., Zhu, D.Z., 2020. Using an integrative research approach to improve fish migrations in regulated rivers: a case study on Pacific salmon in the Fraser River, Canada. Hydrobiologia 849, 385–405.

Bido, A.F., Urbinati, E.C., Makrakis, M.C., Celestino, L.F., Serra, M., Markrakis, S., 2018. Stress indicators for *Prochilodus lineatus* (Characiformes: Procholodontidae) breeders during passage through a fish ladder. Mar. Freshw. Res. 69, 1814–1821.

Black, E.C., 1958. Hyperactivity as a lethal factor in fish. J. Fish. Res. Board Can. 15, 573–586.

Bleckmann, H., 1994. Reception of hydrodynamic stimuli in aquatic and semiaquatic animals. Prog. Zool. 41, 1–115.

Bleckmann, H., 2008. Peripheral and central processing of lateral line information. J. Comp. Physiol. A 194, 145–158.

Bond, M.H., Westley, P.A.H., Dittman, A.H., Holecek, D., Marsh, T., Quinn, T.P., 2017. Combined effects of barge transportation, river environment, and rearing location on straying and migration of adult snake river fall-run Chinook salmon. Trans. Am. Fish. Soc. 146, 60–73.

Bonga, S.E., 1997. The stress response in fish. Physiol. Rev. 77, 591–625.

Bouck, G.R., 1980. Etiology of gas bubble disease. Trans. Am. Fish. Soc. 109, 703–707.

Boys, C.A., Pflugrath, B.D., Mueller, M., Pander, J., Deng, Z.D., Geist, J., 2018. Physical and hydraulic forces experienced by fish passing through three different low-head hydropower turbines. Mar. Freshw. Res. 69, 1934–1944.

Brauner, C.J., Seidelin, M., Madsen, S.S., Jensen, F.B., 2000. Effects of fresh water hyperoxia and hypercapnia exposures and their influences on subsequent seawater transfer in Atlantic salmon (*Salmo salar*) smolts. Can. J. Fish. Aquat. Sci. 57, 2054–2064.

Brett, R., 1995. Energetics. In: Groot, C., Margolis, L., Clarke, W.C. (Eds.), Physiological Ecology of Pacific Salmon. UBC Press, Vancouver, pp. 1–68.

Brown, R.S., Geist, D.R., 2002. Determination of swimming speeds and energetic demands of upriver migrating fall chinook salmon (*Oncorhynchus tshawytscha*) in the Klickitat River, Washington. In: Pacific Northwest Laboratory, Bonneville Power Administration. Project 22063, Contract 42663A. Richland, Washington, p. 52.

Brown, R.S., Carlson, T.J., Welch, A.E., Stephenson, J.R., Abernethy, C.S., Ebberts, B.D., Langeslay, M.J., Ahmann, M.L., Feil, D.H., Skalski, J.R., Townsend, R.L., 2009. Assessment of barotrauma from rapid decompression of depth-acclimated juvenile Chinook salmon bearing radiotelemetry transmitters. Trans. Am. Fish. Soc. 138, 1285–1301.

Brown, R.S., Ahmann, M.L., Trumbo, B.A., Foust, J., 2012a. Fish protection: cooperative research advances fish friendly turbine design. Hyd. Rev. 31, 48–53.

Brown, R.S., Pflugrath, B.D., Colotelo, A.H., Brauner, C.J., Carlson, T.J., Deng, Z.D., Seaburg, A.G., 2012b. Pathways of barotrauma in juvenile salmonids exposed to simulated hydroturbine passage: Boyle's law vs. Henry's law. Fish. Res. 121, 43–50.

Brown, R.S., Colotelo, A.H., Pflugrath, B.D., Boys, C.A., Baumgartner, L.J., Deng, Z.D., Silva, L.G., Brauner, C.J., Mallen-Cooper, M., Phonekhampeng, O., Thorncraft, G., 2014. Understanding barotrauma in fish passing hydro structures: a global strategy for sustainable development of water resources. Fisheries 39, 108–122.

Brown, R.S., Walker, R.W., Stephenson, J.R., 2015. A Preliminary Assessment of Barotrauma Injuries and Acclimation Studies for Three Fish Species (No. PNNL-24720). Pacific Northwest National Laboratory (PNNL), Richland, WA (US), https://doi.org/10.2172/1237809.

Budy, P., Thiede, G.P., Bouwes, N., Petrosky, C.E., Schaller, H., 2002. Evidence linking delayed mortality of Snake River salmon to their earlier hydrosystem experience. N. Am. J. Fish. Manag. 22, 35–51.

Bunt, C.M., Cooke, S.J., McKinley, R.S., 2000. Assessment of the Dunnville Fishway for passage of walleyes from Lake Erie to the Grand River, Ontario. J. Great Lakes Res. 26, 482–488.

Bunt, C., Castro-Santos, T., Haro, A., 2012. Performance of fish passage structures at upstream barriers to migration. River Res. Appl. 28, 457–478.

Burgetz, I.J., Rojas-Vargas, A., Hinch, S.G., Randall, D.J., 1998. Initial recruitment of anaerobic metabolism during submaximal swimming in rainbow trout (*Oncorhynchus mykiss*). J. Exp. Biol. 201, 2711–2721.

Burnett, N.J., Hinch, S.G., Donaldson, M.R., Furey, N.B., Patterson, D.A., Roscoe, D.W., Cooke, S.J., 2014a. Alterations to dam-spill discharge influence sex-specific activity, behaviour and passage success of migrating adult sockeye salmon. Ecohydrology 7, 1094–1104.

Burnett, N.J., Hinch, S.G., Braun, D.C., Casselman, M.T., Middleton, C.T., Wilson, S.M., Cooke, S.J., 2014b. Burst swimming in areas of high flow: delayed consequences of anaerobiosis in wild adult sockeye salmon. Physiol. Biochem. Zool. 87, 587–598.

Burnett, N.J., Hinch, S.G., Bett, N.N., Braun, D.C., Casselman, M.T., Cooke, S.J., Gelchu, A., Lingard, S., Middleton, C.T., Minke-Martin, V., White, C.F.H., 2017. Reducing carryover effects on the migration and spawning success of sockeye salmon through a management experiment of dam flows. River Res. Appl. 33, 3–15.

Byford, G.J., Wagner, C.M., Hume, J.B., Moser, M.L., 2016. Do native Pacific lamprey and invasive sea lamprey share an alarm cue? Implications for use of a natural repellent to guide imperiled Pacific lamprey into fishways. N. Am. J. Fish. Manag. 36, 1090–1096.

Čada, G.F., 1997. Shaken, Not Stirred: The Recipe for a Fish Friendly Turbine. Oak Ridge National Laboratory. Contract No. DE-AC05-96OR22464. https://www.osti.gov/servlets/purl/510550.

Čada, G.F., 1998. Fish passage mitigation at hydroelectric power projects in the United States. In: Jungwirth, M., Schmutz, S., Weiss, S. (Eds.), Fish Migration and Fish Bypasses. Fishing News Books, Oxford, pp. 208–219.

Čada, G.F., 2001. The development of advanced hydroelectric turbines to improve fish passage survival. Fisheries 26, 14–23.

Čada, G.F., Schweizer, P.E., 2012. The Application of Traits-Based Assessment Approaches to Estimate the Effects of Hydroelectric Turbine Passage on Fish Populations. ORNL/TM-2012/110, Oak Ridge National Laboratory, Oak Ridge, Tenn, p. 37, https://doi.org/10.2172/1038082.

Carpenter-Bundhoo, L., Butler, G.L., Bond, N.R., Bunn, S.E., Reinfelds, I.V., Kennard, M.J., 2020. Effects of a low-head weir on multi-scaled movement and behavior of three riverine fish species. Sci. Rep. 10, 6817.

Castro-Santos, T, 2005. Optimal swim speeds for traversing velocity barriers: an analysis of volitional high-speed swimming behavior of migratory fishes. J. Exp. Biol. 208, 421–432.

Castro-Santos, T., Cotel, A., Webb, P., 2009. Fishway evaluations for better bioengineering: an integrative approach. 69, 557-575. In: Haro, A.J., Smith, K.L., Rulifson, R.A., Moffit, C.M., Klauda, R.J., Dadswell, M.J., Avery, T.S. (Eds.), Challenges for Diadromous Fishes in a Dynamic Global Environment. American Fisheries Society Symposium, Bethesda, MD.

Castro-Santos, T., Goerig, E., He, P., Lauder, G.V., 2022. Applied aspects of locomotion and biomechanics. Fish Physiol. 39A, 91–140.

Caudill, C.C., Daigle, W.R., Keefer, M.L., Boggs, C.T., Jepson, M.A., Burke, B.J., Zabel, R.W., Bjornn, T.C., Peery, C.A., 2007. Slow dam passage in adult Columbia River salmonids associate with unsuccessful migration: delayed negative effects of passage obstacles or condition-dependent mortality? Can. J. Fish. Aquat. Sci. 64, 979–995.

Chagnaud, B.P., Bleckmann, H., Hofmann, M.H., 2008. Lateral line nerve fibers do not code bulk water flow direction in turbulent flow. Fortschr. Zool. 111, 204–217.

Chapman, D.C., Carlson, C., Weitkamp, D., Matthews, G., Stevenson, J., Miller, M., 1997. Homing in sockeye and Chinook salmon transported around part of their smolt migration route in the Columbia River. N. Am. J. Fish. Manag. 17, 101–113.

Clark, T.D., Sandblom, E., Cox, G.K., Hinch, S.G., Farrell, A.P., 2008. Circulatory limits to oxygen supply during an acute temperature increase in the Chinook salmon (*Oncorhynchus tshawytscha*). Am. J. Physiol. Regul. Integr. Comp. Physiol. 295, 1631–1639.

Clark, T.D., Furey, N.B., Rechisky, E.L., Gale, M.K., Jeffries, K.M., Porter, A.D., Casselman, M.T., Lotto, A.G., Patterson, D.A., Cooke, S.J., Farrell, A.P., Welch, D.W., Hinch, S.G., 2016. Tracking wild salmon smolts to the ocean reveals distinct regions of nocturnal movement and high mortality. Ecol. Appl. 26, 959–978.

Clay, C.H., 1995. Design of Fishways and Other Fish Facilities, second ed. Lewis Publishers, Boca Raton.

Cocherell, D.E., Kawabata, A., Kratville, D.W., Cocherell, S.A., Kaufman, R.C., Anderson, E.K., Chen, Z.Q., Bandeh, H., Rotondo, M.M., Padilla, R., Churchwell, R., Kavvas, M.L., Cech Jr., J.J., 2011. Passage performance and physiological stress response of adult white sturgeon ascending a laboratory fishway. J. Appl. Ichthyol. 27, 327–334.

Colotelo, A.H., Pflugrath, B.D., Brown, R.S., Brauner, C.J., Mueller, R.P., Carlson, T.J., Deng, D.Z., Ahmann, M.L., Trumbo, B.A., 2012. The effect of rapid and sustained decompression on barotrauma in juvenile brook lamprey and Pacific lamprey: implications for passage at hydroelectric facilities. Fish. Res. 129, 17–20.

Connor, A.R., Elling, C.H., Black, E.C., Collins, G.B., Gauley, J.R., Trevor-Smith, E., 1964. Changes in glycogen and lactate levels in migrating salmonid fishes ascending experimental endless fishways. J. Fish. Res. Board Can. 21, 255–290.

Cook, K.V., McConnachie, S.H., Gilmour, K.M., Hinch, S.G., Cooke, S.J., 2011. Fitness and behavioral correlates of pre-stress and stress-induced plasma cortisol titers in pink salmon (*Oncorhynchus gorbuscha*) upon arrival at spawning grounds. Horm. Behav. 60, 489–497.

Cooke, S.J., Hinch, S.G., 2013. Improving the reliability of fishway attraction and passage efficiency estimates to inform fishway engineering, science, and practice. Ecol. Eng. 58, 123–132.

Cooke, S.J., Crossin, G.T., Patterson, D.A., English, K.K., Hinch, S.G., Young, J.L., Alexander, R., Healey, M.C., Van Der Kraak, G., Farrell, A.P., 2005. Coupling non-invasive physiological assessments with telemetry to understand inter-individual variation in behaviour and survivorship of sockeye salmon: development and validation of a technique. J. Fish Biol. 67, 1–17.

Cooke, S.J., Hinch, S.G., Donaldson, M.R., Clark, T.D., Eliason, E.J., Crossin, G.T., Raby, G.D., Jeffries, K.M., Lapointe, M., Miller, K., Patterson, D.A., Farrell, A.P., 2012. Conservation physiology in practice: how physiological knowledge has improved our ability to sustainably manage Pacific salmon during up-river migration. Philos. Trans. R. Soc. Lond. B 367, 1757–1769.

Cooper, J.C., Hasler, A.D., 1974. Electroencephalographic evidence for retention of olfactory cues in homing coho salmon. Science 183, 336–338.

Cooper, J.C., Scholz, A.T., Horrall, R.M., Hasler, A.D., Madison, D.M., 1976. Experimental confirmation of the olfactory hypothesis with homing, artificially imprinted coho salmon (*Oncorhynchus kisutch*). J. Fish. Res. Board Can. 33, 703–710.

Costa, M.J., Lennox, R.J., Katopodis, C., Cooke, S.J., 2017. Is there evidence for flow variability as an organism-level stressor in fluvial fish? J. Ecohydraul. 2, 68–83.

Crew, A.V., Keatley, B.E., Phelps, A.M., 2017. Literature review: fish mortality risks and international regulations associated with downstream passage through hydroelectric facilities. Can. Tech. Rep. Fish. Aquat. Sci. 3207. iv + 47 p.

Crossin, G.T., Hinch, S.G., Farrell, A.P., Whelly, M.P., Healey, M.C., 2003. Pink salmon (*Oncorhynchus gorbuscha*) migratory energetics: response to migratory difficulty and comparisons with sockeye salmon (*Oncorhynchus nerka*). Can. J. Zool. 81, 1986–1995.

Crossin, G.T., Hinch, S.G., Farrell, A.P., Higgs, D.A., Lotto, A.G., Oakes, J.D., Healey, M.C., 2004. Energetics and morphology of sockeye salmon: effects of upriver migratory distance and elevation. J. Fish Biol. 65, 788–810.

Crossin, G.T., Hinch, S.G., Cooke, S.J., Welch, D.W., Lotto, A.G., Patterson, D.A., Jones, S.R.M., Leggatt, R.A., Mathes, M.T., Shrimpton, J.M., Van Der Kraak, G., Farrell, A.P., 2008. Exposure to high temperature influences the behaviour, physiology, and survival of sockeye salmon during spawning migration. Can. J. Zool. 86, 127–140. https://doi.org/10.1139/Z07-122.

Deng, Z.D., Guensch, G.R., McKinstry, C.A., Mueller, R.P., Dauble, D.D., Richmond, M.C., 2005. Evaluation of fish-injury mechanisms during exposure to turbulent shear flow. Can. J. Fish. Aquat. Sci. 62, 1513–1522.

Deng, Z.D., Carlson, T.J., Duncan, J.P., Richmond, M.C., Dauble, D.D., 2010. Use of an autonomous sensor to evaluate the biological performance of the advanced turbine at Wanapum dam. J. Renew. Sustain. Energy 2, 053104.

Deng, Z.D., Lu, J., Myjak, M.J., Martinez, J.J., Tian, C., Morris, S.J., Carlson, T.J., Zhou, D., Hou, H., 2014. Design and implementation of a new autonomous sensor fish to support advanced hydropower development. Rev. Sci. Instrum. 85, 115001.

Dittman, A.H., Quinn, T.P., 1996. Homing in Pacific salmon: mechanisms and ecological basis. J. Exp. Biol. 199, 83–91.

Dittman, A.H., Quinn, T.P., Nevitt, G.A., Hacker, B., Storm, D.R., 1997. Sensitization of olfactory guanylyl cyclase to a specific imprinted odorant in coho salmon. Neuron 19, 381–389.

Dobson, G.P., Hochachka, P.W., 1987. Role of glycolysis in adenylate depletion and repletion during work and recovery in teleost white muscle. J. Exp. Biol. 129, 125–140.

Dominy, C.L., 1971. Changes in blood lactic acid concentrations in alewives (*Alosa pseudoharengus*) during passage through a pool and weir fishway. J. Fish. Res. Board Can. 28, 1215–1217.

Drenner, S.M., Clark, T.D., Whitney, C.K., Martins, E.G., Cooke, S.J., Hinch, S.G., 2012. A synthesis of tagging studies examining the behaviour and survival of anadromous salmonids in marine environments. PLoS One 7 (3), e31311.

Drenner, S.M., Harrower, W.L., Casselman, M.T., Bett, N.N., Bass, A.L., Middleton, C.T., Hinch, S.G., 2018. Whole-river manipulation of olfactory cues affects upstream migration of sockeye salmon. Fish. Manag. Ecol. 25, 488–500.

Electric Power Research Institute (EPRI), 2011. Fish passage through turbines: application of conventional hydropower data to hydrokinetic technologies. In: Final Report. 2011. Report No. 1024638.

Eliason, E.J., Clark, T.D., Hague, M.J., Hanson, L.M., Gallagher, Z.S., Jeffries, K.M., Gale, M.K., Patterson, D.A., Hinch, S.G., Farrell, A.P., 2011. Differences in thermal tolerance among sockeye salmon populations. Science 332, 109–112.

Eliason, E.J., Clark, T.D., Hinch, S.G., Farrell, A.P., 2013. Cardiorespiratory collapse at high temperature in swimming adult sockeye salmon. Conserv. Physiol. 1, cot008. https://doi.org/10.1093/conphys/cot008.

Elvidge, C.K., Reid, C.H., Ford, M.I., Sills, M., Patrick, P.J., Gibson, D., Backhouse, S., Cooke, S.J., 2019. Ontogeny of light avoidance in juvenile lake sturgeon. J. Appl. Ichthyol. 35, 202–209.

Enders, E.C., Gessel, M.H., Anderson, J.J., Williams, J.G., 2012. Effects of decelerating and accelerating flows on juvenile salmonid behavior. Trans. Am. Fish. Soc. 141, 357–364.

Eros, S.K., Milligan, C.L., 1996. The effect of cortisol on recovery from exhaustive exercise in rainbow trout (*Oncorhynchus mykiss*): potential mechanisms of action. Physiol. Zool. 69, 1196–1214.

Evans, A.F., Hostetter, N.J., Collis, K., Roby, D.D., Loge, F.J., 2014. Relationship between juvenile fish condition and survival to adulthood in steelhead. Trans. Am. Fish. Soc. 14, 899–909.

Farrell, A.P., 2007. Cardiorespiratory performance during prolonged swimming tests with salmonids: a perspective on temperature effects and potential analytical pitfalls. Philos. Trans. R. Soc. B 362, 2017–2030.

Farrell, A.P., 2016. Pragmatic perspective on aerobic scope: peaking, plummeting, pejus and apportioning. J. Fish Biol. 88, 322–343.

Farrell, A.P., Jones, D.R., 1992. The heart. In: Hoar, W.S., Randall, D.J., Farrell, A.P. (Eds.), Fish Physiology. vol. 12A. Academic Press, San Diego, CA, pp. 1–88.

Farrell, A.P., Steffensen, J.F., 1987. An analysis of the energetic cost of the branchial and cardiac pumps during sustained swimming in trout. Fish Physiol. Biochem. 4, 73–79.

Farrell, A.P., Gallaugher, P.E., Routledge, R., 2001a. Rapid recovery of exhausted adult coho salmon after commercial capture by troll fishing. Can. J. Fish. Aquat. Sci. 58, 2319–2324.

Farrell, A.P., Thorarensen, H., Axelsson, M., Crocker, C.E., Gamperl, A.K., Cech, J.J., 2001b. Gut blood flow in fish during exercise and severe hypercapnia. Comp. Biochem. Physiol. A 128, 549–561.

Farrell, A.P., Hinch, S.G., Cooke, S.J., Patterson, D.A., Crossin, G.T., Lapointe, M., Mathes, M.T., 2008. Pacific salmon in hot water: applying metabolic scope models and biotemetry to predict the success of spawning migrations. Physiol. Biochem. Zool. 81, 697–708.

Ferguson, J.W., 2005. The Behavior and Ecology of Downstream Migrating Atlantic salmon (*Salmo salar* L.) and Brown trout (*Salmo trutta* L.) in Regulated Rivers in Northern Sweden. No. 44. Vattenbruksinstitutionen, Umeå. 71 pp.

Ferguson, J.W., Absolon, R.F., Carlson, T.J., Sandford, B.P., 2006. Evidence of delayed mortality on juvenile Pacific salmon passing through turbines at Columbia River dams. Trans. Am. Fish. Soc. 135, 139–150.

Ferguson, J.W., Sandford, B.P., Reagan, R.E., Gilbreath, L.G., Meyer, E.B., Ledgerwood, R.D., Adams, N.S., 2007. Bypass system modification at Bonneville dam on the Columbia River improved the survival of juvenile salmon. Trans. Am. Fish. Soc. 136, 1487–1510.

Ferrari, M.R., Miller, J.R., Russell, G.L., 2007. Modeling changes in summer temperature of the Fraser River during the next century. J. Hydrol. 342, 336–346.

Fidler, L.E., 1988. Gas Bubble Trauma in Fish. University of British Columbia, Vancouver, British Columbia, Doctoral dissertation.

Fjeldstad, H.P., Alfredsen, K., Forseth, T., 2013. Atlantic salmon fishways: the Norwegian experience. Vann 2, 191–204.

Fjeldstad, H.P., Pulg, U., Forseth, T., 2018. Safe two-way migration for salmonids and eel past hydropower structures in Europe: a review and recommendations for best-practice solutions. Mar. Freshw. Res. 69, 1834–1847.

Flett, P.A., Van Der Kraak, G., Munkittrick, K.R., Leatherland, J.F., 1996. Overripening as the cause of low survival to hatch in Lake Erie coho salmon (*Oncorhynchus kisutch*) embryos. Can. J. Zool. 74, 851–857.

Ford, M.I., Elvidge, C.K., Baker, D., Pratt, T.C., Smokorowski, K.E., Patrick, P., Sills, M., Cooke, S.J., 2017. Evaluating a light-louver system for behavioural guidance of age-0 white sturgeon. River Res. Appl. 33, 1286–1294.

Fretwell, M.R., 1989. Homing behaviour of adult sockeye salmon in response to a hydroelectric diversion of homestream waters at Seton Creek. In: International Pacific Salmon Fisheries Commission Progress Report No. 25, p. 38. Vancouver, BC.

Furey, N.B., Hinch, S.G., Lotto, A.G., Beauchamp, D.A., 2015. Extensive feeding on sockeye salmon *Oncorhynchus nerka* smolts by bull trout *Salvelinus confluentus* during initial outmigration into a small, unregulated, and inland British Columbia river. J. Fish Biol. 86, 392–401.

Furey, N.B., Bass, A.L., Miller, K.M., Li, S., Lotto, A.G., Healy, S.J., Drenner, S.M., Hinch, S.G., 2021. Infected juvenile salmon experience increased predation risk during freshwater migration. R. Soc. Open Sci. 8, 201522. https://doi.org/10.1098/rsos.201522.

Gale, M.K., Hinch, S.G., Eliason, E.J., Cooke, S.J., Patterson, D.A., 2011. Physiological impairment of adult sockeye salmon in fresh water after simulated capture-and-release across a range of temperatures. Fish. Res. 112, 85–95.

Ginetz, R.M., Larkin, P.A., 1976. Factors affecting rainbow trout (*Salmo gairdneri*) predation on migrant fry of sockeye salmon (*Oncorhynchus nerka*). J. Fish. Res. Board Can. 33, 19–24.

Goniea, T.M., Keefer, M.L., Bjornn, T.C., Peery, C.A., Bennett, D.H., Stuehrenberg, L.C., 2006. Behavioral thermoregulation and slowed migration by adult fall Chinook salmon in response to high Columbia River water temperatures. Trans. Am. Fish. Soc. 135, 408–419.

Goodwin, R.A., Politano, M., Garvin, J.W., Nestler, J.M., Hay, D., Anderson, J.J., Weber, L.J., Dimperio, E., Smith, D.L., Timko, M., 2014. Fish navigation of large dams emerges from their modulation of flow field experience. Proc. Natl. Acad. Sci. 111, 5277–5282.

Gosselin, J.L., Buhle, E.R., Van Holmes, C., Beer, W.N., Iltis, S., Anderson, J.J., 2021. Role of carryover effects in conservation of wild Pacific salmon migrating regulated rivers. Ecosphere 12. https://doi.org/10.1002/ecs2.3618, e03618.

Gowans, A.R., Armstrong, J.D., Priede, I.G., Mckelvey, S., 2003. Movements of Atlantic salmon migrating upstream through a fish-pass complex in Scotland. Ecol. Freshw. Fish 12, 177–189.

Hammer, C., 1995. Fatigue and exercise tests with fish. Comp. Biochem. Physiol. A 112, 1–20.

Haro, A., Kynard, B., 1997. Video evaluation of passage efficiency of American shad and sea lamprey in a modified Ice Harbor fishway. N. Am. J. Fish. Manag. 17, 981–987.

Horodysky, A.Z., Schweitzer, C.C., Brill, R.W., 2022. Applied sensory physiology and behavior. Fish Physiol. 39A, 33–90.

Harrower, W.L., Bett, N.N., Hinch, S.G., 2020. BRGMON-14: effectiveness of cayoosh flow dilution, dam operation, and fishway passage on delay and survival of upstream migration of Salmon in the Seton-Anderson watershed. In: Final Report. Prepared for St'át'imc Eco-Resources Ltd. and BC Hydro. The University of British Columbia, Vancouver, BC. 154 pp.–1 App.

Hartman, W.L., Strickland, C.W., Hoopes, D.T., 1962. Survival and behaviour of sockeye salmon fry migrating into Brooks Lake, Alaska. Trans. Am. Fish. Soc. 91, 133–139.

Hatry, C., Thiem, J.D., Binder, T.R., Hatin, D., Dumont, P., Stamplecoskie, K.M., Molina, J.M., Smokorowski, K.E., Cooke, S.J., 2014. Comparative physiology and relative swimming performance of three redhorse (*Moxostoma* spp.) species: associations with fishway passage success. Physiol. Biochem. Zool. 87, 148–159.

Hinch, S.G., Bratty, J.M., 2000. Effects of swim speed and activity pattern on success of adult sockeye salmon migration through an area of difficult passage. Trans. Am. Fish. Soc. 129, 604–612.

Hinch, S.G., Rand, P.S., 1998. Swim speeds and energy use of river migrating adult sockeye salmon: role of local environment and fish characteristics. Can. J. Fish. Aquat. Sci. 55, 1821–1831.

Hinch, S.G., Rand, P.S., 2000. Optimal swim speeds and forward assisted propulsion: energy conserving behaviours of up-river migrating salmon. Can. J. Fish. Aquat. Sci. 57, 2470–2478.

Hinch, S.G., Standen, E.M., Healey, M.C., Farrell, A.P., 2002. Swimming patterns and behaviour of upriver migrating adult pink (*Oncorhynchus gorbuscha*) and sockeye (*O. nerka*) salmon as assessed by EMG telemetry in the Fraser River, British Columbia, Canada. Hydrobiologia 165, 147–160.

Hinch, S.G., Cooke, S.J., Healey, M.J., Farrell, A.P., 2006. Behavioural physiology of fish migrations: salmon as a model approach. In: Sloman, K., Balshine, S., Wilson, R. (Eds.), Fish Physiology. Vol. 24. Behaviour and Physiology of Fish. Elsevier Press, pp. 239–295.

Hinch, S.G., Bett, N.N., Eliason, E.J., Farrell, A.P., Cooke, S.J., Patterson, D.A., 2021. Exceptionally high mortality of migrating female salmon: a large-scale emerging trend and a conservation concern. Can. J. Fish. Aquat. Sci. 78, 639–654. https://cdnsciencepub.com/doi/abs/10.1139/cjfas-2020-0385.

Hockersmith, E.E., Muir, W.D., Smith, S.G., Sandford, B.P., Perry, R.W., Adams, N.S., Rondorf, D.W., 2003. Comparison of migration rate and survival between radio-tagged and PIT-tagged migrant yearling Chinook salmon in the Snake and Columbia Rivers. N. Am. J. Fish. Manag. 23, 404–413.

Hou, F., Zhiqun, Z.D., Deng, D., Martinez, J.J., Fu, T., Duncan, J.P., Johnson, G.E., Lu, J., Skalski, J.J., Townsend, R.L., Tan, L., 2018. A hydropower biological evaluation toolset (HBET) for characterizing hydraulic conditions and impacts of hydro-structures on fish. Energies 11, 990. https://doi.org/10.3390/en11040990.

International Commission on Large Dams, 2018. General Synthesis. Retrieved May 9, 2019, from https://www.icold-cigb.org/GB/world_register/general_synthesis.asp.

Jeffries, K.M., Hinch, S.G., Martins, E.G., Clark, T.D., Lotto, A.G., Patterson, D.A., Cooke, S.J., Farrell, A.P., Miller, K.M., 2012. Sex and proximity to reproductive maturity influence the survival, final maturation, and blood physiology of Pacific salmon when exposed to high temperature during a simulated migration. Physiol. Biochem. Zool. 85, 62–73.

Jones, D.R., 1982. Anaerobic exercise in teleost fish. Can. J. Zool. 60, 1131–1134.

Juanes, F., Gephard, S., Beland, K.F., 2004. Long-term changes in migration timing of adult Atlantic salmon (*Salmo salar*) at the southern edge of the species distribution. Can. J. Fish. Aquat. Sci. 61, 2392–2400. https://doi.org/10.1139/f04-207.

Keefer, M.L., Caudill, C.C., 2014. Homing and straying by anadromous salmonids: a review of mechanisms and rates. Rev. Fish Biol. Fish. 24, 333–368.

Keefer, M.L., Caudill, C.C., Peery, C.A., Boggs, C.T., 2008. Non-direct homing behaviours by adult Chinook salmon in a large, multi-stock river system. J. Fish Biol. 72, 27–44.

Kennedy, R., Moffett, I., Allen, M., Dawson, S., 2013. Upstream migratory behaviour of wild and ranched Atlantic salmon *Salmo salar* at a natural obstacle in a coastal spate river. J. Fish Biol. 83, 515–530.

Kramer, D.L., McLaughlin, R.L., 2001. The behavioral ecology of intermittent locomotion. Am. Zool. 41, 137–153.

Langdon, S.A., Collins, A.L., 2000. Quantification of the maximal swimming performance of Australasian glass eels, *Anguilla australis* and *Anguilla reinhardtii*, using a hydraulic flume swimming chamber. N. Z. J. Mar. Freshw. Res. 34, 629–636.

Larinier, M., 2001. Dams, fish and fisheries: opportunities, challenges and conflict resolution. In: FAO Fisheries Technical Paper No. 419. FAO, Rome, pp. 45–90.

Lee, C.G., Farrell, A.P., Lotto, A.G., Hinch, S.G., Healey, M.C., 2003a. Excess post-exercise oxygen consumption in adult sockeye (*Oncorhynchus nerka*) and coho (*O. kisutch*) salmon following critical speed swimming. J. Exp. Biol. 206, 3253–3260.

Lee, C.G., Farrell, A.P., Lotto, A.G., MacNutt, M.J., Hinch, S.G., Healey, M.C., 2003b. Effects of temperature on swimming performance and oxygen consumption in adult sockeye (*Oncorhynchus nerka*) and coho (*O. kisutch*) salmon stocks. J. Exp. Biol. 206, 3239–3251.

Lehner, B., Liermann, C.R., Revenga, C., Vörösmarty, C., Fekete, B., Crouzet, P., Döll, P., Endejan, M., Frenken, K., Magome, J., Nilsson, C., Robertson, J.C., Rödel, R., Sindorf, N., Wisser, D., 2011. High-resolution mapping of the world's reservoirs and dams for sustainable river-flow management. Front. Ecol. Environ. 9, 494–502. https://doi.org/10.1890/100125.

Lema, S.C., Nevitt, G.A., 2004. Evidence that thyroid hormone induces olfactory cellular proliferation in salmon during a sensitive period for imprinting. J. Exp. Biol. 207, 3317–3327.

Li, P., Zhang, W., Burnett, N.J., Zhu, D.Z., Casselman, M., Hinch, S.G., 2021. Evaluating dam water release strategies for migrating adult salmon using computational fluid dynamic modeling and biotelemetry. Water Resour. Res. 57, e2020WR028981. https://doi.org/10.1029/2020WR028981.

Lindmark, E., Gustavsson, L.H., 2008. Field study of an attraction channel as entrance to fishways. River Res. Appl. 24, 564–570. https://doi.org/10.1002/RRA.1145.

Lucas, M.C., Baras, E., 2001. Migration of Freshwater Fishes. Blackwell Science Ltd, Oxford.

Lundqvist, H., Rivinoja, P., Leonardsson, K., McKinnell, S., 2008. Upstream passage problems for wild Atlantic salmon (*Salmo salar* L.) in a regulated river and its effect on the population. Hydrobiologia 602, 111–127.

Mallen-Cooper, M., Brand, D.A., 2007. Non-salmonids in a salmonid fishway: what do 50 years of data tell us about past and future fish passage? Fish. Manag. Ecol. 14, 319–332.

Martinez, J.J., Deng, Z.D., Titzler, P.S., Duncan, J.P., Lu, J., Mueller, R.P., Tian, C., Trumbo, B.A., Ahmann, M.L., Renholds, J.F., 2019. Hydraulic and biological characterization of a large Kaplan turbine. Renew. Energy 131, 240e249.

Martins, E.G., Hinch, S.G., Patterson, D.A., Hague, M.J., Cooke, S.J., Miller, K.M., Lapointe, M.F., English, K.K., Farrell, A.P., 2011. Effects of river temperature and climate warming on stock-specific survival of adult migrating Fraser River sockeye salmon (*Oncorhynchus nerka*). Glob. Change Biol. 17, 99–114.

Mateus, A.P., Anjos, L., Cardoso, J.R., Power, D.M., 2017. Chronic stress impairs the local immune response during cutaneous repair in Gilthead Sea bream (*Sparus aurata*, L.). Mol. Immunol. 87, 267–283.

McCormick, S.D., Hansen, L.P., Quinn, T.P., Saunders, R.L., 1998. Movement, migration, and smolting of Atlantic salmon (*Salmo salar*). Can. J. Fish. Aquat. Sci. 55, 77–92.

McInerney, J.E., 1964. Salinity preference: an orientation mechanism in salmon migration. J. Fish. Res. Board Can. 21, 995–1018.

Mckenzie, D.J., Hale, M.E., Domenici, P., 2007. Locomotion in primitive fishes. In: McKenzie, D.J., Farrell, A.P., Brauner, C.J. (Eds.), Primitive Fishes, Fish Physiology. vol. 26. Elsevier, pp. 319–380.

Mesa, M.G., Bayer, J.M., Seelye, J.G., 2003. Swimming performance and physiological responses to exhaustive exercise in radio-tagged and untagged Pacific lampreys. Trans. Am. Fish. Soc. 132, 483–492.

Middleton, C.T., Hinch, S.G., Martins, E.G., Braun, D.C., Patterson, D.A., Burnett, N.J., Minke-Martin, V., Casselman, M.T., Gelchu, A., 2018. Effects of natal water concentration and temperature on the behavior of up-river migrating sockeye salmon. Can. J. Fish. Aquat. Sci. 75, 2375–2389.

Milligan, C.L., 1996. Metabolic recovery from exhaustive exercise in rainbow trout. Comp. Biochem. Physiol. A 113, 51–60.

Milligan, C.L., Hooke, G.B., Johnson, C., 2000. Sustained swimming at low velocity following a bout of exhaustive exercise enhances metabolic recovery in rainbow trout. J. Exp. Biol. 203, 921–926.

Minke-Martin, V., Hinch, S.G., Braun, D.C., Burnett, N.J., Casselman, M.T., Eliason, E.J., Middleton, C.T., 2018. Physiological condition and migratory experience affect fitness-related outcomes in adult female sockeye salmon. Ecol. Freshw. Fish 27, 296–309.

Morin, P.-P., Anderson, Ø., Haug, E., Døving, K.B., 1994. Changes in serum free thyroxine, prolactin, and olfactory activity during induced smoltification in Atlantic salmon (*Salmo salar*). Can. J. Fish. Aquat. Sci. 51, 1985–1992.

Muir, W.D., Smith, S.G., Williams, J.G., Sandford, B.P., 2001. Survival of juvenile salmonids passing through bypass systems, turbines, and spillways with and without flow deflectors at Snake River dams. N. Am. J. Fish. Manag. 21, 135–146.

Muir, W.D., Marsh, D.M., Sandford, B.P., Smith, S.G., Williams, J.G., 2006. Post-hydropower system delayed mortality of transported Snake River stream-type Chinook salmon: unraveling the mystery. Trans. Am. Fish. Soc. 135, 1523–1534.

Mussen, T.D., Cech Jr., J.J., 2019. Assessing the use of vibrations and strobe lights at fish screens as enhanced deterrents for two estuarine fishes. J. Fish Biol. 95, 238–246.

Naughton, G.P., Caudill, C.C., Keefer, M.L., Bjornn, T.C., Stuehrenberg, L.C., Peery, C.A., 2005. Late-season mortality during migration of radio-tagged adult sockeye salmon (*Oncorhynchus nerka*) in the Columbia River. Can. J. Fish. Aquat. Sci. 62, 30–47.

Nehlsen, W., Williams, J.E., Lichatowich, J.A., 1991. Pacific Salmon at the crossroads: stocks at risk from California, Oregon, Idaho, and Washington. Fisheries 16, 4–21.

Nguyen, V.M., Martins, E.G., Raby, G.D., Donaldson, M.R., Lotto, A.G., Patterson, D.A., Robichaud, D., English, K.K., Farrell, A.P., Willmore, W.G., Rudd, M.A., Hinch, S.G., Cooke, S.J., 2014. Disentangling the roles of air exposure, gillnet injury, and facilitated recovery on the post-capture and release mortality and behavior of adult migratory sockeye salmon (*Oncorhynchus nerka*) in freshwater. Physiol. Biochem. Zool. 87, 125–135.

Noonan, M.J., Grant, J.W.A., Jackson, C.D., 2012. A quantitative assessment of fish passage efficiency. Fish Fish. 13, 450–464.

Nyqvist, D., Calles, O., Bergman, E., Hagelin, A., Greenberg, L., 2016. Post-spawning survival and downstream passage of landlocked Atlantic salmon (*Salmo salar*) in a regulated river: is there potential for repeat spawning? River res. Appl. Ther. 32, 1008–1017.

Nyqvist, D., Greenberg, L., Goerig, E., Calles, O., Bergman, E., Ardren, W.R., Castro-Santos, T., 2017a. Migratory delay leads to reduced passage success of Atlantic salmon smolts at a hydroelectric dam. Ecol. Freshw. Fish. 26, 707–718.

Nyqvist, D., McCormick, S.D., Greenberg, L., Ardren, W.R., Bergman, E., Calles, O., Castro-Santos, T., 2017b. Downstream migration and multiple dam passage by Atlantic Salmon smolts. N. Am. J. Fish. Manag. 37, 816–828.

Nyqvist, D., Nilsson, P.A., Alenäs, P.I., Elghagen, J., Hebrand, M., Karlsson, S., Kläppe, S., Calles, O., 2017c. Upstream and downstream passage of migrating adult Atlantic salmon: remedial measures improve passage performance at a hydropower dam. Ecol. Eng. 102, 331–343.

Nyqvist, D., Bergman, E., Calles, O., Greenberg, L., 2017d. Intake approach and dam passage by downstream-migrating Atlantic Salmon Kelts. River Res. Appl. 33, 697–706.

O'Connor, C.M., Norris, D.R., Crossin, G.T., Cooke, S.J., 2014. Biological carryover effects: linking common concepts and mechanisms in ecology and evolution. Ecosphere 5, 28. https://doi.org/10.1890/ES13-00388.1.

Odeh, M. (Ed.), 1999. Innovations in Fish Passage Technology. American Fisheries Society, Bethesda, MA.

Office of Technology Assessment (OTA), 1995. Fish Passage Technologies: Protection at Hydropower Facilities. OTAENV-641. U.S. Government Printing Office, Washington, DC.

Olsen, R.E., Oppedal, F., Tenningen, M., Vold, A., 2012. Physiological response and mortality caused by scale loss in Atlantic herring. Fish. Res. 129, 21–27.

Parasiewicz, P., Eberstaller, J., Weiss, S., Schmutz, S., 1998. Conceptual guidelines for nature-like bypass channels. In: Jungwirth, M., Schmutz, S., Weiss, S. (Eds.), Fish Migration and Fish Bypasses. Fishing News Books, Oxford, pp. 348–362.

Parsley, M.J., Wright, C.D., van der Leeuw, B.K., Kofoot, E.E., Peery, C.A., Moser, M.L., 2007. White sturgeon (*Acipenser transmontanus*) passage at the Dalles dam, Columbia River, USA. J. Appl. Ichthyol. 23, 627–635.

Peake, S.J., Farrell, A.P., 2004. Locomotory behaviour and post-exercise physiology in relation to swimming speed, gait transition and metabolism in free-swimming smallmouth bass (*Micropterus dolomieu*). J. Exp. Biol. 207, 1563–1575.

Petersen, J.H., 1994. Importance of spatial pattern in estimating predation on juvenile salmonids in the Columbia River. Trans. Am. Fish. Soc. 123, 924–930.

Pickering, A.D., Pottinger, T.G., 1989. Stress responses and disease resistance in salmonid fish: effects of chronic elevation of plasma cortisol. Fish Physiol. Biochem. 7, 253–258.

Pickering, A.D., Pottinger, T.G., 1995. Biochemical effects of stress. In: Hochachka, P.W., Mommsen, T.P. (Eds.), Biochemistry and Molecular Biology of Fishes, Volume 5: Environmental and Ecological Biochemistry. Elsevier, pp. 349–379.

Pleizier, N.K., Algera, D., Cooke, S.J., Brauner, C.J., 2020. A meta-analysis of gas bubble trauma in fish. Fish Fish. 21, 1175–1194.

Pleizier, N.K., Rost-Komiya, B., Cooke, S.J., Brauner, C.J., 2021. The lack of avoidance of total dissolved gas supersaturation in juvenile rainbow trout. Hydrobiologia 848, 4837–4850. https://doi.org/10.1007/s10750-021-04676-w.

Poletto, J.L., Cocherell, D.E., Ho, N., Cech Jr., J.J., Klimley, A.P., Fangue, N.A., 2014. Juvenile green sturgeon (*Acipenser medirostris*) and white sturgeon (*Acipenser transmontanus*) behavior near water-diversion fish screens: experiments in a laboratory swimming flume. Can. J. Fish. Aquat. Sci. 71, 1030–1038.

Pon, L.B., Hinch, S.G., Cooke, S.J., Patterson, D.A., Farrell, A.P., 2009a. Physiological, energetic and behavioural correlates of successful fishway passage of adult sockeye salmon Oncorhynchus nerka in the Seton River, British Columbia. J. Fish. Biol. 74, 1323–1336.

Pon, L.B., Hinch, S.G., Cooke, S.J., Patterson, D.A., Farrell, A.P., 2009b. A comparison of the physiological condition, and fishway passage time and success of migrant adult sockeye salmon at Seton River dam, British Columbia, under three operational water discharge rates. N. Am. J. Fish. Manag. 29, 1195–1205.

Pon, L.B., Hinch, S.G., Suski, C.D., Patterson, D.A., Cooke, S.J., 2012. The effectiveness of tissue biopsy as a means of assessing the physiological consequences of fishway passage. River Res. Appl. 28, 1266–1274.

Pracheil, B.M., DeRolph, C.R., Schramm, M.P., Bevelhimer, M.S., 2016. A fish-eye view of riverine hydropower systems: the current understanding of the biological response to turbine passage. Rev. Fish Biol. Fish. 26, 153–167.

Quinn, T.P., 2018. The Behavior and Ecology of Pacific Salmon and Trout, second ed. University of Washington Press. 520 pp.

Quinn, T.P., Adams, D.J., 1996. Environmental changes affecting the migratory timing of American shad and sockeye salmon. Ecology 77, 1151–1162.

Quinn, T.P., Hodgson, S., Peven, C., 1997. Temperature, flow and the migration of adult sockeye salmon (Oncorhynchus nerka) in the Columbia River. Can. J. Fish. Aquat. Sci. 54, 1349–1360.

Rand, P.S., 2011. Oncorhynchus nerka. In: The IUCN Red List of Threatened Species 2011., e.T135301A4071001. https://doi.org/10.2305/IUCN.UK.2011-2.RLTS.T135301A4071001.en.

Ricklefs, R.E., Wikelski, M., 2002. The physiology/life-history nexus. Trends Ecol. Evol. 17, 462–468.

Rieman, B.E., Beamesderfer, R.C., Vigg, S., Poe, T.P., 1991. Estimated loss of juvenile salmonids to predation by northern squawfish, walleyes, and smallmouth bass in John Day reservoir, Columbia River. Trans. Am. Fish. Soc. 120, 448–458.

Robards, M.D., Quinn, T.P., 2002. The migratory timing of adult summer-run steelhead in the Columbia River over six decades of environmental change. Trans. Am. Fish. Soc. 131, 523–536.

Romão, F., Santos, J.M., Katopodis, C., Pinheiro, A.N., Branco, P., 2018. How does season affect passage performance and fatigue of potamodromous cyprinids? An experimental approach in a vertical slot fishway. Water 10, 395.

Roscoe, D.W., Hinch, S.G., 2010. Effectiveness monitoring of fish passage facilities: historical trends, geographic patterns and future directions. Fish Fish 11, 12–33.

Roscoe, D.W., Hinch, S.G., Cooke, S.J., Patterson, D.A., 2010. Behaviour and thermal experience of adult sockeye salmon migrating through stratified lakes near spawning grounds: the roles of reproductive and energetic states. Ecol. Freshw. Fish 19, 51–62.

Roscoe, D.W., Hinch, S.G., Cooke, S.J., Patterson, D.A., 2011. Fishway passage and post-passage mortality of up-river migrating sockeye salmon in the Seton River, British Columbia. River Res. Appl. 27, 693–705.

Sackville, M., Tang, S., Nendick, L., Farrell, A.P., Brauner, C.J., 2011. Pink salmon (*Oncorhynchus gorbuscha*) osmoregulatory development plays a key role in sea louse (*Lepeophtheirus salmonis*) tolerance. Can. J. Fish. Aquat. Sci. 68, 1077–1086.

Sackville, M., Wilson, J.M., Farrell, A.P., Brauner, C.J., 2012. Water balance trumps ion balance for early marine survival of juvenile pink salmon (*Oncorhynchus gorbuscha*). J. Comp. Physiol. B 182, 781–792.

Sandblom, E., Clark, T.D., Hinch, S.G., Farrell, A.P., 2009. Sex-specific differences in cardiac control and hematology of sockeye salmon (*Oncorhynchus nerka*) approaching their spawning grounds. Am. J. Physiol. Regul. Integr. Comp. Physiol. 297, 1136–1143.

Schiewe, M.H., 1974. Influence of dissolved atmospheric gas on swimming performance of juvenile Chinook Salmon. Trans. Am. Fish. Soc. 103, 717–721.

Schreck, C.B., Tort, L., 2016. The concept of stress in fish. In: Schreck, C.B., Tort, L., Farrell, A.P., Brauner, C.J. (Eds.), Biology of Stress in Fish: Fish Physiology. vol. 35. Elsevier Inc, pp. 1–34.

Schreck, C.B., Solazzi, M.F., Johnson, S.L., Nichelson, T.E., 1989. Transportation stress affects performance of coho salmon, *Oncorhynchus kisutch*. Aquaculture 82, 15–20.

Schreck, C.B., Stahl, T.P., David, L.E., Roby, D.D., Clemens, B.J., 2006. Mortality estimates of juvenile spring–summer Chinook salmon in the lower Columbia River and estuary, 1992–1998: evidence for delayed mortality? Trans. Am. Fish. Soc. 135, 457–475.

Schwalme, K., Mackay, W.C., Lindner, D., 1985. Suitability of vertical slot and Denil fishways for passing north-temperate, nonsalmonid fish. Can. J. Fish. Aquat. Sci. 42, 1815–1822.

Schwevers, U., Adam, B., 2020. Fish-friendly turbines. In: Fish Protection Technologies and Fish Ways for Downstream Migration. Springer, Cham.

Seals-Price, C., Schreck, C.B., 2003. Effects of bacterial kidney disease on saltwater preference of juvenile spring Chinook salmon, Oncorhynchus tshawytscha. Aquaculture 222, 331–341.

Selye, H., 1936. A syndrome produced by diverse nocuous agents. J. Clin. Neuropsychol. 10, 230a–231.

Shively, R.S., Thomas, P.P., Mindi, B.S., Peters, R., 1996. Criteria for reducing predation by northern squawfish near juvenile salmonid bypass outfalls at Columbia River dams. Regul. River 12, 493–500.

Shrimpton, J.M., Randall, D.J., Fidler, L.E., 1990. Factors affecting swim bladder volume in rainbow trout (*Oncorhynchus mykiss*) held in gas supersaturated water. Can. J. Zool. 68, 962–968.

Silva, A.T., Lucas, M.C., Castro-Santos, T., Katopodis, C., Baumgartner, L.J., Thiem, J.D., Aarestrup, K., Pompeu, P.S., O'Brien, G.C., Braun, D.C., Burnett, N.J., Zhu, D.Z., Fjeldstad, H.P., Forseth, T., Rajaratnam, N., Williams, J.G., Cooke, S.J., 2018. The future of fish passage science, engineering and practice. Fish Fish. 19, 340–364.

Slaney, T.L., Hyatt, K.D., Northcote, T.G., Fielden, R.J., 1996. Status of anadromous Salmon and Trout in British Columbia and Yukon. Fisheries 21, 20–35.

Smith, S.G., Muir, W.D., Williams, J.G., Skalski, J.R., 2002. Factors associated with travel time and survival of migrant yearling Chinook salmon and steelhead in the lower Snake River. N. Am. J. Fish. Manag. 22, 385–405.

Standen, E.M., Hinch, S.G., Healey, M.C., Farrell, A.P., 2002. Energetics of upriver migrating adult pink (*Oncorhynchus gorbuscha*) and sockeye (*O. nerka*) salmon as assessed by EMG telemetry in the Fraser River canyon, British Columbia. Can. J. Fish. Aquat. Sci. 59, 1809–1818.

Standen, E.M., Hinch, S.G., Rand, P.S., 2004. Influence of river currents on path selection and swimming efficiency of migrating adult sockeye salmon. Can. J. Fish. Aquat. Sci. 61, 905–912.

Steinhausen, M.F., Sandblom, E., Eliason, E.J., Verhille, C., Farrell, A.P., 2008. The effect of acute temperature increases on the cardiorespiratory performance of resting and swimming sockeye salmon (*Oncorhynchus nerka*). J. Exp. Biol. 211, 3915–3926.

Stephenson, J.R., Gingerich, A.J., Brown, R.S., Pflugrath, B.D., Deng, Z., Carlson, T.J., Langeslay, M.J., Ahmann, M.L., Johnson, R.L., Seaburg, A.G., 2010. Assessing barotrauma in neutrally and negatively buoyant juvenile salmonids exposed to simulated hydro-turbine passage using a mobile aquatic barotrauma laboratory. Fish. Res. 106, 271–278.

Swanson, C., Young, P.S., Cech Jr., J.J., 2005. Close encounters with a fish screen: integrating physiological and behavioral results to protect endangered species in exploited ecosystems. Trans. Am. Fish. Soc. 134, 5.

Tabor, R.A., Brown, G.S., Luiting, V.T., 2004. The effect of light intensity on sockeye salmon fry migratory behavior and predation by cottids in the Cedar River, Washington. N. Am. J. Fish. Manag. 24, 128–145.

Teffer, A.K., Bass, A.L., Miller, K.S., Patterson, D.A., Juanes, F., Hinch, S.G., 2018. Infections, fisheries capture, temperature and host responses: multi-stressor influences on survival and behavior of adult Chinook salmon. Can. J. Fish. Aquat. Sci. 75, 2069–2083.

Teles, M., Santos, M.A., Pacheo, M., 2004. Responses of European eel (*Anguilla Anguilla* L.) in two polluted environments: in situ experiments. Ecotoxicol. Environ. Saf. 58, 373–378.

Thornton, K.W., Kimmel, B.L., Payne, F.E., 1990. Reservoir Limnology: Ecological Perspectives. Wiley, 246 p.

Thorstad, E.B., Okland, F., Kroglund, F., Jepsen, N., 2003. Upstream migration of Atlantic salmon at a power station on the river Nidelva, southern Norway. Fish. Manag. Ecol. 10, 139–146.

Thorstad, E.B., Okland, F., Aarestrup, K., Heggberget, T.G., 2008. Factors affecting the within-river spawning migration of Atlantic salmon, with emphasis on human impacts. Rev. Fish Biol. Fish. 18, 345–371.

Tiemann, J.S., Gillette, D.P., Wildhaber, M.L., Edds, D.R., 2004. Effects of lowhead dams on riffle-dwelling fishes and macroinvertebrates in a midwestern river. Trans. Am. Fish. Soc. 133, 705–717.

Tort, L., 2011. Stress and immune modulation in fish. Dev. Comp. Immunol. 35, 1366–1375.

Travade, F., Larinier, M., 2002. Fish locks and fish lifts. Bull. Fr. Peche Piscic. 364, 102–118.

Trump, C.L., Leggett, W.C., 1980. Optimum swimming speeds in fish: the problem of currents. Can. J. Fish. Aquat. Sci. 37, 1086–1092.

Vergeynst, J., Pauwels, I., Baeyens, R., Coeck, J., Nopens, I., De Mulder, T., Mouton, A., 2019. The impact of intermediate-head navigation locks on downstream fish passage. River Res. Appl. 35, 224–235.

Vezza, P., Libardoni, F., Manes, C., Tsuzaki, T., Bertoldi, W., Kemp, P.S., 2020. Rethinking swimming performance tests for bottom-dwelling fish: the case of European glass eel (*Anguilla anguilla*). Sci. Rep. 10, 1–11.

Videler, J.J., 1993. Fish Swimming. Chapman & Hall, London.

Vogel, S., 1994. Life in moving fluids. In: The Physical Biology of Flow. Princeton University Press, Princeton.

Wagner, G.N., Kuchel, L.J., Lotto, A., Patterson, D.A., Shrimpton, J.M., Hinch, S.G., Farrell, A.P., 2006. Routine and active metabolic rates of migrating adult wild sockeye salmon (*Oncorhynchus nerka* Walbaum) in seawater and freshwater. Physiol. Biochem. Zool. 79, 100–108.

Wang, P., Dong, S., Lassoie, J.P., 2014. A global review of large dam construction. In: The Large Dam Dilemma. Springer, Dordrecht. https://doi.org/10.1007/978-94-007-7630-2_1.

Wang, S., Fu, B., Piao, S., et al., 2016. Reduced sediment transport in the Yellow River due to anthropogenic changes. Nat. Geosci. 9, 38–41.

Wang, Y., Heigenhauser, G.J., Wood, C.M., 1994. Integrated responses to exhaustive exercise and recovery in rainbow trout white muscle: acid-base, phosphogen, carbohydrate, lipid, ammonia, fluid volume and electrolyte metabolism. J. Exp. Biol. 195, 227–258.

Webber, J.D., Chun, S.N., MacColl, T.R., Mirise, L.T., Kawabata, A., Anderson, E.K., Cheong, T.S., Kavvas, L., Rotondo, M.M., Hochgraf, K.L., Churchwell, R., Cech Jr., J.J., 2007. Upstream swimming performance of adult white sturgeon: effects of partial baffles and a ramp. Trans. Am. Fish. Soc. 136 (2), 402–408.

Weihs, D., 1974. Energetic advantages of burst swimming of fish. J. Theor. Biol. 48, 215–229.

Weitkamp, D.E., Katz, M., 1980. A review of dissolved gas supersaturation literature. Trans. Am. Fish. Soc. 109, 659–702.

White, D.K., Swanson, C., Young, P.S., Cech Jr., J.J., Chen, Z.Q., Kavvas, M.L., 2007. Close encounters with a fish screen II: Delta smelt behavior before and during screen contact. Trans. Am. Fish. Soc. 136 (2), 528–538.

Wikelski, M., Cooke, S.J., 2006. Conservation physiology. Trends Ecol. Evol. 21, 38–46. https://doi.org/10.1016/j.tree.2005.10.018.

Wood, C.M., 1991. Acid-base and ion balance, metabolism, and their interactions, after exhaustive exercise in fish. J. Exp. Biol. 160, 285–308.

Wood, C.M., Turner, J.D., Graham, M.S., 1983. Why do fish die after severe exercise? J. Fish Biol. 22, 189–201.

Woodhead, A.D., 1975. Endocrine Physiology of Fish Migration. G. Allen and Ungin.

Wu, G., Yang, Y., Zeng, L., 2007. Kinematics, hydrodynamics and energetic advantages of burst-and-coast swimming of koi carps (*Cyprinus carpio koi*). J. Exp. Biol. 210, 2181–2191.

Yamamoto, Y., Hino, H., Ueda, H., 2010. Olfactory imprinting of amino acids in lacustrine sockeye salmon. PLoS One 5, e8633.

Young, J.L., Hinch, S.G., Cooke, S.J., Crossin, G.T., Patterson, D.A., Farrell, A.P., Van Der Kraak, G., Lotto, A.G., Lister, A., Healey, M.C., English, K.K., 2006. Physiological and energetic correlates of en route mortality for abnormally early migrating adult sockeye salmon (*Oncorhynchus nerka*) in the Thompson River, British Columbia. Can. J. Fish. Aquat. Sci. 63, 1067–1077.

Young, P.S., Swanson, C., Cech Jr., J.J., 2010. Close encounters with a fish screen III: behavior, performance, physiological stress responses, and recovery of adult delta smelt exposed to two-vector flows near a fish screen. Trans. Am. Fish. Soc. 139 (3), 713–726.

Zarfl, C., Lumsdon, A.E., Berlekamp, J., Tydecks, L., Tockner, K., 2015. A global boom in hydropower dam construction. Aquat. Sci. 77, 161–170. https://doi.org/10.1007/s00027-014-0377-0.

Zhang, Y., Claireaux, G., Takle, H., Jørgensen, S.M., Farrell, A.P., 2018. A three-phase excess post-exercise oxygen consumption in Atlantic salmon *Salmo salar* and its response to exercise training. J. Fish Biol. 92, 1385–1403.

Zhang, Y., Gilbert, M.J.H., Farrell, A.P., 2019. Finding the peak of dynamic oxygen uptake during fatiguing exercise in fish. J. Exp. Biol. 222 (12), jeb196568.

Zhang, Y., Gilbert, M.J.H., Farrell, A.P., 2020. Measuring maximum oxygen uptake with an incremental swimming test and by chasing rainbow trout to exhaustion inside a respirometry chamber yield the same results. J. Fish Biol. 97, 28–38.

Chapter 10

# Invasive species control and management: The sea lamprey story

Michael P. Wilkie[a,*], Nicholas S. Johnson[b], and Margaret F. Docker[c]
[a]*Department of Biology and Laurier Institute for Water Science, Wilfrid Laurier University, Waterloo, ON, Canada*
[b]*U.S. Geological Survey, Great Lakes Science Center, Hammond Bay Biological Station, Millersburg, MI, United States*
[c]*Department of Biological Sciences, University of Manitoba, Winnipeg, MB, Canada*
**Corresponding author: e-mail: mwilkie@wlu.ca*

## Chapter Outline



**Fish Physiology, Vol. 39B. https://doi.org/10.1016/bs.fp.2022.09.001**

Control of invasive species is a critical component of conservation biology given the catastrophic damage that they can cause to the ecosystems they invade. This is particularly evident with sea lamprey (*Petromyzon marinus*) in the Laurentian Great Lakes. Native to the Atlantic Ocean, the sea lamprey's ability to osmoregulate in fresh water, its wide thermal tolerance, generalist diet, and high fecundity allowed it to rapidly reach pest proportions in the prey-rich Great Lakes once it gained access through shipping canals. The invasion exacerbated declines in Great Lakes fisheries caused by overharvest, culminating in the crash of lake trout (*Salvelinus namaycush*) and other fish populations. In the last 60 years, however, a highly successful sea lamprey control program has reduced sea lamprey to ~10% of their peak abundance and has been instrumental in enabling the rehabilitation of the Great Lakes ecosystem. In this chapter, we: (1) discuss the likely vectors of the invasion and the physiological attributes of sea lamprey that enabled them to become established in the Great Lakes; (2) review the two cornerstones of the sea lamprey control program—which relies on a combination of pesticides to eradicate multiple generations of larval sea lamprey in their nursey streams, and in-stream barriers to restrict the upstream migration of spawning lamprey—both of which exploit unique physiological vulnerabilities of sea lamprey; (3) describe how sea lamprey control can adversely affect non-target species and how these can be mitigated; (4) show how physiology-based approaches are improving our understanding of the lethal and sublethal effects of sea lamprey on host fishes; and (5) discuss the future of conservation physiology in sea lamprey control. The prime challenge in the next several decades of the Anthropocene will be to further refine the specificity of control tools while maintaining their efficacy, and to adapt to a warming climate and other anthropogenic activities affecting the Great Lakes and their tributaries.

## 1 Introduction

At first glance, a chapter on a notorious invasive fish, the sea lamprey (*Petromyzon marinus*), might seem out of place in a book focused on fish conservation in the Anthropocene. In the Laurentian Great Lakes of North America, as well as the New York Finger Lakes (Cayuga and Seneca lakes), Oneida Lake, and Lake Champlain, parasitism of fishes by hematophagous (blood-feeding) juvenile sea lamprey had contributed to the decline of many native fishes by the mid-1900s, resulting in catastrophic damage to the basin's ecosystem, not to mention severe socioeconomic disruption to communities whose livelihoods relied on tourism, and recreational and commercial fishing (see Marsden and Siefkes, 2019). Although sea lamprey are of conservation concern in their native range in drainages of the Atlantic Ocean, especially in Europe (see Lucas et al., 2021; Maitland et al., 2015), in the Great Lakes, they are, without doubt, a species of conservation significance based on the ongoing threat they pose to fisheries, including many species at risk in the basin.

In the last 60 years, sea lamprey control research has revealed much about the ecology, behavior, and physiology of sea lamprey, providing a foundation for improving conservation efforts aimed at protecting sea lamprey in their native range and other lamprey species of conservation concern. Lamprey conservation initiatives are growing in importance due to ever-growing interest in the biology and evolution of these phylogenetically ancient jawless fishes (see Section 2.1; for recent reviews see Clemens and Wang, 2021; Clemens et al., 2017; Lucas et al., 2021; Maitland et al., 2015). Here, we focus on sea lamprey in the Laurentian Great Lakes (hereafter the Great Lakes), and review how their invasion devastated fisheries in the basin, and how a better understanding of their physiology was ultimately exploited to develop a highly successful invasive species control program.

Worldwide, invasive species constitute one of the biggest threats to the conservation of freshwater fishes, which are also threatened by habitat loss and degradation, pollution, overexploitation (overfishing), and climate change (Arthington et al., 2016). In many cases, the threats posed by invasive species, like some other threats to fisheries, may be exacerbated by climate change, due to changes in the thermal regimes of freshwater environments including extremes in seasonal high and low temperatures, the timing of ice coverage, changes in stratification of lakes in more temperate climates, altered precipitation, and altered water flow regimes caused by more frequent and intense storm events leading to flooding. Each of these environmental factors can lead to niche expansion and access to novel habitat for invasive or potentially invasive species, further threatening resident fish populations (Arthington et al., 2016; Bartolai et al., 2015; Lennox et al., 2020; Reid et al., 2019).

This chapter is not intended to be a comprehensive review of lamprey physiology, which has been thoroughly reviewed elsewhere (e.g., Birceanu et al., 2021; Borowiec et al., 2021; Ferreira-Martins et al., 2021a; Manzon et al., 2015; Siefkes, 2017; Sower, 2015, 2018; Wilkie, 2011a,b; Youson, 1980, 2003). Instead, we will focus on aspects of lamprey physiology that we consider to be most relevant for controlling invasive sea lamprey in the Great Lakes, and physiological aspects that may reveal how to minimize their negative impact on host fishes and the Great Lakes ecosystem. We will first address the physiological and ecological characteristics that made the sea lamprey such a highly successful, and devastating, invader in the Great Lakes. We will then describe how the control of sea lamprey populations is also a success story for fisheries conservation in the Great Lakes (i.e., how the lessons learned about its physiology have contributed to the rehabilitation of fisheries in the basin), and how further knowledge will be needed to deal with ongoing and emerging challenges in the Anthropocene that could undermine sea lamprey control and fisheries conservation in the future. We will end with a discussion of how invasive species control efforts are integral to fisheries conservation efforts (i.e., how control and conservation are mutually compatible), and identify the ongoing and future challenges and potential solutions to such efforts in the Anthropocene.

# 2 Introduction to the "stone sucker"

## 2.1 Scientific and cultural importance of lampreys

Lampreys (order Petromyzontiformes, "stone sucker") are one of only two extant groups of jawless vertebrates. Their ordinal name is derived from their unique and signature nesting behavior, where the oral disc is used to move rocks. Lampreys diverged ~500 million years ago from the jawed vertebrate lineage, and they have survived at least four of the five mass extinction events documented since the Cambrian explosion (see Docker et al., 2015). Given their phylogenetic significance, studies on the anatomy, development, physiology, and genomics of lampreys are providing important insight into the evolution of vertebrates (reviewed by Docker et al., 2015; Grillner, 2021; McCauley et al., 2015; Shimeld and Donoghue, 2012; Smith et al., 2013, 2018; York and McCauley, 2020; York et al., 2019, 2021). This research includes study of programmed genome rearrangement which occurs during embryogenesis, when ~20% of the germline genome (hundreds of millions of base pairs) is lost from somatic cell lineages (Smith et al., 2009), and the evolution of adaptive immunity. Adaptive immunity evolved independently in jawless and jawed vertebrates; unlike jawed vertebrates in which antigens are recognized using immunoglobulin-based B-cell and T-cell receptors, lampreys and hagfishes use variable lymphocyte receptors (Boehm et al., 2012).

Although sea lamprey are a significant pest in the Great Lakes basin, they are held in high regard in their native range, especially in Europe (Hansen et al., 2016); other species of lampreys have likewise been long appreciated for food and ceremonial purposes, including by the Māori in New Zealand and Native Americans in the Pacific Northwest (see Almeida et al., 2021; Close et al., 2002; Docker et al., 2015; Lucas et al., 2021).

Thus, despite the small number of extant species (~45 recognized lamprey species, although the exact number continues to be debated; see Docker and Hume, 2019), lampreys continue to "punch above their weight class" in terms of scientific attention. Searching the database Web of Science (Thompson Reuters) for the time period 1864–2013; Docker et al. (2015) found that, although lampreys comprise only 0.14% of living fishes, they represent ~0.95% of the papers written on fishes. Research related to lamprey biology increased in the 1950s in support of sea lamprey control in the Great Lakes, as evidenced by the proceedings of three Sea Lamprey International Symposia (SLIS) published between 1980 and 2021 (Canadian Journal of Fisheries and Aquatic Sciences, Volume 37, Issue 11; Journal of Great Lakes Research, Volume 29, Supplement 1; and Journal of Great Lakes Research, Volume 41, Supplement 1; see McLaughlin et al., 2021). All of this research is good news for biologists interested in using physiological and other tools to inform the control and conservation of lampreys.

## 2.2 Sea lamprey life cycle

All lampreys undergo indirect development, which is characterized by a larval phase followed by a true metamorphosis into a distinctive juvenile stage (Fig. 1A; Youson, 1980, 2003). The protracted larval phase lasts several years during which the animals live burrowed in the soft, silty sediment of rivers

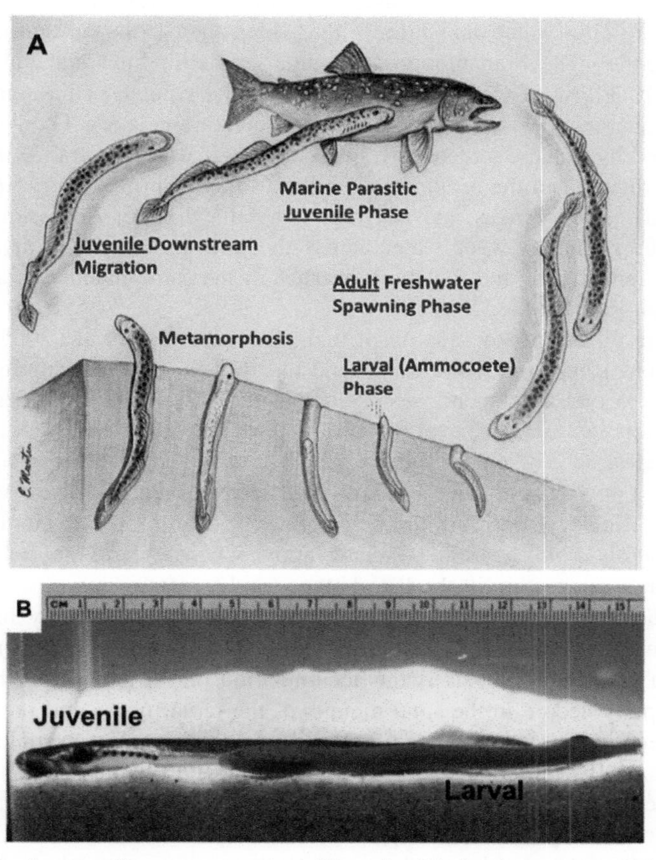


**FIG. 1** Sea lamprey (*Petromyzon marinus*) life cycle. (A) Sea lamprey spend the first the 4–7 years of life as filter-feeding larvae, also known ammocoetes, that live burrowed in the soft silty, substrate of rivers and streams that drain into the Atlantic Ocean, or as is the case with invasive sea lamprey, into the Laurentian Great Lakes. The subsequent metamorphosis is characterized by major changes in physiology and internal and external morphology, preparing sea lamprey for the juvenile, hematophagous (blood-feeding) phase when they prey/parasitize fishes. After the parasitic phase, which last 12–20 months in the Great Lakes but up to 1 year longer in the Atlantic Ocean, the maturing sea lamprey stop feeding and begin the final adult phase during which they migrate up rivers or streams, spawn and then die. (B) Comparison of a larval (foreground) and juvenile sea lamprey (background). *With permission from Wilkie, M.P., 2011a. Lampreys: energetics and development. In: Farrell A.P. (Ed.), Encyclopedia of Fish Physiology: From Genome to Environment, vol. 3. Academic Press, San Diego, pp. 1779–1787.*

and streams as suspension feeders (filter feeders), ingesting mainly organic detritus, along with lesser amounts of diatoms and algae (see reviews by Beamish, 1980; Dawson et al., 2015; Hardisty and Potter, 1971; Sutton and Bowen, 1994). Feeding currents are generated by the contraction of a muscular velum found near the opening of the pharynx, which draws water containing food particles in through the oral cavity and then on to the pharynx. Contraction of the branchial musculature then compresses the pharyngeal chamber, forcing water across the gills, which then exits via one of seven branchiopores. The branchiopores are equipped with branchial valves that prevents backflow of water across the gills as the volume of the pharyngeal chamber increases when the branchial musculature relaxes. Food particles are trapped by mucus, secreted by an endostyle, that lies ventral to the pharynx, and diverted to the esophagus, similar to the set-up found in protochordates such as amphioxus and sea squirts (Mallatt, 1996; Rovainen, 1996). The unidirectional flow of water across the gills promotes $O_2$ uptake and the excretion of $CO_2$ and metabolic wastes, in the same manner as a typical teleost fish gill (Wilkie, 2011b).

As a result of the low quality of their diet (Mundahl et al., 2005; Sutton and Bowen, 1994), larval lampreys tend to grow slowly. Sea lamprey in the Great Lakes typically spend ~3–7 years in the larval phase, but growth is highly variable among streams and individuals (Dawson et al., 2015); Morkert et al. (1998) reported the possibility of metamorphosis occurring at age 2 in a highly productive stream in the Lower Peninsula of Michigan (State of Michigan, U.S.A), while Manion and Smith (1978), monitoring a single year class of sea lamprey isolated above a barrier dam, found that some sea lamprey can remain in the larval stage for 12 years or more.

Sea lamprey metamorphosis typically begins in late June-early July, lasting 3–4 months (Beamish and Potter, 1975; Youson and Potter, 1979). A precondition of metamorphosis is the accumulation of sufficient lipid reserves, which rapidly accrue in the final months of the larval phase and provide the critical energy reserves needed for the non-trophic metamorphic period (Lowe et al., 1973; Manzon et al., 2015; O'Boyle and Beamish, 1977). A continual accumulation of thyroid hormone, followed by a rapid decline at the onset of metamorphosis, also appears to be a key trigger for metamorphosis, although the role of other endocrine signals remains unresolved (Manzon et al., 2015). Environmental cues include fluctuations in water temperature during the preceding larval phase, with elevated water temperature being a prerequisite for the ultimate onset of metamorphosis (Manzon et al., 2015; Youson, 1997).

Metamorphosis is highly coordinated and involves complex changes in the internal and external body plan, including the appearance and complete development of eyes, changes in skin coloration, and more prominent dorsal and caudal fins (see Manzon et al., 2015; Youson, 2003 for reviews; Fig. 1B). Perhaps the most notable change is the replacement of the oral hood with the characteristic multi-toothed oral disc and dagger-like tongue, which are used to attach to and pierce the skin of host fishes during the subsequent juvenile, parasitic phase of their life cycle (Renaud et al., 2009; Rovainen, 1996; Youson and Potter, 1979). Parasitic lampreys also produce a buccal gland secretion called lamphredin, which is known to contain anticoagulant and to have and cytolytic, hemolytic and fibrinolytic properties (Farmer, 1980; Li et al., 2018a; Renaud et al., 2009; Xiao et al., 2007). The lamphredin of the Arctic lamprey (*Lethenteron camtschaticum*) is reported to have sodium ($Na^+$) channel blocker proteins that might serve as a local anaesthetic (Chi et al., 2009; Xiao et al., 2012). In the Laurentian Great Lakes, host fishes frequently die due to blood loss or secondary infections (Farmer, 1980; Farmer et al., 1975; Swink, 2003). Farmer et al. (1975) estimated that juvenile parasitic sea lamprey consumed blood from salmonid fishes at rates ranging between 3% and 30% of their body weight per day, with the rate increasing as the lamprey got larger.

Apart from the eyes and the oral disc, another conspicuous change is the reorganization of the gills. The unidirectionally ventilated gills of the larva and the gastrointestinal tract are completely re-structured and re-organized during metamorphosis (see Manzon et al., 2015). The gills switch to tidal ventilation, allowing the post-metamorphic lampreys to breathe while attached by the oral disc to prey or substrate. A new esophagus also develops that is independent of the pharynx but connects the oral cavity with the intestine (Mallatt, 1996; Manzon et al., 2015; Rovainen, 1996; Wilkie, 2011a).

Fully metamorphosed sea lamprey migrate downstream to the sea (anadromous populations) or to lakes in the fall or spring, which appears to coincide with increased discharge of their natal streams or rivers (reviewed by Dawson et al., 2015; Evans et al., 2021; Manzon et al., 2015). Otherwise, little else is known about the environmental, let alone physiological, factors that trigger downstream migration. Evidence that temperature or photoperiod play any role in triggering downstream migration by sea lamprey is so far equivocal (Evans et al., 2021). This situation is quite unlike that seen with smolting salmonids, in which there is a dynamic interplay between behavioral cues (e.g., lunar phases, schooling), body size, and hormonal status (cortisol, growth hormone, thyroid hormone) that trigger downstream migration to marine environments (see review by McCormick, 2013).

During their parasitic feeding phase, lampreys are called juveniles because they remain sexually immature until they begin their upstream migration (see Docker et al., 2015, 2019). The parasitic juvenile phase of the sea lamprey lasts 12–20 months in the Great Lakes basin (Applegate, 1950; Bergstedt and Swink, 1995), which appears to be 1 year shorter than in the native anadromous population along the Atlantic coast (Beamish, 1980), followed by the migration upstream by the maturing adults to their spawning sites (Beamish and Potter, 1975; Moser et al., 2015; Swink, 2003). All lampreys are semelparous, dying after spawning like Pacific salmon (*Oncorhynchus* spp.) (Beamish and Potter, 1975; Docker et al., 2019). Unlike Pacific salmon, however, sea lamprey do not home to natal streams, but instead select potential spawning habitat first through directed movements toward shorelines based on decreases in depth, possibly by sensing changes in hydrostatic pressure (barokinesis), followed by up and down movements in the water column (vertical casting) to detect river plumes (Meckley et al., 2017). Candidate streams for spawning are then selected based on the presence of chemical odorants, unique bile acids (migratory pheromones) released by the stream-resident larvae (Buchinger et al., 2015; Sorensen and Hoye, 2007; see Section 5.2.1.2). Sea lamprey generally spawn from April to June (see reviews by Johnson et al., 2015; Renaud, 2011). The high fecundity of the females, which deposit 150,000–300,000 eggs in the case of anadromous sea lamprey and ~40,000 to more than 100,000 eggs (average 70,000) deposited by the freshwater-resident populations (Docker et al., 2019; Gambicki and Steinhart, 2017; Manion and Hanson, 1980; Renaud, 2011), was likely a key factor in the sea lamprey's successful invasion of the Great Lakes.

## 3 Invasive species in the Laurentian Great Lakes

### 3.1 Non-native and aquatic invasive species in the Great Lakes

The Laurentian Great Lakes are particularly vulnerable to introductions of new species, with multiple vectors for invasion. The complex connectivity of the Great Lakes to numerous canal systems and diversions linking the Great Lakes basin to the Atlantic Ocean was the cause of some of the earliest invasions, including alewife (*Alosa pseudoharengus*) and sea lamprey (Arthington et al., 2016; Sturtevant et al., 2019). The most common vector in recent years has been via the ballast water of ships, resulting in the introduction of Eurasian ruffe (*Gymnocephalus cernua*), round goby (*Neogobius melanostomus*), and tubenose goby (*Proterorhinus semilunaris*), along with invertebrates including zebra (*Dreissena polymorpha*) and quagga (*Dreissena rostriformis bugensis*) mussels, and the spiny water flea (*Bythotrephes longimanus*). There are now at least 189 non-indigenous fishes, invertebrates, algae, plants, and microbes (bacteria and viruses) in the Great Lakes, and this number appears to be increasing at a rate of 1.3–1.8 species per year (Fig. 2; GLANSIS, 2021; Sturtevant et al., 2019), making it one of the most invaded freshwater basins on Earth.

Although invasive species are non-indigenous, not all non-indigenous species are considered invasive species. Here, we define an invasive species as any non-indigenous species that has become established in its non-native habitat by means other than natural migration, and causes significant socioeconomic and ecological harm (Sturtevant et al., 2019). For instance, salmonid species including Chinook salmon (*Oncorhynchus tshawytscha*), coho salmon

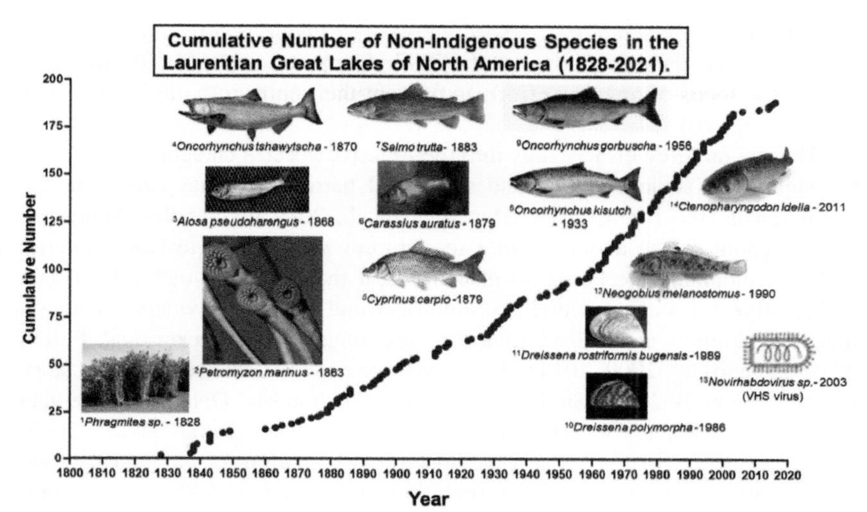


**FIG. 2** Non-indigenous species in the Laurentian Great Lakes since 1828. As of 2021, there were at least 189 non-indigenous fishes, invertebrates, plants, algae, and microbes in the Great Lakes. Data points denote the cumulative number of established non-indigenous species by year. *Insets*: Photos of notable non-indigenous species, with year first reported, and common name in parentheses: 1. *Phragmites* sp. (common reed); 2. *Petromyzon marinus* (sea lamprey); 3. *Alosa pseudoharengus* (alewife); 4. *Oncorhynchus tshawytscha* (Chinook salmon); 5. *Cyprinus carpio* (common carp); 6. *Carassius auratus* (goldfish); 7. *Salmo trutta* (brown trout); 8. *Oncorhynchus kisutch* (coho salmon); 9. *Oncorhynchus gorbuscha* (pink salmon); 10. *Dreissena polymorpha* (zebra mussel); 11. *Dreissena rostriformis bugensis* (quagga mussel); 12. *Neogobius melanostomus* (round goby); 13. *Novirhabdovirus* sp. genotype IV sublineage b (VHS - Viral haemorrhagic septicemia virus); 14. *Ctenopharyngodon idella* (grass carp). Data obtained from the Great Lakes Aquatic Nonindigenous Species Informatin Systems (GLANSIS; https://www.glerl.noaa.gov/glansis/) and Sturtevant et al. (2019). *Image credits and permissions*: [1]J. Gilbert, Ontario Invasive Plant Council (https://www.ontarioinvasiveplants.ca/invasive-plants/species/phragmites/). [2]S.M. Davidson, Wilfrid Laurier University. [3]C. Krueger, Great Lakes Fishery Commission (http://www.glfc.org/photos.php). [5,7,14]U.S. Fish and Wildlife Service, Public domain (https://digitalmedia.fws.gov/digital/collection/natdiglib/search/). [4,8]Under license from iStock.com/Willard. [9]iSttock.com/Krakovski. [6]M.P. Wilkie, Wilfrid Laurier University; [10]A. Benson, USGS, Public domain (https://nas.er.usgs.gov/queries/FactSheet.aspx?speciesID=5). [11]M. Richerson, USGS, Public domain (https://nas.er.usgs.gov/queries/FactSheet.aspx?speciesID=5); [12]P. van der Sluijs, Wikimedia Commons (https://commons.wikimedia.org/wiki/File:Round_goby.jpg). [13]M.P. Wilkie. Created with BioRender.com.

(*Oncorhynchus kisutch*), and rainbow trout (*Oncorhynchus mykiss*) are native to the Pacific Ocean, but they were deliberately introduced into and are now established in the Great Lakes where they are largely beneficial (e.g., because they feed on invasive alewife, whose populations exploded following the near collapse of lake trout) and, for many communities, they continue to have positive socio-economic value as prized gamefishes (Crawford, 2001; Dettmers et al., 2012). On the other hand, invasive carps in the Mississippi drainage (e.g., bighead and silver carps, *Hypophthalmichthys nobilis* and

*Hypophthalmichthys molitrix*, respectively) are causing severe ecological and socioeconomic damage (Irons et al., 2007; Sampson et al., 2009), and they remain the focus of ongoing efforts to prevent their entry into the Great Lakes (Jerde et al., 2011).

The sea lamprey clearly falls into the invasive species category in terms of the significant socioeconomic and ecological harm that it has caused in the Great Lakes (see Brant, 2019; Sturtevant et al., 2019 for details). There has been ongoing debate about whether sea lamprey were native to Lake Ontario, perhaps colonizing the lake post-glacially near the beginning of the Holocene (~11,500 years), but the weight of empirical and historical evidence suggests that sea lamprey are indeed invasive (see reviews by Docker and Potter, 2019; Eshenroder, 2009, 2014). Arguments for native status were genetic studies from the early 2000s. Studies using both mitochondrial DNA sequence data (Waldman et al., 2004, 2009) and microsatellite loci (Bryan et al., 2005) showed evidence for long-term vicariance (i.e., separation) of the Lake Ontario and Atlantic drainage populations, providing support for post-Pleistocene colonization following the last ice age. Although these studies used the genetic markers available at the time, they are relatively limited compared to those now available, and Eshenroder (2014) presented a compelling case for invasion via canals. Eshenroder (2014) concluded that sea lamprey entered the Lake Ontario drainage no earlier than the 1860s, likely following the completion of the Erie Canal which linked the Great Lakes basin to the Atlantic Ocean through the Hudson River or Susquehanna River (see also Docker et al., 2021).

Regardless of the origin of sea lamprey in Lake Ontario, there is no doubt that sea lamprey are non-native and invasive in Lake Erie and the upper Great Lakes (Lakes Michigan, Huron, and Superior). The route of invasion was via the Welland Canal, completed in 1829 and subsequently modified in the early 1900s, which allowed sea lamprey to by-pass Niagara Falls and enter Lake Erie first and then the other Great Lakes either by swimming through the canal, via attachment to the hulls of ships, by attaching to host fishes, or a combination of all three (Brant, 2019; Docker et al., 2021; Eshenroder, 2009, 2014; Smith and Tibbles, 1980). Sea lamprey were first detected in Lake Erie in 1921, followed by Lakes Michigan, Huron, and Superior in 1936, 1937, and 1939, respectively (Brant, 2019; Smith and Tibbles, 1980).

The invasion, coupled with ongoing overfishing, contributed to catastrophic declines in numerous fish species including large, commercially lucrative coldwater fishes such as the lake trout, for which harvest in Lake Huron declined by more than 95% between 1938 and 1954, followed by the complete collapse of its fishery in 1966 (Berst and Spangler, 1972; Morse et al., 2003; Smith and Tibbles, 1980). Along with lake trout, burbot (*Lota lota*) populations collapsed in Lake Huron by the mid-1960s, accompanied by the extinction of four deepwater cisco (*Coregonus* spp.) populations, along with declines in lake whitefish (*Coregonus clupeaformis*), lake herring (*Coregonus artedii*), and walleye (*Sander vitreus*) populations. Similar declines were observed in Lake Michigan, and a few years later Lake Superior, in which the lake trout harvest decreased by almost 90% between 1950 and 1960 (Smith and Tibbles 1980). The decline in top predatory fishes due to sea lamprey parasitism also led to population explosions of invasive alewife and rainbow smelt (*Osmerus mordax*), leading to frequent die-offs of alewives which contributed to poor water quality and fouling of beaches, particularly in Lake Michigan (Dettmers et al., 2012). Sea lamprey continue to threaten the Great Lake's commercial and recreational fisheries, which were valued at more than 7 billion dollars (U.S.) in 2008, including money spent on related activities such as boating, travel, and tourism (Krantzberg and de Boer, 2008), but the fishery is likely worth much more in today's dollars.

## 3.2 Features of a successful invasion

Though not yet a formally recognized geological time unit, the Anthropocene is generally defined as the period during which human activity has been the dominant influence on climate and the environment (Lewis and Maslin, 2015; Steffen et al., 2007). Although preindustrial human activity also had devastating impacts on many species and ecosystems, such damage has greatly accelerated in recent centuries (Lewis and Maslin, 2015; Steffen et al., 2007). Depending on your frame of reference, the Anthropocene started with the Industrial Revolution in the 1800s, when a great expansion of industrial activity led to much greater reliance on fossil fuels, or during the so-called Great Acceleration after World War II when rates of greenhouse gas emissions, including atmospheric carbon dioxide ($CO_2$), increased due to an explosion of economic activity and fossil fuel use (Lewis and Maslin, 2015; Savenije et al., 2014; Steffen et al., 2011; Waters et al., 2016). At the same time, isotopes arising from nuclear detonations and persistent industrial chemicals and compounds (e.g., plastics, cement, metals) began to accumulate in the Earth's (geological) strata, potentially serving as a geological marker of the new epoch (Lewis and Maslin, 2015). Regardless of the date described for the beginning of the Anthropocene, increased human population growth, industrial activity, and expanded global trade have led to pollution, habitat degradation, climate change, and invasive species, which have all contributed to marked declines in biodiversity in terrestrial and aquatic ecosystems (Arthington et al., 2016; Reid et al., 2019).

The rate of invasions has greatly increased in recent decades (Fig. 1; Sturtevant et al. (2019)), mainly due to the introduction of vectors that either directly provided non-native species with access to recipient sites (e.g., in the ballast of ships and on recreational watercraft) or the removal of natural barriers to invasion (due to increased connectivity of waterways). As noted above, however, not all introduced non-native species become invasive. The invasiveness of a species depends on many other factors, often described by the "Tens Rule," which states that ~10% of all introduced species become established in their recipient habitats, with a further 10% subsequently causing serious ecosystem disturbance and economic damage (Williamson and Fitter, 1996). So, what then causes an introduced species to become invasive?

Invasion theory states that the ability of non-native species to invade a novel habitat is influenced by: (1) their life history in their native range or some other invaded site; (2) the presence of vectors to transport them or give them access to the new site; and (3) propagule pressure which considers the number of invaders and invasion events that allow the non-native species to become established and then spread (Leung and Mandrak, 2007; Lockwood et al., 2009; Perkins et al., 2011; Sakai et al., 2001). Numerous studies of aquatic invasive invertebrates and vertebrates have indicated that propagule pressure is a key determinant of invasion success (Kolar and Lodge, 2001, 2002; Leung and Mandrak, 2007; Lockwood et al., 2009; Simberloff, 2009). Thus, the greater the propagule pressure, the more likely the establishment of a non-native species; if propagule pressure is insufficient, there will be too few individuals and limited mating opportunities to effectively reproduce, and/or a lack of genetic diversity that can lead to genetic bottlenecks (Sakai et al., 2001; Simberloff, 2009).

Another key to predicting invasion success is "invasibility," which describes the environmental conditions needed to permit a non-invasive species to become established at its new site and spread (Leung and Mandrak, 2007). The establishment and spread of the invader in their new environment ultimately depend upon their ability to pass through the "ecological filter" characteristic of the new site (Crowl et al., 2008; Kelley, 2014). Kelley (2014) identified two major components of the ecological filter, biotic and abiotic, which have the capacity to block or slow establishment and spread. Hence, invaders must have physiological traits that allow them to maintain homeostasis, performance, and fitness in the new habitat. In the case of freshwater invaders such as the sea lamprey, biotic filters may include predation by resident fishes, competition for resources, or lack of suitable prey. Abiotic filters may include salinity, temperature, environmental disturbance(s), or habitat features required for spawning and reproduction. In other words, the invader must be able to either tolerate and/or adapt to the new environment (Kelley, 2014; Lennox et al., 2015). This can be accomplished by: (1) prior adaptation to a similar environment in their native range; (2) wide tolerance to environmental perturbations; and (3) high genetic diversity and phenotypic plasticity, allowing the invader to rapidly adapt to the new environment.

## 3.3 The success of sea lamprey in the Laurentian Great Lakes

### *3.3.1 Life history*

In general, the prolonged life cycle of the sea lamprey (6–8+ years; see Section 2.2) would not normally be consistent with the establishment and spread of invasive species, many of which have short generation times

(Baker, 1974). In the case of the sea lamprey, however, the prolonged burrowing phase of larval sea lamprey protects them from predation in their natal streams (Dawson et al., 2015), and variation in age at metamorphosis (Dawson et al., 2015; Manzon et al., 2015) helps dampen variance in recruitment to the juvenile population (Schindler et al., 2010). Moreover, the high fecundity of the female adult sea lamprey (Docker et al., 2019) and their polygynous mating habits (Johnson et al., 2015) is expected to provide sufficient genetic diversity as raw material for adaptation. As noted above (Section 2.2), sea lamprey in the Great Lakes produce an average of 70,000 eggs per female (Docker et al., 2019), which is much higher than the smaller-bodied native lampreys that it appears capable of outcompeting (COSEWIC, 2020); for example, at the two extremes, northern brook lamprey (*Ichthyomyzon fossor*) and silver lamprey (*Ichthyomyzon unicuspis*) females produce an average of 1200 and 19,000 eggs, respectively (reviewed by Docker et al., 2019).

Sea lamprey are ecological generalists, able to survive in a range of aquatic habitats, and able to feed on a wide range of teleost and other fishes during their juvenile parasitic stage. Native to both sides of the Atlantic Ocean across a wide latitudinal range (i.e., from Newfoundland to Florida and the Gulf of Mexico and from Scandinavia to the Mediterranean Sea; Potter et al., 2015), sea lamprey readily survived in the Great Lakes following their initial introduction into Lake Ontario, and subsequently Lake Erie and the upper Great Lakes. The non-philopatric and anadromous nature of its life history likely predisposed its colonization, establishment, and spread throughout the Great Lakes. Unlike Pacific salmon, sea lamprey do not home to natal streams (Bergstedt and Seelye, 1995), but instead use a "suitable river" strategy (Waldman et al., 2008), by locating appropriate spawning tributaries by attraction to larval pheromones (Bergstedt and Seelye, 1995; Buchinger et al., 2015; Sorensen and Vrieze, 2003). This strategy is probably related to the nomadic life history of juvenile sea lamprey, whose distribution would be highly dependent on that of their hosts; in their native range in the North Atlantic Ocean, some hosts, depending on the species, may transport sea lamprey over 100–1000s of kilometers (Renaud, 2011; Waldman et al., 2008).

A non-philopatric life history would also facilitate dispersal to new regions (more so than species that exhibit strong site fidelity), particularly given that adult sea lamprey appear to be attracted to the odors of the four lamprey species that are native to the Great Lakes—silver, northern brook, chestnut (*Ichthyomyzon castaneus*), and American brook (*Lethenteron appendix*) lampreys. Larval pheromones may also increase the probability of finding a mate by causing aggregation of adults (Wagner et al., 2009), which would be important to ensure successful reproduction at low density in the early stages of invasion. Lack of site fidelity also likely reduces the evolution of local adaptation (Spice et al., 2012) and increases the likelihood of broad climatic tolerances, suggesting that sea lamprey larvae—whose parents may have been reared in geographically distant streams—are able to survive across a wide temperature range and environmental conditions (Potter and Beamish, 1975; Sutherby, 2019; reviewed by Borowiec et al., 2021).

### *3.3.2 Thermal physiology*

A broad thermal tolerance for larval sea lamprey may have been very important in allowing them to physiologically function over the wide range of temperatures characteristic of tributaries flowing into the Great Lakes. The "thermal niche" of Great Lakes' larval sea lamprey is proposed to fall between 17.8 and 21.8 °C, with a preferred temperature of 20.8 °C in the summer (Holmes and Lin, 1994). However, sea lamprey are surprisingly tolerant of warm waters, with incipient lethal temperature thresholds of 31 °C, which are 1–2 °C higher than native larval American brook lamprey and northern brook lamprey (Potter and Beamish, 1975). Although these differences are slight, they may have been sufficient to allow sea lamprey to outcompete these native lampreys at the edges of their ranges (Kelley, 2014), facilitating their establishment in Great Lakes tributaries. Behavioral experiments, measuring the maximal critical temperature ($CT_{max}$), the temperature at which larval sea lamprey emerge from their burrows, has since confirmed their high thermal tolerance, with $CT_{max}$ values of 32–34 °C (Sutherby, 2019). The high thermal tolerance of sea lamprey is also borne out by their cellular stress or heat shock protein response. Experiments on larval, pre-metamorphic, metamorphosing, and post-metamorphic sea lamprey have demonstrated that changes in temperature approaching 13–16 °C are needed to induce production of heat shock proteins HSP70 and HSP90 in the gills, liver, kidney, and intestine, with thresholds for HSP induction falling between 25 and 29 °C (Wood et al., 1998, 1999).

Notably, warmer acclimation temperatures did not appear to facilitate much greater temperature tolerance in sea lamprey, with $CT_{max}$ only increasing marginally (1.9–2.0 °C) from 32.5 to 34.4 °C when acclimated to 5 and 19 °C, respectively (Sutherby, 2019). Nor do warmer acclimation temperatures appear to increase the threshold needed to induce HSP production (Wood et al., 1999). The acclimation response ratio ($ARR = \Delta CT_{max}/\Delta Acclimation\ T = 0.12$), an index of thermal plasticity, was also very low compared to freshwater fishes, the majority of which had two- to fivefold greater ARR values (Comte and Olden, 2017; Sutherby, 2019). Thermal plasticity, as predicted by relatively high ARR values, has been hypothesized to be a feature common to aquatic invasive species (Kelley, 2014), but this does not appear to have been the case with sea lamprey. Indeed, more recent work has demonstrated that in freshwater fish species, those with higher thermal tolerances tend to have lower thermal acclimation capacity, which may represent an evolutionary trade-off between thermal tolerance and thermal plasticity (Comte and Olden, 2017).

A broad thermal width, the maximal high and low temperatures at which an animal can survive (Zerebecki and Sorte, 2011), also appears to be characteristic of sea lamprey, at least in the larval stages. Using respirometry to determine the maximal metabolic rate and the standard metabolic rate of larval sea lamprey, the thermal width of larval sea lamprey was in fact demonstrated to be very wide, with metabolic aerobic scope steadily increasing from approximately 6 μmol $g^{-1}$ $h^{-1}$ at 6 °C to more than 15 μmol $g^{-1}$ $h^{-1}$ at 28 °C, before dropping markedly as temperature approached their known incipient lethal temperatures (H. Flavio and M.P. Wilkie, Unpublished data; Fig. 3). This broad "thermal width" observed in sea lamprey is thought to be a feature of many aquatic invasive species (Kelley, 2014), and would have been particularly beneficial in the sea lamprey's native range, where juveniles could be translocated great distances whilst parasitizing marine fishes and encounter great variation in the thermal regimes of the rivers ultimately used for spawning and rearing larvae. It is also notable that $CT_{max}$ differs little between Great Lakes sea lamprey and larval anadromous sea lamprey native to the Atlantic coast of Canada, varying by 1 °C or less depending on acclimation temperature (Sutherby, 2019). Thus, the high thermal width and high innate thermal tolerance of sea lamprey, rather than thermal plasticity, may have been the key adaptations that facilitated the sea lamprey's occupation of the Great Lakes.

### 3.3.3 *Freshwater ion and osmoregulation*

Another abiotic filter faced by juvenile parasitic sea lamprey was long-term survival in fresh water. In anadromous sea lamprey populations, most of the juvenile parasitic phase is spent in sea water, with the animals migrating downstream within a few months of completing metamorphosis (Beamish and Potter, 1975; Hansen et al., 2016). The freshwater to seawater transition of the sea lamprey is characterized by a reversal of the ion and osmotic gradients across the gills, necessitating a switch from hyperosmoregulation to hypoosmoregulation strategies. To survive in fresh water, lampreys and other diadromous fishes have evolved strategies to: (1) take up ions from the water using active transport; (2) minimize ion loss across the gills and body surface; and (3) excrete copious amounts of urine to counter the osmotic influx of water (Edwards and Marshall, 2013; Ferreira-Martins et al., 2021a).

As in teleost fishes, $Na^+$ and chloride ($Cl^-$) uptake across the lamprey gill is facilitated by high-affinity, low-capacity $Na^+$ and $Cl^-$ transporters (Morris, 1972; Morris and Bull, 1970; Wilkie et al., 1998) and thought to take place using mitochondria rich cells (MRC), henceforth called ionocytes (Evans et al., 2005; Ferreira-Martins et al., 2021a). Two types of freshwater ionocytes have been characterized in sea lamprey: (1) ammocoete ionocytes that are widely distributed on the gill lamellae and lack the complex tubular network common to other ionocytes (Peek and Youson, 1979); and (2) intercalated ionocytes, which are solitary and present in the freshwater larval and juvenile stages and lost following seawater migration (Reis-Santos et al., 2008), before re-appearing during their freshwater spawning run (Bartels and Potter, 2004;

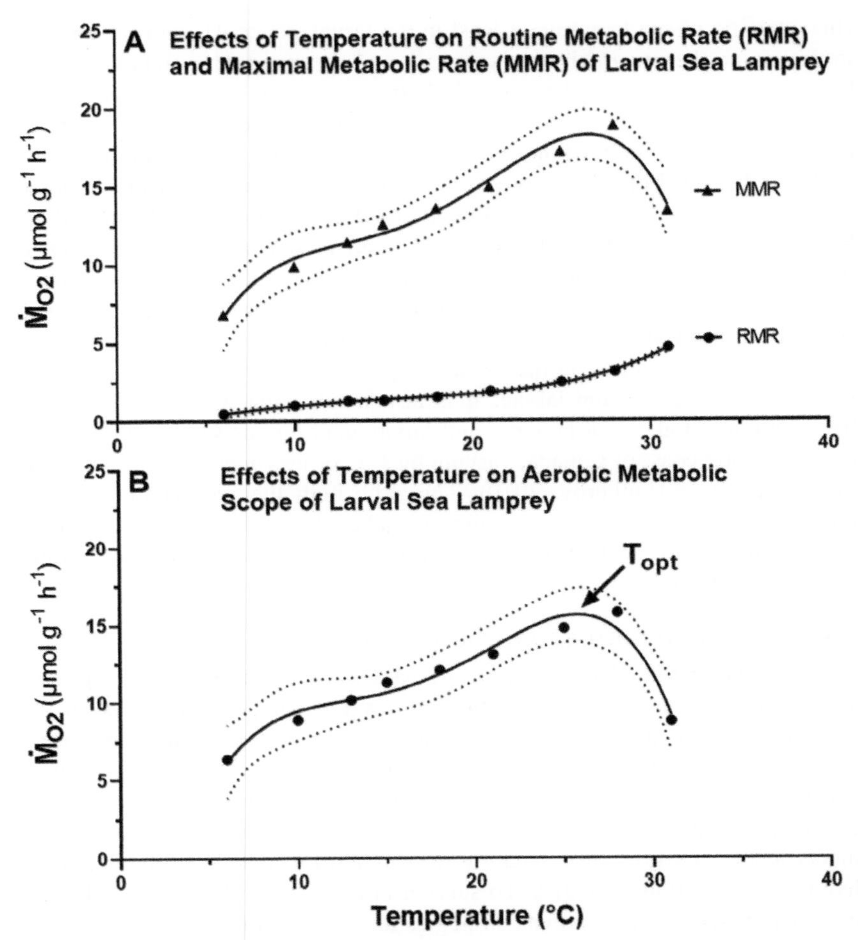


**FIG. 3** Thermal performance of larval sea lamprey (*Petromyzon marinus*) at different temperatures. (A) Larval sea lamprey, acclimated to a water temperature of 13 °C, were left overnight in intermittent flow respirometers to measure resting metabolic rate (RMR) following acute exposure to a range of temperatures (6–31 °C), followed by measurements of maximum metabolic rate (MMR) after exhaustive chasing (5 min) at the same temperature. (B) Metabolic aerobic scope, the difference between MMR–RMR, of the same larval lamprey calculated from the data presented in (A). Arrow denotes the thermal optima ($T_{opt}$). Data represent mean oxygen consumption rates ($\dot{M}_{O2}$) of $N = 12$ animals per temperature. Curves were fitted by non-linear regression using a centred fourth order polynomial function, with corresponding confidence intervals (dotted lines; Graphpad. Prism 8). *H. Flavio and M.P. Wilkie, Dept. Biology, Wilfrid Laurier University, Waterloo, Ontario, Canada. Unpublished data.*

Ferreira-Martins et al., 2016). Each type of ionocyte expresses vacuolar type $H^+$-ATPase (V-ATPase) proteins on the apical membrane and $Na^+/K^+$-ATPase (NKA) on the basolateral membrane (Reis-Santos et al., 2008; Sunga et al., 2020). In freshwater fishes, including sea lamprey, the

V-ATPase is thought to help generate the electrochemical gradient to drive $Na^+$ uptake across the apical epithelium via apical $Na^+$ channels (Ferreira-Martins et al., 2021a). Less is known about $Cl^-$ uptake, but recent messenger RNA (mRNA) evidence suggests that apical chloride transport in freshwater juvenile sea lamprey may take place via a $Na^+$:$Cl^-$ co-transporter (NCC; Barany et al., 2021) and/or a $Cl^-/HCO_3^-$ exchange set-up on the apical membrane, and a basolateral cystic fibrosis transmembrane conductance regulator (CFTR) which conveys the $Cl^-$ into the extracellular fluid (Ferreira-Martins et al. 2021a).

Following metamorphosis, the ammocoete ionocytes are lost, and branchial NKA activity increases many-fold due to the appearance of seawater (SW) ionocytes (Reis-Santos et al., 2008; Sunga et al., 2020). The SW ionocytes are arranged side by side within the interlamellar spaces, and display intense immunoreactivity to NKA due to their widespread distribution along the basolateral membrane and complex tubular system characteristic of these cells (Bartels and Potter, 2004; Reis-Santos et al., 2008; Zydlewski and Wilkie, 2013). This arrangement, very similar to the SW ionocytes of teleosts, generates high concentrations of $Na^+$ in the paracellular pathways between adjacent gill cells, generating the electrochemical gradient needed to drive $Na^+$ secretion (Edwards and Marshall, 2013; Evans et al., 2005; Zydlewski and Wilkie, 2013). A bumetanide sensitive basolateral $Na^+$:$K^+$:2 $Cl^-$ cotransporter (NKCC1) is thought to transport $Cl^-$ into the intracellular space of the SW ionocyte (Shaughnessy and McCormick, 2020), followed by $Cl^-$ diffusion to the water via an apical CFTR which has not yet been identified (Edwards and Marshall, 2013; Ferreira-Martins et al., 2021a). Intercalated ionocytes disappear in seawater acclimated juvenile sea lamprey, with corresponding decreases in the abundance of V-ATPase expressed on the gills (Reis-Santos et al., 2008).

Prolonged survival of juvenile sea lamprey in fresh water would depend upon retaining at least some of the components of their freshwater ionoregulatory machinery, and suppression of the ion secretion processes. This appears to be the case as there is some retention of V-ATPase protein in the juvenile lamprey so long as they remain in fresh water following metamorphosis, suggesting that the intercalated ionocytes have retained some of their capacity to take up $Na^+$ from fresh water (Reis-Santos et al., 2008; Sunga et al., 2020). Based on mRNA expression studies, NCC function is also maintained in fresh water juveniles, but significantly downregulated after acclimation to salt water, further suggesting that additional components of the $Na^+$ and $Cl^-$ uptake machinery remain intact in fresh water but are lost on introduction to salt water (Barany et al., 2021). Flexibility in the function of both the intercalated ionocytes and SW ionocytes may explain why anadromous populations of juvenile sea lamprey have been reported to remain in fresh water for up to 6 months following transformation (Beamish and Potter, 1975; Beamish et al., 1978), and may ultimately explain why juveniles were able to remain in the fresh waters of the Great Lakes for even longer.

Gill ion permeability would also need to be controlled by regulating the expression of tight junctions (TJ), which form selective barriers between pavement cells and ionocytes that regulate solute/ion movements between the extracellular fluid and the water (Chasiotis et al., 2012; Kolosov et al., 2013). Tight junction structure and expression shows great plasticity through the life cycle of lampreys and other fishes, particularly in response to changes in salinity or exposure to very dilute waters (Chasiotis et al., 2012; Edwards and Marshall, 2013; Evans et al., 2005). Ultrastructural analysis revealed that the number of TJ strands (zona occludens) between adjacent SW ionocytes of juvenile pouched lamprey (*Geotria australis*; native to Australia, New Zealand, and Argentina in the southern hemisphere) was reduced following transition between fresh and salt water, leading to increases in $Na^+$ permeability known to promote paracellular $Na^+$ secretion (Bartels and Potter, 1991). More recent work using molecular and immunohistochemical methods demonstrated that the expression and abundance of tricellulin protein and claudin genes increased in response to acclimation to dilute (ion-poor water), consistent with previous observations that tricellulin and claudins were involved in decreasing gill ion permeability (Kolosov et al., 2017, 2020). It was also notable that the claudin TJ protein composition of post-metamorphic sea lamprey gills was similar to those of larval sea lamprey (Kolosov et al., 2020), suggesting that gill ion permeability may be comparable before and after metamorphosis, provided the animals remain in fresh water. If the founding populations of juvenile sea lamprey were able to maintain low gill ion permeability, it too may have been a key to their longer-term survival in fresh water.

The flexible timing of the migration of juvenile sea lamprey to sea water following metamorphosis, along with increasing evidence that their physiological responses to fresh and sea water exposure are highly plastic, suggests that the sea lamprey are facultatively anadromous (Zydlewski and Wilkie, 2013). In other words, they can either migrate downstream to the sea or, in some instances, remain in fresh water. Such strategies are not unprecedented, as facultative anadromy has been reported in brown trout (*Salmo trutta*), and populations of pink salmon (*Oncorhynchus gorbuscha*) and coho salmon, although once thought to be strictly anadromous, have readily populated the Great Lakes (Docker and Potter, 2019; Railsback et al., 2014). If sea lamprey were facultatively anadromous, it would suggest that it was not osmoregulatory constraints that limited their freshwater colonization, but rather access to abundant prey of a suitable size range (see Docker and Potter, 2019). Under such circumstances, sea lamprey would have been able to feed and osmoregulate in fresh water for sufficient time to complete the juvenile phase and proceed to their adult spawning phase, which would have ultimately allowed them to initially colonize Lake Ontario, and other nearby waters.

### 3.3.4 Feeding

A generalist diet was also likely key to the sea lamprey's successful invasion. Even though sea lamprey show host preferences (see Section 5.1.2), sea lamprey have been reported to feed on more than 50 fish species (and even marine mammals) throughout its native and introduced range (see reviews by Quintella et al., 2021; Renaud and Cochran, 2019). This generalist diet likely allowed them to spread throughout the Great Lakes by parasitizing a wide variety of native fishes including lake trout, burbot, and ciscoes (*Coregonus* spp.). This means that they are less likely to be impacted from changes in the distribution and abundance of individual host species. For example, with the extirpation of some species of ciscoes and the near collapse of the lake trout fishery (Brant, 2019), sea lamprey appear to have readily switched to feeding on non-native Pacific salmon (Adams and Jones, 2021; Renaud and Cochran, 2019).

Feeding would also have been important for maintaining ion and osmotic balance in the Great Lakes. As pointed out by Wood and Bucking (2011), feeding plays a known, but often underappreciated, role in maintaining salt and water balance in fishes. For instance, feeding can offset ion losses and stabilize plasma ion balance in rainbow trout exposed to conditions that lower plasma $Na^+$ or $Cl^-$, such as transfer to ion-poor water or exposure to acidic pH or metals (D'Cruz et al., 1998; Kamunde et al., 2005). The role that feeding plays in maintaining ion balance in juvenile sea lamprey could provide a great deal of additional insight into the factors that allowed anadromous sea lamprey to become established in the Great Lakes.

## 4 The sea lamprey control program: Exploiting the unique physiological vulnerabilities of an invader

### 4.1 International cooperation leads to effective sea lamprey control and fish conservation

In response to declining fish stocks and the sea lamprey invasion, the Great Lakes Fishery Commission (GLFC) was formed in 1955 after the signing of the "Convention on Great Lakes Fisheries between the United States of America and Canada" (September 10, 1954. Washington, D.C. see Gaden et al., 2021 for review). The mandate of the GLFC was to establish a "comprehensive program" to eradicate or control sea lamprey populations in the basin, as well as to manage fisheries in the Great Lakes (Gaden et al., 2021). What resulted was a highly successful invasive species control program involving extensive cross-border collaboration between U.S. and Canadian government agencies including Fisheries and Oceans Canada, the U.S. Fish and Wildlife Service, and the U.S. Geological Survey. As described below, the sea lamprey control program exploited unique features of the sea lamprey life history, behavior, and physiology to develop methods to reduce sea lamprey populations by more than 90% following their peak in the mid-1950s and the virtual collapses of fisheries in the basin (Siefkes, 2017).

The key to the success of the sea lamprey control program was the combination of chemical control to eradicate multiple year-classes of larval sea lamprey in their nursey streams, and the continued development and improvement of barriers and traps to prevent the upstream migration of spawning lamprey (Marsden and Siefkes, 2019; Siefkes, 2017; see Section 4.3). While Great Lakes fisheries have not been fully restored, these measures have contributed to a rehabilitation of the Great Lakes ecosystem with the recovery of many of its native species, particularly lake trout and whitefish populations (Grunder et al., 2021; Heinrich et al., 2003; Lavis et al., 2003; Marsden and Siefkes, 2019; Morse et al., 2003; Smith and Tibbles, 1980; Sullivan et al., 2003). Continued exploitation of unique aspects of the sea lamprey's life history, behavior, and physiology (see review by Borowiec et al., 2021) will be critical as the GLFC moves toward a more integrated pest management framework (Section 4.3) and to adapt to future challenges. A particular focus will be on how climate change-induced changes in water temperature and hydrology could undermine the efficacy of these very effective methods of sea lamprey control, and what can be done to meet these challenges (see Lennox et al., 2020).

## 4.2 Chemical control of sea lamprey

Early efforts at chemical control of sea lamprey were led by Vernon Applegate (Applegate and King, 1962; Applegate et al., 1957, 1961), who also provided early reports on the biology of these animals (Applegate, 1950). Due to their relatively sessile, burrow-dwelling life style, the larval stage was the focus of efforts to develop a selective piscicide, or lampricide, that specifically targeted sea lamprey. A total of 4346 chemicals were screened for toxicity to larval sea lamprey and non-target fishes before 3-trifluoromethyl-4-nitrophenol (TFM; Fig. 4) was chosen due to its ease of application and because it was thought to be largely lamprey specific (Applegate et al., 1957). TFM is now used to control sea lamprey in all five of the Great Lakes, Lake Champlain (in New York and Vermont), and the Finger Lakes (New York). To be effective as a control tool, TFM needs to be added to streams at a concentration sufficient to ensure that all sea lamprey in a stream reach will be exposed to the minimal lethal concentration (MLC) for at least 9 h, which is equivalent to the 9-h LC99.9 of TFM (Bills et al., 2003; McDonald and Kolar, 2007; Wilkie et al., 2019). Therefore, at the site of application, TFM is applied at higher concentrations (e.g., 1.3–1.5× the MLC) to compensate for dilution from incoming tributaries, groundwater upwellings, and run-off as the TFM block drifts downstream (Barber and Steeves, 2021).

Niclosamide (Fig. 4), commonly used as a molluscicide to control snail populations and intestinal parasites in pets and livestock, is often co-applied with TFM to increase its toxicity, while still maintaining the TFM's specificity to sea lamprey (Wilkie et al., 2019, 2021). Very little niclosamide, 1–2% of the TFM concentration, is needed to increase the toxicity of TFM by 30–40% (Boogaard et al., 2003). Experiments with rainbow trout have

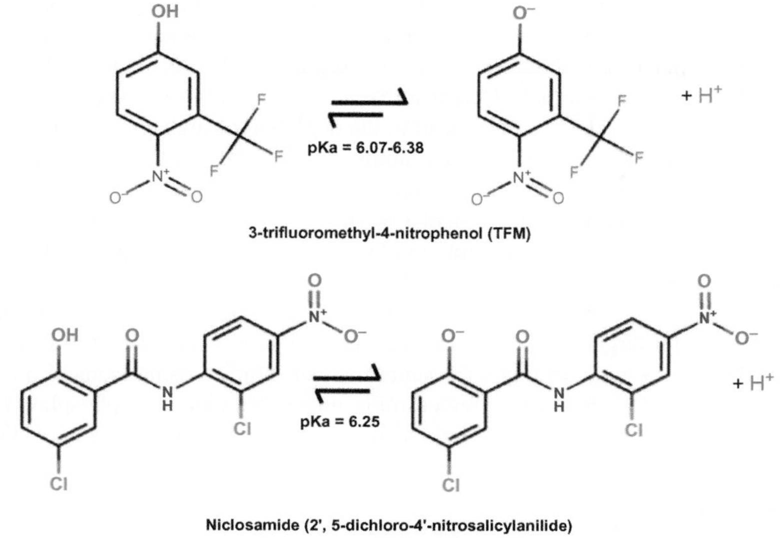

**FIG. 4** The chemical structure and log acid dissociation constants (pKa) of 3-trifluoromethyl-4-nitrophenol (TFM) and niclosamide (2′, 5-dichloro-4′-nitrosalicylanilide), which are applied to streams containing larval sea lamprey (*Petromyzon marinus*) as a means of control in the Laurentian Great Lakes. See text for further details. Chemical structures obtained courtesy of the U.S. National Library of Medicine (https://chem.nlm.nih.gov/chemidplus).

demonstrated that the two chemicals act in a synergistic manner (greater than additive) with one another (Hepditch et al., 2021). This synergistic action of niclosamide reduces TFM requirements for lampricide treatments, and results in considerable cost savings. A granular formulation of niclosamide, called granular Bayluscide™, which sinks and is bottom-acting, is used to target larval sea lamprey in stream mouths and some of the very large, high discharge rivers found in the Great Lakes, including the St. Mary's River, which is the single largest source of sea lamprey in Lake Michigan (Criger et al., 2021).

### *4.2.1 Mechanism of action of lampricides*

More is known about the mode of action of TFM, and the factors affecting its uptake, distribution, and elimination by sea lamprey and non-target fishes, than is known for niclosamide (Wilkie et al., 2019). However, both TFM and niclosamide interfere with aerobic ATP generation by uncoupling oxidative phosphorylation in cellular mitochondria (Birceanu et al., 2011; Borowiec et al., 2022; Huerta et al., 2020; Niblett and Ballantyne, 1976). The resulting mismatch in ATP supply and demand forces sea lamprey and non-target fishes to rely on anaerobic pathways of ATP production, leading to a depletion of anaerobic energy reserves such as glycogen, and corresponding increases in lactate in brain and muscle (Birceanu et al., 2009, 2014; Clifford et al., 2012;

Henry et al., 2015; Ionescu et al., 2021, 2022a,b). Niclosamide also causes a marked elevation in lactate, along with corresponding reductions in muscle intracellular pH in larval sea lamprey (Ionescu et al., 2022a). It is likely that TFM also causes similar acid-base disturbances, but this has not yet been confirmed. In addition, TFM can also lead to the depletion of high energy phosphagens such as phosphocreatine in sea lamprey (Birceanu et al., 2009; Clifford et al., 2012), and phosphoarginine in limpet molluscs (Viant et al., 2001). Death is ultimately thought to be due to depletion of energy reserves in the body and a corresponding inability to maintain cellular homeostasis, particularly in highly sensitive tissues such as the brain (Birceanu et al., 2009; Clifford, 2012; Henry et al., 2015). Recent work by Ionescu et al. (2021, 2022b) has demonstrated that non-target rainbow trout and lake sturgeon (*Acipenser fulvescens*) respond to niclosamide in much the same manner. Studies using isolated mitochondria have also shown that niclosamide uncouples oxidative phosphorylation in lamprey mitochondria, but with much greater potency than TFM (Borowiec et al., 2022).

### 4.2.2 Selectivity of TFM

The greater sensitivity of larval lampreys to TFM compared to other fishes is due to the relative inability of larval lampreys to detoxify TFM (Bussy et al., 2018a,b; Kane et al., 1994; Lawrence et al., 2021; Lech and Statham, 1975). Although TFM is largely lamprey specific, it is not sea lamprey specific. The native lampreys in the Great Lakes are also highly sensitive to TFM (Neave et al., 2021; Schuldt and Goold, 1980). King and Gabel (1985) showed that northern and American brook lampreys were slightly less susceptible to TFM than sea lamprey, but the differences were generally insufficient to allow for selective control of sea lamprey where the species overlap. Northern and American brook lampreys often inhabit upstream reaches from which sea lamprey are now excluded by barriers, but the migratory silver lamprey has an in-stream distribution similar to that of sea lamprey (COSEWIC, 2020). Vulnerability to lampricides is considered a threat to northern brook and silver lampreys in the Great Lakes basin, and the Great Lakes-St. Lawrence population of both species are listed on the Canadian Species at Risk Act (SARA) as Special Concern (COSEWIC, 2020; Maitland et al., 2015; Neave et al., 2021). More work is needed to understand how lampricide treatments can be better targeted to minimize effects to native lampreys. Lamprey sensitivity to TFM also extends to other stages of the sea lamprey cycle, including the adult stage in which the MLC is almost 50% lower than in the larval stage (Henry et al., 2015). This greater TFM sensitivity of the adults could be because they are in the terminal spawning phase of their life cycle.

The greater sensitivity of lampreys to TFM is related to their limited expression of the Phase II biotransformation detoxification enzyme, UDP glucuronyl transferase (UPD-GT), which catalyzes the addition of a glucuronide functional group to form TFM-glucuronide; this transformation makes TFM more water soluble and facilities its excretion via the bile and/or urinary routes (Wilkie et al., 2019). In non-target fishes such as the bluegill (*Lepomis macrochirus*), channel catfish (*Ictalurus punctatus*), and rainbow trout, the maximal activity (two- to threefold) and affinity (1.5- to 4-fold) of UDP-GT is higher than in sea lamprey (Kane et al., 1994). Recent high-throughput transcriptomics analysis has also revealed that there are a far greater number of UDP-GT transcripts (mRNA levels) and higher expression in bluegill compared to sea lamprey, providing a genetic foundation to explain the greater sensitivity of sea lamprey to TFM (M.J. Lawrence, P. Grayson, and K.M. Jeffries, University of Manitoba, Personal communication.).

At present, it is unclear if Phase I biotransformation plays a significant role in TFM metabolism in sea lamprey or non-target fishes. Bussy et al. (2018a,b) provided in vitro evidence indicating that some cytochrome P450 enzymes play a role in TFM metabolism by sea lamprey, including the generation of a unique amino-metabolite; they speculated that this amino-metabolite could contribute to TFM toxicity in sea lamprey. However, the implications of these observations still need to be worked out.

### 4.2.3 Non-target effects of lampricides

Many have argued that TFM is selective, but in fact there are a number of invertebrate and vertebrate species, including some of conservation concern, that are adversely affected by TFM exposure. In addition to native lampreys, mudpuppies (*Necturus maculosus*) and lake sturgeon are also sensitive to TFM (Boogaard et al., 2003; Kaye, 2021; Pratt et al., 2021). Lake sturgeon, assessed as either endangered or threatened throughout the Great Lakes basin (COSEWIC, 2017; COSSARO, 2017; Hayes and Caroffino, 2012), are of concern because young-of-the-year (Age 0) lake sturgeon that are less than 100 mm in length are particularly sensitive to TFM. At this stage, their 12-h TFM LC50s overlap with those of larval sea lamprey, and could lead to potential mortality (McDonald and Kolar, 2007; O'Connor et al., 2017). Hepditch et al. (2019), using $^{14}$C-TFM to measure TFM uptake, demonstrated that these life-stage-dependent increases in TFM sensitivity were size-dependent, with rates of TFM uptake inversely scaling with the body size (reviewed in Wilkie et al., 2021). A relatively low capacity to detoxify TFM using glucuronidation compared to other non-target fishes also appears to be important, but this does not appear to be life-stage-dependent in lake sturgeon (O. Birceanu and M.P. Wilkie, Unpublished observations).

Mitigating or avoiding non-target effects is a priority of the sea lamprey control program, and the GLFC is thus promoting and funding research in this area (Boogaard et al., 2003; Kaye, 2021; O'Connor et al., 2017; Pratt et al., 2021). In the short- and medium-term, lampricide applications are likely to continue to be used because ceasing them could severely undermine sea lamprey population control efforts and fisheries conservation and management in the basin, as was recently demonstrated using an age-structured population model; this model showed that ceasing lampricide applications on certain streams containing nursery habitat for lake sturgeon would improve lake sturgeon recruitment from those streams, but the resulting increase in sea lamprey recruitment could result in greater sea lamprey parasitism on larger, immature lake sturgeon and other large-bodied fishes (Dobiesz et al., 2018). Another option, applying TFM at lower concentrations in waters occupied by both larval sea lamprey and lake sturgeon, lowered sturgeon mortality but at the cost of increased larval sea lamprey survival, recruitment of parasites, and greater damage to lake trout fisheries (Dobiesz et al., 2018; O'Connor et al., 2017; Pratt et al., 2021). Together, field observation and physiological studies indicate that delaying treatments to later in the year, when age-0 lake sturgeon are larger and show markedly lower rates of TFM uptake (Hepditch et al., 2019), is one option that is being used to reduce lake sturgeon mortality without compromising sea lamprey control efforts (Pratt et al., 2021).

The chemical properties of TFM give it high diffusability across the gills of sea lamprey and results in high rates of accumulation (Wilkie et al., 2021), but this also results in its rapid elimination following exposure (Hlina et al., 2017; Tessier et al., 2018). Of particular note is how resilient sea lamprey and non-target fishes are to sublethal exposure to TFM and niclosamide. Work conducted over the last decade or so suggests that both sea lamprey and non-target fishes such as rainbow trout and lake sturgeon that survive TFM and/or niclosamide exposure completely restore their glycogen reserves and phosphocreatine, eliminate lactate, and correct acid-base balance within hours (24h or less) following exposure (Clifford et al., 2012; Ionescu et al., 2022a,b). However, further work is needed to more thoroughly quantify how quickly sturgeon and other non-target fishes completely recover from lampricide exposure including further assessments of their physiological performance (e.g., restoration of energy stores including lipids, protein and carbohydrates; swimming capacity) and behavior.

### 4.2.4 Impacts of climate change on lampricide use

As water temperatures continue to rise due to climate change, the effectiveness and use of TFM could be compromised through its effects on both sea lamprey metabolism and TFM detoxification. For a thorough review of the potential effects of climate change on several aspects of sea lamprey control, readers are referred to Lennox et al. (2020). The effectiveness of TFM, and likely niclosamide, will likely change as water temperatures increase in the Great Lakes where, depending upon how much atmospheric $CO_2$ concentrations increase, air temperatures are expected to rise by 1–6 °C by 2100 (Angel and Kunkel, 2010). Similar changes in the thermal regime, as well as water flow, will also be expected to take place in Great Lakes tributaries (Lennox et al., 2020).

As waters warm in the Great Lakes basin, sea lamprey are expected to become less sensitive to TFM as summertime water temperatures increase (Hlina et al., 2021). Sea lamprey sensitivity was already known to be

1.5- to 2.0-fold greater in the late summer than in the spring (Scholefield et al., 2008), and subsequent work demonstrated that this greater TFM tolerance in the summer is primarily a function of temperature rather than body condition, size, or proximate body composition (e.g., lipid stores) of the animals (Fig. 5) (Hlina et al., 2021; Muhametsafina et al., 2019). Although rate of TFM uptake increases with temperature (Hlina et al., 2021), this rate increase is likely offset by increases in TFM detoxification as has been reported in other fishes exposed to other phenolic compounds similar to TFM such as 4-nitrophenol, 2,4-nitrophenol (Howe et al., 1994; Patra et al., 2015). This interpretation was also supported by experiments using $^{14}$C-TFM, in which larval sea lamprey exposed to the MLC of TFM at different temperatures experienced greater total TFM accumulation at 6 °C than at 12 or 22 °C (Hlina et al., 2021).

The above findings imply that, at warmer water temperatures, there will be a greater risk of residual sea lamprey that survive treatments unless additional amounts of TFM are added to compensate for the higher temperatures. Non-target mortality could also be higher if greater amounts of TFM are needed for treatments, making it necessary to learn more about how other fishes and invertebrates respond to TFM at different temperatures. However, agencies may be able to work around these problems by adding TFM to larger rivers in the spring or fall when water temperatures are cooler, and less total TFM will be needed compared to the warmer summer months. Ultimately, temperature will likely have to be incorporated into treatment protocols that are currently used to more accurately determine the required TFM application amounts based on pH and alkalinity (Bills et al., 2003).

In addition to temperature, water pH is projected to decrease by 0.29–0.49 pH units by 2100 due to the anticipated increases in atmospheric $CO_2$ partial pressure equilibrating with the water, but this would be highly dependent on how local geological factors affect water chemistry, particularly alkalinity which determines water acid buffer capacity (Phillips et al., 2015). It also seems doubtful that lower water pH would adversely affect TFM effectiveness because persistently lower pH would result in a lower MLC, decreasing the concentrations of TFM required to kill larval sea lamprey.

## 4.3 Barriers to migration and trapping

Lampricides are the only tools used in the sea lamprey control program that reduce sea lamprey abundance in predictable ways (Hanson and Swink, 1989; Hodges, 1972; but see Miehls et al. (2020, 2021). Their application is logistically feasible and impacts to non-target species are more likely to be tolerated (Kaye, 2021; Wilkie et al., 2021), because sea lamprey larvae are concentrated into relatively small habitat patches by barriers that block adult sea lamprey migration. Without the presence of barriers restricting sea lamprey access to spawning habitat, the costs of sea lamprey control—and the

A

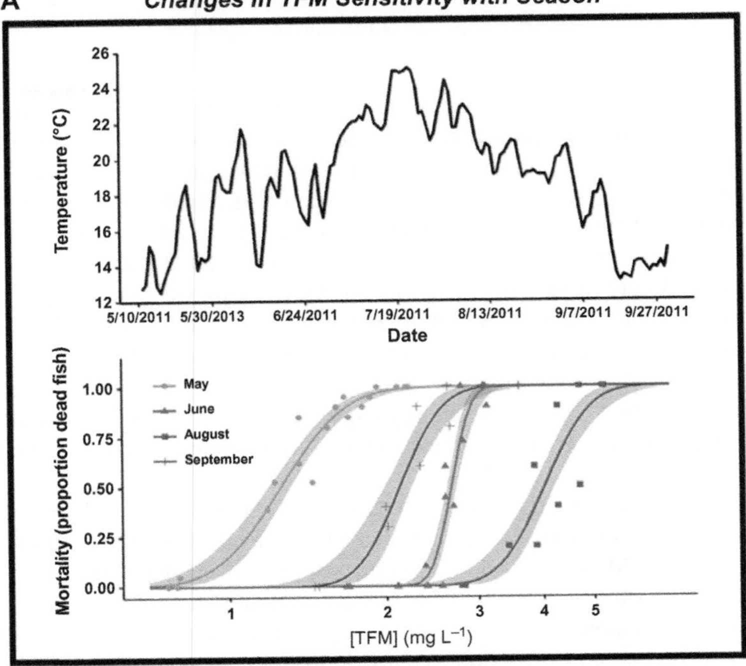


B Higher Acclimation Temperatures Lower TFM Sensitivity

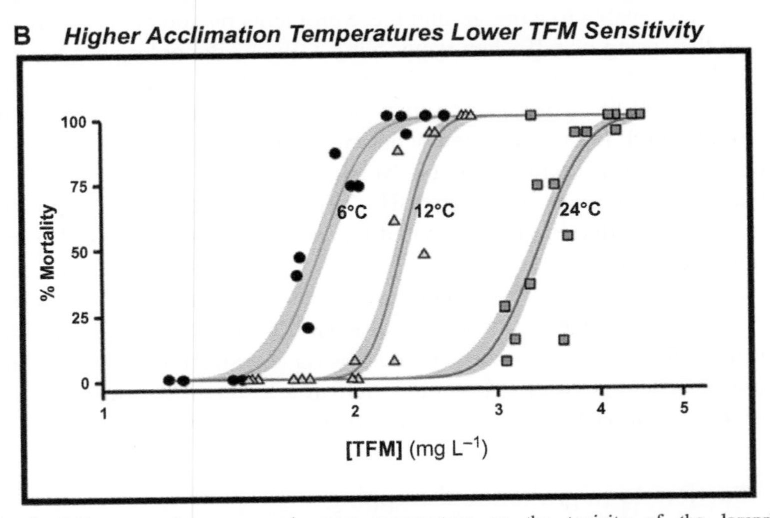


**FIG. 5** Influences of season and water temperature on the toxicity of the lampricide 3-trifluoromethyl-4-nitrophenol (TFM) to larval sea lamprey (*Petromyzon marinus*) (A) captured from Deer Creek, Michigan, or (B) from the Au Sable River, Michigan, and acclimated for ~10 days to cold (6°C), moderate (12°C) or warm temperatures (24°C). Shifts in the dose-response curves (±95% confidence intervals; shaded regions) to the right indicate greater tolerance to TFM and correspondingly higher 12-h $LC_{50}$ values. Note that tolerance is greatest in the summer and at warm temperatures, when the dose-response curves are shifted furthest to the right, and lowest in the spring and at cold temperatures (left-most curves). *With permission from Muhametsafina, A., Birceanu, O., Hlina, B.L., Tessier, L.R., Wilkie, M.P., 2019. Warmer waters increase the larval sea lamprey's (Petromyzon marinus) tolerance to the lampricide 3-trifluoromethyl-4-nitrophenol (TFM). J. Great Lakes Res. 45, 921–933 and Hlina, B.L., Birceanu, O., Robinson, C.S., Dhiyebi, H., Wilkie, M.P., 2021. The relationship between thermal physiology and lampricide sensitivity in larval sea lamprey (Petromyzon marinus). J. Great Lakes Res. 47, S272–S284.*

resultant effects of parasitism on Great Lakes fishes—would rapidly become untenable. Therefore, we review sea lamprey migration barriers, their impacts to non-target species, and the paradox of how continued use of these human-made structures to control sea lamprey is threatened by climate change. Interested readers also are directed to the numerous historical and recent reviews on barriers and trapping (Applegate, 1950; Applegate and Smith, 1951a; Brant, 2019; Hrodey et al., 2021; Hunn and Youngs, 1980; Lavis et al., 2003; McLaughlin et al., 2007, 2013; Miehls et al., 2020; Moser et al., 2015; Walter et al., 2021; Zielinski and Freiburger, 2021; Zielinski et al., 2019).

The first tool to control sea lamprey populations were physical barriers erected in Great Lakes tributaries to block and trap sea lamprey migrating upstream to spawn (Applegate, 1950; Fig. 6A). These barriers exploit the anguilliform body of sea lamprey and their limited leaping ability compared to other fishes in the basin (Webb, 1994), and they remain a critical component of the program (Hrodey et al., 2021). The first sea lamprey barriers built in the 1940s and 1950s were designed with little regard to non-target fishes; little work had been conducted at that time on the effects of the barriers on non-target species, and the ongoing effects that sea lamprey were having on the ecosystem and economy were thought to justify rapid action and deployment (Brant, 2019). Under average streamflow conditions and when cleaned daily, sea lamprey migration could be blocked using mesh screen ($25\,mm^2$ hardware cloth) framed in wood and arrayed across the entire channel (Fig. 6A). These weirs presumably blocked all bidirectional fish movement, although selective removal of sea lamprey could be achieved by adding traps that exploited lamprey shape and behavior. Because upstream-migrating sea lamprey probe barriers extensively in the spring when water temperatures reach 12–16°C (Applegate, 1950; McCann et al., 2018a), weirs were built that were "V-shaped" with a trap placed where the wings of the weir joined at the furthest upstream extent of the barrier. Funnels having entrances of ~35 mm-diameter allowed sea lamprey to enter the traps while excluding larger fishes like suckers (Catostomidae; Miehls et al., 2019). In addition to physically removing sea lamprey so they do not spawn below the barrier or in neighboring streams (Applegate and Smith, 1951a,b), barrier-integrated traps are also important to the control program by providing estimates of abundance by allowing mark-recapture assessment (Robinson et al., 2021). However, the mesh used in the early barriers frequently tore, plugged with debris, and were overtopped during high discharge (Applegate and Smith, 1951a,b). Therefore, the mechanical weirs and traps designed in the 1950s were not considered a scalable long-term solution for controlling sea lamprey throughout Great Lake tributaries.

In efforts to reduce barrier cleaning and maintenance, while also not impounding water upstream of physical structures, alternating electrical current (AC) was tested as the first non-physical barrier for sea lamprey

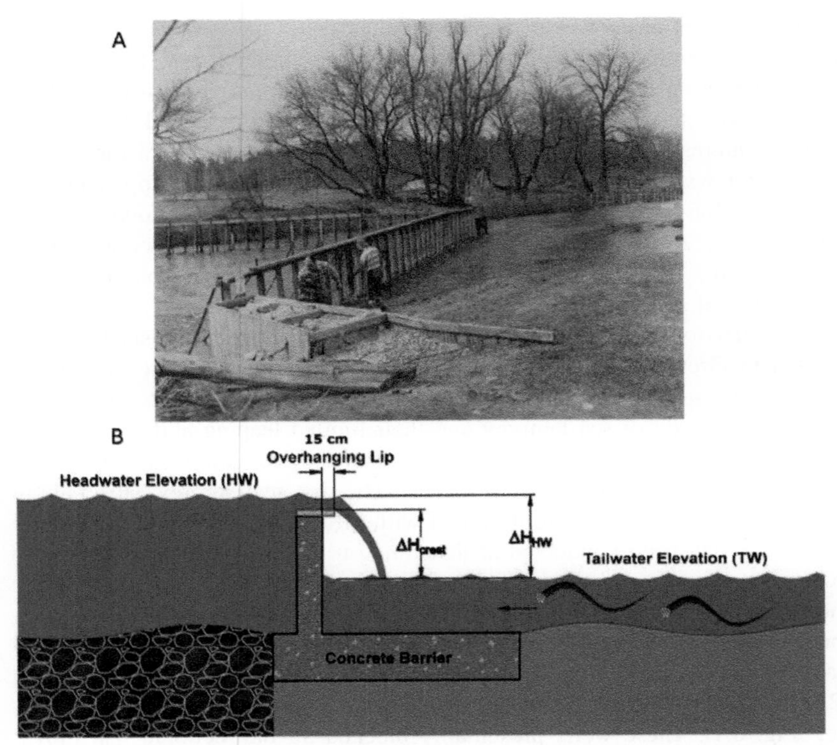


**FIG. 6** Barriers designed to block adult sea lamprey (*Petromyzon marinus*) migration in Great Lakes tributaries. (A) Barrier from the 1950s constructed of angled screens that directed sea lamprey toward a trap in the middle of the river. Upstream of the trap, vertical electrodes produce alternating current electricity to block any sea lamprey that breaches the mechanical barrier and trap (Image in Public Domain of U.S. Government). (B) Modern-day fixed-crest sea lamprey barrier with an overhanging lip. $\Delta H_{crest}$ = crest elevation, which is the difference in elevation between the surface of the tail water and the barrier crest. A crest elevation of 45 cm is recommended to block adult sea lamprey in Great Lakes tributaries. $\Delta H_{HW}$ = hydrologic head and is the difference in elevation between the surface of the tail water and head water. *From Zielinski, D.P., McLaughlin, R., Castro-Santos, T., Paudel, B., Hrodey, P., Muir, A., 2019. Alternative sea lamprey barrier technologies: history as a control tool. Rev. Fish. Sci. Aquac. 27, 438–457; Public Domain of U.S. Government.*

(Applegate et al., 1952; reviewed in McLain et al., 1965). The first electrical systems used 110-V AC actuated through vertical electrodes suspended above the stream (Applegate et al., 1952; Fig. 6A). AC barriers were effective at blocking fishes under normal streamflow conditions, but, like mechanical weirs, could be washed out or allow sea lamprey to bypass the electric field along the stream edges during floods. Large die-offs of non-target fishes (1000 or more fish) were also observed near AC barriers. Because the polarity of the direct current (DC) does not alternate, DC is less injurious to fishes and was first tested during 1956 to reduce non-target fish mortality (McLain, 1957). DC barriers did reduce fish mortality, but they remained unselective and blocked the migration of all fishes (McLain, 1957). Therefore, the greatest weaknesses of mechanical and electrical barriers include blocking and killing non-target fishes, washouts during floods, and continued need for periodic maintenance to clean screens or electrodes. Only one mechanical trap (Bridgeland Creek, Ontario) and two electrical barriers (Ocqueoc and Black Mallard rivers, Michigan) were used for sea lamprey control in 2021 (Zielinski et al., 2019).

Starting in the late 1950s, permanent low-head weirs were designed to semi-selectively block sea lamprey (Stauffer, 1964; Wigley, 1959). Design criteria for a purpose-built fixed-crest sea lamprey barrier are: (1) maintaining greater than a 45-cm vertical differential between the barrier crest and the elevation of the water downstream of the barrier (the tail water); and (2) fitting the crest with a 15-cm overhanging lip (Fig. 6B; Zielinski et al., 2019). The disadvantages of these structures are an increase in cost and time to engineer/build, they impound water upstream of the device, and block migrating of non-target fishes. The advantages are that they can block sea lamprey up to the 25- to 50-year flood event, have a life expectancy of 30–50 years, and do not require daily or weekly maintenance. When desired by management agencies, purpose-built sea lamprey barriers can be engineered with pools downstream of the barrier to allow non-native steelhead (*O. mykiss*) and other Pacific salmon (*Oncorhynchus* spp.) to jump over the barrier, but movement of non-jumping native species is still impeded (see below).

Dams and low-head weirs remain a critical component of the sea lamprey control program, but they face challenges to their continued use and effectiveness. In 2020, the sea lamprey control program identified nearly 600 fixed-crest barriers that block sea lamprey migration in tributaries of the Great Lakes. These barriers limit sea lamprey access to as much as 450,000 km of stream habitat, thus eliminating the need for lampricide treatment in these areas (Hrodey et al., 2021). Only about 10% of these structures were built or modified with the specific goal of blocking sea lamprey; the majority were built for flood control, energy production, or recreation, with an average build date for these structures of 1943 (roughly coinciding with when sea lamprey invaded the upper Great Lakes; Hrodey et al., 2021). However, the aging infrastructure, along with challenges related to ownership, liability, and high costs to rebuild will be a major challenge to maintain their ability to block sea lamprey migration for the next several decades.

Existing sea lamprey barriers face at least two other significant challenges: they are still minimally selective for sea lamprey, and changing temperature and precipitation regimes are increasing the risk that barriers will be unable to maintain blockage (Lennox et al., 2020; Zielinski and Freiburger, 2021). While some low-head weirs allow passage of fishes with strong leaping ability, most native fishes are also blocked by these structures. For example, suckers (*Catostomus* spp.), walleye, and lake sturgeon are unable to pass low-head barriers unless manually passed via a trap and sort fishway (Dodd et al., 2003;

Pratt et al., 2009; Tews et al., 2021). Erecting and operating low-head weirs only when adult sea lamprey are migrating may allow some native fish passage. However, sea lamprey migration overlaps with that of most spring- and summer-migrating fishes (Vélez-Espino et al., 2011) so, to block ~99% of the adult sea lamprey run, as many as 40–100% of non-target species would also be blocked (Vélez-Espino et al., 2011). Sea lamprey begin entering streams in the spring when river temperatures approach 4 °C, and entry peaks at ~12 °C often during high discharge (Applegate, 1950; McCann et al., 2018a); arrival and probing at barriers peaks at ~15 °C (McCann et al., 2018b). Sea lamprey have also been observed entering streams in the late summer (Applegate, 1950; Johnson et al., 2021b), and, although the reproductive success of early and late migrants remains unclear (McLaughlin et al., 2007), recruitment of larval lamprey in years when lampricide treatment occurs from May to July indicate that adult sea lamprey spawning later in the summer can still reproduce after TFM treatment (Jubar et al., 2021; Sullivan et al., 2021). For sea lamprey barriers to be considered effective and successful, they will need to better exploit the unique physiology of sea lamprey to selectively block their upstream migration while passing native fishes (see Section 4.3).

Changes in seasonal temperatures and precipitation in the Great Lakes basin is another challenge to the continued reliance on fixed-crest dams and barriers. Great Lakes tributaries, especially on the leeward side of the Great Lakes, are warming earlier in the spring, and these warming trends correspond to earlier sea lamprey migration (McCann et al., 2018b). In the most extreme circumstances (e.g., the St. Marys River, the connecting channel between Lake Superior and Huron), sea lamprey migration timing has shifted 30 days earlier in the past 30 years. Thus, traps and barriers need to be deployed sooner and remain effective longer to block the entire sea lamprey migration. Larger and more frequent flood events also increase the risk that sea lamprey will migrate upstream of low-head barriers while the barriers are inundated, or that the barrier will wash out completely like those on the Tittabawassee River, Michigan, in spring 2020 (Hrodey et al., 2021). Understanding exactly how climate change will influence the effectiveness of sea lamprey barriers remains a critical research priority of the Great Lakes Fishery Commission (Buckley, 2021).

## 4.4 Movement to integrated pest management

More than 40 years ago, Sawyer (1980) emphasized how the sea lamprey control program could benefit by embracing the ideologies of integrated pest management. The definitions and philosophical components of integrated pest management are numerous (Ehler, 2006; Hubert et al., 2021; Stenberg, 2017), but they mostly emphasize using a combination of biological, chemical, and physical control approaches in a socioeconomic context to keep pest populations below those causing economic injury (Dent, 1995). Sea lamprey control today uses an integrated approach by combining chemical and physical control methods to reduce sea lamprey populations to levels where fishery restoration activities can be successful. Integrated control could be improved if damage caused by sea lamprey to fish stocks would be described more precisely and if additional and species-specific control tools were developed.

Sea lamprey control tools have been integrated since their inception because neither migration barriers (physical controls targeting the adult stage) nor lampricides (chemical controls targeting the larval stage) would be useful without the other (Christie and Goddard, 2003; Hubert et al., 2021). Barriers alone do not necessarily reduce sea lamprey populations; blocked sea lamprey, if not removed through trapping, can still spawn downstream of barriers or in neighboring streams (Applegate and Smith, 1951a,b). Lampricides reduce sea lamprey abundance (Hanson and Swink, 1989; Hodges, 1972), but their application is feasible (Kaye, 2021; Wilkie et al., 2021) only because sea lamprey larvae are concentrated into relatively small habitat patches by migration barriers. Without barriers, lampricide treatments in Great Lakes tributaries would cost ~$200 million USD per year, compared to the current budget ~$13 million USD (Hrodey et al., 2021). Furthermore, barriers that exclude sea lamprey from the upper reaches of tributaries protect sensitive non-target species, such as native non-migratory northern brook and American brook lampreys, from exposure to lampricides (Maitland et al., 2015; Marsden and Siefkes, 2019).

Lampricide treatments are prescribed to individual tributaries of the Great Lakes to meet targets established by management agencies. Lampricide treatment options are ranked based on the cost to kill larvae where electrofishing surveys assess larval abundance, length, and spatial distribution throughout Great Lakes tributaries (Jubar et al., 2021). Targets for the sea lamprey control program are based on the abundance of adult sea lamprey and percentage of lake trout wounded in each of the Great Lakes (Treska et al., 2021). Abundance and wounding-based metrics were set by Great Lakes management agencies, and they explicitly relate acceptable sea lamprey-induced mortality rates to fish stocks of importance (Treska et al., 2021).

Another area for future progress for integrated control is that the adult stage and, especially, the metamorphosing and parasitic juvenile life stages are not currently targeted by the control program in any meaningful or systematic way (Borowiec et al., 2021; Lamsa et al., 1980; Siefkes et al., 2021; but see Lewandoski et al., 2021a). In addition to assessment traps that remove upstream migrants, tools to reduce reproduction could be direct such as catching and killing adults (Johnson et al., 2016) or indirect by releasing sterilized adult males to mate with females (Bravener and Twohey 2016) or disruption of sensory systems that reduce reproductive success (Buchinger et al., 2020). Juveniles could be targeted with traps (Evans et al., 2021; Miehls et al., 2019) or aspects of metamorphosis could be disrupted (Manzon et al., 2015). Future sea lamprey control could further implement integrated pest management where managers have a more precise understanding of the damage sea lamprey cause and can draw on a more diverse set of assessment and control tools based on sea lamprey's unique physiology (Borowiec et al., 2021; Siefkes, 2017; see Section 4.4).

# 5 The future of conservation physiology in sea lamprey control

Conservation physiologists—by including a detailed mechanistic understanding of the physiology of both sea lamprey and their hosts (Wikelski and Cooke, 2006)—are well-equipped to advance the goals of integrated sea lamprey control where combinations of control tools keep sea lamprey populations below those causing ecological and economic injury. Doing so will require two things: (1) refining an acceptable threshold of damage caused by sea lamprey as the fish communities and lakes change; and (2) developing effective and sea lamprey-specific control tools to keep populations below those targets. A better understanding of the impact sea lamprey have on fish populations is helping to define management targets where the level of sea lamprey-induced mortality is reduced to a level compatible with fish community objectives (Treska et al., 2021). Where controls are needed, conservation physiology is likely to be central to developing physical, chemical, and biological methods of sea lamprey control with greater specificity and fewer non-target effects. Other authors have reviewed some of the ways in which the sea lamprey control program exploits the physiological vulnerabilities of sea lamprey (e.g., Borowiec et al., 2021; Siefkes, 2017). Here, we highlight how conservation physiology is shaping the future of sea lamprey control, starting with characterizing lethal and sublethal damage to fish stocks and continuing with a discussion of the next generation of control tools aimed at exploiting physiological characteristics that are unique to each sea lamprey life stage.

## 5.1 Predicting the lethality and stress induced by parasitic sea lamprey to host fishes

### *5.1.1 Assessing the damage caused by parasitic sea lamprey*

As ectoparasites, sea lamprey are a significant pest in the Great Lakes because of the stress they inflict on their hosts, causing direct mortality, reducing growth and reproduction due to loss of blood, or resulting in secondary infection of the wounds. Therefore, a mechanistic understanding of the damage sea lamprey cause is paramount for defining the level of control required to keep populations below those causing ecological injury (Stewart et al., 2003), but directly measuring the damage caused by sea lamprey in the Great Lakes has been largely intractable because of the lakes' enormous size, diversity of habitats, temperatures, and fish communities (Harvey et al., 2008). Furthermore, fishes killed by sea lamprey are difficult to survey because they sink to the bottom (Bergstedt and Schneider, 1988), fishes that survive attack may be wounded in ways that are not easy to quantify (Firkus et al., 2021; Fig. 7A),

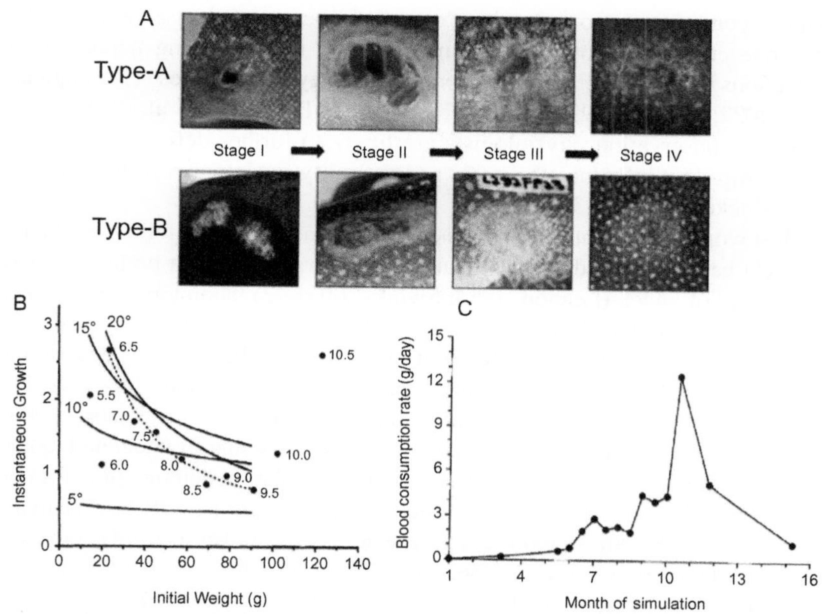


**FIG. 7** Conservation physiologists have played an important role defining sea lamprey (*Petromyzon marinus*)-induced damage caused to fish stocks. (A) Sea lamprey-induced wounding on lake trout classified according to King (1980). Wounds piercing the skin and exposing musculature are Type A and those that abrade the sink are Type B. Stage 1 wounds are those that are recent with no healing, whereas stage 4 wounds are nearly healed. Critical assumptions are that wounds are correctly classified by field staff and wounds heal in a predictable way. (B) Estimated instantaneous growth rates of parasitic sea lamprey as a function of initial weight and water temperature. Solid lines are estimates derived from Farmer et al. (1977) in the laboratory, and the dotted lines are estimates from Bergstedt and Swink (1995) based on wild-captured parasitic lamprey from northern Lake Huron. Points are estimates of growth per half month where "6.5" is June 15th and 10.5 is October 15th. Note that estimated growth rates of wild-captured sea lamprey in October are much higher than predicted by Farmer et al. (1977). (C) Blood consumption of parasitic sea lamprey as a function of month of the year (1 = January) as predicted by an individual-based bioenergetics model calibrated with data from Bergstedt and Swink (1995). *Panel (A) from Firkus, T.J., Murphy, C.A., Adams, J.V., Treska, T.J., Fischer, G., 2021. Assessing the assumptions of classification agreement, accuracy, and predictable healing time of sea lamprey wounds on lake trout. J. Great Lakes Res. 47, S368–S377. under open use Creative Commons License. Panel (B) from Bergstedt, R.A., Swink, W.D., 1995. Seasonal growth and duration of the parasitic life stage of the landlocked sea lamprey (*Petromyzon marinus*). Can. J. Fish. Aquat. Sci. 52, 1257–1264; product of official U.S. Government Work and public domain. Panel (C) from Madenjian, C.P., Cochran, P.A., Bergstedt, R.A., 2003. Seasonal patterns in growth, blood consumption, and effects on hosts by parasitic-phase sea lampreys in the Great Lakes: an individual-based model approach. J. Great Lakes Res. 29, 332–346; product of official U.S. Government Work and public domain.*

and the blood meal sea lamprey consume is not readily identifiable through traditional gut content analysis (Johnson et al., 2021a). Bence et al. (2003) went so far as to suggest that the uncertainties, inconsistencies, and discrepancies surrounding prediction of how fish mortality responds to changes in sea lamprey control were so great that "one could be tempted to give up on the enterprise entirely." Most key advancements in understanding parasite-host interactions have used conservation physiology approaches (Bullingham et al., 2022; Farmer, 1980; Sepúlveda et al., 2012; Smith et al., 2016) and, no doubt, conservation physiologists are likely to further define mechanistic relationships that influence the stress and mortality sea lamprey inflict on their hosts (Cooke et al., 2013).

Most work on sea lamprey parasitism can be placed into the following theme areas: (1) behavioral studies in laboratory tanks evaluating host preference and the lethality of attack (Lennon, 1954; Swink, 2003); (2) laboratory experiments evaluating physiological responses of sea lamprey and their hosts (Farmer, 1980; Farmer et al., 1975; Sepúlveda et al., 2012); (3) field studies describing sea lamprey-induced wounding and mortality (Adams and Jones, 2021; Adams et al., 2021a,b; Hume et al., 2021a; Johnson and Anderson, 1980; Lantry et al., 2015; Simpkins et al., 2021); and (4) modeling to integrate laboratory and field observations in bioenergetics and population demographic frameworks (Kitchell, 1990; Kitchell and Breck, 1980; Irwin et al., 2012; Madenjian et al., 2003, 2008). Since about 2005, researchers have also identified biomarkers in sea lamprey that reveal aspects of their feeding history (stable isotopes, Harvey et al., 2008; fatty acids, Happel et al., 2017; host fish DNA, Johnson et al., 2021a), and they have described biomarkers in host fishes that begin to characterize the sublethal effects of sea lamprey attack (Bullingham et al., 2022; Goetz et al., 2016; Sepúlveda et al., 2012). The feeding behavior of parasitic sea lamprey and responses of host fishes has been reviewed comprehensively in several contexts (Bence et al., 2003; Farmer, 1980; Johnson and Anderson, 1980; Renaud and Cochran, 2019; Stewart et al., 2003; Swink, 2003). Here, we briefly review each of the four identified theme areas, and end by discussing threats and opportunities in the Anthropocene, as well as ideas for how parasitic sea lamprey could be targeted for control.

### 5.1.2 Behavioral studies reveal host selection and lethality of attack

Given the difficulty of studying parasitic stage sea lamprey in the Atlantic Ocean and Great Lakes, biologists gleaned much insight concerning sea lamprey parasitism in laboratory tanks. The first such study was conducted in a bathtub in 1914 (Gage, 1928), and the studies conducted since have revealed relationships regarding host selection and how host size, host species, host strain, and water temperature influence lethality of attack. Swink (2003) provides a succinct and detailed review of laboratory studies prior to 2003, and interested readers are referred there; we provide only a short summary of the key findings highlighted therein.

The first studies describing the feeding of parasitic sea lamprey from the Great Lakes were conducted in the 1950s (Lennon, 1954; Parker and Lennon, 1956). Aquaria, holding roughly 200 L of water supplied from Lake

Huron at ambient temperatures, were used and these early studies coarsely described the blood meal lamprey consumed, a secretion termed lamphredin with anti-coagulant properties and cytolytic effects (Baxter, 1956), a reduction in red cell count in lethally wounded white suckers (*Catostomus commersonii*), and the lethality of attacks. These studies showed that, when presented with 11 fish species from the Great Lakes, sea lamprey exhibited host preferences (see below). They also provided the first estimates of how many kilograms of fish sea lamprey kill in the wild (16.8 kg) by growing recently transformed juveniles to the adult stage over 14–18 months using mostly small hosts (16–870 g).

Interest in laboratory sea lamprey parasitism studies peaked again after the GLFC committed to adopting integrated pest management in the early 1980s (Christie and Goddard, 2003; Sawyer, 1980). Swink and colleagues conducted numerous tank studies to improve estimates of lamprey-induced losses of teleosts (Swink, 2003), mostly with parasitic sea lamprey recently captured by commercial fishers and in ~150 L tanks supplied with water at ambient temperatures from Lake Huron (the exception being during the summer when temperatures were to chilled 10–14 °C). Collectively, studies found that sea lamprey are size selective in that they avoid hosts less than 600 mm when larger hosts are present (Swink, 1991; see also Cochran, 1985; Farmer and Beamish, 1973). Also, while the total biomass of fish killed by sea lamprey in the laboratory can range widely (1.2–36 kg), the kg of fish killed is generally highest when the feeding occurs in warmer water (Swink, 2003), when the sea lamprey is large (September–November), and when sea lamprey are feeding on large hosts (> ~800 mm), the latter indicating that being large does not ensure survival from lamprey attack, at least in the context of these tank environments (Swink, 1990). When confined in tanks and given multiple species to attach to, sea lamprey will attach more frequently to splake ("speckled" or brook trout (*Salvelinus fontinalis*) × lake trout hybrids), common carp (*Cyprinus carpio*), and white sucker than brown bullhead (*Ameiurus nebulosus*), redhorses (*Moxostoma* spp.), lake whitefish, and burbot (Farmer and Beamish, 1973). These laboratory results were not consistent with observations from the Great Lakes because burbot and lake whitefish populations declined in the Great Lakes after sea lamprey invaded and carp did not. Swink (2003) speculated that deviations from observations of wounds on wild vs laboratory fishes was the result of these fishes occupying different habitats in the lakes, highlighting that the spatial and thermal ecology of sea lamprey and potential host fishes is critical for understanding what fishes are attacked and the damage caused (Bergstedt et al., 2003a). More recent studies have specifically focused on species highly prioritized for restoration; for example, defining how lethality and recovery from attacks on lake sturgeon was size-specific and that high abundance of parasitic sea lamprey may also threaten sturgeon restoration goals (Dobiesz et al., 2018; Patrick et al., 2009). The latter is an important reminder of the degree to which sea lamprey control and native fish conservation is intertwined. Although there is evidence that lake sturgeon can be killed or adversely affected by TFM treatment if it is applied when these early life stages are present in streams (Boogaard et al., 2003; see Section 4.2.3), lack of sufficient sea lamprey control would also have negative effects on lake sturgeon growth and survival (Patrick et al., 2009).

### *5.1.3 Physiological studies define sea lamprey growth and mechanisms underlying lethal and sublethal effects to hosts*

Studies by Farmer and colleagues were first in describing mechanistic physiological changes that occurred in sea lamprey and their hosts during parasitism (reviewed in Farmer, 1980). Their innovative study in 1975, where sea lamprey fed on rainbow trout and lake trout with red blood cells labeled with radioactive chromium ($^{51}Cr$) provided estimates of the quantity of blood consumed by sea lamprey (Farmer et al., 1975). Blood consumption of sea lamprey (30–157 g) ranged from 3% to 30% (mean 12%) of the lamprey's body weight per day, a consumption rate greater than that reported for most teleosts. Conversion efficiencies were also higher than most fishes (40–50%), in part, because less than 3% of the available energy was lost as fecal material given the high digestibility of the blood meal. Applying their physiological data, they estimated the number of days it would take for trout to die, assuming a feeding rate 12% of body weight per day across a range of sea lamprey sizes; for example, a 200-g sea lamprey was predicted to kill a 1-kg trout in 3 days whereas the time for a 50-g sea lamprey was 22 days. Interestingly, the caloric content of the blood meal declined after a few days of feeding, and was only half that of fish that were not parasitized. Therefore, the high feeding rate of sea lamprey on host fishes may be necessary to compensate for lower bloodmeal energy densities during prolonged attachments and may be a motivating factor for sea lamprey to detach from their hosts prior to killing them. Farmer et al. (1975) terminated their studies in September and only experimented on sea lamprey up to 160 g; therefore, questions remain about how size-specific consumption changes in sea lamprey >160 g, and especially in the fall of the year when host mortality is expected to be highest (Madenjian et al., 2003). It is also unclear if $^{51}Cr$ may have impacted the health of the trout or sea lamprey.

Follow-up work by Farmer et al. (1977) described positive relationships between water temperature, lethality of sea lamprey attack, and sea lamprey growth rate. Instantaneous growth rates of sea lamprey 10–30 g were greatest at 20 °C and at 15 °C for lamprey 30–90 g. Mortality of hosts (white sucker in this case) increased over the entire temperature range tested (4–20 °C). Instantaneous growth rates of sea lamprey declined as lamprey size increased (Fig. 7B). Based on these data, the authors concluded that host mortality was seasonal and most likely increased as water temperature and sea lamprey size peaked in late summer and fall. Swink (2003), applying these and data from his own experiments, estimated that a single sea lamprey could kill roughly 12kg of fish between May and September.

A limitation of experiments in tank environments is that they likely provided an additional source of stress to sea lamprey and their hosts. Parker and Lennon (1956) noted that sea lamprey reared in the lab from metamorphosis to the adult stage were about half the size (52g) of wild-captured adults (125g), perhaps because sea lamprey were not feeding as aggressively or feeding conversion rates were lower in the lab. Swink (2003) noted that up to 10% of lake trout in his tank experiments died without being attacked by a sea lamprey, and that up to 40% of the mortality observed during parasitism trials occurred post-attachment (many from wound infections; Swink, 1993). A study comparing mortality of rainbow trout and lake trout after a single sea lamprey attachment found that rainbow trout were more likely to survive, but the authors admitted that the higher survival of rainbow trout may have been because that species is better able to adapt to laboratory tanks (Swink and Hanson, 1989). While studies by Farmer et al. (1975, 1977) were conducted in much larger tanks (68,000 and 2000L, respectively), the potential confounding effects of laboratory environments requires that physiological, behavioral, and lethality outcomes be contrasted with field observations.

### *5.1.4 Field observations largely confirm behavioral and physiological studies*

Reviewing the large number of field studies documenting sea lamprey size, sea lamprey-induced mortality, or wounding is beyond the scope of this chapter, and most have been reviewed elsewhere (Hume et al., 2021a; Johnson and Anderson, 1980; Swink 2003). However, several field studies are highlighted below because they corroborate or qualify conclusions made regarding host selection, host mortality, and sea lamprey growth from laboratory studies.

Laboratory observations regarding host selection have largely been corroborated in the Great Lakes. Assessment in Lakes Superior (Pycha and King, 1975), Huron (Rutter and Bence, 2003), and Ontario (Schneider et al., 1996) found evidence that sea lamprey actively avoid small lake trout when larger hosts were available (Swink, 2003). Population modeling frameworks integrating assessments of sea lamprey wounds on fish also conclude that sea lamprey will target other fishes (cf. Farmer and Beamish, 1973) in circumstances when lake trout are not highly abundant (Adams and Jones, 2021; Lantry et al., 2015). An exception is a recent study on Lake Huron that concluded that sea lamprey may feed on Chinook salmon more than lake trout even through lake trout are abundant (Simpkins et al., 2021). Higher overall survival of Seneca strain lake trout in the Great Lakes (Ontario: Schneider et al., 1996; Huron: Scribner et al., 2018; Michigan: Larson et al., 2021) relative to other strains is not because Seneca strain lake trout are better able to survive sea lamprey attack, as revealed in the laboratory by Swink and Hanson (1986). Instead, field studies reporting that Seneca strain lake trout occupy colder water support the alternative hypotheses that they have reduced interactions with sea lamprey (Elrod and Schneider, 1987; Schneider et al., 1996) and lower mortality rates if attacked in colder water (per Farmer et al., 1977).

Laboratory observations describing increases in lethality with increasing water temperature and sea lamprey size have also been mostly confirmed in the Great Lakes. In Lake Ontario, nearly all lake trout killed by sea lamprey were recovered in trawls from early October to mid-November (Schneider et al., 1996); the authors concluded that sea lamprey-induced mortality on lake trout was minimal at other times of the year. Bergstedt and Schneider (1988) likewise documented a pulse in lake trout mortalities in the fall, and they found that most of the dead lake trout were between 600 and 800 mm in length, which is consistent with conclusions by Swink (2003) and Farmer (1977) that even large lake trout are not immune from sea lamprey-induced mortality. In northern Lake Huron, estimated sea lamprey-induced mortality of lake whitefish also peaked in September and October (Spangler et al., 1980). However, higher host mortality in the fall contrasts with observations in the lab, where Swink (2003) observed many sea lamprey-induced lake trout mortalities in August and September. Higher summer water temperatures in the laboratory studies and possible host stress in laboratory tanks may account for this discrepancy.

Rates of sea lamprey growth predicted from laboratory studies have also been corroborated in wild-caught sea lamprey from Lake Huron. Using lengths and weights of ~2300 parasitic sea lamprey collected between 1984 and 1990 to model growth, Bergstedt and Swink (1995) found instantaneous growth rates matched those calculated by Farmer et al. (1977) for lab-reared sea lamprey for June to September. However, a growth spurt in wild sea lamprey was observed in October that was not described in earlier laboratory studies (Fig. 7B). Bergstedt and Swink (1995) speculated that rapid October growth may have been attributed to higher water temperatures occupied by host fishes as warm surface waters mixed with and thickened the epilimnion (later confirmed by Bergstedt et al., 2003a), possible increased appetite to support gonad development, and relative ease of finding hosts when mature lake trout and whitefish aggregate to spawn (Madenjian et al., 2003).

### *5.1.5 Bioenergetics integrates physiology, behavior, and fish community data to yield damage estimates and control targets*

A classic application of conservation physiology has been the use of bioenergetics and economic injury models that incorporate fish community data and lake water temperatures with knowledge of sea lamprey and host fish physiology. Doing so has provided the most descriptive estimates of sea lamprey-induced damage to fishes (Madenjian et al., 2003, 2008), including when it occurs (Kitchell and Breck, 1980), and objective means for setting sea lamprey control targets (Irwin et al., 2012; Treska et al., 2021).

Kitchell and Breck (1980) were the first to piece together these disparate data sources into a modeling framework that match feeding and growth rates described by Farmer et al. (1975). They found that the previously described relationships of sea lamprey respiration and feeding rates that vary with water temperature and size was sufficient to account for observed sea lamprey growth dynamics in the Great Lakes. They also predicted that lethality of attack would be greatest in October when sea lamprey are near maximum size and temperatures can still be near 10 °C (see Section 5.1.4). Farmer et al. (1975) also noted that the most important factor influencing the final sea lamprey size and kilograms of fishes killed was the thermal history of sea lamprey.

The bioenergetics framework was further extended to an individual-based modeling approach where variation in sea lamprey size could be explicitly accounted for (MacKay, 1992; Madenjian et al., 2003) and was updated with more recent data on sea lamprey energy density (Cochran et al., 2003) and respiration (Hanson et al., 1997). The individual-based model was applied to evaluate cumulative sea lamprey-induced mortality in northern Lake Huron under water temperature regimes experienced by Seneca strain lake trout and Marquette strain lake trout (per data collected by Bergstedt et al., 2003a). Modeling results demonstrated that these two strains did not experience differences in total number of lake trout killed or blood consumed, but in both strains, mortality peaked in October when maximum consumption of blood by sea lamprey peaked at around 12 g per day (Madenjian et al., 2003; Fig. 7C).

Bioenergetics has been further used to refine estimates of the probability of lake trout surviving sea lamprey attack and sex-specific differences in blood consumption. Madenjian et al. (2008) used catch curve analyses, mark-recapture techniques, and observed wounding rates from Lake Champlain to estimate that adult lake trout have a 74% probability of surviving a sea lamprey attack. They then used the individual-based bioenergetics model to extend estimates of the probability of lake trout surviving in Lake Champlain to estimate probability of survival in Lake Huron, where sea lamprey are larger. When the bioenergetics model was applied to Lake Huron, the probability of lake trout surviving a sea lamprey attack was 66%, higher than the 55% probability of survival used in fishery management models (Bence et al., 2003). Over the course of the parasitic stage, Madenjian et al. (2008) predicted that ~0.5 lake trout (250 lake trout per 500 sea lamprey) would die directly from a parasitic lamprey. The possible discrepancy between model estimates and laboratory studies (Swink, 2003) was that lake trout in the lab may also have died from stress and post-detachment infections (up to 40% of the mortality reported in the lab). Given these results, Madenjian et al. (2008) concluded that successful rehabilitation of lake trout may not be as simple as further reducing sea lamprey populations and may also require habitat improvements, reductions in fishing mortality, and further mitigation of alewife impacts, another invasive species in the Great lakes that can feed on lake trout fry and, if alewife are in turn consumed by lake trout, can cause thiamine deficiency in the latter (Brant, 2019). Later, Madenjian et al. (2014) found that mercury concentrations in male sea lamprey were 16% higher than in females and, by applying bioenergetics, they concluded that males would need to consume ~10% more blood than females to achieve the same size at spawning because of their greater activity and standard metabolic rate. Therefore, sea lamprey sex and sex ratios may be important to consider when conducting laboratory studies and developing models of sea lamprey-induced mortality.

Despite much progress in describing sea lamprey growth, feeding rate, and induced mortality on hosts, an exceptional amount of uncertainty remains concerning how sea lamprey-induced damage varies through time and space, and as fish communities vary. For example, estimates of the probability of lake trout surviving an attack ranged from 14% to 74% for large hosts (Treska et al., 2021). Therefore, the fishery management agencies have adopted sea lamprey management targets that are direct measures (abundance of sea lamprey, lake trout wounding rates) rather than temporally and spatially explicit model-based economic injury levels (Treska et al., 2021). Conservation physiology still has a major role to play in addressing the many uncertainties, as discussed by Bence et al. (2003) and Stewart et al. (2003). Specifically, despite additional insight into hosts alternative to lake trout (Adams and Jones, 2021; Adams et al., 2021a,b; Lantry et al., 2015; Simpkins et al., 2021), it is still unclear how a diverse fish community (and hence diverse sea lamprey hosts) lessen sea lamprey-induced mortality to lake trout populations. Very little is known about the survival rates of parasitic sea lamprey across the Great Lakes (Madenjian et al., 2008; Swink and Johnson, 2014), especially from the time they exit the stream to the end of summer, the period during which host mortality may be minimal. Additional field and laboratory studies that better describe mechanistic relationships between host blood loss, host preference, host mortality, and wounding on surviving hosts would further help reduce uncertainty. A better description of sea lamprey thermal ecology, potentially collected with archival tags, would also be helpful for fitting more realistic bioenergetics models.

### 5.1.6 *Biomarkers to determine the sublethal damage experienced by hosts and diet of individual sea lamprey*

Biomarkers are commonly used in conservation physiology applications to quantity sublethal stressors to fish (Cooke et al., 2013). Because as many as ~60% of lake trout may survive sea lamprey attack, the sublethal acute stress and energy required to recover likely has important population level impacts (Smith et al., 2016). Morphological deformities are one type of biomarker (Sun et al., 2009), and the wounds caused by sea lamprey that have detached from their hosts has been the primary biomarker used in the sea lamprey control program to assess sea lamprey abundance and damage to lake trout stocks

(Treska et al., 2021; Fig. 7A). Sea lamprey-induced wounding on Great Lakes fishes has been extensively reviewed elsewhere including how to describe wounds (King, 1980), how wounding can vary with fish communities (Adams and Jones, 2021; Lantry et al., 2015; Simpkins et al., 2021), and how wounding rates on lake trout are used as sea lamprey control targets (Treska et al., 2021). Because consistency in how field staff assign wound ratings remains low and wound healing can be highly variable (Ebener et al., 2003; Firkus et al. 2021), physiological biomarkers would be a valuable complement to current wounding assessments.

Biomarkers describing sublethal effects on hosts fishes include hematocrit, sex steroids, protein function and composition, and growth rates. A reduction in red blood cell count in host fishes was first described in the 1950s (Lennon, 1954; Parker and Lennon, 1956) and, more recently, reductions in hematocrit, plasma protein, and hemoglobin due to sea lamprey parasitism were found to be strong predictors of mortality in lake sturgeon (Sepúlveda et al., 2012). For example, a 1% decline in hematocrit corresponded to a 5% increase in probability of death. Unexpectedly, lake sturgeon that survived sea lamprey attack generally did not restore hemoglobin, hematocrit, and plasma protein to control levels within 2 weeks, highlighting that sea lamprey can cause substantial sublethal physiological complications even in lake sturgeon up to 760mm in length.

Only a few studies have examined the sublethal effects and identified biomarkers of sea lamprey attack in fishes. Comparing the sublethal effects of sea lamprey parasitism on lake trout morphotypes (lean vs siscowets), Smith et al. (2016) found that lean lake trout showed the most direct responses to parasitism; steroid-binding function was impaired, and more resources were invested into growth than into reproduction, resulting in larger fish. These results are consistent with the physiology of these strains; lean lake trout experience higher growth rates and younger age-at-maturity than siscowets. Both morphotypes experienced endocrine disruptions with reductions in plasma testosterone concentrations and follicle stimulating hormone, potential indicators of longer-term reproductive impairment. The potential reduction in gamete quantity and quality of those lake trout surviving sea lamprey attack may influence stock-recruitment and could be important to consider harvest and production models (Smith et al., 2016).

Using RNA-sequencing to examine changes in the liver transcriptome of these same lean and siscowet lake trout following sea lamprey parasitism, Goetz et al. (2016) found that genes involved with regulation of inflammation, cellular damage, energy utilization, and osmoregulatory imbalances showed strong transcriptional responses with many genes involved with catalytic processes. Many of the transcriptional responses were shared between the morphotypes, but pathways related to carbohydrate and lipid metabolism were not, and admittedly, these strain differences may have largely been a function of the differences in how these morphotypes store and use carbohydrates and lipids. Interestingly, the kynurenine pathway, which supports immunotolerance in some animals, was downregulated (Goetz et al., 2016). The buccal gland secretions of parasitic lampreys contain immunosuppressants (e.g., L-3-hydroxykynurenine O-sulfate in Arctic lamprey; Odani et al., 2012), and Goetz et al. (2016) suggested that downregulation of the kynurenine pathway might be a mechanism to block immunosuppression.

Recent work on blood clotting may provide the best option yet for characterizing a physiological marker that is feasible to use during fishery assessments (Bullingham et al., 2022). Although the sample size was low ($N=13$) and experiments included only male siscowet lake trout, investigation of the proteome of parasitized lake trout showed that components and regulators of fibrin clot formation were compromised for at least 7 months after severe lamprey attack (Type A wounds penetrating the muscle; King, 1980). With further validation, clotting assays could provide an objective physiological biomarker to complement the wound assessments currently used as morphological markers of parasitism.

Biomarkers in sea lamprey have also been investigated to better understand what hosts sea lamprey have been feeding on. Rather than just inferring what species are parasitized by sea lamprey from wounds or direct observation of attachment, these biomarkers allow for greater insights into the diet of individual sea lamprey. Harvey et al. (2008) collected parasitic sea lamprey from six ecoregions of Lake Superior and compared nitrogen and carbon stable isotopes to those of potential host fishes. Sea lamprey captured from the main basin of the lake and parasites that were larger in size, had high $\delta^{15}N$ and $\delta^{13}C$, implying they were feeding on higher trophic level fishes, presumably lake trout. In contrast, in the semi-enclosed Black Bay on the north shore of Lake Superior, sea lamprey had lower $\delta^{15}N$ and $\delta^{13}C$ and hence were likely feeding on the abundant lower trophic level fishes such as suckers and coregonines. Building on the evaluation of fatty acid composition used to describe aspects of sea lamprey feeding ecology of the southwest coast of Europe (Lança et al., 2011, 2013), fatty acid profiles from adult sea lamprey captured for Lake Michigan tributaries were described and compared to fatty acids of possible hosts fishes (Happel et al., 2017). Although the approach could not be used to draw conclusions on specific host fishes that were parasitized, the overall variability in fatty acid profiles led to the conclusion that sea lamprey feed on a wide variety of host species. This is consistent with observations that sea lamprey (as a species or population) parasitize a wide range of species (e.g., Quintella et al., 2021; Renaud and Cochran, 2019), and confirms that most individuals feed on multiple species. To directly assess what hosts individual sea lamprey have fed on to the species level, DNA-based assessment of fecal material for host DNA using metabarcoding was validated in a proof-of-concept study (Johnson et al., 2021a). Interestingly, host fish DNA was able to be extracted from non-feeding adults captured in streams, and results from adult fecal samples suggested *Catostomus* spp. may be an important sea lamprey host in northern Lake Huron. Because adult sea lamprey do not return to natal streams (Bergstedt and Seelye, 1995) and populations can mix broadly throughout a Great Lake (Docker et al., 2021; Swink and Johnson, 2014), standardized sampling of sea lamprey diet may be less biased during the adult life stage once outstanding uncertainties about DNA degradation and evacuation times can be characterized.

### 5.1.7 Challenges and opportunities in the Anthropocene

In the Anthropocene, the warming of the Great Lakes is potentially the most significant challenge to minimizing damage caused by sea lamprey and may further confound efforts to define damage caused by sea lamprey (Fig. 8A). Because much of the damage caused by sea lamprey to fish stocks is based

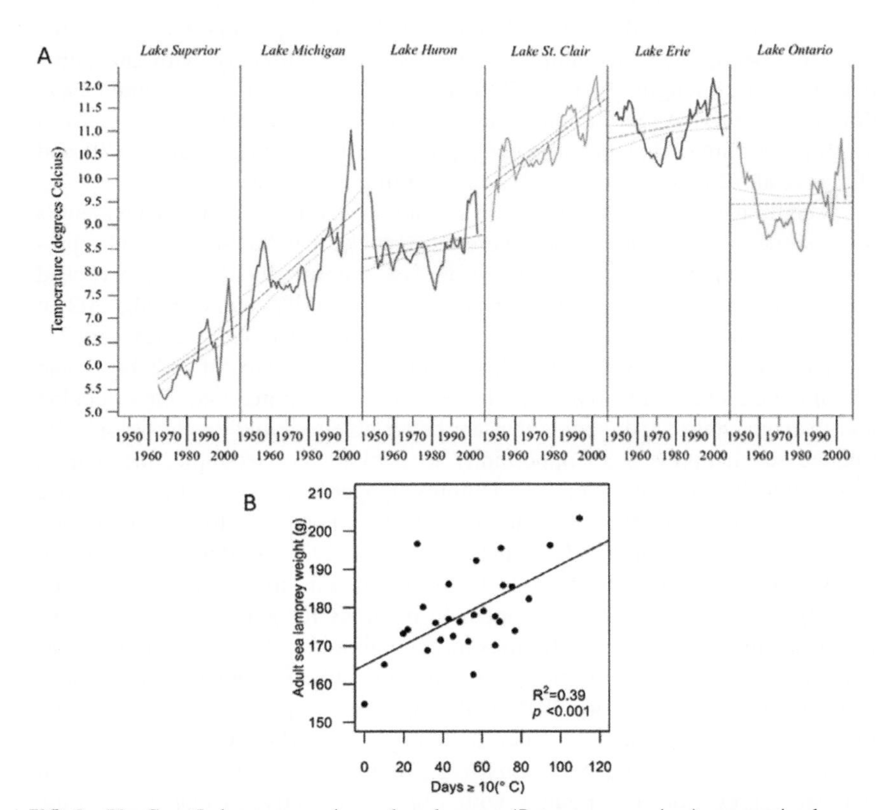


**FIG. 8** The Great Lakes are warming and sea lamprey (*Petromyzon marinus*) are growing larger. (A) Average surface water temperature of the Great Lakes ~1950–2005 with confidence intervals (lines) (B) Correlation between size (g) of adult sea lamprey in Lake Superior and the number of days that surface water temperatures in Lake Superior were above 10°C. *Panel (A) taken from Hume, J.B., Bravener, G.A., Flinn, S., Johnson, N.S., 2021a. What can commercial fishery data in the Great Lakes reveal about juvenile sea lamprey (*Petromyzon marinus*) ecology and management? J. Great Lakes Res. 47, S590–S603 (Creative Commons License); Panel (B) from Cline, T.J., Kitchell, J.F., Bennington, V., McKinley, G.A., Moody, E.K., Weidel, B. C., 2014. Climate impacts on landlocked sea lamprey: implications for host-parasite interactions and invasive species management. Ecosphere 5, 1–13 (Creative Commons License).*

on the temperature of the water they occupy, a warmer Great Lakes may correspond to larger sea lamprey that kill more kilograms of fish (Hume et al., 2021a). Indeed, adult sea lamprey size varies among the Great Lakes and is largest in the warmest lake, Lake Erie (Docker et al., 2021). Lake Superior has been described as the fastest warming lake on Earth, and an analysis of water temperatures and adult sea lamprey size found that adult sea lamprey size increased 12% between 1980 and 2008, which the authors predicted resulted in an estimated 5% increase in blood consumption during that period (Cline et al., 2014; Fig. 8B). Continued warming will likely result in larger sea lamprey that cause more damage to fish stocks. Larger adults may also be more capable of surmounting low-head barriers (Hume et al., 2021b), and larger females will be more fecund (Docker et al., 2019). Given that the total number of eggs in lampreys increases approximately with the cubic power of body length, even a 5% increase in female body size could have dramatic effects on recruitment. The challenges sea lamprey control faces in a changing climate has been recently presented by Lennox et al. (2020) and Hume et al. (2021b); interested readers are directed there.

A significant opportunity in future sea lamprey research and management is to continue building robust and diverse fish stocks that can survive and reproduce despite modest levels of attack from sea lamprey. Field studies reviewed above (Adams and Jones, 2021; Harvey et al., 2008; Hume et al., 2021a; Johnson et al., 2021a; Lantry et al., 2015) suggest that sea lamprey target a diversity of hosts if available and, therefore, the diversity of fish hosts can lessen the direct impacts to any single species. Furthermore, continued stocking and management of fish strains that are better able to avoid or physiologically tolerate sea lamprey is an opportunity to explore. For example, the Seneca strain, which coexisted with sea lamprey longer than Great Lakes strains, may be better able to avoid sea lamprey attack and continue to be more prevalent in Lakes Michigan and Huron (Larson et al., 2021; Scribner et al., 2018).

Although the parasitic juvenile life stage of sea lamprey has never been directly targeted for control, advances in methods for genetically manipulating fishes may provide opportunities in upcoming decades. For example, creating stains of lake trout or other fishes that are lethal to sea lamprey has been considered (Thresher et al., 2019a). Perhaps sea lamprey can be genetically modified to disrupt appetite, alter food digestion (Mathai et al., 2021), or modify the buccal gland secretions (e.g., immunosuppressants, inhibitors of platelet aggregation, compounds with fibrinolytic activity) that aid feeding (Odani et al., 2012; Wang et al. 2010; Xiao et al. 2007). Since sea lamprey-induced mortality appears to be greatest in the fall (see Section 5.1.4) and that is when lake trout and whitefish are in spawning aggregations, perhaps sensory traps could be deployed near critical spawning areas to divert or capture juvenile sea lamprey (see Section 5.2.1). See Evans et al. (2021) for a discussion of potential ideas for targeting parasitic sea lamprey in the Great Lakes proper.

## 5.2 Reducing larval recruitment by removing and redirecting adult sea lamprey and disrupting reproduction

The adult life stage of sea lamprey was first targeted for control in the 1940s and 1950s (Applegate, 1950). Barriers to limit access to spawning habitat (Hrodey et al., 2021) and traps to census adult populations (Adams et al., 2021a,b; see Section 4.3) have remained largely unchanged since then. However, most sea lamprey barriers also block native fishes (Zielinski et al., 2019), and traps do not capture enough adult sea lamprey to consistently reduce recruitment of larvae. Therefore, traps, although essential for assessment, are not considered a control tool (Miehls et al., 2021). Recognizing this lost opportunity, nearly half of the research funds spent on sea lamprey control by the GLFC in recent decades have been directed toward the goal of developing more effective and selective means to target the adult life stage (Siefkes et al., 2021). However, much progress remains to be made, and conservation physiology can contribute.

The management goals of targeting the adult life stage are to reduce recruitment of larvae throughout the entire watershed or to concentrate recruitment into areas of watersheds that are smaller, have low quality habitat, or are easier to treat with lampricide. Doing so may function as an alternative to lampricide treatment (if treatment is no longer needed) or as a supplement to lampricide treatment (if treatments occur less often and fewer larvae survive treatment; Lamsa et al., 1980; Siefkes et al., 2021). These goals can be accomplished by removing eggs from the watershed (removing females in traps), reducing the probability of eggs being fertilized and hatching (mating disruption and sterilization), or concentrating where sea lamprey deposit their eggs (redirection into or away from specific tributaries with physical or non-physical cues; Noatch and Suski, 2012).

Indeed, the sensory systems, swimming performance, and reproductive endocrinology of adult sea lamprey have and continue to be researched to support the development of next generation control tools. The sensory modalities garnering the most interest since the 1980s include electroreception, olfaction, vision, and audition (e.g., see reviews by Borowiec et al., 2021; Johnson et al., 2015; Siefkes, 2017). The ability of sea lamprey to swim, climb, and attach to surfaces with their oral disc also continues to be explored (e.g., Moser et al., 2015). The reproductive endocrinology and behavior of sea lamprey has also been studied extensively to inform approaches and population level consequences of sterilizing and releasing adult male sea lamprey (e.g., Docker et al., 2019; Sower, 2018). Taken together, because the reproductive success of sea lamprey can be high at low adult densities (Dawson and Jones, 2009) and no single behaviorally- or physiologically-based control tool is likely to be effective on 100% of the adult population (Zielinski et al., 2021), the integration of tactics exploiting multiple aspects of sea lamprey sensory physiology, morphology, and endocrinology is likely to be most successful for reducing recruitment consistently within and across streams (Siefkes et al., 2021; Zielinski et al., 2021). In the subsections following, we discuss how sensory systems, swimming performance, and endocrinology of adult sea lamprey could potentially be exploited in an integrated control program.

## 5.2.1 Sensory physiology of adult sea lamprey

The sensory physiology of adult sea lamprey was reviewed in the context of control in several recent publications (Borowiec et al., 2021; Elmer et al., 2021; Miehls et al., 2020; Siefkes, 2017). The focus here will be to briefly highlight the state of the science, possible applications in sea lamprey control, potential limitations, and outstanding research questions to move the ideas toward selective and effective sea lamprey control. Sensory modalities are presented in roughly the chronological order they were first considered for sea lamprey control.

### 5.2.1.1 Electroreception and galvanotaxis

Although electricity was developed and used as a non-physical sea lamprey barrier in the 1950s (McLain et al., 1965), the physiological responses of the animals were largely unknown at the time the first electrical barriers were developed. In the 1980s, the electroreceptive abilities of sea lamprey were confirmed and found to be like other fishes (0.1–10 μV/cm; Bodznick and Northcutt, 1981; Chung-Davidson et al., 2004). Weak electrical fields were also found to elicit reproductive endocrine responses in adult males (Chung-Davidson et al., 2008), but this potential role of electroreception in maturation and reproduction remains mostly uninvestigated to date.

Despite advancements in understanding the physiological mechanisms of electroreception in the 1980s and 2000s, this knowledge has not yet been translated toward the design of more effective and selective systems that use electricity to block and guide adult sea lamprey by galvanotaxis (Zielinski et al., 2019). The most significant advancement occurred in the 1950s when pulsed direct current (DC) was used instead of alternating current (AC), which reduced non-target mortality of upstream migrating fishes to less than 2% and guided up to 90% of sea lamprey to a trap (McLain, 1957). Variations on pulsed DC electrical barriers were tested again starting in the 1990s with the electrical field generated using electrodes mounted to the bottom (Swink, 1999; Tews et al., 2021) or electrodes suspended vertically in the water column (Johnson et al., 2014, 2016, 2021b). Success of newer deployments have been mixed. The systems described by Swink (1999) and Tews et al. (2021) that used electrodes mounted on the bottom have since been decommissioned because larvae were found upstream of the devices and due to a lack of public support (Tews et al., 2021). Use of vertical electrode systems in small rivers has offered more portability, higher percentages

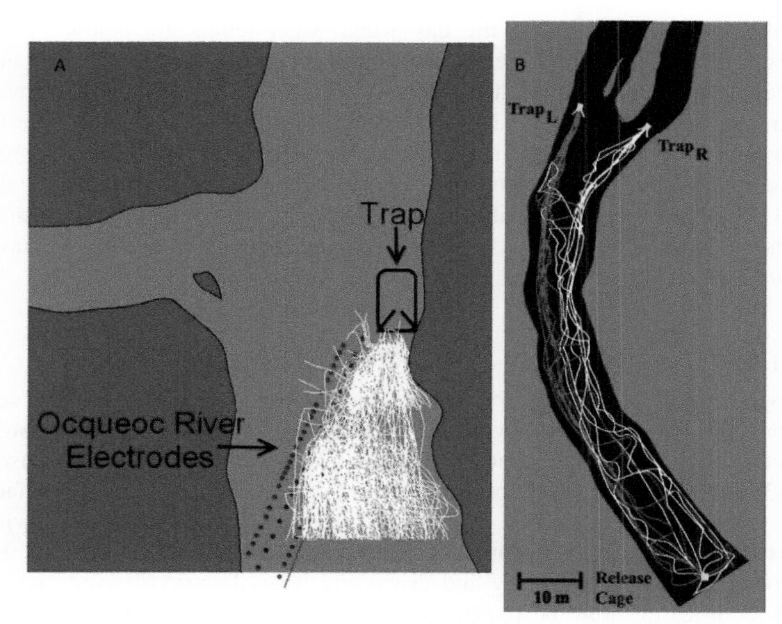


**FIG. 9** Examples of behavioral guidance of adult sea lamprey (*Petromyzon marinus*) moving upstream in the Ocqueoc River, Michigan. (A) Sexually immature females moving upstream (white lines) are guided toward a trap by vertical electrodes (red and blue dots), while a few escape through the field. Trap entrance is 1.5m wide. (B) Sexually mature females moving upstream toward traps baited with a synthesized sex pheromone component. Red movement tracks are when the left trap ($Trap_L$) was baited with pheromone and the other trap was not baited. White movement tracks are when the right trap ($Trap_R$) was baited with pheromone. *Panel (A) from Johnson, N.S., Thompson, H.T., Holbrook, C., Tix, J.A., 2014. Blocking and guiding adult sea lamprey with pulsed direct current from vertical electrodes. Fish. Res., 150, 38–48.; U.S.A. public domain. Panel (B) from Johnson, N.S., Yun, S.S., Thompson, H.T., Brant, C.O., Li, W., 2009. A synthesized pheromone induces upstream movement in female sea lamprey and summons them into traps. Proc. Natl. Acad. Sci. U. S. A. 106, 1021-1026.; U.S.A. public domain.*

of adult sea lamprey being guided into traps (Johnson et al., 2014, 2016; Fig. 9A) and reduced recruitment of larvae when used as a seasonal barrier to migration (Johnson et al., 2021b).

The most significant disadvantage of using electric fields to guide and block sea lamprey remains its non-selectivity and impacts to non-target species. The intensity of electric fields deployed for sea lamprey control generally varies from 0.4 to 3.0 V/cm with the duty cycle of pulsed DC around 10% (percent of time electricity is pulsed; Johnson et al., 2021b; Tews et al., 2021), which is many magnitudes of order above the detection threshold of adult sea lamprey (Bodznick and Northcutt, 1981). At these intensities, the electric field immobilizes fishes (electronarcosis; Geddes, 1965; Johnson et al., 2014), and fish die if they are immobilized longer than several minutes. These impacts have been reduced by turning the system off when sea lamprey are not present (i.e., during the day when water temps are below 12 °C: Johnson et al., 2016; after the migratory season: Johnson et al., 2021b) and by designing electric fields that gradually increase in voltage gradient over several meters (Tews et al., 2021). Thus far, research has not been able to uncover lower voltage gradients that consistently block or guide sea lamprey without influencing the behavior of non-target fishes (see Johnson et al., 2014), and this remains an outstanding opportunity for conservation physiologists to improve upon the vertical electrode systems that recently have shown promise (Johnson et al., 2021b).

### 5.2.1.2 Olfaction

Recognizing its potential for control, biologists and scientists with the sea lamprey control program have shown exceptional interest in sea lamprey olfaction and olfactory-induced behaviors. Kleerekoper and Mogensen (1963) were the first to describe in detail the potency of the sea lamprey olfactory system to detect odors of host fishes. Teeter (1980) followed with focused laboratory experiments describing olfactory cues that influence the behavior of adult sea lamprey and discussing how olfactory-derived behavioral modification may be useful for reducing sea lamprey recruitment. At the same time, given the observed large declines in adult spawning runs the year immediately after lampricide treatment eliminated most larval sea lamprey from a stream, field biologists were hypothesizing that adult sea lamprey use chemical cues from larval lampreys to evaluate the suitability of tributaries for spawning (Moore and Schleen, 1980).

An exceptionally broad and novel body of work has been reported since the 1980s characterizing the potency of the adult olfactory system to detect chemosensory cues (Aurangzeb et al., 2021) and how chemosensory cues can be prominent motivators of their migratory and mating behavior (Fissette et al., 2021). The influence of chemosensory cues on sea lamprey behavior and potential use in control have been reviewed frequently and recently (Aurangzeb et al., 2021; Buchinger et al., 2015; Fissette et al., 2021; Li et al., 2007; Sorensen and Hoye, 2007; Sorensen and Johnson 2016; Twohey et al., 2003a). Therefore, the subsequent text serves instead as a primer and to highlight topics receiving less focus in these previous reviews and where opportunities for conservation physiologists still exist.

Instead of homing to natal streams to spawn (Bergstedt and Seelye, 1995), sea lamprey enter streams that contain larval lampreys (Moore and Schleen, 1980; Fine et al., 2004) as determined by smelling compounds emitted by larvae (Teeter, 1980) at concentrations as low as $10^{-13}$ M (Twohey et al., 2003a). If these migratory pheromones could be identified and replicated, they could be exceptionally useful for redirecting adult sea lamprey into tributaries containing efficient traps or where spawning and larval habitat yields poor recruitment. To date, roughly 13 compounds isolated from water containing larval lamprey have been identified as putative migratory pheromones, but researchers have only had the means to test a few of the compounds identified; none alone or in combination have elicited the same robust preference response migratory sea lamprey exhibit toward extracts of natural larval lamprey odors (Fissette et al., 2021; Li et al., 2018b; Meckley et al., 2012, 2014; Sorensen et al., 2005). Additional obstacles for widescale use in sea lamprey control are that: (1) most compounds identified are expensive to synthesize because they are novel or have stereoisomers; (2) outcompeting natural sources of migratory pheromone is challenging because sea lamprey only swim toward higher concentrations of migratory pheromone when differences in concentration are 10-fold or more (Wagner et al., 2009); and (3) larval lampreys native to the Great Lakes release similar migratory pheromones (Fine et al., 2004), so pheromone sources from native lampreys are common and are not eliminated by extirpating sea lamprey from the watershed. Key research opportunities for conservation physiologists therefore include describing the entire pheromone mixture, developing less expensive ways to synthesize compounds that have been identified, and envisioning places and times migratory pheromones from native lampreys can be leveraged to manipulate sea lamprey behavior for control purposes.

Before becoming sexually mature and when migrating upstream, sea lamprey actively avoid odors of other dead lampreys and predators. These predation-related semiochemicals, if identified and synthesized, could be used to push sea lamprey out of stream sections that are challenging to trap or treat with lampricides (Imre et al., 2010; Wagner et al., 2011). Field trials conducted with extracted odors from dead sea lamprey found that sea lamprey avoid areas of a natural stream activated with alarm cue (Bals and Wagner, 2012; Di Rocco et al., 2016), and that application of alarm cues can hasten arrival at traps (Hume et al., 2015) and upstream movement (Luhring et al., 2016). Like the efforts to identify the migratory pheromone, chemically identifying alarm cues is not complete; the cues are likely not lipids and may be water-soluble nitrogenous compounds (Dissanayake et al., 2016, 2019). Other obstacles for current widescale use in sea lamprey control are that: (1) in some contexts, sea lamprey may habituate and stop responding to the cue after 4 h (Hanson, 2021; Imre et al., 2016); (2) the presence of migratory pheromone may reduce avoidance responses to alarm cues (Byford, 2016); and (3) sea lamprey alarm cues may not be completely selective for sea lamprey, as other lampreys in the Great Lakes likely release and response to the same cues (Byford et al., 2016; Hume and Wagner, 2018). Conservation physiologists continue to explore the potential use of alarm cues with areas of interest focusing on identifying active compounds, testing approaches to minimize habituation (pulsed; varying concentration), uses in fishways, and synergizing applications with migratory pheromone (Fissette et al., 2021).

After sexually maturing and arriving at riffle spawning habitats, sea lamprey are less likely to behaviorally respond to migratory pheromones and alarm cues, but they do begin responding to sex pheromones released by male sea lamprey via gills (Li et al., 2002) and their semen (Scott et al., 2019). Ovulated females are attracted over several kilometers to these pheromones (Johnson et al., 2006) and at concentrations low as $10^{-14}$ M (Johnson et al., 2009). In a natural stream, when no other sources of pheromone were present, a trap baited with sex pheromones released by spermiated males captured 74% of ovulated females (Johnson et al., 2005). Given early successes in simple contexts (Johnson et al., 2009; Fig. 9B), the concept of pheromone-baiting traps was embraced as having the potential to be a highly selective and effective option for removing eggs from the spawning population (Li et al., 2007; Twohey et al., 2003a). Like migratory pheromones and predation-related alarm cues, sex pheromones consist of multiple components, all of which have yet to be identified (~12 putative pheromones identified to date). However, a major component of the pheromone that induces upstream movement to spawning riffles has been identified (Li et al., 2002; 3-keto petromyzonol sulfate), and the synthesized pheromone component increases trapping efficiencies of females and disrupts female orientation to natural pheromones plumes (Johnson et al., 2009). When this component was applied to traps used for sea lamprey assessment in the Great Lakes (Adams et al., 2021a,b), trap efficiencies increased modestly between 0% and 30% (Johnson et al., 2013, 2020). This pheromone component was registered in the U.S.A. and Canada and was the first vertebrate pheromone registered for use in public waters in these countries (Fredricks et al., 2021). However, as of 2022, the registered pheromone has not been actively used in sea lamprey control. The primary obstacle to the widespread use for trapping of the registered sex pheromone component is that modestly increasing the removal rate of female sea lamprey may not be sufficient to consistently overcome compensatory effects and reduce recruitment (Dawson et al., 2016; Jones et al., 2003). If used broadly, the pheromone could also have some negative consequences for chestnut lamprey and other native lampreys because they use the same sex pheromone component (Buchinger et al., 2021); however, these negative consequences may be less than those caused by lampricides and barriers. Sex pheromone antagonists (Buchinger et al., 2020) that block female reception of male sex pheromones are a promising area of research and may be useful for disrupting mate finding and fertilization success.

Semiochemicals that are attractive (migratory pheromone) and repulsive (alarm cues; pheromone antagonists) could be used in combination to pull (attract) and push (repel) sea lamprey toward traps or tributaries (Hume et al., 2020a; Miller and Cowles, 1990). An ideal combination of bipolar semiochemicals for push-pull applications would be the use of migratory pheromones to pull sexually immature sea lamprey to traps and alarm cues to push sexually immature sea lamprey away from areas of the stream without traps (Byford, 2016; Hume et al., 2020a). However, given challenges synthesizing the migratory pheromone, recent tests have targeted migratory sea lamprey with synthesized sex pheromones as a pull and alarm cue as the push (Hume et al., 2020a). In these tests, the alarm cue was a strong push for migratory adults, but the sex pheromone was not a strong pull (Hume et al., 2020a). Another combination of bipolar semiochemicals may be to target ovulated females using sea pheromone antagonists as the push and synthesized sex pheromones as the pull; this combination awaits testing.

Because many of the sea lamprey migratory and sex pheromones are not released by teleost fishes, quantification of their concentration in stream water may be a promising avenue for assessing sea lamprey abundance or biomass (Fissette et al., 2021). Accordingly, methods have been established to extract and quantify sea lamprey pheromones in water with high specificity and at biologically relevant concentrations ($10^{-12}$ to $10^{-13}$M; Stewart et al., 2011; Wang et al., 2013; Xi et al., 2011). Before pheromone quantification could be broadly useful for assessing sea lamprey populations, uncertainties about how release and degradation of pheromones varies across seasons, the larval and adult life stages, and different lampreys need to be resolved (Fissette et al., 2021). Simultaneous detection of pheromones and environmental DNA (eDNA) have the potential to provide complementary alternatives or supplements to traditional sampling methods; eDNA could identify the lamprey species present (i.e., distinguishing sea lamprey from native lampreys; Gingera et al., 2016; Schloesser et al., 2018), while the concentration of particular pheromones in the water sample could provide insights into the life history stage(s) present (see Docker and Hume, 2019).

#### 5.2.1.3 Photosensory and vision

Applegate (1950) reported that the eyes of adult sea lamprey degrade and become opaque as they mature, and there is little evidence to suggest that they use vision to avoid predators. Perhaps accordingly, photosensory systems and vision in adult sea lamprey has received less attention by researchers and the sea lamprey control program. An early field experiment evaluating white light (9–100 lx) as a non-physical attractant or repellent of adult sea lamprey found that catch rates in lit traps were about 5× higher than an adjacent unlit trap (Purvis et al., 1985). Subsequent work by Binder and McDonald (2008) found that nocturnal migratory behavior of adults is mediated by dermal photoreceptors, not the eyes (Binder and McDonald, 2007), and that the sensitivity of dermal photoreceptors decreases as lampreys become sexually mature, explaining the observation that sea lamprey will spawn during the day (Applegate, 1950). A lab and field study by Stamplecoskie et al. (2012) confirmed that when lit and unlit traps are adjacent to each other, lit traps catch more sea lamprey; however, if traps are separated by $\geq$10 m, no increase in trap catch was observed relative to the unlit trap. These results are consistent with the physiology of the adult sea lamprey photosensory system in that dermal photoreceptors are unlikely to be able to discriminate between light sources several meters away. Stamplecoskie et al. (2012) also documented that non-target fishes were more likely to be captured in lit traps compared to those with no light, so the approach would not be selective for sea lamprey. Future research could evaluate if sea lamprey and non-target fishes have varied behavioral responses to different portions of the light spectrum (see Borowiec et al., 2021 for a longer review).

#### 5.2.1.4 Auditory

Auditory systems in sea lamprey have also received relatively little attention by researchers and the sea lamprey control program until recently. A physiological analysis of hearing capability reports that adult sea lamprey likely hear only low frequency sounds ranging between 50 and 200 Hz (Mickle et al., 2019). Behavioral studies in the lab (Mickle et al., 2019) and field (Heath et al., 2021) report that low frequency sound (70–90 Hz) increased movement (lab) and induced avoidance upstream migrating adult sea lamprey (in-stream). Since most fishes also detect low frequency sounds (Popper and Fay, 1999), it remains unclear if sound can selectively repel sea lamprey. How sea lamprey respond to auditory cues changes with increases in ambient noise (increasing discharge) are yet to be described.

### *5.2.2 Morphology and swimming physiology of upstream migrating sea lamprey*

#### 5.2.2.1 Anguilliform morphology

In the Great Lakes, the anguilliform body plan is unique to lampreys and American eel (*Anguilla rostrata*) and could serve as a selective trait to target for control. The anguilliform body form is among the least efficient swimming forms in fishes (Lighthill, 1969), and it establishes physical limits to propulsion and jumping (reviewed in Hoover and Murphy, 2018; Zielinski et al., 2020). Therefore, the manipulation of hydrologic conditions to create combinations of velocity and turbulence that selectively overcome the physiological and morphological limits of sea lamprey swimming could be a selective non-physical means to guide sea lamprey to traps or block access to tributaries.

Beamish (1974), working with newly metamorphosed sea lamprey, was first to report that swimming performance of sea lamprey was lower than most teleosts and that, like other fishes, sea lamprey could swim further if larger (100 g vs 2 g) and in warmer water (5 vs 15 °C). Swimming evaluations of adult sea lamprey were conducted by Bergstedt et al. (1981) and McAuley (1996), and both data sets were analyzed by Hoover and Murphy (2018) with the conclusion that conservative estimates of maximum swim speeds were 4.0–4.5 m/s. Given the constraints that the experimental apparatus imposed on locomotion in past studies (Bergstedt et al., 1981; Hanson, 1980; McAuley, 1996), swim tests in larger flumes could be useful. The design of velocity barriers that are selective for sea lamprey is complicated by needing to know the swimming ability of native fishes and how swimming varies with fish length and water temperature (McAuley, 1996). Zielinski et al. (2020) suggest that sea lamprey may fatigue at slower water velocities than several migratory fishes in the Great Lakes. Adult sea lamprey may also be unable to navigate in high turbulence environments with similar efficiency as fishes with paired fins. Indeed, in a fishway located on a Lake Superior tributary, white sucker were found to navigate turbulent areas of the fishway better than sea lamprey (Lewandoski et al., 2021b).

The morphological distinction of the anguilliform body form has also been used to distinguish sea lamprey from other fishes in hydroacoustic and video analyses. Automated, real-time sorting based on morphology could therefore be useful in next generation selective traps and barriers. Trained observers of high-resolution underwater acoustic videos were able to consistently distinguish sea lamprey from other fishes (McCann et al., 2018a; Swanson et al., 2021), and open access data sets may be useful for training machine learning tools for high-resolution hydroacoustic data (McCann et al., 2018c). When images of dewatered fishes (13 species; dewatered for a few seconds) captured from a high-speed camera array were evaluated with deep convolutional neural networks, nearly 100% of the images were correctly classified as lamprey vs non-lamprey (Eickholt et al., 2020). Additional work will be needed to distinguish small sea lamprey from large native lampreys, particularly silver lamprey, the largest of the native lampreys and a migratory species whose distribution overlaps most with that of sea lamprey (Schuldt and Goold, 1980). Silver lamprey typically average 224–248 mm at maturity, although they can reach up to ~400 mm (COSEWIC, 2020); sea lamprey in the Great Lakes average 450–500 mm, but can range from ~300 to 650 mm (see Docker et al., 2021). Sorting based on image recognition could prove to be a highly effective and efficient (in terms of labor) means for selectively removing sea lamprey from valued fishes at dams and fishways if a device can be designed to capture and introduce fishes into the dewatered camera scanning system tested by Eickholt et al. (2020).

#### 5.2.2.2 Oral disc

The unique lamprey suctorial oral disc is used for feeding, nest construction, and locomotion (Section 2.2) and could be exploited for selective trapping and surveillance. In terms of developing a velocity barrier, sea lamprey have a considerable advantage relative to other fishes to overcome deficiencies in swim performance; they attach to surfaces with their oral disc (Shi et al., 2021), allowing them to remain stationary (Quintella et al., 2004) and recover (Wilkie et al., 1998) when in high velocities (Quintella et al., 2004). Therefore, any hydraulic-based control tool for adult sea lamprey also needs to incorporate non-attachment surfaces that disrupt the oral disc from achieving suction (described in McAuley, 1996). Interestingly, the suction dynamics of the oral disc have being described in detail (Shi et al., 2021), and sensors that detect sea lamprey attachment (González-Afanador et al., 2021) could be used against sea lamprey, for example, to deliver an electric shock, trigger the deployment of a capture device, or as an assessment tool to quantify how many sea lamprey are present. Traps have also been designed to exploit the ability of adult sea lamprey to attach and climb inclined surfaces with studs (eel ladders; Hume et al., 2020b; Reinhardt and Hrodey, 2019). Hume et al. (2020b) reported a 98% capture rate of sea lamprey that approached the base of a studded ramp in an artificial channel that was placed in a river, suggesting that such traps represent an opportunity for nearly 100% selective removal of sea lamprey if they can be directed to these inclined surfaces.

### 5.2.3 Sterilization of adult male sea lamprey

Sterilization and release of adult male sea lamprey has been developed and tested to reduce sea lamprey recruitment by reducing embryo survival. The topic has been reviewed extensively elsewhere (Bergstedt et al., 2003b; Bergstedt and Twohey, 2007; Bravener and Twohey, 2016; Hanson and Manion, 1980; Siefkes, 2017; Twohey et al., 2003b), so here, only an overview of the method is highlighted with a short discussion of ways conservation physiologists may be able to help develop new and selective sterilizing agents for integrated sea lamprey control.

Conceptually, if a sterilizing agent can be administered to adult sea lamprey that does not otherwise influence their fitness, the release of sterilized males to mate with (and waste the reproductive effort of) females will result in a larger net reduction than simply killing the male after it is captured (Knipling, 1959; Li et al., 2021). Because adult sea lamprey no longer feed during upstream migration, the release of adult males also does not inflict additional damage to the fishery (compared to sterilization of larvae or juveniles). As such, several sterilizing agents have been investigated for sea lamprey (Ciereszko et al., 2004; Hanson, 1990; Hanson and Manion, 1980; Rinchard et al., 2000; Sower, 2003), but to date only P,P-bis(1-aziridinyl)-N-methylphosphinothioic amide (bisazir) was found to result in complete sterilization of males (Hanson and Manion, 1980; Twohey et al., 2003b) with minimal influence on migratory or mating behavior in the laboratory and field (Bergstedt and Twohey, 2007; Bergstedt et al., 2003b; Bravener and Twohey, 2016).

Bisazir achieves sterilization by damaging spermatozoa DNA (Ciereszko et al., 2005). Because bisazir is highly mutagenic to all organisms, it must be administered to sea lamprey in a contained facility by a highly specialized injecting machine (Twohey et al., 2003b). This injection method greatly limits the mobility of the technique around the Great Lakes, increases operational costs, and increases handling time and therefore the likelihood that sterilized males are not maturing at the same time as wild males (Bergstedt and Twohey, 2007). Field tests in tributaries of Lake Superior (Kaye et al., 2003), the St. Marys River (Bravener and Twohey, 2016; Criger et al., 2021), and in the Cheboygan River in the Lake Huron basin (Johnson et al., 2021c) suggest that ratios of sterile males to normal males may need to be ~50:1 to achieve large and consistent reductions in larval recruitment. Such a high sterile-to-normal male ratio was recently found to be effective in insect control programs (Mastrangelo et al., 2018). These high ratios can only be achieved by importing large numbers of sterile males into select tributaries (Johnson et al., 2021c) or by deploying highly efficient trapping that is able to remove 98% of the wild males to be sterilized and released (Siefkes et al., 2021). Either way, large numbers of adult male sea lamprey need to be captured. Therefore, an effective sterilization program also needs an effective trapping program—in the same way that the lampricide program needs an effective barrier program, highlighting the importance of integrated pest management. At present, sterilization is only used in four tributaries of Lake Huron close to the injection facility. In these streams, ratios of 40:1 can be achieved by sourcing males captured from nearby streams (Johnson et al., 2021c).

Sterilization of male sea lamprey using bisazir is not ideal given its human health hazards (Twohey et al., 2003b). Further research into the reproductive physiology of lampreys could help identify and develop sterilizing agents that could be used streamside with minimal adverse effects to other fishes or humans. A large body of work on the reproductive endocrinology of sea lamprey by Sower and Bryan (reviewed in Bryan et al., 2008; Docker et al., 2019; Sower, 2003, 2015, 2018) identified novel reproductive hormones. Three unique gonadotropin-releasing hormones (GnRHs) have been characterized in lampreys (Sower, 2015, 2018), and lampreys appear to use a mix of classical and non-classical reproductive steroids, including steroids that differ from those of other vertebrates in possessing an additional hydroxyl group at the C15 position (see reviews by Borowiec et al., 2021; Bryan et al., 2008; Docker et al., 2019). Exploiting these differences, hormonal antagonists (e.g., analogs of GnRH) could potentially be developed as agents to sterilize male sea lamprey streamside while working to achieve minimal risk to human health.

RNA interference (RNAi) technology has the potential to offer a safe and inexpensive means of sterilizing male or even female sea lamprey. With the capacity for species-specificity, unlike bisazir which is mutagenic to all species, streamside application could be possible, thus considerably increasing the geographic scope of a sterile adult release program compared to bisazir. RNAi, or post-transcriptional gene silencing, is already being used to develop species-specific insecticides for the control of agricultural pests or disease vectors (see reviews by Christiaens et al., 2020; Koch and Kogel, 2014; Xue et al., 2012). In June 2017, the U.S. Environmental Protection Agency registered four RNAi-based products for the control of corn rootworm (*Diabrotica virgifera virgifera*), which has developed resistance to several other pesticides. RNAi is triggered by the introduction of double-stranded RNA (dsRNA) into the cell cytoplasm (Campbell et al., 2015). When the dsRNA is designed specifically to target the pest species' genes (with deliberate mismatches with gene sequences in non-target species), RNAi can result in the directed destruction of the target species' messenger RNA (mRNA), thus blocking the synthesis of specific proteins critical to survival and reproduction (Campbell et al., 2015; Horak, 2020). Unlike genetic control mechanisms (e.g., gene-drive based approaches) that are being considered and/or developed for pest control (see Ferreira-Martins et al., 2021b; Thresher et al. 2019a,b; Section 5.4), RNAi does not genetically modify organisms, and risk evaluation for RNAi-based tools is generally similar to that used for chemical pesticides (Horak, 2020).

Most RNAi research and development to date has focused on invertebrate pest control and use in vertebrate animal models to treat human diseases; progress has been slower in applying RNAi to vertebrate pests (Campbell et al., 2015; Horak, 2020; Murphy et al., 2019). However, in a proof-of-concept study, Heath et al. (2014) showed that injection of short interfering, double-stranded RNAs (siRNAs) into sea lamprey embryos could reduce the mRNA levels of targeted genes by more than 50%; transcript levels were also reduced by oral administration of siRNAs complexed with liposomes fed to larval lamprey. One of the siRNA treatments also caused increased mortality. The genes targeted by Heath et al. (2014) were highly conserved across vertebrates (i.e., 75–99% identity with zebrafish (*Danio rerio*) gene sequences), but this study demonstrated that RNAi can be used in lampreys. Species specificity would require design of siRNAs that selectively target gene sequences that are unique to sea lamprey (Lantz et al., 2022; see Section 5.4).

Efforts to use RNAi as a non-surgical method to induce sterility in domestic cats and dogs and to disrupt fertility in feral horses and dogs are relevant to development of sterilization in sea lamprey (e.g., Dissen et al., 2012, 2017). Delivering dsRNA by injection is far more targeted than release into the environment, and it is not prone to degradation like oral delivery, which requires a carrier to facilitate transfer across the gut (He et al., 2013). Dissen et al. (2012) identified the following necessary conditions for success of sterilization in dogs and cats: (1) the target gene must be essential for fertility; (2) the method must be able to effectively and specifically silence the gene of interest (i.e., so that disruption is non-lethal); (3) the method of delivering the silencing agent must be minimally invasive; and (4) the silencing effect must be sustained for the lifespan of the target species. Using these criteria, they designed siRNAs against the Kiss1 gene, which encodes for one of the kisspeptin peptides that plays a major role in reproduction (Dissen et al., 2012; Ogawa and Parhar, 2018). The Kiss1 gene is expressed in the hypothalamic region of the brain, and animals lacking Kiss1 gene products are infertile (Dissen et al. 2012, 2017). Two kisspeptin genes have been identified in sea lamprey (Felip et al., 2009; Lee et al., 2009), and Sobrido-Cameán et al. (2021) showed that Kiss1 is expressed in the hypothalamus of post-metamorphic juvenile and upstream migrating adult sea lamprey. Sequencing of the sea lamprey genome and other genomic resources are helping to improve our understanding of genes and gene networks that control many aspects of sea lamprey biology (Smith et al., 2013, 2018; Yasmin et al., 2022; see review by York et al. 2019). Working in close collaboration with physiologists to ensure that gene expression patterns are interpreted in the context of the whole organism, it is possible that other potential targets for RNAi-induced sterility in adult sea lamprey can be identified (see Section 5.4).

### 5.2.4 Integrating knowledge of sensory physiology, swimming performance, and sterilization to develop robust systems of integrated adult controls

Despite numerous potential approaches to reduce sea lamprey recruitment through enhanced trapping or selective barriers, none are currently deployed in the sea lamprey control program (Siefkes et al., 2021; only non-selective physical barriers are used). Each tool has tradeoffs related to effectiveness, non-target impacts, and cost; therefore, options exist to tailor these tools to fit the physical, biological, and social context of the deployment location. However, when new control tactics have been deployed, they were often tested alone and not integrated with other complementary techniques to reduce reproduction, with the notable exception of the push-pull approaches in Hume et al. (2020a). While testing individual tools provides the best scientific evaluation of that tool, it does not address the larger management desire to maximize reductions in reproduction (Dawson and Jones, 2009). Realizing this need, the GLFC has funded two long-term initiatives to explicitly test combinations of adult sea lamprey control tools that complement each other to maximize effectiveness and selectivity in open stream channels and fishways.

FishPass is an initiative being funded by the GLFC on the Ottaway River, Michigan, to develop completely automated and selective passage of valued fishes at a dam while blocking invasive fishes like adult sea lamprey (Zielinski et al., 2020). Realizing that dam removal can benefit both desirable and undesirable species (the connectivity conundrum), FishPass aims to develop selective connectivity by leveraging unique aspects of fish physiology, morphology, and ecology with engineering solutions. Modeling technological approaches using single stream recycling, FishPass aims to integrate multiple sorting mechanisms targeting specific attributes of species to ensure that invasive species (sea lamprey) are blocked while native species are passed. Concepts and methods developed through the FishPass initiative may be applicable at other dam sites in the Great Lakes and throughout the world.

SupCon (supplemental sea lamprey control initiative), a second initiative funded by the GLFC, aims to deploy combinations of adult sea lamprey control tools that synergize with each other in 12 streams for 12 years (Siefkes et al., 2021). The difference from FishPass is that SupCon will apply solutions in natural stream channels utilizing infrastructure that is deployed seasonally rather than at existing dam sites. Rates of recruitment, larval production, and juvenile production relative to status quo lampricide treatments will be evaluated within the framework of an active adaptive assessment plan (Lewandoski et al., 2021a) where family groups of larvae are genetically tagged (Sard et al., 2020) and siblings are sampled in subsequent years to track their growth, movement, survival, sex, and age at metamorphosis using close kin mark recapture (Bravington et al., 2016). Combinations of traps, electrical barriers, and sterile male release have already resulted in reductions in larval recruitment in the four study streams that adult controls have been deployed in thus far (Siefkes et al., 2021) and supplemental control deployments are planned in other experiment streams starting in 2023.

## 5.3 Exploiting the physiology of metamorphosis and outmigration

Metamorphosing and outmigrating juvenile sea lamprey still in the stream at time of treatment are killed by lampricides, but otherwise this life stage is not specifically targeted by the sea lamprey control program. Although outmigrating juvenile sea lamprey have been trapped throughout the history of sea lamprey control with drift nets, screw traps, and inclined screen traps, these efforts were not viewed as large-scale programmatic control tools that could reduce populations throughout the Great Lakes. Instead, they were useful for assessing the effectiveness of lampricide control (Hanson and Swink, 1989) and collecting biological and behavioral data on this life stage including their size (Applegate, 1950), sex ratios (Applegate and Thomas, 1965), timing of outmigration (Applegate, 1961; Applegate and Brynildsen, 1952; Miehls et al., 2019), and location in the stream channel while moving downstream (Sotola et al., 2018).

However, as Evans et al. (2021) notes in his review of the pre-feeding juvenile life stage, understanding and targeting metamorphosing and outmigrating sea lamprey offers several advantages for sea lamprey control. Metamorphosing sea lamprey: (1) are still in or near their natal stream and are thus aggregated in space, unlike feeding juveniles in the open lake; and (2) have not yet inflicted damage to the fishery (unlike adult sea lamprey which will already have killed, on average, greater than 15 kg of fish by the time they are captured on their upstream migration; see Section 5.1.2). If the natural mortality rate of metamorphosing and outmigrating juvenile sea lamprey varies predictably among streams and that variability can be described, the control program could also benefit by investing more resources in barriers, lampricides, or trapping outmigrating juveniles in the streams where natural mortality is low. Furthermore, if the physiological processes of metamorphosis can be disrupted, it could prove to be a highly selective means to target sea lamprey for control.

### 5.3.1 Metamorphosis

The environmental and physiological triggers of metamorphosis have been reviewed thoroughly elsewhere (e.g., Manzon et al., 2015; Youson, 2003). Most notably, metamorphosis depends on a rise in spring water temperature (with peak rates of metamorphosis in the laboratory occurring at ~21 °C), the accumulation of sufficient lipid reserves to fuel for this non-trophic process, and thyroid hormones. In lampreys, high levels of thyroid hormones appear to promote larval growth and a subsequent sharp decline is important for metamorphosis (see Manzon et al., 2015; Youson, 2003; and references therein). Precocious—albeit incomplete—metamorphosis (e.g., in small larvae and out of season) has been induced with goitrogen treatments that depress thyroid hormone levels (Holmes and Youson, 1993; Manzon et al., 2001; Manzon and Youson, 1997).

Other reviews have suggested ways that lamprey metamorphosis could be disrupted for control purposes, including suppression of metamorphosis so that sea lamprey never become parasitic juveniles or early initiation of metamorphosis (before sea lamprey have acquired sufficient lipid reserves) that could lead to death due to starvation (Borowiec et al., 2021; Docker et al., 2003; Youson, 2003). However, much is still not known about the regulation of metamorphosis in lampreys, and nuanced control would be needed for specificity. Even though metamorphosis is rare in other freshwater fishes, some—most notably other Great Lakes lampreys and American eel—also exhibit true metamorphosis, as do most amphibians (Manzon et al., 2015), so it would be difficult to disrupt sea lamprey metamorphosis without having effects on non-target species. Identifying species-specific targets for control will require considerable collaboration between physiologists and molecular biologists (see Section 5.4).

Perhaps more achievable on the short term is using knowledge of the physiology of metamorphosis to better predict the onset of metamorphosis. Because the goal of lampricide treatments is to kill the larvae before they metamorphose into parasitic juveniles, detecting impending metamorphosis is critical. A combination of length and condition factor ($W/L^3 \times 10^6$, where W is weight in g and L is length in mm) is useful for defining a potential pool of presumptively metamorphic larvae (Youson et al., 1993), and models that also include latitude, longitude, drainage area, average larval density, and productivity can improve these predictions (Treble et al., 2008). However, more work is needed to refine the accuracy of these predictions, especially since the criteria for predicting metamorphosis appear to vary among streams and years (Henson et al., 2003). Current larval sea lamprey assessment protocols largely focus on estimating the density of larvae >100 mm in length to provide an index of the relative abundance of larvae with the potential to metamorphose the following year, and streams with the lowest cost per kill ratio are given the highest priority for treatment

(Christie et al., 2003; Hansen and Jones, 2008; Jubar et al., 2021). More accurate predictions of impending metamorphosis could finesse the sea lamprey control program's ability to rank streams and permit more "just in time" treatments (i.e., rather than pre-emptively treating high-density streams sooner or more often than needed). Interestingly, novel omega-shaped muscle fibers in metamorphosing and large northern brook lamprey larvae may serve as a potential marker for metamorphosis before other morphological features are evident (Anderson et al., 2019). These "omega muscle units" may identify a growth spurt of myotomes elongating axially to increase swimming efficiency as lampreys transform from relatively sedentary filter feeders to more mobile juveniles. The timing of the appearance of these fibers in sea lamprey has not been investigated, but further study could help improve the schedule of lampricide treatments in the Great Lakes. More refined scheduling will be particularly important given concerns that climate change may reduce the time to metamorphosis because of possible faster larval growth, higher larval condition, and more rapid spikes in spring water temperatures as an environmental trigger of metamorphosis (Lennox et al., 2020).

### 5.3.2 Outmigration

Understanding the timing of outmigration is critical if the control program is going to target juveniles with trapping approaches that exploit their sensory physiology. The environmental triggers and sensory physiology of outmigration were recently reviewed (Evans et al., 2021), so what follows is a brief review followed by speculation about how environmental changes in the future may influence the timing and sensory physiology of outmigration.

Outmigration occurs after metamorphosis is complete and coincides with specific environmental conditions (Potter, 1980). Increases in stream discharge (flood pulses) seem to be the single most important environmental trigger for outmigration (Applegate, 1950; Quintella et al., 2005). In Europe, where snowpack and river ice are not common, the outmigration is unimodal and varies across regions depending on the timing of the rainy season (January–April in Spain, Silva et al., 2013; December in France, Taverny and Elie, 2009; Germany March–May, Baer et al., 2018). In the Great Lakes, where snowpack is common, the outmigration is bimodal occurring from October to December with autumnal rain events, and from February to April with spring snowmelt (Applegate, 1961; Swink and Johnson, 2014). While observation data strongly support that outmigration coincides with flood pulses, it remains unclear what sensory cues sea lamprey are using to determine that discharge has increased. Floods are associated with increases in turbidity (visual and olfactory cues), increases in depth and velocity (mechanoreception), increases in ambient low frequency sounds (auditory), and changes in water temperature. If conservation physiologists can define what cues associated with floods trigger migration, practitioners may devise ways to manipulate those cues to trigger outmigration when streams are not flooding which is when traps are easier to operate and when sea lamprey may be better able to swim toward stimuli guiding them to traps.

Juvenile sea lamprey migrate mostly at night (Miehls et al., 2019; Potter, 1980; but see Applegate, 1950; Baer et al., 2018) and can occur anywhere in the stream channel (bank to bank; bottom to surface; Sotola et al., 2018). However, outmigrants show a slight preference for higher water velocities occurring in the middle of the water column where channel depth is greatest (Miehls et al., 2019; Sotola et al., 2018). Because juveniles are negatively buoyant (due to their lack of a swim bladder), their nonrandom positioning in high velocity areas suggests they actively swim to maintain body orientation (Evans et al., 2021; Sotola et al., 2018). Therefore, manipulation of cues that juvenile sea lamprey sense to maintain position in the stream channel may offer opportunities to guide migrants toward trap entrances.

Physiological and behavioral responses of outmigrating juveniles to auditory, chemosensory, visual, and electrical cues have recently been reviewed (Borowiec et al., 2021; Evans et al., 2021). Auditory cues have received the least attention, but a recent laboratory study reported that recently metamorphosed juveniles detect sounds in the 50–300 Hz range and increase activity when exposed to those frequencies (Mickle et al., 2018). During metamorphosis, the olfactory organ enlarges, and juveniles become sensitive to odors emitted by potential hosts (Kleerekoper and Mogensen, 1963) and of dead conspecifics (Johnson et al., 2019). In a large laboratory flume, outmigrating juveniles exposed to odors extracted from dead larvae (putative alarm cue) increased the speed of downstream movement, but they did not move laterally across the stream channel to avoid the odor (Johnson et al., 2019; Fig. 10A). When the odor from dead larvae was applied in combination with light stimuli that attracted sea lamprey laterally across the stream channel (100 lx at a distance of 1 m from the light source; Haro et al., 2020), more sea lamprey were captured in the lighted trap relative to when only exposed to the light stimuli (push with alarm cue, pull with light cue; Johnson et al., 2019; Fig. 10B). Sea lamprey may respond more intensely to wavelengths of light that they are physiologically most sensitive to (500–550 nm; Fain, 2020), but this has yet to be tested in a behavioral context. Juvenile sea lamprey are electroreceptive (Miehls et al., 2017) and vertical fields of pulsed DC with voltage gradients ranging from 0.3 to 0.8 V/cm have pushed downstream moving juveniles toward traps (Johnson and Miehls, 2014; Fig. 10B), but only when velocities were less than 30 cm/s (Miehls et al., 2017). The relatively poor ability of outmigrating juveniles to swim against high velocity currents (Beamish, 1974) may be a limitation of any behavioral guidance strategy unless the tool can be deployed in pools or estuaries or over a long reach of stream (such as a light array 100 m long). To date, no instream experiments have been conducted to evaluate the efficacy of these sensory cues to manipulate behavior; doing so is the next logical step.

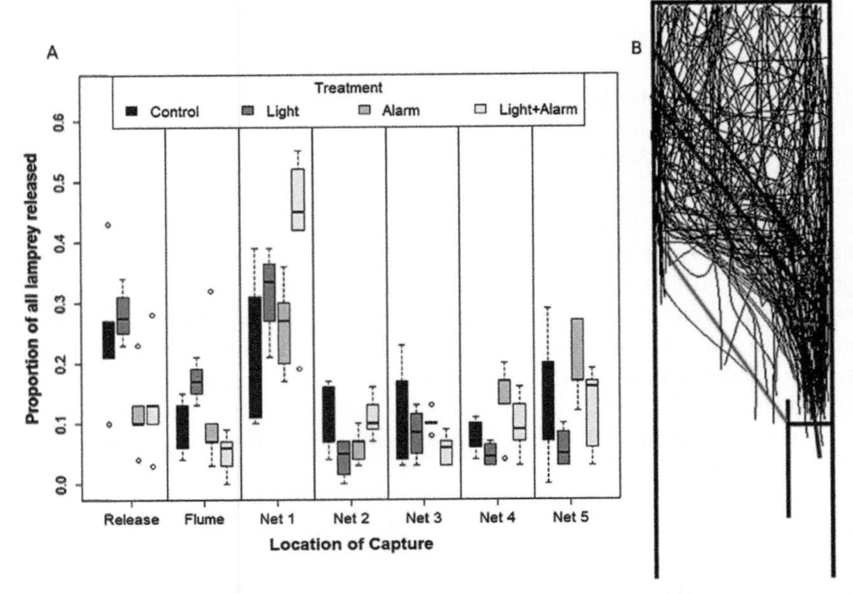


**FIG. 10** Examples of behavioral guidance of downstream migrating juvenile sea lamprey (*Petromyzon marinus*). (A) The distribution of downstream moving juvenile sea lamprey in an experimental flume when light was administered upstream of Net 1 and alarm cue extract from dead larval sea lamprey was applied in the middle of the channel (upstream of Net 3). Nets were arrayed across the channel at the downstream end with Net 1 adjacent to the left wall and Net 5 adjacent to the right wall (looking downstream). "Release" means lamprey did not leave the release area. "Flume" means lamprey left the release area but did not move downstream to the nets. Alarm cue increased the number of sea lamprey moving downstream (push) and light attracted sea lamprey (pull). (B) Movement of downstream migrating juvenile sea lamprey in an experimental flume when pulsed direct current was applied at the blue (negative electrodes) and red (positive electrodes) lines. The black horizonal line was the target for guidance and was 0.46m wide. *Panel (A) from Johnson, N.S., Miehls, S.M., Haro, A.J., Wagner, C.M., 2019. Push and pull of downstream moving juvenile sea lamprey (Petromyzon marinus) exposed to chemosensory and light cues. Conserv. Physiol. 7, coz080.; U.S.A. Public Domain. Panel (B) from Johnson, N.S., Miehls, S., 2014. Guiding out-migrating juvenile Sea Lamprey (Petromyzon marinus) with pulsed direct current. River Res. Appl. 30, 1146–1156; U.S.A. public domain.*

In the future, changes in the timing of when precipitation events occur could result in changes in outmigration timing; perhaps changes are already occurring, as has been observed in the timing of adult sea lamprey upstream migration (McCann et al., 2018b; see Section 4.3). Outmigration dynamics may also be expected to be modified by light pollution (Haro et al., 2020), sound pollution (Mickle et al., 2019), and regulated stream flows (Quintella et al., 2005). These changes may hinder conservation goals where sea lamprey are native, and they need to be understood and accounted for in the context of sea lamprey control.

## 5.4 Genetic control options

Because it is now possible to edit the genomes of organisms with great precision (Piaggio et al., 2017), genetic control tools appear promising for species-specific and cost-effective control of pest species (e.g., Alphey, 2014; Galizi et al., 2014; Jacobs-Lorena, 2006; Teem et al., 2020). Clustered Regularly Interspaced Short Palindromic Repeats/Cas9 (CRISPR/Cas9) gene editing has been shown to be effective in the sea lamprey (e.g., Square et al., 2015), and genetic options are being evaluated for sea lamprey control (Ferreira-Martins et al., 2021b; Thresher et al., 2014, 2019a,b). However, considerable efforts will be needed for the basic research required to: (1) develop and improve existing tools for genome editing and transformation specific to sea lamprey and the control program; (2) identify suitable candidate genes to serve as targets for modification—that is, gene sequences or products that, when disrupted, enhanced, or ingested, result in species-specific physiological or metabolic changes in sea lamprey that impair survival or fertility or distort population sex ratios; and (3) develop strategies to evaluate and mitigate risk (e.g., to prevent spread to non-target populations or species).

To identify suitable candidate genes to serve as targets for modification, close collaboration between physiologists and molecular biologists will again be needed. A better understanding of the molecular basis of key physiological processes in lampreys will be key to developing effective and species-specific alternatives or supplements to current sea lamprey control methods. Possible targets include genes involved in embryological development, sex determination, sex differentiation, metamorphosis, or sexual maturation. Sequencing of the sea lamprey genome (Smith et al., 2013, 2018) and subsequent genomic and transcriptomic studies are helping to improve our understanding of genes and gene networks that control many aspects of sea lamprey biology (Ajmani et al., 2021; Hockman et al., 2019; Jones et al., 2019; McCauley et al., 2015; Parker et al., 2019; Yasmin et al., 2022; see reviews by York and McCauley, 2020; York et al. 2019, 2021). Transcriptomic studies can also be used to deduce cellular mechanisms of lampricide toxicity, uptake, and elimination that could potentially be exploited to increase the potency and specificity of lampricides (Lantz et al., 2022). Even a cursory review of potential target genes is not possible here, but a physiological genomic approach could be most effective. An integrative approach could be used, for example, to interpret transcriptomic responses in a physiological context (e.g., when comparing gene expression between stages, tissues, or environmental conditions), avoiding a pitfall of many transcriptomic studies which may be "short on biology and long on [gene] lists" (DeWoody et al., 2013). Understanding the details of the functioning of gene products can help provide context to our understanding of the whole organism and its environment (Wikelski and Cooke, 2006).

Exploration of genetic options for the control of sea lamprey could address many ethical and societal concerns regarding the genetic manipulation of organisms released into the environment (Thresher et al., 2019a,b). Given the very real danger that deleterious alleles could spread out of the Great Lakes to the anadromous sea lamprey population, self-limiting genetic control options and confined gene drives will likely be preferred over unconfined gene drive options for sea lamprey control (Ferreira-Martins et al., 2021b; Thresher et al., 2019a,b), but evaluations could provide information on the extent to which there is social license to develop and use such technologies A survey of professional fishery managers, recreational anglers, and other stakeholders showed strong support for initiation of research and development related to genetic options to control sea lamprey (Thresher et al., 2019b). Further exploration of the feasibility and management applications of such technologies could involve public consultation and careful and ongoing evaluation of the actual and perceived risks and benefits of genetic control.

## 6 Conclusions

Invasive fishes disrupt fishery conservation efforts in aquatic ecosystems worldwide; therefore, invasive species control is a critical component of conservation biology. Conservation physiology-based approaches are well positioned to describe the stressors induced by invasive species and develop selective and effective methods of population control through mechanism-based evidence. The importance of conservation physiology-based approaches is highly evident in our case study of invasive sea lamprey in the Great Lakes—the most heavily invaded aquatic system in the world—where a pre-existing physiological capacity to survive and thrive in fresh water, along with a broad range of fish species on which it will prey and its high fecundity, facilitated a rapid invasion of sea lamprey through shipping canals. Conservation physiology approaches have been paramount for providing a mechanistic understanding of the lethal and sublethal effects sea lamprey cause to host fishes, defining the mechanisms of lampricide toxicity, and designing effective migration barriers and traps that exploit weaknesses (poor swimming and jumping abilities) and strengths (olfactory) in their physiology. As a result, sea lamprey populations have been reduced to 10% of their historic highs for nearly 50 years, representing one of the longest and most successful invasive species control campaigns in the world (Robinson et al., 2021).

Lessons learned for sea lamprey control do not occur in a vacuum and are applicable to efforts to conserve valued populations of lampreys worldwide and control other invasive fishes (Hume et al., 2021c; Sorensen, 2021). An understanding of population demographics, habitat requirements, and anthropogenic stressors are critical regardless of whether the goal is to restore or control a fish population. This understanding is evident in lampreys where information gleaned by efforts to better control their populations has been applicable and more impactful when also used for conservation (and visa versa;

Docker and Hume, 2019; Hume et al., 2021c). The lessons learned in sea lamprey control have been broadly informative to other fishery conservation efforts to control invasive carps, round goby, and sometimes even non-native trout (brook trout and rainbow trout; Sorensen, 2021). Clearly, the lessons learned controlling other invasive species have applicability to the sea lamprey story, like how carbon dioxide barriers could be effective for carps and sea lamprey (Suski, 2020).

Continued success in controlling sea lamprey relies on an ability to maintain support from the public to continue to operate (social licence) and a key aspect of maintaining that licence is minimizing or eliminating negative impacts to non-target species (Gaden et al., 2021). Therefore, the prime challenge for conservation physiologists in subsequent decades will be to develop the next generation of control tools that are specific to sea lamprey while maintaining their efficacy. Work in this regard has already begun, and key areas of focus in subsequent decades are likely to include next generation lampricides (e.g., Lantz et al., 2022), selective barriers and traps that exploit sensory physiology (Section 5.2.1), rapid stream-side sterilization of male sea lamprey using RNAi (Section 5.2.3), and exploration of genetic control tools (Section 5.4).

Continued success in controlling sea lamprey will also rely on the ability to adapt as anthropogenic activities continue to change the Great Lakes and their tributaries (Burkett et al., 2021; Lennox et al., 2020). A warming climate is likely to have multiple cascading negative impacts for sea lamprey control including: (1) increased survival and growth of larval sea lamprey in streams; (2) shortened life cycle by reducing the years to metamorphosis; (3) increased damage to fish stocks as larger sea lamprey are feeding at higher water temperatures; (4) increased fecundity as a function of increasing adult size; (5) increased ability of larvae to survive exposure to TFM at higher stream temperature and pH; and (6) increased ability of larger sea lamprey to breach migration barriers given that swimming performance is size and temperature dependent and destructive floods may become more frequent. Each issue identified above has a physiological basis, and a mechanistic understanding of physiology will be key to addressing these and other challenges as biologists continue to strive to conserve fisheries through the deepening of the Anthropocene.

## References

Adams, J.V., Jones, M.L., 2021. Evidence of host switching: sea lampreys disproportionately attack Chinook salmon when lake trout abundance is low in Lake Ontario. J. Great Lakes Res. 47, S604–S611.

Adams, J.V., Jones, M.L., Bence, J.R., 2021a. Using simulation to understand annual sea lamprey marking rates on lake trout. J. Great Lakes Res. 47, S628–S638.

Adams, J.V., Barber, J.M., Bravener, G.A., Lewandoski, S.A., 2021b. Quantifying Great Lakes sea lamprey populations using an index of adults. J. Great Lakes Res. 47, S335–S346.

Ajmani, N., Yasmin, T., Docker, M.F., Good, S.V., 2021. Transcriptomic analysis of gonadal development in parasitic and non-parasitic lampreys (*Ichthyomyzon* spp.), with a comparison of genomic resources in these non-model species. Genes Genomes Genet. 11, jkab030.

Almeida, P.R., Arakawa, H., Aronsuu, K., et al., 2021. Lamprey fisheries: history, trends and management. J. Great Lakes Res. 47, S159–S185.

Alphey, L., 2014. Genetic control of mosquitoes. Annu. Rev. Entomol. 59, 205–224.

Anderson, J.E., Cunha, A., Docker, M.F., 2019. Novel "omega fibers" in superficial body-wall myotomes during metamorphosis in the northern brook lamprey, *Ichthyomyzon fossor*. Can. J. Zool. 97, 1218–1224.

Angel, J.R., Kunkel, K.E., 2010. The response of Great Lakes water levels to future climate scenarios with an emphasis on Lake Michigan-Huron. J. Great Lakes Res. 36, 51–58.

Applegate, V.C., 1950. Natural history of the sea lamprey, *Petromyzon marinus*, in Michigan. In: U.S. Fish Wildl. Serv. Spec. Sci. Rep. Fish. No. 55. 235 pp.

Applegate, V.C., 1961. Downstream movement of lampreys and fishes in the Carp Lake River, Michigan. In: U.S. Fish Wildl. Serv. Spec. Sci. Rep. Fish. No. 387. 71 pp.

Applegate, V.C., Brynildsen, C.L., 1952. Downstream movement of recently transformed sea lampreys, *Petromyzon marinus*, in the Carp Lake River, Michigan. Trans. Am. Fish. Soc. 81, 275–290.

Applegate, V.C., King, E.L., 1962. Comparative toxicity of 3-trifluormethyl-4-nitrophenol (TFM) to larval lampreys and eleven species of fishes. Trans. Am. Fish. Soc. 91, 342–345.

Applegate, V.C., Smith, B.R., 1951a. Movement and dispersion of a blocked spawning run of sea lampreys in the Great Lakes. N. Am. Wildl. Conf. 16, 243–251.

Applegate, V.C., Smith, B.R., 1951b. Sea lamprey spawning runs in the Great Lakes, 1950. In: U.S. Fish Wildl. Serv. Spec. Sci. Rep. Fish. No. 61. 49 pp.

Applegate, V.C., Thomas, M.L.H., 1965. Sex ratios and sexual dimorphism among recently transformed sea lampreys, *Petromyzon marinus* Linnaeus. J. Fish. Res. Brd. Can. 22 (3), 695–711.

Applegate, V.C., Smith, B.R., Nielsen, W.L., 1952. Use of electricity in the control of sea lampreys: electromechanical weirs and traps and electrical barriers. In: U.S. Fish Wildl. Serv. Spec. Sci. Rep. Fish. No. 92. 52 pp.

Applegate, V.C., Howell, J.H., Hall, A.E., Smith, M.A., 1957. Toxicity of 4,346 chemicals to larval lampreys and fishes. In: U.S. Fish Wildl. Serv. Spec. Sci. Rep. Fish. vol. 207, pp. 1–157.

Applegate, V.C., Howell, J.H., Moffett, J.W., Johnson, B.G.H., Smith, M.A., 1961. Use of 3-trifluormethyl-4-nitrophenol as a selective sea lamprey larvicide. In: Great Lakes Fish. Comm. Tech. Rep. vol. 1, pp. 3–35.

Arthington, A.H., Dulvy, N.K., Gladstone, W., Winfield, I.J., 2016. Fish conservation in freshwater and marine realms: status, threats and management. Aquat. Conserv. 26, 838–857.

Aurangzeb, Z., Daghfous, G., Innes, L., Dubuc, R., Zielinski, B., 2021. Current understanding of lamprey chemosensory systems. J. Great Lakes Res. 47, S650–S659.

Baer, J., Hartmann, F., Brinker, A., 2018. Abiotic triggers for sea and river lamprey spawning migration and juvenile outmigration in the River Rhine, Germany. Ecol. Freshw. Fish 27, 988–998.

Baker, H., 1974. The evolution of weeds. Annu. Rev. Ecol. Syst. 5, 1–24.

Bals, J.D., Wagner, C.M., 2012. Behavioral responses of sea lamprey (*Petromyzon marinus*) to a putative alarm cue derived from conspecific and heterospecific sources. Behaviour 149, 901–923.

Barany, A., Shaughnessy, C.A., Pelis, R.M., Fuentes, J., Mancera, J.M., McCormick, S.D., 2021. Tissue and salinity specific Na+/Cl− cotransporter (NCC) orthologues involved in the adaptive osmoregulation of sea lamprey (*Petromyzon marinus*). Sci. Rep. 11, 22698.

Barber, J., Steeves, T.B., 2021. Sea Lamprey Control in the Great Lakes 2020 - Annual Report to the Great Lakes Fishery Commission. Great Lakes Fishery Commission, Ann Arbor, MI.

Bartels, H., Potter, I.C., 1991. Structural changes in the zonulae occludentes of the chloride cells of young adult lampreys following acclimation to sea water. Cell Tissue Res. 265, 447–457.

Bartels, H., Potter, I.C., 2004. Cellular composition and ultrastructure of the gill epithelium of larval and adult lampreys - Implications for osmoregulation in fresh and seawater. J. Exp. Biol. 207, 3447–3462.

Bartolai, A.M., He, L., Hurst, A.E., Mortsch, L., Paehlke, R., Scavia, D., 2015. Climate change as a driver of change in the Great Lakes St. Lawrence River Basin. J. Great Lakes Res. 41, 45–58.

Baxter, E.W., 1956. Observations on the buccal glands of lampreys (Petromyzonidae). Proc. Zool. Soc. 127, 95–118.

Beamish, F., 1974. Swimming performance of adult sea lamprey, *Petromyzon marinus*, in relation to weight and temperature. Trans. Am. Fish. Soc. 103 (2), 355–358.

Beamish, F.W.H., 1980. Biology of the North American anadromous sea lamprey, *Petromyzon marinus*. Can. J. Fish. Aquat. Sci. 37, 1924–1943.

Beamish, F.W.H., Potter, I.C., 1975. Biology of anadromous sea lamprey (*Petromyzon marinus*) in New Brunswick. J. Zool. 177, 57–72.

Beamish, F.W.H., Strachan, P.D., Thomas, E., 1978. Osmotic and ionic performance of the anadromous sea lamprey, *Petromyzon marinus*. Comp. Biochem. Physiol. 60A, 435–443.

Bence, J.R., Bergstedt, R.A., Christie, G.C., Cochran, P.A., Ebener, M.P., Koonce, J.F., Rutter, M.-A., Swink, W.D., 2003. Sea lamprey (*Petromyzon marinus*) parasite-host interactions in the Great Lakes. J. Great Lakes Res. 29, 253–282.

Bergstedt, R.A., Schneider, C.P., 1988. Assessment of sea lamprey (*Petromyzon marinus*) predation by recovery of dead lake trout (*Salvelinus namaycush*) from Lake Ontario, 1982–85. Can. J. Fish. Aquat. Sci. 45, 1406–1410.

Bergstedt, R.A., Seelye, J.G., 1995. Evidence for lack of homing by sea lampreys. Trans. Am. Fish. Soc. 124, 235–239.

Bergstedt, R.A., Swink, W.D., 1995. Seasonal growth and duration of the parasitic life stage of the landlocked sea lamprey (*Petromyzon marinus*). Can. J. Fish. Aquat. Sci. 52, 1257–1264.

Bergstedt, R.A., Twohey, M.B., 2007. Research to support sterile-male-release and genetic alteration techniques for sea lamprey control. J. Great Lakes Res. 33, 48–69.

Bergstedt, R.A., Rottiers, D.V., Foster, N.R., 1981. Laboratory determination of maximum swimming speed of migrating sea lampreys: a feasibility study. Administrative Report 81-3, Great Lakes Fishery Laboratory, US Fish and Wildlife Service, Ann Arbor, MI.

Bergstedt, R.A., Argyle, R.L., Seelye, J.G., Scribner, K.T., Curtis, G.L., 2003a. In situ determination of the annual thermal habitat use by lake trout (*Salvelinus namaycush*) in Lake Huron. J. Great Lakes Res. 29, 347–361.

Bergstedt, R.A., McDonald, R.B., Twohey, M.B., Mullett, K.M., Young, R.J., Heinrich, J.W., 2003b. Reduction in sea lamprey hatching success due to release of sterilized males. J. Great Lakes Res. 29, 435–444.

Berst, A.H., Spangler, G.R., 1972. Lake Huron – Effects of exploitation, introductions, and eutrophication on salmonid community. J. Fish. Res. Board Can. 29, 877–887.

Bills, T.D., Boogaard, M.A., Johnson, D.A., Brege, D.C., Scholefield, R.J., Westman, R.W., Stephens, B.E., 2003. Development of a pH/alkalinity treatment model for applications of the lampricide TFM to streams tributary to the Great Lakes. J. Great Lakes Res. 29 (Suppl. 1), 510–520.

Binder, T.R., McDonald, D.G., 2007. Is there a role for vision in the behaviour of sea lampreys (*Petromyzon marinus*) during their upstream spawning migration? Can. J. Fish. Aquat. Sci. 64, 1403–1412.

Binder, T.R., McDonald, D.G., 2008. The role of dermal photoreceptors during the sea lamprey (*Petromyzon marinus*) spawning migration. J. Comp. Physiol. A 194, 921–928.

Birceanu, O., McClelland, G.B., Wang, Y.X.S., Wilkie, M.P., 2009. Failure of ATP supply to match ATP demand: the mechanism of toxicity of the lampricide, 3-trifluoromethyl-4-nitrophenol (TFM), used to control sea lamprey (*Petromyzon marinus*) populations in the Great Lakes. Aquat. Toxicol. 94, 265–274.

Birceanu, O., McClelland, G.B., Wang, Y.S., Brown, J.C.L., Wilkie, M.P., 2011. The lampricide 3-trifluoromethyl-4-nitrophenol (TFM) uncouples mitochondrial oxidative phosphorylation in both sea lamprey (*Petromyzon marinus*) and TFM-tolerant rainbow trout (*Oncorhynchus mykiss*). Comp. Biochem. Physiol. C Toxicol. Pharmacol. 153, 342–349.

Birceanu, O., Sorensen, L.A., Henry, M., McClelland, G.B., Wang, Y., Wilkie, M.P., 2014. The effects of the lampricide 3-trifluoromethy1-4-nitrophenol (TFM) on fuel stores and ion balance in a non-target fish, the rainbow trout (*Oncorhynchus mykiss*). Comp. Biochem. Physiol. C Toxicol. Pharmacol. 160, 30–41.

Birceanu, O., Tessier, L.R., Huerta, B., Li, W.M., McDonald, A., Wilkie, M.P., 2021. At the intersection between toxicology and physiology: what we have learned about sea lampreys and bony fish physiology from studying the mode of action of lampricides. J. Great Lakes Res. 47, S673–S689.

Bodznick, D., Northcutt, R.G., 1981. Electroreception in lampreys: evidence that the earliest vertebrates were electroreceptive. Science 212, 465–467.

Boehm, T., Iwanami, N., Hess, I., 2012. Evolution of the immune system in the lower vertebrates. Annu. Rev. Genomics Hum. Genet. 13, 127–149.

Boogaard, M.A., Bills, T.D., Johnson, D.A., 2003. Acute toxicity of TFM and a TFM/niclosamide mixture to selected species of fish, including lake sturgeon (*Acipenser fulvescens*) and mudpuppies (*Necturus maculosus*), in laboratory and field exposures. J. Great Lakes Res. 29, 529–541.

Borowiec, B.G., Docker, M.F., Johnson, N.S., Moser, M.L., Zielinski, B., Wilkie, M.P., 2021. Exploiting the physiology of lampreys to refine methods of control and conservation. J. Great Lakes Res. 47, S723–S741.

Borowiec, B., Birceanu, O., Wilson, J.M., McDonald, A., Wilkie, M.P., 2022. Niclosamide is a much more potent inhibitor of mitochondrial respiration than TFM in invasive sea lamprey (*Petromyzon marinus*). Environ. Sci. Technol. 56, 4970–4979.

Brant, C., 2019. Great Lakes sea lamprey: the 70 year war on a biological invader. University of Michigan Press, Ann Arbor.

Bravener, G., Twohey, M., 2016. Evaluation of a sterile-male release technique: a case study of invasive sea lamprey control in a tributary of the Laurentian Great Lakes. N. Am. J. Fish. Manag. 36, 1125–1138.

Bravington, M.V., Skaug, H.J., Anderson, E.C., 2016. Close-kin mark-recapture. Stat. Sci. 31, 259–274.

Bryan, M.B., Zalinski, D., Filcek, K.B., Libants, S., Li, W., Scribner, K.T., 2005. Patterns of invasion and colonization of the sea lamprey (*Petromyzon marinus*) in North America as revealed by microsatellite genotypes. Mol. Ecol. 14, 3757–3773.

Bryan, M.B., Scott, A.P., Li, W., 2008. Sex steroids and their receptors in lamprey. Steroids 73, 1–12.

Buchinger, T.J., Siefkes, M.J., Zielinski, B.S., Brant, C.O., Li, W., 2015. Chemical cues and pheromones in the sea lamprey (*Petromyzon marinus*). Front. Zool. 12, 1–11.

Buchinger, T.J., Scott, A.M., Fissette, S.D., Brant, C.O., Huertas, M., Li, K., Johnson, N.S., Li, W., 2020. A pheromone antagonist liberates female sea lamprey from a sensory trap to enable reliable communication. Proc. Natl. Acad. Sci. U. S. A. 117, 7284–7289.

Buchinger, T.J., Fissette, S.D., Huerta, B., Li, K., Johnson, N.S., Li, W., 2021. Pheromone pollution from invasive sea lamprey misguides a native confamilial. Curr. Zool. 67, 333–335.

Buckley, C.M., 2021. Evaluation of Great Lakes Sea Lamprey Control Barrier Effectiveness Under Climate Change. Master's thesis, Michigan State University, Lansing.

Bullingham, O.M.N., Firkus, T.J., Goetz, F.W., Murphy, C.A., Alderman, S.L., 2022. Lake charr (*Salvelinus namaycush*) clotting response may act as a plasma biomarker of sea lamprey (*Petromyzon marinus*) parasitism: implications for management and wound assessment. J. Great Lakes Res. 48, 207–218.

Burkett, D.P., Barber, J.M., Steeves, T.B., Siefkes, M.J., 2021. Sea lamprey control 2020–2040: charting a course through dynamic waters. J. Great Lakes Res. 47, S809–S814.

Bussy, U., Chung-Davidson, Y.-W., Buchinger, T., Li, K., Smith, S.A., Jones, A.D., Li, W., 2018a. Metabolism of a sea lamprey pesticide by fish liver enzymes. Part A: identification and synthesis of TFM metabolites. Anal. Bioanal. Chem. 410, 1749–1761.

Bussy, U., Chung-Davidson, Y.-W., Buchinger, T., Li, K., Smith, S.A., Jones, A.D., Li, W., 2018b. Metabolism of a sea lamprey pesticide by fish liver enzymes. Part B: method development and application in quantification of TFM metabolites formed *in vivo*. Anal. Bioanal. Chem. 410, 1763–1774.

Byford, G.J., 2016. Altering the migratory routes of sea lamprey through the application of semiochemicals. Master's thesis, Michigan State University, East Lansing, MI.

Byford, G.J., Wagner, C.M., Hume, J.B., Moser, M.L., 2016. Do native Pacific lamprey and invasive sea lamprey share an alarm cue? Implications for use of a natural repellent to guide imperiled Pacific lamprey into fishways. N. Am. J. Fish. Manag. 36 (5), 1090–1096.

Campbell, K.J., Beek, J., Eason, C.T., et al., 2015. The next generation of rodent eradications: innovative technologies and tools to improve species specificity and increase their feasibility on islands. Biol. Conserv. 185, 47–58.

Chasiotis, H., Kolosov, D., Bui, P., Kelly, S.P., 2012. Tight junctions, tight junction proteins and paracellular permeability across the gill epithelium of fishes: a review. Respir. Physiol. Neurobiol. 84, 269–281.

Chi, S., Xiao, R., Li, Q., Zhou, L., He, R., Qi, Z., 2009. Suppression of neuronal excitability by the secretion of the lamprey (*Lampetra japonica*) provides a mechanism for its evolutionary stability. Pflugers Archiv Eur. J. Physiol. 458, 537–545.

Christiaens, O., Whyard, S., Vélez, A.M., Smagghe, G., 2020. Double-stranded RNA technology to control insect pests: current status and challenges. Front. Plant Sci. 11, 451.

Christie, G.C., Goddard, C.I., 2003. Sea Lamprey International Symposium (SLIS II): advances in the integrated management of sea lamprey in the Great Lakes. J. Great Lakes Res. 29, 1–14.

Christie, G.C., Adams, J.V., Steeves, T.B., et al., 2003. Selecting Great Lakes streams for lampricide treatment based on larval sea lamprey surveys. J. Great Lakes Res. 29 (Suppl. 1), 152–160.

Chung-Davidson, Y.W., Yun, S.S., Teeter, J., Li, W., 2004. Brain pathways and behavioral responses to weak electric fields in parasitic sea lampreys (*Petromyzon marinus*). Behav. Neurosci. 118, 611.

Chung-Davidson, Y.W., Bryan, M.B., Teeter, J., Bedore, C.N., Li, W., 2008. Neuroendocrine and behavioral responses to weak electric fields in adult sea lampreys (*Petromyzon marinus*). Horm. Behav. 54, 34–40.

Ciereszko, A., Babiak, I., Dabrowski, K., 2004. Efficacy of animal anti-fertility compounds against sea lamprey (*Petromyzon marinus*) spermatozoa. Theriogenology 61 (6), 1039–1050.

Ciereszko, A., Wolfe, T.D., Dabrowski, K., 2005. Analysis of DNA damage in sea lamprey (*Petromyzon marinus*) spermatozoa by UV, hydrogen peroxide, and the toxicant bisazir. Aquat. Toxicol. 73, 128–138.

Clemens, B.J., Wang, C.J., 2021. Dispelling misperceptions of native lampreys (*Entosphenus* and *Lampetra* spp.) in the Pacific northwest (USA). Conserv. Sci. Pract. 2021, e402.

Clemens, B.J., Beamish, R.J., Kelly, C.C., et al., 2017. Conservation challenges and research needs for Pacific lamprey in the Columbia River basin. Fisheries 42, 268–280.

Clifford, A.M., Henry, M., Bergstedt, R., McDonald, D.G., Smits, A.S., Wilkie, M.P., 2012. Recovery of larval sea lampreys from short-term exposure to the pesticide 3-trifluoromethyl-4-nitrophenol: implications for sea lamprey control in the Great Lakes. Trans. Am. Fish. Soc. 141, 1697–1710.

Cline, T.J., Kitchell, J.F., Bennington, V., McKinley, G.A., Moody, E.K., Weidel, B.C., 2014. Climate impacts on landlocked sea lamprey: implications for host-parasite interactions and invasive species management. Ecosphere 5, 1–13.

Close, D.A., Fitzpatrick, M.S., Li, H.W., 2002. The ecological and cultural importance of a species at risk of extinction, Pacific lamprey. Fisheries, 2719–2725.

Cochran, P.A., 1985. Size-selective attack by parasitic lampreys: consideration of alternate null hypotheses. Oecologia 67, 137–141.

Cochran, P.A., Hodgson, J.Y., Kinziger, A.P., 2003. Change in energy density of the sea lamprey (*Petromyzon marinus*) during its parasitic phase: implications for modeling food consumption and growth. J. Great Lakes Res. 29 (Suppl. 1), 297–306.

Comte, L., Olden, J.D., 2017. Evolutionary and environmental determinants of freshwater fish thermal tolerance and plasticity. Glob. Chang. Biol. 23, 728–736.

Cooke, S.J., Sack, L., Franklin, C.E., Farrell, A.P., Beardall, J., Wikelski, M., Chown, S.L., 2013. What is conservation physiology? Perspectives on an increasingly integrated and essential science. Conserv. Physiol. 1, cot001.

COSEWIC, 2017. COSEWIC Assessment and status report on the lake sturgeon (*Acipenser fulvescens*), Western Hudson Bay populations, Saskatchewan-Nelson River population, Southern Hudson Bay-James Bay population, Great Lakes-Upper St. Lawrence populations. Committee on the Status of Endangered Wildlife in Canada, Ottawa. xxx +153 pp (Species at Risk Public registry website. Accessed February 17, 2021).

COSEWIC, 2020. COSEWIC assessment and status report on the Northern Brook Lamprey *Ichthyomyzon fossor* (Great Lakes—Upper St. Lawrence populations and Saskatchewan—Nelson River populations) and the Silver Lamprey *Ichthyomyzon unicuspis* (Great Lakes—Upper St. Lawrence populations, Saskatchewan—Nelson River populations and Southern Hudson Bay—James Bay populations) in Canada. Committee on the Status of Endangered Wildlife in Canada, Ottawa. xxiv + 156 pp. (https://www.canada.ca/en/environment-climate-change/services/species-risk-public-registry.html).

COSSARO, 2017. Ontario Species at Risk Evaluation Report for Lake Sturgeon (*Acipenser fulvescens*) Saskatchewan-Nelson River populations Southern Hudson Bay-James Bay populations Great Lakes-Upper St. Lawrence populations. Committee on the Status of Species at Risk in Ontario (COSSARO). COSSARO Candidate Species at Risk Evaluation for Lake Sturgeon. Committee on the Status of Species at Risk in Ontario (COSSARO). Publication 671/92. http:/cossaroagency.ca/.

Crawford, S., 2001. Salmonine introductions to the Laurentian Great Lakes: an historical review and evaluation of ecological effects. Can. Spec. Publ. Fish. Aquat. Sci. 132, 1–205.

Criger, L.A., Barber, J.M., Bravener, G.A., Brenden, T.O., Neave, F.B., 2021. The evolution of sea lamprey control in the St. Marys River: 1997–2019. J. Great Lakes Res. 47, S479–S491.

Crowl, T.A., Crist, T.O., Parmenter, R.R., Belovsky, G., Lugo, A.E., 2008. The spread of invasive species and infectious disease as drivers of ecosystem change. Front. Ecol. Environ. 6, 238–246.

Dawson, H.A., Jones, M.L., 2009. Factors affecting recruitment dynamics of Great Lakes sea lamprey (*Petromyzon marinus*) populations. J. Great Lakes Res. 35, 353–360.

Dawson, H.A., Quintella, B.R., Almeida, P.R., Treble, A.J., Jolley, J.C., 2015. The ecology of larval and metamorphosing lampreys. In: Docker, M.F. (Ed.), Lampreys: Biology, Conservation and Control. 1. Springer, Dordrecht, pp. 75–137.

Dawson, H.A., Jones, M.L., Irwin, B.J., Johnson, N.S., Wagner, M.C., Szymanski, M.D., 2016. Management strategy evaluation of pheromone-baited trapping techniques to improve management of invasive sea lamprey. Nat. Resour. Model. 29, 448–469.

D'Cruz, L.M., Dockray, J.J., Morgan, I.J., Wood, C.M., 1998. Physiological effects of sublethal acid exposure in juvenile rainbow trout on a limited or unlimited ration during a simulated global warming scenario. Physiol. Zool. 71, 359–376.

Dent, D.R. (Ed.), 1995. Integrated Pest Management. Chapman and Hall, London, p. 356.

Dettmers, J.M., Goddard, C.I., Smith, K.D., 2012. Management of alewife using Pacific salmon in the Great Lakes: whether to manage for economics or the ecosystem? Fisheries 37, 495–501.

DeWoody, J., Abts, K.C., Fahey, A.L., et al., 2013. Of contigs and quagmires: next-generation sequencing pitfalls associated with transcriptomic studies. Mol. Ecol. Resour. 13, 551–558.

Di Rocco, R.T., Johnson, N.S., Brege, L., Imre, I., Brown, G.E., 2016. Sea lamprey avoid areas scented with conspecific tissue extract in Michigan streams. Fish. Manag. Ecol. 23, 548–560.

Dissanayake, A.A., Wagner, C.M., Nair, M.G., 2016. Chemical characterization of lipophilic constituents in the skin of migratory adult sea lamprey from the Great Lakes region. PLoS One 11, e0168609.

Dissanayake, A.A., Wagner, C.M., Nair, M.G., 2019. Nitrogenous compounds characterized in the deterrent skin extract of migratory adult sea lamprey from the Great Lakes region. PLoS One 14, e0217417.

Dissen, G.A., Lomniczi, A., Boudreau, R.L., et al., 2012. Applying gene silencing technology to contraception. Reprod. Domest. Anim. 47, 381–386.

Dissen, G.A., Adachi, K., Lomniczi, A., et al., 2017. Engineering a gene silencing viral construct that targets the cat hypothalamus to induce permanent sterility: an update. Reprod. Domest. Anim. 52, 354–358.

Dobiesz, N.E., Bence, J.R., Sutton, T., Ebener, M., Pratt, T.C., O'Connor, L.M., Steeves, T.B., 2018. Evaluation of sea lamprey-associated mortality sources on a generalized lake sturgeon population in the Great Lakes. J. Great Lakes Res. 44, 319–329.

Docker, M.F., Hume, J.B., 2019. There and back again: lampreys in the 21st century and beyond. In: Docker, M.F. (Ed.), Lampreys: Biology, Conservation and Control. 2. Springer, Dordrecht, pp. 527–570.

Docker, M.F., Potter, I.C., 2019. Life history evolution in lampreys: alternative migratory and feeding types. In: Docker, M.F. (Ed.), Lampreys: Biology, Conservation and Control. 2. Springer, Dordrecht, pp. 287–409.

Docker, M.F., Sower, S.A., Youson, J.H., Beamish, F.W.H., 2003. Future sea lamprey control through regulation of metamorphosis and reproduction: a report from the SLIS II New Science and Control workgroup. J. Great Lakes Res. 29 (Suppl. 1), 801–807.

Docker, M.F., Hume, J.B., Clemens, B.J., 2015. Introduction: a surfeit of lampreys. In: Docker, M.F. (Ed.), Lampreys: Biology, Conservation and Control. 1. Springer, Dordrecht, pp. 1–34.

Docker, M.F., Beamish, F.W.H., Yasmin, T., Bryan, M.B., Khan, A., 2019. The lamprey gonad. In: Docker, M.F. (Ed.), Lampreys: Biology, Conservation and Control. 2. Springer, Dordrecht, pp. 1–186.

Docker, M.F., Bravener, G.A., Garroway, C.J., Hrodey, P.J., Hume, J.B., Johnson, N.S., Lewandoski, S., Ogden, J.L., Zollweg-Horan, E., 2021. A review of sea lamprey dispersal and population structure in the Great Lakes and the implications for control. J. Great Lakes Res. 47, S549–S569.

Dodd, H.R., Hayes, D.B., Baylis, J.R., et al., 2003. Low-head sea lamprey barrier effects on stream habitat and fish communities in the Great Lakes Basin. J. Great Lakes Res. 29, 386–402.

Ebener, M.P., Bence, J.R., Bergstedt, R.A., Mullett, K.M., 2003. Classifying sea lamprey marks on Great Lakes lake trout: observer agreement, evidence on healing times between classes, and recommendations for reporting of marking statistics. J. Great Lakes Res. 29, 283–296.

Edwards, S.L., Marshall, W.S., 2013. Principles and patterns of osmoregulation and euryhalinity in fishes. In: McCormick, S.D., Farrell, A.P., Brauner, C.J. (Eds.), Fish Physiology: Euryhaline Fishes. In: Fish Physiology, 32. Elsevier, Amsterdam, pp. 1–44.

Ehler, L.E., 2006. Integrated pest management (IPM): definition, historical development and implementation, and the other IPM. Pest Manag. Sci. 62, 787–789.

Eickholt, J., Kelly, D., Bryan, J., Miehls, S., Zielinski, D., 2020. Advancements towards selective barrier passage by automatic species identification: applications of deep convolutional neural networks on images of dewatered fish. ICES J. Mar. Sci. 77, 2804–2813.

Elmer, L.K., Madliger, C.L., Blumstein, D.T., Elvidge, C.K., Fernández-Juricic, E., Horodysky, A.Z., Johnson, N.S., McGuire, L.P., Swaisgood, R.R., Cooke, S.J., 2021. Exploiting common senses: sensory ecology meets wildlife conservation and management. Conserv. Physiol. 9, coab002.

Elrod, J.H., Schneider, C.P., 1987. Seasonal bathythermal distribution of juvenile lake trout in Lake Ontario. J. Great Lakes Res. 13, 121–134.

Eshenroder, R.L., 2009. Comment: mitochondrial DNA analysis indicates sea lampreys are indigenous to Lake Ontario. Trans. Am. Fish. Soc. 138, 1178–1189.

Eshenroder, R.L., 2014. The role of the Champlain Canal and Erie Canal as putative corridors for colonization of Lake Champlain and Lake Ontario by Sea Lampreys. Trans. Am. Fish. Soc. 143, 634–649.

Evans, D.H., Piermarini, P.M., Choe, K.P., 2005. The multifunctional fish gill: dominant site of gas exchange, osmoregulation, acid-base regulation, and excretion of nitrogenous waste. Physiol. Rev. 85, 97–177.

Evans, T.M., Wagner, C.M., Miehls, S.M., Johnson, N.S., Haas, T.F., Dunlop, E., Manzon, R.G., 2021. Before the first meal: the elusive pre-feeding juvenile stage of the sea lamprey. J. Great Lakes Res. 47, S580–S589.

Fain, G.L., 2020. Lamprey vision: photoreceptors and organization of the retina. Semin. Cell Dev. Biol. 106, 5–11.

Farmer, G.J., 1980. Biology and physiology of feeding in adult lampreys. Can. J. Fish. Aquat. Sci. 37, 1751–1761.

Farmer, G.J., Beamish, F.W.H., 1973. Sea lamprey (*Petromyzon marinus*) predation on freshwater teleosts. J. Fish. Res. Board Can. 30, 601–605.

Farmer, G.J., Beamish, F.W.H., Robinson, G.A., 1975. Food consumption of the adult landlocked sea lamprey, *Petromyzon marinus*, L. Comp. Biochem. Physiol. A Comp. 50, 753–757.

Farmer, G.J., Beamish, F.W.H., Lett, P.F., 1977. Influence of water temperature on the growth rate of the landlocked sea lamprey (*Petromyzon marinus*) and the associated rate of host mortality. J. Fish. Board Can. 34, 1373–1378.

Felip, A., Zanuy, S., Pineda, R., Pinilla, L., Carrillo, M., Tena-Sempere, M., Gómez, A., 2009. Evidence for two distinct KiSS genes in non-placental vertebrates that encode kisspeptins with different gonadotropin-releasing activities in fish and mammals. Mol. Cell. Endocrinol. 312, 61–71.

Ferreira-Martins, D., Coimbra, J., Antunes, C., Wilson, J.M., 2016. Effects of salinity on upstream-migrating, spawning sea lamprey, *Petromyzon marinus*. Conserv. Physiol. 4, cov064.

Ferreira-Martins, D., Wilson, J.M., Kelly, S.P., Kolosov, D., McCormick, S.D., 2021a. A review of osmoregulation in lamprey. J. Great Lakes Res. 47, S51–S79.

Ferreira-Martins, D., Champer, J., McCauley, D.W., Zhang, Z., Docker, M.F., 2021b. Genetic control of invasive sea lamprey in the Great Lakes. J. Great Lakes Res. 47, S764–S775.

Fine, J.M., Vrieze, L.A., Sorensen, P.W., 2004. Evidence that petromyzontid lampreys employ a common migratory pheromone that is partially comprised of bile acids. J. Chem. Ecol. 30, 2091–2110.

Firkus, T.J., Murphy, C.A., Adams, J.V., Treska, T.J., Fischer, G., 2021. Assessing the assumptions of classification agreement, accuracy, and predictable healing time of sea lamprey wounds on lake trout. J. Great Lakes Res. 47, S368–S377.

Fissette, S.D., Buchinger, T.J., Wagner, C.M., Johnson, N.S., Scott, A.M., Li, W., 2021. Progress towards integrating an understanding of chemical ecology into sea lamprey control. J. Great Lakes Res. 47, S660–S672.

Fredricks, K., Johnson, N., Hubert, T., Siefkes, M., 2021. Registration and application of sea lamprey pheromones for sea lamprey control in the United States and Canada. J. Great Lakes Res. 47, S448–S454.

Gaden, M., Brant, C., Stedman, R.C., Cooke, S.J., Young, N., Lauber, T.B., Nguyen, V.M., Connelly, N.A., Knuth, B., 2021. Shifting baselines and social license to operate: challenges in communicating sea lamprey control. J. Great Lakes Res. 47, S800–S808.

Gage, S.H., 1928. The lampreys of New York State—life history and economics. In: Biological Survey of the Oswego River System. New York Conservation Department Supplement to the 17th Annual Report, 1927. New York Conservation Department, pp. 158–191.

Galizi, R., Doyle, L.A., Menichelli, M., Bernardini, F., Deredec, A., Burt, A., Stoddard, B.L., Windbichler, N., Crisanti, A., 2014. A synthetic sex ratio distortion system for the control of the human malaria mosquito. Nat. Commun. 5, 3977.

Gambicki, S., Steinhart, G.B., 2017. Changes in sea lamprey size and fecundity through time in the Great Lakes. J. Great Lakes Res. 43, 209–214.

Geddes, L.A., 1965. Electronarcosis. Med. Electron. Biol. Eng. 3, 11–26.

Gingera, T.D., Steeves, T.B., Boguski, D.A., Whyard, S., Li, W., Docker, M.F., 2016. Detection and identification of lampreys in Great Lakes streams using environmental DNA. J. Great Lakes Res. 42, 649–659.

GLANSIS, 2021. Great Lakes Aquatic Nonindigenous Species Information System. National Oceanic and Atmospheric Administration (NOAA). U.S. Department of Commerce. https:/www.glerl.noaa.gov/glansis/ accessed December 6, 2021.

Goetz, F., Smith, S.E., Goetz, G., Murphy, C.A., 2016. Sea lampreys elicit strong transcriptomic responses in the lake trout liver during parasitism. BMC Genomics 17, 1–16.

González-Afanador, I., Shi, H., Holbrook, C., Tan, X., Sepúlveda, N., 2021. Sea lamprey detection and characterization using interdigitated electrode (IDE) contact sensor. IEEE Sens. J. 21, 27947–27956.

Grillner, S., 2021. Evolution of the vertebrate motor system — from forebrain to spinal cord. Curr. Opin. Neurobiol. 71, 11–18.

Grunder, S.A., Markham, J.L., Sullivan, W.P., Eilers, C., Tallon, K., McGarry, D., 2021. A review of sea lamprey control in Lake Erie, 2000–2019. J. Great Lakes Res. 47, S506–S522.

Hansen, G.J.A., Jones, M.L., 2008. A rapid assessment approach to prioritizing streams for control of Great Lakes sea lampreys (*Petromyzon marinus*): a case study in adaptive management. Can. J. Fish. Aquat. Sci. 65, 2471–2484.

Hansen, M.J., Madenjian, C.P., Slade, J.W., Steeves, T.B., Almeida, P.R., Quintella, B.R., 2016. Population ecology of the sea lamprey (*Petromyzon marinus*) as an invasive species in the Laurentian Great Lakes and an imperiled species in Europe. Rev. Fish Biol. Fish. 26, 509–535.

Hanson, L.H., 1980. 1980 Study to Determine Burst Swimming Speed Of Spawning-Run Sea Lampreys (*Petromyzon marinus*). Research Completion Report, Hammond Bay Biological Station. US Fish and Wildlife Service, Millersburg, MI.

Hanson, L.H., 1990. Sterilizing effects of cobalt-60 and cesium-137 radiation on male sea lampreys. N. Am. J. Fish. Manag. 10, 352–361.

Hanson, M.E., 2021. Making a Repellent: Overcoming Physiological Impediments to Guiding Migratory Sea Lamprey (*Petromyzon marinus*) With an Alarm Cue. Doctoral dissertation, Michigan State University, Lansing.

Hanson, L.H., Manion, P.J., 1980. Sterility method of pest control and its potential role in an integrated sea lamprey (*Petromyzon marinus*) control program. Can. J. Fish. Aquat. Sci. 37, 2108–2117.

Hanson, L.H., Swink, W.D., 1989. Downstream migration of recently metamorphosed sea lampreys in the Ocqueoc River, Michigan, before and after treatment with lampricides. N. Am. J. Fish. Manag. 9, 327–331.

Hanson, P.C., Johnson, T.B., Schindler, D.E., Kitchell, J.F., 1997. Fish Bioenergetics 3.0 for Windows. University of Wisconsin Center for Limnology, Sea Grant Institute, Madison, Wisconsin.

Happel, A., Rinchard, J., Czesny, S., 2017. Variability in sea lamprey fatty acid profiles indicates a range of host species utilization in Lake Michigan. J. Great Lakes Res. 43, 182–188.

Hardisty, M.W., Potter, I.C., 1971. The behaviour, ecology and growth of larval lampreys. In: Hardisty, M.W., Potter, I.C., I.C. (Eds.), The Biology of Lampreys. 1. Academic Press, London, pp. 85–125.

Haro, A., Miehls, S., Johnson, N.S., Wagner, C.M., 2020. Evaluation of visible light as a cue for guiding downstream migrant juvenile sea lamprey. Trans. Am. Fish. Soc. 149, 635–647.

Harvey, C.J., Ebener, M.P., White, C.K., 2008. Spatial and ontogenetic variability of sea lamprey diets in Lake Superior. J. Great Lakes Res. 34, 434–449.

Hayes, D.B., Caroffino, D.C. (Eds.), 2012. Michigan's Lake Sturgeon Rehabilitation Strategy. Michigan Department of Natural Resources, Fisheries Special Report 62, Lansing, MI.

He, C., Yin, L., Tang, C., Yin, C., 2013. Multifunctional polymeric nanoparticles for oral delivery of TNF-a siRNA to macrophages. Biomaterials 34, 2843–2854.

Heath, G., Childs, D., Docker, M.F., McCauley, D.W., Whyard, S., 2014. RNA interference technology to control pest sea lampreys—A proof-of-concept. PLoS One 9, e88387.

Heath, V.L., Miehls, S., Johnson, N., Higgs, D.M., 2021. Behavioural response of sea lamprey (*Petromyzon marinus*) to acoustic stimuli in a small stream. Can. J. Fish. Aquat. Sci. 78, 341–348.

Heinrich, J.W., Mullett, K.M., Hansen, M.J., Adams, J.V., Klar, G.T., Johnson, D.A., Christie, G. C., Young, R.J., 2003. Sea lamprey abundance and management in Lake Superior, 1957 to 1999. J. Great Lakes Res. 29, 566–583.

Henry, M., Birceanu, O., Clifford, A.M., McClelland, G.B., Wang, Y.S., Wilkie, M.P., 2015. Life stage dependent responses to the lampricide, 3-trifluoromethy1-4-nitrophenol (TFM), provide insight into glucose homeostasis and metabolism in the sea lamprey (*Petromyzon marinus*). COMP. Biochem. Physiol. C Toxicol. Pharmacol. 169, 35–45.

Henson, M.P., Bergstedt, R.A., Adams, J.V., 2003. Comparison of spring measures of length, weight, and condition factor for predicting metamorphosis in two populations of sea lamprey (*Petromyzon marinus*) larvae. J. Great Lakes Res. 29 (Suppl. 1), 204–213.

Hepditch, S.L.J., Tessier, L.R., Wilson, J.M., Birceanu, O., O'Connor, L.M., Wilkie, M.P., 2019. Mitigation of lampricide toxicity to juvenile lake sturgeon: the importance of water alkalinity and life stage. Conserv. Physiol. 7, coz089.

Hepditch, S.L.J., Birceanu, O., Wilkie, M.P., 2021. A toxic unit and additive index approach to understanding the interactions of 2 piscicides, 3-trifluoromethyl-4-nitrophenol and niclosamide, in rainbow trout. Environ. Toxicol. Chem. 40, 1419–1430.

Hlina, B.L., Tessier, L.R., Wilkie, M.P., 2017. Effects of water pH on the uptake and elimination of the piscicide, 3-trifluoromethyl-4-nitrophenol (TFM), by larval sea lamprey. Comp. Biochem. Physiol. C Toxicol. Pharmacol. 200, 9–16.

Hlina, B.L., Birceanu, O., Robinson, C.S., Dhiyebi, H., Wilkie, M.P., 2021. The relationship between thermal physiology and lampricide sensitivity in larval sea lamprey (*Petromyzon marinus*). J. Great Lakes Res. 47, S272–S284.

Hockman, D., Chong-Morrison, V., Green, S.A., et al., 2019. A genome-wide assessment of the ancestral neural crest gene regulatory network. Nat. Commun. 10, 4689.

Hodges, J.W., 1972. Downstream migration of recently transformed sea lampreys before and after treatment of a Lake Michigan tributary with a lampricide. J. Fish. Res. Board Can. 29, 1237–1240.

Holmes, J.A., Lin, P., 1994. Thermal niche of larval sea lamprey, *Petromyzon marinus*. Can. J. Fish. Aquat. Sci. 51, 253–262.

Holmes, J.A., Youson, J.H., 1993. Induction of metamorphosis in landlocked sea lampreys, *Petromyzon marinus*. J. Exp. Zool. 267, 598–604.

Hoover, J.J., Murphy, C.E., 2018. Maximum Swim Speed of Migrating Sea Lamprey (*Petromyzon marinus*): Reanalysis of Data From a Prior Study. Technical Report ERDC/TN ANSRP-18-1. U.S. Army Engineer Research and Development Center, Vicksburg, MS.

Horak, K.E., 2020. RNAi: applications in vertebrate pest management. Trends Biotechnol. 38, 1200–1202.

Howe, G.E., Marking, L.L., Bills, T.D., Rach, J.J., Mayer, F.L., 1994. Effects of water temperature and pH on toxicity of terbufos, trichlorfon, 4-nitrophenol and 2,4-dinitrophenol to the amphipod *Gammarus pseudolimnaeus* and rainbow trout (*Oncorhynchus mykiss*). Environ. Toxicol. Chem. 13, 51–66.

Hrodey, P.J., Lewandoski, S.A., Sullivan, W.P., Barber, J.M., Mann, K.A., Paudel, B., Symbal, M.J., 2021. Evolution of the Sea Lamprey Control Barrier Program: the importance of lowermost barriers. J. Great Lakes Res. 47, S285–S296.

Hubert, T., Miller, J., Burkett, D., 2021. A brief introduction to integrated pest management for aquatic systems. N. Am. J. Fish. Manag. 41, 264–275.

Huerta, B., Chung-Davidson, Y.-W., Bussy, U., Zhang, Y., Bazil, J.N., Li, W., 2020. Sea lamprey cardiac mitochondrial bioenergetics after exposure to TFM and its metabolites. Aquat. Toxicol. 219, 105380.

Hume, J.B., Wagner, M., 2018. A death in the family: sea lamprey (*Petromyzon marinus*) avoidance of confamilial alarm cues diminishes with phylogenetic distance. Ecol. Evol. 8, 3751–3762.

Hume, J.B., Meckley, T.D., Johnson, N.S., Luhring, T.M., Siefkes, M.J., Wagner, C.M., 2015. Application of a putative alarm cue hastens the arrival of invasive sea lamprey (*Petromyzon marinus*) at a trapping location. Can. J. Fish. Aquat. Sci. 72, 1799–1806.

Hume, J.B., Luhring, T.M., Wagner, C.M., 2020a. Push, pull, or push–pull? An alarm cue better guides sea lamprey towards capture devices than a mating pheromone during the reproductive migration. Biol. Invasions 22, 2129–2142.

Hume, J.B., Lucas, M.C., Reinhardt, U., Hrodey, P.J., Wagner, C.M., 2020b. Sea lamprey (*Petromyzon marinus*) transit of a ramp equipped with studded substrate: implications for fish passage and invasive species control. Ecol. Eng. 155, 105957.

Hume, J.B., Bravener, G.A., Flinn, S., Johnson, N.S., 2021a. What can commercial fishery data in the Great Lakes reveal about juvenile sea lamprey (*Petromyzon marinus*) ecology and management? J. Great Lakes Res. 47, S590–S603.

Hume, J.B., Almeida, P.R., Buckley, C.M., Criger, L.A., Madenjian, C.P., Robinson, K.F., Wang, C.J., Muir, A.M., 2021b. Managing native and non-native sea lamprey (*Petromyzon marinus*) through anthropogenic change: a prospective assessment of key threats and uncertainties. J. Great Lakes Res. 47, S704–S722.

Hume, J.B., Bracken, F.S., Mateus, C.S., Brant, C.O., 2021c. Synergizing basic and applied scientific approaches to help understand lamprey biology and support management actions. J. Great Lakes Res. 47, S24–S37.

Hunn, J.B., Youngs, W.D., 1980. Role of physical barriers in the control of sea lamprey (*Petromyzon marinus*). Can. J. Fish. Aquat. Sci. 37, 2118–2122.

Imre, I., Brown, G.E., Bergstedt, R.A., McDonald, R., 2010. Use of chemosensory cues as repellents for sea lamprey: potential directions for population management. J. Great Lakes Res. 36, 790–793.

Imre, I., Di Rocco, R.T., Brown, G.E., Johnson, N.S., 2016. Habituation of adult sea lamprey repeatedly exposed to damage-released alarm and predator cues. Environ. Biol. Fishes 99, 613–620.

Ionescu, R., Hepditch, S., Wilkie, M., 2021. The lampricide 3-trifluoromethyl-4-nitrophenol (TFM) causes temporary metabolic disturbances in juvenile lake sturgeon (*Acipenser fulvescens*): implications for sea lamprey control and fish conservation. Conserv. Physiol. 9, coab069.

Ionescu, R., Mitrovic, D., Wilkie, M., 2022a. Reversible disruptions to energy supply and acid-base balance in larval sea lamprey exposed to the pesticide: niclosamide (2'5-dichloro-4'-nitrosalicylanilide). Aquat. Toxicol. 242, 106006.

Ionescu, R., Mitrovic, D., Wilkie, M., 2022b. Disturbances to energy metabolism in juvenile lake sturgeon (*Acipenser fulvescens*) following exposure to niclosamide. Ecotoxicol. Environ. Saf. 229, 112969.

Irons, K.S., Sass, G.G., McClelland, M.A., Stafford, J.D., 2007. Reduced condition factor of two native fish species coincident with invasion of non-native Asian carps in the Illinois River, USA—is this evidence for competition and reduced fitness? J. Fish Biol. 71, 258–273.

Irwin, B.J., Liu, W., Bence, J.R., Jones, M.L., 2012. Defining economic injury levels for sea lamprey control in the Great Lakes basin. N. Am. J. Fish. Manag. 32, 760–771.

Jacobs-Lorena, M., 2006. Genetic approaches for malaria control. In: Knols, B.G.J., Louis, C. (Eds.), Bridging Laboratory and Field Research for Genetic Control of Disease Vectors. Springer, Dordrecht, pp. 53–65.

Jerde, C.L., Mahon, A.R., Chadderton, W.L., Lodge, D.M., 2011. "Sight-unseen" detection of rare aquatic species using environmental DNA. Conserv. Lett. 4, 150–157.

Johnson, B.G.H., Anderson, W.C., 1980. Predatory-phase sea lampreys (*Petromyzon marinus*) in the Great Lakes. Can. J. Fish. Aquat. Sci. 37, 2007–2020.

Johnson, N.S., Miehls, S., 2014. Guiding out-migrating juvenile Sea Lamprey (*Petromyzon marinus*) with pulsed direct current. River Res. Appl. 30, 1146–1156.

Johnson, N., Siefkes, M., Li, W., 2005. Capture of ovulating female sea lampreys in traps baited with spermiating male sea lampreys. N. Am. J. Fish. Manag. 25, 67–72.

Johnson, N.S., Luehring, M.A., Siefkes, M.J., Li, W., 2006. Mating pheromone reception and induced behavior in ovulating female sea lampreys. N. Am. J. Fish Manag. 26 (1), 88–96.

Johnson, N.S., Yun, S.S., Thompson, H.T., Brant, C.O., Li, W., 2009. A synthesized pheromone induces upstream movement in female sea lamprey and summons them into traps. Proc. Natl. Acad. Sci. U. S. A. 106, 1021–1026.

Johnson, N.S., Thompson, H.T., Holbrook, C., Tix, J., 2014. Blocking and guiding adult sea lamprey with pulsed direct current from vertical electrodes. Fish. Res. 150, 38–48.

Johnson, N.S., Siefkes, M.J., Wagner, C.M., Dawson, H., Wang, H., Steeves, T., Twohey, M., Li, W., 2013. A synthesized mating pheromone component increases adult sea lamprey (*Petromyzon marinus*) trap capture in management scenarios. Can. J. Fish. Aquat. Sci. 70, 1101–1108.

Johnson, N.S., Buchinger, T.J., Li, W., 2015. Reproductive ecology of lampreys. In: Docker, M.F. (Ed.), Lampreys: Biology, Conservation and Conservation and Control. 1. Springer, Dordrecht, pp. 265–303.

Johnson, N.S., Miehls, S., O'Connor, L.M., Bravener, G., Barber, J., Thompson, H., Tix, J.A., Bruning, T., 2016. A portable trap with electric lead catches up to 75% of an invasive fish species. Sci. Rep. 6, 1–8.

Johnson, N.S., Miehls, S.M., Haro, A.J., Wagner, C.M., 2019. Push and pull of downstream moving juvenile sea lamprey (*Petromyzon marinus*) exposed to chemosensory and light cues. Conserv. Physiol. 7, coz080.

Johnson, N.S., Lewandoski, S.A., Alger, B.J., O'Connor, L., Bravener, G., Hrodey, P., Huerta, B., Barber, J., Li, W., Wagner, C.M., Siefkes, M.J., 2020. Behavioral responses of sea lamprey to varying application rates of a synthesized pheromone in diverse trapping scenarios. J. Chem. Ecol. 46, 233–249.

Johnson, N.S., Lewandoski, S.A., Merkes, C., 2021a. Assessment of sea lamprey (*Petromyzon marinus*) diet using DNA metabarcoding of feces. Ecol. Indic. 125, 107605.

Johnson, N.S., Snow, B., Bruning, T., Jubar, A., 2021b. A seasonal electric barrier blocks invasive adult sea lamprey (*Petromyzon marinus*) and reduces production of larvae. J. Great Lakes Res. 47, S310–S319.

Johnson, N.S., Jubar, A.K., Keffer, D.A., Hrodey, P.J., Bravener, G.A., Freitas, L.E., McCarter, J.T., Siefkes, M.J., 2021c. A case study of sea lamprey (*Petromyzon marinus*) control and ecology in a microcosm of the Great Lakes. J. Great Lakes Res. 47, S492–S505.

Jones, M.L., Bergstedt, R.A., Twohey, M.B., Fodale, M.F., Cuddy, D.W., Slade, J.W., 2003. Compensatory mechanisms in Great Lakes sea lamprey populations: implications for alternative control strategies. J. Great Lakes Res. 29, 113–129.

Jones, J., Wellband, K., Zielinski, B., Heath, D.D., 2019. Transcriptional basis of copper-induced olfactory impairment in the sea lamprey, a primitive invasive fish. Gen. Genom. Genet. 9, 933–941.

Jubar, A.K., Frank, R.J., Keffer, D.A., Neave, F.B., Symbal, M.J., Steeves, T.B., 2021. Prioritizing lampricide treatments in Great Lakes tributaries and lentic areas during 2000–2017. J. Great Lakes Res. 47, S238–S246.

Kamunde, C.N., Niyogi, S., Wood, C.M., 2005. Interaction of dietary sodium chloride and waterborne copper in rainbow trout (*Oncorhynchus mykiss*): copper toxicity and sodium and chloride homeostasis. Can. J. Fish Aquat. Sci. 62, 390–399.

Kane, A.S., Kahng, M.W., Reimschuessel, R., Nhamburo, P.T., Lipsky, M.M., 1994. UDP-glucuronosyltransferase kinetics for 3-trifluoromethyl-4-nitrophenol (TFM) in fish. Trans. Am. Fish. Soc. 123, 217–222.

Kaye, C.A., 2021. Addressing the risk of lampricide exposure to three federally listed species in US streams. J. Great Lakes Res. 47, S388–S406.

Kaye, C.A., Heinrich, J.W., Hanson, L.H., McDonald, R.B., Slade, J.W., Genovese, J.H., Swink, W.D., 2003. Evaluation of strategies for the release of male sea lampreys (*Petromyzon marinus*) in Lake Superior for a proposed sterile-male-release program. J. Great Lakes Res. 29, 424–434.

Kelley, A.L., 2014. The role thermal physiology plays in species invasion. Conserv. Physiol. 2, cou045.

King Jr., E.L., 1980. Classification of sea lamprey (*Petromyzon marinus*) attack marks on Great Lakes lake trout (*Salvelinus namaycush*). Can. J. Fish. Aquat. Sci. 37 (11), 1989–2006.

King Jr., E.L., Gabel, J.A., 1985. Comparative toxicity of the lampricide 3-trifluoromethyl-4-nitrophenol to ammocoetes of three species of lampreys. Great Lakes Fish. Comm. Tech. Rep. 47, 1–5.

Kitchell, J.F., 1990. The scope for mortality caused by sea lamprey. Trans. Am. Fish. Soc. 119, 642–648.

Kitchell, J.F., Breck, J.E., 1980. Bioenergetics model and foraging hypothesis for sea lamprey (*Petromyzon marinus*). Can. J. Fish. Aquat. Sci. 37, 2159–2168.

Kleerekoper, H., Mogensen, J., 1963. Role of olfaction in the orientation of *Petromyzon marinus*. I. Response to a single amine in prey's body odor. Physiol. Zool. 36, 347–360.

Knipling, E.F., 1959. Sterile-male method of population control: successful with some insects, the method may also be effective when applied to other noxious animals. Science 130, 902–904.

Koch, A., Kogel, K.H., 2014. New wind in the sails: improving the agronomic value of crop plants through RNAi-mediated gene silencing. Plant Biotechnol. J. 12, 821–831.

Kolar, C.S., Lodge, D.M., 2001. Progress in invasion biology: predicting invaders. Trends Ecol. Evol. 16, 199–204.

Kolar, C.S., Lodge, D.M., 2002. Ecological predictions and risk assessment for alien fishes in North America. Science 298, 1233–1236.

Kolosov, D., Bui, P., Chasiotis, H., Kelly, S.P., 2013. Claudins in teleost fishes. Tiss. Barriers 1 (3), e25391-1–e25391-15.

Kolosov, D., Bui, P., Donini, A., Wilkie, M.P., Kelly, S.P., 2017. A role for tight junction-associated MARVEL proteins in larval sea lamprey (*Petromyzon marinus*) osmoregulation. J. Exp. Biol. 220, 3657–3670.

Kolosov, D., Bui, P., Wilkie, M.P., Kelly, S.P., 2020. Claudins of sea lamprey (*Petromyzon marinus*) - organ-specific expression and transcriptional responses to water of varying ion content. J. Fish Biol. 96, 768–781.

Krantzberg, G., de Boer, C., 2008. A valuation of ecological services in the Laurentian Great Lakes Basin with an emphasis on Canada. J. Am. Water Works Assoc. 100, 100–111.

Lamsa, A.K., Rovainen, C.M., Kolenosky, D.P., Hanson, L.H., 1980. Sea lamprey (*Petromyzon marinus*) control—where to from here? Report of the SLIS control theory task force. Can. J. Fish. Aquat. Sci. 37, 2175–2192.

Lança, M.J., Rosado, C., Machado, M., Ferreira, R., Alves-Pereira, I., Quintella, B.R., Almeida, P.R., 2011. Can muscle fatty acid signature be used to distinguish diets during the marine trophic phase of sea lamprey (*Petromyzon marinus*, L.). Comp. Biochem. Physiol. B Biochem. Mol. Biol. 159, 26–39.

Lança, M.J., Machado, M., Ferreira, R., Alves-Pereira, I., Quintella, B.R., De Almeida, P.R., 2013. Feeding strategy assessment through fatty acid profiles in muscles of adult sea lampreys from the western Iberian coast. Sci. Mar. 77, 281–291.

Lantry, B., Adams, J., Christie, G., Schaner, T., Bowlby, J., Keir, M., Lantry, J., Sullivan, P., Bishop, D., Treska, T., Morrison, B., 2015. Sea lamprey mark type, marking rate, and parasite–host relationships for lake trout and other species in Lake Ontario. J. Great Lakes Res. 41, 266–279.

Lantz, S.R., Adair, R.A., Amberg, J.J., et al., 2022. Next generation lampricides: a three-stage process to develop improved control tools for invasive sea lamprey. Can. J. Fish. Aquat. Sci. 79, 692–702.

Larson, W.A., Kornis, M.S., Turnquist, K.N., Bronte, C.R., Holey, M.E., Hanson, S.D., Treska, T.J., Stott, W., Sloss, B.L., 2021. The genetic composition of wild recruits in a recovering lake trout population in Lake Michigan. Can. J. Fish. Aquat. Sci. 78, 286–300.

Lavis, D.S., Henson, M.P., Johnson, D.A., Koon, E.M., Ollila, D.J., 2003. A case history of sea lamprey control in Lake Michigan: 1979 to 1999. J. Great Lakes Res. 29, 584–598.

Lawrence, M.J., Mitrovic, D., Foubister, D., Bragg, L.M., Sutherby, J., Docker, M.F., Servos, M.R., Wilkie, M.P., Jeffries, K.M., 2021. Contrasting physiological responses between invasive sea lamprey and non-target bluegill in response to acute lampricide exposure. Aquat. Toxicol. 237, 105848.

Lech, J.J., Statham, C.N., 1975. Role of glucuronide formation in selective toxicity of 3-trifluoromethyl-4-nitrophenol (TFM) for sea lamprey—comparative aspects of TFM uptake and conjugation in sea lamprey and rainbow trout. Toxicol. Appl. Pharmacol. 31, 150–158.

Lee, Y.R., Tsunekawa, K., Moon, M.J., et al., 2009. Molecular evolution of multiple forms of kisspeptins and GPR54 receptors in vertebrates. Endocrinology 150, 2837–2846.

Lennon, R.E., 1954. Feeding mechanism of the sea lamprey and its effect on host fishes. Fishery Report 98. In: US Fish Wildl. Serv. Fish. Bull. 56, pp. 247–293.

Lennox, R., Choi, K., Harrison, P.M., Paterson, J.E., Peat, T.B., Ward, T.D., Cook, S.J., 2015. Improving science-based invasive species management with physiological knowledge, concepts, and tools. Biol. Invasions 17 (8), 2213–2227.

Lennox, R.J., Bravener, G.A., Lin, H.Y., Madenjian, C.P., Muir, A.M., Remucal, C.K., Robinson, K.F., Rous, A.M., Siefkes, M.J., Wilkie, M.P., Zielinski, D.P., 2020. Potential changes to the biology and challenges to the management of invasive sea lamprey *Petromyzon marinus* in the Laurentian Great Lakes due to climate change. Glob. Chang. Biol. 26, 1118–1137.

Leung, B., Mandrak, N.E., 2007. The risk of establishment of aquatic invasive species: joining invasibility and propagule pressure. Proc. R. Soc. B 274, 2603–2609.

Lewandoski, S.A., Brenden, T.O., Siefkes, M.J., Johnson, N.S., 2021a. An adaptive management implementation framework for evaluating supplemental sea lamprey (*Petromyzon marinus*) controls in the Laurentian Great Lakes. J. Great Lakes Res. 47, S753–S763.

Lewandoski, S.A., Hrodey, P., Miehls, S., Piszczek, P.P., Zielinski, D.P., 2021b. Behavioral responses of sea lamprey (*Petromyzon marinus*) and white sucker (*Catostomus commersonii*) to turbulent flow during fishway passage attempts. Can. J. Fish. Aquat. Sci. 78, 409–421.

Lewis, S.L., Maslin, M.A., 2015. Defining the anthropocene. Nature 519, 171–180.

Li, W., Scott, A.P., Siefkes, M.J., Yan, H., Liu, Q., Yun, S.S., Gage, D.A., 2002. Bile acid secreted by male sea lamprey that acts as a sex pheromone. Science 296, 138–141.

Li, W., Twohey, M., Jones, M., Wagner, M., 2007. Research to guide use of pheromones to control sea lamprey. J. Great Lakes Res. 33, 70–86.

Li, B.W., Gou, M., Han, J.M., Yuan, X.F., Li, Y.Y., Li, T.S., Jiang, Q., Xiao, R., Li, Q.W., 2018a. Proteomic analysis of buccal gland secretion from fasting and feeding lampreys (*Lampetra morii*). Proteome Sci. 16 (1), 1–9.

Li, K., Brant, C.O., Huertas, M., Hessler, E.J., Mezei, G., Scott, A.M., Hoye, T.R., Li, W., 2018b. Fatty-acid derivative acts as a sea lamprey migratory pheromone. Proc. Natl. Acad. Sci. U. S. A. 115, 8603–8608.

Li, M., Yang, T., Bui, M., et al., 2021. Suppressing mosquito populations with precision guided sterile males. Nat. Commun. 12, 1–10.

Lighthill, M.J., 1969. Hydromechanics of aquatic animal propulsion. Annu. Rev. Fluid Mech. 1, 413–446.

Lockwood, J.L., Cassey, P., Blackburn, T.M., 2009. The more you introduce the more you get: the role of colonization pressure and propagule pressure in invasion ecology. Divers. Distrib. 15, 904–910.

Lowe, D.R., Beamish, F.W.H., Potter, I.C., 1973. Changes in proximate body composition of landlocked sea lamprey *Petromyzon marinus* (L) during larval life and metamorphosis. J. Fish Biol. 5, 673–682.

Lucas, M.C., Hume, J.B., Almeida, P.R., Aronsuu, K., Habit, E., Silva, S., Wang, C.J., Zampatti, B., 2021. Emerging conservation initiatives for lampreys: research challenges and opportunities. J. Great Lakes Res. 47, S690–S703.

Luhring, T.M., Meckley, T.D., Johnson, N.S., Siefkes, M.J., Hume, J.B., Wagner, C.M., 2016. A semelparous fish continues upstream migration when exposed to alarm cue, but adjusts movement speed and timing. Anim. Behav. 121, 41–51.

MacKay, N.A., 1992. Evaluating the size effects of lampreys and their hosts: application of an individual based model. In: DeAngelis, D., Gross, L. (Eds.), Individual-Based Models and Approaches in Ecology: Populations, Communities, and Ecosystems. Chapman & Hall, New York, pp. 278–294.

Madenjian, C.P., Cochran, P.A., Bergstedt, R.A., 2003. Seasonal patterns in growth, blood consumption, and effects on hosts by parasitic-phase sea lampreys in the Great Lakes: an individual-based model approach. J. Great Lakes Res. 29, 332–346.

Madenjian, C.P., Chipman, B.D., Marsden, J.E., 2008. New estimates of lethality of sea lamprey (*Petromyzon marinus*) attacks on lake trout (*Salvelinus namaycush*): implications for fisheries management. Can. J. Fish. Aquat. Sci. 65, 535–542.

Madenjian, C.P., Johnson, N.S., Siefkes, M.J., Dettmers, J.M., Blum, J.D., Johnson, M.W., 2014. Mercury accumulation in sea lamprey (*Petromyzon marinus*) from Lake Huron. Sci. Total Environ. 470, 1313–1319.

Maitland, P.S., Renaud, C.B., Quintella, B.R., Close, D.A., Docker, M.F., 2015. Conservation of native lampreys. In: Docker, M.F. (Ed.), Lampreys: Biology, Conservation and Control. 1. Springer, Dordrecht, pp. 375–428.

Mallatt, J., 1996. Ventilation and the origin of jawed vertebrates: a new mouth. Zool. J. Linn. Soc. 117, 329–404.

Manion, P.J., Hanson, L.H., 1980. Spawning behaviour and fecundity of lampreys from the upper 3 Great Lakes. Can. J. Fish. Aquat. Sci. 37, 1635–1640.

Manion, P.J., Smith, B.R., 1978. Biology of larval and metamorphosing sea lampreys, *Petromyzon marinus*, of the 1960 year class in the Big Garlic River, Michigan, Part II, 1966–72. Great Lakes Fish. Comm. Tech. Rep. 30, 1–35.

Manzon, R.G., Youson, J.H., 1997. The effects of exogenous thyroxine ($T_4$) or triiodothyronine ($T_3$), in the presence and absence of potassium perchlorate, on the incidence of metamorphosis and on serum $T_4$ and $T_3$ concentrations in larval sea lampreys (*Petromyzon marinus* L.). Gen. Comp. Endocrinol. 106, 211–220.

Manzon, R.G., Holmes, J.A., Youson, J.H., 2001. Variable effects of goitrogens in inducing precocious metamorphosis in sea lampreys (*Petromyzon marinus*). J. Exp. Zool. 289, 290–303.

Manzon, R.G., Youson, J.H., Holmes, J.A., 2015. Lamprey metamorphosis. In: Docker, M. (Ed.), Lampreys: Biology, Conservation and Control. 1. Springer, Dordrecht, pp. 139–214.

Marsden, J.E., Siefkes, M.J., 2019. Control of invasive sea lamprey in the Great Lakes, Lake Champlain, and Finger Lakes of New York. In: Docker, M.F. (Ed.), Lampreys: Biology, Conservation and Control. 2. Springer, Dordrecht, pp. 441–479.

Mastrangelo, T., Kovaleski, A., Botteon, V., Scopel, W., Costa, M.D.L.Z., 2018. Optimization of the sterilizing doses and overflooding ratios for the South American fruit fly. PLoS One 13, e0201026.

Mathai, P.P., Byappanahalli, M.N., Johnson, N.S., Sadowsky, M.J., 2021. Gut microbiota associated with different sea lamprey (*Petromyzon marinus*) life stages. Front. Microbiol. 12, 706683.

McAuley, T.C., 1996. Development of an Instream Velocity Barrier to Stop Sea Lamprey (*Petromyzon marinus*) Migrations in Great Lakes Streams. Master's thesis, University of Manitoba, Winnipeg, Manitoba.

McCann, E.L., Johnson, N.S., Hrodey, P.J., Pangle, K.L., 2018a. Characterization of sea lamprey stream entry using dual-frequency identification sonar. Trans. Am. Fish. Soc. 147, 514–524.

McCann, E.L., Johnson, N.S., Pangle, K.L., 2018b. Corresponding long-term shifts in stream temperature and invasive fish migration. Can. J. Fish. Aquat. Sci. 75, 772–778.

McCann, E., Li, L., Pangle, K., Johnson, N., Eickholt, J., 2018c. An underwater observation dataset for fish classification and fishery assessment. Sci. Data 5, 180190.

McCauley, D.W., Docker, M.F., Whyard, S., Li, W., 2015. Lampreys as diverse model organisms in the genomics era. BioScience 65, 1046–1056.

McCormick, S.D., 2013. Smolt physiology and endocrinology. In: McCormick, S.D., Farrell, A.P., Brauner, C.J. (Eds.), Fish Physiology: Euryhaline Fishes. 32. Elsevier, Amsterdam, pp. 199–251.

McDonald, D.G., Kolar, C.S., 2007. Research to guide the use of lampricides for controlling sea lamprey. J. Great Lakes Res. 33, 20–34.

McLain, A.L., 1957. The control of the upstream movement of fish with pulsated direct current. Trans. Am. Fish. Soc. 86, 269–284.

McLain, A.L., Smith, B.R., Moore, H.H., 1965. Experimental control of sea lampreys with electricity on the south shore of Lake Superior, 1953-1960. In: Great Lakes Fishery Commission Tech. Rep. 10. Ann Arbor, Michigan.

McLaughlin, R.L., Hallett, A., Pratt, T.C., O'Connor, L.M., McDonald, D.G., 2007. Research to guide use of barriers, traps, and fishways to control sea lamprey. J. Great Lakes Res. 33, 7–19.

McLaughlin, R.L., Smyth, E.R.B., Castro-Santos, T., Jones, M.L., Koops, M.A., Pratt, T.C., Veliz-Espino, L.A., 2013. Unintended consequences and trade-offs of fish passage. Fish Fish. 14, 580–604.

McLaughlin, R., Adams, J.V., Almeida, P.R., et al., 2021. Foreword: control and conservation of lampreys beyond 2020—proceedings from the 3rd Sea Lamprey International Symposium (SLIS III). J. Great Lakes Res. 47, S1–S10.

Meckley, T.D., Wagner, C.M., Luehring, M.A., 2012. Field evaluation of larval odor and mixtures of synthetic pheromone components for attracting migrating sea lampreys in rivers. J. Chem. Ecol. 38, 1062–1069.

Meckley, T.D., Wagner, C.M., Gurarie, E., 2014. Coastal movements of migrating sea lamprey (*Petromyzon marinus*) in response to a partial pheromone added to river water: implications for management of invasive populations. Can. J. Fish. Aquat. Sci. 71, 533–544.

Meckley, T.D., Gurarie, E., Miller, J.R., Wagner, C.M., 2017. How fishes find the shore: evidence for orientation to bathymetry from the non-homing sea lamprey. Can. J. Fish. Aquat. Sci. 74, 2045–2058.

Miehls, S.M., Johnson, N.S., Haro, A., 2017. Electrical guidance efficiency of downstream-migrating juvenile sea lampreys decreases with increasing water velocity. Trans. Am. Fish. Soc. 146, 299–307.

Mickle, M.F., Miehls, S.M., Johnson, N.S., Higgs, D.M., 2019. Hearing capabilities and behavioural response of sea lamprey (*Petromyzon marinus*) to low-frequency sounds. Can. J. Fish. Aquat. Sci. 76, 1541–1548.

Miehls, S., Sullivan, P., Twohey, M., Barber, J., McDonald, R., 2020. The future of barriers and trapping methods in the sea lamprey (*Petromyzon marinus*) control program in the Laurentian Great Lakes. Rev. Fish Biol. Fish. 30, 1–24.

Miehls, S.M., Dawson, H.A., Maguffee, A.C., Johnson, N.S., Jones, M.L., Dobiesz, N., 2021. Where you trap matters: Implications for integrated sea lamprey management. J. Great Lakes Res. 47, S320–S327.

Miehls, S.M., Holbrook, C.M., Marsden, J.E., 2019. Diel activity of newly metamorphosed juvenile sea lamprey (*Petromyzon marinus*). PLoS One 14, e0211687.

Miller, J.R., Cowles, R.S., 1990. Stimulo-deterrent diversion: a concept and its possible application to onion maggot control. J. Chem. Ecol. 16, 3197–3212.

Moore, H.H., Schleen, L.P., 1980. Changes in spawning runs of sea lamprey (*Petromyzon marinus*) in selected streams of Lake Superior after chemical control. Can. J. Fish. Aquat. Sci. 37, 1851–1860.

Morkert, S.B., Swink, W.D., Seelye, J.G., 1998. Evidence for early metamorphosis of sea lampreys in the Chippewa River, Michigan. N. Am. J. Fish. Manag. 18, 966–971.

Morris, R., 1972. Osmoregulation. In: Potter C., L., Hardisty M., W. (Eds.), The Biology of Lampreys. 2. Academic Press, London, pp. 192–239.

Morris, R., Bull, J.M., 1970. Studies on freshwater osmoregulation in ammocoete larva of *Lampeta planeri* (Bloch). 3. Effect of external and internal sodium on sodium transport. J. Exp. Biol. 52, 275–290.

Morse, T.J., Ebener, M.P., Koon, E.M., Morkert, S.B., Johnson, D.A., Cuddy, D.W., Weisser, J.W., Mullett, K.M., Genovese, J.H., 2003. A case history of sea lamprey control in Lake Huron: 1979 to 1999. J. Great Lakes Res. 29, 599–614.

Moser, M.L., Almeida, P.R., Kemp, P.S., Sorensen, P.W., 2015. Lamprey spawning migration. In: Docker, M.F. (Ed.), Lampreys: Biology, Conservation and Conservation and Control. 1. Springer, Dordrecht, pp. 215–263.

Muhametsafina, A., Birceanu, O., Hlina, B.L., Tessier, L.R., Wilkie, M.P., 2019. Warmer waters increase the larval sea lamprey's (*Petromyzon marinus*) tolerance to the lampricide 3-trifluoromethyl-4-nitrophenol (TFM). J. Great Lakes Res. 45, 921–933.

Mundahl, N.D., Erickson, C., Johnston, M.R., Sayeed, G.A., Taubel, S., 2005. Diet, feeding rate, and assimilation efficiency of American brook lamprey. Environ. Biol. Fishes 72, 67–72.

Murphy, E.C., Russell, J.C., Broome, K.G., Ryan, G.J., Dowding, J.E., 2019. Conserving New Zealand's native fauna: a review of tools being developed for the Predator Free 2050 programme. J. Ornithol. 160, 883–892.

Neave, F.B., Booth, R.M.W., Philipps, R.R., Keffer, D.A., Bravener, G.A., Coombs, N., 2021. Changes in native lamprey populations in the Great Lakes since the onset of sea lamprey (*Petromyzon marinus*) control. J. Great Lakes Res. 47, S378–S387.

Niblett, P.D., Ballantyne, J.S., 1976. Uncoupling of oxidative-phosphorylation in rat-liver mitochondria by lamprey larvicide TFM (3-trifluoromethyl-4-nitrophenol). Pestic. Biochem. Physiol. 6, 363–366.

Noatch, M.R., Suski, C.D., 2012. Non-physical barriers to deter fish movements. Environ. Rev. 20, 71–82.

O'Boyle, R.N., Beamish, F.W.H., 1977. Growth and intermediary metabolism of larval and metamorphosing stages of the landlocked sea lamprey *Petromyzon marinus*. Environ. Biol. Fishes 2, 103–120.

O'Connor, L.M., Pratt, T.C., Steeves, T.B., Stephens, B., Boogaard, M., Kaye, C., 2017. In situ assessment of lampricide toxicity to age-0 lake sturgeon. J. Great Lakes Res. 43, 189–198.

Odani, S., Ito, N., Hasegawa, M., Uchiumi, T., Hase, S., 2012. Identification of L-3-hydroxykynurenine O-sulfate in the buccal gland secretion of the parasitic lamprey *Lethenteron japonicum*. Amino Acids 43, 2505–2512.

Ogawa, S., Parhar, I.S., 2018. Biological significance of kisspeptin-Kiss 1 receptor signaling in the habenula of teleost species. Front. Endocrinol. 9, 222.

Parker, P.S., Lennon, R.E., 1956. Biology of the sea lamprey in its parasitic phase. Research report 44. In: US Fish Wildl. Serv. Fish. Bull. vol. 57. 32 pp.

Parker, H.J., Bronner, M.E., Krumlauf, R., 2019. An atlas of anterior *hox* gene expression in the embryonic sea lamprey head: *hox*-code evolution in vertebrates. Dev. Biol. 453, 19–33.

Patra, R.W., Chapman, J.C., Lim, R.P., Gehrke, P.C., Sunderam, R.M., 2015. Interactions between water temperature and contaminant toxicity to freshwater fish. Environ. Toxicol. Chem. 34, 1809–1817.

Patrick, H.K., Sutton, T.M., Swink, W.D., 2009. Lethality of sea lamprey parasitism on lake sturgeon. Trans. Am. Fish. Soc. 138, 1065–1075.

Peek, W.D., Youson, J.H., 1979. Transformation of the interlamellar epithelium of the gills of the anadromous sea lamprey, *Petromyzon marinus*, during metamorphosis. Can. J. Zool. 57, 1318–1332.

Perkins, L.B., Leger, E.A., Nowak, R.S., 2011. Invasion triangle: an organizational framework for species invasion. Ecol. Evol. 1, 610–625.

Phillips, J.C., McKinley, G.A., Bennington, V., Bootsma, H.A., Pilcher, D.J., Sterner, R.W., Urban, N.R., 2015. The potential for $CO_2$-induced acidification in freshwater: a Great Lakes case study. Oceanography 28, 136–145.

Piaggio, A.J., Segelbacher, G., Seddon, P.J., et al., 2017. Is it time for synthetic biodiversity conservation? Trends Ecol. Evol. 32, 97–107.

Popper, A.N., Fay, R.R., 1999. The auditory periphery in fishes. In: Fay, R.R., Popper, A.N. (Eds.), Comparative Hearing: Fish and Amphibians. Springer Handbook of Auditory Research. 11. Springer, New York, pp. 43–100.

Potter, I.C., 1980. Ecology of larval and metamorphosing lampreys. Can. J. Fish. Aquat. Sci. 37, 1641–1657.

Potter, I., Beamish, F., 1975. Lethal temperatures in ammocoetes of four species of lampreys. Acta Zool. 56, 85–91.

Potter, I.C., Gill, H.S., Renaud, C.B., Haoucher, D., 2015. The taxonomy, phylogeny, and distribution of lampreys. In: Docker, M.F. (Ed.), Lampreys: Biology, Conservation and Control. 1. Springer, Dordrecht, pp. 35–74.

Pratt, T.C., Morrison, B.J., Quinlan, H.R., Elliott, R.F., Grunder, S.A., Chiotti, J.A., Young, B.A., 2021. Implications of the sea lamprey control program for lake sturgeon conservation and rehabilitation efforts. J. Great Lakes Res. 47, S421–S429.

Pratt, T.C., O'Connor, L.M., Hallett, A.G., McLaughlin, R.L., Katopodis, C., Hayes, D.B., Bergstedt, R.A., 2009. Balancing aquatic habitat fragmentation and control of invasive species: enhancing selective fish passage at sea lamprey control barriers. Trans. Am. Fish. Soc. 138, 652–665.

Purvis, H, Chudy, C.L., King Jr., E.L., Dawson, V.K., 1985. Response of spawning-phase sea lampreys (*Petromyzon marinus*) to a lighted trap. Technical Report No. 42. Great Lakes Fishery Commission, Ann Arbor, MI, pp. 15–25.

Pycha, R.L., King, G.R., 1975. Changes in the lake trout population of southern Lake Superior in relation to the fishery, the sea lamprey, and stocking, 1950-70. In: Great Lakes Fishery Commission Tech. Rep. 28. Ann Arbor, Michigan.

Quintella, B.R., Andrade, N.O., Koed, A., Almeida, P.R., 2004. Behavioural patterns of sea lampreys' spawning migration through difficult passage areas, studied by electromyogram telemetry. J. Fish Biol. 65, 961–972.

Quintella, B.R., Andrade, N.O., Espanhol, R., Almeida, P.R., 2005. The use of PIT telemetry to study movements of ammocoetes and metamorphosing sea lampreys in river beds. J. Fish Biol. 66, 97–106.

Quintella, B.R., Clemens, B.J., Sutton, T., Lança, M.J., Madenjian, C.P., Happel, A., Harvey, C.J., 2021. At-sea feeding ecology of parasitic lampreys. J. Great Lakes Res. 47, S72–S89.

Railsback, S.F., Harvey, B.C., White, J.L., 2014. Facultative anadromy in salmonids: linking habitat, individual life history decisions, and population-level consequences. Can. J. Fish. Aquat. Sci. 71, 1270–1278.

Reid, A.J., Carlson, A.K., Creed, I.F., et al., 2019. Emerging threats and persistent conservation challenges for freshwater biodiversity. Biol. Rev. 94, 849–873.

Reinhardt, U.G., Hrodey, P.J., 2019. Trap happiness and catch bias in sea lamprey traps. Fishes 4, 34.

Reis-Santos, P., McCormick, S.D., Wilson, J.M., 2008. Ionoregulatory changes during metamorphosis and salinity exposure of juvenile sea lamprey (*Petromyzon marinus* L.). J. Exp. Biol. 211, 978–988.

Renaud, C.B., 2011. Lampreys of the world. An annotated and illustrated catalogue of lamprey species known to date. In: FAO Species Catalogue for Fishery Purposes. No. 5. FAO, Rome. 109 pp.

Renaud, C.B., Cochran, P.A., 2019. Post-metamorphic feeding in lampreys. In: Docker, M. (Ed.), Lampreys: Biology, Conservation and Control. vol. 2. Springer, Dordrecht, pp. 247–285.

Renaud, C.B., Gill, H.S., Potter, I.C., 2009. Relationships between the diets and characteristics of the dentition, buccal glands and velar tentacles of the adults of the parasitic species of lamprey. J. Zool. 278, 231–242.

Rinchard, J., Ciereszko, A., Dabrowski, K., Ottobre, J., 2000. Effects of gossypol on sperm viability and plasma sex steroid hormones in male sea lamprey, *Petromyzon marinus*. Toxicol. Lett. 111, 189–198.

Robinson, K.F., Miehls, S.M., Siefkes, M.J., 2021. Understanding sea lamprey abundances in the Great Lakes prior to broad implementation of sea lamprey control. J. Great Lakes Res. 47, S328–S334.

Rovainen, C.M., 1996. Feeding and breathing in lampreys. Brain Behav. Evol. 48, 297–305.

Rutter, M.A., Bence, J.R., 2003. An improved method to estimate sea lamprey wounding rate on hosts with application to lake trout in Lake Huron. J. Great Lakes Res. 29, 320–331.

Sakai, A.K., Allendorf, F.W., Holt, J.S., et al., 2001. The population biology of invasive species. Annu. Rev. Ecol. Evol. Syst. 32, 305–332.

Sampson, S.J., Chick, J.H., Pegg, M.A., 2009. Diet overlap among two Asian carp and three native fishes in backwater lakes on the Illinois and Mississippi rivers. Biol. Invasions 11, 483–496.

Sard, N.M., Smith, S.R., Homola, J.J., Kanefsky, J., Bravener, G., Adams, J.V., Holbrook, C.M., Hrodey, P.J., Tallon, K., Scribner, K.T., 2020. RAPTURE (RAD capture) panel facilitates analyses characterizing sea lamprey reproductive ecology and movement dynamics. Ecol. Evol. 10, 1469–1488.

Savenije, H.H.G., Hoekstra, A.Y., van der Zaag, P., 2014. Evolving water science in the Anthropocene. Hydrol. Earth Syst. Sci. 18, 319–332.

Sawyer, A.J., 1980. Prospects for integrated pest management of the sea lamprey (*Petromyzon marinus*). Can. J. Fish. Aquat. Sci. 37, 2081–2092.

Schindler, D.E., Hilborn, R., Chasco, B., et al., 2010. Population diversity and the portfolio effect in an exploited species. Nature 465, 609–612.

Schloesser, N.A., Merkes, C.M., Rees, C.B., Amberg, J.J., Steeves, T.B., Docker, M.F., 2018. Correlating sea lamprey density with environmental DNA detections in the lab. Manag. Biol. Invasions 9, 483–495.

Schneider, C.P., Owens, R.W., Bergstedt, R.A., O'Gorman, R., 1996. Predation by sea lamprey (*Petromyzon marinus*) on lake trout (*Salvelinus namaycush*) in southern Lake Ontario, 1982-1992. Can. J. Fish. Aquat. Sci. 53, 1921–1932.

Scholefield, R.J., Slaght, K.S., Stephens, B.E., 2008. Seasonal variation in sensitivity of larval sea lampreys to the lampricide 3-trifluoromethyl-4-nitrophenol. N. Am. J. Fish. Manag. 28, 1609–1617.

Schuldt, R.J., Goold, R., 1980. Changes in the distribution of native lampreys in Lake Superior tributaries in response to sea lamprey (*Petromyzon marinus*) control, 1953–77. Can. J. Fish. Aquat. Sci. 37, 1872–1885.

Scott, A.M., Zhang, Z., Jia, L., Li, K., Zhang, Q., Dexheimer, T., Ellsworth, E., Ren, J., Chung-Davidson, Y.W., Zu, Y., Neubig, R.R., 2019. Spermine in semen of male sea lamprey acts as a sex pheromone. PLoS Biol. 17, e3000332.

Scribner, K., Tsehaye, I., Brenden, T., Stott, W., Kanefsky, J., Bence, J., 2018. Hatchery strain contributions to emerging wild lake trout populations in Lake Huron. J. Hered. 109, 675–688.

Sepúlveda, M.S., Patrick, H.K., Sutton, T.M., 2012. A single sea lamprey attack causes acute anemia and mortality in lake sturgeon. J. Aquat. Anim. Health 24, 91–99.

Shaughnessy, C.A., McCormick, S.D., 2020. Functional characterization and osmoregulatory role of the $Na^{+}$-$K^{+}$-$2Cl^{-}$ cotransporter in the gill of sea lamprey (*Petromyzon marinus*), a basal vertebrate. Am. J. Physiol. Regul. Integr. Comp. Physiol. 318, R17–R29.

Shi, H., Holbrook, C.M., Cao, Y., Sepúlveda, N., Tan, X., 2021. Measurement of suction pressure dynamics of sea lampreys, *Petromyzon marinus*. PLoS One 16, e0247884.

Shimeld, S.M., Donoghue, P.C.J., 2012. Evolutionary crossroads in developmental biology: cyclostomes (lamprey and hagfish). Development 139, 2091–2099.

Siefkes, M.J., 2017. Use of physiological knowledge to control the invasive sea lamprey (*Petromyzon marinus*) in the Laurentian Great Lakes. Conserv. Physiol. 5, cox031.

Siefkes, M.J., Johnson, N.S., Muir, A.M., 2021. A renewed philosophy about supplemental sea lamprey controls. J. Great Lakes Res. 47, S742–S752.

Silva, S., Servis, M.J., Vieira-Lanero, R., Cobo, F., 2013. Downstream migration and hematophagous feeding of newly metamorphosed sea lampreys (*Petromyzon marinus* Linnaeus, 1758). Hydrobiologia 700, 277–286.

Simberloff, D., 2009. The role of propagule pressure in biological invasions. Annu. Rev. Ecol. Evol. Syst. 40, 81–102.

Simpkins, D.G., Kornis, M.S., Maguffee, A.C., Bence, J.R., Pankow, K.W., Bronte, C.R., 2021. Spatial and temporal variation in marking rates and severity of sea lamprey attacks on salmonines in Lakes Michigan and Huron. J. Great Lakes Res. 47, S612–S627.

Smith, B.R., Tibbles, J.J., 1980. Sea lamprey (*Petromyzon marinus*) in Lakes Huron, Michigan, and Superior - History of invasion and control, 1936-78. Can. J. Fish. Aquat. Sci. 37, 1780–1801.

Smith, J.J., Antonacci, F., Eichler, E.E., Amemiya, C.T., 2009. Programmed loss of millions of base pairs from a vertebrate genome. Proc. Natl. Acad. Sci. U. S. A. 106, 11212–11217.

Smith, J.J., Kuraku, S., Holt, C., et al., 2013. Sequencing of the sea lamprey (*Petromyzon marinus*) genome provides insights into vertebrate evolution. Nat. Genet. 45, 415–421.

Smith, S.E., Sitar, S.P., Goetz, F.W., Huertas, M., Armstrong, B.M., Murphy, C.A., 2016. Differential physiological response to sea lamprey parasitism between lake trout (*Salvelinus namaycush*) morphotypes from Lake Superior. Can. J. Fish. Aquat. Sci. 73, 1815–1829.

Smith, J.J., Timoshevskaya, N., Ye, C., et al., 2018. The sea lamprey germline genome provides insights into programmed genome rearrangement and vertebrate evolution. Nat. Genet. 50, 270–277.

Sobrido-Cameán, D., Yanez-Guerra, L.A., Deber, A., Rodicio, M.C., Barreiro-Iglesias, A., 2021. Expression of kisspeptin 1 in the brain of the adult sea lamprey *Petromyzon marinus*. Life 11, 1174.

Sorensen, P.W., 2021. Introduction to the biology and control of invasive fishes and a special issue on this topic. Fishes 6, 69.

Sorensen, P.W., Hoye, T.R., 2007. A critical review of the discovery and application of a migratory pheromone in an invasive fish, the sea lamprey *Petromyzon marinus* L. J. Fish Biol. 71, 100–114.

Sorensen, P.W., Johnson, N.S., 2016. Theory and application of semiochemicals in nuisance fish control. J. Chem. Ecol. 42, 698–715.

Sorensen, P.W., Vrieze, L.A., 2003. The chemical ecology and potential application of the sea lamprey migratory pheromone. J. Great Lakes Res. 29 (Suppl. 1), 66–84.

Sorensen, P.W., Fine, J.M., Dvornikovs, V., Jeffrey, C.S., Shao, F., Wang, J., Vrieze, L.A., Anderson, K.R., Hoye, T.R., 2005. Mixture of new sulfated steroids functions as a migratory pheromone in the sea lamprey. Nat. Chem. Biol. 1, 324–328.

Sotola, V.A., Miehls, S.M., Simard, L.G., Marsden, J.E., 2018. Lateral and vertical distribution of downstream migrating juvenile sea lamprey. J. Great Lakes Res. 44, 491–496.

Sower, S.A., 2003. The endocrinology of reproduction in lampreys and applications for male lamprey sterilization. J. Great Lakes Res. 29, 50–65.

Sower, S.A., 2015. The reproductive hypothalamic-pituitary axis in lampreys. In: Docker, M.F. (Ed.), Lampreys: Biology, Conservation and Control. 1. Springer, Dordrecht, pp. 305–373.

Sower, S.A., 2018. Landmark discoveries in elucidating the origins of the hypothalamic-pituitary system from the perspective of a basal vertebrate, sea lamprey. Gen. Comp. Endocrinol. 264, 3–15.

Spangler, G.R., Robson, D.S., Regier, H.A., 1980. Estimates of lamprey-induced mortality in whitefish, *Coregonus clupeaformis*. Can. J. Fish. Aquat. Sci. 37, 2146–2150.

Spice, E.K., Goodman, D.H., Reid, S.B., Docker, M.F., 2012. Neither philopatric nor panmictic: Microsatellite and mtDNA evidence suggests lack of natal homing but limits to dispersal in Pacific lamprey. Mol. Ecol. 21, 2916–2930.

Square, T., Romášek, M., Jandzik, D., Cattell, M.V., Klymkowsky, M., Medeiros, D.M., 2015. CRISPR/Cas9-mediated mutagenesis in the sea lamprey *Petromyzon marinus*: a powerful tool for understanding ancestral gene functions in vertebrates. Development 142, 4180–4187.

Stamplecoskie, K.M., Binder, T.R., Lower, N., Cottenie, K., McLaughlin, R.L., McDonald, D.G., 2012. Response of migratory sea lampreys to artificial lighting in portable traps. N. Am. J. Fish. Manag. 32, 563–572.

Stauffer, T.M., 1964. An experimental sea lamprey barrier. Prog. Fish Cult. 26, 80–83.

Steffen, W., Crutzen, P.J., McNeill, J.R., 2007. The anthropocene: are humans now overwhelming the great forces of nature? Ambio 36, 614–621.

Steffen, W., Grinevald, J., Crutzen, P., McNeill, J., 2011. The anthropocene: conceptual and historical perspectives. Philos. Trans. A Math. Phys. Eng. Sci. 369, 842–867.

Stenberg, J.A., 2017. A conceptual framework for integrated pest management. Trends Plant Sci. 22, 759–769.

Stewart, T.J., Bence, J.R., Bergstedt, R.A., Ebener, M.P., Lupi, F., Rutter, M.A., 2003. Recommendations for assessing sea lamprey damages: toward optimizing the control program in the Great Lakes. J. Great Lakes Res. 29, 783–793.

Stewart, M., Baker, C.F., Cooney, T., 2011. A rapid, sensitive, and selective method for quantitation of lamprey migratory pheromones in river water. J. Chem. Ecol. 37, 1203–1207.

Sturtevant, R.A., Mason, D.M., Rutherford, E.S., Elgin, A., Lower, E., Martinez, F., 2019. Recent history of nonindigenous species in the Laurentian Great Lakes; An update to Mills et al., 1993 (25 years later). J. Great Lakes Res. 45, 1011–1035.

Sullivan, W.P., Christie, G.C., Cornelius, F.C., Fodale, M.F., Johnson, D.A., Koonces, J.F., Larson, G.L., McDonald, R.B., Mullett, K.M., Murray, C.K., Ryan, P.A., 2003. The sea lamprey in Lake Erie: a case history. J. Great Lakes Res. 29, 615–636.

Sullivan, W.P., Burkett, D.P., Boogaard, M.A., Criger, L.A., Freiburger, C.E., Hubert, T.D., Leistner, K.G., Morrison, B.J., Nowicki, S.M., Robertson, S.N., Rowlinson, A.K., 2021. Advances in the use of lampricides to control sea lampreys in the Laurentian Great Lakes, 2000–2019. J. Great Lakes Res. 47, S216–S237.

Sun, P.L., Hawkins, W.E., Overstreet, R.M., Brown-Peterson, N.J., 2009. Morphological deformities as biomarkers in fish from contaminated rivers in Taiwan. Int. J. Environ. Res. Public Health 6, 2307–2331.

Sunga, J., Wilson, J.M., Wilkie, M.P., 2020. Functional re-organization of the gills of metamorphosing sea lamprey (*Petromyzon marinus*): preparation for a blood diet and the freshwater to seawater transition. J. Comp. Physiol. B 190, 701–715.

Suski, C.D., 2020. Development of carbon dioxide barriers to deter invasive fishes: insights and lessons learned from bigheaded carp. Fishes 5, 3.

Sutherby, J., 2019. The Effect of Temperature on Sea Lamprey (*Petromyzon marinus*): Ecological and Cellular Implications (M.Sc. Thesis). University of Manitoba, Winnipeg, Manitoba, Canada.

Sutton, T.M., Bowen, S.H., 1994. Significance of organic detritus in the diet of larval lampreys in the Great Lakes basin. Can. J. Fish. Aquat. Sci. 51, 2380–2387.

Swanson, R.G., McCann, E., Johnson, N., Zielinski, D., 2021. Environmental factors influencing annual sucker (*Catostomus* sp.) migration into a Great Lakes tributary. J. Great Lakes Res. 47, 1159–1170.

Swink, W.D., 1990. Effect of lake trout size on survival after a single sea lamprey attack. Trans. Am. Fish. Soc. 119, 996–1002.

Swink, W.D., 1991. Host-size selection by parasitic sea lampreys. Trans. Am. Fish. Soc. 120, 637–643.

Swink, W.D., 1993. Effect of water temperature on sea lamprey growth and lake trout survival. Trans. Am. Fish. Soc. 122, 1161–1166.

Swink, W.D., 1999. Effectiveness of an electrical barrier in blocking a sea lamprey spawning migration on the Jordan River, Michigan. N. Am. J. Fish. Manag. 19, 397–405.

Swink, W.D., 2003. Host selection and lethality of attacks by sea lampreys (*Petromyzon marinus*) in laboratory studies. J. Great Lakes Res. 29, 307–319.

Swink, W.D., Hanson, L.H., 1986. Survival from sea lamprey (*Petromyzon marinus*) predation by two strains of lake trout (*Salvelinus namaycush*). Can. J. Fish. Aquat. Sci. 43, 2528–2531.

Swink, W.D., Hanson, L.H., 1989. Survival of rainbow trout and lake trout after sea lamprey attack. N. Am. J. Fish. Manag. 9, 35–40.

Swink, W.D., Johnson, N.S., 2014. Growth and survival of sea lampreys from metamorphosis to spawning in Lake Huron. Trans. Am. Fish. Soc. 143, 380–386.

Taverny, C., Elie, P., 2009. Assessment of biological knowledge and the state of migratory lamprey habitats in the Gironde basin—Proposals for priority actions (in French). Final Report No. 123, Groupement de Bordeaux, Bordeaux, France.

Teem, J.L., Alphey, L., Descamps, S., et al., 2020. Genetic biocontrol for invasive species. Front. Bioeng. Biotechnol. 8, 452.

Teeter, J., 1980. Pheromone communication in sea lampreys (*Petromyzon marinus*): implications for population management. Can. J. Fish. Aquat. Sci. 37, 2123–2132.

Tessier, L.R., Long, T.A.F., Wilkie, M.P., 2018. Influence of body size, metabolic rate and life history stage on the uptake and excretion of the lampricide 3-trifluoromethyl-4-nitrophenol (TFM) by invasive sea lampreys (*Petromyzon marinus*). Aquat. Toxicol. 194, 27–36.

Tews, J.M., Adams, J.V., Mann, K.A., Koon, E.M., Heinrich, J.W., 2021. A review of an electric weir and fishway in a Great Lakes tributary from conception to termination. J. Great Lakes Res. 47, S297–S309.

Thresher, R.E., Hayes, K., Bax, N.J., Teem, J., Benfey, T.J., Gould, F., 2014. Genetic control of invasive fish: technological options and its role in integrated pest management. Biol. Invasions 16, 1201–1216.

Thresher, R.E., Jones, M., Drake, D.A.R., 2019a. Evaluating active genetic options for the control of sea lamprey (*Petromyzon marinus*) in the Laurentian Great Lakes. Can. J. Fish. Aquat. Sci. 76, 1186–1202.

Thresher, R.E., Jones, M., Drake, D.A.R., 2019b. Stakeholder attitudes towards the use of recombinant technology to manage the impact of an invasive species: sea lamprey in the North American Great Lakes. Biol. Invasions 21, 575–586.

Treble, A.J., Jones, M.L., Steeves, T.B., 2008. Development and evaluation of a new predictive model for metamorphosis of Great Lakes larval sea lamprey (*Petromyzon marinus*) populations. J. Great Lakes Res. 34, 404–417.

Treska, T.J., Ebener, M.P., Christie, G.C., Adams, J.V., Siefkes, M.J., 2021. Setting and tracking suppression targets for sea lampreys in the Great Lakes. J. Great Lakes Res. 47, S357–S367.

Twohey, M.B., Sorensen, P.W., Li, W., 2003a. Possible applications of pheromones in an integrated sea lamprey management program. J. Great Lakes Res. 29, 794–800.

Twohey, M.B., Heinrich, J.W., Seelye, J.G., Fredricks, K.T., Bergstedt, R.A., Kaye, C.A., Scholefield, R.J., McDonald, R.B., Christie, G.C., 2003b. The sterile-male-release technique in Great Lakes sea lamprey management. J. Great Lakes Res. 29, 410–423.

Vélez-Espino, L.A., McLaughlin, R.L., Jones, M.L., Pratt, T.C., 2011. Demographic analysis of trade-offs with deliberate fragmentation of streams: control of invasive species versus protection of native species. Biol. Conserv. 144, 1068–1080.

Viant, M.R., Walton, J.H., Tjeerdema, R.S., 2001. Comparative sublethal actions of 3-trifluoromethyl-4-nitrophenol in marine molluscs, as measured by in vivo P-31 NMR. Pestic. Biochem. Physiol. 71, 40–47.

Wagner, C.M., Twohey, M.B., Fine, J.M., 2009. Conspecific cueing in the sea lamprey: do reproductive migrations consistently follow the most intense larval odour? Anim. Behav. 78, 593–599.

Wagner, C.M., Stroud, E.M., Meckley, T.D., 2011. A deathly odor suggests a new sustainable tool for controlling a costly invasive species. Can. J. Fish. Aquat. Sci. 68, 1157–1160.

Waldman, J.R., Grunwald, C., Roy, N.K., Wirgin, I.I., 2004. Mitochondrial DNA analysis indicates sea lampreys are indigenous to Lake Ontario. Trans. Am. Fish. Soc. 133, 950–960.

Waldman, J., Grunwald, C., Wirgin, I., 2008. Sea lamprey *Petromyzon marinus*: an exception to the rule of homing in anadromous fishes. Biol. Lett. 4, 659–662.

Waldman, J., Daniels, R., Hickerson, M., Wirgin, I., 2009. Mitochondrial DNA analysis indicates sea lampreys are indigenous to Lake Ontario: response to comment. Trans. Am. Fish. Soc. 138, 1190–1197.

Walter, L.M., Dettmers, J.M., Tyson, J.T., 2021. Considering aquatic connectivity trade-offs in Great Lakes barrier removal decisions. J. Great Lakes Res. 47, S430–S438.

Wang, J., Han, X., Yang, H., Lu, L., Wu, Y., Liu, X., Wu, Y., Liu, X., Guo, R., Zhang, Y., Zhang, Y., Li, Q., 2010. A novel RGD-toxin protein, Lj-RGD3, from the buccal gland secretion of *Lampetra japonica* impacts diverse biological activities. Biochimie 92, 1387–1396.

Wang, H., Johnson, N., Bernardy, J., Hubert, T., Li, W., 2013. Monitoring sea lamprey pheromones and their degradation using rapid stream-side extraction coupled with UPLC-MS/MS. J. Sep. Sci. 36, 1612–1620.

Waters, C.N., Zalasiewicz, J., Summerhayes, C., et al., 2016. The Anthropocene is functionally and stratigraphically distinct from the Holocene. Science 351, aad2622.

Webb, P.W., 1994. Mechanics and physiology of animal swimming. In: Maddock, L., Bone, Q., Rayner, J.M.V. (Eds.), The Biology of Fish Swimming. Cambridge University Press, pp. 45–62.

Wigley, R.L., 1959. Life history of the sea lamprey of Cayuga Lake, New York. US Fish Wildl. Serv. Fish. Bull. 59, 561–617.

Wikelski, M., Cooke, S.J., 2006. Conservation physiology. Trends Ecol. Evol. 21, 38–46.

Wilkie, M.P., 2011a. Lampreys: energetics and development. In: Farrell, A.P. (Ed.), Encyclopedia of Fish Physiology: From Genome to Environment. vol. 3. Academic Press, San Diego, pp. 1779–1787.

Wilkie, M.P., 2011b. Lampreys: environmental physiology. In: Farrell, A.P. (Ed.), Encyclopedia of Fish Physiology: From Genome to Environment. vol. 3. Academic Press, San Diego, pp. 1788–1799.

Wilkie, M.P., Couturier, J.E., Tufts, B., 1998. Mechanisms of acid-base regulation in migrant sea lampreys (*Petromyzon marinus*) following exhaustive exercise. J. Exp. Biol. 201, 1473–1482.

Wilkie, M.P., Hubert, T.D., Boogaard, M.A., Birceanu, O., 2019. Control of invasive sea lampreys using the piscicides TFM and niclosamide: Toxicology, successes & future prospects. Aquat. Toxicol. 211, 235–252.

Wilkie, M.P., Tessier, L.R., Boogaard, M., O'Connor, L., Birceanu, O., Steeves, T.B., Sullivan, W.P., 2021. Lampricide bioavailability and toxicity to invasive sea lamprey and non-target fishes: the importance of alkalinity, pH, and the gill microenvironment. J. Great Lakes Res. 47, S407–S420.

Williamson, M., Fitter, A., 1996. The varying success of invaders. Ecology 77, 1661–1666.

Wood, C.M., Bucking, C., 2011. The role of feeding in salt and water balance. In: Grosell, M., Farrell, A.P., Brauner, C.J. (Eds.), Fish Physiology: The Multifunctional Gut of Fish. 30. Elsevier, Amsterdam, pp. 166–212.

Wood, L.A., Brown, I.R., Youson, J.H., 1998. Characterization of the heat shock response in the gills of sea lampreys and a brook lamprey at different intervals of their life cycles. Comp. Biochem. Physiol. 120A, 509–518.

Wood, L.A., Brown, I.R., Youson, J.H., 1999. Tissue and developmental variations in the heat shock response of sea lampreys (*Petromyzon marinus*): effects of an increase in acclimation temperature. Comp. Biochem. Physiol. 123A, 35–42.

Xi, X., Johnson, N.S., Brant, C.O., Yun, S.S., Chambers, K.L., Jones, A.D., Li, W., 2011. Quantification of a male sea lamprey pheromone in tributaries of Laurentian Great Lakes by liquid chromatography–tandem mass spectrometry. Environ. Sci. Technol. 45, 6437–6443.

Xiao, R., Li, Q.W., Perrett, S., He, R.Q., 2007. Characterization of the fibrinogenolytic properties of the buccal gland secretion from *Lampetra japonica*. Biochimie 89, 383–392.

Xiao, R., Pang, Y., Li, Q.W., 2012. The buccal gland of *Lampetra japonica* is a source of diverse bioactive proteins. Biochimie 94, 1075–1079.

Xue, X.-Y., Mao, Y.-B., Tao, X.-Y., Huang, Y.-P., Chen, X.-Y., 2012. New approaches to agricultural insect pest control based on RNA interference. Adv. Insect Physiol. 42, 73–117.

Yasmin, T., Grayson, P., Docker, M.F., Good, S.V., 2022. Pervasive male-biased expression throughout the germline-specific regions of the sea lamprey genome supports key roles in sex differentiation and spermatogenesis. Commun. Biol. 5, 434.

York, J.R., McCauley, D.W., 2020. Functional genetic analysis in a jawless vertebrate, the sea lamprey: insights into the developmental evolution of early vertebrates. J. Exp. Biol. 223, jeb206433.

York, J.R., Lee, E.M.-J., McCauley, D.W., 2019. The lamprey as a model vertebrate in evolutionary developmental biology. In: Docker, M.F. (Ed.), Lampreys: Biology, Conservation and Control. 2. Springer, Dordrecht, pp. 481–526.

York, J.R., Thresher, R.E., McCauley, D.W., 2021. Applying functional genomics to the study of lamprey development and sea lamprey population control. J. Great Lakes Res. 47, S639–S649.

Youson, J.H., 1980. Morphology and physiology of lamprey metamorphosis. Can. J. Fish. Aquat. Sci. 37, 1687–1710.

Youson, J.H., 1997. Is lamprey metamorphosis regulated by thyroid hormones? Am. Zool. 37, 441–460.

Youson, J.H., 2003. The biology of metamorphosis in sea lampreys: endocrine, environmental, and physiological cues and events, and their potential application to lamprey control. J. Great Lakes Res. 29 (Suppl. 1), 26–49.

Youson, J.H., Potter, I.C., 1979. Description of the stages in metamorphosis of the anadromous sea lamprey, *Petromyzon marinus* L. Can. J. Zool. 57, 1808–1817.

Youson, J.H., Holmes, J.A., Guchardi, J.A., et al., 1993. Importance of condition factor and the influence of water temperature and photoperiod on metamorphosis of sea lamprey, *Petromyzon marinus*. Can. J. Fish. Aquat. Sci. 50, 2448–2456.

Zerebecki, R.A., Sorte, C.J.B., 2011. Temperature tolerance and stress proteins as mechanisms of invasive species success. Plos One 6, e14806.

Zielinski, D.P., Freiburger, C., 2021. Advances in fish passage in the Great Lakes basin. J. Great Lakes Res. 47, S439–S447.

Zielinski, D.P., McLaughlin, R., Castro-Santos, T., Paudel, B., Hrodey, P., Muir, A., 2019. Alternative sea lamprey barrier technologies: history as a control tool. Rev. Fish. Sci. Aquac. 27, 438–457.

Zielinski, D.P., McLaughlin, R.L., Pratt, T.C., Goodwin, R.A., Muir, A.M., 2020. Single-stream recycling inspires selective fish passage solutions for the connectivity conundrum in aquatic ecosystems. BioScience 70, 871–886.

Zielinski, D.P., Miehls, S., Burns, G., Coutant, C., 2021. Adult sea lamprey respond to induced turbulence in a low current system. TJoE 6, 82–90.

Zydlewski, J., Wilkie, M.P., 2013. Freshwater to seawater transitions in migratory fishes. In: McCormick, S.D., Farrell, A.P., Brauner, C.J. (Eds.), Fish Physiology: Euryhaline Fishes. 32. Elsevier, Amsterdam, pp. 253–326.

Chapter 11

# Conservation Physiology of fishes for tomorrow: Successful conservation in a changing world and priority actions for the field

**Lisa M. Komoroske**[a,*] **and Kim Birnie-Gauvin**[b]

[a]*Department of Environmental Conservation, University of Massachusetts Amherst, Amherst, MA, United States*

[b]*National Institute of Aquatic Resources, Technical University of Denmark, Silkeborg, Denmark*

[*]*Corresponding author: e-mail: lkomoroske@umass.edu*

## Chapter Outline



**Fish Physiology, Vol. 39B. https://doi.org/10.1016/bs.fp.2022.04.016**

Marine and freshwater fishes are currently facing complex and extensive challenges, from warming waters and habitat degradation to direct human-wildlife interactions. Within this context, Conservation Physiology has made some important contributions to advance our understanding of the underlying mechanisms leading to these problems, as well as in offering practical solutions. However, there remains much space for the field to grow and significantly expand its impact on real-world conservation of fish biodiversity. As the planet continues to change, so do the problems that fish encounter, and so must the field. Importantly, systemic changes must occur to better represent the diversity of peoples and knowledges who have been historically systemically excluded in this field (and others). In this chapter, we discuss some of the remaining key challenges that conservation physiologists need to overcome to protect and effectively manage fish species around the globe in the coming decades, including those related to diversity, equity, inclusivity and justice. We make suggestions for overcoming these challenges, highlighting examples of how physiological knowledge has been used to conserve fishes and other taxa and providing key resources for understanding and addressing inequities in the field, which may serve as guidance for scientists and practitioners seeking to advance these goals. We finish with a list of priority actions needed to ensure that the field of Conservation Physiology remains relevant and successful in its quest to promote long-term sustainable and equitable conservation solutions.

## 1 Introduction

Evidence-based conservation is necessary to mitigate and reverse global declines in the diversity and abundance of fishes. As the urgency for adequate conservation and management actions is increasing, demands for a better understanding of the physiological ecology of species, particularly for those at risk, have also increased. To do so, Conservation Physiology—which broadly aims to uncover the physiological mechanisms that drive population declines and biodiversity loss in the face of environmental change—uses a range of tools (i.e., the Conservation Physiology toolbox; Madliger et al., 2018) broadly originating from the field of physiology. Over the last decade, these tools have been expanded, validated and refined. The field has slowly moved away from the narrow definition relating to stress physiology (i.e., understanding how humans induce "stress" in animals, Wikelski and Cooke, 2006), with the addition of many sub-disciplines spanning cardiorespiratory physiology, chemical communication, immunology, neurophysiology, and genomics, to name a few (Madliger et al., 2018; see also Volume A). This expansion now allows Conservation Physiology to contribute to and offer solutions more broadly for conservation, restoration and management (Madliger et al., 2021). The previous chapters have highlighted diverse examples of how Conservation Physiology can help to assess and manage the recovery of endangered species (see Chapter 1, Volume 39B: Yanagitsuru et al., 2022; Chapter 2, Volume 39B: Gary Anderson et al., 2022), to evaluate the effects of pollutants and stressors (see Chapter 3, Volume 39B: De Boeck et al., 2022; Chapter 6, Volume 39B: Chapman et al., 2022; Chapter 7, Volume 39B: Rummer and Illing, 2022; Chapter 8, Volume 39B: Strailey and Suski, 2022), to assess the impacts of anthropogenic activities (see Chapter 4, Volume 39B: Killen et al., 2022; Chapter 9, Volume 39B: Hinch et al., 2022; Chapter 10, Volume 39B: Wilkie, 2022), as well as changes in environmental conditions (see Chapter 5, Volume 39B: Luis Val et al., 2022), and to control invasive species (see Chapter 11, Volume 39B: Komoroske and Birnie-Gauvin, 2022). These chapters have underscored how our understanding of the physiological mechanisms that may protect species or make them more vulnerable plays an important role for fish conservation.

That said, Conservation Physiology is an applied discipline, and its usefulness to conservation relies on the scientific community generating the necessary knowledge needed by practitioners to make informed decisions (Cooke et al., 2020). Failure to generate relevant knowledge that advises on the most pressing conservation issues means that Conservation Physiology is failing in its primary objective (Ames et al., 2020). As the field continues to grow and evolve, we are faced with some outstanding and emerging challenges including (1) linking physiological mechanisms to management-relevant scales, (2) contextualizing physiological results into real-world scenarios, (3) using syntheses to understand emergent patterns, (4) broadening phylogenomic and ecological diversity representation, and (5) creating an inclusive field that values the perspectives and knowledges of all peoples. In this final chapter we look forward, discussing these challenges and how conservation physiologists may go about making their science more relevant within these contexts. We then consider what defines successful Conservation Physiology and discuss several success stories that may be used as guidance for tackling these challenges more broadly as the field advances in the next decade and into the future.

## 2 Linking physiological mechanisms to management-relevant scales

Though it has been long appreciated that environmental and anthropogenic stressor impacts can resonate across biological scales, linking physiological measures to population and ecological level effects has been a key challenge of Conservation Physiology since its inception (Ames et al., 2020; Cooke and O'Connor, 2010). In addition to the classic "biological hierarchy" (i.e., genes to ecosystems), linking across other types of scales, such as connecting the spatial and temporal scales measured in physiological studies with those typically employed in management frameworks, is critical to conservation strategies (Cooke et al., 2014). Robust studies to understand these connections, particularly in complex, natural systems, have always been essential to the incorporation of physiological knowledge into conservation management, but recent focus on ecosystem-based management approaches and predicting future ecosystems under global change scenarios reinforces this need. Physiological tools can play especially important roles in these contexts if they can promote early detection of environmental-population linkages before it is too late to mitigate impacts and reverse population declines (Bergman et al., 2019; Ellis et al., 2012). Ideally, studies aiming to make population-level assessments would collect physiological measures explicitly paired with fitness and population monitoring data, but this is frequently not possible due to resources, life histories and other constraints. Without this complementary data, it can be difficult to assess if observed physiological perturbations will resonate up the biological hierarchy, or if organisms will be able to compensate and avoid fitness reductions. Moreover, without population trend data, even robust studies connecting physiological impacts to changes in key vital rates expected to affect population dynamics can fall short of demonstrating population impacts in situ due to emergent properties and complexities of natural systems (Ames et al., 2020; Chown et al., 2010).

Despite these challenges continuing to hinder progress in many systems (Ames et al., 2020), there has been a steady increase in interdisciplinary and creative approaches linking physiological measures with population trajectories and forecasts in recent years. Perhaps unsurprisingly, research has been highly uneven across taxa and approaches. For example, only three fish studies ($\sim$10%) were reported in a recent review of 29 studies linking environmental, physiological and demography measures (termed "EPD," Bergman et al., 2019), though the strict criteria of inclusion of this study should be noted. Much more work in this realm is urgently needed and is one of the top priorities for Conservation Physiology over the next decade and into the future. To facilitate this goal, we highlight some advances and discuss remaining barriers and opportunities for their broader adoption in the conservation of fishes. These diverse examples also emphasize that there are many possible ways to achieve this, and the "best" approach varies by the system, species, study context, as well as relevant research and management questions.

## 2.1 Linking stress biomarkers with environmental conditions and demographic trends

Studies that associate physiological biomarkers with demographic trends have been limited (Bergman et al., 2019), but serve as some of the best examples for linking to population viability, so we draw from these here. These examples also demonstrate the importance of physiological measures for early detection of threats to inform conservation actions before it is too late. Indeed, some biological indices can serve to highlight the effects of some stressors *before* any significant changes in abundance or other population-level metrics become apparent. This is crucial given that many population abundance estimates have significant associated uncertainties; in some instances, large changes in abundance would need to occur before being detected if more sensitive biological health metrics were not available. While physiological markers employed extend beyond traditional "stress" measures (including immune function, nutritional physiology, toxicology and reproductive physiology), the majority of studies to date use some type of biomarker of stress in a broad sense of the term. As mentioned above, fish are under-represented in the EPD literature. Given the extensive nature of stress physiology research utilizing biomarkers in a fisheries relevant context, as well as population monitoring programs and stock assessments, this is perhaps surprising and could represent prime opportunities for interdisciplinary collaboration to build on existing datasets and frameworks to address this gap (Brosset et al., 2021). Though there will inherently be system-specific differences and needs, researchers can draw from the successes in other taxa where integrative EPD approaches have been more broadly applied, such as birds and mammals. For example, coupling measures of corticosterone and survival in common murres identified linkages between food availability and population trends (Kitaysky et al., 2007), and associations of glucocorticoid and androgen levels with population growth rates in Cape mountain zebras identified stress and reproductive physiology mechanisms underlying population declines (Lea et al., 2018). Additionally, accumulating a critical mass of these studies within a taxon or technique is key to advancing the field by enabling syntheses to reveal broader patterns, such as a recent meta-analysis demonstrating that glucocorticoids are negatively associated with fitness, but strongly context dependent (Schoenle et al., 2021).

Interpreting biomarkers can be challenging, particularly in complex natural environments (Wingfield and Kitaysky, 2002), and is most robust when using multiple physiological measures (Madliger and Love, 2014). Combining physiological approaches can also assist with data interpretation when some traits may also respond to other intrinsic or external factors, such as hormone levels that fluctuate naturally over seasons (Bonier et al., 2009). This can be done within and across sub-disciplines, however currently the overwhelming majority of studies employing multiple measures do so within one sub-discipline (Bergman et al., 2019). Moreover, many metrics with high potential to complement traditional stress physiology measures have been largely underrepresented in studies linking environmental disturbance to demographic trends. However, in recent years the availability of molecular and whole-organism physiology tools has rapidly advanced, and there have been an impressive amount of studies employing techniques from multiple sub-disciplines and across biological levels to understand physiological impacts on fishes of concern (e.g., Bowen et al., 2020; Jeffries et al., 2014; see Chapter 8, Volume 39A: Jeffries et al., 2022). This includes new markers (e.g., genomic tools; Connon et al., 2018), as well as optimization of existing ones to be reliably measured rapidly and affordably in non-model systems (e.g., point-of-care devices that can provide reliable measurements in the field, etc.). Importantly, the endpoints chosen must be compatible with reliably addressing the research questions at hand as well as logistical constraints; for example it is challenging to collect certain types of samples from some species, or properly preserve them in remote field settings. Brosset et al. (2021) recently published a review of physiology biomarkers in fisheries management that includes a summary table of the utility and constraints of different biomarkers and decision flow chart that is a valuable resource for researchers conducting these studies. They also highlight the largely underutilized potential of conducting biomarker studies in conjunction with fisheries surveys to link biomarker with population level data. Looking forward, if more of these types of studies can additionally incorporate analyses of demographic and environmental data from in situ monitoring programs, we expect they will play a critical role in advancing our understanding of the linkages between physiological mechanisms and population viability.

## 2.2 Leveraging advances in macrophysiology and landscape physiology

Conservation Physiology studies are often conducted on one or a few populations or species, aiming to assess anthropogenic effects in a specific organism or group of concern. However, practitioners frequently need to predict impacts for many species in a community or at other broadscale levels (e.g., Ecosystem approach to fisheries management) to assess relative risks and inform management decisions. The fields of macrophysiology (the investigation of physiological traits over large geographical, temporal and phylogenetic scales; Chown and Gaston, 2008, 2016) and landscape physiology (incorporation of physiological measures with spatial habitat data to assess how landscape patterns affect species persistence; Ellis et al., 2012) offer appealing approaches to identify how physiological responses link to broader patterns at higher levels, such as population growth, species interactions, and spatial dynamics (Chown and Gaston, 2016). Additionally, by establishing "rules" that emerge from large-scale analyses across taxa and systems, these studies can potentially offer predictive insight for species without direct empirical data (Chown and Gaston, 2008; Chown et al., 2002). For example, macrophysiology studies of thermal tolerances across broad spatial and phylogenetic scales have been used to forecast species responses to climate warming and assess adaptive capacity within and across taxa (Bennett et al., 2021; Chown and Gaston, 2016). Such results can inform trait-based risk assessments at regional and broader scales that are needed in ecosystem-based management approaches (Bremner, 2008; Chown, 2012). Additionally, landscape physiology studies coupling physiological metrics with demographic and spatial data in human-altered landscapes can shed light on the mechanisms through which habitat fragmentation and other forms of degradation affect population persistence (Ellis et al., 2012; Ziv and Davidowitz, 2019).

These disciplines also have limitations (e.g., see table 1 of key challenges in Chown and Gaston, 2016). Particularly relevant for this context, the reliance on trait and biomarkers data from many studies with different methodologies and assumptions can create confounding issues in the datasets. Additionally, traits can vary for a multitude of reasons, and correlative analyses do not typically assess underlying mechanisms (Ames et al., 2020), or consider the influences of other factors not captured at broad scales (e.g., microclimates, phenotypic plasticity and/or local adaptation; Chevin et al., 2013; Woods et al., 2015). In these ways, macrophysiology and landscape physiology are complementary to Conservation Physiology, such that combining techniques across these disciplines may be one avenue to improve our ability to link physiological mechanisms with population and ecosystem impacts (Ellis et al., 2012). Yet, despite several papers making this point over the past decade, to our knowledge, there have been relatively few studies to date that employ such an integrative approach in fishes (but see Jeffrey et al., 2015; King et al., 2016). Perhaps this is because it is often not clear how these contrasting techniques can be effectively combined. Looking forward, hopefully increased efforts by conservation physiologists to overcome this challenge and incorporate advances in these and other disciplines into their studies will contribute to better connections between physiological mechanisms and management-relevant biological impacts.

## 2.3 Integrating physiology into mechanistic models

A variety of modeling approaches have long been important tools to examine and forecast impacts of natural and anthropogenic activities on fish population and species' viability (Koenigstein et al., 2016). Many early models were correlative in nature with limited capacity to integrate physiological knowledge. Examples include species distribution models (SDMs) that examined overlaps with current species' ranges and environmental conditions to predict future distributions with forecasted environmental changes (Araújo and Peterson, 2012; Jeschke and Strayer, 2008). However, in the past decade significant advances have been made in developing process-based, or mechanistic, models that can incorporate individual-level processes and link them to population dynamics. Process-based models have the advantage of being able to explicitly connect patterns to underlying mechanisms, whereas correlative models cannot. These advances not only address some of the common shortcomings of correlative models, but in taking a more "bottom-up" approach, open the door to greater integration of physiological metrics that are measured at the individual level to inform conservation (Boult and Evans, 2021; Evans et al., 2015; Johnston et al., 2019). Moreover, the advances described in Sections 2.1 and 2.2 above in biomarkers and macrophysiology can provide empirical data to greatly inform and improve mechanistic modeling efforts (Chown and Gaston, 2016), especially when aims are to predict responses under novel conditions such as forecasted climate change scenarios.

A few types of mechanistic models that have particular promise in this realm include agent-based models (also often referred to as individual based models; IBMs McLane et al., 2011), process-based SDMs (a.k.a. ecological niche models; ENMs), and dynamic energy budgets (DEB; Lavaud et al., 2021; Sousa et al., 2010; see Chapter 6, Volume 39A: Bernier and Alderman, 2022). See table 1 in Johnston et al. (2019) for a summary of approaches, and Lavaud et al. (2021) for discussion of their common foundation in DEB theory and potential applications in Conservation Physiology. For example, in Atlantic mackerel (*Scomber scombrus*), Boyd et al. (2018) demonstrated the power of IBMs to integrate physiological and ecological factors with spatial and temporal variation to understand fish population dynamics. Some studies in fishes and other taxa have drawn on components of multiple of these approaches, such as Buckley (2008) that integrated individual bioenergetics into a process-based SDM to better predict range dynamics in a North American lizard (*Sceloporus undulates*). In another example, Fordham et al. (2013) demonstrated that integrating physiological metrics and demographic processes into biophysical-SDMs substantially improved predictions of abundance and range shifts in two species of abalone because they were better able to capture the complex, non-linear relationships between physiology, the environment and demography that frequently exist in natural populations. These and other adaptations where physiological data are combined with simulations of demographic effects may also be useful in applying the EPD approach described above (Bergman et al., 2019). Finally, because energetic demands and trade-offs compose a fundamental connection between physiological, demographic and ecological processes (Boult and Evans, 2021; Chipps and Wahl, 2008; Cooke et al., 2014; Horodysky et al., 2015) DEB models can be powerful tools to evaluate linkages between bioenergetics and population dynamics (for further discussion and examples, see the diverse case studies highlighted using these approaches in Chapter 6, Volume 39A: Bernier and Alderman, 2022).

An on-going hurdle precluding widespread application of mechanistic models and larger-scale comparisons is their need for species- and system-specific data for parameterization (reviewed in Boult and Evans, 2021; Evans et al., 2015; Johnston et al., 2019). For example, Freitas et al. (2010) conducted DEB analyses to compare across a number of North Atlantic fish and invertebrate marine species, but were limited by a lack of species-specific thermal tolerance and energetic data. Further, because of this specificity, models developed for one system can be difficult to apply to others, further limiting their general use. Recent advances in technologies (e.g., biologgers and remote sensing satellites; Metcalfe et al., 2012) and leveraging existing monitoring programs, local knowledge and citizen science approaches may offer avenues to gather the system-specific data needed for parameterization (Dickinson et al., 2010; Johnston et al., 2019; Jørgensen et al., 2012), while continued advances in the field of modeling will hopefully lead to more standardized, broadly applicable models (Boult and Evans, 2021; Railsback and Grimm, 2019). Many additional issues need to be addressed, and continued cross-talk between physiologists and modelers will be essential to realize the potential of mechanistic models to link across biological levels in fishes of conservation concern (Jørgensen et al., 2012). For example, determining the best types of physiological metrics to employ, how to account for differences in experimental methodologies and traits measured, and the scales (spatial, temporal, phylogenetic, etc.) over which data are needed to capture intra- and inter-species variability and robustly parameterize models are all complex issues that will require careful consideration (Evans et al., 2015; Johnston et al., 2019). However, if accomplished, the broader application of mechanistic models will facilitate syntheses to evaluate whether common patterns or contributing factors emerge to inform more generally across species and ecosystems.

## 3 Contextualizing physiological results into real-world scenarios

Mechanistic studies are frequently conducted by necessity via isolating one (or few) factor(s) to examine their effects and related physiological processes. However, fishes exist in complex ecosystems with environmental heterogeneity, and thus experience a combination of abiotic and biotic stressors that vary spatiotemporally. For example, we may be interested in understanding the effects of increased temperature on aerobic scope in a reef fish of conservation concern, but increased temperature may also be accompanied by ocean acidification or hypoxia (see Chapter 7, Volume 39B: Rummer and Illing, 2022). To forecast the impacts of, and resilience to, abiotic and biotic stressors, particularly under different global change and management scenarios, there is a strong need for studies to make robust links to environmental and ecological realism (Box 1; Todgham and Stillman, 2013). This means we need to understand how multiple stressors may act independently, cumulatively and/or sequentially to affect fishes. We further need to understand how these stressors may vary in space and time, across life stages, and how this variation influences fish physiology. Moreover, behavior has seldom been included as a metric in physiology studies, but provides a way in which fish may cope or be further impacted by stressors, and needs to become a salient consideration.

## 3.1 The need for environmental and ecological realism

Environments are not static but dynamic, and conditions vary in space and time (see Chapter 4, Volume 39B: Killen et al., 2022; Chapter 5, Volume 39B: Luis Val et al., 2022; Chapter 7, Volume 39B: Rummer and Illing, 2022; Chapter 8, Volume 39B: Strailey and Suski, 2022). Changing climates are often discussed in terms of averages, but extreme events, including increased variation in environmental factors, are already becoming common and are predicted to become further exacerbated in the future (IPCC, 2021). Furthermore, though there has been much focus on temperature, oxygen content and carbon dioxide content, these are not the only changing environmental factors; habitats are becoming increasingly degraded and fragmented, and many species are faced with overexploitation and other anthropogenic disturbances that can alter adaptive capacity, species interactions, and ecological dynamics (McNeil and Sasse, 2016; Pauly and Zeller, 2016; Perry et al., 2005; see previous chapters in Vol 39B: Cooke et al., 2022). In most real-world scenarios, fish are not exposed to a single stressor, hence the importance of exploring the impacts of multiple (simultaneous) stressors. This issue was highlighted in a study by (Faleiro et al., 2015), where adult seahorses (*Hippocampus guttulatus*) were exposed to ocean warming and acidification. The seahorses were found to perform relatively well with increased temperature, but did not perform well when faced with the combined effect of warming and acidification. Similarly, Antarctic dragonfish were found to survive best and develop fastest in the presence of acidification, but a combined warming and acidification treatment caused a dose-dependent, synergistic decrease in survival and a slower development (Flynn et al., 2015). This is also the case for a range of other fish species (see previous chapters in Vol 39B: Cooke et al., 2022), which have shown some capacity to adjust physiological processes to cope, or acclimate to single stressors, but not necessarily or as successfully to multiple simultaneous stressors (Leeuwis et al., 2021). The

**BOX 1 Contextualizing physiological data from Fraser River sockeye salmon with environmental realism provides scientific advice for stock management.**

In many ways, the success of physiology as a tool in conservation relies on its integration in management to affect changes in real life. The challenge then, is for researchers to transmit the plethora of available physiological information based on individuals to predictions about populations desired by managers (Coristine et al., 2014). One such success story pertains to the Fraser River sockeye salmon (*Oncorhynchus nerka*) (Cooke et al., 2012; Patterson et al., 2016). Concern over the decline in the species resulted in major political attention, specifically related to the causes of mortality during upriver migration. Starting in the early to mid-1900s, particularly following a series of major rockslides resulting from railroad construction in the Fraser Canyon (Hell's Gate), the need to understand the impacts of water discharge on sockeye was recognized (see Patterson et al., 2016). In the 1990s, swimming performance, variation in swimming behavior and migration survival at Hell's Gate was directly evaluated using electronic tagging (reviewed in Patterson et al., 2016), and provided a physiological (and mechanistic) explanation for why hydraulically challenging stretches of the Fraser were more difficult to ascend and could result in mortality. Subsequent modeling of energy expenditures in relation to flow helped guide management efforts, though growing concern over increasing river temperatures and its potential consequences for salmon led to more research on the individual and combined impacts of water temperature on salmon migration.

Hell's Gate, Fraser River, British Columbia, Canada (Image by Kim Birnie-Gauvin ©)

The realization that predictive modeling on such management targets could not possibly be attained without consideration for environmental realism reflecting changing conditions and the presence of multiple stressors in the wild has encouraged the use of Conservation Physiology for managing Fraser River sockeye salmon. Conservation Physiology research has provided the scientific evidence to enact changes in decisions made at the management level related to water temperature, high discharge, pathogens, sex, and interactions with fisheries. For example, the fate of Pacific salmon released after fisheries interactions and capture has long been of interest, and has been studied within the context of physiological

*Continued*

**BOX 1 Contextualizing physiological data from Fraser River sockeye salmon with environmental realism provides scientific advice for stock management.—cont'd**

impacts and post-release mortality extensively. The effects of gear type, levels of physiological stress, benefits of recovery techniques, variation among populations, and combined effects of capture and high temperatures have been studied. It is now well established, for example, that fish may recover from capture, but not necessarily from capture when temperatures are too high (Patterson et al., 2016 and references therein). Again, this information can be, and has been, integrated into management to reflect environmental realism, and continues to be used to adjust harvest. This integration of physiological knowledge into management decisions has been facilitated by the legal background laid out by the Pacific Salmon Treaty, which mandates both Canada and the USA to prevent overfishing and to ensure specific spawning escapement goals.

Spawning sockeye salmon (Image by Kim Birnie-Gauvin ©)

interaction between nitrate (a pervasive aquatic pollutant), hypoxia and heat stress was recently studied in the European grayling (*Thymallus thymallus*), with findings showing that nitrate-exposed fish were far more susceptible to both hypoxia and heat stress (Rodgers et al., 2021). Conversely, low oxygen levels were found to help prevent the negative consequences of acute warming on mitochondrial efficiency in the European seabass (*Dicentrarchus labrax*, Thoral et al., 2021), suggesting that exposure to some stressors may have protective effects on other stressors. In the meager (*Argyrosomus regius*), warming caused greater accumulation of mercury, but this effect was offset by ocean acidification, again suggesting a type of antagonistic effect of warming and ocean acidification (Sampaio et al., 2018). Still, other studies have shown that elevated acidification, but not increased temperature, can hinder behavior, while increased temperature, but not acidification, decreases physiological performance (Laubenstein et al., 2019). Collectively, these studies demonstrate that stressor impacts can vary widely across species or realms, reinforcing the critical importance of ecologically relevant multi-stressor experiments to understand how stressors may interact, and to identify the most important stressor(s) to mitigate.

Stressors may occur acutely, chronically or repeatedly, and as such their effects are likely to vary depending on the time course over which exposure occurs. For example, multiple studies have demonstrated that motorboat noise affects the cardiac physiology of fishes in a negative manner (e.g., Graham and Cooke, 2008; Jain-Schlaepfer et al., 2018; see Chapter 7, Volume 39B: Rummer and Illing, 2022), but others have shown repeated exposure to motorboat noise led to increased tolerance to this stressor (e.g., threespot dascyllus, *Dascyllus trimaculatus*; Nedelec et al., 2016). At initial exposure, fish showed increased ventilation rate and hiding behavior, but this response disappeared after 2 weeks of exposure. In a series of other studies, fish were acclimated to fluctuating temperatures rather than constant temperatures, to better reflect environmental conditions in the wild (e.g., Salinas et al., 2019). The time scale over which exposure occurs must become a more salient consideration as the field of Conservation Physiology moves forward.

The above examples have integrated ecological and environmental realism, but the field has largely been dominated by the study of single stressors. Of course, this may represent an evolution of the field, as a minimal understanding of the impacts of single stressors may have led researchers to better appreciate that the combined effect of multiple stressors may be different than the sum of its parts. Future studies should strive for their experimental design to reflect real-world scenarios. This includes multi-stressor exposure over variable exposure timescales (e.g., hours, days, weeks, months), as well as a better representation of environmental fluctuations observed in nature (e.g., static vs cyclic exposures). Meso- or megacosms may provide one such approach to study the impacts of stressors under more controlled, but still relevant, conditions (e.g., Nagrodski et al., 2013). Advances in environmental monitoring (e.g., remotely sensed data, instrumenting aquatic systems) can greatly assist in guiding experimental design to mimic environmentally realistic contexts. Such environmental monitoring is not always accessible, and can be costly to operate, but coupling such real-life environmental data to biological sampling can facilitate the interpretation and applicability of findings and ensure that these findings are environmentally and ecologically relevant. In doing so, conservation physiologists will increase the predictive capacity of fish responses to changes in environmental conditions and relevance to management decisions (Todgham and Stillman, 2013).

## 3.2 The need for studies across life stages and populations

Animal life histories can be finely tuned to their environments via genetic adaptation, though most have at least some capacity to cope with environmental change through phenotypic plasticity. In addition to local adaptation among populations of the same species, individuals may also experience different environmental conditions across life stages (e.g., Gardiner et al., 2010; and see Chapter 9, Volume 39A: Schulte and Healy, 2022). Consequently, thresholds and responses to abiotic and biotic factors may vary within- and across individuals, populations and species. This can make it challenging to extrapolate results from a particular life stage or population to others that are unstudied or conducted under different conditions. In some instances, embryos and larvae may be the most at risk to environmental stressors (Killen et al., 2007), while in other cases adults may be most at risk (Komoroske et al., 2014), given that selection pressures differ across life stages (e.g., Crozier et al., 2008). Some life stages may be particularly sensitive to some stressors, causing population bottlenecks and a reduction in population size (Petitgas et al., 2013). Therefore, future research conducted across life stages will provide crucial information on the presence of developmental plasticity, while research across populations will shed light on the presence of local adaptation and adaptive capacity, both of which need to be assessed thoroughly if we are to accurately predict the biological impacts of global climate change.

Physiological indices are often measured at shorter timescales such that they give us a "glimpse" at a particular time, but there can be downstream physiological, behavioral and ecological consequences in later life stages or later seasons. Particularly for species with complex life histories, such as migratory species, experiences from one life stage can have significant repercussions on later life stages, a phenomenon termed carryover effect (O'Connor and Cooke, 2015). Thus, studying the same individuals over time may provide insights into behavioral repeatability and the heritability of traits, as well as phenotypic plasticity and the underpinning mechanisms that may affect or dictate ecological outcomes (e.g., Birnie-Gauvin et al., 2021a; Eldøy et al., 2019). Assessing the link between previous experiences with current animal performance is also critical to understand the underlying mechanisms that govern fitness as well as to predict the effects of environmental stressors on individuals and populations. Failure to consider carryover effects may lead to flawed conclusions about specific conservation measures (O'Connor and Cooke, 2015). For example, sockeye salmon are relatively capable of overcoming the Seton Dam fishway in the Fraser River basin, but use anaerobic metabolism to do so (Burnett et al., 2014). Following a successful ascent, a large proportion of the fish died in the upstream reservoir and did not successfully reproduce, likely due to an inability to recover. If only passage success was evaluated, managers may think that the fishway is adequate because most fish passed it. But, reverting to anaerobic metabolism had effects that carried over into the subsequent phase of the salmon's migration. Such carryover effects are likely common but have tended to be overlooked and understudied with research largely focused on more immediate impacts (O'Connor and Cooke, 2015). While many carryover effects are negative, they could be exploited positively for conservation benefits. For example, protecting key life stages from stressors known to have carryover effects could have positive effects on physiological processes, making individuals more resistant to later stressors (O'Connor and Cooke, 2015). One example from Bordeleau et al. (2018) suggests that Atlantic salmon brought in hatcheries to presumably boost local populations via stocking have higher stress levels and lower survival to repeat spawning than wild-spawned salmon. Thus, hatchery operations may be causing consequences at the population-level by reducing rates of iteroparity. Studies on carryover effects may require long-term experiments that can be challenging to conduct, but developments in tracking technology, like the previous example, will likely make such studies possible and feasible.

## 3.3 The need to integrate behavior into physiological experiments and field studies

Physiological studies tend to focus on the effects of various factors on physiological indices, but do not necessarily integrate behavior into experiments. However, it is well known that behavior is often the first line of defense for animals to cope with challenging environmental conditions. As such, an important consideration for understanding the effects of environmental and anthropogenic challenges is the assessment of animal movement and behavior. Behavior may change as a result of a stressor, such as hiding in response to a predator or moving out of a hypoxic environment to avoid unfavorable conditions. Cortese et al. (2021) studied both the physiological and behavioral effects of anemone bleaching on orange-fin anemonefish (*Amphiprion chrysopterus*) and found that anemonefish living in bleached anemones had reduced metabolic costs compared to anemonefish living in healthy anemones, but that they had lower growth rates and spent more time foraging. This type of information is important as it highlights how relatively short-term thermal anomalies (which cause bleaching) can have long-term impacts, not only on the anemones themselves but also on the organisms that rely on them.

Historically, our ability to resolve animal behavior, particularly in the wild, was constrained by a lack of appropriate tracking and monitoring technologies. The continued development of telemetry and biologging devices is now making the assessment of movements and behavior easier, and doable on a greater range of species with miniaturization (Hussey et al., 2015). These devices can be coupled with sensors that monitor heart rate, salinity, temperature, depth or acceleration, and thus provide information on how fish respond to environmental change (Cooke et al., 2004).

# 4 Broadening phylogenomic and ecological diversity representation

To determine the physiological mechanisms that drive population declines and biodiversity loss in the face of environmental change, we must understand how populations or species across the tree of life will respond to these changes. We further need to be able to assess whether some patterns can be generalized, and under what circumstances, if we are to achieve broad scale biodiversity conservation. As such, studies need to be conducted in species with varied evolutionary histories and ecological diversity, rather than extrapolating findings from one well studied group to others. Fishes are the most diverse group of vertebrates, inhabiting nearly every aquatic system on the planet, yet much of our understanding of their physiology comes disproportionately from a small number of species (see for example the focus on salmonids in fish passage studies, see Chapter 10, Volume 39B: Wilkie, 2022). In part, this could be due to greater available funding for species with stronger management and economic implications (though see Section 6.3 for further discussion of bias in these determinations). Regardless of why, a limited knowledge base was recently identified as a major constraint in the application of Conservation Physiology in marine fishes (McKenzie et al., 2016). As one example, we extracted the studied fish species from all primary research articles published in *Conservation Physiology* and grouped them by similarity of Order. We identified that perch-like fishes (including coral reef fishes and icefishes), salmonids, and sharks (including flat sharks) were the most studied groups of fishes, highlighting a bias in the species generally considered in the context of Conservation Physiology (Fig. 1).

## 4.1 Comparing and contrasting species: Questioning physiological paradigms

Conducting parallel studies to compare and contrast responses across groups of fishes is key to understanding the diversity that exists in nature and forecast future ecological communities under global change. While some findings may be somewhat constant across diverse fish groups, many are not, and thus a broad knowledge base is required to challenge existing paradigms (e.g., Lefevre, 2016). For example, the oxygen- and capacity-limited thermal tolerance (OCLTT) suggests that aerobic scope declines outside the optimal temperature range due to impaired oxygen uptake (Pörtner, 2002). Therefore, environmental conditions set the limits within which energetic demands can be met (Farrell, 2011). There is convincing evidence that failures in upriver spawning migrations of Pacific salmon (*Oncorhynchus* species) are a result of abnormally high river temperatures, which impair swimming performance due to a reduced aerobic scope (Eliason et al., 2011; Patterson et al., 2016), supporting the OCLTT hypothesis. However, other studies have failed to find evidence that aerobic scope declines when reaching upper lethal temperatures

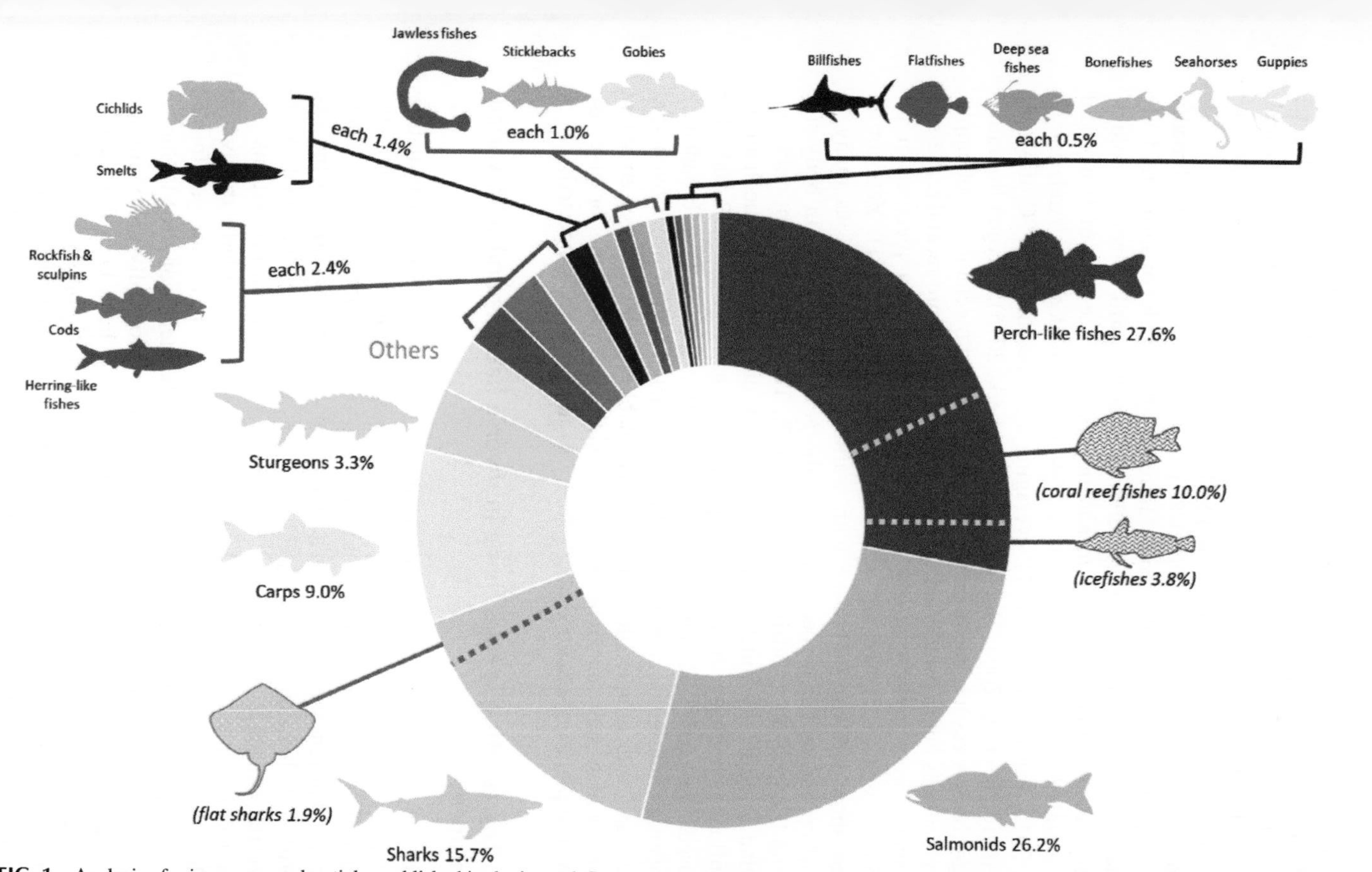


**FIG. 1** Analysis of primary research articles published in the journal *Conservation Physiology* on fish species. Results demonstrate that perch-like fishes (including coral reef fishes and icefishes), salmonids, and sharks (including flat sharks) are the most studied groups of fishes in Conservation Physiology. Here, "Others" includes charachins, ricefishes, silversides and toothcarps. Search was done 25 January 2021, with 150 relevant studies identified (491 initial hits, scoped at the title and abstract level for relevance). Note that percentages represent percentages of all studies included in this analysis, even those in brackets.

(Claësson et al., 2016; Norin et al., 2014). The broader applicability of the OCLTT has thus been questioned, and remains the subject of an active debate, but the diverse results, particularly across species, from OCLTT research highlight the need to expand phylogenomic and ecological diversity representation in these Conservation Physiology studies.

Another example of a physiological paradigm that may not be broadly applicable across species is the temperature-size rule (TSR) or shrinking body size paradigm, which has also been referred to as the third universal ecological response to global warming in aquatic systems. The size of fishes and other ectotherms have declined by 5–20% in recent decades (Audzijonyte et al., 2016; van Rijn et al., 2017), averaging a 3% decline per 1 °C of warming (Angilletta et al., 2004; Hoefnagel and Verberk, 2015). Oxygen limitation to larger body sizes has been posited as the underlying mechanism, but empirical evidence is conflicting (Audzijonyte et al., 2019). More recently, the aerobic scope protection hypothesis has been suggested as a mechanism for reduced growth under warmer conditions ( Jutfelt et al., 2021). Yet, not all species show reduced body size with warming. For example, as many as 45% of coastal reef fish species in Australia actually increased in body size over the last decades (Audzijonyte et al., 2020). These findings not only suggest that there is still much to uncover to fully understand the mechanisms that influence body size under warming, but also challenge the TSR and other proposed "universal" ecological responses. Within this context, broadening phylogenetic and ecological representation to include often overlooked species or taxa with unique life histories may enable the detection of emergent patterns, which may reinforce existing physiological paradigms or give rise to novel hypotheses when existing paradigms are not fully supported.

## 4.2 Representing species living in diverse habitats and with varied life histories

Part of gaining a better understanding of the diversity that exists in nature, particularly within the context of Conservation Physiology, is studying species that live in diverse habitats and those that display a range of life history strategies. Currently, we lack physiological data and information for the vast majority of fish species and geographical areas, though it is important to recognize the existence of traditional knowledge of species in ecosystems (see Section 6.4). This lack of knowledge seems to be particularly true where developed fish ecophysiology research communities have not yet formed (Killen et al., 2016; McKenzie et al., 2016; Seebacher et al., 2014; Sunday et al., 2012). In these locations, economically or ecologically important species, and those easy to obtain or house in captivity, have been the focus of most research. McKenzie et al. (2016) called for physiologists to collaborate effectively to fill knowledge gaps: to study more species, perhaps those harder to obtain or keep in captivity. Several meta-analyses (see Sections 5 and 4.1) have demonstrated that findings are rarely applicable across all species and habitats, underscoring the importance of reflecting the diversity of fish and the systems they inhabit in Conservation Physiology research. This work is complementary to within species assessments across life stages and populations discussed in Section 3.2, which collectively is essential to advance our understanding of broadscale patterns. In an analysis on hypoxia tolerance from 96 published studies, Rogers et al. (2016) found tolerance was highly variable among species, and influenced both by abiotic (temperature, $CO_2$, acidification) and biotic (feeding) factors. These results suggest that fishes adapted to dissimilar habitats and displaying diverse life histories can vary greatly in their physiological processes and responses, and such species are therefore important to include when assessing broadscale trends and patterns in Conservation Physiology.

## 4.3 Thinking outside the box

Broadening species representation is challenging because more funds are often available for certain groups with high direct economic values (e.g., salmonids, tuna, or coral reef fishes), some species are difficult to obtain, and many species are not amenable to traditional physiological experiments (e.g., pelagic fishes such as delta smelt and sardines, or large species such as white sharks). A salient consideration within this context is of course the fact that less accessible species may prevent the consideration of sex, life-stage and other relevant factors in studies due to small sample sizes and other limitations. We acknowledge that this will likely continue to be a challenge, but nonetheless encourage future work to expand species representation, especially as our understanding of these species within the Conservation Physiology lens is currently limited (if existent at all). Recent emerging techniques play a key role here to help researchers think outside the box to address this need; particularly as technologies are becoming more cost-effective for application in situ, studies can be expanded to non-model species with greater ease. The advancement of sequencing and bioinformatics methods is now making genome and transcriptome-wide approaches possible to study Conservation Physiology questions in non-model species with limited existing genomic resources (Connon et al., 2018; see Chapter 8, Volume 39A: Jeffries et al., 2022). The miniaturization of tracking technologies has been critical in expanding the range and life stages of species that scientists can track (Hussey et al., 2015), and as technological innovation further advances in this realm it will likely continue to open new opportunities and applications. Other important advancements include the development of portable equipment that scientists can easily bring in the field with them. Examples include respirometers, and portable clinical analyzers to measure physiological indices like lactate, glucose and pH (also see Chapter 4, Volume 39A: Brownscombe et al., 2022). Being able to easily transport such devices into the field reduces overall research costs and stress on the fish themselves, by avoiding having to house animals in the lab. Of course the accuracy, precision and power of these tools must be tested, but evidence suggests that (at least some) field methods are highly comparable to laboratory methods (e.g., inducing maximum metabolic rate in the field, Little et al., 2020; iSTAT portable analyzer, Harter et al., 2014). Recently, portable respirometers have even been developed for large-bodied fishes (bull sharks, *Carcharhinus leucas*; Byrnes et al., 2020), once more expanding the range of species from which physiological data can be obtained. As tools continue to be developed and/or improved, and as costs continue to diminish, researchers will be able to study physiology in ways and species never-before investigated. Nonetheless, it is important to remind ourselves that ultimately conservation needs should drive research priorities using these technologies, not the reverse.

## 5 Using syntheses to understand emergent patterns

The expansion of scientific research (approximately 2.5 million articles published annually, Plume and Van Weijen, 2014) reflects an important increase in knowledge generation, but much of this knowledge is scattered across (increasingly) diverse sources. This makes it challenging for knowledge users and decision-makers to find and digest the relevant information, or even make sense of sometimes contradictory results (Pullin et al., 2020). Evidence synthesis—the process of searching for and summarizing a body of research on a topic to inform decisions—is a particularly important tool within this context that has been largely neglected until more recently (Pullin et al., 2020). Given the wealth of ecophysiological data that already exists, advanced approaches in data synthesis (i.e., meta-analyses and systematic reviews) will be important tools to leverage the large amount of existing data to assess generalized patterns and identify key outcome modifiers, particularly to forecast future population projections. Below, we highlight a few different types of syntheses with examples of relevant questions that have been tackled to date to demonstrate the utility of these approaches and stimulate ideas for their application in other priority topics in Conservation Physiology.

### 5.1 Stressor-specific syntheses

One way to use meta-analyses in the context of Conservation Physiology is by performing stressor-specific meta-analyses to explore patterns in the effects of specific stressors across a wide range of species. One such meta-analysis investigated the effects of warming and ocean acidification on the respiratory physiology of marine ectotherms, including fishes (Lefevre, 2016), and found that there were no clear overall combined effects of temperature and $CO_2$. $CO_2$ caused an increase in resting metabolic rate in only 18 of 125 datasets and a decrease in 25 of 125 datasets. When $CO_2$ had an effect, synergistic effects with temperature were most common, as opposed to additive effects. The author suggests that responses to elevated temperatures and $CO_2$ vary across species, with no single unifying principle (Lefevre, 2016).

As a second example, as mentioned in Section 4.1, a meta-analysis investigating the effect of temperature on the body size of 335 coastal reef fish species in Australia found that temperature drove spatial and temporal changes in body size, but not necessarily in the manner expected (Audzijonyte et al., 2020). Approximately 55% of species were smaller with increased water temperature, but 45% were larger. Thus, although several studies, and even other meta-analyses, have found support for reduced body size as a universal response to global warming (e.g., Daufresne et al., 2009; Gardner et al., 2011), this work suggests that trends may actually be somewhat dependent on the species and geographical areas considered (see Section 5.1).

Conservation physiologists can also learn new statistical tools or partner with colleagues with complimentary quantitative expertise to take advantage of recent statistical advances to shed new light on emergent patterns of existing datasets. In a different type of large-scale analysis, Bigman et al. (2021) explored the relative importance of respiratory capacity (measured as respiratory surface area) and temperature on metabolic rate across the vertebrate tree of life. To do so, the authors used a novel phylogenetic Bayesian multilevel modeling framework coupled with species-specific metabolic rate and respiratory surface area data from published datasets for 109 species. Their robust analysis revealed that respiratory capacity explained twice as much of the variation in metabolic rate as temperature. The authors suggest that understanding the processes and capacity for oxygen acquisition and transport would provide an opportunity to understand the history of metabolism over evolutionary time, and improve our predictive models for climate change (Bigman et al., 2021).

## 5.2 Species-specific syntheses

Another way of using meta-analyses to better integrate the use of physiological data in management schemes is by performing single species (or single taxon) meta-analyses for particularly vulnerable or at-risk species (or groups of species). Such an analysis was recently conducted for the imperiled green sturgeon (*Acipenser medirostris*) and summarized the mean effects of important stressors (i.e., increased temperatures, salinity, low food availability and presence of contaminants) on important physiological traits (e.g., growth, swimming performance, thermal tolerance and heat shock protein expression) (Rodgers et al., 2019). The authors found that all stressors impaired growth and noted other costs associated with specific stressors, but they were also able to suggest specific management actions to mitigate those impacts, including reducing salt intrusion in nursery habitats and maintaining optimal water temperatures during peak spawning periods. The meta-analysis further identified gaps in the data, demonstrating that research had been biased toward juvenile and adult life-history stages, overlooking embryonic, yolk-sac larvae and post yolk-sac larvae life stages (Rodgers et al., 2019). In another study, Santos et al. (2021) combined the findings of 368 control-treatment contrasts obtained from 53 studies on elasmobranchs to better understand the effects of warming, acidification and deoxygenation (and their various combinations).

The authors further compared these effects across life stages, lineages, reproductive modes, lifestyles (benthic, benthopelagic or pelagic) and climates. Their findings suggest that elasmobranchs are most impacted by warming, which had clear directional effects, including decreased development time, as well as increased metabolism, feeding and thermal tolerance. Acidification, in contrast, had the most context-dependent effects, with no directional trends. Still, deoxygenation was the least studied, along with combinations of treatments; thus, data paucity precluded meaningful conclusions (Santos et al., 2021, and see Section 3). Such analyses can help guide future work on the species directly to areas of research where important knowledge gaps exist. Of course, meta-analyses and syntheses are statistically powerful, but literature reviews may also be useful in this realm (e.g., Birnie-Gauvin et al., 2021b; Hinch et al., 2021; Jordan et al., 2013).

## 5.3 Making data accessible and standardized

One current critical limitation for robust data synthesis is inconsistent reporting of data and methods in empirical studies (Whitlock, 2011). Thus, not only does data synthesis research need to be prioritized, but there is a strong need for increased training and requirement of standardized data and methods that facilitate use in meta-analyses and syntheses. Making data publicly available and accessible is only one requirement of public data archiving (PDA) policies, generally to allow reproducibility of results (Moore et al., 2010; Whitlock et al., 2010), but there are currently no requirements for how data are presented. An analysis of 100 datasets from journals with strong PDA standards in the fields of ecology and evolution found that 56% of datasets were incomplete, and 64% were archived such that reuse was impossible (Roche et al., 2015). These findings highlight the need for guidelines to make data standardized, and not just accessible.

# 6 Creating an inclusive field that values the perspectives and knowledges of all peoples[a]

The majority of STEM fields including physiology and fisheries sciences, as well as conservation and management organizations, have historically lacked equal opportunity and representation across race, gender, cultures and other

---

[a]*Statement of position*: The authors of this chapter self-identify as white, cis-gendered women, and thus have only lived experience through this lens. For this section, we have integrated and cited the work of BIPOC, LGTBQIA+, and other diverse scientists and conservation practitioners to draw on their knowledge to provide examples and suggestions for how to advance DEIJ (diversity, equity, inclusion and justice) in our field. However, we recognize that we cannot speak for these groups directly, nor are they monolithic (e.g., the perspectives of some BIPOC scientists cited do not necessarily represent everyone who identifies in this group), and strongly encourage all readers seeking to contribute to DEIJ in STEM and conservation practice to further examine the literature cited here directly, as well as the rich works of scholars in these topics more broadly.

forms of diversity (Li and Koedel, 2017; National Science Foundation, 2019). This lack of diversity has especially been the case for positions of leadership, where largely white male perspectives and western knowledge structures have dominated over others. Though some recent progress has been made, much more work is needed to achieve diverse and inclusive fields of study, governance structures, and real-world equitable management practices. For Conservation Physiology in particular, an inherently applied field, the continued lack of inclusivity is not only a problem from a moral and equity standpoint, but it also strongly limits the advances and impacts the field can achieve. For example, many Indigenous peoples and local communities (IPLCs; Dawson et al., 2021) have extensive knowledge of species of concern that can guide studies and contribute to development of best practices for conservation. Further, historically marginalized groups are often most affected by conservation crises and environmental degradation. Thus, not only is the science strengthened from a diversity of perspectives and knowledge sources (Hofstra et al., 2020), if a key goal of Conservation Physiology is to maintain and restore biodiversity and healthy fish populations around the globe, engagement of communities sharing their habitats and utilizing ecological resources is essential.

Both systemic change and individual actions are needed to accomplish diversity, equity, inclusion and justice (DEIJ) goals in Conservation Physiology and beyond. It is critical to recognize that inaction to change the status quo is not a neutral stance, but rather a choice that allows current inequitable and discriminatory practices and culture to persist (Kendi, 2019). Thus, we call upon all individuals in our field to meaningfully engage in DEIJ efforts, particularly those in positions of privilege and/or authority who currently hold more power to influence larger-scale institutional change. Here, we discuss several common barriers to realizing DEIJ goals in academic and conservation spheres relevant to Conservation Physiology and highlight suggested strategies for dismantling them. Importantly, this is not an exhaustive list and the specific issues, identities and histories of marginalized groups, and strategies for success inherently will vary among countries, sub-fields and other contexts.

## 6.1 Dismantling colonial and racist legacies

Many STEM fields and conservation practices have deep roots in colonial, racist, and sexist ideologies (Chaudhury and Colla, 2020; Cronin et al., 2021; Halsey et al., 2020). Dismantling these practices to move into a future of equitable biodiversity sustainability first requires understanding and acknowledging the breadth and depth of these legacies. Doing so includes acknowledging the many ways they have shaped our views of the natural world, dictated who is entitled to produce scientific knowledge, and how this has strongly influenced past and current conservation management policies. For example, for centuries Western scientists have widely engaged in colonial or parachute science, in which they travel to and conduct research in low-income countries, excluding local and Indigenous communities, and not engaging ethically with the traditional owners of the lands and seas on which they work (de Vos, 2020; Stefanoudis et al., 2021). A recent study found that 89% of authors publishing on coral reef science come from developed countries, despite most coral reef habitats occurring in developing nations (Ahmadia et al., 2021). Not only have these practices set false narratives for Westerners being the sole purveyors of scientific knowledge (i.e., many "discoveries" published by Western scientists were already well known by IPLCs; Das and Lowe, 2018), but biodiversity management plans enacted without integration of IPLC stakeholder knowledge and engagement often fail to accomplish conservation goals (Dawson et al., 2021). Additionally, many foundational figures in fields adjacent to Conservation Physiology (e.g., ecology, evolution, and conservation biology) also promoted problematic racist and other discriminatory views that still pervade our teaching of these subjects (Müller-Wille, 2014; Shields and Bhatia, 2009). This not only minimizes past contributions of BIPOC and other marginalized scholars, but perpetuates false perceptions and non-inclusive cultures that discourages students from historically underrepresented groups from pursuing studies in these fields (*though in-depth analysis of these issues is outside the scope of this chapter, we strongly encourage readers to see, Cronin et al. (2021), Chaudhury and Colla (2020), and the many sources cited within for robust background and discussion of these topics*). A final example particularly relevant to the Conservation Physiology of fishes is how historical paradigms have shaped management policies of native freshwater fishes in North America, largely prioritizing a small proportion of species (e.g., Salmonidae and Micropterus spp.) valued by white male anglers over most others. Many of these diverse species continue to be deemed as useless (i.e., "rough fish") and ignored in fisheries management plans, despite their long important roles in sustenance fisheries for IPLCs and immigrant populations, and demonstrated ecological values (Rypel et al., 2021). In turn, these biases connect to issues discussed previously, such as disproportionate research funding availability and the consequent over-representation of a few groups largely composing our physiological knowledge in fishes (see Section 4).

These highlighted examples are but a few of the many ways in which discrimination and inequity is deeply embedded in our STEM and conservation cultures (Kendi, 2019). We do not list specific recommendations here to address these long-standing, complex issues because the "right" solutions depend heavily on the geographic, cultural, and other contexts (e.g., in the classroom or laboratory research setting vs stakeholder engagement in conservation policy development). Instead, we emphasize the need for all to meaningfully engage with the literature and other resources from experts on these topics that are most relevant to the avenues in which they can influence individual and institutional change (see the literature cited throughout this section). Taking time to learn and understand these histories is a critical component for all those practicing Conservation Physiology and adjacent fields to engage meaningfully in effective DEIJ efforts. Without taking these contexts into account, many well-meaning efforts will fail to achieve their intended goals, or even cause greater harm to marginalized peoples and biodiversity. Rather, dismantling these problematic legacies and creating equitable solutions will take concerted, long-term efforts tailored to different situations, but ultimately are necessary to achieve an inclusive, sustainable future for biodiversity and all peoples.

## 6.2 Promoting equitable opportunity and inclusive practices

As discussed above, the best avenues for conservation physiologists to influence change will vary depending on institution, geographical location, position (e.g., academic professor, government researcher, or NGO scientist) and other factors, but undoubtedly there are opportunities for every person to take on active roles in promoting diversity and inclusion. As one simple example, the majority of biology textbooks, including most curricula for animal physiology, ecology and evolution, are heavily skewed toward emphasizing the scientific contributions of heterosexual cis white men over those of BIPOC (Black people, Indigenous people, and people of color), women, and LGBTQIA+ (lesbian, gay, bisexual, transgender, queer, intersex, asexual and other gender identities) scholars, as well as sources of traditional knowledge (Kimmerer, 2002; Wood et al., 2020). Thus, for those engaging with students in the classroom, a simple first step can be to reshape course curricula to be more inclusive. Collated resources of recent initiatives such as Project Biodiversify (https:/projectbiodiversify.org/) can serve as a good starting point. Such changes are well worth the time investment—simultaneously disrupting the false narrative that these fields have been solely built by white men, while providing role models for diverse students to foster a sense of belonging in STEM (O'Brien et al., 2020). Additionally, students from underrepresented groups are often less exposed to knowledge of how to pursue careers in science, so those in teaching or early mentorship roles (e.g., via K-12 engagement or undergraduate research advising) can incorporate modules demystifying the process into their programs (i.e., explaining hidden curriculum).

For those in mentorship and leadership roles at the graduate level and beyond, there are many actions that can positively impact the advancement of individual mentees as well as broadscale institutional change. Several recent excellent articles discussing these topics can guide allies[b] in understanding and adopting equitable practices in research laboratories, academic departments, conservation organizations and professional societies (e.g., see Applewhite, 2021; Cronin et al., 2021; Davies et al., 2021; Keisling et al., 2020; Schell et al., 2020). Engaging with these resources directly will help

[b]There are multiple definitions of the term "ally"; here we refer to the term as defined by UW College of Environment DEIJ glossary: *individuals not belonging to a marginalized group who "acknowledge oppression and actively commit to reducing their own complicity, investing in strengthening their own knowledge and awareness of oppression"* (https:/environment.uw.edu/about/diversity-equity-inclusion/tools-and-additional-resources/glossary-dei-concepts/).

determine the best approaches for specific circumstances, but here we highlight a few examples that are broadly applicable across many contexts. First, the recruitment of mentees and employees from historically systemically excluded groups is an important initial step toward DEIJ goals, but needs to be coupled with systematic changes that create inclusive work cultures for these students and scholars to thrive and advance to leadership positions (Halsey et al., 2020; Tseng et al., 2020). While it is critical to include and listen to the perspectives and needs of marginalized scholars throughout processes to achieve these goals, the work to transform our field and institutions cannot be placed on their shoulders (Halsey et al., 2020; Jimenez et al., 2019; Keisling et al., 2020). Secondly, although actions are needed at all levels of organizations, individual research groups can contribute by taking actions such as creating inclusive codes of conduct, implementing strategies for inclusive lab meetings, and embracing and normalizing thoughtful discussions of DEIJ topics as part of professional development activities (see examples and strategies discussed in Forrester, 2020; Golden et al., 2021). Finally, all individuals (including allies and those belonging to marginalized groups) across career levels and positions can elevate the work of historically systemically excluded scientists and conservationists by nominating them for awards at their institutions and professional societies, citing their work in their publications, and inviting them to present their work at seminars and conferences (Applewhite, 2021; Schell et al., 2020).

## 6.3 Field work safety and support systems

Like many other disciplines in the natural sciences, conducting fieldwork can be an important component of Conservation Physiology research and can provide new students with the practical experience and inspiration to pursue further study (Morales et al., 2020). Early exposure and access to fieldwork opportunities are often lower among students from historically underrepresented groups, and when they do participate, discriminatory conditions and non-inclusive cultures can lead to negative experiences (Anadu et al., 2020; Demery and Pipkin, 2021). This can range from microaggressions that marginalize and isolate individuals to feel as though they do not belong in these outdoor spaces, to sexual and other forms of harassment that create dangerous conditions for individuals with certain identities (Chaudhury and Colla, 2020; Estrada et al., 2016; Finney, 2014; Morales et al., 2020). While safety and inclusion in the field is an important consideration for all researchers that should be robustly planned for in advance, it is critical to recognize that students and scholars identifying as BIPOC, LGBTQIA+, women and other marginalized groups often experience differential and heightened risks and barriers (Anadu et al., 2020). This is especially true for those with intersectional identities, such as women of color. Moreover, the consequences of these negative experiences can reverberate far beyond the field setting, significantly impacting individuals' mental health, productivity and professional advancement (O'Brien et al., 2020).

Recently, several excellent articles have been published outlining specific actions that team leaders and organizations can take to overcome these barriers in different fieldwork and educational settings (see Anadu et al., 2020; Demery and Pipkin, 2021; Morales et al., 2020). Importantly, these include prioritizing time and resources for proactive measures, such as conducting racial and gender risk assessments and anti-discrimination trainings for all team members before going into the field. The knowledge gained from these can then be used to create shared codes of conduct and plans to address and de-escalate situations that may arise (e.g., 5D strategy for bystander interventions: direct, distract, delegate, delay, document; https://www.ihollaback.org/bystander-resources/). Proactive planning by supervisors and institutions for safety support can be particularly important for situations where at-risk students and scholars may need to conduct field work alone and/or in remote settings (see list of suggested specific actions for various contexts in table 1 in Demery and Pipkin, 2021). Finally, although such preventative actions will hopefully reduce the occurrences of negative interactions and dangerous conditions, robust planning also needs to include the establishment of transparent mechanisms for reporting incidents and supporting at-risk individuals when they do occur.

## 6.4 Embracing multiple knowledge systems in research and conservation practices

There has been increasing recognition in STEM and conservation fields that understanding of the natural world is not limited to knowledge generated by Western science (Jessen et al., 2021). Indigenous peoples and local communities have deep knowledge of their surrounding species and ecosystems, often spanning decades and centuries before "discovery" or study by Western scientists (Das and Lowe, 2018; Dawson et al., 2021). Of particular importance to the conservation of fishes is that many IPLCs have rich, long-standing cultural connections to and robust knowledge of the ecology and sustainable management of local fish populations (Reid et al., 2021). Though integrating traditional ecological knowledge (TEK; Kimmerer, 2002) and Native science (Bang et al., 2018) into Western scientific frameworks is an increasingly common approach, this still takes a colonial perspective of how IPLC knowledge can be "used" by Western science, rather than viewing them as complementary, coexisting knowledge sources (following the framework of *Etuaptmumk*, "Two-Eyed Seeing" in Mi'kmaw, as proposed by Reid et al., 2021 in the context of fisheries research and management). Not only can the methodologies and knowledge types be different in Indigenous science relative to Western science, but they are often grounded in dichotomous cultural perspectives and value systems (e.g., taxonomically vs ecologically, or competitively vs collaboratively; see Bang et al., 2018 for a deeper examination of this topic). Additionally, it is well documented that positive outcomes for both conservation goals and community well-being are significantly higher when IPLC knowledge and perspectives are valued in policy development and resource governance (Dawson et al., 2021). Therefore, approaches based on the Two-Eyed Seeing framework can be particularly transformative as the knowledge should transform the holder, and consequently the holder bears responsibility to act on that knowledge. Such frameworks not only aid in recognizing the inherent, independent value of IPLC knowledge, but can also be critical for the co-production of equitable, effective natural resources policy and decision-making (Reid et al., 2021).

One critical first step is often for Western scientists to educate themselves on the history of land they (and their institutions) currently occupy, the existing IPLC knowledge of their ecosystem/species of study, and broader historical and contemporary contexts that may factor into the best next steps for engaging with Native scholars and communities (e.g., have colonial or other discriminatory actions and policies marginalized their knowledge and voices?). Next steps may include working with IPLCs to build partnerships for research studies and/or policy development. It is important to enter into such collaborative endeavors recognizing the significant commitment, communication and learning required for success, especially among groups that may have just reason from the past for mistrust (Jessen et al., 2021; Reid et al., 2021). Encouragingly, there have been increasing examples of successful coupling of Indigenous Knowledge and Western science in fisheries management as well as broader ecology, evolution and physiology contexts that can serve as case studies for those seeking to develop partnerships and approaches based on knowledge co-existence and co-production (e.g., see thorough reviews across a multitude of systems in Jessen et al., 2021 and Reid et al., 2021).

## 7 What is "successful" Conservation Physiology?

In any applied field of study, to be truly successful research ultimately must result in positive outcomes in real-world contexts. For Conservation Physiology, this can take multiple forms; but overall success means measurable contributions to species recovery and persistence, ecological stability and restoration, and other conservation applications, including threat assessments of endangered species (Birnie-Gauvin et al., 2017). Achieving this requires critical additional steps outside of the basic process of scientific inquiry, which are most effective when undertaken not only once the research is concluded, but via continuous engagement with practitioners and stakeholders, before, during and after studies are conducted. There are already excellent examples of such work in the discipline undertaking and accomplishing these goals (Cooke et al., 2012), but to advance the field in the next decade and beyond these need to become the rule rather than the exception. The specific actions and additional knowledge required can be context-dependent depending on the topic, governance structures, and stakeholder communities but typically fall into a few categories that entail some common themes and activities that researchers can identify appropriate analogs in their systems.

## 7.1 Improving integration of physiological data into management frameworks

Despite a wealth of ecophysiological data relevant to key issues in the conservation of fishes and continued calls for better integration of physiological information into fish and fisheries management (Horodysky et al., 2015; Wikelski and Cooke, 2006), this information is still underutilized (Mahoney et al., 2018). For example, an assessment of the US Endangered Species Act found that fish and birds had some of the lowest proportion of recovery plans that include physiological data beyond the description of natural history (Mahoney et al., 2018). In some cases this is due to gaps in the relevant physiological data (e.g., sparse data on the sensory biology of pelagic species due to logistical constraints; Horodysky et al., 2016), or fundamental incompatibilities between the types of information generated by physiological studies and management goals. However, in many instances physiological knowledge is relevant for conservation, but a lack of engagement between physiologists, practitioners and policymakers plays a dominant role in preventing its translation into management plans—and as such, are missed opportunities. Many well-intentioned scientists who genuinely would like their work to have conservation impacts do not have training in understanding management and legal frameworks or policy processes. Consequently, study designs are often poorly matched with management goals and/or are published in peer-reviewed journals stating they "will be useful for conservation" without any engagement with practitioners in direct, productive ways to understand their findings and how they can be integrated into management plans (e.g., highlighted for coral reef fishes in Illing and Rummer, 2017). Conversely, many practitioners do not have access to papers or well-aligned training to evaluate the relevance to conservation goals from the peer-reviewed literature alone. These problems are not unique to Conservation Physiology, but pose a particular challenge for this newer discipline that is not already well understood by decision makers (Cooke et al., 2021).

Physiologists, practitioners and policymakers are often speaking different "languages" and operating within very different institutional frameworks. For example, while conservation physiologists often focus on generating specific mechanistic understanding of a species or system, managers and policymakers typically need to integrate many types of information and priorities when developing and implementing species management plans (Cooke et al., 2012). It is unrealistic, and unnecessary, for everyone to become experts in all realms, but awareness of these differences, some effort to understand the fundamentals of each, and open communication can go a long way in facilitating productive relationships. Engaging in early discussions with practitioners is key to understanding management priorities to ensure study designs are well-aligned with the information needed (see extended discussion of this topic and examples in Laubenstein and Rummer, 2021), while continued communication can help managers better understand what types of data physiology and behavioral studies can deliver (particularly important as technologies advance). Such partnerships have led to successful design and implementation of alternate fishing gear informed by sensory biology (see Chapter 2, Volume 39A: Horodysky et al., 2022), behavior, and other physiological studies to reduce bycatch and/or post-release mortality in large pelagic fishes (reviewed in Horodysky et al., 2016; Jordan et al., 2013). Further, physiological studies evaluating efficacy of different gear coupled with social science quantifying compliance can inform the rankings of fisheries for certification programs (e.g., Marine Stewardship Council) that aim to influence consumer behavior toward increasing demand for more sustainable fisheries (Horodysky et al., 2016).

These and other approaches to understand differences in knowledge structures ultimately help physiologists better give scientific advice and understand how their findings connect (or not) to real-world management "levers" that can be adjusted to support conservation goals. For example, Patterson et al. (2016) describes how physiological studies and extensive engagement between scientists and managers have informed sockeye salmon recovery efforts in the Fraser River (see Box 1). These efforts resulted in evidence-based advice to help reduce fishing mortality related to handling and temperature stress to integrate into management plans and policies. However, it is instructive to note that despite extensive efforts and being one of the most studied species of Pacific salmon in Canada, uncertainty in multiple aspects of how study findings relate to complex real-world contexts still poses a challenge to developing clear directives for management actions. Additionally, relationship-building and continued engagement can greatly assist in finding the right "home" for different types of physiological information within management frameworks. For example, during the study design phase of research on Pacific sardine thermal physiology and energetics, stakeholder meetings identified that current stock assessment models determining harvest allowances are not compatible with including physiological parameters, and thus would be unlikely to be informed by results no matter the specific experimental choices (N. Wegner, pers. comm.). In contrast, since Pacific sardine are a primary forage fish in the highly economically and ecologically valuable California Current ecosystem, discussions also revealed that ecosystem and climate modeling efforts are currently hampered by a lack of this empirical physiological knowledge. These analyses, in turn, are increasingly important elements of identifying and forecasting critical habitat and integrating food web dynamics into ecosystem-based management.

## 7.2 Engaging directly with public and stakeholder communities

Conservation practitioners routinely engage with stakeholders and the public to communicate science informed management plans and policies as described above, where the scientists may be one or more steps removed. However, in some contexts, alternative avenues where scientists can engage directly or through other translational partners can also be highly productive for positive conservation outcomes. One realm perhaps well-suited for such interactions is recreational fishing communities because the stakeholders can be enthusiastic, open to engagement, and have established platforms for direct communication (e.g., fishing club meetings, active online presences, etc.). Additionally, many conservation fish physiologists are also anglers themselves and have at least some exposure to and understanding of those communities. In some cases, scientists have formed direct collaborative relationships with anglers to study and inform the community on best practices (e.g., rockfish barotrauma, Box 2), while in others, partners like non-governmental organizations can serve as critical links to translate the science and work with stakeholders to shift cultural behavioral norms (e.g., implementing science-based best practices for survival in catch and release recreational fisheries, Box 3). While the specific actions in these case studies are not necessarily applicable to all situations, common themes such as willingness to try new communication platforms and methods, approaching stakeholder engagement as multi-directional idea and knowledge transfer, and partnering with science communication experts may be useful tools for other systems and conservation contexts.

## 7.3 Common themes of success to inform effective conservation in the future

Given the complexities and context-dependencies of management across species, governance structures and systems, there will never be a prescriptive one-size-fits-all approach to successful science-informed conservation. However, many of the examples above highlight that building collaborative teams with complementary expertise and perspectives is a key component of success. This takes effort and continued engagement from all sides, which is no small feat, especially when there are often very different incentive structures and priority directives among parties. Yet, for physiologists motivated to have their work truly affect conservation goals, these approaches not only alleviate the pressure for scientists to feel like they need to become an expert in all realms on their own, but also can lead to more creative, innovative science itself. It can also help to avoid missteps or unintended consequences, for example by including social scientists on the teams to evaluate how human perceptions and behaviors may alter the real-world outcomes of biologically-based recommendations (Guckian et al., 2018). Exposing scientists early in their career to perspectives and frameworks of practitioners and policymakers can jumpstart their ability to engage in productive interactions. Existing programs

**BOX 2 Leveraging stakeholder-scientist-management partnerships for successful conservation.**

Fish caught at depth and brought quickly to the surface frequently suffer from barotrauma (i.e., tissue damage and other physical negative impacts due to pressure differences in gaseous spaces within compared to outside of the body; especially problematic in physoclistous fishes), which can substantially impact survival upon release (Rummer and Bennett, 2005). This is of particular concern for rockfishes on the West Coast of the U.S. that can be sensitive to barotrauma (Jarvis and Lowe, 2008) and have suffered population declines due to overfishing. Regulations for bag limits and no-take for some hard-hit species to help restore populations results in individuals frequently being caught and released by recreational anglers. As these restrictions were put in place, the recreational fishing community recognized that releasing fish that cannot properly descend and succumb to post-release mortality is futile and would not help rebuild populations, and could result in additional restrictions on anglers. Recognizing this mismatch in practices with policy intentions, representatives from the Sportfishing Association of California (SAC, particularly president Ken Franke) approached NOAA suggesting they work together to evaluate different descending devices that had begun to be developed and marketed to address this issue.

Working in partnership, NOAA scientists and SAC first conducted studies on Commercial Passenger Fishing Vessels across California to evaluate the impact of different devices on short-term rockfish survival (Bellquist et al., 2019; Fig. 2). Importantly, they incorporated factors such as ease of use and preference to ensure results would be relevant for real-world scenarios, including use by fishermen with varying levels of experience. In addition to informing regional best practices, results of this work have led to the adoption of required practices in other regions (e.g., the Canada Department of Fisheries and Oceans in British Columbia). At the time of study, Bellquist et al. (2019) documented less than half of fishermen participating in the study owned and regularly used descending devices. Thus, it was critical that they did not just publish their findings, but also actively engaged with stakeholders and regulators directly to share the results and encourage adoption of best practices (e.g., conducting demonstrations at sportfishing meetings and presenting to the Pacific Fisheries Management Council). In these engagement and outreach realms, it has been particularly helpful to leverage recent technology such as underwater videos that show fish descending, recovering, and swimming away after release at depth to demonstrate that the devices truly work (N. Wegner, pers. comm.), and working with local artists, NGOs and science communicators to create educational media like the well-known Rockfish Recompression video (https://www.youtube.com/watch?v=EiZFghwVOyI). Though formal survey studies are needed to robustly assess if these and other outreach efforts are shifting behavioral norms in recreational fishing communities, they represent critical components of truly successful Conservation Physiology.

*Continued*

**BOX 2 Leveraging stakeholder-scientist-management partnerships for successful conservation.**

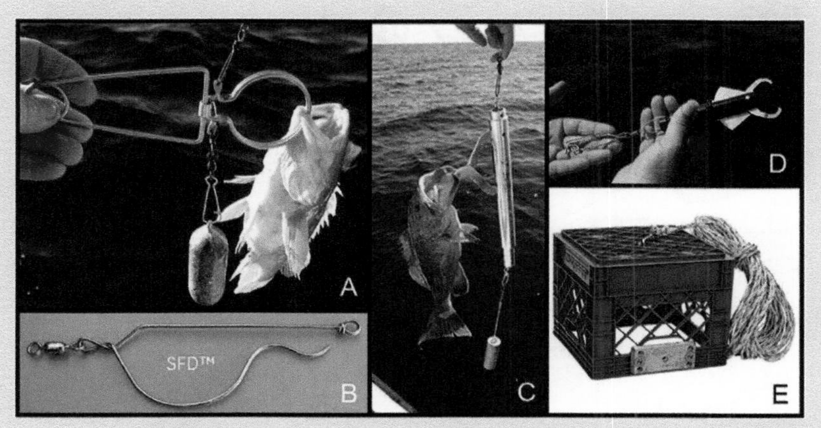


**FIG. 2** Different descending devices assessed in a study with recreational anglers by Bellquist et al. (2019). *Reprinted with permissions.*

Finally, NOAA scientists have continued their collaborative work with recreational fishermen to better understand sublethal impacts and delayed mortality. Leveraging acoustic technologies to track fish released from recreational fishing vessels using descending devices over longer periods is shedding light on important differential long-term survival among rockfish species and sizes (Wegner et al., 2021). Given the implications for population viability if these impacts are not accounted for, these results may play an important role in decisions to amend time-area closures to maintain recent population recovery successes and support healthy fisheries.

**BOX 3 Non-governmental organizations as bridges for translational science.**

The popularity of catch and release fishing has continued to boom in recent years, yet some common handling practices can have detrimental effects on fish survival post-release related to physiological impacts (Brownscombe et al., 2017). Collating and translating the science into best practices that anglers can follow under real-world conditions is a critical step to mitigate the impacts on fish populations, but is often a missing link when scientists lack the time, incentives and training to do this directly. Additionally, anglers often can't access scientific articles, and when they do it can be difficult to translate the findings into their angling practices. *Keep Fish Wet* (https:/www.keepfishwet.org/) is an NGO that recognized this missed opportunity and has spear-headed efforts to fill this gap through the development of several innovative translational science and outreach programs (Fig. 3).

*Continued*

**BOX 3 Non-governmental organizations as bridges for translational science.—cont'd**

**FIG. 3** Infographic developed by *Keep Fish Wet* to communicate best handling practices for striped bass catch and release fisheries, based on results of Conservation Physiology studies.

The model that *Keep Fish Wet* uses is grounded in the idea of building community with complementary expertise and perspectives. Led by executive director Sascha Clark Danylchuk, the team is not only composed of science advisors and board members directly involved in related Conservation Physiology research, but also education and engagement specialists, science ambassadors, and a network of partnering angling, conservation, and other organizations. This allows the strengths of each to

*Continued*

**BOX 3 Non-governmental organizations as bridges for translational science.**

shine and leads to better conservation outcomes; researchers who know the science best may not always be well-versed in effective science communication, and crafting messaging that are both accessible to general public audiences and true to the nuances and uncertainties of what is known can be particularly challenging. Including team members with extensive expertise in science communication and social science helps to ensure not only that the research will be understood by the target audience, but also increases the likelihood of affecting positive behavioral changes toward best practices. Similarly, their model emphasizes the importance of multi-directional communication where anglers and fishing organizations can provide feedback on the research or science-based suggested practices, as well as bring questions and ideas to the team to inform research priorities and stimulate new studies.

Key elements of success include using diverse modalities to engage stakeholders and not being afraid to branch out and try new approaches (S. Clark Danylchuk, pers. comm.). For example, the organization realized the importance of translating scientific articles and demonstrating applications of them via multiple platforms to reach a broad audience (e.g., their science blog, *Finsights,* newsletter and social media). In particular, they have homed in on identifying critical "pinchpoints" for recreational fishing communities (i.e., where a large proportion of the community interacts) such as shops, guides and lodges and partnered with them to disseminate best practices information and resources. Also, as a complement to their ambassador program that consists of fishing guides and anglers who promote implementation of best practices within their local and regional communities, *Keep Fish Wet* also launched a science ambassador program to facilitate the angling communities getting to know the people behind the research. They again use multi-modalities to highlight the work of these scientists, such as blog posts and live-streaming short interviews on social media at conferences. These efforts have helped to begin to break down barriers that are common between recreational anglers and scientists or managers.

*Keep Fish Wet* is making important contributions to fish conservation but also faces challenges common to many other conservation organizations. First and foremost, measuring impact and success are instrumental to achieving their mission and directing their efforts and resources. Some measures like views or alt-metrics are easy to quantify but deciphering whether these are truly correlated with widespread behavioral changes and shifts in community norms to positively affect population viability of fishes is more difficult. These outcomes can be much more challenging to assess, but ultimately are required to determine true conservation success. Social science approaches are helping to fill this gap, and as we look to the next decade of Conservation Physiology, we anticipate continued need for development and implementation of these techniques. Additionally, communicating the best available science for complex issues can be tricky, particularly when there is nuance or uncertainty combined with misconceptions of how science works and/or misinformation. Like almost all aspects of society, in an increasing digital age where information—both true and false—is online and at everyone's fingertips, the broader importance of science and political literacy will continue to grow and be vital for conservation success. Creating open-source resources and lowering barriers to access (e.g., *Keep Fish Wet* is not a membership based nonprofit and all their information is available for free) can help create a more informed and science literate community.

that partner fellows with mentors in these realms and provide associated professional development training can serve as excellent models for achieving this (e.g., the U.S. Sea Grant Graduate Fellow and Mitacs Canadian Policy Science Fellow programs).

Ultimately, some physiology research will not be directly useful to management, and in many contexts, more biological data with the latest technologies is not necessarily what is needed to solve conservation problems. This does not mean that such research should not be conducted, as it still will contribute to our fundamental knowledge of the world and may be useful in the future or in other contexts (e.g., generating public support and elevating awareness of issues by sharing discoveries that highlight how amazing the natural world is). But through gaining an understanding of the different perspectives and priorities of practitioners and policymakers, scientists can more authentically and productively engage with their counterparts on the ways that their work can help fuel conservation successes.

## 8 Looking forward: Priorities for the next decade and beyond

The application of physiological knowledge, tools and concepts to understand and solve conservation problems is the primary goal of Conservation Physiology. Though the field has come a long way from the stress physiology focus and has helped solve some impressive fish conservation problems, there remains ample opportunities for the discipline to mature and improve. Within this chapter, we focused on some of the remaining broad conceptual challenges facing the conservation physiologists, and provided guidance on priority actions to help overcome them (Table 1). A recent effort from a team of experts in the field of Conservation Physiology created a list of 100 pressing questions related to the field, spanning 10 diverse thematic areas (Cooke et al., 2021). This list provided specific research areas to guide future work that can inform the conservation and management of biodiversity as effectively as possible, and the challenges and conceptual priorities discussed within this chapter complement this list. It is our hope that addressing the priorities identified here *within* the context of each of these pressing research themes will strengthen the applicability and outcome of the research agenda proposed by Cooke et al. (2021).

Conservation Physiology is only one of the many branches of conservation science, but its suitability to understand the underlying mechanisms that drive species declines make it particularly powerful (Madliger et al., 2021). As we enter a new era of conservation science ("Decade of Ecosystem Restoration 2021–2030"), we call on current and future fish physiologists to think creatively to contribute to solving today's most pressing conservation issues. This includes integrating across scales and disciplines, collaborating with conservation practitioners and stakeholders to solve complex problems, and using emerging technological and other developments that offer exciting

**TABLE 1 Top priority actions for the conservation physiologists of tomorrow.**

| Priority actions | Suggestions to implement priority actions |
|---|---|
| Link physiological mechanisms to management-relevant scales | • Pair physiological studies with environmental and demographic data<br>• Draw on approaches from complementary fields for metrics and analyses, such as macrophysiology and landscape physiology<br>• Integrate physiological data into mechanistic models |
| Reflect environmental & ecological realism | • As best as possible, directly reflect conditions experienced in the wild, including the presence of multiple stressors over variable time scales |
| Conduct research across life stages & populations | • To identify the most vulnerable life stages within species or groups of species, conduct research studies across multiple life stages<br>• Examine the role of local adaptation by conducting studies on populations across species' ranges<br>• Explore potential carryover effects (and their underlying mechanisms) of stressors across seasons and life stages |
| Examine trends and patterns in existing data | • Collate, analyze and synthesize existing data to explore trends and patterns within species or within specific themes of conservation relevance |
| OA Make data publicly accessible | • Collect and report data and all supporting information such that it can be functionally used in syntheses<br>• Make raw data available whenever possible to make syntheses of large datasets easier |

*Continued*

**TABLE 1 Top priority actions for the conservation physiologists of tomorrow.—Cont'd**

| | | |
|---|---|---|
|  | Broaden phylogenomic & ecological representation | • Conduct and support studies beyond the 'popular' species across fish groups, habitats, trophic levels and life histories<br>• Analyze patterns across these contrasting species to examine physiological paradigms |
|  | Create an inclusive field of research that values the perspectives and knowledges of all peoples | • Educate yourself on the history of relevant sub-disciplines, occupied land, and institutions to determine the best actions to promote the dismantling of racist and colonial structures<br>• Create inclusive codes of conduct, implement strategies for inclusive meetings & research teams, and normalize thoughtful discussions of DEIJ topics as part of professional development activities<br>• Plan for and allocate ample resources to ensure safety in the field for historically marginalized team members<br>• Co-develop questions, documents and knowledge to co-produce learned insights, outcomes and decisions with Native scientists |
| 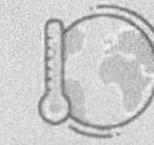 | Make explicit links to the real world | • Every research paper should have an explicit link to the real world that is easily identifiable, and simple enough to understand and distill by the readership |
| | Work to make science actionable for conservation | • Engage with managers, policymakers and stakeholders throughout research project development and implementation<br>• Cross-train the next generation of scientists in policy and conservation frameworks |

opportunities to strengthen the field, including (but not limited to) telemetry, community science, genomics and environmental sensing (see Chapter 4, Volume 39A: Brownscombe et al., 2022; Chapter 8, Volume 39A: Jeffries et al., 2022). Perhaps most importantly, this work needs to be conducted in concert with greater efforts to produce systemic changes for a more inclusive and equitable culture. This includes supporting efforts for largescale, structural changes in our field and society, as well as individuals' responsibility to educate themselves on the history and context in which they work. Conservation physiologists have some grand tasks ahead, and our hope is that this chapter provides the foundation to meet these demands in the coming years and beyond.

## Acknowledgments

We would like to thank Amy Teffer for contributions to the conceptual framework and original outline of this chapter. We also wish to thank the editors and four reviewers for their thoughtful feedback.

## References

Ahmadia, G., et al., 2021. Limited progress in improving gender and geographic representation in coral reef science. Front. Mar. Sci. 8. https://doi.org/10.3389/fmars.2021.731037.

Ames, E.M., et al., 2020. Striving for population-level conservation: integrating physiology across the biological hierarchy. Conserv. Physiol. 8 (1), coaa019.

Anadu, J., Ali, H., Jackson, C., 2020. Ten steps to protect BIPOC scholars in the field. Eos 101 (November). https://doi.org/10.1029/2020EO150525.

Angilletta Jr., M.J., Steury, T.D., Sears, M.W., 2004. Temperature, growth rate, and body size in ectotherms: fitting pieces of a life-history puzzle. Integr. Comp. Biol. 44 (6), 498–509.

Applewhite, D.A., 2021. A year since George Floyd. Mol. Biol. Cell 32 (19), 1797–1799.

Araújo, M.B., Peterson, A.T., 2012. Uses and misuses of bioclimatic envelope modeling. Ecology 93 (7), 1527–1539.

Audzijonyte, A., et al., 2016. Trends and management implications of human-influenced life-history changes in marine ectotherms. Fish Fish. 17 (4), 1005–1028.

Audzijonyte, A., et al., 2019. Is oxygen limitation in warming waters a valid mechanism to explain decreased body sizes in aquatic ectotherms? Glob. Ecol. Biogeogr. 28 (2), 64–77.

Audzijonyte, A., et al., 2020. Fish body sizes change with temperature but not all species shrink with warming. Nat. Ecol. Evol. 4 (6), 809–814.

Bang, M., Marin, A., Medin, D., 2018. If indigenous peoples stand with the sciences, will scientists stand with us? Daedalus 147 (2), 148–159.

Bellquist, L., et al., 2019. Effectiveness of descending devices to mitigate the effects of barotrauma among rockfishes (*Sebastes* spp.) in California recreational fisheries. Fish. Res. 215, 44–52.

Bennett, J.M., et al., 2021. The evolution of critical thermal limits of life on earth. Nat. Commun. 12, 1198. https://doi.org/10.1038/s41467-021-21263-8.

Bergman, J.N., et al., 2019. Scaling from individual physiological measures to population-level demographic change: case studies and future directions for conservation management. Biol. Conserv. 238, 108242.

Bernier, N.J., Alderman, S.L., 2022. Applied aspects of fish endocrinology. Fish Physiol. 39A, 253–320.

Bigman, J.S., et al., 2021. Respiratory capacity is twice as important as temperature in explaining patterns of metabolic rate across the vertebrate tree of life. Sci. Adv. 7 (19), eabe5163. https://doi.org/10.1126/sciadv.abe5163.

Birnie-Gauvin, K., et al., 2017. Conservation physiology can inform threat assessment and recovery planning processes for threatened species. Endanger. Species Res. 32, 507–513.

Birnie-Gauvin, K., Bordeleau, X., Cooke, S.J., Davidsen, J.D., Eldøy, S.H., Eliason, E.J., Moore, A., Aarestrup, K., 2021a. Life-history strategies in salmonids: the role of physiology and its consequences. Biol. Rev. 96, 2304–2320.

Birnie-Gauvin, K., Koed, A., Aarestrup, K., 2021b. Repeatability of migratory behavior suggests trade-off between size and survival in a wild iteroparous salmonid. Funct. Ecol. 35 (12), 2717–2727.

Bonier, F., et al., 2009. The relationship between fitness and baseline glucocorticoids in a passerine bird. Gen. Comp. Endocrinol. 163 (1–2), 208–213.

Bordeleau, X., Hatcher, B.G., Denny, S., Fast, M.D., Whoriskey, F.G., Patterson, D.A., Crossin, G.T., 2018. Consequences of captive breeding: fitness implications for wild-origin, hatchery-spawned Atlantic salmon kelts upon their return to the wild. Biol. Conserv. 225, 144–153.

Boult, V.L., Evans, L.C., 2021. Mechanisms matter: predicting the ecological impacts of global change. Glob. Chang. Biol. 27 (9), 1689–1691.

Bowen, L., et al., 2020. Transcriptomic response to elevated water temperatures in adult migrating Yukon River Chinook salmon (*Oncorhynchus tshawytscha*). Conserv. Physiol. 8 (1), coaa084.

Boyd, R., et al., 2018. A general approach to incorporating spatial and temporal variation in individual-based models of fish populations with application to Atlantic mackerel. Ecol. Model. 382, 9–17.

Bremner, J., 2008. 'Species' traits and ecological functioning in marine conservation and management. J. Exp. Mar. Biol. Ecol. 366 (1), 37–47.

Brosset, P., et al., 2021. Physiological biomarkers and fisheries management. Rev. Fish Biol. Fish. 31, 797–819. https://doi.org/10.1007/s11160-021-09677-5.

Brownscombe, J.W., et al., 2017. Best practices for catch-and-release recreational fisheries—angling tools and tactics. Fish. Res. 186, 693–705.

Brownscombe, J.W., Lawrence, M.J., Deslauriers, D., Filgueira, R., Boyd, R.J., Cooke, S.J., 2022. Applied fish bioenergetics. Fish Physiol. 39A, 141–188.

Buckley, L.B., 2008. Linking traits to energetics and population dynamics to predict lizard ranges in changing environments. Am. Nat. 171 (1), E1–E19.

Burnett, N.J., et al., 2014. Burst swimming in areas of high flow: delayed consequences of anaerobiosis in wild adult sockeye salmon. Physiol. Biochem. Zool. 87 (5), 587–598.

Byrnes, E.E., et al., 2020. Respirometer in a box: development and use of a portable field respirometer for estimating oxygen consumption of large-bodied fishes. J. Fish Biol. 96 (4), 1045–1050.

Chapman, L.J., Nyboer, E.A., Fugère, V., 2022. Fish response to environmental stressors in the Lake Victoria Basin ecoregion. Fish Physiol. 39B, 273–324.

Chaudhury, A., Colla, S., 2020. Next steps in dismantling discrimination: lessons from ecology and conservation science. Conserv. Lett. 14, e12774. https://doi.org/10.1111/conl.12774.

Chevin, L.-M., Collins, S., Lefèvre, F., 2013. Phenotypic plasticity and evolutionary demographic responses to climate change: taking theory out to the field. Funct. Ecol. 27, 967–979. https://doi.org/10.1111/j.1365-2435.2012.02043.x.

Chipps, S.R., Wahl, D.H., 2008. Bioenergetics modeling in the 21st century: reviewing new insights and revisiting old constraints. Trans. Am. Fish. Soc. 137 (1), 298–313.

Chown, S.L., 2012. Trait-based approaches to conservation physiology: forecasting environmental change risks from the bottom up. Philos. Trans. R. Soc. B 367, 1615–1627. https://doi.org/10.1098/rstb.2011.0422.

Chown, S.L., Gaston, K.J., 2008. Macrophysiology for a changing world. Proc. Biol. Sci. 275 (1642), 1469–1478.

Chown, S.L., Gaston, K.J., 2016. Macrophysiology—progress and prospects. Funct. Ecol. 30, 330–344. https://doi.org/10.1111/1365-2435.12510.

Chown, S.L., Addo-Bediako, A., Gaston, K.J., 2002. Physiological variation in insects: large-scale patterns and their implications. Comp. Biochem. Physiol. B Biochem. Mol. Biol. 131 (4), 587–602.

Chown, S.L., et al., 2010. Population responses within a landscape matrix: a macrophysiological approach to understanding climate change impacts. Evol. Ecol. 24 (3), 601–616.

Claësson, D., Wang, T., Malte, H., 2016. Maximal oxygen consumption increases with temperature in the European eel (*Anguilla anguilla*) through increased heart rate and arteriovenous extraction. Conserv. Physiol. 4 (1), cow027.

Connon, R.E., et al., 2018. The utility of transcriptomics in fish conservation. J. Exp. Biol. 221 (2). https://doi.org/10.1242/jeb.148833.

Cooke, S.J., O'Connor, C.M., 2010. Making conservation physiology relevant to policy makers and conservation practitioners: making conservation physiology relevant. Conserv. Lett. 3 (3), 159–166.

Cooke, S.J., et al., 2004. Biotelemetry: a mechanistic approach to ecology. Trends Ecol. Evol. 19 (6), 334–343. https://doi.org/10.1016/j.tree.2004.04.003.

Cooke, S.J., et al., 2012. Conservation physiology in practice: how physiological knowledge has improved our ability to sustainably manage Pacific salmon during up-river migration. Philos. Trans. R. Soc. B 367 (1596), 1757–1769. https://doi.org/10.1098/rstb.2012.0022.

Cooke, S.J., et al., 2014. Conservation physiology across scales: insights from the marine realm. Conserv. Physiol. 2 (1), cou024.

Cooke, S.J., et al., 2020. Reframing conservation physiology to be more inclusive, integrative, relevant and forward-looking: reflections and a horizon scan. Conserv. Physiol. 8 (1), coaa016.

Cooke, S.J., et al., 2021. One hundred research questions in conservation physiology for generating actionable evidence to inform conservation policy and practice. Conserv. Physiol. 9 (1), coab009.

Coristine, L.E., et al., 2014. A conceptual framework for the emerging discipline of conservation physiology. Conserv. Physiol. 2 (1), cou033.

Cortese, D., et al., 2021. Physiological and behavioral effects of anemone bleaching on symbiont anemonefish in the wild. Funct. Ecol. 35, 663–674. https://doi.org/10.1111/1365-2435.13729 (1365-2435.13729).

Cronin, M.R., et al., 2021. Anti-racist interventions to transform ecology, evolution and conservation biology departments. Nat. Ecol. Evol. 5, 1213–1223. https://doi.org/10.1038/s41559-021-01522-z.

Crozier, L.G., et al., 2008. Potential responses to climate change in organisms with complex life histories: evolution and plasticity in Pacific salmon. Evol. Appl. 1 (2), 252–270.

Das, S., Lowe, M., 2018. Nature read in black and white: decolonial approaches to interpreting natural history collections. J. Nat. Sci. Collect. 6, 4–14.

Daufresne, M., Lengfellner, K., Sommer, U., 2009. Global warming benefits the small in aquatic ecosystems. Proc. Natl. Acad. Sci. U. S. A. 106 (31), 12788–12793.

Davies, S.W., et al., 2021. Promoting inclusive metrics of success and impact to dismantle a discriminatory reward system in science. PLoS Biol. 19 (6), e3001282.

Dawson, N., et al., 2021. The role of indigenous peoples and local communities in effective and equitable conservation. Ecol. Soc. 26 (3), 19. https://doi.org/10.5751/ES-12625-260319.

de Vos, A., 2020. The Problem of "Colonial Science" Conservation Projects in the Developing World Should Invest in Local Scientific Talent and Infrastructure. Scientific American (1 July).

De Boeck, G., Rodgers, E., Town, R.M., 2022. Using ecotoxicology for conservation: From biomarkers to modeling. Fish Physiol. 39B, 111–174.

Demery, A.J.C., Pipkin, M.A., 2021. Safe fieldwork strategies for at-risk individuals, their supervisors and institutions. Nat. Ecol. Evol. 5, 5–9. https://doi.org/10.1038/s41559-020-01328-5.

Dickinson, J.L., Zuckerberg, B., Bonter, D.N., 2010. Citizen science as an ecological research tool: challenges and benefits. Annu. Rev. Ecol. Evol. Syst. 41, 149–172. https://doi.org/10.1146/annurev-ecolsys-102209-144636.

Eldøy, S.H., Bordeleau, X., Crossin, G.T., Davidsen, J.G., 2019. Individual repeatability in marine migratory behavior: a multi-population assessment of anadromous brown trout tracked through consecutive feeding migrations. Front. Ecol. Evol. 7. https://doi.org/10.3389/fevo.2019.00420.

Eliason, E.J., et al., 2011. Differences in thermal tolerance among sockeye Salmon populations. Science 332 (6025), 109–112. https://doi.org/10.1126/science.1198767.

Ellis, R.D., McWhorter, T.J., Maron, M., 2012. Integrating landscape ecology and conservation physiology. Landsc. Ecol. 27 (1), 1–12.

Estrada, M., et al., 2016. Improving underrepresented minority student persistence in STEM. CBE Life Sci. Educ. 15 (3), es5. https://doi.org/10.1187/cbe.16-01-0038.

Evans, T.G., Diamond, S.E., Kelly, M.W., 2015. Mechanistic species distribution modelling as a link between physiology and conservation. Conserv. Physiol. 3 (1), cov056.

Faleiro, F., et al., 2015. Seahorses under a changing ocean: the impact of warming and acidification on the behavior and physiology of a poor-swimming bony-armoured fish. Conserv. Physiol. 3 (1), cov009.

Farrell, A.P., 2011. Encyclopedia of Fish Physiology: From Genome to Environment. Academic Press, United States.

Finney, C., 2014. Black Faces, White Spaces: Reimagining the Relationship of African Americans to the Great Outdoors. UNC Press Books.

Flynn, E.E., Bjelde, B.E., Miller, N.A., Todgham, A.E., 2015. Ocean acidification exerts negative effects during warming conditions in a developing Antarctic fish. Conserv. Physiol. 3 (1), cov033.

Fordham, D.A., et al., 2013. Population dynamics can be more important than physiological limits for determining range shifts under climate change. Glob. Chang. Biol. 19 (10), 3224–3237.

Forrester, N., 2020. Diversity in science: next steps for research group leaders. Nature 585 (7826), S65–S67.

Freitas, V., et al., 2010. Temperature tolerance and energetics: a dynamic energy budget-based comparison of North Atlantic marine species. Philos. Trans. R. Soc. B 365, 3553–3565. https://doi.org/10.1098/rstb.2010.0049.

Gardiner, N.M., Munday, P.L., Nilsson, G.E., 2010. Counter-gradient variation in respiratory performance of coral reef fishes at elevated temperatures. PLoS One 5 (10), e13299.

Gardner, J.L., et al., 2011. Declining body size: a third universal response to warming? Trends Ecol. Evol. 26 (6), 285–291.

Gary Anderson, W., Schreier, A., Crossman, J.A., 2022. Conservation aquaculture—A sturgeon story. Fish Physiol. 39B, 39–109.

Golden, N., et al., 2021. Ten simple rules for productive lab meetings. PLoS Comput. Biol. 17 (5), e1008953.

Graham, A.L., Cooke, S.J., 2008. The effects of noise disturbance from various recreational boating activities common to inland waters on the cardiac physiology of a freshwater fish, the largemouth bass (*Micropterus salmoides*). Aquat. Conserv. Mar. Freshwat. Ecosyst. 18 (7), 1315–1324.

Guckian, M., et al., 2018. Assessing anglers' intentions to communicate best practices in a catch-and-release fishery for steelhead in British Columbia, Canada. J. Environ. Manag. 219, 252–259.

Halsey, S.J., et al., 2020. Elevate, don't assimilate, to revolutionize the experience of scientists who are black, indigenous and people of colour. Nat. Ecol. Evol. 4 (10), 1291–1293.

Harter, T.S., Shartau, R.B., Brauner, C.J., Farrell, A.P., 2014. Validation of the i-STAT system for the analysis of blood parameters in fish. Conserv. Physiol. 2 (1), cou037.

Hinch, S.G., Bett, N.N., Eliason, E.J., Farrell, A.P., Cooke, S.J., Patterson, D.A., 2021. Exceptionally high mortality of adult female salmon: a large-scale emerging trend and a conservation concern. Can. J. Fish. Aquat. Sci. 99 (999), 1–16.

Hinch, S.G., Bett, N.N., Farrell, A.P., 2022. A conservation physiological perspective on dam passage by fishes. Fish Physiol. 39B, 429–487.

Hoefnagel, K.N., Verberk, W.C.E.P., 2015. Is the temperature-size rule mediated by oxygen in aquatic ectotherms? J. Therm. Biol. 54, 56–65.

Hofstra, B., Kulkarni, V.V., Galvez, S.M.N., 2020. The diversity–innovation paradox in science. PNAS 117 (17), 9284–9291. https://doi.org/10.1073/pnas.191537811.

Horodysky, A.Z., Cooke, S.J., Brill, R.W., 2015. Physiology in the service of fisheries science: why thinking mechanistically matters. Rev. Fish Biol. Fish. 25 (3), 425–447.

Horodysky, A.Z., et al., 2016. Fisheries conservation on the high seas: linking conservation physiology and fisheries ecology for the management of large pelagic fishes. Conserv. Physiol. 4 (1), cov059.

Horodysky, A.Z., Schweitzer, C.C., Brill, R.W., 2022. Applied sensory physiology and behavior. Fish Physiol. 39A, 33–90.

Hussey, N.E., et al., 2015. Aquatic animal telemetry: a panoramic window into the underwater world. Science 348 (6240), 1255642.

Illing, B., Rummer, J.L., 2017. Physiology can contribute to better understanding, managing, and conserving coral reef fishes. Conserv. Physiol. 5 (1), cox005. https://doi.org/10.1093/conphys/cox005.

IPCC, 2021. Climate Change 2021—The Physical Science Basis. Cambridge University Press, Cambridge, UK.

Jain-Schlaepfer, S., et al., 2018. Impact of motorboats on fish embryos depends on engine type. Conserv. Physiol. 6 (1), coy014.

Jarvis, E.T., Lowe, C.G., 2008. The effects of barotrauma on the catch-and-release survival of southern California nearshore and shelf rockfish (Scorpaenidae, *Sebastes* spp.). Can. J. Fish. Aquat. Sci. 65 (7), 1286–1296. https://doi.org/10.1139/f08-071.

Jeffrey, J.D., et al., 2015. Linking landscape-scale disturbances to stress and condition of fish: implications for restoration and conservation. Integr. Comp. Biol. 55 (4), 618–630.

Jeffries, K.M., et al., 2014. Immune response genes and pathogen presence predict migration survival in wild salmon smolts. Mol. Ecol. 23 (23), 5803–5815.

Jeffries, K.M., Jeffrey, J.D., Holland, E.B., 2022. Applied aspects of gene function for the conservation of fishes. Fish Physiol. 39A, 389–433.

Jeschke, J.M., Strayer, D.L., 2008. Usefulness of bioclimatic models for studying climate change and invasive species. Ann. N. Y. Acad. Sci. 1134, 1–24.

Jessen, T.D., et al., 2021. Contributions of indigenous knowledge to ecological and evolutionary understanding. Front. Ecol. Environ. 20 (2), 93–101. https://doi.org/10.1002/fee.2435.

Jimenez, M.F., et al., 2019. Underrepresented faculty play a disproportionate role in advancing diversity and inclusion. Nat. Ecol. Evol. 3, 1030–1033. https://doi.org/10.1038/s41559-019-0911-5.

Johnston, A.S.A., et al., 2019. Predicting population responses to environmental change from individual-level mechanisms: towards a standardized mechanistic approach. Proc. Biol. Sci. 286 (1913), 20191916.

Jordan, L.K., et al., 2013. Linking sensory biology and fisheries bycatch reduction in elasmobranch fishes: a review with new directions for research. Conserv. Physiol. 1 (1), cot002.

Jørgensen, C., et al., 2012. Conservation physiology of marine fishes: advancing the predictive capacity of models. Biol. Lett. 8 (6), 900–903.

Jutfelt, F., et al., 2021. Aerobic scope protection reduces ectotherm growth under warming. Funct. Ecol. 35, 1397–1407. https://doi.org/10.1111/1365-2435.13811.

Keisling, B., et al., 2020. Does our vision of diversity reduce harm and promote justice? GSA Today. https://doi.org/10.1130/GSATG429GW.1.

Kendi, I.X., 2019. How to be an Antiracist. Random House Publishing Group.

Killen, S.S., et al., 2007. Little left in the tank: metabolic scaling in marine teleosts and its implications for aerobic scope. Proc. Biol. Sci. 274 (1608), 431–438.

Killen, S.S., et al., 2016. Ecological influences and morphological correlates of resting and maximal metabolic rates across teleost fish species. Am. Nat. 187 (5), 592–606.

Killen, S.S., Hollins, J., Koeck, B., Lennox, R.J., Cooke, S.J., 2022. Consequences for fisheries in a multi-stressor world. Fish Physiol. 39B, 175–207.

Kimmerer, R.W., 2002. Weaving traditional ecological knowledge into biological education: a call to action. Bioscience 52 (5), 432–438.

King, G.D., et al., 2016. Watershed-scale land use activities influence the physiological condition of stream fish. Physiol. Biochem. Zool. 89 (1), 10–25.

Kitaysky, A.S., Piatt, J.F., Wingfield, J.C., 2007. Stress hormones link food availability and population processes in seabirds. Mar. Ecol. Prog. Ser. 352, 245–258.

Koenigstein, S., et al., 2016. Modelling climate change impacts on marine fish populations: process-based integration of ocean warming, acidification and other environmental drivers. Fish Fish. 17 (4), 972–1004.

Komoroske, L.M., et al., 2014. Ontogeny influences sensitivity to climate change stressors in an endangered fish. Conserv. Physiol. 2 (1), cou008.

Komoroske, L.M., Birnie-Gauvin, K., 2022. Conservation Physiology of fishes for tomorrow: Successful conservation in a changing world and priority actions for the field. Fish Physiol. 39B, 581–628.

Laubenstein, T.D., Rummer, J.L., 2021. Communication in conservation physiology: linking diverse stakeholders, promoting public engagement, and encouraging application. In: Conservation Physiology—Applications for Wildlife Conservation and Management. Oxford University Press, Oxford.

Laubenstein, T.D., Rummer, J.L., McCormick, M.I., Munday, P.L., 2019. A negative correlation between behavioural and physiological performance under ocean acidification and warming. Sci. Rep. 9 (1), 1–10.

Lavaud, R., Filgueira, R., Augustine, S., 2021. The role of dynamic energy budgets in conservation physiology. Conserv. Physiol. 9 (1), coab083.

Lea, J.M.D., et al., 2018. Non-invasive physiological markers demonstrate link between habitat quality, adult sex ratio and poor population growth rate in a vulnerable species, the cape mountain zebra. Funct. Ecol. 32 (2), 300–312.

Leeuwis, R.H.J., et al., 2021. Research on sablefish (*Anoplopoma fimbria*) suggests that limited capacity to increase heart function leaves hypoxic fish susceptible to heat waves. Proc. Biol. Sci. 288 (1946), 20202340.

Lefevre, S., 2016. Are global warming and ocean acidification conspiring against marine ectotherms? A meta-analysis of the respiratory effects of elevated temperature, high CO2 and their interaction. Conserv. Physiol. 4 (1), cow009. https://doi.org/10.1093/conphys/cow009.

Li, D., Koedel, C., 2017. Representation and salary gaps by race-ethnicity and gender at selective public universities. Educ. Res. 46, 343–354.

Little, A.G., et al., 2020. Maxed out: optimizing accuracy, precision, and power for field measures of maximum metabolic rate in fishes. Physiol. Biochem. Zool. 93 (3), 243–254.

Luis Val, A.L., Duarte, R.M., Campos, D., de Almeida-Val, V.M.F., 2022. Environmental stressors in Amazonian riverine systems. Fish Physiol. 39B, 209–271.

Madliger, C.L., Love, O.P., 2014. The need for a predictive, context-dependent approach to the application of stress hormones in conservation: stress hormones and conservation. Conserv. Biol. 28 (1), 283–287.

Madliger, C.L., et al., 2018. The conservation physiology toolbox: status and opportunities. Conserv. Physiol. 6 (1), coy029.

Madliger, C.L., et al., 2021. The second warning to humanity: contributions and solutions from conservation physiology. Conserv. Physiol. 9 (1). https://doi.org/10.1093/conphys/coab038.

Mahoney, J.L., Klug, P.E., Reed, W.L., 2018. An assessment of the US endangered species act recovery plans: using physiology to support conservation. Conserv. Physiol. 6 (1), coy036.

McKenzie, D.J., et al., 2016. Conservation physiology of marine fishes: state of the art and prospects for policy. Conserv. Physiol. 4 (1), cow046.

McLane, A.J., et al., 2011. The role of agent-based models in wildlife ecology and management. Ecol. Model. 222 (8), 1544–1556.

McNeil, B.I., Sasse, T.P., 2016. Future ocean hypercapnia driven by anthropogenic amplification of the natural CO2 cycle. Nature 529 (7586), 383–386.

Metcalfe, J.D., et al., 2012. Conservation physiology for applied management of marine fish: an overview with perspectives on the role and value of telemetry. Philos. Trans. R. Soc. B 367, 1746–1756. https://doi.org/10.1098/rstb.2012.0017.

Moore, A.J., et al., 2010. The need for archiving data in evolutionary biology. J. Evol. Biol. 23 (4), 659–660.

Morales, N., et al., 2020. Promoting inclusion in ecological field experiences: examining and overcoming barriers to a professional rite of passage. Bull. Ecol. Soc. Am. 101 (4), e01742. https://doi.org/10.1002/bes2.1742.

Müller-Wille, S., 2014. Race and history: comments from an epistemological point of view. Sci. Technol. Hum. Values 39 (4), 597–606.

Nagrodski, A., Murchie, K.J., Stamplecoskie, K.M., Suski, C.D., Cooke, S.J., 2013. Effects of an experimental short-term cortisol challenge on the behavior of wild creek chub *Semotilus atromaculatus* in mesocosm and stream environments. J. Fish Biol. 82 (4), 1138–1158.

National Science Foundation, 2019. Women, Minorities, and Persons With Disabilities in Science and Engineering. Special report, NSF.

Nedelec, S.L., et al., 2016. Repeated exposure to noise increases tolerance in a coral reef fish. Environ. Pollut. 216, 428–436. https://doi.org/10.1016/j.envpol.2016.05.058.

Norin, T., Malte, H., Clark, T.D., 2014. Aerobic scope does not predict the performance of a tropical eurythermal fish at elevated temperatures. J. Exp. Biol. 217 (Pt 2), 244–251.

O'Brien, L.T., Bart, H.L., Garcia, D.M., 2020. Why are there so few ethnic minorities in ecology and evolutionary biology? Challenges to inclusion and the role of sense of belonging. Soc. Psychol. Educ. 23 (2), 449–477.

O'Connor, C.M., Cooke, S.J., 2015. Ecological carryover effects complicate conservation. Ambio 44 (6), 582–591.

Patterson, D.A., et al., 2016. A perspective on physiological studies supporting the provision of scientific advice for the management of Fraser River sockeye salmon (*Oncorhynchus nerka*). Conserv. Physiol. 4 (1), cow026.

Pauly, D., Zeller, D., 2016. Catch reconstructions reveal that global marine fisheries catches are higher than reported and declining. Nat. Commun. 7, 10244. https://doi.org/10.1038/ncomms10244.

Perry, A.L., et al., 2005. Climate change and distribution shifts in marine fishes. Science 308 (5730), 1912–1915.

Petitgas, P., et al., 2013. Impacts of climate change on the complex life cycles of fish. Fish. Oceanogr. 22 (2), 121–139.

Plume, A., Van Weijen, D., 2014. Publish or perish? The rise of the fractional author. Res. Trends 38 (3), 16–18.

Pörtner, H.O., 2002. Climate variations and the physiological basis of temperature dependent biogeography: systemic to molecular hierarchy of thermal tolerance in animals. Comp. Biochem. Physiol. A Mol. Integr. Physiol. 132 (4), 739–761.

Pullin, A.S., et al., 2020. Informing conservation decisions through evidence synthesis and communication. In: Sutherland, W.J., et al. (Eds.), Conservation Research, Policy and Practice. Cambridge University Press, United Kingdom, pp. 114–123.

Railsback, S.F., Grimm, V., 2019. Agent-Based and Individual-Based Modeling: A Practical Introduction, second ed. Princeton University Press.

Reid, A.J., et al., 2021. "Two-eyed seeing": an indigenous framework to transform fisheries research and management. Fish Fish. 22 (2), 243–261.

Roche, D.G., et al., 2015. Public data archiving in ecology and evolution: how well are we doing? PLoS Biol. 13 (11), e1002295.

Rodgers, E.M., et al., 2019. Integrating physiological data with the conservation and management of fishes: a meta-analytical review using the threatened green sturgeon (*Acipenser medirostris*). Conserv. Physiol. 7 (1), coz035.

Rodgers, E.M., et al., 2021. Double whammy: nitrate pollution heightens susceptibility to both hypoxia and heat in a freshwater salmonid. Sci. Total Environ. 765, 142777.

Rogers, N.J., et al., 2016. A new analysis of hypoxia tolerance in fishes using a database of critical oxygen level (Pcrit). Conserv. Physiol. 4 (1), cow012. https://doi.org/10.1093/conphys/cow012.

Rummer, J.L., Bennett, W.A., 2005. Physiological effects of swim bladder overexpansion and catastrophic decompression on red snapper, *Lutjanus campechanus*. Trans. Am. Fish. Soc. 134 (6), 1457–1470. https://doi.org/10.1577/T04-235.1.

Rummer, J.L., Illing, B., 2022. Coral reef fishes in a multi-stressor world. Fish Physiol. 39B, 325–391.

Rypel, A.L., et al., 2021. Goodbye to "rough fish": paradigm shift in the conservation of native fishes. Fisheries 46, 605–616. https://doi.org/10.1002/fsh.10660.

Salinas, S., et al., 2019. Trait variation in extreme thermal environments under constant and fluctuating temperatures. Philos. Trans. R. Soc. Lond. Ser. B Biol. Sci. 374 (1768), 20180177.

Sampaio, E., Lopes, A.R., Francisco, S., Paula, J.R., Pimentel, M., Maulvault, A.L., Rosa, R., 2018. Ocean acidification dampens physiological stress response to warming and contamination in a commercially-important fish (Argyrosomus regius). Sci. Total Environ. 618, 388–398.

Santos, C.P., et al., 2021. Elasmobranch responses to experimental warming, acidification, and oxygen loss-a meta-analysis. Front. Mar. Sci. 8, 1380.

Schell, C.J., et al., 2020. Recreating Wakanda by promoting black excellence in ecology and evolution. Nat. Ecol. Evol. 4 (10), 1285–1287.

Schoenle, L.A., et al., 2021. Does variation in glucocorticoid concentrations predict fitness? A phylogenetic meta-analysis. Gen. Comp. Endocrinol. 300, 113611.

Schulte, P.M., Healy, T.M., 2022. Physiological diversity and its importance for fish conservation and management in the Anthropocene. Fish Physiol. 39A, 435–477.

Seebacher, F., White, C.R., Franklin, C.E., 2014. Physiological plasticity increases resilience of ectothermic animals to climate change. Nat. Clim. Chang. 5 (1), 61–66.

Shields, S.A., Bhatia, S., 2009. Darwin on race, gender, and culture. Am. Psychol. 64 (2), 111–119.

Sousa, T., et al., 2010. Dynamic energy budget theory restores coherence in biology. Philos. Trans. R. Soc. Lond. Ser. B Biol. Sci. 365 (1557), 3413–3428.

Stefanoudis, P.V., et al., 2021. Turning the tide of parachute science. Curr. Biol. 31 (4), R184–R185.

Strailey, K.K., Suski, C.D., 2022. Restoration physiology of fishes: Frontiers old and new for aquatic restoration. Fish Physiol. 39B, 393–428.

Sunday, J.M., Bates, A.E., Dulvy, N.K., 2012. Thermal tolerance and the global redistribution of animals. Nat. Clim. Chang. 2 (9), 686–690.

Thoral, E., et al., 2021. Low oxygen levels can help to prevent the detrimental effect of acute warming on mitochondrial efficiency in fish. Biol. Lett. 17 (2), 20200759.

Todgham, A.E., Stillman, J.H., 2013. Physiological responses to shifts in multiple environmental stressors: relevance in a changing world. Integr. Comp. Biol. 53 (4), 539–544.

Tseng, M., et al., 2020. Strategies and support for black, indigenous, and people of colour in ecology and evolutionary biology. Nat. Ecol. Evol. 4 (10), 1288–1290.

van Rijn, I., et al., 2017. Large but uneven reduction in fish size across species in relation to changing sea temperatures. Glob. Chang. Biol. 23 (9), 3667–3674.

Wegner, N.C., et al., 2021. Post-release survival and prolonged sublethal effects of capture and barotrauma on deep-dwelling rockfishes (genus Sebastes): implications for fish management and conservation. ICES J. Mar. Sci. 78 (9), 3230–3244. https://doi.org/10.1093/icesjms/fsab188.

Whitlock, M.C., 2011. Data archiving in ecology and evolution: best practices. Trends Ecol. Evol. 26 (2), 61–65.

Whitlock, M.C., et al., 2010. Data archiving. Am. Nat. 175 (2), 145–146.

Wikelski, M., Cooke, S.J., 2006. Conservation physiology. Trends Ecol. Evol. 21 (1), 38–46.

Wilkie, M.P., Johnson, N.S., Docker, M.F., 2022. Invasive species control and management: The sea lamprey story. Fish Physiol. 39B, 489–579.

Wingfield, J.C., Kitaysky, A.S., 2002. Endocrine responses to unpredictable environmental events: stress or anti-stress hormones? Integr. Comp. Biol. 42 (3), 600–609.

Wood, S., et al., 2020. A scientist like me: demographic analysis of biology textbooks reveals both progress and long-term lags. Proc. Biol. Sci. 287 (1929), 20200877.

Woods, H.A., Dillon, M.E., Pincebourde, S., 2015. The roles of microclimatic diversity and of behavior in mediating the responses of ectotherms to climate change. J. Therm. Biol. 54, 86–97.

Yanagitsuru, Y.R., Davis, B.E., Baerwald, M.R., Sommer, T.R., Fangue, N.A., 2022. Using physiology to recover imperiled smelt species. Fish Physiol. 39B, 1–37.

Ziv, Y., Davidowitz, G., 2019. When landscape ecology meets physiology: effects of habitat fragmentation on resource allocation trade-offs. Front. Ecol. Evol. 7, 137.

# Other volumes in the Fish Physiology series

# Index

Note: Page numbers followed by "*f*" indicate figures and "*t*" indicate tables.

## B

# C

## F

## J



## K



## L



## M

## Q



## R



## S

## T

Printed and bound by CPI Group (UK) Ltd, Croydon, CR0 4YY
08/06/2025
01896870-0011